The
Role of Glia
in
Neurotoxicity

Second Edition

The
Role of Glia
in
Neurotoxicity

Second Edition

Edited by
Michael Aschner
Lucio G. Costa

CRC Press
Taylor & Francis Group
Boca Raton London New York

CRC Press is an imprint of the
Taylor & Francis Group, an **informa** business

CRC Press
Taylor & Francis Group
6000 Broken Sound Parkway NW, Suite 300
Boca Raton, FL 33487-2742

First issued in paperback 2019

© 2005 by Taylor & Francis Group, LLC
CRC Press is an imprint of Taylor & Francis Group, an Informa business

No claim to original U.S. Government works

ISBN-13: 978-0-8493-1794-1 (hbk)
ISBN-13: 978-0-367-39338-0 (pbk)

Library of Congress Card Number 2004054600

Library of Congress Cataloging-in-Publication Data

The role of glia in neurotoxicity/edited by Michael Aschner & Lucio Costa.–2nd ed.
    p. cm.
    Includes bibliographical references and index.
    ISBN 0-8493-1794-0 (alk. paper)
    1. Neurotoxicology. 2. Neuroglia. I. Aschner, Michael. II. Costa, Lucio (Lucio G.)

RC347.5.R65 2004
616.8'047–dc22   2004054600

Visit the Taylor & Francis Web site at
http://www.taylorandfrancis.com

and the CRC Press Web site at
http://www.crcpress.com

Preface

In contrast to the previously emphasized support roles of glia for neurons, the new information coming from the application of novel techniques provides evidence that glia play critical roles in the maintenance of CNS homeostasis throughout life. As illustrated in this book, the role of glial cells extends well beyond passive structural support and sensitivity to axon commands. In fact, neuroglia and neurons establish a highly dynamic reciprocal relationship that influences subsequent nervous tissue growth, morphology, behavior, and repair. Neuron-centric definitions of the nervous system are, therefore, inadequate and should incorporate the obvious pervasive involvement of neuroglia. Neuroglial interactions with other cell types and between themselves, and the complexity of these interactions, provide numerous strategic sites of action for neurotoxic chemicals or neuropathic disease processes. This reciprocity between neurons and neuroglia suggests that the morphological and physiological attributes of neurons are a product of this cell–cell interaction and vice versa. Their potential in modulating damage and repair in the nervous system is reflected in this book.

This publication is intended to fill a missing gap in the literature. With the increasing importance of glia, substantial publications, including a number of textbooks, have been published over the last decade, and a journal (*Glia*) is dedicated to studies on these cells. Many of the published textbooks are multifaceted and multidisciplinary, encompassing neurophysiology, neuroanatomy, neuroscience, neurochemistry, and neuropharmacology. None, however, includes the role of glia in neurotoxicity, a timely topic, and a subject in its own right. This second volume represents the combined efforts of internationally recognized scientists who believe that we are at the threshold of comprehending the role afforded by neuroglia in brain function and injury, in general. We are thankful for their contribution to this project, some new and some who have agreed to update their

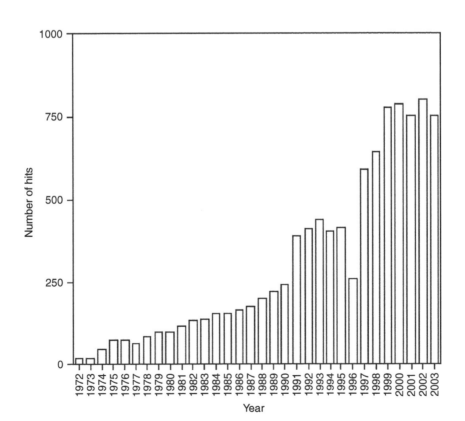

chapters from the original publication in 1996. The "coming of age" of glia has come to bear since the pioneering work by Kuffler and his colleagues (Kuffler et al., 1966) on the K^+ buffering capacity of these cells in invertebrate models. A CRISP (Computer Retrieval of Information on Scientific Projects) search of database of federally funded biomedical research projects conducted at universities, hospitals, and other research institutions conducted on May 3, 2004, utilizing the keywords *astrocyte*, *glia*, *microglia*, *oligodendrocyte*, or *Schwann* identified 9975 hits for the query for the years 1972 (first year for which the data are available) to the present. While in 1972 only 16 biomedical research projects included the above keywords, the number of research projects has mushroomed to 752 in 2003. A PubMed search also vindicates the Rodney Dangerfield cells of the CNS (those seeking "respect") with 50,292 publications between 1972 and present identified for the keyword *glia*.

When we first thought about this book in the early 1990s, we imagined an audience made up of established researchers in fields connected in different ways to neurotoxicology. We also hoped that clinicians and students could learn about the many different aspects of neurotoxicity related to neuroglia. While everyone contributing to the book is a true authority in some domain of knowledge, there are a number of different types of knowledge that are represented among them. We urged the authors to provide all of the necessary details but to also take the time to explain concepts and terms that are of a specialized nature. We hope that these efforts materialized in the final format of the second edition of the book, making it attractive to students, researchers, and clinicians alike.

The Editors

Pierre Morell Tribute for
The Role of Glia in Neurotoxicology

The field of neurotoxicology and the neuroscience community at large lost a valued member, and a great many neuroscientists lost a treasured friend and colleague, when Pierre Morell passed away on July 15, 2003, after a brief illness. Pierre was a brilliant scientist, a distinguished and highly respected university professor, a mentor to countless students and colleagues alike, a tireless supporter of numerous scientific endeavors outside his own lab, but most importantly, a very kind, generous, and caring human being. We are honored and humbled by the request to provide a memorial tribute to Pierre.

Our reflections on Pierre's life fittingly begin with his parents, who were very important to him from the moment of his birth until his last breath. In the late 1930s, the newlywed Anatol and Halina (Leszczynska) Mosewicki escaped the prewar horrors of Nazi-occupied Poland, crossing the frozen Siberian wastelands of Stalin's Russia and gaining ship's passage to Japan. After a brief stay in Japan and passage through the United States, they eventually arrived at the Sosua refugee community in the Dominican Republic where, on December 10, 1941, Pierre was born. After several years, the young family realized their dream of coming to the United States, settling in New York City, where Anatol soon began his long and successful career as an academic research scientist in the Biochemistry Department of the Albert Einstein College of Medicine. Anatol's collaborative efforts with Gilbert Ashwell on hepatic glycoprotein receptors and the clearance of circulating glycoproteins represent some of the early seminal work demonstrating the importance of carbohydrate moieties on glycoproteins in defining biological function and metabolism. Pierre's mother, Helena, enjoyed two successful careers, first as a biochemist and then as a social worker. Both are now retired to Stony Brook, New York, adjacent to the Stony Brook campus where they remain in the midst of a caring community of scholars and thinkers.

The young Pierre attended the prestigious Bronx High School of Science, and upon graduation entered Columbia University, majoring in chemistry and competing on the varsity swim team. Following graduation in 1963, Pierre began his graduate studies in the Biochemistry Department at Einstein, working in the laboratory of Julius Marmur on early studies of RNA sequences in bacteria. The first of his many publications appeared in P.N.A.S. USA in 1965. For his postdoctoral training, Pierre joined Norm Radin's lab at the University of Michigan, where his early work on sphingolipid biosynthesis began the path that led to his long and distinguished research career examining various aspects of myelin metabolism in both normal and abnormal conditions.

Pierre returned to the Albert Einstein College of Medicine in 1968 for his first faculty position in the Department of Neurology, where he continued his studies on metabolism of myelin glycolipids. Well before the transgenic and gene knockout era, Pierre was among the first to realize the potential of mutant mice for understanding the underlying mechanisms responsible for myelin function and dysfunction. Indeed, he remained active in this area until his death. In 1973, Pierre joined the Department of Biochemistry at the University of North Carolina at Chapel Hill, where he remained for the remainder of his life. Pierre assembled and edited the first edition of his *Myelin* book when he was only 35 years old, and for many years, this and a subsequent edition were the definitive reference on myelin for both beginners and established investigators. This accomplishment showed not only his precocious development as a neuroscientist, but also his organizational skills and underlying drive to get things done right and done well, traits he put to good use in many areas of science and academia.

A recurring theme in Pierre's work was the use of neurotoxicants and other perturbed systems to learn more about the metabolism and function of myelin and about the underlying mechanisms of action of directed insults. Notable among his early studies in this area were those in collaboration with Martin Krigman on developmental effects of lead on myelination. Later studies included elucidation of the underlying metabolic defect in tellurium-induced demyelinating neuropathy (with

Maria Wagner, Jean Harry, Arrel Toews, and Tom Bouldin as collaborators) and, most recently, studies on cuprizone-induced demyelination and subsequent remyelination in the CNS, in collaboration with Glenn Matsushima, Jeffrey Mason, and others.

A hallmark of his approach to neurotoxicology was his view of the field not as a specific discipline, but as an amalgam of various disciplines or approaches, each contributing vital input and complementing the others to provide much more powerful insights than would have been possible by isolated investigations. His insistence that each individual contribution to this multi-faceted approach be exacting and solid basic science was a cornerstone of his own research programs, and a view he also tirelessly advanced as a member of countless study sections, review panels, and editorial review boards. Pierre was the joy of journal editors and executive secretaries, and the bane of submitting authors and grant applicants, further demonstration of his uncompromising insistence on excellence and integrity in scientific research. He served on numerous NIH and National Multiple Sclerosis Society study sections and was a member of the National Research Council/National Academy of Sciences Committee on Neurotoxicology and Risk Assessment. He served on the editorial boards of the *Journal of Neurochemistry and Brain Research* and as a reviewer for numerous journals.

Pierre was very active in many aspects of the neuroscience and general academic community, both at UNC and on national and international levels. He was an early, tireless, and perpetual supporter of the American and International Societies for Neurochemistry and the Society for Neuroscience, working hard behind the scenes to make them successful. For 10 years, he was Director of the UNC Curriculum in Neurobiology, the second-oldest neuroscience training program in the United States. He presided over unprecedented growth in neuroscience graduate training and established UNC as a leader in this field. Not only students but also postdocs, junior faculty, staff, and many others benefited from his leadership and guidance through the years, because their development and professional growth was always an important priority for Pierre. He introduced the potentials of molecular biology to the UNC neuroscience community (via a several-day workshop) some years before the impact of such techniques upon neuroscience was generally appreciated. In addition, he was a driving force in the maturation of the university's neuroscience research effort, playing a major role in forming what is now the freestanding Neuroscience Center. He was a wise mentor and counselor to more than a generation of neuroscientists, and his willingness to share his knowledge, advice, and guidance extended far beyond the boundaries of his university. On numerous occasions, he was willing to note that the emperor wore no clothes, and he was always guided by what was right, not by what was expedient. Yet no matter the issue or ultimate outcome, he could always be depended on to give a wistful smile, reach into his desk drawer, and say, "Here everyone, have a Snickers bar!" His humanity, even more than his considerable scientific accomplishments, is his most lasting legacy to those who knew him.

Despite his seemingly constant attention to scientific and academic matters, Pierre was also a loving and devoted husband and father. Married to Bonnie Jean Brown, whom he met during a summer school interlude at UC-Berkeley (Pierre from Columbia, Bonnie from Cornell), theirs was a happy and lasting relationship blessed with two children, Sharon and David. Pierre's granddaughter Katy, born in 2002, brought him great joy and happiness during his final year.

A realistic review of the life and times of Pierre Morell cannot ignore a very important aspect of his life outside the lab (but not outside the realm of science!). Pierre was an avid scuba diver, and his budding love for this activity in the early 1990s quickly (and typically) evolved into a role as distinguished teacher. Rapidly becoming certified at all levels of instruction, he taught several wildly popular courses of scuba diving at UNC and was locally famous for taking his students to the Florida Keys for their final checkout/certification dives. Although he held academic ranks in several departments, he seemed most proud of his appointment as Adjunct Professor of Physical Education and Exercise Sport Science, and many students who knew not of his scientific acumen will miss his energy, enthusiasm, and guidance too.

Pierre's premature passing has left a void in the neuroscience community and in many individual lives that will be difficult, if not impossible, to fill. His sharp wit, his lively and enthusiastic approach to science and to life in general, his incisive (and sometimes wildly inappropriate) remarks about all aspects of life, and his generous assistance and advice will all be sorely missed. Although we will all miss Pierre very much, we are better for having known this truly remarkable and cherished colleague, teacher, mentor, and most importantly, friend.

Respectfully submitted
Arrel Toews, Jean Harry, and Richard Mailman

Editors

Michael Aschner received a Ph.D. from the University of Rochester, School of Medicine and Dentistry, Department of Anatomy and Neurobiology in 1985. He is currently the Gray E. B. Stahlman Professor of Neuroscience in the Department of Pediatrics at the Vanderbilt University Medical Center in Nashville, Tennessee. He also holds joint appointments at the Kennedy Center and in the Department of Pharmacology. Dr. Aschner's research interests are in the neurobiology and physiology of astrocytes and the mechanisms of central nervous system injury. Dr. Aschner has been particularly interested in metal uptake and distribution in the brain, devoting the last 20 years of his research to the mechanisms of transport of methylmercury, manganese, and uranium across the endothelium composing the blood–brain barrier, as well as their cellular and molecular mechanisms of neurotoxicity. Dr. Aschner has served on numerous national and international toxicology panels, and he has authored 200 peer-reviewed manuscripts and chapters in the area of neurotoxicology.

Lucio G. Costa is Professor of Toxicology in the Department of Environmental and Occupational Health Sciences at the University of Washington in Seattle, and Professor of Pharmacology in the Department of Pharmacology and Human Physiology at the University of Bari Medical School. He completed graduate studies in pharmacology at the University of Milano and did postdoctoral work at the University of Texas in Houston. Dr. Costa's research interests are in the area of neurotoxicology and developmental neurotoxicology, in particular the biochemical and molecular mechanisms of neurotoxicity and the role of gene–environment interactions in neurotoxicity and neurodegenerative diseases. He has authored more than 250 peer-reviewed manuscripts, book chapters, and books in the area of neurotoxicology. He has been funded over the years particularly by NIEHS and NIAAA; the latter project deals with the role of glia in the developmental neurotoxicity of alcohol. Dr. Costa has served on numerous national and international panels, committees, and review boards dealing with toxicological issues. He was a founding member of the International Neurotoxicology Association and served as its first president. He has also received various awards including the Achievement Award from the Society of Toxicology. He currently serves as Councilor of the Italian Society of Toxicology.

Contributors

Jan Albrecht
Department of Neurotoxicology
Medical Research Centre
Polish Academy of Sciences
Warsaw, Poland

J. Steven Alexander
Department of Molecular and Cellular
  Physiology
Louisiana State University Health Sciences
  Center
Shreveport, Louisiana

Takao Asano
Department of Neurosurgery
Institute of Laboratory Animal Science
Saitama Medical Center/School
Kawagoe, Saitama, Japan

Judy L. Aschner
Department of Pediatrics
Wake Forest University School of Medicine
Winston-Salem, North Carolina

Michael Aschner
Department of Physiology and Pharmacology
  and Interdisciplinary Program in
  Neuroscience
Wake Forest University School of Medicine
Winston-Salem, North Carolina

Stefan Bröer
School of Biochemistry and Molecular Biology
Australian National University
Canberra, Australia

Poonlarp Cheepsunthorn
Faculty of Medicine
Chulalongkorn University
Bangkok, Thailand

Valerie Chock
Department of Anesthesia and Department
  of Neonatology
Stanford University School of Medicine
Stanford, California

James R. Connor
Department of Neural and Behavioral
  Sciences
Pennsylvania State University
M.S. Hershey Medical Center
Hershey, Pennsylvania

Lucio G. Costa
Department of Environmental and
  Occupational Health Sciences
University of Washington
Seattle, Washington

Gabriele Dini
Cerebrovascular Research
Cleveland Clinic Foundation
Cleveland, Ohio

Stefano Ferroni
Department of Human and General
  Physiology
University of Bologna
Bologna, Italy

Vanessa A. Fitsanakis
Department of Physiology and Pharmacology
Wake Forest University School
  of Medicine
Winston-Salem, North Carolina

Rodrigo Franco
Departamento de Biofísica, Instituto
  de Fisiología Celular
Universidad Nacional Autónoma de México
México

Stephanie J. Garcia
Department of Physiology and Pharmacology
Wake Forest University School of Medicine
Winston-Salem, North Carolina

Rona Giffard
Department of Anesthesia and Department
  of Neurosurgery
Stanford University School of Medicine
Stanford, California

Tomás R. Guilarte
Department of Environmental Health Sciences
The Johns Hopkins University
Bloomberg School of Public Health
Baltimore, Maryland

Marina Guizzetti
Department of Environmental and Occupational Health Sciences
University of Washington
Seattle, Washington

Kerri L. Hallene
Cerebrovascular Research
Cleveland Clinic Foundation
Cleveland, Ohio

G. Jean Harry
Laboratory of Molecular Toxicology
National Institute of Environmental Health Sciences
National Institutes of Health
Department of Health and Human Services
Research Triangle Park, NC

Juan Hidalgo
Institute of Neurosciences and Department of Cellular Biology, Physiology and Immunology
Animal Physiology Unit
Faculty of Sciences
Autonomous University of Barcelona
Bellaterra, Barcelona, Spain

Damir Janigro
Cerebrovascular Research
Cleveland Clinic Foundation
Cleveland, Ohio

Kelly M. Kight
Cerebrovascular Research
Cleveland Clinic Foundation
Cleveland, Ohio

Rolf Knoth
Pathological Institute
Department of Neuropathology
Freiburg, Germany

Marina Marinovich
Department of Pharmacological Sciences
University of Milan
Milan, Italy

Ralf Peter Meyer
Pathological Institute
Department of Neuropathology
Freiburg, Germany

Alireza Minagar
Department of Neurology
Louisiana State University Health Sciences Center
Shreveport, Louisiana

Pierre Morell
Department of Biochemistry and Biophysics and Neuroscience Center
University of North Carolina
Chapel Hill, North Carolina

Takashi Mori
Discovery Research Laboratories III
Minase Research Institute
Ono Pharmaceutical Co. Ltd.
Osaka, Japan

Mark Noble
Department of Biomedical Genetics
University of Rochester Medical Center
Rochester, New York

Michael D. Norenberg
Department of Pathology and Department of Biochemistry and Molecular Biology
University of Miami School of Medicine and Veterans Affairs Medical Center
Miami, Florida

James P. O'Callaghan
Molecular Neurotoxicology Team
Toxicology and Molecular Biology Branch
Centers for Disease Control and Prevention-NIOSH
Morgantown, West Virginia

Herminia Pasantes-Morales
Departamento de Biofísica
Instituto de Fisiología Celular
Universidad Nacional Autónoma de México
México

Amanda D. Phelka
Toxicology Program
University of Michigan
Ann Arbor, Michigan

Martin A. Philbert
Toxicology Program
University of Michigan
Ann Arbor, Michigan

Yongchang Qian
Department of Veterinary Anatomy
  and Public Health
Texas A&M University
College of Veterinary Medicine and Center
  for Environmental and Rural Health
College Station, Texas

David E. Ray
MRC Applied Neuroscience Group
School of Biomedical Science
University of Nottingham Medical School
Nottingham, England

Arne Schousboe
Department of Pharmacology
Danish University of Pharmaceutical Sciences
Copenhagen, Denmark

Peter Schubert
Max Planck Institute of Neurobiology
Martinsried, Germany

Gouri Shanker
Department of Physiology and Pharmacology
Wake Forest University School of Medicine
Winston-Salem, North Carolina

Paul Shapshak
Department of Psychiatry and Behavioral
  Sciences, Department of Neurology,
  Department of Pathology, Department of
Pediatrics McDonald Foundation GeneTeam,
  and Comprehensive Drug Research Center
University of Miami School of Medicine
Miami, Florida

William F. Silverman
Department of Morphology
The Zlotowski Center for Neuroscience
Faculty of Health Sciences
Ben-Gurion University of the Negev
Beer Sheva, Israel

Offie P. Soldin
Department of Oncology
Division of Cancer Genetics and Epidemiology
Lombardi Cancer Center and Department of
  Medicine
Division of Endocrinology and Metabolism
Georgetown University School of Medicine
Washington, D.C.

Ursula Sonnewald
Department of Neuroscience
Norwegian University of Science and
  Technology
Trondheim, Norway

Krishnan Sriram
Molecular Neurotoxicology Team
Toxicology and Molecular Biology Branch
Centers for Disease Control and
  Prevention-NIOSH
Morgantown, West Virginia

Wolfgang J. Streit
Department of Neuroscience
University of Florida College of Medicine
Gainesville, Florida

Tore Syversen
Department of Clinical Neuroscience
Norwegian University of Science and
  Technology
Trondheim, Norway

Narito Tateishi
Discovery Research Laboratories III
Minase Research Institute
Ono Pharmaceutical Co. Ltd.
Osaka, Japan

Evelyn Tiffany-Castiglioni
Department of Veterinary Anatomy
  and Public Health
Texas A&M University
College of Veterinary Medicine and Center
  for Environmental and Rural Health
College Station, Texas

Arrel D. Toews
Department of Biochemistry and Biophysics
  and Neuroscience Center
University of North Carolina
Chapel Hill, North Carolina

Annabella Vitalone
Department of Pharmacology and Human
  Physiology
University of Bari Medical School
Bari, Italy

Barbara Viviani
Department of Pharmacological Sciences
University of Milan
Milan, Italy

Benedikt Volk
Pathological Institute
Department of Neuropathology
Freiburg, Germany

Helle S. Waagepetersen
Department of Pharmacology
Danish University of Pharmaceutical Sciences
Copenhagen, Denmark

E. Spencer Williams
Toxicology Program
University of Michigan
Ann Arbor, Michigan

Kevin Yagle
Department of Radiology
University of Washington
Seattle, Washington

Xuesheng Zhang
Department of Neural and Behavioral Sciences
Pennsylvania State University
M.S. Hershey Medical Center
Hershey, Pennsylvania

Contents

Neural Precursor Cells and Toxicant Action: Addressing the Analysis of Low-Level Toxicant Exposure

Mark Noble

CONTENTS

0-8493-1749-0/05/$0.00+$1.50

1.1 INTRODUCTION

It is becoming increasingly recognized that the understanding of developmental toxicology lies, at least in part, in identifying the adverse developmental effects of toxicant exposure on the stem cells and progenitor cells responsible for generating differentiated cell types. Such studies may have considerable relevance to understanding the mechanisms by which toxicants perturb normal development.

In this review, the analytical systems currently available for studying the biology of toxicant action on neural precursor cells (a generic term used to include both stem cells and lineage-restricted progenitor cells) will be presented, along with examples where the utilization of such systems has provided potentially useful insights. The study of the mechanisms that underlie normal and abnormal development of the CNS has been of major interest to a large number of laboratories, and this endeavor has been associated with the elaboration of a variety of systems for studying such problems at the cellular biological level.

A problem of particular concern in this review is that quite subtle effects on precursor cell function can have enormous developmental implications. The analysis of subtle effects may be particularly important to the study of low-level toxicant exposure, but it is nonetheless particularly problematic for *in vivo* research. For example, as will be discussed in more detail later, one of the central points at which metabolic processes, extracellular signaling molecules, and toxicants appear to converge is in modulating the balance between self-renewing division and differentiation in dividing precursor cells. Our studies on clonal cell families *in vitro* have demonstrated that a change from 0.5 to 0.65 in the probability of exiting the cell cycle (i.e., turning into a nondividing terminally differentiated cell) is associated with >60% reductions in progenitor cell number over 6 days. Were similar levels of change in this vital balance to underlie, for example, the reduced birth size that is often associated with toxicant exposure, this would be very difficult to study *in vivo*. For example, consider a rapidly growing tissue in which 5% of cells normally are labeled with a 2-h pulse of 5-bromodeoxyuridine (BrdU), and 60% of the cells in a tissue undergo division over a 24-h time period. In experimental animals, in which the probability of exiting the cell cycle has been increased from 0.5 to 0.65 by exposure to low levels of a chemical of interest, labeling with BrdU over a 2-hour time period would lead to a difference in labeling of 5% of cells versus 4.25% of cells. Even over 24 h of labeling, this difference would only improve to 60% versus 51%. Such differences would be lost in the background of variation between animals. Extending the *in vivo* labeling period to several days, if this could be done without compromising normal development, would not lead to greater insight, as the prolonged labeling would lead to the labeling of all dividing cells. As another example, similarly dramatic changes could result from simply increasing by 5–10% the levels of programmed cell death that occur during normal development. A graphic demonstration of the profound effect on cell generation associated with increasing the amount of cell death at any given division from just 25% of cells to 27.5% (a change that could not be detected by any existing analytical approaches) is shown in Figure 1.1. Yet, as for the analysis of cell division, it is not clear whether any existing analytical approaches would allow such alterations to be detected by *in vivo* analysis.

This chapter first reviews our current understanding of CNS development from the neuroepithelial stem cell to the differentiated cells of the CNS, with particular emphasis on the generation of oligodendrocytes (the myelinating cells of the CNS). Generation of oligodendrocytes is compromised by a wide variety of physiological insults, leading to defects in myelination *in vivo*; such defects prevent normal impulse conduction and can be neurologically as devastating as neuronal loss. Moreover, the generation of oligodendrocytes is understood in more detail than is the case for any CNS cell type, and provides an important paradigm system for analyzing multiple biological problems. Our current understanding of the effects of intracellular redox state on modulating cellular function will then be considered as an example of the ability of the approaches described to analyze relatively subtle changes in cell function. We next will consider the importance of syndromes in

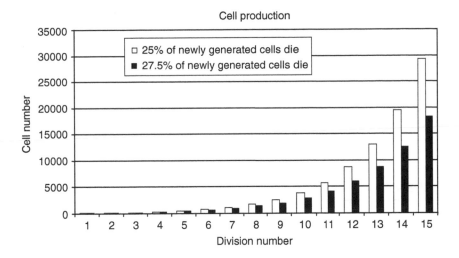

Figure 1.1 A 10% increase in cell death for 15 cell generations is associated with a 38% reduction in cell number. This graph is provided to illustrate the profound effects associated with relatively small changes in cell generation. To generate this graph, the assumption first was made that at every cell division 25% of newly generated cells die, and the total number of cells produced over 15 generations was determined (white bars). The cell death profile was then raised by just 10%, so that 27.5% of cells would die over 15 cell generations (black bars). Thus, this would be an example of the theoretical consequences of exposure to "subclinical" toxicant levels for a period of less than 2 weeks (even allowing for a cell cycle time of close to 24 hours, which is considerably longer than that thought to apply during embryogenesis). Similarly large changes can occur by lengthening the cell cycle, increasing the probability of differentiation occurring at each division (thus removing dividing precursor cells from the population). None of these changes could be detected by acute analysis of markers of cell death (TUNEL staining) or cell division (BrdU incorporation).

which sublethal effects on precursor cell function have profound developmental effects, after which effects of toxicants on cells of the oligodendrocyte lineage — and also on neuronal lineages — will be considered. Finally, consideration will be given to the biological and mathematical problems involved in bridging the gap between *in vitro* and *in vivo* studies, so as to improve our abilities to study developmental problems associated with exposure to levels of toxicants encountered by many millions of people around the world.

1.2 GENERAL CONCEPTS: PURIFIED POPULATIONS AND CLONAL ANALYSES

Two of the most important tools in studying precursor cell function are purified populations of primary precursor cells and the ability to grow and analyze cells at clonal density. Each of these tools provides different advantages, with the combination of the two being particularly powerful.

The study of purified populations of primary precursor cells allows ready determination of whether particular effects are mediated directly on the population of interest, or instead represent toxicities mediated through effects on a separate cell type. For example, a toxicant that causes death of neurons in the brain, or in mixed cultures of brain cells, could exert its activity by killing the neurons themselves. It is also possible, however, that the toxicant might kill neurons by causing other cells of the CNS to release neurotoxic substances. That such a possibility can occur is indicated by studies demonstrating that exposure of astrocytes to methylmercury can cause the release of glutamate,[5,138] thus leading to neuronal death. In the case where neuron is the direct target of action, the goal would be to discover means of protecting neurons from the toxicant. In the case of indirect action, in contrast, the goal instead would be to block the effects of glutamate, or to protect the astrocytes from the effects of methylmercury. Indirect toxicity could also be mediated, for example, by activation of macrophages or microglia, causing them to release the agents of destruction that typify their biology. Determining the target of toxicant action requires the analysis of purified cell populations, with mixed or reconstituted cultures then being used to bring the experimental system closer to the complexity of the CNS itself.

Clonal analysis is of great importance for multiple reasons. In the analysis of precursor cell differentiation, it is clonal analysis that allows one to determine the full differentiation potential of individual cells and to determine whether a truly homogeneous cell population has been isolated. As shall be shown, clonal analysis also enables the detailed analysis of subtle changes in precursor cell division and differentiation, thus facilitating the study of toxicant effects at very low exposure levels.

1.3 CURRENT KNOWLEDGE OF CNS PRECURSOR CELL BIOLOGY ALLOWS A DETAILED ANALYSIS OF NEUROTOXICITY OF TOXICANTS

1.3.1 Overview

Recent advances in identifying the multiple precursor cell populations (Figure 1.2) that contribute to development of the nervous system have greatly altered our abilities to dissect the biology of

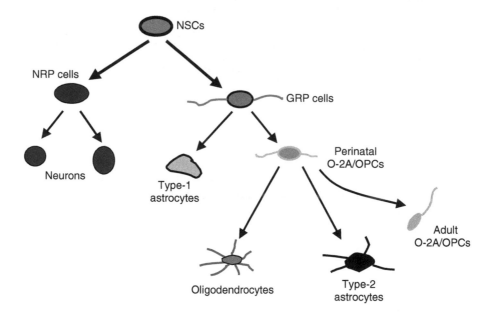

Figure 1.2 Diagrammatic representation of the lineages involved in CNS development. Neuroepithelial stem cells (NSCs) give rise to neuron-restricted precursor (NRP) cells and glial-restricted precursor (GRP) cells shortly after neural tube formation.[122,168–170] At later developmental stages, GRP cells give rise to oligodendrocyte-type-2 astrocyte progenitor cells (also referred to as oligodendrocyte precursor cells, and here abbreviated as O-2A/OPCs).[81,167] Perinatal O-2A/OPCs give rise to a further specialized progenitor cell, the O-2A/OPC of the adult CNS.[214] As discussed in the text, *adult* O-2A/OPCs differ from their perinatal counterparts in multiple aspects consistent with the reduced generation or replacement of myelin associated with aging. This diagram includes only cells that have been isolated directly from the CNS, have been studied at the clonal level *in vitro* to determine their differentiation potential, and (except for the *adult* O-2A/OPC) have been transplanted back in the animal and found to exhibit the same differentiation potentials *in vitro* and *in vivo*. The literature also suggests the existence of a specialized astrocyte progenitor cell, which may be derived from GRP cells.[110] Some studies also suggest the existence of specialized neuron-oligodendrocyte progenitors in the spinal cord and cortex, but whether such cells actually exist or may instead represent NSCs is still a matter of discussion.[143,144,179]

developmental neurotoxicity. Multiple cellular lineages can be isolated as purified populations, expanded *in vitro*, and used to determine whether they are directly sensitive to the action of any given compound of interest. Synergy between toxicants and other compounds can also be readily studied, as can protective strategies. In addition, availability of cell-type specific markers suitable for staining sectioned material enable *in vivo* testing of many of the hypotheses formulated on the basis of studies initiated *in vitro*.

The developmental history of the oligodendrocyte is particularly well understood and provides an important paradigm system for identifying the multiple points at which toxicants may act to alter normal development. The key turning point in cellular biological analysis of the origin of oligodendrocytes came with the identification of a bipotential progenitor cell in cultures of optic nerve[167] that can differentiate *in vitro* into an oligodendrocyte or into a particular kind of astrocyte (called the type-2 astrocyte[166]). Numerous studies also have confirmed the ability of these precursor cells to generate oligodendrocytes following transplantation,[83,208,212] although it remains unknown whether there are circumstances in which the potential of these cells to generate astrocytes is utilized *in vivo*.[62,99] This uncertainty led to the use of two names for these cells, one being the oligodendrocyte-type-2 astrocyte (O-2A) progenitor cell (to reflect the cells' developmental potential) and the other name being the oligodendrocyte precursor cell (OPC, to reflect the clear importance of this cell in oligodendrocyte generation). As both names are reflective of important properties of this cell, we use the combined abbreviation of O-2A/OPC. O-2A/OPCs have been isolated from many different regions of the CNS and from multiple species.

Cellular biological studies on the earliest events in spinal cord development have led to the identification of a second glial precursor cell that may be a crucial ancestor of the oligodendrocyte. These studies originated with observations that neuroepithelial stem cells generate two antigenically distinct populations of lineage-restricted precursor cells *in vitro*, each restricted to the generation of either neurons or glia. Glial-restricted precursor (GRP) cells[168] were labeled with the A2B5 monoclonal antibody and were shown to give rise to oligodendrocytes and two antigenically distinct populations of astrocytes (corresponding to previous descriptions of type-1 and type-2 astrocytes). Neuron-restricted precursor (NRP) cells, in contrast, expressed the polysialylated form of the neural cell adhesion molecule (PSA-NCAM) and were shown to give rise to multiple different kinds of neurons and not to glia.[122]

Both GRP cells and NRP cells can be directly isolated from the developing rat spinal cord and grown as purified populations.[122,169] Freshly isolated cells exhibit the same lineage restrictions as those cells derived from neuroepithelial stem cells *in vitro*. Clonal studies have demonstrated that GRP cells retain their tripotential nature even after weeks of *in vitro* expansion and several serial reclonings[169] and also exhibit these same restrictions following transplantation *in vivo*. GRP cells generate both oligodendrocytes and astrocytes following transplantation into brain or spinal cord and do not generate neurons even when they migrate into such neurogenic zones as the rostral migratory stream and olfactory bulb.[89] NRP cells, in contrast, generate only neurons (including motor neurons), even upon transplantation into such CNS regions as the adult spinal cord.[28,87,105] This generation of neurons by NRP cells following transplantation into the adult CNS is particularly striking in light of multiple studies indicating that NSCs in similar transplants generate astrocytes or do not differentiate at all[29] unless transplantation is delayed for a week or more after injury.[146]

GRP cells differ from O-2A/OPCs in multiple ways.[144,169] Freshly isolated GRP cells from the E13.5 rat spinal cord are dependent upon exposure to FGF-2 for both their survival and their division, while division and survival of O-2A/OPCs can be promoted by PDGF and other chemokines. Consistent with this difference in chemokine-response patterns, GRP cells freshly isolated from the E13.5 spinal cord do not express receptors for PDGF (although they do express such receptors with continued growth *in vitro* or *in vivo*, as discussed later). These populations also differ in their response to inducers of differentiation. For example, exposure of GRP cells to the combination of FGF-2 and ciliary neurotrophic factor (CNTF) induces these cells to differentiate into astrocytes (primarily expressing the antigenic phenotype of type-2 astrocytes[169]). In contrast, exposure of O-2A/OPCs to FGF-2 + CNTF promotes the generation of oligodendrocytes.[17,118,120] Moreover, the behavior of these two precursor cell populations following transplantation is strikingly different. The ability of GRP cells to readily generate astrocytes following transplantation into the adult CNS[89] stands in striking contrast to the behavior of primary O-2A/OPCs, which thus far only generate oligodendrocytes in such transplantations[62] (although it has been reported that O-2A/OPC cell lines will generate astrocytes if transplanted in similar circumstances[69]).

Antigenic and *in situ* analysis of development *in vivo* has confirmed that cells with the A2B5+ antigenic phenotype of GRP cells arise in spinal development several days prior to the appearance of GFAP-expressing astrocytes, and also prior to the appearance of cells expressing markers of radial glia.[110] Thus, these cells can be isolated directly from the developing spinal cord, and cells with the appropriate antigenic phenotype have been found to exist *in vivo* at appropriate ages to play important roles in gliogenesis.

Thus far, analysis of A2B5+ cells isolated from the early embryonic spinal cord reveals a great degree of homogeneity in their ability to generate oligodendrocytes and type-1 and type-2 astrocytes *in vitro*.[81,169] In addition, GRP cells have been isolated from multiple species and by multiple means. For example, such cells have been isolated from the rat spinal cord, the mouse spinal cord, and from murine embryonic stem cells.[137] In addition, A2B5+ precursor cells restricted to the generation of astrocytes and oligodendrocytes have been derived from cultures of human embryonic brain cells.[54] Both mouse and human cells share the ability of rat GRP cells to generate oligodendrocytes and more than one antigenically defined population of astrocytes.

1.3.2 The GRP Cell as an Ancestor of the O-2A/OPC

A number of questions arise from the fact that it is possible to isolate two distinct precursor cell populations (i.e., GRP cells and O-2A/OPCs) from the developing animal, each of which can generate oligodendrocytes. Is the relationship between these two populations one of lineage restriction or lineage convergence? If GRP cells and O-2A/OPCs are related, what signals promote the generation of one from the other and how can the existence of both populations be integrated with existing studies on the generation of oligodendrocytes during spinal cord development?

In vitro studies have demonstrated that GRP cells can give rise to O-2A/OPCs if exposed to particular signaling molecules. In these experiments, cultures of GRP cells derived from E13.5 rats were grown in conditions known to induce the generation of oligodendrocytes.[81] At the initiation of these experiments, no cells in the GRP cell cultures were labeled with the O4 antibody, which can be used to recognize O-2A/OPCs at a stage of development at which the generation of both oligodendrocytes and type-2 astrocytes *in vitro* is possible.[14,84,207] When GRP cells that originally had been grown in the presence of FGF for several days were exposed to a combination of PDGF and thyroid hormone (TH), however, O4+ cells were generated in the cultures. Purification of cells that were O4+ but did not express galactocerebroside (GalC, a marker of oligodendrocytes), and subsequent examination of the differentiation potential of these cells at the clonal level, confirmed that they behaved like O-2A/OPCs rather than like GRP cells.[81] When O4+GalC- cells were grown in conditions that induced the generation of astrocytes, the resulting clones contained only type-2 astrocytes. In contrast, O4-negative cells derived from the GRP cell cultures cells behaved as did freshly isolated GRP cells in these conditions and generated clones containing both type-1 and type-2 astrocytes. Moreover, we could find no GalC+O4- oligodendrocytes in any conditions, which would have at least raised the possibility that oligodendrocytes might be generated directly from GRP cells. Such results are consistent with previous observations that passage through an O4+/GalC- stage of development is required for oligodendrocyte generation from bipotential O-2A/OPCs.[72–74]

At present, the simplest model of oligodendrocyte generation that appears to be consistent with the data discussed thus far would be that production of these cells requires the initial generation of GRP cells from NSCs followed by the generation of O-2A/OPCs from GRP cells.[81,144] Previous studies[168,169] indicated strongly that GRP cells are a necessary intermediate between NSCs and differentiated glia, and our subsequent experiments raise the possibility that O-2A/OPCs are a necessary intermediate between GRP cells and oligodendrocytes, at least in the developing spinal cord.[81] It is important to note, however, that some investigators have suggested that there may also (or instead) be another oligodendrocyte precursor cell that has the ability to generate motor neurons rather than astrocytes. The idea that motor neurons and oligodendrocytes might be developmentally related was particularly strongly suggested by findings that compromising the function of members of the *Olig* gene family can prevent the generation of both motor neurons and oligodendrocytes.[111,199,220] These studies have been discussed in detail in a number of detailed recent reviews.[144,174,179,186] The results of these studies are also consistent with the existence of two separate precursor pools, such as NRP cells and GRP cells, in each of which *Olig* gene expression plays distinct roles in the promotion of particular cell fates.[144,174,179,199,220]

1.3.3 Regional Differences in Precursor Cell Biology

One of the important lessons learned from studies of oligodendrocyte development is that O-2A/OPCs from different regions of the CNS may have very different properties, which appear to be associated with the physiological requirements of the tissue from which they are inhibited. This problem has been most thoroughly studied in the optic nerve and the cortex, in studies that initially were designed to elucidate the biological mechanisms underlying the differing developmental time courses that characterize different regions of the CNS. For example, neuron production in the rat spinal cord is largely complete by the time of birth, is still ongoing in the rat cerebellum for at least several days after birth, and continues in the olfactory system and in some regions of the hippocampus of multiple species throughout life. Similarly, myelination has long been known to progress in a rostral-caudal direction, beginning in the spinal cord significantly earlier than in the brain.[68,98,113] Even within a single CNS region, myelination is not synchronous. In the rat optic nerve, for example, myelinogenesis occurs with a retinal-to-chiasmal gradient, with regions of the nerve nearest the retina becoming myelinated first.[68,187] The cortex itself shows the widest range of timing for myelination, both initiating later than many other CNS regions[68,98,113] and exhibiting an ongoing myelinogenesis that can extend over long periods of time. This latter characteristic is seen perhaps most dramatically in the human brain, for which it has been suggested that myelination may not be complete until after several decades of life.[21,217]

Variant time courses of development in different CNS regions could be due to two fundamentally different reasons. One possibility is that precursor cells are sufficiently plastic in their developmental programs that local differences in exposure to modulators of division and differentiation may account for these variances. Alternatively, it may be that the precursor cells resident in particular tissues express differing biological properties related to the timing of development in the tissue to which they contribute.

1.3.3.1 *Cortical O-2A/OPCs Appear to Be Specialized to Undergo Extended Self-Renewal*

Comparison of A2B5[+] populations derived from cortex, optic nerve, and optic chiasm of P7 rats revealed that they all were similar in their lineage restriction but differed strikingly in their relative probabilities of undergoing division or differentiation in a variety of conditions. All of these populations shared the well-characterized differentiation phenotype of the optic nerve-derived cells, which is the ability to generate oligodendrocytes and type-2 astrocytes (but not type-1 astrocytes) *in vivo*.[161] These O-2A/OPCs are referred to in the following sections of this review by indicating the tissue of derivation by adding (CX) for cortex and (ON) for optic nerve.

Analysis of clonal growth of cells grown in chemically defined medium supplemented with platelet-derived growth factor (PDGF), but lacking thyroid hormone (TH), revealed that cortex-derived

O-2A/OPCs exhibited a much greater tendency to undergo continued self-renewal than did their counterparts from optic nerve or optic chiasm. For example, more than half of clones of O-2A/OPCs(ON) contained at least one oligodendrocyte after 3 days of *in vivo* growth in these conditions, and this proportion increased to include almost all clones after 7 days of *in vitro* growth. The percentage of cells in these cultures that were oligodendrocytes increased from $21 \pm 11\%$ on day 3 to more than half on day 7. In contrast, in clones of O-2A/OPCs(CX) grown in identical conditions, only very few clones contained at least one oligodendrocyte after 3 days of *in vitro* growth, and this proportion increased only marginally after 7 days of *in vitro* growth. Even after 10 days of *in vitro* growth, less than $20 \pm 1\%$ of O-2A/OPC(CX) clones grown in PDGF contained one or more oligodendrocytes. Similarly, the overall percentage of oligodendrocytes was markedly lower in O-2A/OPC(CX) clones than in clones of O-2A/OPCs(ON) at all time points analyzed. The percentage of oligodendrocytes seen in clonal cultures of O-2A/OPCs(CX) was < 2% on days 3, 7, and 10 in these basal division conditions. Moreover, these clones rarely contained more than one to two oligodendrocytes regardless of the number of O-2A/OPCs found within the clone.

The ability of O-2A/OPCs(CX) to undergo extended self-renewal when exposed to PDGF was associated with the generation of large clonal sizes and with division that continued for several weeks of *in vitro* growth. O-2A/OPCs(CX) isolated as single cells and grown in the presence of PDGF for 10 days generated clones with average sizes of 100 ± 16 cells/clone, a threefold expansion in clonal size over that observed on day 7. In contrast, clones derived from optic nerve did not exhibit significant expansion in their numbers after day 7, at which time the average clonal size was 7 ± 4 cells. Moreover, O-2A/OPCs(CX) cells were capable of dividing for more than 6 weeks when exposed continuously to PDGF, a continuation of division that we have never observed in cultures of O-2A/OPCs(ON) grown in these conditions.

The remarkable ability of cortex-derived O-2A/OPCs to undergo continued self-renewal when exposed only to PDGF as a mitogen is also strikingly different from observations in the spinal cord. In this latter tissue, it appears that promotion of extended division of perinatal-derived O-2A/OPCs requires exposure not just to PDGF but also to additional cytokines. In particular, the combination of PDGF and the chemokine growth-regulated oncogene-alpha (GRO-α; CXCL1) has been found to promote division of spinal cord-derived O-2A/OPCs, while PDGF applied on its own is relatively ineffective[175] (unpublished observations).

1.3.3.2 Cortex-Derived O-2A/OPCs Exhibit Reduced Responsiveness to Inducers of Oligodendrocyte Generation

The enhanced self-renewal capacity of cortex-derived O-2A/OPCs was mirrored by a lessened responsiveness of these cells to two well-characterized inducers of oligodendrocyte generation, triiodothyronine (T3) and ciliary neurotrophic factor (CNTF).[18,91,120] Even when grown in the presence of T3, cultures of O-2A/OPCs(CX) were far less likely to exhibit extensive differentiation than was seen in the optic nerve–derived cultures.[161] For example, after 7 days of growth in the presence of T3, only a small minority of O-2A/OPC lineage cells in O-2A/OPC(CX) cultures were oligodendrocytes, compared with the great majority being oligodendrocytes in comparable O-2A/OPC(ON) cultures. It was not, however, that O-2A/OPCs(CX) were incapable of responding to T3. In fact, exposure of O-2A/OPCs(CX) to T3 for 7 days was associated with an eightfold increase in the percentage of clones containing only oligodendrocytes (2% vs. 16%) and a 3.5-fold increase in the number of clones containing at least one oligodendrocyte (16% vs. 58%). After 10 days of *in vitro* growth in the presence of T3, $93 \pm 2\%$ of the clones contained at least one oligodendrocyte, thus confirming that the overwhelming majority of these clones were capable of generating oligodendrocytes. Nonetheless, when compared directly with the extent of differentiation seen in cultures of O-2A/OPCs(ON) grown in these conditions, it was apparent that cortex-derived cells were much more refractory to the induction of differentiation. A similarly reduced induction of differentiation in cortex-derived O-2A/ OPCs was observed in response to CNTF.

1.3.3.3 Cortex-Derived O-2A/OPCs Appear to Be Intrinsically Different from Cells from Optic Nerve or Chiasm

Differences between cortex-, optic nerve–, and optic chiasm–derived O-2A/OPCs appear to be cell-intrinsic.[161] We first tested whether O-2A/OPCs(CX) were secreting soluble factors that promoted self-renewal by growing O-2A/OPCs(ON) in the presence of O-2A/OPCs(CX) and examining the generation of oligodendrocytes by the O-2A/OPCs(ON). In converse experiments, we determined whether soluble factor(s) secreted by O-2A/OPCs(ON) could enhance oligodendrocyte generation by O-2A/OPCs(CX). Cells were plated in ratios of 1:20, with the population to be examined in a smaller dot surrounded by a larger ring of the putative producers of modulatory factors. As controls, we cultured both the O-2A/OPCs(CX) and the O-2A/OPCs(ON) in the presence of other cells of the same type. To determine whether coculture might alter the ability of one population to influence the behavior of the other population, we also grew the cells in medium conditioned by either progenitor population in the absence of coculture. We found that the likelihood of O-2A/OPCs(CX) and O-2A/OPCs(ON) undergoing either self-renewal or differentiation into oligodendrocytes was unchanged by exposure to medium conditioned by the other cell type. Similar results to those in the above coculture experiments were obtained when O-2A/OPCs(ON) or O-2A/OPCs(OC) were exposed to conditioned medium from either cells from the same or the comparative tissue.

1.3.3.4 Developmental Relevance of Regional and Age-Related Differences of O-2A/OPCs

The varied properties we observed in different O-2A/OPC populations could theoretically represent a developmental progression, for which the phenotype of O-2A/OPC(ON) cells represents the most mature pattern of behavior. While it is difficult to rule out this possibility, some observations suggest that, at least for O-2A/OPC(CX) cells, this may not be true. As all cells were isolated from animals of the same age, invoking a developmental progression would require positing a different timing of this progression in each tissue, which would still make these populations biologically different from each other. Indeed, the fact that O-2A/OPCs(CX) continue to express their characteristic potential for continuous and extended self-renewal even after 6 weeks of *in vitro* growth suggests that if such a transition occurs, it may occur over quite a long time frame. In addition, O-2A/OPCs(CX) derived from P13 cortex were still far more prone to undergo self-renewal than O-2A/OPCs(ON) isolated from P7 animals. For example, less than 20% of O-2A/OPC(CX) clones derived from P13 rats contained one or more oligodendrocytes after 7 days of *in vitro* growth in the basal division conditions as compared with a value of 94% for O-2A/OPC(ON) clones from P7 rats. The proportion of P13-derived O-2A/OPC(CX) clones containing at least one oligodendrocyte was increased to 67% in the presence of T3, but even in these conditions only 25% of the cells in the cultures actually differentiated into oligodendrocytes after 7 days (compared with a value of 56% for O-2A/OPC[ON] cells from P7 rats). Thus, it is possible that the cortex-derived cells are continuously different from their counterparts isolated from other regions of the CNS.

The differences that distinguish O-2A/OPCs (ON), O-2A/OPCs (OC), and O-2A/OPCs (CX) appear to be reflective of differing physiological requirements of the tissues to which these cells contribute. For example, a variety of experiments have indicated that the O-2A/OPC population of the optic nerve arises from a germinal zone located in or near the optic chiasm and enters the nerve by migration.[147,189] Thus, it would not be surprising if progenitor cells of the optic chiasm expressed properties expected of cells at a potentially earlier developmental stage than those cells that are isolated from optic nerve of the same physiological age. Such properties would be expected to include the capacity to undergo a greater extent of self-renewal, much as has been seen when the properties of O-2A/OPCs from optic nerves of embryonic rats and postnatal rats have been compared.[71] With respect to the properties of cortical progenitor cells, physiological considerations also appear to be consistent with our observations. The cortex is one of the last regions of the CNS in which myelination is initiated, and the process of myelination also can continue for extended periods in this region[68,98,113] If the biology of a precursor cell population is reflective of the developmental characteristics of the tissue in which it resides, then one might expect that O-2A/OPCs isolated from this tissue would not initiate oligodendrocyte generation until a later time than occurs with O-2A/OPCs isolated from structures in which myelination occurs earlier. In addition, cortical O-2A/OPCs might be physiologically required to make oligodendrocytes for a longer time due to the long period of continued development in this tissue, at least as this has been defined in the human CNS.[21,217]

1.4 INTRACELLULAR REDOX STATE AS A MODULATOR OF PRECURSOR CELL FUNCTION

One of the outcomes of developmental studies that sheds light on the regional differences between O-2A/OPCs, and which is perhaps the greatest relevance to the analysis of toxicant action, is the sensitivity of precursor cell function to relatively metabolic changes. Of particular concern in this regard are the effects of intracellular redox state on modulating the balance between division and differentiation.

Several lines of evidence indicate that regulation of intracellular redox state is a critical modulator of precursor cell function. There is a direct correlation between the redox state of precursor cells at the time of their isolation from the organism and their tendency to either divide or differentiate when induced to divide *in vitro*.[145,161,190] O-2A/OPCs that are more reduced *in vivo* undergo a great deal more self-renewal *in vitro* than those that are oxidized at the time of their isolation.[190] Indeed, it appears that these differences in intracellular redox state may be mechanistically relevant to the already discussed differences between O-2A/OPCs isolated from different regions of the developing CNS, such as the cortex and optic nerve. O-2A/OPCs isolated from the cortex, which exhibit much greater tendencies to undergo continued division *in vitro* than optic nerve cells, are more reduced at the time of isolation from the animal than are cells isolated from optic nerve.[161]

Not only does it appear that the organism utilizes intracellular redox state to modulate precursor cell function, but it also seems that modulation of this state is a point of convergence for the action of extracellular signaling molecules. In our own studies, we found that growth factors that enhance self-renewal made O-2A/OPCs more reduced, while those that promote differentiation make cells more oxidized. Moreover these changes in intracellular redox state appear to be essential components of the means by which such factors alter the balance between self-renewal and differentiation,[190] and pharmacological interference with the redox alterations induced by these signaling molecules effectively abrogated their effects. Pharmacological manipulation of redox state also alters the balance between division and differentiation, with agents that make cells more reduced enhancing self-renewing division and sublethal doses of pro-oxidants inducing differentiation.

Intracellular redox balance also plays a critical role in modulating the responsiveness of cells to a variety of signaling molecules. For example, growth of oligodendrocytes in the presence of antioxidants or cysteine prodrugs that can enhance production of glutathione (the major antioxidant found within cells) dramatically enhances responsiveness to promoters of cell survival.[121] In contrast, reducing glutathione content of lymphocytes (for example, by growth in the presence of buthionine sulfoximine, which inhibits glutathione production) reduces their response to mitogens and other activators.[65,66,195]

It is well established that exposure to many substances of interest in this review is associated with changes in intracellular redox state, with oxidative stress being associated with exposure to, e.g., alcohol,[124] glutamate,[7] lead,[56,90] mercury,[183,218] and multiple other environmental toxicants.[10,200] The redox challenges associated with many physiological stressors may be of particular importance in understanding the effects of low levels of toxicant exposure, as even small changes in intracellular redox state may dramatically alter cellular function. For example, as little as a 10% decrease in intracellular levels of glutathione (a major regulator of intracellular redox balance) significantly decreases calcium influx in peripheral blood lymphocytes stimulated with anti-CD3 antibody.[194] Similarly, levels of pharmacological agents that cause 10–15% changes in average redox-based fluorescence values are sufficient to alter the balance between self-renewal and differentiation in dividing O-2A/OPCs.[190] These small changes are readily achieved with exposure of O-2A/OPCs to MeHg in concentrations of less than 5 parts per billion (MN et al., unpublished observations) and are consistent with the later discussed effects of MeHg and Pb on O-2A/OPC function.

1.5 THE IMPORTANCE OF SUBLETHAL CHANGES IN PRECURSOR CELL FUNCTION IN DEVELOPMENTAL MALADIES

One of the critical realizations to emerge from developmental biology has been the recognition that reductions in the number of differentiated cells may occur for other reasons than cell death. Induction of premature differentiation of precursor cells (thus removing them from the mitotic cycle), or the absence of a substance required for their differentiation, both may lead to a relative paucity of differentiated cells. Such processes appear to be of profound importance in multiple developmental maladies.

One example of a syndrome in which sublethal alterations in precursor cell function appears to be of central importance is fetal and early postnatal hypothyroidism. Deficiencies in thyroid hormone levels, usually associated with iodine deficiency, are a major cause of mental retardation, myelination failure, and other developmental disorders.[51,103] Children born to mothers with reduced thyroid hormone (TH) levels during early stages of pregnancy perform poorly on later tests of neurological and cognitive function.[85,116,160] In addition, the thyroid status of neonates and children has a significant long-term impact on their behavior, locomotor ability, speech, hearing, and cognition.[104] Hypothyroidism-associated deficiencies of myelination have been observed in the cerebral cortex, visual and auditory cortex, hippocampus, and cerebellum.[11,12,177] In addition, TH can modulate such processes as dendritic and axonal growth, synaptogenesis, and neuronal migration.[58,60,61] Hypothyroidism in developing rats causes a retarded elaboration of the neuropil in the cerebral cortex and adversely affects cerebellar Purkinje cells. Neuronal bodies are smaller and more densely packed, with diminished dendritic branching and elongation and altered distribution of dendritic spines.[142] Axonal density is decreased to such an extent as to reduce by as much as 80% the probability of appropriate axodendritic interaction,[59] a reduction that would greatly reduce the complexity of developing neural networks.

Hypothyroidism provides an example of a syndrome in which the ability of precursor cells to differentiate normally is compromised. O-2A/OPCs cultured in the absence of TH generate only a fraction of the oligodendrocytes found when TH is present.[18,70,91,190] In vivo, the number of oligodendrocytes found in optic nerves of young postnatal hypothyroid rats is reduced by about 80%, the same level of reduction observed in vitro when cells are grown in the absence of TH.[3,91] Earlier points in the developmental sequence leading to oligodendrocytes may also be compromised, as TH also promotes the generation of O-2A/OPCs from GRP cells, at least in vitro.[81] Reductions in oligodendrocyte number associated with a failure of oligodendrocyte generation in hypothyroid animals may be further compounded by effects of TH on oligodendrocytes themselves, as TH promotes oligodendrocyte-specific gene expression[92,158,159,205] and possibly also survival.[95] The clearest case for the generation of myelination defects involves, however, the inhibition of normal oligodendrocyte generation.

Studies on the effects of TH on neuronal precursor cells are less detailed than studies on glial precursors, but studies in the cerebellum indicate that TH deficiency is associated with abnormalities in migration and differentiation of granule neuron precursor cells.[148] In hypothyroid animals, the external granular layer (the habitat of migrating and dividing granule neuron precursor cells) persists significantly longer than it does in normal animals, consistent with a failure to induce differentiation in an appropriately timely manner.

1.6 TOXICANT EXPOSURE AND THE DEVELOPING CNS

The information discussed thus far sets the stage for toxicant analysis by providing analytical systems and two important paradigms. The first paradigm is that sublethal alterations in precursor cell function may be as dramatic in their effects on development as the induction of cell death. This is not to say that cell death is unimportant. The salient point rather is that it is critical to consider whether conditions in which cell death occur represent only the tip of the iceberg of important damage, with the physiological relevance of more lesser exposure levels being found in sublethal events. The second critical paradigm is that sublethal exposure to compounds that make cells more oxidized can induce premature differentiation. As discussed in the introduction, inducing only slight changes in the probability that a precursor cell will exit the cell cycle to generate a nondividing differentiated cell type can dramatically alter the total number of cells produced in a population. While most studies on toxicant-associated alterations have focused on the induction of cell death, there is already sufficient evidence to indicate that sublethal effects are frequent and are deserving of much greater attention than has thus far been the case.

1.6.1   Alcohol and Nicotine

Two syndromes in which both sublethal and lethal damage to immature cells and precursor cells appear to be important are caused by perinatal exposure to alcohol and nicotine. Fetal alcohol syndrome is the leading known cause of mental retardation in the United States,[2] while maternal smoking substantially increases the risk of learning disabilities, behavioral problems, and attention deficit/hyperactivity disorder in the offspring.[19,55,141,188] Fetal exposure to ethanol in rats and humans is associated with myelination defects;[150,155,157,191] with neuron loss in the hippocampus, the principal sensory nucleus of the trigeminal nerve,[130,131,206] and among Purkinje cells of the cerebellum;[108,117] and with smaller brain size and thickness of the cerebral cortex.[1,129] Exposure to nicotine during prenatal development causes extensive cytotoxicity, as evidenced by cytoplasmic vacuolation, enlargement of intercellular spaces, and a sharply increased incidence of pyknotic/apoptotic cells in the forebrain, midbrain, and hindbrain.[188]

Ethanol and nicotine disrupt multiple developmental processes involving both precursor cells and newly generated differentiated cells. Ethanol inhibits proliferation and migration of neuronal precursor cells and disturbs several critical steps in neocortical gliogenesis.[82,93,135] Ethanol also blocks the function of signaling pathways involved in neuroepithelial and neuronal precursor cell division, leading to both reduced cell division and cell death.[106,112] Prenatal animals exposed to alcohol show altered cell generation, proliferation, migration, and also altered development of dendritic arbors.[127,128,132–134,165] In addition, ethanol reduces oligodendrocyte generation, may alter synthesis of myelin lipids,[164] and is associated with delays in oligodendrocyte maturation *in vitro*.[33,50] Ethanol exposure during the prenatal and early postnatal period also triggers widespread apoptosis of early differentiating neurons in the developing rat forebrain,[93] with associated cell loss and synaptic abnormalities.[188] Effects of nicotine on neuroepithelial cell development, and killing of hippocampal progenitor cells *in vitro*, have been reported at nicotine dosages not greatly different from the plasma concentration of nicotine in smokers or transdermal patch users.[22,96,107]

1.6.2   Toxicant Exposure

Environmental chemicals have long been known to disrupt normal CNS development.[196] For example, long-term exposure to lead (Pb) during neurodevelopmental periods may cause substantial neurological-cognitive deficits, with the developing CNS being more sensitive than the mature CNS to this metal.[13,38,41,76,77,172] Mercury also is capable of adversely affecting development,[196] and recent studies suggest that at least 7% of U.S. women of reproductive age, as well as 20%

of 3- to 6-year-old children, exceed the exposure limit currently thought to be safe.[39,40] Exposure to polychlorinated biphenyls (PCBs) is another important cause of such neurological problems as cognitive impairment, developmental delays, and adverse behavioral and emotional effects.[196] It has been a consistent theme in toxicant research that new studies demonstrate that exposure levels previously thought to be safe are in fact associated with serious adverse effects.[196] Thus, it is of great importance to understand the ways in which exposure to low doses of toxicants may perturb normal development.

Recent studies indicate that low-dose exposure to toxicants may be another cause of sublethal alterations in precursor cell function. For example, concentrations of Pb as low as 1 μM compromise both survival and division of O-2A/OPCs and also compromise oligodendrocyte maturation *in vitro*.[52] Although the hypo- and demyelination associated with exposure to high levels of Pb[76,101,201,204] has been thought to be a secondary consequence of neuronal damage,[201] the sensitivity of O-2A/OPCs and oligodendrocytes to this metal should force reevaluation of such hypotheses.

Our own ongoing studies have focused thus far on mercury exposure, with an interest in both methylmercury and ethylmercury. It is clear from unfortunate experiences with contaminated wheat in Iraq and contaminated fish in Japan that high levels of exposure to MeHg is associated with severe abnormalities in the developing brain, including neuronal migration disorders and diffuse gliosis of the periventricular white matter.[34] Studies in the Faroe Islands, the Seychelles Islands, New Zealand, and the Amazon Basin have further found that children born from mothers exposed during pregnancy to moderate doses of MeHg showed significantly reduced performance on several neuropsychological tests.[46,47,57,79,80] Children exposed to mercury during development may exhibit a range of neurological problems, including cerebral palsy, developmental delay, and white matter astrocytosis.[31,126]

The developing nervous system is more sensitive to MeHg neurotoxicity than the adult nervous system.[37,140] MeHg appears to have a wide range of toxic effects on the developing CNS. For example, developmental exposure to MeHg is associated with decreases in cell survival, myelination, and cerebral dysgenesis,[15,26,32] decreased expression and activity of proteins involved in neurotrophic factor signaling,[16,88,139] and changes in neurotrophic factor expression.[102] MeHg has a high affinity for thiol groups, thus making proteins and peptides bearing cysteines susceptible to structural and functional modification. One such protein target has been suggested to be microtubules, which are disrupted by MeHg exposure in human fibroblasts,[181] mouse splenic lymphocytes,[180] neuroblastoma,[162] glioma cells,[136,162] and embryonal carcinoma cells.[27,78,213] In cerebellar granule neurons, microtubule fragmentation has been observed at MeHg concentrations in the range of 0.5–1 μM. As MeHg dosages of 1 μM or less have been observed to induce apoptosis of cerebellar granule neurons,[30] however, it is not clear whether the microtubule fragmentation that has been observed represents a process distinct from the action of apoptosis-related proteases.

MeHg also accumulates in astrocytes and inhibits both glutamate and aspartate uptake[25,49,138] as well as causing a calcium-dependent release of excitatory amino acids from these cells.[5,138] These actions would lead to an increase in glutamate concentrations in the extracellular milieu. Stimulation of glutamate release is of particular interest, as glutamate has been suggested to play a contributory role in MeHg damage. In particular, it has been reported that antagonists of the *N*-methyl-*D*-aspartate (NMDA) receptor can block MeHg-induced toxic effects in cerebral neuron culture,[151] raising questions as to the extent to which toxic effects are due to direct actions of MeHg, to glutamate, or to a combination of the two.

Alterations in extracellular glutamate concentrations are of great concern in the developing CNS, especially in light of extensive evidence that it is glutamate release that may be a major contributor to the extensive white matter damage that occurs with disturbing frequency in premature infants. Such damage is caused by hypoxic-ischemic insults, particularly during the period from 24–32 weeks postconception. These insults can cause extensive cerebral white matter damage and oligodendrocyte loss, associated with permanent areas of hypomyelination. This disorder, termed periventricular leukomalacia (PVL), is the leading cause of neurological disability in premature infants, including spastic motor deficits (cerebral palsy) and cognitive deficits.[115,171,173,209]

A critical contributor to PVL appears to be glutamate-induced death of oligodendrocytes and their precursor cells.[109,125,178] During the developmental period of greatest risk for PVL, human cerebral white matter is populated primarily by cells thought to be human O-2A/OPCs.[9] Studies in the rat have shown that O-2A/OPCs and immature oligodendrocytes are killed by glutamate-activated calcium fluxes through α–amino-3-hydroxy-5-methyl-4-isoxazolepropionic acid (AMPA) and kainite receptors, with release of glutamate being caused both by traumatic damage and as a consequence of oxygen-glucose deprivation.[8,53,64,67,94,97,119,123,154,176,182,210] Differentiated oligodendrocytes are also vulnerable to killing by glutamate, in this case through activation of a glutamate-cysteine exchange pump that causes oxidative stress as a consequence of intracellular cysteine depletion.[8] The combined killing of both oligodendrocytes and O-2A/OPCs creates a demyelinated lesion in which the precursor cells necessary for repair are eliminated. Whether increases in glutamate concentration, as might be associated with mercury exposure, would have adverse affects of a more subtle nature is not known.

Another organic mercury compound that has become of considerable recent interest is thimerosal, a vaccine preservative that contains 49.6% ethylmercury (by weight) as its active ingredient. Concern has been raised that apparent increases in the prevalence of autism (from 1 in 2000 prior to 1970 up to 1 in 500 in 1996[75]) have paralleled the increased mercury intake induced by mandatory inoculations. In 1999, the FDA recorded thimerosal usage in over 30 vaccine products.[63] According to the classification of thimerosal-containing vaccines provided by the Massachusetts Department of Public Health, as of June 2002 thimerosal was still in use as a preservative in a significant number of vaccines, including diphtheria/tetanus, hep B, influenza, meningococcus, and rabies vaccines. The World Health Organization, the American Academy of Pediatrics, and the U.S. Public Health Service have all voiced support for phasing out thimerosal usage as a vaccine preservative, but the WHO has stressed that this may not be an option for developing countries. While a recent Danish study[114] failed to find a link between autism and vaccination with the MMR (measles, mumps, rubella) vaccine, this is not a thimerosal-containing vaccine and thus did not shed light on controversies related to autism and mercury exposure. The hypothesis that mercury exposure and autism are linked is discussed extensively in Bernard et al.,[23] including information on the multiple similarities between the neurological symptoms seen in mercury poisoning and those considered to typify autism.

It is not the intent of this review to offer a stance on any link between thimerosal and autism, but instead to raise questions about the toxicities of both MeHg and ethylmercury (thimerosal) for neural precursor cells. This is important in understanding general principles related to mercury toxicity. It also is striking how little is known about toxicity of thimerosal. A PubMed search of the literature linking *methylmercury* with *neurotoxicity* reveals 168 publications, of which over 120 appear to be primary research papers. In striking contrast, a similar search using *ethylmercury* (or *thimerosal*) and *neurotoxicity* as search terms reveals five publications, none of which are primary research papers. Thus, despite the extensive exposure of individuals to thimerosal, little appears to be known about the effects of this preservative on CNS development.

The amount of mercury that would have been delivered to a child born in the 1990s in association with vaccination over the first 2 years of life is not small and was delivered in bolus form (as part of a vaccination). The amount of mercury injected at birth was 12.5 µg, followed by 62.5 µg at 2 months, 50 µg at 4 months, another 62.5 µg during the infant's 6-month immunizations, and a final 50 µg at about 15 months.[86] Concerns exist that infants under 6 months may be inefficient at mercury excretion, most likely due to their inability to produce bile, the main excretion route for organic mercury.[36,100] More recent studies have challenged these concerns, reporting that blood mercury in thimerosal-exposed 2-month-olds ranged from less than 3.75 to 20.55 parts per billion; in 6-month-olds all values were lower than 7.50 parts per billion.[156] As we shall discuss, such concentrations are very much in the range that can modify precursor cell function, at least *in vitro*.

1.6.3 Mercury Toxicity for Precursor Cells Occurs with Exposure Levels as Low as 4–5 Parts per Billion

Given the aforementioned considerations, we became interested in examining the effects of organic mercury compounds on CNS precursor cell function. This work is ongoing, and this review will only summarize those areas of research where experiments have been repeated multiple times. Progress to date appears to give significant concern that exposure levels currently deemed to be safe are in fact far more biologically active than was anticipated.

indicate that methylmercury, another important organic heavy metal toxicant, compromises O-2A/OPC division and differentiation at exposure dosages as low as 20–30 nM (MN et al., unpublished observations). Such dosages, which are equivalent to exposures of 4–6 parts per billion,[153] are within the range of blood mercury levels found within the bloodstream of approximately 8% of women of childbearing age in the United States.[163]

Exposure of cells to MeHg is generally described using two terminologies, that of molarity and that of parts per million (ppm) or parts per billion (ppb) of MeHg in the fluid or tissue examined. According to the literature,[153] an exposure concentration of MeHg of 100 nM is equivalent to 0.02 ppm (i.e., 20 ppb), based on the molecular weight of MeHg. In order to facilitate discussion of relevant literature, both terminologies are used in this application. This is done because *in vivo* studies generally analyze information using ppm or ppb scales, while *in vitro* studies tend to refer to molar concentrations of MeHg. As will be seen throughout Section C, in the great majority of our experiments we are observing effects on oligodendrocyte survival and progenitor cell function at exposure doses of $\leq$ 20 nM, which would be equivalent to 5 parts per billion,[153] revealing an unexpectedly low threshold for mercury toxicity in these populations.

Our first concern was to establish the dosages at which glial precursor cells and oligodendrocytes might be affected by organic mercury compounds. We have found that the LD_{50} values for both methylmercury and ethylmercury (thimerosal) are in the range of 10 parts per billion for oligodendrocytes and extend down to 4–5 parts per billion for O-2A/OPCs and GRP cells. Thus, in contrast with studies published by multiple previous investigators on cell lines, astrocytes, and neurons, in which sublethal dosages have been routinely observed to be in the range of 250–500 nM (i.e., ~ 50–100 parts per billion), studies on precursor cells and oligodendrocytes reveal sensitivities at concentrations an order of magnitude lower.

Sublethal exposure levels of 25 nM (5 parts per billion) and lower reduce O-2A/OPC division and increase the oxidative state of cells to a level that we previously have seen reduces cell division and interferes with intracellular signaling processes. In agreement with such findings, we find that these low exposure levels also reduce efficacy of growth factors in causing erk2 phosphorylation. That mercury exposure would have such effects is not surprising, although the dosage at which we see such effects is lower than generally studied. There are, however, multiple indications that at least some portion of MeHg toxicity may be related to its pro-oxidant activity. MeHg triggers production of reactive oxygen species (ROS).[4,35,184,185,193] Additionally, given that ROS adversely affect glutamate transport into astrocytes,[192] altered glutathione homeostasis likely leads to impaired astrocytic handling of glutamate, and as a result altered astrocytic energy metabolism. Multiple studies indicate that such antioxidants/oxygen radical scavengers such as vitamin E, GSH, selenium, catalase, etc., can provide some protection against the toxic effects of MeHg *in vitro* and *in vivo*.[48,152,219] The toxicity of MeHg also can be enhanced by decreasing intracellular GSH levels (Aschner et al. 1994), or lessened by increasing intracellular GSH levels (Mullaney et al. 1994; Park et al. 1996). Potential mechanisms involved in this protective effect of GSH include direct conjugation of GSH with MeHg followed by the efflux of the conjugate, GSH acting as an intracellular buffer for MeHg, thereby limiting the net MeHg concentration available for interaction with sensitive macromolecules, and its acting as a ROS scavenger. However, it appears that there are also other components to MeHg toxicity (at least in some cell types), as some studies have indicated that N-acetyl-L-cysteine (NAC) does not protect cerebellar granule neurons from MeHg, nor does treatment with *L*-buthionine-(*S,R*)-sulfoximine increase toxicity.[149]

Further evidence that exposure to doses of mercury currently thought to be safe have come from studies of the effects of thimerosal exposure on the regulation of insulin-like growth factor-1 (IGF-1) and dopamine-stimulated methionine synthase (MS) activity and folate-dependent methylation of phospholipids in SH-SY5Y human neuroblastoma cells, via a PI3-kinase- and MAP-kinase-dependent mechanism. Stimulation of this pathway increases DNA methylation, a process of critical importance in the regulation of normal development by growth factor exposure. The ethylmercury-containing preservative thimerosal, and mercury itself, inhibited both IGF-1- and dopamine-stimulated methylation and eliminated MS activity.[211] The IC50 for thimerosal was 1 n*M*, demonstrating a remarkable sensitivity of this critical regulatory pathway to a compound in which blood levels above this range are routine in infants receiving thimerosal-containing vaccines. These findings thus describe a novel growth factor signaling pathway that regulates MS activity and thereby modulates methylation reactions, including DNA methylation. The potent inhibition of this pathway by mercury and thimerosal (as well as by low concentrations of ethanol, lead, and aluminum) suggests that it may be an important target of neurodevelopmental toxins.

1.7 CHALLENGES FOR THE TOXICOLOGY OF THE FUTURE: ANALYSIS OF LOW-DOSE EXPOSURE AND COMBINATORIAL INSULTS

The approaches described in this review hopefully offer some useful beginnings in the analysis of the effects of low-dose toxicant exposure, a problem of critical importance throughout the world. *In vitro* approaches, particularly those involving the study of those precursor cells essential to the building of the brain and spinal cord, clearly provide levels of sensitivity not offered by previously studied cell populations. Moreover, if it is generally correct that low-dose exposure to toxicants can interfere with the normal methylation of DNA,[211] then the sensitive techniques being applied to the study of DNA methylation in cancer[42–45] will rapidly have relevance to the field of toxicology.

One set of tools that would be extremely useful is the development of techniques that allow the generally narrative analysis of biology to be conducted in a more quantitative manner. *In vitro*, we and our colleagues have been working to create such tools by developing algorithms for the analysis of differentiation within clonal populations.[24,197,198,215,216,221] Such approaches have helped us to obtain a better understanding of the details of oligodendrocyte generation and are now being applied to the analysis of toxicant function.

The understanding of regional differences in precursor cell function seems likely to be of value in elucidating the biological basis of the ability of particular toxicants to preferentially alter function in particular regions of the CNS. For example, the more reduced O-2A/OPCs of the cortex[161] might show a different vulnerability to toxicant action than the more oxidized O-2A/OPCs isolated from the optic nerve. As another example, differing sensitivities to particular toxicants and differentiation signals of early embryonic GRP cells and of the O-2A/OPCs that characterize later developmental stages may be relevant to understanding how toxicants exert different effects depending on the age of the animal exposed to them.

One of the greatest challenges that lie ahead is the development of approaches that allows the study of subtle effects of toxicant action *in vivo*. As discussed in the introduction, fairly minor variations in the probability of differentiation or cell survival can have enormous effects on the number of differentiated cells generated during development. Some research teams are providing mathematical approaches to the analysis of alterations in such parameters as cell division,[133] but we are far from having the tools required to effectively analyze such a problem. It may be that the application of such approaches as analysis of developmental changes in DNA methylation will prove useful in such analyses.

It also seems likely that one of the most important contributions that will be made to the study of low-dose exposure to toxicants will come from the study of the effects of two or more insults. It is only in the highly artificial situation of the laboratory that an organism is subjected to only a single insult. In the real world, we are all subjected to a large range of toxicants — be they heavy metals, pesticides, industrial pollutants, and so forth. The importance of examining conjoined physiological insults in understanding the action of toxicants is supported by multiple studies. For example, glutamate and MeHg appear to interact to exacerbate cell damage, as indicated by studies showing that coapplication of nontoxic concentrations of MeHg and glutamate leads to the appearance of neuronal lesions typical of excitotoxic damage.[6] Coapplication of combinations of toxicants also is associated with striking increases in tissue damage. For example, exposure of pregnant mice to lead nitrate and sodium arsenite, at doses that show limited maternal toxicity, concurrently with MeHg at a toxic dose is associated with supra-additive levels of maternal toxicity.[20] Another example of CNS damage associated with toxicant coadministration is seen in the effects of the combination of 1,1'-dimethyl-4,4'-bipyridinium (paraquat) with manganese ethylenebisdithiocarbamate (maneb) on striatal dopaminergic systems. In these experiments, it was found that the combination of paraquat and maneb produces a far greater effect on striatal dopaminergic systems than the additive effects of each compound given alone, and is even capable of producing permanent and progressive lesions resembling those seen in Parkinson's disease.[202,203] In our own ongoing studies, it is clear that exposure of precursor cells and oligodendrocytes to a variety of toxicants at doses that do not cause cell death greatly increases the vulnerability of these cells to other insults. If such changes in vulnerability are associated with low-dose toxicant exposure *in vivo*, such studies may provide important insights into the greatly variable responses to injury and recovery seen in virtually all clinical conditions. Perhaps low-dose toxicant exposure will prove as, or more, important than genetics in understanding such variability.

REFERENCES

1.   Aase, J.M., Clinical recognition of FAS, *Alcohol World*, 18, 5, 1994.
2.   Abel, E.L. and Sokol, R.J., Incidence of fetal alcohol syndrome and economic impact of FAS-related anomalies, *Drug Alcohol Depend.*, 19, 51–70, 1987.
3.   Ahlgren, S., Wallace, H., Bishop, J., Neophytou, C., and Raff, M., Effects of thyroid hormone on embryonic oligodendrocyte precursor cell development *in vivo* and *in vitro*, *Mol. Cell Neurosci.*, 9, 420, 1997.
4.   Ali, S.F., LeBel, C.P., and Bondy, S.C., Reactive oxygen species formation as a biomarker of methyl-mercury and trimethyltin neurotoxicity, *Neurotoxicology*, 13, 637– 648, 1992.
5.   Aschner, M., Du, Y.L., Gannon, M., and Kimelberg, H.K., Methyl-mercury induced alterations in excitatory amino acid transport in rat primary astrocyte cultures, *Brain Res.*, 602, 181, 1993.
6.   Aschner, M., Yao, C.P., Allen, J.W., and Tan, K.H., Methylmercury alters glutamate transport in astrocytes, *Neurochem. Int.*, 37, 199–206, 2000.
7.   Atlante, A., Calissano, P., Bobba, A., Giannattasio, S., Marra, E., and Passarella, S., Glutamate neurotoxicity, oxidative stress and mitochondria, *FEBS Lett.*, 497, 1–5, 2001.
8.   Back, S.A., Gan, X., Li, Y., Rosenberg, P.A., and Volpe, J.J., Maturation-dependent vulnerability of oligodendrocytes to oxidative stress-induced death caused by glutathione depletion, *J. Neurosci.*, 18, 6241–6253, 1998.
9.   Back, S.A., Luo, N.L., Borenstein, N.S., Levine, J.M., Volpe, J.J., and Kinney, H.C., *J. Neurosci.*, 21, 1302–1312, 2001.
10.   Bagchi, D., Balmoori, J., Bagchi, M., Ye, X., Williams, C.B., and Stohs, S.J., Comparative effects of TCDD, endrin, naphthalene and chromium (VI) on oxidative stress and tissue damage in the liver and brain tissues of mice, *Toxicology*, 175, 73.
11.   Balazs, R., Brooksbank, B.W., Davison, A.N., Eayrs, J.T., and Wilson, D.A., The effect of neonatal thyroidectomy on myelination in the rat brain, *Brain Res.*, 15, 219–232, 1969.

12. Balazs, R., Kovacs, S., Cocks, W.A., Johnson, A.L., and Eayrs, J.T., Effect of thyroid hormone on the biochemical maturation of rat brain: postnatal cell formation, *Brain Res.*, 25, 555–570, 1971.

13. Banks, E.C., Ferretti, L.E., and Shucard, D.W., Effects of low level lead exposure on cognitive function in children: A review of behavioral, neuropsychological and biological evidence, *Neurotoxicology*, 18, 237, 1997.

14. Barnett, S.C., Hutchins, A.M., and Noble, M., Purification of olfactory nerve ensheathing cells from the olfactory bulb, *Dev. Biol.*, 155, 337, 1993.

15. Barone, S., Jr., Haykal-Coates, N., Parran, D.K., and Tilson, H.A., Gestational exposure to methylmercury alters the developmental pattern of trk-like immunoreactivity in the rat brain and results in cortical dysmorphology, *Brain Res. Dev. Brain Res.*, 109, 13–31, 1998.

16. Barone, S., Jr., Haykal-Coates, N., Parran, D.K., and Tilson, H.A., Gestational exposure to methylmercury alters the developmental pattern of trk-like immunoreactivity in the rat brain and results in cortical dysmorphology, *Brain Res. Dev. Brain Res.*, 109, 13–31, 1998.

17. Barres, B., Burne, J., Holtmann, B., Thoenen, H., Sendtner, M., and Raff, M., Ciliary neurotrophic factor enhances the rate of oligodendrocyte generation, *Mol. Cell. Neurosci.*, 8, 146, 1996.

18. Barres, B.A., Lazar, M.A., and Raff, M.C., A novel role for thyroid hormone, glucocorticoids and retinoic acid in timing oligodendrocyte development, *Development*, 120, 1097, 1994.

19. Bell, G.L. and Lau, K., Perinatal and neonatal issues of substance abuse, *Pediatr. Clin. North Am.*, 42, 261–281, 1995.

20. Belles, M., Albina, M.L., Sanchez, D.J., Corbella, J., and Domingo, J.L., Interactions in developmental toxicology: effects of concurrent exposure to lead, organic mercury, and arsenic in pregnant mice, *Arch. Environ. Contam. Toxicol.*, 42, 93, 2002.

21. Benes, F.M., Turtle, M., Khan, Y., and Farol, P., Myelination of a key relay zone in the hippocampal formation occurs in the human brain during childhood, adolescence and adulthood, *Arch. Gen. Psychiat.*, 51, 477, 1994.

22. Berger, F., Gage, F.H., and Vijayaraghavan, S., Nicotinic receptor-induced apoptotic cell death of hippocampal progenitor cells, *J. Neurosci.*, 8, 6871, 1998.

23. Bernard, S., Enayati, A., Redwood, L., Roger, H., and Binstock, T., Autism: a novel form of mercury poisoning, *Med. Hypotheses*, 56, 462, 2001.

24. Boucher, K., Yakovlev, A., Mayer-Proschel, M., and Noble, M., A stochastic model of temporally regulated generation of oligodendrocytes in cell culture, *Math. Biosci.*, 159, 47, 1999.

25. Brookes, N. and Kristt, D.A., Inhibition of amino acid transport and protein synthesis by HgCl2 and methylmercury in astrocytes: selectivity and reversibility, *J. Neurochem.*, 53, 1228–1237, 1989.

26. Burbacher, T.M., Rodier, P.M., and Weiss, B., Methylmercury developmental neurotoxicity: a comparison of effects in humans and animals, *Neurotoxicol. Teratol.*, 12, 191–202, 1990.

27. Cadrin, M., Wasteneys, G.O., Jones-Villeneuve, E.M., Brown, D.L., and Reuhl, K.R., Effects of methylmercury on retinoic acid-induced neuroectodermal derivatives of embryonal carcinoma cells, *Toxicol. Cell Biol.*, 4, 61–80, 1988.

28. Cao, Q.L., Howard, R.M., Dennison, J.B., and Whittemore, S.R., Differentiation of engrafted neuronal-restricted precursor cells is inhibited in the traumatically injured spinal cord, *Exp. Neurol.*, 177, 349, 2002.

29. Cao, Q.L., Zhang, Y.P., Howard, R.M., Walters, W.M., Tsoulfas, P., and Whittemore, S.R., Pluripotent stem cells engrafted into the normal or lesioned adult rat spinal cord are restricted to a glial lineage, *Exp. Neurol.*, 167, 48, 2001.

30. Castoldi, A.F., Barni, S., Turin, I., Gandini, C., and Manzo, L., Early acute necrosis, delayed apoptosis and cytoskeletal breakdown in cultured cerebellar granule neurons exposed to methylmercury, *J. Neurosci. Res.*, 60, 775–787, 2000.

31. Castoldi, A.F., Coccini, T., Ceccatelli, S., and Manzo, L., Neurotoxicity and molecular effects of methylmercury, *Brain Res. Bull.*, 55, 197, 2001.

32. Chang, L.W., Reuhl, K.R., and Lee, G.W., Degenerative changes in the developing nervous system as a result of *in utero* exposure to methylmercury, *Environ. Res.*, 14, 414, 1977.

33. Chiappelli, F., Taylor, A.N., Espinosa de los Monteros, A., and de Vellis, J., Fetal alcohol delays the developmental expression of myelin basic protein and transferrin in rat primary oligodendrocyte cultures, *Int. J. Dev. Neurosci.*, 9, 67, 1991.

34. Cho, B.H., The effects of methylmercury on the developing brain, *Prog. Neurobiol.*, 32, 447–470, 1989.

35. Choi, B.H., Yee, S., and Robles, M., The effects of glutathione glycoside in methyl mercury poisoning, *Toxicol. Appl. Pharmacol.*, 141, 357–364, 1996.

36. Clarkson, T.W., Molecular and ionic mimicry of toxic metals, *Annu. Rev. Pharmacol. Toxicol.*, 32, 545, 1993.

37. Clarkson, T.W., The toxicology of mercury, *Crit. Rev. Clin. Lab. Sci.*, 34, 369–403, 1997.

38. Cohen-Hubal, E.A., Sheldon, L.S., Burke, J.M., McCurdy, T.R., Berry, M.R., Rigas, M.L., Zartarian, V.G., and Freeman, N.C.G., Children's exposure assessment: A review of factors influencing children's exposure, and the data available to characterize and assess that exposure, *Environ. Health Perspect.*, 108, 475, 2000.

39. U.S. Environmental Protection Agency, Mercury Study Report to Congress, Vol. I, 1997, pp. 3–39. Available at http://www.epa.gov/ttn/atw/112nmerc/volume1.pdf

40. U.S. Environmental Protection Agency, Mercury Study Report to Congress, Vol. VII, 1997. Available at http://www.epa.gov/ttn/atw/112nmerc/volume7.pdf

41. Cory-Slechta, D.A., Relationships between Pb-induced changes in neurotransmitter system function and behavioral toxicity, *Neurotoxicology*, 18, 673, 1997.

42. Costello, J.F., DNA methylation in brain development and gliomagenesis, *Front. Biosci.*, 8, S175, 2003.

43. Costello, J.F., Berger, M.S., Huang, H.S., and Cavenee, W.K., Silencing of p16/CDKN2 expression in human gliomas by methylation and chromatin condensation, *Cancer Res.*, 56, 2405, 1996.

44. Costello, J.F., Fruhwald, M.C., Smiraglia, D.J., Rush, L.J., Robertson, G.P., Gao, X., Wright, F.A., Feramisco, J.D., Peltomaki, P., Lang, J.C., Schuller, D.E., Yu, L., Bloomfield, C.D., Caligiuri, M.A., Yates, A., Nishikawa, R., Su Huang, H., Petrelli, N.J., Zhang, X., O'Dorisio, M.S., Held, W.A., Cavenee, W.K., and Plass, C., Aberrant CpG-island methylation has non-random and tumour-type-specific patterns, *Nat. Genet.*, 132, 2000.

45. Costello, J.F., Plass, C., and Cavenee, W.K., Aberrant methylation of genes in low-grade astrocytomas, *Brain Tumor Pathol.*, 17, 49, 2000.

46. Crump, K.S., Kjellstrom, T., Shipp, A.M., Silvers, A., and Stewart, A., Influence of prenatal mercury exposure upon scholastic and psychological test performance: Benchmark analysis of a New Zealand cohort, *Risk Anal.*, 18, 701, 1998.

47. Crump, K.S., Van Landingham, C., Shamlaye, C., Cox, C., Davidson, P.W., Myers, G.J., and Clarkson, T.W., Benchmark concentrations for methylmercury obtained from the Seychelles Island Development Study, *Environ. Health Perspect.*, 108, 257, 2000.

48. Dare, E., Goetz, M.E., Zhivotovsky, B., Manzo, L., and Ceccatelli, S., The antioxidants J811 and 17 estradiol protect cerebellar granule cells from methylmercury induced apoptotic cell death, *J. Neurosci. Res.*, 62, 557–565, 2000.

49. Dave, V., Mullaney, K.J., Goderie, S., Kimelberg, H.K., and Aschner, M., Astrocytes as mediators of methylmercury neurotoxicity: effects on D-aspartate and serotonin uptake, *Dev. Neurosci.*, 16, 222–231, 1994.

50. Davies, D.L. and Ross, T.M., Long-term ethanol-exposure markedly changes the cellular composition of cerebral glial cultures, *Dev. Brain Res.*, 62, 151, 1991.

51. Delange, F., The disorders induced by iodine deficiency, *Thyroid*, 4, 107, 1994.

52. Deng, W., McKinnon, R.D., and Poretz, R.D., Lead exposure delays the differentiation of oligodendroglial progenitors *in vitro*, and at higher doses induces cell death, *Toxicol. Appl. Pharmacol.*, 174, 235, 2001.

53. Deng, W., Rosenberg, P.A., Volpe, J.J., and Jensen, F.E., Calcium-permeable AMPA/kainate receptors mediate toxicity and preconditioning by oxygen-glucose deprivation in oligodendrocyte precursors, *Proc. Natl. Acad. Sci. U.S.A.*, 100, 6801, 2003.

54. Dietrich, J., Noble, M., and Mayer-Proschel, M., Characterization of A2B5+ glial precursor cells from cryopreserved human fetal brain progenitor cells, *Glia*, 40, 65, 2002.

55. DiFranza, J.R. and Lew, R.A., Effect of maternal cigarette smoking on pregnancy complications and Sudden Infant Death Syndrome, *J. Fam. Pract.*, 40, 385–394, 1995.

56. Ding, Y., Gonick, H.C., and Vaziri, N.D., Lead promotes hydroxyl radical generation and lipid peroxidation in cultured aortic endothelial cells, *Am. J. Hypertens.*, 13, 552–555, 2000.

57. Dolbec, J., Mergler, D., Sousa-Passos, C.J., Sousa de Morais, S., and Lebel, J., Methylmercury exposure affects motor performance of a riverine population of the Tapajos River, Brazilian Amazon, *Int. Arch. Occup. Environ. Health*, 73, 195, 2000.

58. Eayrs, J.T., The cerebral cortex of normal and hypothyroid rats, *Acta Anat.*, 25, 160–183, 1955.

59. Eayrs, J.T., Thyroid and developing brain: anatomical and behavioural effects, in *Hormones in Development*, Hamburgh, M. and Barrington, E., Eds., Appleton–Century–Crofts, New York, 1971.

60. Eayrs, J.T. and Horne, G., The development of cerebral cortex in hypothyroid and starved rats, *Anat. Rec.*, 121, 53, 1955.

61. Eayrs, J.T. and Taylor, S.H., The effect of thyroid deficiency induced by methylthiouracil on the maturation of the central nervous system, *J. Anat.*, 85, 350–358, 1951.

62. Espinosa de los Monteros, A., Zhang, M., and De Vellis, J., O2A progenitor cells transplanted into the neonatal rat brain develop into oligodendrocytes but not astrocytes, *Proc. Natl. Acad. Sci. U.S.A.*, 90, 50, 1993.

63. FDA, Mercury Compounds in Drugs and Food, 98N-1109, November 16, 1999.

64. Fern, R. and Möller, T., Rapid ischemic cell death in immature oligodendrocytes: a fatal glutamate release feedback loop, *J. Neurosci.*, 20, 34–42, 2000.

65. Fidelus, R.K., Ginouves, P., Lawrence, D., and Tsan, M.F., Modulation of intracellular glutathione concentrations alters lymphocyte activation and proliferation, *Exp. Cell Res.*, 170, 269, 1987.

66. Fidelus, R.K. and Tsan, M.F., Enhancement of intracellular glutathione promotes lymphocyte activation by mitogen, *Cell. Immunol.*, 97, 155, 1986.

67. Follett, P.L., Rosenberg, P.A., Volpe, J.J., and Jensen, F.E., NBQX attenuates excitotoxic injury in developing white matter, *J. Neurosci.*, 20, 9235, 2000.

68. Foran, D.R. and Peterson, A.C., Myelin acquisition in the central nervous system of the mouse revealed by an MBP-LacZ transgene, *J. Neurosci.*, 12, 4890, 1992.

69. Franklin, R.J. and Blakemore, W.F., Glial-cell transplantation and plasticity in the O-2A lineage— implications for CNS repair, *Trends Neurosci.*, 18, 151, 1995.

70. Gao, F., Apperly, J., and Raff, M., Cell-intrinsic timers and thyroid hormone regulate the probability of cell-cycle withdrawal and differentiation of oligodendrocyte precursor cells, *Dev. Biol.*, 197, 54, 1998.

71. Gao, F. and Raff, M., Cell size control and a cell-intrinsic maturation program in proliferating oligodendrocyte precursor cells, *J. Cell Biol.*, 138, 1367, 1997.

72. Gard, A.L. and Pfeiffer, S.E., Two proliferative stages of the oligodendrocyte lineage (A2B5 + O4- and O4 + GalC-) under different mitogenic control, *Neuron*, 5, 615, 1990.

73. Gard, A.L. and Pfeiffer, S.E., Glial cell mitogens bFGF and PDGF differentially regulate development of O4 + GalC-oligodendrocyte progenitors, *Dev. Biol.*, 159, 618, 1993.

74. Gard, A.L., Williams, W.C.n., and Burrell, M.R., Oligodendroblasts distinguished from O-2A glial progenitors by surface phenotype (O4 + GalC-) and response to cytokines using signal transducer LIFR beta, *Dev. Biol.*, 167, 596, 1995.

75. Gillberg, C. and Wing, L., Autism: not an extremely rare disorder, *Acta Psychiatr. Scand.*, 99, 399, 1999.

76. Goyer, R.A., Lead toxicity: Current concerns, *Environ. Health Perspect.*, 100, 177, 1993.

77. Goyer, R.A. and Clarkson, T.W., Toxic effects of metals, in *Casarett and Doull's Toxicology: The Basic Science of Poisons*, Klaassen, C.D., Ed., McGraw-Hill, New York, 2001, pp. 811–867.

78. Graff, R.D., Philbert, M.A., Lowndes, H.E., and Reuhl, K.R., The effect of glutathione depletion on methyl mercury-induced microtubule disassembly in cultured embryonal carcinoma cells, *Toxicol. Appl. Pharmacol.*, 120, 20–28, 1993.

79. Grandjean, P., Weihe, P., White, R.F., and Debes, F., Cognitive performance of children prenatally exposed to "safe" levels of methylmercury, *Environ. Res.*, 77, 165, 1998.

80. Grandjean, P., White, R.F., Nielsen, A., Cleary, D., and de Oliveira Santos, E.C., Methylmercury neurotoxicity in Amazonian children downstream from gold mining, *Environ. Health Perspect.*, 107, 587, 1999.

81. Gregori, N., Proschel, C., Noble, M., and Mayer-Pröschel, M., The tripotential glial-restricted precursor (GRP) cell and glial development in the spinal cord: Generation of bipotential oligodendrocyte-type-2 astrocyte progenitor cells and dorsal-ventral differences in GRP cell function, *J. Neurosci.*, 22, 248, 2002.

82. Gressens, P., Lammens, M., Picard, J.J., and Evrard, P., Ethanol-induced disturbances of gliogenesis and neurogenesis in the developing murine brain. An *in vitro* and *in vivo* immunohistochemical, morphological, and ultrastructural study, *Alcohol Alcohol.*, 27, 219–226, 1992.

83. Groves, A.K., Barnett, S.C., Franklin, R.J., Crang, A.J., Mayer, M., Blakemore, W.F., and Noble, M., Repair of demyelinated lesions by transplantation of purified O-2A progenitor cells, *Nature*, 362, 453, 1993.

84. Grzenkowski, M., Niehaus, A., and Trotter, J., Monoclonal antibody detects oligodendroglial cell surface protein exhibiting temporal regulation during development, *Glia*, 28, 128–137, 1999.

85. Haddow, J.E., Glenn, E., Palomaki, B.S., Walter, C., Allan, M.D., Williams, J.R., Knight, G.J., Gagnon, J., O'Heir, C.E., Mitchell, M.L., Hermos, R.J., Waisbren, S.E., Faix, J.D., and Klein, R.Z., Maternal thyroid deficiency during pregnancy and subsequent neuropsychological development of the child, *N. Engl. J. Med.*, 341, 549–555, 1999.

86. Halsey, N.A., Limiting infant exposure to thimerosal in vaccines and other sources of mercury, *JAMA*, 282, 1999.

87. Han, S.S., Kang, D.Y., Mujtaba, T., Rao, M.S., and Fischer, I., Grafted lineage-restricted precursors differentiate exclusively into neurons in the adult spinal cord, *Exp. Neurol.*, 177, 360, 2002.

88. Haykal-Coates, N., Shafer, T.J., Mundy, W.R., and Barone, S., Jr., Effects of gestational methylmercury exposure on immunoreactivity of specific isoforms of PKC and enzyme activity during postnatal development of the rat brain, *Brain Res. Dev. Brain Res.*, 109, 33, 1998.

89. Herrera, J., Yang, H., Zhang, S.C., Proschel, C., Tresco, P., Duncan, I.D., Luskin, M., and Mayer-Proschel, M., Embryonic-derived glial-restricted precursor cells (GRP cells) can differentiate into astrocytes and oligodendrocytes *in vivo*, *Exp. Neurol.*, 171, 11, 2001.

90. Hsu, P., Liu, M., Hsu, C., Chen, L., and Guo, Y., Lead exposure causes generation of reactive oxygen species and functional impairment in rat sperm, *Toxicology*, 122, 133–143, 1997.

91. Ibarrola, N., Mayer-Proschel, M., Rodriguez-Pena, A., and Noble, M., Evidence for the existence of at least two timing mechanisms that contribute to oligodendrocyte generation *in vitro*, *Dev. Biol.*, 180, 1, 1996.

92. Ibarrola, N. and Rodriguez-Pena, A., Hypothyroidism coordinately and transiently affects myelin protein gene expression in most rat brain regions during postnatal development, *Brain Res.*, 752, 285, 1997.

93. Ikonomidou, C., Bittigau, P., Ishimaru, M.J., Wozniak, D.F., Koch, C., Genz, K., Price, M.T., Stefovska, V., Horster, F., Tenkova, T., Dikranian, K., and Olney, J.W., Ethanol-induced apoptotic neurodegeneration and fetal alcohol syndrome, *Science*, 287, 1056–1060, 2000.

94. Itoh, T., Beesley, J., Itoh, A., Cohen, A.S., Kavanaugh, B., Coulter, D.A., Grinspan, J.B., and Pleasure, D., AMPA glutamate receptor-mediated calcium signaling is transiently enhanced during development of oligodendrocytes, *J. Neurochem.*, 81, 390, 2002.

95. Jones, S.A., Jolson, D.M., Cuta, K.K., Mariash, C.N., and Anderson, G.W., Triiodothyronine is a survival factor for developing oligodendrocytes, *Mol. Cell. Endocrin.*, 199, 49, 2003.

96. Joschko, M.A., Dreosti, I.E., and Tulsi, R.S., The teratogenic effects of nicotine *in vitro* in rats: A light and electron microscope study, *Neurotoxicol. Teratol.*, 13, 307–316, 1991.

97. Kelland, E.E. and Toms, N.J., Group I metabotropic glutamate receptors limit AMPA receptor-mediated oligodendrocyte progenitor cell death, *Eur. J. Pharmacol.*, 424, R3, 2001.

98. Kinney, H.C., Brody, B.A., Kloman, A.S., and Gilles, F.H., Sequence of central nervous system myelination in human infancy. II. Patterns of myelination in autopsied infants, *J. Neuropath. Exp. Neurol.*, 47, 217, 1988.

99. Knapp, P.E., Studies of glial lineage and proliferation *in vitro* using an early marker for committed oligodendrocytes, *J. Neurosci. Res.*, 30, 336, 1991.

100. Koos, B.J. and Longo, L.D., Mercury toxicity in the pregnant woman, fetus, and newborn infant, *Am. J. Obst. Gyn.*, 126, 390, 1976.

101. Krigman, M.R., Druse, M.J., Traylor, T.D., Wilson, M.H., Newell, L.R., and Hogan, E.L., Lead encephalopathy in the developing rat: Effect on myelination, *J. Neuropathol. Exp. Neurol.*, 33, 58, 1974.

102. Lärkfors, L., Oskarsson, A., Sundberg, J., and Ebendal, T., Methylmercury induced alterations in the nerve growth factor leel in the developing brain, *Brain Res. Dev.*, 62, 287, 1991.

103. Lazarus, J.H., Thyroid hormone and intellectual development: a clinician's view, *Thyroid*, 9, 659, 1999.

104. Legrand, J., Thyroid hormone effects on growth and development, in *Thyroid Hormone Metabolism*, Henneman, G., Ed., M. Dekker, Inc., New York, 1986, p. 503.

105. Li, R., Thode, S., Zhou, J., Richard, N., Pardinas, J., Rao, M.S., and Sah, D.W., Motoneuron differentiation of immortalized human spinal cord cell lines, *J. Neurosci. Res.*, 59, 342, 2000.

106. Li, Z., Lin, H., Zhu, Y., Wang, M.W., and Luo, J., Disruption of cell cycle kinetics and cyclin-dependent kinase system by ethanol in cultured cerebellar granule progenitors, *Dev. Brain Res.*, 132, 47, 2001.

107. Lichtensteiger, W., Ribary, U., Schlumpf, M., Odermatt, B., and Widmer, H.R., Prenatal adverse effects of nicotine on the developing brain, *Prog. Brain Res.*, 73, 137–157, 1988.

108. Light, K.E., Belcher, S.M., and Pierce, D.R., Time course and manner of purkinje neuron death following a single ethanol exposure on postnatal day 4 in the developing rat, *Neuroscience*, 114, 327, 2002.

109. Lipton, S.A. and Rosenberg, P.A., Excitatory amino acids as a final common pathway for neurologic disorders, *N. Engl. J. Med.*, 330, 613– 622, 1994.

110. Liu, Y., Wu, Y., Lee, J.C., Xue, H., Pevny, L.H., Kaprielian, Z., and Rao, M.S., Oligodendrocyte and astrocyte development in rodents: an *in situ* and immunohistological analysis during embryonic development, *Glia*, 40, 25, 2002.

111. Lu, Q.R., Sun, T., Zhu, Z., Ma, N., Garcia, M., Stiles, C.D., and Rowitch, D.H., Common developmental requirement for Olig function indicates a motor neuron/oligodendrocyte connection, *Cell*, 109, 75, 2002.

112. Ma, W., Li, B.S., Maric, D., Zhao, W.Q., Lin, H.J., Zhang, I., Pant, H.C., and Barker, J.L., Ethanol blocks both basic fibroblast growth factor- and carbachol-mediated neuroepithelial cell expansion with differential effects on carbachol-activated signaling pathways, *Neuroscience*, 118, 37, 2002.

113. Macklin, W.B. and Weill, C.L., Appearance of myelin proteins during development in the chick central nervous system, *Dev. Neurosci.*, 7, 170, 1985.

114. Madsen, K.M., Hviid, A., Vestergaard, M., Schendel, D., Wohlfahrt, J., Thorsen, P., Olsen, J., and Melbye, M., A population-based study of measles, mumps and rubella vaccination and autism, *N. Eng. J. Med.*, 347, 1477, 2002.

115. Mallard, E.C., Rees, S., Stringer, M., Cock, M.L., and Harding, R., Effects of chronic placental insufficiency on brain development in fetal sheep, *Pediatr. Res.*, 43, 262–270, 1998.

116. Man, E.B., Brown, J.F., and Scrunian, S.A., Maternal hypothyroxinemia: psychoneurological deficits of progeny, *Ann. Clin. Lab. Sci.*, 21, 227–239, 1991.

117. Marcus, J.C., Neurological findings in the fetal alcohol syndrome, *Neuropediatrics*, 18, 158, 1987.

118. Marmur, R., Kessler, J.A., Zhu, G., Gokhan, S., and Mehler, M.F., Differentiation of oligodendroglial progenitors derived from cortical multipotent cells requires extrinsic signals including activation of gp130/LIFbeta receptors, *J. Neurosci.*, 18, 9800, 1998.

119. Matute, C., Properties of acute and chronic kainate excitotoxic damage to the optic nerve, *Proc. Natl. Acad. Sci. U.S.A.*, 95, 10229–10234, 1998.

120. Mayer, M., Bhakoo, K., and Noble, M., Ciliary neurotrophic factor and leukemia inhibitory factor promote the generation, maturation and survival of oligodendrocytes *in vitro*, *Development*, 120, 142, 1994.

121. Mayer, M. and Noble, M., N-acetyl-L-cysteine is a pluripotent protector against cell death and enhancer of trophic factor-mediated cell survival *in vitro*, *Proc. Natl. Acad. Sci. U.S.A.*, 91, 7496, 1994.

122. Mayer-Pröschel, M., Kalyani, A., Mujtaba, T., and Rao, M.S., Isolation of lineage-restricted neuronal precursors from multipotent neuroepithelial stem cells, *Neuron*, 19, 773, 1997.

123. McDonald, J.W., Althomsons, S.P., Hyrc, K.L., Choi, D.W., and Goldberg, M.P., Oligodendrocytes from forebrain are highly vulnerable to AMPA/kainate receptor-mediated excitotoxicity, *Nat. Med.*, 4, 291, 1998.

124. McDonough, K.H., Antioxidant nutrients and alcohol, *Toxicology*, 189, 89, 2003.

125. Meldrum, B. and Garthwaite, J., Excitatory amino acid neurotoxicity and neurodegenerative disease, *Trends Pharmacol. Sci.*, 11, 379–387, 1990.

126. Mendola, P., Selevan, S.G., Gutter, S., and Rice, D., Environmental factors associated with a spectrum of neurodevelopmental deficits, *Ment. Retard. Dev. Disabil. Res. Rev.*, 8, 188, 2002.

127. Miller, M.W., Effect of prenatal exposure to ethanol on the development of cerebral cortex: I. Neuronal generation, *Alcohol. Clin. Exp. Res.*, 12, 440–449, 1988.

128. Miller, M.W., Effect of prenatal exposure to ethanol on neocortical development: II. Cell proliferation in the ventricular and subventricular zones of the rat, *J. Comp. Neurol.*, 278, 326–338, 1989.

129. Miller, M.W., Effects of prenatal exposure to ethanol on cell proliferation and neuronal migration, in *Development of the Central Nervous System: Effects of Alcohol and Opiates*, Miller, M.W., Ed., Wiley-Liss, Inc., New York, 1992, p. 47.

130. Miller, M.W., Effect of pre- or postnatal exposure to ethanol on the total number or neurons in the principal sensory nucleus of the trigeminal nerve: Cell proliferation versus neuronal death, *Alcohol. Clin. Exp. Res.*, 19, 1359–1364, 1995.

131. Miller, M.W., Generation of neurons in the rat denate gyrus and hippocampus: Effects of prenatal exposure to ethanol treatment with ethanol, *Alcohol. Clin. Exp. Res.*, 19, 1500–1509, 1995.

132. Miller, M.W., Chiaia, N.L., and Rhoades, R.W., Intracellular recording and injection study of corticospinal neurons in the rat somatosensory cortex: Effect of prenatal exposure to ethanol, *J. Comp. Neurol.*, 297, 91–105, 1990.

133. Miller, M.W. and Nowakowski, R.S., Effect of prenatal exposure to ethanol on the cell cycle kinetics and growth fraction in the proliferative zones of fetal rat cerebral, *Alcohol. Clin. Exp. Res.*, 15, 229–232, 1991.

134. Miller, M.W. and Potempa, G., Numbers of neurons and glia in mature rat somatosensory cortex: Effect of prenatal exposure to ethanol, *J. Comp. Neurol.*, 293, 92–102, 1990.

135. Miller, M.W. and Robertson, S., Prenatal exposure to ethanol alters the postnatal development and transformation of radial glia to astrocytes in the cortex, *J. Comp. Neurol.*, 337, 253–266, 1993.

136. Miura, K., Inokawa M., and Imura, N., Effects of methylmercury and some metal ions on microtubule networks in mouse glioma cells and *in vitro* tubulin polymerization, *Toxicol. Appl. Pharmacol.*, 73, 218–231, 1984.

137. Mujtaba, J., Piper, D., Groves, A., Kalyani, A., Lucero, M., and Rao, M.S., Lineage restricted precursors can be isolated from both the mouse neural tube and cultures ES cells, *Dev. Biol.*, 214, 113, 1999.

138. Mullaney, K.J., Fehm, M.N., Vitarella, D.E., Wagoner, D.E.J., and Aschner, M., The role of -SH groups in methylmercuric chloride-induced D-aspartate, and rubidium release from rat primary astrocyte cultures, *Brain Res.*, 641, 1, 1994.

139. Mundy, W.R., Parran, D.K., and Barone, S., Jr., Gestational exposure to methylmercury alters the developmental pattern of neurotrophin- and neurotransmitter-induced phosphoinositide (PI) hydrolysis, *Neurotoxicity Res.*, 1, 271, 2000.

140. Myers, G.J. and Davidson, P.W., Prenatal methylmercury exposure and children: neurologic, developmental, and behavioral research, *Environ. Health Perspect.*, 106 (Suppl. 3), 841–847, 1998.

141. Naeye, R.L., Cognitive and behavioral abnormalities in children whose mothers smoked cigarettes during pregnancy, *J. Dev. Behav. Pediatr.*, 13, 425–428, 1992.

142. Nicholson, J.L. and Altman, J., The effects of early hypo- and hyperthyroidism on the development of the rat cerebellar cortex. I. Cell proliferation and differentiation, *Brain Res.*, 44, 13–23, 1972.

143. Noble, M., Arhin, A., Gass, D., and Mayer-Proschel, M., The cortical ancestry of oligodendrocytes: Common principles and novel features, *Dev. Neurosci.*, 25, 217, 2003.

144. Noble, M., Pröschel, C., and Mayer-Proschel, M., Getting a GR(i)P on oligodendrocyte development, *Dev. Biol.*, 265, 33, 2004.

145. Noble, M., Smith, J., Power, J., and Mayer-Pröschel, M., Redox state as a central modulator of precursor cell function, *Ann. N.Y. Acad. Sci.*, 991, 251, 2003.

146. Ogawa, Y., Sawamoto, K., Miyata, T., Miyao, S., Watanabe, M., Nakamura, M., Bregman, B.S., Koike, M., Uchiyama, Y., Toyama, Y., and Okano, H., Transplantation of *in vitro*-expanded fetal neural progenitor cells results in neurogenesis and functional recovery after spinal cord contusion injury in adult rats, *J. Neurosci. Res.*, 69, 925, 2002.

147. Ono, K., Bansal, R., Payne, J., Rutishauser, U., and Miller, R.H., Early development and dispersal of oligodendrocyte precursors in the embryonic chick spinal cord, *Development*, 121, 1743, 1995.

148. Oppenheimer, J.H. and Schwartz, H.L., Molecular basis of thyroid hormone-dependent brain development, *Endocrine Rev.*, 18, 462, 1997.

149. Ou, Y.C., White, C.C., Krejsa, C.M., Ponce, R.A., Kavanagh, T.J., and Faustman, E.M., The role of intracellular glutathione in methylmercury-induced toxicity in embryonic neuronal cells, *Neurotoxicology*, 20, 793–804, 1999.

150. Özer, E., Saraioglu, S., and Güre, A., Effect of prenatal ethanol exposure on neuronal migration, neurogenesis and brain myelination in the mice brain, *Clin. Neuropathol.*, 19, 21, 2000.

151. Park, S.T., Lim, K.T., Chung, Y.T., and Kim, S.U., Methylmercury induced neurotoxicity in cerebral neuron culture is blocked by antioxidants and NMDA receptor antagonists, *Neurotoxicology*, 17, 37–46, 1996.

152. Park, S.T., Lim, K.T., Chung, Y.T., and Kim, S.U., Methylmercury-induced neurotoxicity in cerebral neuron culture is blocked by antioxidants and NMDA receptor antagonists, *Neurotoxicology*, 17, 37–46, 1996.

153. Parran, D.K., Mundy, W.R., and Barone, S., Jr., Effects of methylmercury and mercuric chloride on differentiation and cell viability in PC12 cells, *Toxicol. Sci.*, 59, 278, 2002.

154. Patneau, D.K., Wright, P.W., Winters, C., Mayer, M.L., and Gallo, V., Glial cells of the oligodendrocyte lineage express both kainate- and AMPA-preferring subtypes of glutamate receptor, *Neuron*, 12, 357, 1994.

155. Phillips, D.E. and Kreuger, S.K., Effects of ethanol exposure on glial cell development in the optic nerve, *Exp. Neurol.*, 107, 97, 1990.

156. Pichichero, M.E., Cernichiari, E., Lopreiato, J., and Treanor, J., Mercury concentrations and metabolism in infants receiving vaccines containing thimerosal: a descriptive study, *Lancet*, 360, 1711, 2002.

157. Pinzo-Duran, M.D., Renau-Piqueras, J., and Guerri, C., Developmental changes in the optic nerve related to ethanol consumption in pregnant rats: analysis of the ethanol-exposed optic nerve, *Teratology*, 48, 305, 1993.

158. Pombo, P.M., Barettino, D., Ibarrola, N., Vega, S., and Rodriguez-Pena, A., Stimulation of the myelin basic protein gene expression by 9-cis-retinoic acid and thyroid hormone: activation in the context of its native promoter, *Mol. Brain Res.*, 64, 92, 1999.

159. Pombo, P.M., Ibarrola, N., Alonso, M.A., and Rodriguez-Pena, A., Thyroid hormone regulates the expression of the MAL proteolipid, a component of glycolipid-enriched membranes, in neonatal rat brain, *J. Neurosci. Res.*, 52, 584, 1998.

160. Pop, V.J., Kuijpens, J.L., van Baar, A.L., Verkerk, G., van Son, M.M., de Vijlder, J.J., Vulsma, T., Wiersinga, W.M., Drexhage, H.A., and Vader, H.L., Low maternal free thyroxine concentrations during early pregnancy are associated with impaired psychomotor development in infancy, *Clin. Endoc.*, 50, 149, 1999.

161. Power, J., Mayer-Proschel, M., Smith, J., and Noble, M., Oligodendrocyte precursor cells from different brain regions express divergent properties consistent with the differing time courses of myelination in these regions, *Dev. Biol.*, 245, 362, 2002.

162. Prasad, K.N., Nobles, E., and Ramanujam, M., Differential sensitivity of glioma cells and neuroblastoma cells to methylmercury toxicity in cultures, *Environ. Res.*, 19, 189–201, 1979.

163. Centers for Disease Control and Prevention, *Second National Report on Human Exposure to Environmental Chemicals*, Department of Health and Human Services, 2003. Retrieved from http://www.cdc.gov/exposurereport/

164. Putzke, J., De Beun, R., Schreiber, R., De Vry, J., Tolle, T.R., Zieglgansberger, W., and Spanagel, R., Long-term alcohol self-administration and alcohol withdrawal differentially modulate microtubule-associated protein 2 (MAP2) gene expression in the rat brain, *Brain Res. Mol. Brain Res.*, 62, 196–205, 1988.

165. Qiang, M., Wang, M.W., and Elberger, A.J., Second trimester prenatal alcohol exposure alters development of rat corpus callosum, *Neurotox. Teratol.*, 24, 719, 2002.

166. Raff, M.C., Abney, E.R., Cohen, J., Lindsay, R., and Noble, M., Two types of astrocytes in cultures of developing rat white matter: differences in morphology, surface gangliosides, and growth characteristics, *J. Neurosci.*, 3, 1289, 1983.

167. Raff, M.C., Miller, R.H., and Noble, M., A glial progenitor cell that develops *in vitro* into an astrocyte or an oligodendrocyte depending on the culture medium, *Nature*, 303, 390, 1983.

168. Rao, M. and Mayer-Pröschel, M., Glial restricted precursors are derived from multipotent neuroepithelial stem cells, *Dev. Biol.*, 188, 48, 1997.

169. Rao, M., Noble, M., and Mayer-Pröschel, M., A tripotential glial precursor cell is present in the developing spinal cord, *Proc. Natl. Acad. Sci. U.S.A.*, 95, 3996, 1998.

170. Rao, M.S., Multipotent and restricted precursors in the central nervous system, *Anat. Rec.*, 257, 137, 1999.

171. Rees, S., Stringer, M., Just, Y., Hooper, S.B., and Harding, R., The vulnerability of the fetal sheep brain to hypoxemia at mid-gestation, *Brain Res. Dev. Brain Res.*, 103, 103, 1997.

172. Rice, D.C., Parallels between attention deficit hyperactivity disorder and behavioral effects produced by neurotoxic exposure in monkeys, *Environ. Health Perspect.*, 108 (Suppl. 3), 405, 2001.

173. Rice, J.E., Vannucci, R.C., and Brierley, J.B., The influence of immaturity on hypoxia-ischemia brain damage in the rat, *Ann. Neurol.*, 9, 131, 1981.

174. Richardson, W.D., Smith, J.K., Sun, T., Pringle, N.P., Hall, A.C., and Woodruff, R., Oligodendrocyte lineage and the motor neuron connection, *Glia*, 29, 136, 2000.

175. Robinson, S., Tani, M., Strieter, R., Ransohoff, R., and Miller, R.H., The chemokine growth-regulated oncogene-alpha promotes spinal cord oligodendrocyte precursor proliferation, *J. Neurosci.*, 18, 10457, 1998.

176. Rosenberg, P.A., Dai, W., Gan, X.D., Ali, S., Fu, J., Back, S.A., Sanchez, R.M., Segal, M.M., Follett, P.L., Jensen, F.E., and Volpe, J.J., Mature myelin basic protein-expressing oligodendrocytes are insensitive to kainate toxicity, *J. Neurosci. Res.*, 71, 237, 2003.

177. Rosman, N.P., Malone, M.J., Helfenstein, M., and Kraft, E., The effect of thyroid deficiency on myelination of brain, *Neurology*, 22, 99–106, 1972.

178. Rothman, S.M. and Olney, J.W., Glutamate and the pathophysiology of hypoxic-ischemic brain damage, *Ann. Neurol.*, 19, 105, 1986.

179. Rowitch, D.H., Lu, R.Q., Kessaris, N., and Richardson, W.D., An 'oligarchy' rules neural development, *Trends Neurosci.*, 25, 417, 2002.

180. Roy, C., Prasad, K.V.S., Reuhl, K.R., Little, J.E., Valentine, B.K., and Brown, D.L., Taxol protects the microtubules of Concanavalin A-activated lymphocytes from disassembly by methylmercury, but DNA synthesis is still inhibited, *Exp. Cell Res.*, 195, 345–352, 1991.

181. Sager, P.R., Doherty, R.A., and Olmsted, J.B., Interaction of methylmercury with microtubules in cultured cells and *in vitro*, *Exp. Cell Res.*, 146, 127–137, 1983.

182. Sanchez-Gomez, M.V. and Matute, C., *Neurobiol. Dis.*, 6, 475–485, 1999.

183. Sarafian, T. and Verity, M.A., Oxidative mechanisms underlying methyl mercury neurotoxicity, *Int. J. Dev. Neurosci.*, 9, 147–153, 1991.

184. Sarafian, T.A., Bredesen, D.E., and Verity, M.A., Cellular resistance to methylmercury. *Neurotoxicology*, 17, 27–36, 1996.

185. Sarafian, T.A., Vartavarian, L., Kane, D.J., Bredesen, D.E., and Verity, M.A., bcl-2 expression decreases methylmercury-induced free-radical generation and cell killing in a neural cell line, *Toxicol. Lett.*, 74, 149–155, 1994.

186. Sauvageot, C.M. and Stiles, C.D., Molecular mechanisms controlling cortical gliogenesis, *Curr. Opin. Neurobiol.*, 12, 244, 2002.

187. Skoff, R.P., Toland, D., and Nast, E., Pattern of myelination and distribution of neuroglial cells along the developing optic system of the rat and rabbit, *J. Comp. Neurol.*, 191, 237, 1980.

188. Slotkin, T.A., Prenatal exposure to nicotine: What can we learn from animal models?, in *Maternal Substance Abuse and the Developing Nervous System*, Zagon, I.S. and Slotkin, T.A., Eds., Academic Press, San Diego, CA, 1992, p. 97.

189. Small, R.K., Riddle, P., and Noble, M., Evidence for migration of oligodendrocyte-type-2 astrocyte progenitor cells into the developing rat optic nerve, *Nature*, 328, 155, 1987.

190. Smith, J., Ladi, E., Mayer-Pröschel, M., and Noble, M., Redox state is a central modulator of the balance between self-renewal and differentiation in a dividing glial precursor cell, *Proc. Natl. Acad. Sci. U.S.A.*, 97, 10032, 2000.

191. Snyder, A.K., Response of glia to alcohol, in *The Role of Glia in Neurotoxicity*, Aschner, M. and Kimelberg, H.K., Eds., CRC Press, Boca Raton, FL, 1996, p. 111.

192. Sorg, O., Horn, T.F., Yu, N., Gruol, D.L., and Bloom, F.E., Inhibition of astrocyte glutamate uptake by reactive oxygen species: role of antioxidant enzymes, *Mol. Med.*, 3, 431–440, 1997.

193. Sorg, O., Schilter, B., Honnegger, P., and Monnet-Tschudi, F., Increased vulnerability of neurons and glial cells to low concentrations of methylmercury in a pro-oxidant situation, *Acta Neuropathol.*, 96, 621–627, 1998.

194. Staal, F., Anderson, M., Staal, G., Herzenberg, L., Gitler, C., and Herzenberg, L., Redox regulation of signal transduction: tyrosine phosphorylation and calcium influx, *Proc. Natl. Acad. Sci. U.S.A.*, 91, 3619, 1994.

195. Staal, F.J., Anderson, M.T., Staal, G.E., Herzenberg, L.A., Gitler, C., and Herzenberg, L.A., Redox regulation of signal transduction: tyrosine phosphorylation and calcium influx, *Proc. Natl. Acad. Sci. U.S.A.*, 91, 3619, 1994.

196. Stein, J., Schettler, T., Wallinga, D., and Valenti, M., In harm's way: Toxic threats to child development, *Dev. Behav. Ped.*, 23, S13, 2002.

197. Szabo, A., Boucher, K., Carroll, W., Klebanov, L., Tsodikov, A., and Yakovlev, A., Variable selection and pattern recognition with gene expression data generated by the microarray technology, *Math. Biosci.*, 176, 71–98, 2002.

198. Szabo, A., Boucher, K., Jones, D., Klebanov, L., Tsodikov, A., and Yakovlev, A., Multivariate exploratory tools for microarray data analysis, *Biostatistics*, 4, 555, 2003.

199. Takebayashi, H., Nabeshima, Y., Yoshida, S., Chisaka, O., Ikenaka, K., and Nabeshima, Y., The basic helix-loop-helix factor olig2 is essential for the development of motoneuron and oligodendrocyte lineages, *Curr. Biol.*, 12, 1157, 2002.

200. Tandon, S.K., Singh, S., Prasad, S., Khandekar, K., Dwivedi, V.K., Chatterjee, M., and Mathur, N., Reversal of cadmium induced oxidative stress by chelating agent, antioxidant or their combination in rat, *Toxicol. Lett.*, 145, 211, 2003.

201. Tennekoon, G., Aitchison, C.S., Frangia, J., Price, D.L., and Goldberg, A.M., Chronic lead intoxication: Effects on developing optic nerve, *Ann. Neurol.*, 5, 558, 1979.

202. Thiruchelvam, M., Richfield, E.K., Baggs, R.B., Tank, A.W., and Cory-Slechta, D.A., The nigrostriatal dopaminergic system as a preferential target of repeated exposures to combined paraquat and maneb: implications for Parkinson's disease, *J. Neurosci.*, 20, 2000.

203. Thiruchelvam, M., Richfield, E.K., Goodman, B.M., Baggs, R.B., and Cory-Slechta, D.A., Developmental exposure to the pesticides paraquat and maneb and the Parkinson's disease phenotype, *Neurotoxicology*, 23, 621, 2002.

204. Toews, A.D., Krigman, M.R., Thomas, D.J., and Morell, P., Effect of inorganic lead exposure on myelination in rat, *Neurochem. Res.*, 5, 605, 1980.

205. Tosic, M., Torch, S., Comte, V., Dolivo, M., Honegger, P., and Matthieu, J.M., Triiodothyronine has diverse and multiple stimulating effects on expression of the major myelin protein genes, *J. Neurochem.*, 59, 1770, 1992.

206. Trana, T.D. and Kelly, S.J., Critical periods for ethanol-induced cell loss in the hippocampal formation, *Neurotox. Teratol.*, 25, 519, 2003.

207. Trotter, J. and Schachner, M., Cells positive for the O4 surface antigen isolated by cell sorting are able to differentiate into astrocytes or oligodendrocytes, *Brain Res. Dev. Brain Res.*, 46, 115, 1989.

208. Utzschneider, D.A., Archer, D.R., Kocsis, J.D., Waxman, S.G., and Duncan, I.D., Transplantation of glial cells enhances action potential conduction of amyelinated spinal cord axons in the myelin-deficient rat, *Proc. Natl. Acad. Sci. U.S.A.*, 53, 1994.

209. Volpe, J.J., Hypoxic-ischemic encephalopathy: neuropathology and pathogenesis, in *Neurology of the Newborn*, W.B. Saunders, Philadelphia, 2001, pp. 296–330.

210. Volpe, J.J., Neurobiology of periventricular leukomalacia in the premature infant, *Pediatr. Res.*, 50, 553–562, 2001.

211. Waly, M., Olteanu, H., Bamerjee, R., Choi, S.-W., Mason, J.B., Parker, B.S., Sukumar, S., Shim, S., Sharmal, A., Benzecryl, J.M., Power-Charnitsky, V.-A., and Deth, R.C., Activation of methionine synthase by insulin-like growth factor-1 and dopamine: a target for neurodevelopmental toxins and thimerosal, *Molec. Psychiat.*, 1, 2002.

212. Warrington, A.E., Barbarese, E., and Pfeiffer, S.E., Differential myelinogenic capacity of specific developmental stages of the oligodendrocyte lineage upon transplantation into hypomyelinating hosts, *J. Neurosci. Res.*, 34, 1, 1993.

213. Wasteneys, G.O., Cadrin, M., Reuhl, K.R., and Brown, D.L., The effects of methylmercury on the cytoskeleton of murine embryonal carcinoma cells, *Cell Biol. Toxicol.*, 4, 41–60, 1988.

214. Wren, D., Wolswijk, G., and Noble, M., *In vitro* analysis of the origin and maintenance of O-2A^{adult} progenitor cells, *J. Cell Biol.*, 116, 167, 1992.

215. Yakovlev, A., Mayer-Proschel, M., and Noble, M., A stochastic model of brain cell differentiation in tissue culture, *J. Math. Biol.*, 37, 49, 1998.

216. Yakovlev, A.Y., Boucher, K., Mayer-Pröschel, M., and Noble, M., Quantitative insight into proliferation and differentiation of O-2A progenitor cells *in vitro*: The clock model revisited, *Proc. Natl. Acad. Sci. U.S.A.*, 95, 14164, 1998.

217. Yakovlev, P.L. and Lecours, A.R., The myelogenetic cycles of regional maturation of the brain, in *Regional Development of the Brain in Early Life*, Minkowski, A. et al., Eds., Blackwell, Oxford, 1967, p. 3.

218. Yonaha, M., Saito, M., and Sagai, M., Stimulation of lipid peroxidation by methyl mercury in rats, *Life Sci.*, 32, 1507–1514, 1983.
219. Yonaha, M., Saitoh, M., and Sagai, M., Stimulation of lipid peroxidation by methyl mercury in rats, *Life Sci.*, 32, 1507–1514, 1983.
220. Zhou, Q. and Anderson, D.J., The bHLH transcription factors olig2 and olig1 couple neuronal and glial subtype specification, *Cell*, 109, 61, 2002.
221. Zorin, A., Mayer-Proschel, M., Noble, M., and Yakovlev, A.Y., Estimation problems associated with stochastic modeling of proliferation and differentiation of O-2A progenitor cells *in vitro*, *Math Biosci.*, 167, 109, 2000.

CHAPTER 2

The Role of Microglia in Neurotoxicity

Wolfgang J. Streit

CONTENTS

2.1 INTRODUCTION

Two basic, but distinct, issues come to mind when considering a role of microglia in neurotoxicity. First, microglia respond rapidly to neuron injury whether it is induced by a neurotoxic agent, physical trauma, or by some other means. The microglial response to neuronal damage is characterized by a number of different events, which include mitosis/proliferation, changes in microglial morphology and phenotype, and increased production of growth factors and cytokines. All of these changes are collectively referred to as microglial activation, and they reflect an acute tissue response to injury. The biological significance of a microglial reaction to neuron injury is likely to be the same as in other tissues, namely, the engagement of cellular mechanisms that initiate the wound-healing process. The second issue concerns the possibility that microglia themselves may exert some type of cellular cytotoxicity that could be detrimental to neurons. The key questions here are, when do microglia become neurotoxic, what are the triggers, and how is microglial cytotoxicity regulated? Obviously, if microglia were neurotoxic constitutively, or could be easily induced to become neurotoxic, given their ubiquitous presence throughout the CNS all neurons would be in constant danger of falling victim to microglia-mediated cytotoxicity. Therefore, microglial neurotoxicity *in vivo* likely represents a process that is under strict control and becomes unleashed only when the need for it arises, or when control mechanisms go awry.

2.2 HISTORICAL OVERVIEW

Microglial cells have moved to the center stage of neuroscience research fairly recently. This is despite the fact that they were recognized as a distinct glial cell type by pioneer neuropathologists, such as Nissl and Spielmeyer, as early as 100 years ago. However, the credit for discovering microglial cells is usually given to the Spanish neuroanatomist del Rio-Hortega, and rightfully so, because del Rio-Hortega provided the first detailed account of this third major glial cell type (del Rio-Hortega, 1932). In the decades that followed Hortega's groundbreaking work, only a handful of neuroscientists continued to study microglia. These investigators were largely concerned with the origin of microglia both in the adult CNS under pathological conditions and in the immature CNS during development. This preoccupation with the origin of microglial cells was due to a claim made by del Rio-Hortega that microglia were derived from the mesodermal germ layer, rather than from the neuroectoderm like other macroglial cells, such as oligodendrocytes and astrocytes. A longstanding controversy over the origin of microglia was the consequence. Today, this controversy can be considered largely resolved (Streit, 2001), and most researchers in the field would agree with del Rio-Hortega's original concept for reasons that will become clear in the subsequent paragraphs. However, most recently, *in vitro* data within the context of stem cell biology have suggested a possible new role for microglia during ontogeny as multipotential stem cells (Yokoyama et al., 2004). Clearly, this represents an exciting possibility that is supported also by the fact that microglia retain robust mitotic potential throughout life (Flanary and Streit, 2004), as well as by studies that demonstrate proregenerative effects of microglial transplants following CNS injury (Rabchevsky and Streit, 1997).

A critical event that brought about the revival of the microglial cell in the mid-1980s was the advent of monoclonal antibodies and lectins that could be used in histochemical protocols to visualize microglia in sections of brain tissue. Although del Rio-Hortega had developed a silver carbonate method for staining microglia, cell localization with this technique was not always successful and depended on many different variables, including animal species. Enzyme histochemical approaches that were originally developed to visualize other cells in different organs but could also stain microglia were put to use only sparingly for reasons unknown. Once immunohistochemistry and lectin histochemistry were available, they quickly became the methods of choice for studying microglial cells, causing a precipitous increase in publications and interest in microglial cells over the past 10 to 15 years.

A second methodological advancement which also came about in the mid-1980s was the isolation of microglial cells from primary mixed cultures of neonatal brain. This made it possible to study purified microglia in the culture dish and opened the door for a large number of possible experiments designed to examine the response of microglia to a variety of stimuli. One problem associated with the tissue culture experiments is the difficulty in establishing relevance of the *in vitro* findings to microglia in the CNS. Nevertheless, it was fortunate that two lines of new and completely different methodologies converged at a point in time when almost nothing was known about microglial functions. Based on the advancements made in microglial research over the past 10 years using both *in vivo* and *in vitro* approaches, microglia are now widely viewed as specially adapted immune cells within the CNS. This novel conceptual development has led to a reevaluation of the brain as an immunologically privileged organ and has given a new meaning to the discipline of neuroimmunology.

2.3 CURRENT QUESTIONS

2.3.1 Microglial Responses to Neuronal Injury

One of the earliest known cellular changes that occur in the microenvironment surrounding injured neurons is a microglial reaction. Typically, the microglial reaction is spatially confined to the region where the afflicted neuron cell bodies are located, although it often spreads to involve white matter areas containing the axons of injured neurons, especially if there is neuron loss due to acute degeneration and subsequent Wallerian degeneration. The time of onset of the microglial reaction, as well as its extent and duration, may vary with the severity of the injury.

2.3.1.1 Axotomy of Motor Neurons

One of the best-studied experimental injury models in terms of microglial reactions is the rat facial nerve paradigm. In this model, the facial nerve is transected or crushed near its exit from the skull, and sections of the brainstem containing the facial nucleus are examined histologically. Facial motor neurons undergo a classical chromatolytic response and subsequently regenerate their peripheral axons. Therefore, axotomy of motor neurons represents an example of reversible neuron injury accompanied by minimal, if any, neuronal loss. The earliest changes in microglial cells can be detected 24 hours after axotomy when microglia change their morphology and increase expression of CR3 complement receptors, surface molecules which are constitutively expressed by microglia in normal brain and which can be detected with a monoclonal antibody designated OX-42 (Robinson et al., 1986; Graeber et al., 1988a). Microglia begin to proliferate about 2 days post-axotomy (Graeber et al., 1988b), and their numbers increase to such an extent that by day 4 or 5 the entire area of the facial nucleus is inundated with microglial cells. The cells are now called reactive (activated) microglia, i.e., they have undergone a transformation from resting to reactive cells. Morphologically, this transformation involves contraction of the long and finely branched processes of resting cells into short and stout cell processes that are characteristic of activated microglia. Many of the activated microglia assume perineuronal positions ensheathing injured motor neurons with their cell processes. Concomitant with this change in morphology, reactive microglia also begin to change their phenotype, that is, they express surface molecules on their membrane which were undetectable prior to activation. Most notably, the surface proteins that are being expressed *de novo* include antigens of the major histocompatibility complex (MHC). While both class I and class II MHC antigens appear on the microglial surface, the time course of expression is different for each of the two classes of molecules, and it is likely that separate microglial subpopulations are responsible for class I and class II expression (Streit et al., 1989a,b; Streit and Graeber, 1993). The appearance of MHC, as well as other immunomolecules, has led to the concept that microglia are immunologically competent cells, and more specifically, that they function as indigenous antigen-presenting cells of the CNS (Hickey and Kimura, 1988; Streit et al., 1988).

It is likely that the activation of microglia as a consequence of axotomy also involves changes in their secretory activity. Studies by Kiefer et al. (1993) and by Streit et al. (2000) have shown that substantial increases in the levels of mRNA encoding transforming growth factor-β1 (TGF-β1) and interleukin-6 (IL-6) occur in the facial nucleus after nerve section. *In situ* hybridization studies suggest that the increase in TGF-β1 occurs in reactive microglial cells (Lehrmann et al., 1998), whereas upregulation of IL-6 mRNA occurs in neurons (Murphy et al., 1999). Interestingly, the levels of interleukin-1 (IL-1) message were not found to be upregulated after motor neuron injury, which is a somewhat unexpected finding, since activated microglia in culture produce IL-1 (Giulian et al., 1986; Hetier et al., 1988), and IL-1 mRNA and protein are known to increase after traumatic brain injuries (Higgins and Olschowka, 1991; Woodroofe et al., 1991; Yan et al., 1992). This finding on one hand supports the view that activated microglia in culture are not to be equated with activated microglia *in vivo* (Hurley et al., 1999), and on the other, that traumatic CNS injury is different from an axotomy lesion in that it involves a massive breakdown of the blood–brain barrier. The blood–brain barrier in the facial nucleus remains intact after facial nerve transection. In view of these differences, it seems likely that cytokine production and secretion by microglial cells is dependent on the experimental circumstances. Accordingly, observations regarding microglial cytokine production are to be considered cautiously and always in the context of the particular lesion paradigm employed.

2.3.1.2 *Cerebral Ischemia*

Transient global forebrain ischemia induced by temporary occlusion of the carotid arteries causes delayed neuronal death in the pyramidal cell layer of the CA1 region in the rat hippocampus (Kirino et al., 1984). The ischemia-induced neuron cell loss which becomes apparent 2-4 days after reperfusion is not only accompanied by a microglial response, as expected, but preceded by microglial activation, which may occur as early as 20 minutes after reperfusion (Morioka et al., 1991; Gehrmann et al., 1992). The appearance of activated microglial cells during the first 48 hours after ischemia, as demonstrated by immunohistochemistry and lectin histochemistry, is not limited to the region that eventually shows neuron loss, i.e., the stratum pyramidale of CA1, nor is it limited to the hippocampus. Activated microglia become widespread throughout the somatosensory cortex and striatum, regions that do not exhibit the blatant neurodegeneration seen in the CA1 pyramidal cell layer. These observations show, similar to the axotomy studies described previously, that microglial activation occurs regardless of whether there is acute neurodegeneration or not. Thus, microglial activation might be precipitated by a transient disruption of neuronal metabolism due to the ischemic insult, or, generally speaking, by neurons in distress. The distressed neurons may or may not undergo degeneration, depending on the nature of injury and the selective vulnerability of the neuronal populations affected. Supporting a neuronal distress theory is the fact that a microglial reaction subsides sooner in those regions that do not experience neurodegeneration. For example, activated microglia are seen in the stratum pyramidale of CA1, as well as in the dentate hilus, within 24 hours after global ischemia, and while the microglial response is sustained in CA1 with the cells eventually transforming into phagocytes and displaying pronounced immunoreactivity for MHC class II antigens, this is not the case in the dentate hilus (Morioka et al., 1992; Gehrmann et al., 1992; Streit, 1993). Moreover, when pyramidal neurons in CA1 are rescued from ischemic damage through administration of MK-801, an NMDA receptor antagonist, the microglial reaction is substantially attenuated (Streit et al., 1992). As in traumatic brain injury, an increase in the levels of IL-1β mRNA has been demonstrated after both transient and permanent ischemia, suggesting a possible role for activated microglial cells in the production of this proinflammatory cytokine (Minami et al., 1991; Liu et al., 1993). Immunohistological studies after direct glutamate injection into the brain are supportive in this regard, as they have reported the presence of IL-1β immunoreactivity on both microglia and astrocytes (Pearson et al., 1999).

2.3.1.3 *Neurotoxic Injury*

A number of neurotoxic substances that cause neurodegeneration and subsequent loss of neurons have been employed for producing selective and localized lesions in the CNS. These lesions are accompanied by a microglial reaction that can be visualized histochemically *in situ*, delineating areas of toxin-induced neurodegeneration. For the following discussion, three neurotoxins with three different mechanisms of action have been chosen, and these are listed in Table 2.1.

2.3.1.3.1 *Ricinus communis* Agglutinin (RCA$_{60}$)

This protein of the castor bean with a molecular weight of 60 kD is a very potent cytotoxin, and minute amounts of toxic ricin are sufficient to kill any cell (Olsnes et al., 1974). The value of

Table 2.1   Some Neurotoxic Substances Eliciting Microglial Activation

Name (Type) of Neurotoxin	Route of Administration	Mechanism of Action
Ricinus communis lectin (protein)	Intraneural injection (peripheral nerve)	Inhibits protein synthesis
Kainic acid (excitatory amino acid)	Intracerebral, systemic, or intraventricular injection; epidural application	Excitotoxicity
Trimethyltin (organotin)	Systemic injection	Unknown

RCA_{60} for neurobiology lies in the fact that it is a protein that is axonally transported (Wiley et al., 1982). After intra-axonal application, RCA_{60} is transported retrogradely to the cell bodies of parent neurons where it inhibits protein synthesis, resulting in rapid neuronal cell death. Using injections of RCA_{60} into a peripheral nerve, it is possible to delete selectively those neurons that contribute their axons to the nerve. This is advantageous because it allows selective obliteration of specific neuronal populations without having to use intracerebral injections that might compromise the blood–brain barrier or cause collateral damage.

In order to differentially characterize microglial responses under conditions of neuronal regeneration versus neuronal degeneration, toxic ricin was injected into the facial nerve followed by crushing of the injection site to facilitate axonal uptake of RCA_{60} (Streit and Kreutzberg, 1988). The contralateral facial nerve was crushed only, and served as an axotomy control. The two key findings from these studies were as follows: (1) Microglia in the degenerating (ricin-injected) facial nucleus respond faster and with greater intensity, as measured by their proliferative index, and (2) microglia undergo morphological transformation into brain macrophages as a result of neurodegeneration. Conceptually, the ricin experiments have established a role for microglia as a source of brain macrophages emphasizing the cells' potential for plasticity, which is expressed in at least three morphologically defined states: resting, reactive (activated), and phagocytic (Streit et al., 1988). Complete transition to the phagocytic state occurs only in the presence of motoneuron degeneration (Figure 2.1B), and microglia-derived brain macrophages are not observed after simple axotomy. Phenotypically, microglia-derived brain macrophages express the same surface molecules as activated microglia, including MHC antigens, as well as various lymphocyte and macrophage markers. With regard to their secretory activity, it remains to be determined if, and how, microglia-derived brain macrophages exhibit a profile of secretory activity that is different from that of activated microglia. This question becomes of paramount importance when considering potential neurotoxic effects of brain macrophages (see next section).

2.3.1.3.2 Kainic Acid

This excitotoxic amino acid has been employed in numerous studies that were undertaken to clarify questions regarding glial responses to neurodegeneration (Akiyama et al., 1988; Andersson et al., 1991; Kaur and Ling, 1992; Pasinetti et al., 1992; Finsen et al., 1993; Mitchell et al., 1993; Morgan et al., 1993; Yabuuchi et al., 1993). Unfortunately, different investigators prefer to use different routes of administering kainic acid, which results in data sets that are difficult to compare directly with each other. Nevertheless, certain parallels regarding microglial activation are quite conspicuously illustrated in all of these studies, and these are similar to the features of microglial activation shown after facial nerve lesions. Microglia take on either activated or phagocytic phenotypes depending on how severely brain regions are affected. Typically, the plump phagocytic forms are seen in areas where neuron cell bodies undergo acute excitotoxic neurodegeneration, while the bushy, activated forms appear to be more numerous in areas of fiber degeneration. MHC antigen expression is increased on most microglia, regardless of whether they are of the activated or phagocytic kind. Some studies have investigated cytokine production, and these show that increased production of IL-1β and TGF-β1 mRNAs is likely to occur in microglial cells, although based on

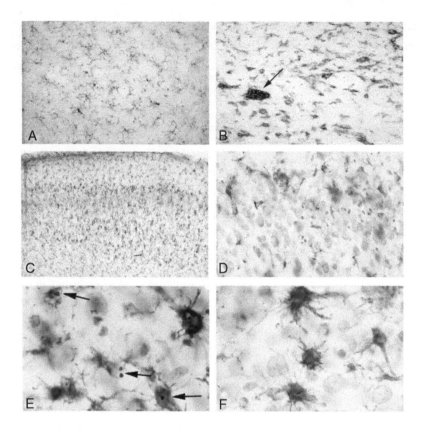

Figure 2.1 Activation of rat microglial cells after intraneural injection of toxic ricin (B) and after systemic intoxication with trimethyltin (C–F). Panel A shows nonactivated, ramified (resting) microglia in the normal
rat facial nucleus stained with OX-42 antibody. In B, degeneration of facial motoneurons has occurred
4 days after injecting ricin into the facial nerve. Microglial cells have become activated and formed
a large phagocytic cluster (arrow) around a degenerating motoneuron, as seen with OX-42 immunostaining. Panels C–F show activated microglia stained with GSA I-B$_4$ isolectin, as described
(Streit, 1990). C and D show the piriform cortex, and E and F show the olfactory bulb at 7 days
after TMT intoxication. Note the presence of nuclear fragments in E (arrows) indicating apoptotis
of neurons and subsequent phagocytosis by microglia. In F, note the presence of activated microglia
among normal-looking neurons. Magnifications: ×110 (A, B); ×55 (C); ×220 (D); ×550 (E, F).

the data shown, other cellular sources cannot be excluded (Morgan et al., 1993; Yabuuchi et al.,
1993). The cytokine data are consistent with what has been shown to occur after traumatic brain
injury, as well as after cerebral ischemia, and the observed increases in IL-1β and TGFβ-1 reflect
a scenario similar to other brain injuries where there is widespread neuronal damage.

2.3.1.3.3 Trimethyltin (TMT)

Systemic administration of TMT causes selective neuron damage primarily in regions of the
limbic system, as shown by cupric-silver impregnation studies (Balaban et al., 1988). Studies
by McCann et al. (1996) have confirmed the regional selectivity of TMT intoxication by using
lectin staining of microglial cells. Following a single systemic dose of TMT, activated microglial
cells are present in the septum, piriform cortex, olfactory bulb, and hippocampus. While some
of the microglial cells display the typical phagocytic morphology with a rounded, ameboid cell
shape, others have stout processes and their cell bodies are considerably enlarged, indicative
of their activated state. Since these activated microglia are frequently seen to coexist side by
side with healthy-looking neurons (Figure 2.1D–F), there is little reason to think that activated

microglia are detrimental to neuronal survival. Evidence for extensive phagocytosis at the light microscopic level, such as the formation of large clusters of microglial cells, as seen after degeneration of facial motor neurons (cf. Figure 2.1B) is rare. The density of neurons in the piriform cortex, as assessed by Nissl staining, is not obviously diminished (Figure 2.1C, D). These observations together with ultrastructural studies showing that TMT causes degeneration of isolated neurons of the hippocampus (Chang et al., 1982) show that the extent of neurodegeneration after TMT poisoning is perhaps not as widespread as suggested by silver impregnation techniques, which can be capricious and difficult to interpret (Walberg, 1971). In summary, TMT does induce selective neurodegeneration limited to specific CNS regions within days after administration. This acute neurodegeneration is delineated by the presence of reactive or phagocytic microglial cells. By carefully studying microglial morphology in conjunction with Nissl counterstaining, it is possible to obtain a qualitative assessment of the extent of the damage. It is unlikely that activated or phagocytic microglia, which are often seen next to healthy neurons, are themselves a source of neurotoxicity. Instead the observed neurodegeneration is due to a direct neurotoxicity of TMT, and TMT-induced neuronal damage is what causes a microglial response.

2.3.1.4 Significance of Microglial Reaction

One of the regrettable shortcomings of *in vivo* studies is that it is often difficult to draw definitive conclusions about the functional significance of a sequence of events captured by a series of static images. Nevertheless, observations made *in vivo* do provide an excellent basis for formulating hypotheses since *in vivo* findings are actually made in the brain rather than on isolated brain cells. The latter may exhibit dramatically altered properties as a result of having been maintained in cell culture. Based on the observations described in the preceding paragraphs regarding the microglial responses to various types of acute neuron injury, perhaps two features can be singled out:

1. Regardless of the way neuronal injury is induced, microglia respond in a rapid, sensitive, and predictable fashion. While the speed with which this response occurs may vary from one paradigm to the next, in each case microglial cells seem to go through a similar sequence of events involving characteristic morphological and phenotypic changes. Eventually, as neurons recover from the injury and debris is cleared by brain macrophages, the cells revert back to their normal resting state.
2. A consistent feature in models of neurodegeneration is the coexistence of activated or phagocytic microglia side by side with normal, healthy-looking neurons. This implies that activated/phagocytic microglia are not damaging to neurons, because if they were, one would expect to find them surrounded by neurons that show structural indications of damage, such as swelling, nuclear fragmentation, or even lysis.

What then might be the purpose of microglial activation other than a stepping stone towards providing macrophages for clearing out possible debris? It is conceivable that microglial activation serves to provide support to injured neurons, and this idea of a neurotrophic role for microglia becomes particularly attractive when considering closely what is happening in the facial nucleus after axotomy. Locally present microglial cells increase dramatically in number and engulf axotomized motor neurons with their processes, stripping away axosomatic synapses in the process (for a review, see Kreutzberg et al., 1989). This accomplishes at least two things: first, afferent and possibly excitatory input to motor neurons is reduced through synaptic stripping, and second, microglial cells make direct and intimate contact with motor neurons. The functional consequence of this cellular reaction, which occurs within the first week after axotomy, is regeneration of the peripheral motor axons. Thus, it is possible that the large number of microglial cells present in the facial nucleus and their close proximity to the motor neurons effectively facilitates production and delivery of neurotrophic support from activated microglia to injured motor neurons. The production of neurotrophic substances, such as transforming growth factor-β, nerve growth factor, and basic fibroblast growth factor by microglia in tissue culture and *in vivo* has been demonstrated (Mallat et al., 1989; Shimojo et al., 1991; Elkabes et al., 1996).

2.3.2 Microglial Neurotoxicity

Recently, much attention has been focused on the possible role of microglia in causing secondary neurodegeneration through the elaboration of neurotoxins during traumatic or ischemic brain injuries. Such a neurotoxic role of microglial cells is based primarily on evidence from *in vitro* studies which show that the supernatants obtained from microglial cell cultures kill cultured neurons. Apparently, such supernatants contain a variety of neurotoxic substances, which includes glutamate, nitric oxide, and reactive oxygen species, as well as yet unidentified neurotoxins with molecular weights smaller than 500 Daltons (Piani et al., 1991; Boje and Arora, 1992; Chao et al., 1992; Giulian et al., 1993). The production of neurotoxins occurs constitutively in cultures of purified microglia but is enhanced substantially by treatment of the cells with bacterial lipopolysaccharide (LPS), interferon-γ (IFN-γ), or zymosan A particles. Paradoxically, other investigators have shown that microglia-conditioned media promote neuronal survival *in vitro* (Nagata et al., 1993; Chamak et al., 1994; Nakajima and Kohsaka, 2002). From these, as well as from numerous other *in vitro* studies, it is clear that once microglial cells are maintained *in vitro*, they exhibit profuse secretory activity, especially if they are subject to stimulation with LPS, IFN-γ, or zymosan. This raises the question of whether the enhanced secretory activity of cultured microglia is a result of taking the cells into culture or a true reflection of microglial secretory activity *in vivo*. If the latter is true, then it becomes essential to confirm microglial secretory activity *in vivo* under normal and pathological conditions. Some studies have begun to investigate this problem. Popovich et al. (1994) have shown strikingly increased levels of quinolinic acid in spinal cord tissue 3 days after a contusion injury. They suggest that the increased production of this neurotoxin is due to increased synthesis by macrophages rather than through seepage from the serum. Giulian et al. (1993) found high levels of a low-molecular-weight neurotoxin in the conditioned media of macrophages cultured from neonatal rat brain, as well as in the conditioned media obtained from injured CNS tissue fragments that were cultured for 24 hours.

The evidence provided by these and other studies suggesting a role of microglia in mediating secondary tissue damage after traumatic injury *in situ* remains circumstantial. Together with the *in vivo* and *in vitro* findings suggestive of a neurotrophic role, one can say that the balance of neurotrophic and neurotoxic effects of microglia *in vivo* probably depends very much on the nature of the experimental paradigm used. In traumatic brain injuries where regeneration does occur, albeit in most instances only minimally, microglial secretory products might help to promote regenerative efforts by injured, but surviving, neurons.

A situation different from traumatic injuries is perhaps represented by disorders involving idiopathic neurodegeneration, such as Alzheimer's disease (AD) or encephalopathy associated with human immunodeficiency virus type 1 (HIV-1) infection. Microglia are the target cell of HIV-1 and produce neurotoxins upon infection (Giulian et al., 1990). It is possible that substances released from HIV-infected, and presumably, "sick" microglia could exert neurotoxic effects in patients and cause neuron loss. Similarly, in AD, if the primary disease mechanism somehow compromised normal microglial cell functions, one could envision the production of neurotoxins by unhealthy microglia followed by secondary neuron damage. There is now evidence from postmortem examinations of human brain that microglia undergo structural abnormalities with normal aging (Streit et al., 2004), suggesting, of course, that such dystrophic microglia also experience deteriorating cell functions, which could be a contributing factor in AD neurodegeneration. Further studies are indicated to investigate the idea of dysfunctional microglial cells.

2.4 CONCLUSIONS AND FUTURE DIRECTIONS

To summarize the preceding paragraphs, neurotoxicity by activated or phagocytic microglia *in vivo* has not been demonstrated. To further explore this question, it will be necessary to determine the composition of potentially neurotoxic, secretory products elaborated by activated and phagocytic microglia *in vivo*. How are activated microglia *in vivo* different from activated microglia *in vitro*? Does the removal of microglia from the normal tissue environment and subsequent transfer into an *in vitro* environment alter their secretory activity to such an extent that they become neurotoxic? In other words, is neurotoxicity of microglia an *in vitro* artifact? Is there an *in vivo* equivalent of cultured microglial cells that have been activated *in vitro* with LPS, IFN-γ, or zymosan A? Another fundamental issue concerns the definition of neuronal damage, as opposed to neuronal degeneration. As this chapter has shown, neurons may be reversibly injured and can recover from an insult. What is the role of microglia with regard to those recovering neurons? Why and how does neuronal damage induce microglial activation? Can we devise better ways to distinguish between reversibly and irreversibly injured neurons *in situ*?

ACKNOWLEDGMENTS

Parts of the work presented in this chapter were supported by a research grant from the Procter & Gamble Company.

LIST OF ABBREVIATIONS

MHC — Major histocompatibility complex
TGF-β — Transforming growth factor-β
IL-6 — Interleukin-6
IL-1 — Interleukin-1
NMDA — *N*-Methyl-*D*-aspartate
CNS — Central nervous system
RCA — *Ricinus communis* agglutinin
TMT — Trimethyltin
LPS — Lipopolysaccharide
IFN-γ — Interferon-gamma
AD — Alzheimer's disease
HIV — Human immunodeficiency virus

REFERENCES

Akiyama, H., Itagaki, S., and McGeer, P.L. (1988). Major histocompatibility complex expression on rat microglia following epidural kainic acid lesions, *J. Neurosci. Res.*, 20, 147–157.

Andersson, P.B., Perry, V.H., and Gordon, S. (1991). The kinetics and morphological characteristics of the macrophage-microglial response to kainic acid-induced neuronal degeneration, *Neuroscience*, 42, 201–214.

Balaban, C.D., O'Callaghan, J.P., and Billingsley, M.L. (1988). Trimethyltin-induced neuronal damage in the rat brain: comparative studies using silver degeneration stains, immunocytochemistry and immunoassay for neuronotypic and gliotypic proteins, *Neuroscience*, 26, 337–361.

Boje, K.M. and Arora, P.K. (1992). Microglial-produced nitric oxide and reactive nitrogen oxides mediate neuronal cell death, *Brain Res.*, 587, 250–256.

Chang, L.W., Tiemeyer, T.M., Wenger, G.R., McMillan, D.E., and Reubl, K.R. (1982). Neuropathology of trimethyltin intoxication. II. Electron microscopy study of the hippocampus, *Environ. Res.*, 29, 445–458.

Chamak, B., Morandi, V., and Mallat, M. (1994). Brain macrophages stimulate neurite growth and regeneration by secreting thrombospondin, *J. Neurosci. Res.*, 38, 221–233.

Chao, C.C., Hu, S., Molitor, T.W., Shaskan, E.G., and Peterson, P.K. (1992). Activated microglia mediate neuronal cell injury via a nitric oxide mechanism, *J. Immunol.*, 149, 2736–2741.

Elkabes, S., DiCicco-Bloom, E.M., and Black, I.B. (1996). Brain microglia/macrophages express neurotrophins that selectively regulate microglial proliferation and function, *J. Neurosci.*, 16, 2508–2521.

Finsen, B.R., Jørgensen, M.B., Diemer, N.H., and Zimmer, J. (1993). Microglial MHC antigen expression after ischemic and kainic acid lesions of the adult rat hippocampus, *Glia*, 7, 41–49.

Flanary, B.E. and Streit, W.J. (2004). Progressive telomere shortening occurs in cultured rat microglia, but not astrocytes, *Glia*, 45, 75–88.

Gehrmann, J., Bonnekoh, P., Miyazawa, T., Hossmann, K.A., and Kreutzberg, G.W. (1992). Immunocytochemical study of an early microglial activation in ischemia, *J. Cereb. Blood Flow Metab.*, 12, 257–269.

Giulian, D., Baker, T.J., Shih, L.N., and Lachman, L.B. (1986). Interleukin-1 of the central nervous system is produced by ameboid microglia, *J. Exp. Med.*, 164, 594–604.

Giulian, D., Vaca, K., and Noonan, C. (1990). Secretion of neurotoxins by mononuclear phagocytes infected with HIV-1, *Science*, 250, 1593–1596.

Giulian, D., Corpuz, M., Chapman, S., Mansouri, M., and Robertson, C. (1993). Reactive mononuclear phagocytes release neurotoxins after ischemic and traumatic injury to the central nervous system, *J. Neurosci. Res.*, 36, 681–693.

Graeber, M.B., Streit, W.J., and Kreutzberg, G.W. (1988a). Axotomy of the rat facial nerve leads to increased CR3 complement receptor expression by activated microglial cells, *J. Neurosci. Res.*, 21, 18-24.

Graeber, M.B., Tetzlaff, W., Streit, W.J., and Kreutzberg, G.W. (1988b). Microglial cells but not astrocytes undergo mitosis following facial nerve axotomy, *Neurosci. Lett.*, 85, 317-321.

Hetier, E., Ayala, J., Denèfle, P., Bousseau, A., Rouget, P., Mallat, M., and Prochiantz, A. (1988). Brain macrophages synthesize interleukin-1 and interleukin-1 mRNAs *in vitro*, *J. Neurosci. Res.*, 21, 391–397.

Hickey, W.F. and Kimura, H. (1988). Perivascular microglial cells of the CNS are bone marrow-derived and present antigen *in vivo*, *Science*, 239, 290–292.

Higgins, G.A. and Olschowka, J.A. (1991). Induction of interleukin-1β mRNA in adult rat brain, *Mol. Brain Res.*, 9, 143–148.

Hurley, S.D., Walter, S.A., Semple-Rowland, S.L., and Streit, W.J. (1999). Cytokine transcripts expressed by microglia *in vitro* are not expressed by ameboid microglia of the developing rat central nervous system, *Glia*, 25, 304–309.

Kaur, C. and Ling, E.A. (1992). Activation and re-expression of surface antigen in microglia following an epidural application of kainic acid in the rat brain, *J. Anat.*, 180, 333–342.

Kiefer, R., Lindholm, D., and Kreutzberg, G.W. (1993). Interleukin-6 and transforming growth factor-β1 mRNAs are induced in rat facial nucleus following motoneuron axotomy, *Eur. J. Neurosci.*, 5, 775–781.

Kirino, T., Tamura, A., and Sano, K. (1984). Delayed neuronal death in the rat hippocampus following transient forebrain ischemia, *Acta Neuropathol.*, 64, 139–147.

Kreutzberg, G.W., Graeber, M.B., and Streit, W.J. (1989). Neuron-glial relationship during regeneration of motoneurons, *Metab. Brain Dis.*, 4, 81–86.

Lehrmann, E., Kiefer, R., Christensen, T., Toyka, K.V., Zimmer, J., Diemer, N.H., Hartung, H.P., and Finsen, B. (1998). Microglia and macrophages are major sources of locally produced transforming growth factor-β1 after transient middle cerebral artery occlusion in rats, *Glia*, 24, 437–448.

Liu, T., McDonnell, P.C., Young, P.R., White, B.A., Siren, A.L., Hallenbeck, J.M., Barone, F.C., and Feuerstein, G.Z. (1993). Interleukin-1β mRNA expression in ischemic rat cortex, *Stroke*, 24, 1746–1751.

Mallat, M., Houlgatte, R., Brachet, P., and Prochiantz, A. (1989). Lipopolysaccharide-stimulated rat brain macrophages release NGF *in vitro*, *Dev. Biol.*, 133, 309–311.

McCann, M.J., O'Callaghan, J.P., Martin, P.M., Bertram, T., and Streit, W.J. (1996). Differential activation of microglia and astrocytes following trimethyltin-induced brain injury, *Neuroscience*, 72, 273–281.

Minami, M., Kuraishi, Y., Yabuuchi, K., Yamazaki, A., and Satoh, M. (1991). Induction of interleukin-1β mRNA in rat brain after transient forebrain ischemia, *J. Neurochem.*, 58, 390–392.

Mitchell, J., Sundstrom, L.E., and Wheal, H.V. (1993). Microglial and astrocytic cell responses in the rat hippocampus after an intracerebroventricular kainic acid injection, *Exp. Neurol.*, 121, 224–230.

Morgan, T.E., Nichols, N.R., Pasinetti, G.M., and Finch, C.E. (1993). TGF-β1 mRNA increases in macrophage/microglial cells of the hippocampus in response to deafferentation and kainic acid-induced neurodegeneration, *Exp. Neurol.*, 120, 291–301.

Morioka, T., Kalehua, A.N., and Streit, W.J. (1991). The microglial reaction in the rat dorsal hippocampus following transient forebrain ischemia, *J. Cereb. Blood Flow Metab.*, 11, 966–973.

Morioka, T., Kalehua, A.N., and Streit, W.J. (1992). Progressive expression of immunomolecules on microglial cells in rat dorsal hippocampus following transient forebrain ischemia, *Acta Neuropathol.*, 83, 149–157.

Murphy, P.G., Borthwick, L.S., Johnston, R.S., Kuchel, G., and Richardson, P.M. (1999). Nature of the retrograde signal from injured nerves that induces interleukin-6 mRNA in neurons, *J. Neurosci.*, 19, 3791–3800.

Nagata, K., Takei, N., Nakajima, K., Saito, H., and Kohsaka, S. (1993). Microglia conditioned medium promotes survival and development of cultured mesencephalic neurons from embryonic brain, *J. Neurosci. Res.*, 34, 357–363.

Nakajima, K. and Kohsaka, S. (2002). Neuroprotective roles of microglia in the central nervous system, in *Microglia in the Regenerating and Degenerating Central Nervous System*, Streit, W.J., Ed., Springer, New York, pp. 188–208.

Olsnes, S., Refsnes, K., and Pihl, A. (1974). Mechanism of action of the toxic lectins abrin and ricin, *Nature*, 249, 627–631.

Pasinetti, G.M., Johnson, S.A., Rozovsky, I., Lampert-Etchells, M., Morgan, D.G., Gordon, M.N., Morgan, T.E., Willoughby, D., and Finch, C.E. (1992). Complement C1qB and C4 mRNAs responses to lesioning in rat brain, *Exp. Neurol.*, 118, 117–125.

Pearson, V.L., Rothwell, N.J., and Toulmond, S. (1999). Excitotoxic brain damage in the rat induces interleukin-1 protein in microglia and astrocytes: Correlation with the progression of cell death, *Glia*, 25, 311–323.

Piani, D., Frei, K., Do, K.Q., Cuénod, M., Fontana, A. (1991). Murine brain macrophages induce NMDA receptor mediated neurotoxicity *in vitro* by secreting glutamate, *Neurosci. Lett.*, 133, 159–162.

Popovich, P.G., Reinhard, J.F., Jr., Flanagan, E.M., and Stokes, B.T. (1994). Elevation of the neurotoxin quinolinic acid occurs following spinal cord trauma, *Brain Res.*, 633, 348–352.

Rabchevsky, A.G. and Streit, W.J. (1997). Grafting of cultured microglial cells into the lesioned spinal cord of adult rats enhances neurite outgrowth, *J. Neurosci. Res.*, 47, 34–48.

del Rio-Hortega, P. (1932). Microglia, in *Cytology and Cellular Pathology of the Nervous System*, Penfield, W., Ed., Hoeber, New York, pp. 481–534.

Robinson, A.P., White, T.M., and Mason, D.W. (1986). Macrophage heterogeneity in the rat as delineated by two monoclonal antibodies MRC OX-41 and MRC OX-42, the latter recognizing complement receptor type 3, *Immunology*, 57, 239–247.

Shimojo, M., Nakajima, K., Takei, N., Hamanoue, M., and Kohsaka, S. (1991). Production of basic fibroblast growth factor in cultured rat brain microglia, *Neurosci. Lett.*, 123, 229–231.

Streit, W.J., Graeber, M.B., and Kreutzberg, G.W. (1988). Functional plasticity of microglia: a review, *Glia*, 1, 301-307.

Streit, W.J. and Kreutzberg, G.W. (1988). The response of endogenous glial cells to motor neuron degeneration induced by toxic ricin, *J. Comp. Neurol.*, 268, 248-263.

Streit, W.J., Graeber, M.B., and Kreutzberg, G.W. (1989a). Peripheral nerve lesion produces increased levels of MHC antigens in the CNS, *J. Neuroimmunol.*, 21, 117-123.

Streit, W.J., Graeber, M.B., and Kreutzberg, G.W. (1989b). Expression of Ia antigens on perivascular and microglial cells after sublethal and lethal neuronal injury, *Exp. Neurol.*, 105, 115–126.

Streit, W.J. (1990). An improved staining method for rat microglial cells using the lectin from *Griffonia simplicifolia* (GSA I-B$_4$), *J. Histochem. Cytochem.*, 38, 1683–1686.

Streit, W.J., Morioka, T., and Kalehua, A.N. (1992). MK-801 prevents microglial reaction in rat hippocampus after forebrain ischemia, *NeuroReport*, 3, 146–148.

Streit, W.J. and Graeber, M.B. (1993). Heterogeneity of microglial and perivascular cell populations — Insights gained from the facial nucleus paradigm, *Glia*, 7, 68–74.

Streit, W.J. (1993). Microglial-neuronal interactions, *J. Chem. Neuroanat.*, 6, 261–266.

Streit, W.J., Hurley, S.D., McGraw, T.S. and Semple-Rowland, S.L. (2000). Comparative evaluation of cytokine profiles and reactive gliosis supports a critical role for interleukin-6 in neuron-glia signaling during regeneration, *J. Neurosci. Res.*, 61, 10–20.

Streit, W.J. (2001). Microglia and macrophages in the developing CNS, *Neurotoxicology*, 22, 619–624.

Streit, W.J., Sammons, N.W., Kuhns, A.J., and Sparks, D.L. (2004). Dystrophic microglia in the aging human brain, *Glia*, 45, 208–212.

Walberg, F. (1971). Does silver impregnate normal and degenerating boutons? A study based on light and electron microscopic observations of the inferior olive, *Brain Res.*, 31, 47–65.

Wiley, R,G,, Blessing, W.W., and Reis, D.J. (1982). Suicide transport: destruction of neurons by retrograde transport of ricin, abrin, and modeccin, *Science*, 216, 889–890.

Woodroofe, M.N., Sarna, G.S., Wadhwa, M., Hayes, G.M., Loughlin, A.J., Tinker, A., and Cuzner, M.L. (1991). Detection of interleukin-1 and interleukin-6 in adult rat brain, following mechanical injury, by *in vivo* microdialysis: evidence of a role for microglia in cytokine production, *J. Neuroimmunol.*, 33, 227–236.

Yabuuchi, K., Minami, M., Katsumata, S., and Satoh, M. (1993). *In situ* hybridization study of interleukin-1β mRNA induced by kainic acid in the rat brain, *Mol. Brain Res.*, 20, 153–161.

Yan, H.Q., Banos, M.A., Herregodts, P., Hooghe, R., and Hooghe-Peters, E.L. (1992). Expression of interleukin (IL)-1β, IL-6 and their respective receptors in the normal rat brain and after injury, *Eur. J. Immunol.*, 22, 2963–2971.

Yokoyama, A., Yang, L., Itoh, S., Mori, K., and Tanaka, J. (2004). Microglia, a potential source of neurons, astrocytes, and oligodendrocytes, *Glia*, 45, 96–104.

CHAPTER 3

Schwann Cell Neurotoxicity

Arrel D. Toews, G. Jean Harry, and Pierre Morell

CONTENTS

3.1 INTRODUCTION

Schwann cells are the glial cells of the peripheral nervous system (PNS). The Schwann cell received its name from the seminal work by Professor Theodor Schwann[1] on the cellular construction of animal tissues. Using teased nerve fibers from fetal pig sciatic nerve, he recognized that each fiber contained rows of closely aligned nuclei and correctly concluded that the mature nerve was produced by the growth and lengthening of these cells during development.[2] We now know that during this process, the Schwann cells ensheathe all axons in the PNS, either by wrapping them with the greatly extended and highly compacted multilamellar extension of their plasma membrane, the myelin sheath,[3] or by sequestering "nonmyelinated" axons within invaginations of the Schwann cell plasma membrane. Most of what we know about these cells and their relationship to nervous system function is with regards to myelin and its role in the rapid efficient conduction of impulses down axons. There are periodic interruptions between adjacent segments of myelin, termed nodes of Ranvier, where axonal membranes are exposed. The high resistance of myelin serves as an electrical insulator, and this property, coupled with the high concentration of voltage-gated sodium channels localized at the nodes, allows impulse conduction to jump from node to node; this is termed saltatory conduction (*L. saltare*, to leap). In contrast, conduction in unmyelinated axons involves a continuous wave of membrane depolarization that moves down the axon. The presence of the myelin sheath thus greatly increases the efficiency of nervous system operation, facilitating conduction while conserving metabolic energy and space.[4]

In addition, Schwann cells influence neuronal and axonal development.[5,6] Schwann cell–derived signals regulate neuronal numbers and axonal caliber during development. They are involved in defining the structural and functional organization of the node of Ranvier and adjacent axonal structures, including the spatial orientation of sodium channels and voltage-dependent potassium channels.[7,8] Disruption of myelin/axon contact results in an arbitrary localization of ion channels and loss of efficient saltatory conduction. In addition, axonal caliber is reduced at the node of Ranvier, possibly a consequence of altered neurofilament phosphorylation and density in this specialized region.[9,10] It is easy to appreciate how even minor loss of myelin from axons, perturbations in its structure or function, or alterations in normal axon/myelin relationships might have deleterious effects on normal nervous system function.

Although largely ignored by neurotoxicologists and neuroscientists in general, the Schwann cells that segregate and sequester the smaller "nonmyelinated" axons in the PNS presumably also play a vital functional role. The response of these nonmyelinating Schwann cells to insults such as those involved in painful neuropathies sometimes associated with some AIDS therapies remains unknown and is an area of possible future investigation.

3.2 ONTOGENIC DEVELOPMENT AND DIFFERENTIATION OF SCHWANN CELLS

The ontogenic development of the myelinating Schwann cell lineage has been extensively studied, mostly in cell cultures, but also *in vivo* in both normal and various gene knockout and transgenic mice.[11–13] Both myelinating and nonmyelinating Schwann cells develop from a common primitive precursor cell derived from the developing neural crest.[14] These proliferative multipotential cells are also capable of differentiating into sensory and autonomic neurons or melanocytes[11,12,14,15] (Figure 3.1). During development, Schwann cell precursors migrate along axonal tracts, proliferate, and then progress through distinct developmental stages.[16,17]

Mitogenesis and differentiation of primitive pluripotent precursor cells into myelinating or nonmyelinating phenotypes is regulated in part by multiple axon-associated signals.[18–23] Schwann cell mitogens active during early development include the neuregulins[24] (glial growth factor isoforms). Inappropriate later neuregulin expression inhibits myelination by preventing axonal segregation and ensheathment.[25] Overexpression of glial growth factor β3 in a transgenic mouse model leads to a hypertrophic demyelinating neuropathy and malignant PNS tumors.[26] These studies suggest that neuronal mitogens, including the neuregulins, may be involved in control of myelination during development, and that inappropriate later activation of these mitogen-signaling pathways may contribute to the initial demyelination and subsequent Schwann cell proliferation seen is some peripheral neuropathies.

During the first few days of postnatal development in rodents, Schwann cells destined to form myelin display a "promyelinating" morphology, establishing a 1:1 relationship with an axon. At this time, the Schwann cells have synthesized a basal lamina but have not initiated myelin membrane biosynthesis.[27–29] The number of these cells declines as development continues,[27] a process involving apoptosis.[30–32] Promyelinating Schwann cell apoptosis is controlled by axonal contact and is in part regulated by neuregulin-1 supplied by the ensheathed axons.[33] Changes in mitotic activity and susceptibility to apoptosis act in concert to match the required number of Schwann cells to existing axons. In addition, the interaction between axon and Schwann cell determines that large-caliber axons (>1 μm) are myelinated, while the small-caliber axons become segregated in cytoplasmic cuffs of nonmyelinating Schwann cells.[34] Although larger axons have thicker myelin sheaths, the presumed close correlation between axon diameter and the number of myelin lamellae has been

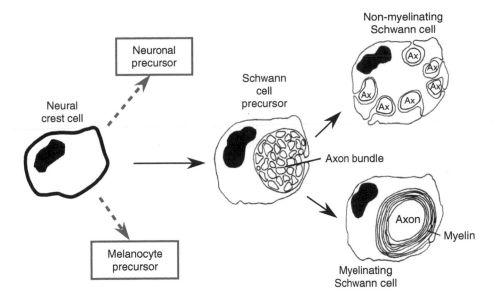

Figure 3.1 Ontogenic development of Schwann cells. Schwann cells of the PNS originate from primitive neural crest cells, proliferative pluripotent cells that can also develop into neurons or melanocytes. The Schwann cell precursor, initially becomes associated with many axons, becomes associated with progressively fewer axons as it matures. "Committed" Schwann cells develop into either nonmyelinating Schwann cells, which remain associated with several axons, individually segregated within invaginations of the Schwann cell plasma membrane, or into myelinating Schwann cells, which are associated with a single axon. The final differentiation into a myelinating phenotype involves upregulation of mRNA expression for myelin proteins and enzymes involved in synthesis of myelin lipids, and the synthesis, elaboration, and compaction of the myelin sheath. Because of the high degree of plasticity of Schwann cells, most of the steps involved in their differentiation and maturation are reversible.

recently challenged.[35] Physical contact by Schwann cells or their secreted factors can also influence synapse formation by neurons.[36–38]

In addition to contact-related signals, factors such as insulin-like growth factor-1 (IGF-1) and transforming growth factor β (TGF-β) have been shown to promote the expression of different phenotypes.[5,22] More recent work has also identified a number of relevant transcription factors, including the zinc-finger transcription factors Krox-20,[39,40] SCIP,[12] Pax3,[41] c-jun,[42] Sry box protein Sox10,[43] and Brn-2.[44] Sox10 is required for the establishment and maintenance of Schwann cell precursors,[43] while the Pou-domain proteins Oct-6, Krox-20, and SCIP (Oct-6/Tst-1) are important for the differentiation of myelinating Schwann cells.[39,45–48] Oct-6 is expressed during fetal development in immature Schwann cells and peaks in promyelinating and early myelinating cells in the first week following birth.[17,49,50] Oct-6 regulates Krox-20, which in turn regulates expression of myelin protein genes and those involved in myelin lipid metabolism.[51] A transient expression of SCIP is induced in Schwann cells by contact with axons; maximal levels are found in promyelinating cells with nearly undetectable levels by the second postnatal week.[12] *In vitro*, this protein is upregulated in cultured Schwann cells by compounds that elevate intracellular cAMP.[52] Krox-20 is present in promyelinating cells about a day after these cells first express SCIP, and its expression continues into adulthood.[12,17] Krox-20 is restricted to the myelinating lineage, while SCIP is expressed developmentally in both myelinating and nonmyelinating Schwann cells. In SCIP knock-out mice, a transient Schwann cell phenotype characterized by a delay in differentiation is present.[46] Mice homozygous for a loss-of-function Krox-20 mutation show a complete permanent block in myelinating Schwann cell differentiation at the promyelinating stage, resulting in drastically decreased expression of myelin-specific gene products.[39,53]

3.3 PNS MYELIN FORMATION

The actual process of myelination in the PNS, which occurs earlier than in CNS, is initiated by contact with the underlying axon. Myelination can be assessed by both anatomical (morphometric analyses of myelin area and thickness, nodal length, etc.) and biochemical (expression of myelin-related genes, synthesis rates of myelin-specific components) techniques. Each of these approaches has provided useful information. Control of the synthesis of myelin components and their subsequent assembly into mature compact myelin is complex and is regulated at multiple levels, including promoter choice, transcription, mRNA splicing and stability, translation, and post-translational processing.[54] In the rodent PNS, initial accumulation of myelin begins with a loose sheath of axons around birth, and this is followed by a period of rapid myelin accumulation during the first month of life.[29] Maximal expression of mRNA for P0 and myelin basic proteins, major structural components of PNS myelin, occurs at 21 days of age,[55] in good agreement with earlier morphometric studies.[27,29]

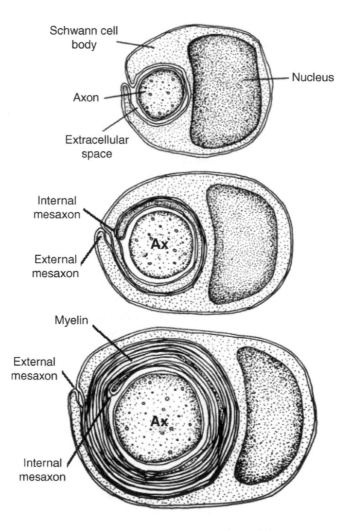

Figure 3.2   Schematic diagram illustrating the formation of PNS myelin by a Schwann cell. The Schwann cell plasma membrane surrounds and contacts the axon, with the apposing layers of the plasma membrane fusing to form the mesaxon, which then wraps itself spirally around the axon as more myelin membrane is laid down. Most of the Schwann cell cytoplasm is extruded as the myelin sheath matures and becomes compacted, although there are still numerous cytoplasmic inclusions in the compact myelin sheath, all continuous with cytoplasm of the Schwann cell perikaryon.

During PNS myelination, Schwann cells extend their plasma membrane to engulf the axon and then, as a double-layered membrane unit, wrap spirally around it (Figure 3.2). As the number of wraps increases, cytoplasm is gradually extruded until the compact multilamellar structure of mature myelin is present. Despite the overall compact fused nature of the myelin sheath, there are still many cytoplasmic inclusions, all continuous with the cytoplasm in the Schwann cell body. These include the lateral loops adjacent to nodes of Ranvier as well as Schmidt-Lanterman clefts and longitudinal incisures (Figure 3.3), potentially dynamic structures relevant to turnover of myelin membrane components (see the following section).

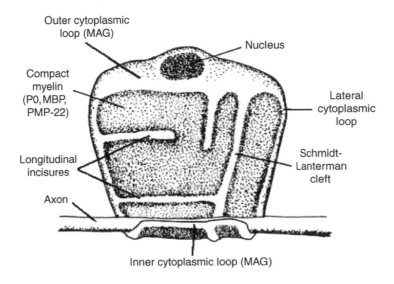

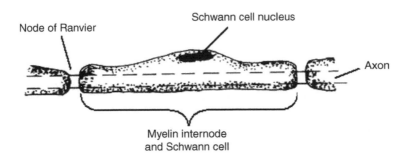

Figure 3.3   Myelin unwrapped. The upper panel shows a single myelinated internode as it would appear if the myelin formed by its associated Schwann cell was unwrapped from around the axon. The area of the extended plasma membrane of the Schwann cell that comprises the myelin sheath is actually much larger in relation to the size of the Schwann cell body than shown in the diagram. Note the many cytoplasmic structures, including the inner, outer, and lateral loops, the longitudinal incisures, and the Schmidt-Lanterman clefts, all of which are continuous with the cytoplasm of the Schwann cell body. These potentially dynamic mobile structures may be involved in replacing individual molecular components of myelin and in maintaining myelin structure and stability. The major structural proteins P0, MBP, and PMP-22 are present in the compact regions of the myelin, while MAG is localized in the regions of myelin associated with the axon. In compact myelin, the myelin membrane is spiraled tightly around the axon, with apposing inner membrane leaflets fusing to produce the major dense line and the outer membrane leaflets fusing to produce the less electron-dense intraperiod line. The lower panel shows a single internode of myelin maintained by a Schwann cell, as well as the nodes of Ranvier, periodic interruptions between adjacent myelinating Schwann cells where the axon is exposed, allowing for the rapid, metabolically efficient salutatory conduction of nervous impulses. (Modified from Morell, P. and Norton, W.T., *Sci. Am.*, 242, 74, 1980. With permission.)

3.4 MYELIN COMPOSITION AND METABOLISM

Myelin has a relatively simple composition.[56-58] It has a very high lipid content (70–80%, higher than any other biological membrane) that contributes to the insulating properties of myelin critical to its physiological function. Because of its high lipid content, myelin has a buoyant density less than that of any other membranes, allowing its isolation with a high yield and degree of purity.[59] The lipid composition of PNS myelin is characterized by a high content of cholesterol and the galactolipids cerebroside (galactosylceramide) and sulfatide (sulfated cerebroside). Phospholipids, including a relatively high level of ethanolamine plasmalogens (phospholipids with a fatty aldehyde linked to the C1 of glycerol instead of a fatty acid), account for most of the remainder of PNS myelin lipids. Gangliosides are also present in low concentrations.

P0, the major protein of PNS myelin accounting for about half the total protein, is a transmembrane protein thought to be involved in myelin compaction.[60,61] P0 may also play a role in the delivery of other myelin proteins to their proper sites and in neuronal–glial interactions.[62] This protein, as well as the P2 protein and PMP-22 (peripheral myelin protein of 22 kDa molecular weight) are unique to PNS myelin. The myelin proteolipid protein (PLP), the major component of CNS myelin, is present in PNS myelin at only very low levels, if at all.[63] While its function in the PNS remains unknown,[64] a human PLP[null] mutation is characterized by a demyelinating peripheral neuropathy not seen in other types of PLP mutations.[65] Other PNS myelin proteins, also present in CNS myelin, include the myelin basic proteins (MBP) and myelin-associated glycoprotein (MAG). MBPs are extrinsic membrane proteins localized to the cytoplasmic membrane surface of myelin. While they account for only about 18% of total myelin protein, their specific association with sulfatides and gangliosides has been suggested as a major contributing factor for stabilization and maintenance of the myelin structure.[66-69] MAG functions include membrane–membrane interactions during myelin formation and maintenance[58,70,71]; it has been implicated in various peripheral neuropathies.[72,73] There are also a number of minor protein components of PNS myelin, including structural proteins and those involved in cell–cell interactions, as well as a number of enzymes, receptors, and second messenger-related proteins. All presumably play roles in maintaining the complex structure and function of PNS myelin. More detailed discussion of these proteins can be found in recent reviews.[13,58,74]

Myelinating Schwann cells must synthesize enormous quantities of required myelin components during a relatively short, tightly programmed window during development, rendering them particularly vulnerable to metabolic insults during this time period (see the following section). Insults occurring during this "vulnerable developmental window" may result in irreversible myelin deficits. Newly synthesized components must also be correctly targeted and transported to their intended destination and then properly assembled into nascent myelin membranes being deposited. The compact structure of mature multilamellar myelin, located a considerable distance from its supporting glial cell body, would seem to dictate a relatively inert metabolism of myelin once deposited. However, it is now known that molecular components of mature myelin do indeed turn over at significant rates. Half-lives are on the order of days to weeks for different myelin lipids,[75] and of several weeks for the peptide backbone of myelin proteins.[58,76] Additionally, some components, such as phosphate groups on polyphosphoinositides[77] and on myelin basic protein,[78] turn over with half-lives of the order of minutes or faster. Much of the available data for metabolic turnover relate to the CNS, but PNS myelin is similar, albeit with a somewhat more rapid overall metabolism. The question of how such turnover of membrane components of compact myelin might be accomplished remains unanswered, although it is possible that transient dynamic invasion of fingers of cytoplasm, visualized ultrastructurally as Schmidt-Lanterman clefts and longitudinal incisures (see Figure 3.3), might be involved.

Another aspect of myelin metabolism relates to possible energy-dependent mechanisms needed to maintain the compact structure of myelin. Water must be excluded from both the cytoplasmic space (major dense line) and the extracellular space (intraperiod line) of compact myelin, and this most likely involves active pumping of ions, with water following osmotically. Compatible with such processes is the demonstrated presence of significant levels of Na$^+$,K$^+$-ATPase,[79] and carbonic anhydrase[80] in compact myelin. These metabolic processes have been proposed as likely targets of certain neurotoxicants that damage myelin by producing intramyelinic edema (see later sections).

3.5 GENERALIZED RESPONSES OF SCHWANN CELLS TO INJURY

Mature myelinating Schwann cells retain a remarkable degree of plasticity. Following a demyelinating insult, these cells dedifferentiate into a more primitive precursor form, but generally do not die. The ability of mature Schwann cells to survive in the absence of axonal contact (a survival prerequisite during early development) is an absolute requirement for nerve repair following toxic insult or injury and may comprise part of a primal survival mechanism. In response to a nerve transection, Schwann cells proliferate and migrate from the proximal and distal stumps to form a continuous tissue cable that serves to guide the regenerating axons, a process influenced by neurotrophin 3 (NT3) via the tyrosine kinase (Trk) C signaling pathway.[81] The program for myelination is inhibited while the Schwann cell is actively migrating. As repair processes are initiated, these Schwann cells proliferate, reestablish contact with axons, and then redifferentiate into the myelinating phenotype, ultimately remyelinating the affected axons and thereby restoring function.

As noted some time ago by Ramon y Cajal,[82] Schwann cells also function as macrophage-like phagocytic cells under conditions where myelin (and axons) are damaged or degenerating. The ability of the Schwann cell to phagocytose its own myelin and retain some of this material for use during remyelination has been examined in a number of experimental models.[83] Schwann cells also possess other macrophage-like characteristics, including the ability to regulate the inflammatory response by inducing T-lymphocyte apoptosis in inflammatory peripheral neuropathies such as experimental autoimmune neuritis.[84,85] Schwann cells also synthesize and secrete proinflammatory cytokines, including those that function as monocyte chemoattractants.[86,87] A sensory neuropathy is the most common neurological complication of HIV infection, and an integral role for chemokines released by Schwann cells in the pathogenesis of this neuropathy has recently been demonstrated.[88] Binding of the HIV envelope glycoprotein gp120 to the CXCR4 chemokine receptor on Schwann cells releases the β-chemokine, RANTES (regulated upon activation, normal T-cell expressed and secreted), which leads to TNFα-mediated neuronal apoptosis and neuritic degeneration. This newly described Schwann cell–neuron interaction may be relevant to the pathophysiology of other peripheral neuropathies as well.

Osteopontin, a cytokine-like molecule expressed by macrophages, has been implicated in posttraumatic tissue repair outside the nervous system. In the PNS, it is constitutively expressed at high levels in normal myelinated Schwann cells, and it is differentially regulated following axonal injury.[89] In addition, it is markedly downregulated in human sural nerves in severe axonal polyneuropathies. The role of osteopontin in degenerative and regenerative events in the PNS remains to be explored.

3.6 PERTURBATIONS INVOLVING PNS MYELIN AND SCHWANN CELLS

Damage to the myelin sheath can be "primary," resulting from some type of insult to myelin or its supporting Schwann cell, or "secondary," wherein it is the axon or neuron that appears initially affected. These distinctions are based almost entirely on morphological observations, with the actual specific underlying insult often remaining unknown. A special case of primary demyelination in the PNS involves "segmental demyelination," a situation in which axons have some internodes with myelin absent, while adjacent segments retain intact myelin. A second-order distinction sometimes encountered in the literature and again based largely on morphological criteria involves damage directly to myelin (myelinopathy) or to the Schwann cell (Schwannopathy).

Mutant mice have proven useful as tools for examining insults to Schwann cells. A differential susceptibility of myelinating cells in the CNS and PNS has been clearly demonstrated in various genetic mutant mice models of dysmyelination. For example, the shiverer mouse (shi; mouse chromosome 18) lacks MBP and fails to make normal CNS myelin but has normal-appearing PNS myelin.[90–92] The lack of any effect in the PNS is perhaps due to substitution of P0 for the structural function of MBP.[60] Various mutations in the myelin proteolipid protein (PLP), known as jimpy (jp), myelin synthesis–deficient (jp msd), and rumpshaker (rsh), have been extensively characterized. In each, there is severe dysmyelination or death of oligodendrocytes in CNS, but the PNS remains relatively normal, presumably a reflection of the lack of PLP in PNS myelin. In the P0-deficient mouse, the selective loss of PNS myelin results in a full body tremor and skeletal muscle atrophy of the hindlimbs.[93] The quaking mouse (qk; mouse chromosome 17) shows signs of dysmyelination predominantly in the CNS, but with some PNS alterations.[94]

The trembler mouse (Tr; mouse chromosome 11) contains a missense point mutation in the PMP-22 gene[95] that results in PNS segmental demyelination, supernumerary Schwann cells forming onion bulb structures,[96] and disorganized neuromuscular junctions.[97] A phenotype characterized by a coarse-action tremor occurs around 14 days of age, resulting in moderate quadriparesis and a waddling gait.[98] Transgenic mice with altered copy numbers of the PMP-22 gene develop a pronounced distal axonopathy in addition to the classical segmental demyelination and onion bulb formation.[99] The pathophysiology of both the Tr mice and the PMP-22 mice suggest that the axonal damage occurs secondary to alterations affecting myelinating Schwann cells.[95,100,101] Charcot-Marie-Tooth disease (CMT),[102,103] also termed hereditary motor and sensory neuropathy, includes a number of disorders affecting the peripheral nervous system. Within this disease group, the different classes are identified by clinical and electrophysiological endpoints as demyelinating and axonal forms. Myelinopathies are characterized by a decreased nerve conduction velocities, while axonopathies have a slightly reduced or normal nerve conduction velocity but a decreased compound muscle action potential.[104] For additional information, the interested reader is referred to recent more-detailed reviews.[105–107]

3.7 CLASSIFICATION OF NEUROTOXIC INSULTS TO SCHWANN CELLS AND MYELIN

A useful classification system for categorizing and understanding various toxic insults to myelin and Schwann cells involves the concept that selective targeting of the toxicant to myelin involves some specialized aspect of this system.[108–110] Some general factors are discussed below, and specific examples follow in subsequent sections.

1. The highly hydrophobic nature of myelin, derived from its high lipid/protein ratio (see above) accounts for the preferential accumulation of certain lipophilic toxicants. This relatively nonspecific physiochemical feature can have relatively specific effects on the lipid-rich myelin sheath.
2. The specialized chemical composition of myelin requires a high level of synthetic capacity in Schwann cells for those membrane components enriched in myelin (the blood–nerve barrier prohibits entry of generally expressed exogenous lipid components, such as cholesterol). Toxicants that preferentially perturb biosynthetic pathways for myelin-enriched components may result in myelinopathies.
3. Because the large accumulation of myelin membrane occurs during a somewhat restricted window of time during development, Schwann cells must operate at or very near their synthetic capacity during this time. Even a relatively generalized metabolic insult, if present during the time of maximal synthetic activity in myelinating Schwann cells, may have a preferentially deleterious effect on myelin accumulation. The mandatory strict temporal sequence of metabolic events

involved in nervous system development, with only a narrow "window of opportunity" for each required developmental event, means that such insults may produce permanent deficits.

4. The highly organized, compact, multilamellar structure of myelin suggests that correct amounts and forms of structural components are critical to myelin function. Even slight changes in composition may alter membrane fluidity and membrane dimensions. Alterations in fatty acid compositions of membrane lipids or cholesterol content may have preferential pathological consequences for myelin, relative to other membranes.

5. Prevention of myelin edema appears to involve some active aspect of metabolism, and the processes involved may be a target for certain neurotoxicants. The accumulation of fluid at the intraperiod line of myelin following exposure to a toxicant presumably involves perturbation of these processes.

6. Some metabolic insults appear to preferentially damage "white matter" without specifically targeting myelinating cells. It may be that the highly interdependent axon/myelin/Schwann cell unit has some feature that renders it preferentially susceptible to some neurotoxicants.

7. The blood–nerve barrier (and the blood–brain barrier of the CNS) separates elements of the circulation from the PNS and may be an important factor in development of PNS neurotoxicity. However, it is often difficult to establish whether compromised blood–nerve barrier function is a factor in demyelinating neuropathies or is rather a consequence of the demyelination.

3.8 SURVEY OF SELECTED SCHWANN CELL/MYELIN TOXICANTS

Triethyltin and hexachlorophene each produce demyelination in both the CNS and PNS, and although a majority of studies have concentrated on the CNS, findings related to mechanisms of action are presumably relevant to the PNS as well. Triethyltin (TET) neurotoxicity in humans became evident when a large number of people suffered severe consequences, including over 100 deaths, as a result of TET contamination of a medication.[111] There is intramyelinic edema caused by splitting of the intraperiod line of myelin with subsequent fluid accumulation.[112] Although the CNS is preferentially affected, rather severe chronic exposure regimens in animals produces PNS myelin edema as well.[113] Neurotoxic specificity of TET for myelin is consistent with its relative hydrophobicity, which suggests it should accumulate in lipid-rich myelin. Specificity of organotin compounds for myelin does not extend to the closely related, but less hydrophobic trimethyltin (TMT), which selectively damages neurons.[114] It appears that TET somehow interferes with the maintenance of osmotic regulation. Impaired oxidative phosphorylation[115] or mitochondrial Mg^{++}-ATPase,[116] action as a Cl^-/OH^- ionophore,[117] or a direct action of TET on myelin involving facilitated Cl (and therefore water) entry[118] have been suggested as mechanisms of toxicity. TET edema can be blocked by acetazolamide, a carbonic anhydrase inhibitor,[119] and the demonstration of carbonic anhydrase in myelin[80] suggests its possible involvement in aspects of water transport involved in maintaining compact myelin.

Hexachlorophene is an antimicrobial agent that was commonly used in operating rooms before evidence of its neurotoxicity emerged. It can be absorbed through the skin, in part due to its relative hydrophobicity, which also presumably accounts for its targeting to myelin. There is myelin edema, again due to splitting of the intraperiod line. Both the CNS and PNS are affected, although CNS is affected at lower concentrations.[120] As with TET, hexachlorophene is neurotoxic in young rats, but the characteristic edema is not evident until after about 15 days of age, when myelin has accumulated in sufficient quantity to serve as a hydrophobic reservoir.[121]

One of the earliest investigations of a toxic demyelinating neuropathy involved inorganic lead intoxication in guinea pigs[122]; segmental demyelination was noted in teased PNS nerve fibers.[123] The nucleoside analog, 2′,3′-dideoxycytidine is clinically relevant as an inhibitor of human immunodeficiency virus, and in a rabbit model, there is preferential damage to myelin. Diphtheric neuropathy, characterized by vacuolation and fragmentation of PNS myelin, is an infrequent occurrence following *Corynebactyeriium diphtheriae* infection. In the animal model produced by intraneural injection of diphtheria toxin, there is primary demyelination, possibly related to preferential inhibition of PNS myelin protein synthesis.[124] With all of the above toxicants, early signs of myelin pathology include edema — the accumulation of fluid between normally compact myelin lamellae. Splitting of the myelin sheath usually occurs at the intraperiod line, where the two external leaflets of the Schwann cell plasma membrane appose one another, although with lead, some splitting also occurs at the major dense line, where the cytoplasmic faces of the myelin membrane meet. In all of the aforementioned disorders, a common theme seems to be that some active pumping mechanism required to exclude water, and therefore required for maintenance of the compact multilamellar structure of myelin, is impaired. There are a whole host of underlying possibilities.[110,118,125]

3.9 TELLURIUM NEUROPATHY AS MODEL FOR PRIMARY TOXICANT-INDUCED DEMYELINATION

Although it has proved difficult to identify and characterize specific neurotoxic insults that selectively target Schwann cells, a considerable amount is known regarding the mechanism of action by which exposure to tellurium produces a primary demyelinating neuropathy in rats.[110,126] The tellurium neuropathy model illustrates how a specific metabolic block in a myelin-enriched component can bring about demyelination. Feeding elemental tellurium (element 52) to weanling rats rapidly produces a peripheral neuropathy; within 3 days of exposure, there is severe hindlimb paralysis. The morphological correlate of this paralysis is a primary segmental demyelination of the sciatic nerve (Figure 3.4), with few observable degenerative changes in axons or nonmyelinating Schwann cells.[127–129] Because of the highly synchronous demyelination, which is followed by rapid and nearly complete remyelination, it has proved useful in examining the temporal sequence of events involved in demyelination and remyelination.

The underlying metabolic basis for this tellurium-induced demyelination is a profound inhibition of cholesterol synthesis by specific inhibition of squalene epoxidase,[130–132] an enzyme near the end of the cholesterol synthesis pathway. Inhibition of cholesterol synthesis occurs within hours of initiating Te exposure and continues for some time after exposure is terminated. Although the block in cholesterol synthesis is systemic (there is massive accumulation of squalene in the liver, accounting for up to 10% of dry weight after 5 days of Te exposure), myelin of the PNS is preferentially affected for several reasons. First, the quantitative significance of cholesterol in myelin means that its synthesis is especially prominent in Schwann cells, and Te exposure is initiated during the period of most rapid myelin accumulation, when the metabolic synthetic capacity of Schwann cells for cholesterol biosynthesis is maximally stressed. The deficit in the supply of a required major structural component of myelin membranes rapidly leads to a coordinate downregulation of expression of mRNA for a whole host of myelin proteins and enzymes involved in synthesis of myelin lipids,[133–136] with subsequent shutdown of the program for myelin synthesis and maintenance. A second factor is the inability of the PNS to utilize circulating cholesterol[137]; the only source of cholesterol available for peripheral nerve is that synthesized locally by the nerve.[138] Furthermore, the compensatory upregulation of mRNA expression for HMG-CoA reductase, the rate-limiting enzyme in cholesterol biosynthesis, seen in liver subsequent to inhibition of cholesterol biosynthesis is not seen in sciatic nerve.[134,139] Rather, control of cholesterol synthesis in Schwann cells appears to be coordinately regulated with other required myelin membrane components.[140]

The lack of cholesterol required to form stable myelin, coupled with the subsequent downregulation of myelin synthesis and maintenance, results in the highly synchronous demyelination of up to one-fourth of the myelin internodes of the PNS, with the largest internodes (and possibly most metabolically stressed Schwann cells) being preferentially vulnerable.[129] When tellurium exposure is discontinued, there is a rapid and synchronous remyelination.

Because there is little axonal degeneration, this model has proven very useful in isolating and examining the specific changes Schwann cells undergo during the processes of demyelination and remyelination. Events occurring secondary to axonal degeneration and regeneration are not

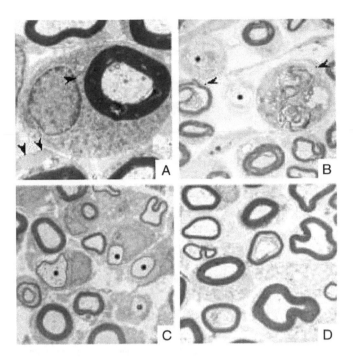

Figure 3.4 Electron micrographs of sciatic nerves from rats at various times after beginning a 7-day exposure to a 1.25% tellurium diet. (A) One day after beginning exposure, lipid droplets (presumably containing squalene, the substrate of the enzyme inhibited by tellurium) are present in the cytoplasm of myelinating Schwann cells (arrows). (B) After 4 days of tellurium exposure, demyelinated axons (*) are present. Nerve fibers with marked degeneration of their myelin sheaths are also present (arrows). (C) Eleven days after beginning a 7-day exposure to tellurium, numerous very thin myelin sheaths are present (*), indicating remyelination by Schwann cells. A still demyelinated axon is present in the lower right corner of this panel (A). (D) Thirty days after beginning a 7-day exposure to tellurium, myelin sheaths of remyelinating axons are now sufficiently thick that it is difficult to distinguish this sciatic nerve section from that obtained from an unexposed age-matched control. (From Harry, G.J. et al., *J. Neurochem.*, 52, 938, 1989. With permission.)

complicating factors in this model. The temporal relationships of various cell biological events and pathophysiological parameters associated with tellurium-induced neuropathy are shown in Figure 3.5. Inhibition of cholesterol synthesis occurs very soon after initiating Te exposure. There is a rapid breakdown of blood–nerve barrier function, which persists for a prolonged period of time.[141] There is rapid upregulation of mRNA expression for the low affinity nerve growth factor receptor (NGF-R),[135] with peak expression of the NGF-R protein peaking during the period of initial remyelination, and the protein product is specifically targeted to local areas of myelin loss.[136] Schwann cells undergoing demyelination revert to a more primitive precursor-cell phenotype, while those associated with adjacent nondemyelinating internodes do not.

Proliferation of Schwann cells is an early stereotypical response to primary segmental demyelination induced by a variety of insults, including tellurium,[128] intraneural injection of lysolecithin,[142] or antibodies against galactocerebroside.[143] There is overproduction of new Schwann cells; however, only about one of every four to six Schwann cells eventually goes on to remyelinate a demyelinated internode.[128] The possible sources of Schwann cells involved in remyelination include those originally associated with unmyelinated axons, newly formed supernumerary Schwann cells in the endoneurial space, and Schwann cells associated with myelinated fibers.[144,145]

An additional component of the PNS response to toxic insult or injury is the invasion of macrophages. These cells play critical roles in both degenerative and regenerative processes.[146,147] During Te-induced demyelination, there is a robust recruitment of macrophages, assessed morphologically

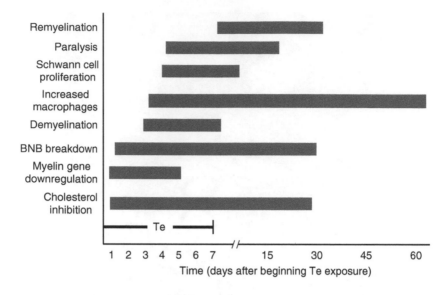

Figure 3.5   Tellurium-induced primary demyelination and remyelination. The temporal sequence and approximate duration of the various processes and events involved are shown in the diagram. In this model of primary demyelination with subsequent remyelination, weanling rats at 20 days of age are placed on a diet containing 1.5% elemental tellurium (Te) for 7 days, after which they are returned to normal rat chow. Although there is complete loss of myelin from about one-fourth of the myelinated internodes in sciatic nerve, there is robust remyelination with near-complete recovery of function within 3 weeks of removing the tellurium insult.

and by mRNA levels for lysozyme, a reliable marker for active phagocytic macrophages.[148] Despite the fact that only about one quarter of the myelinated internodes degenerate, more macrophages are recruited during tellurium-induced demyelination than during a nerve cut or crush, where all myelin as well as axons degenerate. This suggests that Schwann cells recruit macrophages soon after a metabolic insult, prior to any actual demyelination, and that the signal for macrophage recruitment is not directly related to the amount of debris produced. Indeed, a dose- and time-dependent relationship between Schwann cell secretion of monocyte chemoattractant protein-1 (MCP-1), a powerful chemokine relatively specific for monocytes/macrophages, and the subsequent appearance and activity of monocytes/macrophages has been demonstrated.[86] Cholesterol (and perhaps other lipids) derived from degenerating myelin is retained within peripheral nerve for later reutilization during remyelination.[83,149]

As with many other toxicant-induced demyelinating neuropathies, remyelination rapidly follows demyelination induced by exposure to tellurium. Demyelinated axons are present within 3 days of exposure, and remyelinating segments are already apparent within 7 days of beginning exposure (Figure 3.4). Remyelinating axons are initially apparent by their abnormally thin dimensions, but there is rapid deposition of myelin so that by 3 to 4 weeks of recovery, it is difficult to distinguish remyelinated axons from adjacent unaffected internodes.[130]

3.10 THE FUTURE

Modern tools for assessing neurotoxic insult to Schwann cells—The use of morphological and functional/behavioral approaches for assessing neurotoxic insults will always be relevant and useful, as will targeted metabolic and molecular biological studies involving measurements of specific metabolites or steady-state levels of mRNA for selected genes of interest. However, recent advances in the development of very powerful tools of "global" cellular/molecular analyses, specifically genomics,[150] proteomics,[151–153] and metabolomics,[154–157] have the potential to provide a wealth of additional information regarding the actual molecular bases of these insults.[158] Because of the relatively comprehensive nature of the data collected, previously undetected interactions among various genes, their protein products, and levels of related metabolites can be dissected and analyzed. Once the basic underlying biology of the Schwann cell has been characterized with respect to molecular markers for development and maturation, this information can then be applied to better understand the underlying molecular mechanisms involved in Schwann cell reaction to disease or toxic insult. With respect to neurotoxicology, these data will prove very useful not only for elucidating the actual molecular mechanisms and sites of actions of various toxicants, but also in providing potential "biomarkers" for perturbations of the Schwann cell–axon relationship critical for normal nervous system function. This information will hopefully also prove to be relevant to development of therapeutic approaches to minimize damage and promote repair of injured nervous tissue. Currently, there is somewhat limited data with regards to the molecular profile of the peripheral nerve and its various components. Work involving DNA microarray analyses is beginning to establish these profiles with regard to the process of myelination during normal PNS development[51,159] and with respect to PNS injury.[160–162] These studies add to the already existing database of genes known to be regulated during myelin formation and offer the opportunity of "mining" to identify likely candidates for study in toxicant-induced models of PNS/Schwann cell disruption. While gene expression profiles have considerable potential, it is obvious that follow-up studies examining alterations in the encoded protein products (including possible posttranslational processing), as well as products of these proteins, such as membrane lipids, will also often be necessary. Such a multifaceted approach of the full spectrum of metabolism should bring power to our ability to examine the critical features of the Schwann cell, its vulnerability to insult, and the consequences of any such perturbations to the axon and the functioning of the nervous system.

A better understanding of the molecular bases of toxic insults to Schwann cells is of course directly relevant to these cells and the myelin they maintain. However, the usefulness of this knowledge extends beyond the boundaries of the peripheral nervous system. The responses of the PNS and CNS to insult and injury are generally very different, with a much greater potential for regeneration and recovery of function in the PNS. Hopefully, advances in understanding degenerative and regenerative events in the PNS will prove useful in developing strategies to promote remyelination and recovery of function in the CNS as well.

REFERENCES

1. Schwann, T., Mikroskopische untersuchungen uber die uebereinstimmung in der struktur und dem wachsthum der thiere und pflanzen, *Verlag der Sander'schen Buchhandlung*, Berlin, 1, 1839.
2. Gould, R.M. et al., The cell of Schwann: an update, in *Myelin: Biology and Chemistry*, Martenson, R.E., Ed., CRC Press, Boca Raton, 1992, chap. 3, pp. 123–171.
3. Bunge, R.P., Bunge, M.B., and Bates, M., Movements of the Schwann cell nucleus implicate progression of the inner (axon-related) Schwann cell process during myelination, *J. Cell. Biol.*, 109, 273, 1989.
4. Ritchie, J.M., Physiological basis of conduction in myelinated nerve fibers, in *Myelin*, Morell, P., Ed., Plenum Press, New York, 1984, pp. 117–141.
5. Mirsky, R. et al., Schwann cells as regulators of nerve development, *J. Physiol. Paris*, 96, 17, 2002.
6. Martini, R., The effect of myelinating Schwann cells on axons, *Muscle Nerve*, 24, 456, 2001.
7. Arroyo, E.J. and Scherer, S.S., On the molecular architecture of myelinated fibers, *Histochem. Cell. Biol.*, 113, 1, 2000.
8. Peles, E. and Salzer, J.L., Molecular domains of myelinated axons, *Curr. Opin. Neurobiol.*, 10, 558, 2000.
9. De Waegh, S.M., Lee, V.M., and Brady, S.T., Local modulation of neurofilament phosphorylation, axonal caliber, and slow axonal transport by myelinating Schwann cells, *Cell*, 68, 451, 1992.

10. Mata, M., Kupina, N., and Fink, D., Phosphorylation-dependent neurofilament epitopes are reduced at the node of Ranvier, *J. Neurocytol.*, 21, 199, 1992.

11. Mirsky, R. and Jessen, K.R., Schwann cell development, differentiation, and myelination, *Curr. Opin. Neurobiol.*, 6, 89, 1996.

12. Zorick, T.S. and Lemke, G., Schwann cell differentiation, *Curr. Opin. Cell. Biol.*, 8, 870, 1996.

13. Harry, G.J. and Toews, A.D., Myelination, dysmyelination, and demyelination, in *Handbook of Developmental Neurotoxicology*, Slikker, W. and Chang, L., Eds., Academic Press, New York, 1998, pp. 87–115.

14. Le Douarin, N.M. and Kalcheim, C., *The Neural Crest*, Cambridge University Press, Cambridge, U.K., 1999, pp. 328–335.

15. Taylor, V. and Suter, U., Molecular biology of axon-glia interactions in the peripheral nervous system, *Prog. Nucleic Acid Res. Mol. Biol.*, 56, 225, 1997.

16. Jessen, K.R. et al., The Schwann cell precursor and its fate: a study of cell death and differentiation during gliogenesis in rat embyronic nerves, *Neuron*, 12, 509, 1994.

17. Blanchard, A.D. et al., Oct-6 (SCIP, Tst-1) is expressed in Schwann cell precursors, embryonic Schwann cells, and postnatal myelinating Schwann cells: Comparison with Oct-1, Krox-20, and Pax-3, *J. Neurosci. Res.*, 46, 630, 1996.

18. Black, J.A. et al., Effects of delayed myelination by oligodendrocytes and Schwann cells on the macromolecular structure of axonal membrane in rat spinal cord, *J. Neurocytol.*, 15, 745, 1986.

19. Waxman, S.G., Molecular organization of the cell membrane in normal and pathological axons: relation to glial contact, in *Glial-Neuronal Communication in Development and Regeneration*, Althaus, H. and Seifert, W., Eds., Springer Verlag, New York, 1987, p. 711.

20. Waxman, S.G., Rules governing membrane reorganization and axon-glial interactions during the development of myelinated fibers, in *Progress in Brain Research, Neural Regeneration*, Vol. 71, Seil, E.J., Herbert, E., and Carlson, B., Eds., Raven Press, New York, 1987, pp 121–142.

21. Chen, S.J. and DeVries, G.H., Mitogenic effect of axolemma-enriched fraction on cultured oligodendrocytes, *J. Neurochem.*, 52, 325, 1989.

22. Jessen, K.R. and Mirsky, R., Signals that determine Schwann cell identity, *J. Anat.*, 200, 367, 2002.

23. Lobsiger, C.S., Taylor, V., and Suter, U., The early life of a Schwann cell, *Biol. Chem.*, 383, 245, 2002.

24. Buonanno, A. and Fischback, G.D., Neuregulin and ErbB receptor signaling pathways in the nervous system, *Curr. Opin. Neurobiol.*, 11, 287, 2001.

25. Zanazzi, G. et al., Glial growth factor/neuregulin inhibits Schwann cell myelination and induces demyelination, *J. Cell Biol.*, 152, 1289, 2001.

26. Huijbregts, R.P., Roth, K.A., Schmidt, R.E., and Carroll, S.L., Hypertrophic neuropathies and malignant peripheral nerve sheath tumors in transgenic mice overexpressing glial growth factor beta3 in myelinating Schwann cells, *J. Neurosci.*, 23, 7269, 2003.

27. Friede, R.L. and Samorajski, T., Myelin formation in the sciatic nerve of the rat. A quantitative electron microscopic, histochemical and radioautographic study, *J. Neuropathol. Exp. Neurol.*, 27, 546, 1968.

28. Martin, J.R. and Webster, H., Mitotic Schwann cells in developing nerve: their changes in shape, fine structure, and axon relationships, *Dev. Biol.*, 32, 417, 1973.

29. Webster, H. de F., The geometry of peripheral myelin sheaths during their formation and growth in rat sciatic nerves, *J. Cell. Biol.*, 48, 348, 1971.

30. Grinspan, J.B. et al., Axonal interactions regulate Schwann cell apoptosis in developing peripheral nerve: neuregulin receptors and the role of neuregulins, *J. Neurosci.*, 16, 6107, 1996.

31. Syroid, D.E. et al., Cell death in the Schwann cell lineage and its regulation by neuregulin, *Proc. Natl. Acad. Sci. U.S.A.*, 93, 9229, 1996.

32. Nakao, J. et al., Apoptosis regulates the number of Schwann cells at the premyelinating stage, *J. Neurochem.*, 68, 1853, 1997.

33. Garratt, A.N., Britsch, S., and Birchmeier, C., Neuregulin, a factor with many functions in the life of a Schwann cell, *Bioessays*, 22, 987, 2000.

34. Webster, H.D., Development of peripheral nerve fibers, in *Peripheral Neuropathy*, Dyck, P.J., Thomas, P.K., Griffin, J.W., Low, P.A., and Poduslo, J.F., Eds., WB Saunders, Philadelphia, 1993, pp. 23–266.

35. Elder, G.A., Friedrich, V.L. Jr., and Lazzarini, R., Schwann cells and oligodendrocytes read distinct signals in establishing myelin sheath thickness, *J. Neurosci. Res.*, 65, 493, 2001.

36. Herrera, A.A., Qiang, H., and Ko, C.P., The role of perisynaptic Schwann cells in development of neuromuscular junctions in the frog (*Xenopus laevis*), *J. Neurobiol.*, 45, 237, 2000.

37. Yang, J.F. et al., Schwann cells express active agrin and enhance aggregation of acetylcholine receptors on muscle fibers, *J. Neurosci.*, 21, 9572, 2001.

38. Peng, H.B. et al., Differential effects of neurotrophins and Schwann cell-derived signals on neuronal survival/growth and synaptogenesis, *J. Neurosci.*, 23, 5050, 2003.

39. Topilko, P. et al., Krox-20 controls myelination in the peripheral nervous system, *Nature*, 371, 796, 1994.

40. Zorick, T.S. et al., Krox-20 controls SCIP expression, cell cycle exit and susceptibility to apoptosis in developing myelinating Schwann cells, *Development*, 126, 1397, 1999.

41. Kioussi, C., Gross, M.K., and Gruss, P., Pax3; a paired domain gene as a regulator in PNS myelination, *Neuron*, 15, 553, 1995.

42. Stewart, J.H., Expression of c-Jun, Jun B, Jun D and cAMP response element binding protein by Schwann cells and their precursors *in vivo* and *in vitro*, *Eur. J. Neurosci.*, 7, 1366, 1995.

43. Britsch, S. et al., The transcription factor SOX 10 is a key regulator of peripheral glial development, *Genes Dev.*, 15, 66, 2001.

44. Jaegle, M. et al., The POU proteins Brn-2 and Oct-6 share important functions in Schwann cell development, *Genes Dev.*, 17, 1380, 2003.

45. Bermingham, J.R., Jr. et al., Tst-1/Oct-6/SCIP regulates a unique step in peripheral myelination and is required for normal respiration, *Genes Dev.*, 10, 1751, 1996.

46. Jaegle, M. et al., The POU factor Oct-6 and Schwann cell differentiation, *Science*, 273, 507, 1996.

47. Topilko, P. and Meijer, D., Transcription factors that control Schwann cell development and myelination, in *Glial Cell Development*, Jessen, K.R. and Richardson, W.D., Eds., Oxford University Press, Oxford, 2001, pp. 223–244.

48. Ghislain, J. et al., Characterisation of cis-acting sequences reveals a biphasic, axon-dependent regulation of Krox20 during Schwann cell development, *Development*, 129, 155, 2002.

49. Scherer, S.S. et al., Axons regulate Schwann cell expression of the POU transcription factor SCIP, *J. Neurosci.*, 14, 1930, 1994.

50. Arroyo, E.J. et al., Promyelinating Schwann cells express Tst-1/SCIP/Oct-6, *J. Neurosci.*, 18, 7891, 1998.

51. Nagarajan, R. et al., Deciphering peripheral nerve myelination by using Schwann cell expression profiling, *Proc. Natl. Acad. Sci. U.S.A.*, 99, 8998, 2002.

52. Monuki, E.S. et al., Expression and activity of the POU transcription factor SCIP, *Science*, 249, 1300, 1990.

53. Murphy, P. et al., The regulation of Krox-20 expression revels important steps in the control of peripheral glial cell development, *Development*, 122, 2847, 1996.

54. Campagnoni, A.T., Molecular biology of myelination, in *Neuroglia*, Kettenmann, H. and Ransom, B.R., Eds., Oxford University Press, New York, 1995, pp. 555–570.

55. Stahl, N., Harry, J., and Popko, B., Quantitative analysis of myelin protein gene expression during development in the rat sciatic nerve, *Mol. Brain Res.*, 8, 209, 1990.

56. Norton, W.T. and Cammer, W., Isolation and characterization of myelin, in *Myelin*, Morell, P., Ed., Plenum Press, New York, 1984, pp. 147–180.

57. DeWille, J.W. and Horrocks, L.A., Synthesis and turnover of myelin phospholipids and cholesterol, in *Myelin: Biology and Chemistry*, Martenson, R.E., Ed., CRC Press, Boca Raton, 1992, pp. 213–234.

58. Morell, P. and Quarles, R.H., Myelin formation, structure, and biochemistry, in *Basic Neurochem*, 6th ed., Siegel, G.W., Agranoff, B., Albers, R.W., Fisher, S.K., and Uhler, M.D., Eds., Raven Press, New York, 1998, pp. 69–93.

59. Norton, W.T. and Poduslo, S.E., Myelination in rat brain: Method of myelin isolation, *J. Neurochem.*, 21, 749, 1973.

60. Lemke, G. and Axel, R., Isolation and sequence of a cDNA encoding the major structural protein of peripheral myelin, *Cell*, 40, 501, 1985.

61. Lemke, G., Unwrapping the genes of myelin, *Neuron*, 1, 535, 1988.

62. Eichberg, J., Myelin P0: new knowledge and new roles, *Neurochem. Res.*, 27, 1331, 2002.

63. Lemke, G., Myelin and myelination, in *An Introduction to Molecular Neurobiology*, Hall, Z., Ed., Sinauer Associates, Sunderland, MA, 1992, pp. 281–309.

64. Kamholz, J. et al., Structure and expression of proteolipid protein in the peripheral nervous system, *J. Neurosci. Res.*, 31, 231, 1992.

65. Garbern, J.Y. et al., Proteolipid protein is necessary in peripheral as well as central myelin, *Neuron*, 19, 205, 1997.

66. Ong, R.L. and Yu, R.K., Interaction of ganglioside GM1 and myelin basic protein studied by ^{12}C and ^{1}H nuclear magnetic resonance, *J. Neurosci. Res.*, 12, 377, 1984.

67. Maggio, B. and Yu, R.K., Interaction and fusion of unilamellar vesicles containing cerebrosides and sulfatides induced by myelin basic protein, *Chem. Phys. Lipids*, 51, 127, 1989.

68. Maggio, B. and Yu, R.K., Modulation by glycosphingolipids of membrane-membrane interaction induced by myelin basic protein and mellitin, *Biochim. Biophys. Acta*, 112, 105, 1992.

69. Smith, R., The basic protein of CNS myelin: its structure and ligand binding, *J. Neurochem.*, 59, 1589, 1992.

70. Trapp, B.D., The myelin-associated glycoprotein: Location and potential functions, in *Myelination and Dysmyelination*, Duncan, I.D., Skoff, R.P., and Colman, D.R., Eds., New York Academy of Science, New York, 1990, p. 29.

71. Quarles, R.H. et al., Myelin-associated glycoprotein: structure-function relationships and involvement in neurological diseases, in *Myelin: Biology and Chemistry*, Martenson, R.E., Ed., CRC Press, Boca Raton, FL, 1992, pp. 413–448.

72. Mendell, J.R. et al., Polyneuropathy and IgM monoclonal gammopathy: studies on the pathogenetic role of anti-myelin-associated glycoprotein antibody, *Ann. Neurol.*, 17, 243, 1985.

73. Tatum, A.H., Experimental paraprotein neuropathy, demyelination by passive transfer of IgM anti-myelin-associated glycoprotein, *Ann. Neurol.*, 33, 502, 1993.

74. Newman, S., Saito, M., and Yu, R.K., Biochemistry of myelin proteins and enzymes, in *Neuroglia*, Kettenman, H. and Ransom, B.R., Eds., Oxford University Press, New York, 1995, pp. 535–554.

75. Ousley, A.H. and Morell, P., Individual molecular species of phosphatidylcholine and phosphatidyle-thanolamine in myelin turn over at different rates, *J. Biol. Chem.*, 267, 10362, 1992.

76. Benjamins, J.A. and Smith, M.E., Metabolism of myelin, in *Myelin*, 2nd ed., Morell, P., Ed., Plenum Press, New York, 1984, pp. 225–249.

77. Deshmukh, D.S. et al., Rapid incorporation *in vivo* of intracerebrally injected ^{32}Pi into polyphospho-inositides of three subfractions of rat brain myelin, *J. Neurochem.*, 36, 594, 1981.

78. DesJardins, K.C. and Morell, P., Phosphate groups modifying myelin basic proteins are metabolically labile; methyl groups are stable, *J. Cell Biol.*, 97, 438, 1983.

79. Reiss, D.S., Lees, M.B., and Sapirstein, V.S., Is Na+,K+-ATPaase a myelin associated enzyme?, *J. Neurochem.*, 36, 1418, 1981.

80. Cammer, W., Carbonic anhydrase activity in myelin from sciatic nerves of adult and young rats quantitation and inhibitor sensitivity, *J. Neurochem.*, 32, 651, 1979.

81. Yamauchi, J., Chan, J.R., and Shooter, E.M., Neurotrophin 3 activation of TrkC induces Schwann cell migration through the c-Jun N-terminal kinase pathway, *Proc. Natl. Acad. Sci. U.S.A.*, 100, 14421, 2003.

82. Ramon y Cajal, S., *Degeneration and Regeneration of the Nervous System* (translated by May, R.M.), Oxford University Press, London, 1928, pp. 92–98.

83. Goodrum, G.F. and Bouldin, T.W., The cell biology of myelin degeneration and regeneration in the peripheral nervous system, *J. Neuropathol. Exp. Neurol.*, 55, 943, 1996.

84. Wohlleben, G. et al., Regulation of Fas and FasL expression on rat Schwann cells, *Glia*, 30, 373, 2000.

85. Bonetti, B. et al., T-cell cytotoxicity of human Schwann cells: TNFα promotes fasL-mediated apoptosis and IFNγ perforin-mediated lysis, *Glia*, 43, 141, 2003.

86. Toews, A.D., Barrett, C., and Morell, P., Monocyte chemoattractant protein 1 is responsible for macrophage recruitment following injury to sciatic nerve, *J. Neurosci. Res.*, 53, 260, 1998.

87. Rutkowski, J.L. et al., Signals for proinflammatory cytokine secretion by human Schwann cells, *J. Neuroimmunol.*, 101, 47, 1999.

88. Keswani, S.C. et al., Schwann cell chemokine receptors mediate HIV-1 gp120 toxicity to sensory neurons, *Ann. Neurol.*, 54, 287, 2003.

89. Jander, S. et al., Osteopontin: a novel axon-regulated Schwann cell gene, *J. Neurosci. Res.*, 67, 156, 2002.

90. Privat, A. et al., Absence of the major dense line in myelin of the mutant mouse 'shiverer,' *Neurosci. Lett.*, 12, 107, 1979.

91. Kirschner, D.A., and Ganser, A.L., Compact myelin exists in the absence of basic protein in the shiverer mutant mouse, *Nature*, 283, 207, 1980.

92. Rosenbluth, J., Central myelin in the mouse mutant shiverer, *J. Comp. Neurol.*, 194, 639, 1980.

93. Messing, A. et al., P_0 promoter directs expression of reporter and toxin genes to Schwann cells of transgenic mice, *Neuron*, 8, 507, 1992.

94. Trapp, B., Distribution of the myelin-associated glycoprotein and Po protein during myelin compaction in quaking mouse peripheral nerve, *J. Cell Biol.*, 107, 675, 1988.

95. Suter, U. et al., A leucine-to-proline mutation in the putative first transmembrane domain of the 22-kDa peripheral myelin in the trembler-J mouse, *Proc. Natl. Acad. Sci. U.S.A.*, 89, 4382, 1992.

96. Sereda, M. et al., A rat transgenic model for Charcot-Marie-Tooth disease, *Neuron*, 16, 1049, 1996.

97. Gale, A.N., Gomez, S., and Duchen, L.W., Changes produced by a hypomyelinating neuropathy in muscle and its innervation. Morphological and physiological studies in the Trembler mouse, *Brain*, 105, 373, 1982.

98. Henry, E.W. and Sidman, R.L., Long lives for homozygous trembler mutant mice despite virtual absence of peripheral nerve myelin, *Science*, 241, 344, 1988.

99. Sancho, S. et al., Distal axonopathy in peripheral neves of PMP22-mutant mice, *Brain*, 122, 1563, 1999.

100. Watson, D.F. et al., Altered neurofilament phosphorylation and beta tubulin isotypes in Charcot-Marie-Tooth disease type 1, *Neurology*, 44, 2383, 1994.

101. Kamholz, J. et al., Charcot-Marie-Tooth disease type 1: molecular pathogenesis to gene therapy, *Brain*, 123, 222, 2000.

102. Naef, R. and Suter, U., Many facets of the peripheral myelin protein PMP22 in myelination and disease, *Microsc. Res. Tech.*, 41, 359, 1998.

103. Berger, P., Young, P., and Suter, U., Molecular cell biology of Charcot-Marie-Tooth disease, *Neurogenetics*, 4, 1, 2002.

104. Dyck, P.J. et al., Hereditary motor and sensory neuropathies, in *Peripheral Neuropathy*, Dyck, P.J., Thomas, P.K., Griffin, J.W., Low, P.A., and Poduslo, J.F., Eds., WB Saunders, Philadelphia, 1993, pp. 1094–1136.

105. Maier, M., Berger, P., and Suter, U., Understanding Schwann cell-neurone interactions: the key to Charcot-Marie-Tooth disease, *J. Anat.*, 357, 2002.

106. Suter, U. and Scherer, S.S., Disease mechanisms in inherited neuropathies, *Nat. Rev. Neurosci.*, 4, 714, 2003.

107. Zhou, L. and Griffin, J.W., Demyelinating neuropathies, *Curr. Opin. Neurol.*, 16, 307, 2003.

108. Morell, P., Biochemical and molecular bases of myelinopathy, in *Principles of Neurotoxicology*, Chang, L.W., Ed., Marcel Dekker, New York, 1994, chap. 21.

109. Morell, P., Goodrum, J.F., and Bouldin, T.W., Biochemical toxicology of the peripheral nervous system, in *Introduction to Biochemical Toxicology*, 2nd ed., Hodgson, E. and Levi, P.E., Eds., Appleton & Lange, Norwalk, CT, 1994, chap. 18.

110. Morell, P. and Toews, A.D., Schwann cells as targets for neurotoxicants, *NeuroToxicology*, 17, 685–696, 1996.

111. Watanabe, I., Organotins (Triethyltin), in *Experimental and Clinical Neurotoxicology*, Spencer, P.S. and Schaumburg, H.H., Eds., Williams & Wilkins, Baltimore, 1980, chap. 37.

112. Aleu, F.P., Katzman, R., and Terry, R.D., Fine structure and electrolyte analysis of cerebral edema induced by alkyl tin intoxication, *J. Neuropathol. Exp. Neurol.*, 22, 403–413, 1963.

113. Graham, D.I. and Gonatas, N.K., Triethyltin sulfate-induced splitting of peripheral myelin in rats, *Lab. Invest.*, 29, 628, 1973.

114. Bouldin, T.W. et al., Pathogenesis of trimethyltin neuronal toxicity, *Am. J. Pathol.*, 104, 237–249, 1981.

115. Aldridge, W.N. and Cremer, J.E., The biochemistry of organo-tin compounds; diethyltin dichloride and triethyltin sulfate, *Biochem. J.*, 61, 406, 1955.

116. Wassenar, J.S., and Kroon, A.M., Effects of triethyltin on different ATPases, 5'-nucleotidase and phosphodiesterases in grey and white matter of rabbit brain and their relation with brain edema, *Eur. Neurol.*, 10, 349, 1973.

117. Selwyn, M.J. et al., Chloride-hydroxide exchange across mitochondrial, erythrocyte, and artificial lipid membranes mediated by trialkyl and triphenlytin compounds, *Eur. J. Biochem.*, 14, 120, 1970.

118. Cammer, W., Toxic demyelination: biochemical studies and hypothetical mechanisms, in *Experimental and Clinical Neurotoxicology*, Spencer, P.S. and Schaumburg, H.H., Eds., Williams & Wilkins, Baltimore, 1980, chap. 17.

119. Yanagisawa, K. et al., Acetazolamide inhibits the recovery from triethyl tin intoxication: putative role of carbonic anhydrase in dehydration of central myelin, *Neurochem. Res.*, 15, 483, 1990.

120. Towfighi, J., Gonatas, N.K., and McCree, L., Hexachlorophene-induced changes in central and peripheral myelinated axons of developing and adult rats, *Lab. Invest.*, 31, 712, 1974.

121. Nieminen, L., Bjondahl, K., and Mottonen, M., Effect of hexachlorophene on the rat brain during organogenesis, *Food Cosmet. Toxicol.*, 11, 635, 1973.

122. Gombault, A., Contribution à l'étude anatomique de la névrite parenchymateuse subaiguë et chronique: névritre segmentaire périaxile, *Arch. Neurol. (Paris)*, 1, 177, 1880–81.

123. Krigman, M.R., Bouldin, T.W., and Mushak, P., Lead, in *Experimental and Clinical Neurotoxicology*, Spencer, P.S. and Schaumburg, H.H., Eds., Williams & Wilkins, Baltimore, 1980, chap. 34.

124. Pleasure, D.E., Feldman, B., and Prockop, D.J., Diphtheria toxin inhibits the synthesis of myelin proteolipid and basic proteins by peripheral nerve *in vitro*, *J. Neurochem.*, 20, 81, 1980.

125. Cammer, W. and Brion, L.P., Carbonic anhydrase in the nervous system, in *The Carbonic Anhydrases: New Horizons*, Chegwidden, W.R., Carter, N.D., and Edwards, Y.H., Eds., Birkhauser Verlag, Basel, 2000, pp. 475–489.

126. Toews, A.D. et al., Alterations in gene expression associated with primary demyelination and remyelination in the peripheral nervous system, *Neurochem. Res.*, 22, 1271, 1997.

127. Lampert, P.W., Garro, F., and Pentschew, A., Tellurium neuropathy, *Acta. Pathol. (Berlin)*, 15, 308, 1970.

128. Said, G., Duckett, S., and Sauron, B., Proliferation of Schwann cells in tellurium-induced demyelination in young rats, *Acta Neuropathol.*, 53, 173, 1981.

129. Boudlin, T.W. et al., Schwann-cell vulnerability to demyelination is associated with internodal length in tellurium neuropathy, *J. Neuropathol. Exp. Neurol.*, 47, 41, 1988.

130. Harry, G.J. et al., Tellurium-induced neuropathy: Metabolic alterations associated with demyelination and remyelination in rat sciatic nerve, *J. Neurochem.*, 52, 938, 1989.

131. Wagner-Recio, M., Toews, A.D., and Morell, P., Tellurium blocks cholesterol synthesis by inhibiting squalene metabolism: Preferential vulnerability to this metabolic block leads to peripheral nervous system demyelination, *J. Neurochem.*, 57, 1891, 1991.

132. Wagner, M., Toews, A.D., and Morell, P., Tellurite specifically affects squalene epoxidase: Investigations examining the mechanism of tellurium-induced neuropathy, *J. Neurochem.*, 64, 2169, 1995.

133. Toews, A.D. et al., Tellurium-induced neuropathy: A model for reversible reductions in myelin protein gene expression, *J. Neurosci. Res.*, 26, 501, 1990.

134. Toews, A.D. et al., Tellurium-induced alterations in HMG-CoA reductase gene expression and enzyme activity: Differential effects in sciatic nerve and liver suggest tissue-specific regulation of cholesterol synthesis, *J. Neurochem.*, 57, 1902, 1991.

135. Toews, A.D. et al., Primary demyelination induced by exposure to tellurium alters mRNA levels for nerve growth factor receptor, SCIP, 2',3'-cyclic nucleotide 3'-phosphodiesterase, and myelin proteolipid protein in rat sciatic nerve, *Mol. Brain Res.*, 11, 321, 1991.

136. Toews, A.D. et al., Primary demyelination induced by exposure to tellurium alters Schwann-cell gene expression: A model for intracellular targeting of NGF-receptor, *J. Neurosci.*, 12, 3676, 1992.

137. Jurevics, H. et al., Regenerating sciatic nerve does not utilize circulating cholesterol, *Neurochem. Res.*, 23, 401, 1998.

138. Jurevics, H.A. and Morell, P., Sources of cholesterol for kidney and nerve during development, *J. Lipid Res.*, 35, 112, 1994.

139. Toews, A.D. et al., Tissue-specific coordinate regulation of enzymes of cholesterol biosynthesis: sciatic nerve vs liver, *J. Lipid Res.*, 37, 2502, 1996.

140. Toews, A.D. et al., Tellurium causes dose-dependent coordinate downregulation of myelin gene expression, *Mol. Brain Res.*, 49, 113, 1997.

141. Bouldin, T.W. et al., Temporal relationship of blood-nerve barrier breakdown to the metabolic and morphologic alterations of tellurium neuropathy, *Neurotoxicology*, 10, 79, 1989.

142. Hall, S.M. and Gregson, N.A., The effects of mitomycin C on the process of remyelination in mammalian peripheral nervous system, *Neuropathol. Appl. Neurobiol.*, 1, 149, 1975.

143. Saida, K. and Saida T., Proliferation of Schwann cells in demyelinated rat sciatic nerve, *Acta. Neuropathol.*, 71, 251, 1986.

144. Griffin, J.W. et al., Schwann cell proliferation and migration during paranodal demyelination, *J. Neurosci.*, 7, 682, 1987.

145. Griffin, J.W. et al., Schwann cell proliferation following lysolecithin-induced demyelination, *J. Neurocytol.*, 19, 367, 1990.

146. Perry, V.H. and Brown, M.C., Role of macrophages in peripheral nerve degeneration and repair, *Bioessays*, 14, 401, 1992.

147. Griffin, J.W., George, R., and Ho, T., Macrophage systems in peripheral nerves. A review, *J. Neuropathol. Exp. Neurol.*, 52, 553, 1993.

148. Venezie, R.D., Toews, A.D., and Morell, P., Macrophage recruitment in different models of nerve injury: lysozyme as a marker for active phagocytosis, *J. Neurosci. Res.*, 40, 99, 1995.

149. Bouldin, T.W. and Goodrum, J.F., Toxicant-induced demyelinating neuropathy, in *Principles of Neurotoxicology*, Chang, L.W., Ed., Marcel Dekker, New York, 1994, chap. 8.

150. Luo, Z. and Geschwind, D.H., Microarray applications in neuroscience, *Neurobiol. Dis.*, 8, 183, 2001.

151. MacBeath, G., Protein microarrays and proteomics, *Nat. Genet.*, 32 Suppl, 526, 2002.

152. Michaud, G.A. and Snyder, M., Proteomic approaches for the global analysis of proteins, *Biotechniques*, 33, 1308, 2002.

153. LoPachin, R.M. et al., Application of proteomics to the study of molecular mechanisms in neurotoxicology, *Neurotoxicology*, 24, 761, 2003.

154. Phelps, T.J., Palumbo, A.V., and Beliaev, A.S., Metabolomics and microarrays for improved understanding of phenotypic characteristics controlled by both genomics and environmental constraints, *Curr. Opin. Biotechnol.*, 13, 20, 2002.

155. Robosky, L.C. et al., *In vivo* toxicity screening programs using metabonomics, *Comb. Chem. High Throughput Screen.*, 5, 651, 2002.

156. Weckwerth, W., Metabolomics in systems biology, *Annu. Rev. Plant Biol.*, 54, 669, 2003.

157. Viant, M.R., Rosenblum, E.S., and Tieerdema, R.S., NMR-based metabolomics: a powerful approach for characterizing the effects of environmental stressors on organism health, *Environ. Sci. Technol.*, 37, 4982, 2003.

158. Waters, M.D., Olden, K., and Tennant, R.W., Toxicogenomic approach for assessing toxicant-related disease, *Mutat. Res.*, 544, 415, 2003.

159. Verheijen, M.H. et al., Local regulation of fat metabolism in peripheral nerves, *Genes Dev.*, 17, 2450, 2003.

160. Fan, M. et al., Analysis of gene expression following sciatic nerve crush and spinal cord hemisection in the mouse by microarray expression profiling, *Cell. Mol. Neurobiol.*, 21, 497, 2001.

161. Costigan, M. et al., Replicate high-density rat genome oligonucleotide microarrays reveal hundreds of regulated genes in the dorsal root ganglion after peripheral nerve injury, *BMC Neurosci.*, 3, 16, 2002.

162. Xiao, H.S. et al., Identification of gene expression profile of dorsal root ganglion in the rat peripheral axotomy model of neuropathic pain, *Proc. Natl. Acad. Sci. U.S.A.*, 99, 8360, 2002.

163. Morell, P. and Norton, W.T., Myelin, *Sci. Am.*, 242, 74, 1980.

CHAPTER 4

Function of Astrocyte Cytochrome P450 in Control of Xenobiotic Metabolism

Ralf Peter Meyer, Benedikt Volk, and Rolf Knoth

CONTENTS

4.1 INTRODUCTION

Cytochrome P450 is the collective term for a large superfamily of heme-containing proteins that play an important role in the oxidative metabolism of numerous endogenous and foreign compounds.[1] The CYP families 1 to 4 are involved in drug metabolism and are preferentially expressed in the liver. Drug metabolizing CYP isoforms also occur in brain and include CYP2B1, CYP2C29, CYP2D4, CYP2E1, CYP4F, CYP3A9, CYP3A11, and CYP3A13.[2,3] The CYP isoforms in brain are localized in defined neuronal or glial cell populations.[4,5] Several of the CYPs expressed in the brain are inducible by alcohol, neuroleptics, anticonvulsants, and endocrine factors.[6–8]

Important questions regarding CYPs expressed in brain are whether these enzymes are active in drug and xenobiotic metabolism.[9] Pharmaceuticals used in treatment of neurological and mental diseases (e.g., haloperidol, diazepam, carbamazepine, phenytoin) exert their function in specific brain regions, and their local concentration may depend on their biotransformation via internal brain CYP pathways.[4,10,11] A characteristic issue of all of these compounds is that they cross the blood–brain and blood–cerebrospinal fluid barrier mainly by diffusion and therefore penetrate into CYP containing astrocytes, which are a major part of these permeability barriers.[12,13]

The function of such CYP containing astrocytes in response to xenobiotic exposure is a topic of great interest. Pharmaceuticals like those previously mentioned are potentially effective in specified neurons. Therefore their entrance into brain needs to be effectively controlled. Astrocytes present at the brain border lines might regulate drug influx by CYP dependent biotransformation and therefore balance neuronal exposition to the respective drugs. Astrocytes have intermediary and regulative functions especially in the control of neuronal outgrowth and neuronal immune functions.[14–16] The regulation of drug or xenobiotic availability for the neurons may be an important part of these intermediary functions.

4.2 CYP IN THE BRAIN—AN OVERVIEW

CYPs are well known as membrane-bound phase I monooxygenases preferentially expressed in the liver to metabolize xenobiotics and several endogenous compounds. Liver CYP can be regarded as the major part of the body's overall drug control and clearance system.[17,18] Brain CYP, however, is expressed approximately only 1–10% of that in liver and is therefore too low to significantly influence overall pharmacokinetics (Table 4.1). Instead the brain CYPs appear to have specific functions according to the CYP subfamily they belong to and the brain region they are expressed at. The subfamilies 1–4 are involved in drug, steroid, and fatty acid metabolism.[5,9] The families 7, 11, 19, 21, and 27 (mammalian CYPs) take part in cholesterol homeostasis and neurosteroid biosynthesis of the brain.[19]

Furthermore, the regulation of brain CYP function is different from liver. The availability of heme seems to be a crucial factor for CYP enzymatic activity, as the brain does not contain a regulatory heme pool similar to liver.[20,21] The availability of heme reveals to be a limiting factor of CYP function in brain.[22] Moreover, heme appears to affect the incorporation of CYP1A1 into the membrane of endoplasmic reticulum (ER). The process of insertion of heme into the apocytochromes P450 and the mechanism of incorporation into the ER is incompletely known. We speculate that reduced availability of heme in brain may limit the incorporation of CYPs into ER membranes and therefore impair their function.[22]

The brain consists of around 100 billion neurons and 10-fold more glial cells.[23] Neuronal CYPs are often expressed in defined neuronal cell populations in specific regions, such as CYP1A, 1B, 2A, 2G in the olfactory mucosa[24,25] and CYP1A1 in olfactory bulb, striatum, hypothalamus, and cortex.[22,26] CYP3A is present in steroid hormone–sensitive areas such as hippocampus, hypothalamus, olfactory bulb, and cerebellum.[2] These neuronal P450s are discussed to metabolize odorants,

Table 4.1   P450 Content of Liver and Brain Cell Populations of C57Bl/6J Mice

	Liver Total[a]	Brain Total[a]	Brain Neurons[b]	Brain Astrocytes[c]
		pmol P450/mg protein		
Untreated	188	32	15	39
Treated with phenytoin[d]	958	32	30	36

[a] Measurement of CYP content was performed by CO difference spectrum[69] using cell homogenates from total organs.
[b] Neuronal hippocampal cell line from postnatal day 21 was used for investigations.[70]
[c] Astrocyte primary cultures generated from pubs of postnatal day 1 (P1) were used for investigations.[6]
[d] Phenytoin is a potent inducer of CYP2B9, CYP2C29, and CYP3A4, -11, and -13 in mouse tissue.[2,4] Phenytoin was applied in a concentration of 80 mg/kg/day in solid nutrition, causing a drug serum level of 15 mg/ml. This is comparable to the human situation.[71]

xenobiotics, and steroid hormones specifically attending the affected cell populations and therefore maintain cell sensitivity to certain xenobiotics.

Drugs invading the brain are usually trapped and forwarded by astrocytes before they reach their neuronal targets. Due to the much greater number of astrocytes, glial CYPs are more abundantly expressed than the neuronal isoforms. Unlike neurons, astrocytes do not show stringent regiodependent CYP expression pattern.[27] The total CYP content of astrocytes, expressed as pmol P450/mg cellular protein, is about 2.7-fold higher than that of neurons (Table 4.1). This indicates that astroglial CYPs do not have the same function as their neuronal counterparts. They might act as a specific filter for drug influx into the brain. In the present chapter, we discuss first localization of CYP expression, then induction of specific isoforms, and finally the resulting drug biotransformation to elucidate the function of glial CYPs in xenobiotic metabolism.

4.3 LOCALIZATION OF CYP-CONTAINING ASTROCYTES *IN VIVO*

4.3.1 CYP in Astrocytes of Brain Border Lines

4.3.1.1 Blood–Brain Barrier

The blood–brain barrier is the brain's most powerful penetration barrier. Besides the capillary endothelial cells, astrocytes are a major component of it.[13,28] The penetration of drugs into the brain is necessarily highly regulated.[29] According to Seelig et al., three groups of drugs can be distinguished in their ability to cross the blood–brain barrier[30]: a region of very hydrophobic drugs that fail to enter the central nervous system because they remain adsorbed to the membrane, a central area of less hydrophobic drugs that can cross the blood–brain barrier, and a region of relatively hydrophilic drugs that do not cross the blood–brain barrier unless applied at high concentrations. Most of the anticonvulsant or neuroleptic drugs used in therapy today belong to the second group, which cross the blood–brain barrier mainly by diffusion or facilitated transport systems.[6,31,32] Examples for such drugs are carbamazepine, phenytoin, phenobarbital, diazepam, and haloperidol. Interestingly these drugs are either substrates or inducers of specific CYPs such as CYP2C19, 2D6, and 3A in humans.[33] Their counterparts in rat brain are CYP2B1/2 (phenytoin, phenobarbital)[12,34] and in mouse brain are CYP2B9, CYP2C29, and CYP3A11/13 (phenytoin, phenobarbital, carbamazepine).[4,7,35]

Investigations of rat and mouse blood–brain barrier demonstrate that astrocytes surrounding the capillary vessels express either CYP2B1 (rat) (Figure 4.1A) or CYP2C29 (mouse) (Figure 4.1B, C) to high extent. Interestingly, the processes of some astrocytes contact the capillary vessels directly. CYP2B1 immunoreaction clearly occurs just at the junction between vessel and glial cell (Figure 4.1A). Therefore, pharmaceuticals penetrating through the blood–brain barrier directly enter the astrocytes and could easily be trapped by drug metabolizing CYPs. This "metabolic trapping" was observed investigating phenytoin uptake into cultured astrocytes.[6] The CYPs expressed in astrocytes of the blood–brain barrier can be assumed as the brain's first line of metabolic defense against penetrating xenobiotics.

4.3.1.2 Cerebrospinal Fluid–Brain Barrier

The CSF is an important determinant of brain's extracellular fluid that bathes neurons and glia. CSF is secreted mainly by the *choroid plexus* and flows primarily one way from the lateral ventricles through the intraventricular foramina into the third ventricle and from there to the fourth ventricle and the *subarachnoid space*.[36] The composition of the CSF is in a steady state with brain extracellular fluid and is a major way to transport potentially harmful brain metabolites and xenobiotics through the brain. Astrocytes are located between neurons and the cerebral

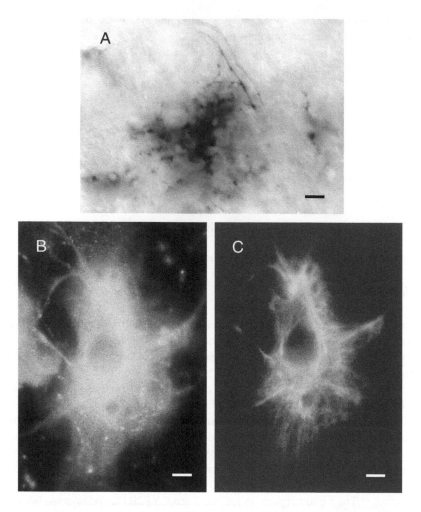

Figure 4.1   Expression of CYP2B1 and CYP2C29 in fibrillary astrocytes of the blood–brain barrier. (A) Expression of CYP2B1 in astrocytes connecting to a capillary vessel in pons region of phenytoin-treated rats. The processes of astrocytes have direct junction toward the capillary vessel. Cryosections of 40 μm were fixed using periodate-lysine-paraformaldehyde and incubated free-floating with monoclonal antibody mab 204 against rat CYP2B1. As chromogen for the visualization of the immunoreaction, 3,3'-diaminobenzidine was used[27] (mab 204 was a generous gift from U.A. Meyer, Biozentrum of the University of Basel, Switzerland) (the bar represents 10 μm). (B, C) Double-labeling of an astrocyte in primary culture (10 days *in vitro*) illustrating colocalization of murine CYP2C29 (B) and glial fibrillary acidic protein (GFAP) (C) after treatment of the cells with phenytoin. Astrocytes originate from mouse cortical hemisphere, cultured at postnatal day 1 (P1). CYP2C29 is highly enriched (the bar represents 10 μm). (B) Indirect immunofluorescence using rhodamin-labeled secondary antibodies coupled to polyclonal antibodies against CYP2C29 (TRITC-CYP2C29, dilution 1:500). (C) Indirect immunofluorescence using fluorescein-labeled secondary antibodies coupled to monoclonal antibody against GFAP (FITC-GFAP, dilution 1:2000).

microcirculation and form a linkage between them.[37,38] This location at the CSF–brain barrier indicates a function of astrocytes in regulation of permeability of xenobiotics to neuronal structures of brain regions along the pathway of CSF flow. We investigated CYP expression of brain–CSF barrier at two characteristic areas of drug penetration within the brain: first, the area of the lateral ventricle neighboring hippocampus and, second, the pons–pia mater region. Glial CYP function in both, hippocampus and pons, is of high interest, because handed penetration of drugs such as phenytoin and ethanol exerts neurotoxic effects especially in hippocampus and cerebellum.[39–41] The pons relays information from cerebral cortex to the cerebellum and is therefore a region of

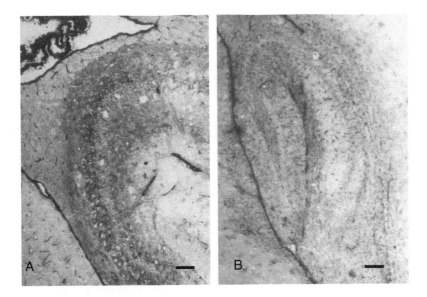

Figure 4.2   Expression of CYP2C29 (A) and GFAP (B) in mouse hippocampus neighboring the lateral ventricle. (A) Fibrillary astrocytes of stratum oriens, the ependymal cells of the lateral ventricle, and the pyramidal neurons of the CA3 region show strong CYP2C29 immunoreactivity. The ependymal cells and the fibrillary astrocytes are part of the CSF–brain barrier (the bar represents 100 µm). (B) Strong hippocampal GFAP expression is located at stratum oriens and the septo-hippocampal structures. These structures also show CYP2C29 immunoreaction. GFAP expression decreases continuously from striatum radiatum to the molecular layer of CA1 region where only very few astrocytes could be detected. These astrocytes obviously do not express CYP2C29 (the bar represents 200 µm). Cryosections of 10–12 µm were stained with polyclonal antibodies against CYP2C29 or monoclonal antibody against GFAP according to the staining instructions for avidin-biotin-peroxidase system (ABC-Kit, Vector, Burlingame, U.S.).

high metabolic exchange originating from endogenous or exogenous compounds circulating in the CSF.[42,43]

The expression profile of CYP2C29 at the CSF–brain barrier supports our hypothesis (Figure 4.2). After treatment with phenytoin glial cells in stratum oriens of hippocampus connecting from the ependymal cell layer of the lateral ventricle to CA3 pyramidal neurons and their axonal extensions through alveus clearly show CYP2C29 expression (Figure 4.2A). Conversely, other fibrillary astrocytes within the hippocampus, located at striatum radiatum, striatum lucidum, or the outer molecular layer of fascia dentata, obviously do not express CYP2C29 to a significant level. Astrocytes present in fimbria and the septohippocampal area also show only moderate expression of CYP2C29 (Figure 4.2). The strong immunoreaction in the pyramidal cell layers is, however, predominantly neuronal. These findings are in accordance with data reported on rat CYP isoforms. Rat CYP2D is predominantly expressed in choroid plexus and pyramidal neurons of hippocampus, while glial expression of CYP2D within hippocampus is on a more moderate level and restricted to molecular and polymorphic layers.[44] Further CYPs that are expressed to a certain amount in hippocampal astrocytes are CYP2C11 (rat) and CYP3A (mouse).[7,45] The CYP2C11 is reported to act in glutamate receptor activation in hippocampus, while the function of glial CYP3A remains to be discussed.

Additionally, the exchange of xenobiotics or endogenous metabolites in the pons is expected to be under efficient control. The pons is covered by the CSF circulation system, ventral by the meningeal layers, dura mater, arachnoid, and pia mater, and dorsal by the fourth ventricle. Consistent to this, astrocytes located at the pons–pia mater junction express CYPs active in drug metabolism to a sufficient extent (Figure 4.3). This is also reported by Ghersi-Egea et al.[46] A tight band of CYP2C29 immunoreactive astrocytes contacts the pia mater in phenytoin-treated mouse brain. With

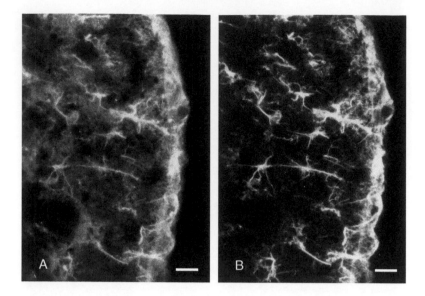

Figure 4.3 Double-labeling of astrocytes located at the penetration barrier of the pons and pia mater region to demonstrate colocalization of CYP2C29 (A) and GFAP (B) in mouse brain. Glial expression of CYP2C29 is diminished with increasing distance from the pia mater. (A) Immunofluorescence with TRITC-CYP2C29. (B) Immunofluorescence with FITC-GFAP. Most of the fibrillary astrocytes are located within the pons–pia mater connections. Cryosections of 10–12 μm were used (the bar represents 20 μm).

increasing distance from this barrier, the number of fibrillary astrocytes and concomitantly the expression of CYP2C29 within the pons is diminished (Figure 4.3A). Interestingly, protoplasmic astrocytes are located to some extent within pons, but contrary to their fibrillary counterparts they do not show significant CYP isoform expression.

4.3.2 Glial CYPs in Parenchymal Structures of the Brain: Cortex and Cerebellum

We next investigated whether astrocytes located in the parenchymal structures of the brain, such as the interior regions of cerebellum and cortex, express CYPs and how these might response to xenobiotic metabolism. Glial cells of molecular layer of cerebellum are mainly of Bergmann glia type, facilitating neuronal outgrowth and migration. However, even after phenytoin exposure these glial cells do not contain CYP2C29 (Figure 4.4). This is an interesting finding, as the Purkinje cells of the cerebellum are highly vulnerable to phenytoin exposure.[41,47] Nevertheless, cerebellum is a region of strong CYP expression of various isoforms due to xenobiotic exposure, but their expression is predominantly neuronal. CYP2B1*, CYP2C29, CYP3A11, and CYP3A13 are found in granule cells or Purkinje neurons after phenytoin or phenobarbital exposure[2,6,27] (Figure 4.4A). CYP2D and CYP2E1 were detected in Purkinje cells.[44,48,49] The Bergmann glia, however, which maintain cerebellar structural organization and therefore are the cerebellum's main glial cell type, obviously do not fulfill a function in xenobiotic control or metabolism.

In cortex, some glial CYP expression of different function is reported. Very moderate glial expression of CYP3A is found in various regions of the cortex, but its function remains unclear.[7] CYP2C11 is located in astrocytes surrounding cortical capillary vessels in rat brain, possibly acting in regulation of vasodilation.[50] Steroidogenic CYP17 is reported to metabolize pregnenolone to DHEA in cortical astrocytes.[51] A main function of cortical astrocytes is the response to inflammation and the subsequent expression of cytokines such as IL-6 and IL-1β. Concomitant to this, a set of

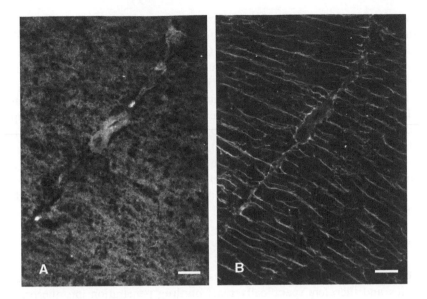

Figure 4.4   Molecular layer of mouse cerebellum. Double-labeling demonstrates that CYP2C29 is expressed to a considerable amount only around a small capillary vessel and in some neuronal structures of Purkinje cells (A). The GFAP-positive fibrillary structures of the Bergmann glia in the molecular layer (B) do not show any CYP2C29 expression. Cryosections of 10–12 μm were incubated with TRITC-CYP2C29 and FITC-GFAP (the bar represents 20 μm).

various CYP isoforms is expressed. This includes CYP1A1/2 (rat), CYP1B1 (human), and CYP2E1 (rat).[52–54] CYPs expressed in cortical astrocytes are obviously an important part of glial cells' response to cellular stress elicited by cerebral injury and cytokines, whereas their function in drug metabolism seems to be of subordinate importance.

4.4 FUNCTIONAL ASPECTS OF CYPs IN ASTROCYTES

4.4.1   Induction of Specific CYP Isoforms

There is clear agreement that induction of brain CYP isoforms takes place in specified regions or cell types of the brain and is not an overall effect similar to liver.[5,9] Drugs penetrating into liver cause induction of a great fraction of CYP isoforms in almost all sinusoidal cells by activation of the specific nuclear receptor signal transduction cascade.[55] As a result, total CYP content of liver is substantially increased (Table 4.1). In brain, the situation turns out to be completely different. Xenobiotics penetrate into the brain at specified regions and enter endothelial and glial cells representing the blood–brain barrier. Expression and activity of a specific CYP isoform may therefore be relatively high in one region or even cell type but lacking in another. To fully understand the significance of glial CYP in brain's xenobiotic metabolism, it is necessary to establish the regiospecific expression profile, regulation, and metabolic activity of each CYP. As shown in Table 4.1, total CYP content is not affected by treatment of murine astrocytes with phenytoin. Investigations of rat astroglial cultures showed only a 1.2- to 1.5-fold increase of total CYP, which is not consistent with general induction of CYP.[12] Measurement of total CYP content seems to be an insufficient tool to elucidate brain's response to xenobiotics.

Indeed the combination of brain-specific subfractionation methodology[56] and isoform-specific molecular and biochemical techniques such as quantitative real time–PCR (RT-PCR), *in situ* hybridization, Northern blots and immunoblots, and specific enzymatic assays revealed induction of several

CYP isoforms present in astrocytes in response to xenobiotic treatment. Phenytoin enhances CYP2C29 and CYP2B1 expression two- to threefold in glial cells.[6,12,34] Rats treated with phenobarbital demonstrated markedly induced levels of CYP2B1, CYP2B2, and CYP3A1 mRNA in the striatum and cerebellum.[57] CYP1B1 is upregulated nearly twofold in oxidative toxic conditions, which implicates participation of glial CYP1B1 in inflammatory or oxidative stress.[53] CYP2E1 is induced nearly sixfold by phorbol-12,13-dibutyrate, a phorbol ester that has tumor-promoting effects in the cell.[58] Investigations of C6 glioma cells revealed the presence of primary transcripts of various isoforms of the families CYP1A, 2A, 2B, 2C, 2D, and 2E, while CYP1A1 and CYP2B1 have been induced by benzo[a]anthracene or phenobarbital, respectively.[59] However, CYP1A1 and 1A2 are in some way depressed under conditions where cytokines such as TNF-α, IL-1β, and IL-6 are released from microglia or astrocytes.[60] In conclusion, astrocytes of the brain show induction of various CYP isoforms, depending on the drug, the region where it penetrates into the brain, and the target area where it works specifically.

4.4.2   Metabolic Activity of CYPs

It clearly follows from the aforementioned facts that glial CYPs are inducible by several drugs and xenobiotics and therefore specifically react on drug penetration into the brain. However, investigations of brain tissue revealed very often no or only marginal metabolism of these drugs.[61] Keeping in mind that brain CYPs acting in xenobiotic metabolism are expressed in a very limited number of brain cells and regions and are differently regulated than those from liver—e.g., by the availability of heme[22]—it becomes obvious that elucidation of CYP mediated metabolism needs a target-oriented methodology. Sufficient tools to overcome these problems and to demonstrate glial CYP mediated metabolism are the use of specific astrocyte cell culture systems or molecular (cDNA-derived) vector-based protein expression in neuronal or glial cell line systems.

The uptake of drugs penetrating into astrocytes of the blood–brain barrier provides evidence for two different consecutive processes, as shown using anticonvulsive drug phenytoin invading in glial cells.[6] First, rapid uptake of the drug by diffusion is observed, followed by a moderate influx caused by the CYP2C29-dependent metabolism of the accumulated intracellular phenytoin. This so-called metabolic trapping phenomenon[62,63] demonstrates that astrocytes are able to convert phenytoin to its main metabolites dihydrodiol, para-, and meta-hydroxyphenyl-phenylhydantoin[64] to a sufficient extent. Twenty-five percent of the invaded phenyotin was degraded to its metabolites within 1 h.[6]

Other studies are consistent with our findings and indicate a prominent function of astrocytes on CYP mediated drug metabolism. Chlorzoxazone, a centrally acting muscle relaxant, is efficiently metabolized by CYP2E1 in astrocytes. This metabolism is sixfold increased by induction of CYP2E1 by phorbol esters.[58] CYP dependent endothelial or glial quinone metabolism at the blood–brain barrier is suspected to be a source of superoxide formation, which may endanger adjacent neurons and astrocytes.[65] An important function of astrocytes surrounding the capillary vessels is the control of hyperemia and angiogenesis. Metabolism of arachidonic acid (AA) to epoxyeicosatrienoic acids (EETs) by glial CYP2C11 is considered to be a major step in glutamate-initiated vasodilation leading to increased blood flow and angiogenesis.[38,45] This seems to be a very old phylogenetic mechanism, as it is also found in the brain of early vertebrate *Fundulus heteroclitus*. Recently detected CYP2N converts AA to EETs in this teleost.[66] Metabolism of AA in astrocytes also occurs during inflammation process and is part of astrocytes' immune response.[60] During infection, astrocytes and microglia become activated and release cytokines such as TNF-α, IL-1β, and IL-6. These cytokines regulate the survival of astrocytes themselves and neurons.[67] In addition to release of cytokines, glial CYP expression is modulated. CYP1A1 and CYP1A2 are depressed and their enzymatic activity is altered, which may contribute to drug toxicity.[68]

4.5 CONCLUSIONS

We conclude that astrocytes play a vital role in the control of drug uptake into the brain and therefore protect highly vulnerable neurons from toxicity evoked by handed local concentration of xenobiotics. The glial CYP isoforms previously mentioned are active in drug metabolism predominantly at the brain border lines and seem to fulfill a key function in this process. They show remarkable inducibility and metabolic activity in response to drug influx. They function in blood-flow regulation and as signaling enzymes in inflammation. It becomes obvious that glial CYPs have different functions than the neuronal ones. CYPs in neurons fulfill specific tasks in specified brain regions such as the hypothalamus, hippocampus, and striatum by providing signaling molecules such as steroids and fatty acids necessary for neuronal cell growth and maintenance, whereas glial CYPs have mainly protective barrier functions, safeguarding the neurons. Many efforts have been made in methodology to elucidate the function of glial CYPs in control of xenobiotic drugs invading the brain. This has substantially expanded our view of brain CYPs. Nevertheless, functional approaches such as reportergene or transactivation assays in combination with CYP isoform protein analysis are necessary to get further insight in brain CYP function.

ACKNOWLEDGMENTS

Work presented in this chapter was supported by grants from the Deutsche Forschungsgemeinschaft (DFG Me 1544/1, DFG Me 1544/4, and DFG Vo 272/5).

REFERENCES

1. Nelson, D.R., Koymans, L., Kamataki, T., Stegeman, J.J., Feyereisen, R., Waxman, D.J., Waterman, M.R., Gotoh, O., Coon, M.J., Estabrook, R.W., Gunsalus, I.C., and Nebert, D.W., P450 superfamily: update on new sequences, gene mapping, accession numbers and nomenclature, *Pharmacogenetics,* 6, 1, 1996.
2. Hagemeyer, C.E., Rosenbrock, H., Ditter, M., Knoth, R., and Volk, B., Predominantly neuronal expression of cytochrome P450 isoforms CYP3A11 and CYP3A13 in mouse brain, *Neuroscience,* 117, 521, 2003.
3. Strobel, H.W., Geng, J., Kawashima, H., and Wang, H., Cytochrome P450-dependent biotransformation of drugs and other xenobiotic substrates in neural tissue, *Drug Metab. Rev.,* 29, 1079, 1997.
4. Volk, B., Meyer, R.P., von Lintig, F., Ibach, B., and Knoth, R., Localization and characterization of cytochrome P450 in the brain. *In vivo* and *in vitro* investigations on phenytoin- and phenobarbital-inducible isoforms, *Toxicol. Lett.,* 83, 655, 1995.
5. Hedlund, E., Gustafsson, J.A., and Warner, M., Cytochrome P450 in the brain; a review, *Curr. Drug Metab.* 2, 245, 2001.
6. Meyer, R.P., Knoth, R., Schiltz, E., and Volk, B., Possible function of astrocyte cytochrome P450 in control of xenobiotic phenytoin in the brain: *in vitro* studies on murine astrocyte primary cultures, *Exp. Neurol.,* 167, 376, 2001.
7. Rosenbrock, H., Hagemeyer, C.E., Ditter, M., Knoth, R., and Volk, B., Identification, induction and localization of cytochrome P450s of the 3A-subfamily in mouse brain, *Neurotox. Res.,* 3, 339, 2001.
8. Hedlund, E., Wyss, A., Kainu, T., Backlund, M., Kohler, C., Pelto-Huikko, M., Gustafsson, J.A., and Warner, M., Cytochrome P4502D4 in the brain: specific neuronal regulation by clozapine and toluene, *Mol. Pharmacol.,* 50, 342, 1996.
9. Strobel, H.W., Thompson, C.M., and Antonovic, L., Cytochromes P450 in brain: function and significance, *Curr. Drug Metab.,* 2, 199, 2001.
10. Riedl, A.G., Watts, P.M., Jenner, P., and Marsden, C.D., P450 enzymes and Parkinson's disease: the story so far, *Mov. Disord.,* 13, 212, 1998.

11. Shahi, G.S., Das, N.P., and Moochhala, S.M., 1-Methyl-4-phenyl-1,2,3,6-tetrahydropyridine-induced neurotoxicity: partial protection against striato-nigral dopamine depletion in C57BL/6J mice by cigarette smoke exposure and by beta-naphthoflavone-pretreatment, *Neurosci. Lett.*, 127, 247, 1991.

12. Kempermann, G., Knoth, R., Gebicke-Haerter, P.J., Stolz, B.J., and Volk, B., Cytochrome P450 in rat astrocytes *in vivo* and *in vitro*: intracellular localization and induction by phenytoin, *J. Neurosci. Res.*, 39, 576, 1994.

13. Bertram, S., Bonitz, D., Gassen, G., Papandrikopoulou, A., Weber, M., Weiler, H., and Wollny, E., Die Blut-Hirn-Schranke — eine Organbarriere, *Kontakte (Darmstadt)*, 1, 17, 1988.

14. Huell, M., Strauss, S., Volk, B., Berger, M., and Bauer, J., Interleukin-6 is present in early stages of plaque formation and is restricted to the brains of Alzheimer's disease patients, *Acta Neuropathol. (Berl.)*, 89, 544, 1995.

15. Strauss, S., Otten, U., Joggerst, B., Pluss, K., and Volk, B., Increased levels of nerve growth factor (NGF) protein and mRNA and reactive gliosis following kainic acid injection into the rat striatum, *Neurosci. Lett.*, 168, 193, 1994.

16. Hamprecht, B., Astroglia cells in culture: receptors and cyclic nucleotides, in *Astrocytes — Biochemistry, Physiology, and Pharmacology of Astrocytes*, Fedoroff, S. and Vernadakis, A., Eds., Academic Press, Orlando, FL, 1986, p. 77.

17. Schenkman, J.B., Historical background and description of the cytochrome P450 monooxygenase system, in *Cytochrome P450*, Schenkman, J.B. and Greim, H., Eds., Springer Verlag, Berlin, 1993, p. 3.

18. Zimniak, P., Waxman, D.J., and Greim, H., Liver cytochrome P450 metabolism of endogenous steroid hormones, bile acids, and fatty acids, in *Cytochrome P450*, Schenkman, J.B., Ed., Springer Verlag, Berlin, 1993, pp. 123–144.

19. Stoffel-Wagner, B., Neurosteroid metabolism in the human brain, *Eur. J. Endocrinol.*, 145, 669, 2001.

20. Meyer, U.A., Schuurmans, M.M., and Lindberg, R.L., Acute porphyrias: pathogenesis of neurological manifestations, *Semin. Liver Dis.*, 18, 43, 1998.

21. Jover, R., Lindberg, R.L., and Meyer, U.A., Role of heme in cytochrome P450 transcription and function in mice treated with lead acetate, *Mol. Pharmacol.*, 50, 474, 1996.

22. Meyer, R.P., Podvinec, M., and Meyer, U.A., Cytochrome P450 CYP1A1 accumulates in the cytosol of kidney and brain and is activated by heme, *Mol. Pharmacol.*, 62, 1061, 2002.

23. Kold, B. and Whishaw, J.Q., *Fundamentals of Human Neuropsychology*, 4th ed., Worth Publishers, New York, U.S., 1996.

24. Huang, P., Rannug, A., Ahlbom, E., Hakansson, H., and Ceccatelli, S., Effect of 2,3,7,8-tetrachlorodibenzo-p-dioxin on the expression of cytochrome P450 1A1, the aryl hydrocarbon receptor, and the aryl hydrocarbon receptor nuclear translocator in rat brain and pituitary, *Toxicol. Appl. Pharmacol.*, 169, 159, 2000.

25. Gu, J., Zhang, Q.Y., Genter, M.B., Lipinskas, T.W., Negishi, M., Nebert, D.W., and Ding, X., Purification and characterization of heterologously expressed mouse CYP2A5 and CYP2G1: role in metabolic activation of acetaminophen and 2,6-dichlorobenzonitrile in mouse olfactory mucosal microsomes, *J. Pharmacol. Exp. Ther.*, 285, 1287, 1998.

26. Schilter, B. and Omiecinski, C.J., Regional distribution and expression modulation of cytochrome P-450 and epoxide hydrolase mRNAs in the rat brain, *Mol. Pharmacol.*, 44, 990, 1993.

27. Volk, B., Hettmannsperger, U., Papp, T., Amelizad, Z., Oesch, F., and Knoth, R., Mapping of phenytoin-inducible cytochrome P450 immunoreactivity in the mouse central nervous system, *Neuroscience*, 42, 215, 1991.

28. Joó, F., The blood-brain barrier *in vitro*: The second decade, *Neurochem. Int.*, 23, 499, 1993.

29. Gragera, R.R., Muñiz, E., and Martínez-Rodriguez, R., Molecular and ultrastructural basis of the blood-brain barrier function. Immunohistochemical demonstration of Na+/K+ ATPase, α-actin, phosphocreatine and clathrin in the capillary wall and its microenvironment, *Cell. Mol. Biol.*, 39, 819, 1993.

30. Seelig, A., Gottschlich, R., and Devant, R.M., A method to determine the ability of drugs to diffuse through the blood-brain barrier, *Proc. Natl. Acad. Sci. U.S.A.*, 91, 68, 1994.

31. Lolin, Y.I., Ratnaraj, N., Hjelm, M., and Patsalos, P.N., Antiepileptic drug pharmacokinetics and neuropharmacokinetics in individual rats by repetitive withdrawal of blood and cerebrospinal fluid: Phenytoin, *Epilepsy Res.*, 19, 99, 1994.

32. Sakaeda, T., Siahaan, T.J., Audus, K.L., and Stella, V.J., Enhancement of transport of D-melphalan analogue by conjugation with L-glutamate across bovine brain microvessel endothelial cell monolayers, *J. Drug Target*, 8, 195, 2000.

33. Flockhart, D.A., P450-drug interaction table. Retrieved in 2004 from http://medicine.iupui.edu/flockhart

34. Ibach, B., Appel, K., Gebicke-Haerter, P., Meyer, R.P., Friedberg, T., Knoth, R., and Volk, B., Effect of phenytoin on cytochrome P450 2B mRNA expression in primary rat astrocyte cultures, *J. Neurosci. Res.*, 54, 402, 1998.

35. Volk, B., Amelizad, Z., Anagnostopoulos, J., Knoth, R., and Oesch, F., First evidence of cytochrome P-450 induction in the mouse brain by phenytoin, *Neurosci. Lett.*, 84, 219, 1988.

36. Rowland, L.P., Fink, M.E., and Rubin, L., Cerebrospinal fluid: Blood-brain-barrier, brain edema, and hydrocephalus, in *Principles of Neural Science*, 3rd ed., Kandel, E.R., Schwartz, J.H., and Jessell, T.M., Eds., Appleton & Lange, East Norwalk, CT, 1991, p. 1050.

37. Reis, D.J. and Iadecola, C., Regulation by the brain of its blood flow and metabolism: role of intrinsic neuronal networks and circulating catecholamines, in *Neuronal Regulation of Brain Circulation*, 1st ed., Owman, C. and Hardebo, J.E., Eds., Elsevier, Amsterdam, 1986, p. 129.

38. Harder, D.R., Zhang, C., and Gebremedhin, D., Astrocytes function in matching blood flow to metabolic activity, *News Physiol. Sci.*, 17, 27, 2002.

39. Ogura, H., Yasuda, M., Nakamura, S., Yamashita, H., Mikoshiba, K., and Ohmori, H., Neurotoxic damage of granule cells in the dentate gyrus and the cerebellum and cognitive deficit following neonatal administration of phenytoin in mice, *J. Neuropathol. Exp. Neurol.*, 61, 956, 2002.

40. Lundqvist, C., Alling, C., Knoth, R., and Volk, B., Intermittent ethanol exposure of adult rats: hippocampal cell loss after one month of treatment, *Alcohol Alcohol*, 30, 737, 1995.

41. Kiefer, R., Knoth, R., Anagnostopoulos, J., and Volk, B., Cerebellar injury due to phenytoin. Identification and evolution of Purkinje cell axonal swellings in deep cerebellar nuclei of mice, *Acta Neuropathol. (Berl.)*, 77, 289, 1989.

42. Roda, F., Pio, J., Bianchi, A.L., and Gestreau, C., Effects of anesthetics on hypoglossal nerve discharge and c-fos expression in brainstem hypoglossal premotor neurons, *J. Comp. Neurol.*, 468, 571, 2004.

43. Hornung, J.P., The human raphe nuclei and the serotonergic system, *J. Chem. Neuroanat.*, 26, 331, 2003.

44. Miksys, S., Rao, Y., Sellers, E.M., Kwan, M., Mendis, D., and Tyndale, R.F., Regional and cellular distribution of CYP2D subfamily members in rat brain, *Xenobiotica*, 30, 547, 2000.

45. Gebremedhin, D., Yamaura, K., Zhang, C., Bylund, J., Koehler, R.C., and Harder, D.R., Metabotropic glutamate receptor activation enhances the activities of two types of Ca2+-activated K+ channels in rat hippocampal astrocytes, *J. Neurosci.*, 23, 1678, 2003.

46. Ghersi-Egea, J.F., Leninger-Muller, B., Suleman, G., Siest, G., and Minn, A., Localization of drug-metabolizing enzyme activities to blood-brain interfaces and circumventricular organs, *J. Neurochem.*, 62, 1089, 1994.

47. Tauer, U., Knoth, R., and Volk, B., Phenytoin alters Purkinje cell axon morphology and targeting *in vitro*, *Acta Neuropathol. (Berl.)*, 95, 583, 1998.

48. Miksys, S., Rao, Y., Hoffmann, E., Mash, D.C., and Tyndale, R.F., Regional and cellular expression of CYP2D6 in human brain: higher levels in alcoholics, *J. Neurochem.*, 82, 1376, 2002.

49. Warner, M. and Gustafsson, J.A., Effect of ethanol on cytochrome P450 in the rat brain, *Proc. Natl. Acad. Sci. U.S.A.*, 91, 1019, 1994.

50. Alkayed, N.J., Narayanan, J., Gebremedhin, D., Medhora, M., Roman, R.J., and Harder, D.R., Molecular characterization of an arachidonic acid epoxygenase in rat brain astrocytes, *Stroke*, 27, 971, 1996.

51. Zwain, I.H. and Yen, S.S., Dehydroepiandrosterone: biosynthesis and metabolism in the brain, *Endocrinology*, 140, 880, 1999.

52. Nicholson, T.E. and Renton, K.W., Modulation of cytochrome P450 by inflammation in astrocytes, *Brain Res.*, 827, 12, 1999.

53. Malaplate-Armand, C., Ferrari, L., Masson, C., Siest, G., and Batt, A.M., Astroglial CYP1B1 up-regulation in inflammatory/oxidative toxic conditions: IL-1beta effect and protection by N-acetylcysteine, *Toxicol. Lett.*, 138, 243, 2003.

54. Tindberg, N., Baldwin, H.A., Cross, A.J., and Ingelman-Sundberg, M., Induction of cytochrome P450 2E1 expression in rat and gerbil astrocytes by inflammatory factors and ischemic injury, *Mol. Pharmacol.*, 50, 1065, 1996.

55. Handschin, C. and Meyer, U.A., Induction of drug metabolism: the role of nuclear receptors, *Pharmacol. Rev.*, 55, 649, 2003.

56. Ghersi-Egea, J.F., Perrin, R., Leininger-Muller, B., Grassiot, M.C., Jeandel, C., Floquet, J., Cuny, G., Siest, G., and Minn, A., Subcellular localization of cytochrome P450, and activities of several enzymes responsible for drug metabolism in the human brain, *Biochem. Pharmacol.*, 45, 647, 1993.

57. Schilter, B., Andersen, M.R., Acharya, C., and Omiecinski, C.J., Activation of cytochrome P450 gene expression in the rat brain by phenobarbital-like inducers, *J. Pharmacol. Exp. Ther.*, 294, 916, 2000.

58. Tindberg, N., Phorbol ester induces CYP2E1 in astrocytes, through a protein kinase C- and tyrosine kinase-dependent mechanism, *J. Neurochem.*, 86, 888, 2003.

59. Geng, J. and Strobel, H.W., Expression, induction and regulation of the cytochrome P450 monooxygenase system in the rat glioma C6 cell line, *Brain Res.*, 784, 276, 1998.

60. Nicholson, T.E. and Renton, K.W., Role of cytokines in the lipopolysaccharide-evoked depression of cytochrome P450 in the brain and liver, *Biochem. Pharmacol.*, 62, 1709, 2001.

61. Warner, M., Hellmold, H., Magnusson, M., Rylander, T., Hedlund, E., and Gustafsson, J.A., Extrahepatic cytochrome P450: role in *in situ* toxicity and cell-specific hormone sensitivity, *Arch. Toxicol. Suppl.*, 20, 455, 1998.

62. Kearfott, K.J., Elmaleh, D.R., Goodman, M., Correira, J.A., Alpert, N.M., Ackerman, R.H., Brownell, G., and Strauss, W.H., Comparison of 2- and 3-18F-fluoro-deoxy-D-glucose for studies of tissue metabolism, *Intern. J. Nucl. Med. Biol.*, 11, 15, 1984.

63. Heichal, O., Ish-Shalom, D., Koren, R., and Stein, W.D., The kinetic dissection of transport from metabolic trapping during substrate uptake by intact cells. Uridine uptake by quiescent and serum-activated Nil 8 hamster cells and their murine sarcoma virus-transformed counterparts, *Biochim. Biophys. Acta*, 551, 169, 1979.

64. Dudley, K.H., Phenytoin metabolism, in *Phenytoin-Induced Teratology and Gingival Pathology*, 1st ed., Dudley, K.H., Hassell, T.M., and Johnston, M.C., Eds., Raven Press, New York, 1980, p. 13.

65. Bayol-Denizot, C., Daval, J.L., Netter, P., and Minn, A., Xenobiotic-mediated production of superoxide by primary cultures of rat cerebral endothelial cells, astrocytes, and neurones, *Biochim. Biophys. Acta*, 1497, 115, 2000.

66. Oleksiak, M.F., Wu, S., Parker, C., Karchner, S.I., Stegeman, J.J., and Zeldin, D.C., Identification, functional characterization, and regulation of a new cytochrome P450 subfamily, the CYP2Ns, *J. Biol. Chem.*, 275, 2312, 2000.

67. Giulian, D., Li, J., Leara, B., and Keenen, C., Phagocytic microglia release cytokines and cytotoxins that regulate the survival of astrocytes and neurons in culture, *Neurochem. Int.*, 25, 227, 1994.

68. Delaporte, E. and Renton, K.W., Cytochrome P4501A1 and cytochrome P4501A2 are downregulated at both transcriptional and post-transcriptional levels by conditions resulting in interferon-alpha/beta induction, *Life Sci.*, 60, 787, 1997.

69. Matsubara, T., Koike, M., Touchi, A., Tochino, Y., and Sugeno, K., Quantitative determination of cytochrome P450 in rat liver homogenate, *Anal. Biochem.*, 75, 596, 1976.

70. Thuerl, C., Otten, U., Knoth, R., Meyer, R.P., and Volk, B., Possible role of cytochrome P450 in inactivation of testosterone in immortalized hippocampal neurons, *Brain Res.*, 762, 47, 1997.

71. Rall, T.W. and Schleifer, L.S., Drugs effective in the therapy of the epilepsies, in *The Pharmacological Basis of Therapeutics*, 7th ed., Gilman, Goodman, and Gilman, Eds., Macmillan Publishing Co., New York, 1985, p. 446.

CHAPTER 5

The Reactive Astrocyte

Michael D. Norenberg

CONTENTS

5.1 INTRODUCTION

Astrocytes are the most abundant cells in the central nervous system (CNS). They are involved in the regulation of the brain microenvironment, in particular neurotransmitter and ionic homeostasis,[1] metabolic support of neurons,[2,3] regulation of energy metabolism,[4,5] synaptic transmission and neuronal excitability,[6] synaptic generation and efficacy,[7,8] neurotrophism,[9] detoxification,[10,11] free-radical scavenging,[12] metal sequestration,[13] development and maintenance of the blood–brain barrier,[14] guidance of neuronal migration during development,[15] and immune/inflammatory functions.[16]

Astrocytes commonly undergo a dramatic transformation referred to as reactive astrocytosis or astrogliosis, which is arguably the most prominent cellular response to diverse forms of CNS injury. It is found in all destructive CNS lesions, including trauma, ischemia, infections, demyelinative disorders (e.g., multiple sclerosis), or degenerative conditions (e.g., Alzheimer's disease, Parkinson's disease). They are also seen following exposure to toxins, such as quinolinic acid, kainic acid, MPTP, 6-hydroxydopamine, methamphetamine, 5,7-dihydrotryptamine, trimethyltin, alcohol, and heavy metals such as iron, aluminum, lead, cobalt, and mercury. For reviews on gliotic responses to various types of tissue injury, see References 17–20.

Reactive astrocytosis may also occur in physiological states. Steward and colleagues[21] have described increased GFAP immunoreactivity (a marker for astrocytosis) following excessive neuronal activity generated by electrically induced seizures. On the other hand, depression in neuronal activity has also been associated with astrogliosis.[22] Astrogliosis has likewise been described in spreading depression,[23] a condition not associated with pathological changes, and also in a learning paradigm.[24]

It should be emphasized that in severe CNS injuries, astrocytes undergo necrosis (along with neurons and other neural elements). Astrogliosis in most instances occurs at the margin rather than in the center of the lesion. As there are a limited number of fibroblasts in the CNS, this leads to the formation of a cavity rather than dense fibrous scars, as occurs in the rest of the body. Smaller lesions may be completely filled by reactive astrocytes. Demyelinating lesions such as plaques of multiple sclerosis (where axons are spared) and Wallerian degeneration do not have a cavitary component and thus are predominantly composed of reactive astrocytes.

Various terms have been applied to the reactive astrocyte: gemistocyte, gemistocytic astrocyte, and Nissl plump astrocyte. The process is referred to as astrocytosis, reactive astrocytosis, fibrous astrocytosis, astrogliosis, reactive astrogliosis, or fibrous gliosis. It is also commonly referred to as gliosis or reactive gliosis. This, however, is not a good practice as cellular elements besides astrocytes (microglia, macrophages, oligodendroglial progenitor cells, meningeal cells, and mesenchymal elements) may contribute to gliosis. Regrettably, the indiscriminate reference to reactive astrocytosis as reactive gliosis has led to erroneous interpretations about the significance of reactive astrocytes in CNS injuries. This unfortunate interchange of terminology has ascribed detrimental effects to reactive astrocytes, when in reality the problem was reactive gliosis or the "glial scar."

The purpose of this chapter is to review critical aspects of astrogliosis, its morphologic features, mechanism(s) of its induction, potential signal transduction pathways mediating this response, and more importantly the significance and potential consequences of astrogliosis.

5.2 MORPHOLOGIC FINDINGS

Cytoplasmic hypertrophy, the dominant light microscopic feature of reactive astrocytes, is best evaluated with conventional hematoxylin and eosin stain (H&E). Whereas in the normal state cytoplasm cannot be identified with H&E, it is readily visible in reactive astrocytes (Figure 5.1B, C). Additionally, the cytoplasmic processes become more numerous, longer, and thickened. Such changes can also be detected with special stains such as the Cajal gold sublimate stain and by immunohistochemistry for the astrocyte-specific intermediate filament, glial fibrillary acidic protein (GFAP) (Figure 5.2). Nuclear changes are prominent and characterized by increased size, irregular outline, dispersion of chromatin, hyperchromaticity, and occasional multinucleation. Nuclear changes are not detectable with GFAP or with gold sublimate stains. Intracytoplasmic beaded eosinophilic inclusions, referred to as Rosenthal fibers, are found around chronic conditions (e.g., arteriovenous malformations, syringomyelia). These inclusions consist of "αB-crystallin,[25] which may represent a heat shock protein.[26]

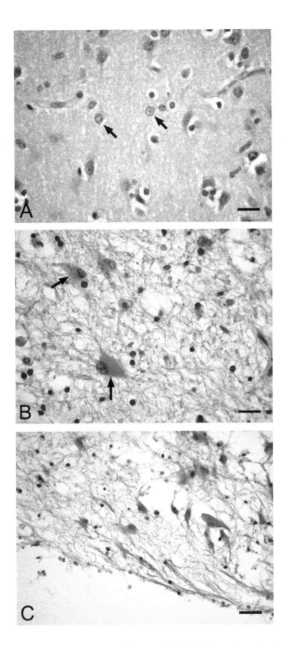

Figure 5.1 Hematoxylin and eosin (H&E) photomicrographs. (A) Normal cerebral cortex. Astrocytes (arrows) are characterized by "naked" nuclei as the cytoplasm is not discernible. Scale bar, 40 μm. (B) Photograph adjacent to a 6-week-old infarct in cerebral cortex. Reactive astrocytes have prominent cytoplasm with several processes and large hyperchromatic nuclei. Scale bar, 40 μm. (C) Six-month-old infarct showing the wall of a cyst (at the bottom). The cyst wall is composed largely of reactive astrocytes, but the degree of cytoplasmic enlargement is much less that in part B. Scale bar, 60 μm.

Electron microscopic findings are consistent with enhanced metabolic activity, i.e., increased numbers of mitochondria and ribosomes, enlarged Golgi complexes, and increased amounts of glycogen.[27] There is occasional evidence of phagocytosis.[27] However, the most striking feature of reactive astrocytes is the accumulation of cytoplasmic bundles of 10-nm intermediate glial filaments.[27] As noted, the principal intermediate filament in the adult resident astrocyte is GFAP.[28–30] In addition, reactive astrocytes also contain the intermediate filaments vimentin[31] and nestin,[32]

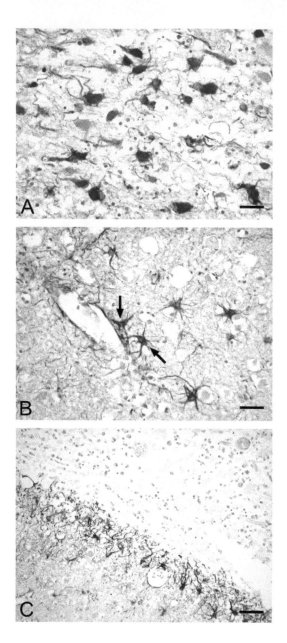

Figure 5.2   GFAP immunostains. (A) Two-month-old infarct illustrating prominent GFAP positivity in reactive astrocytes. Cytoplasmic processes are better depicted in this immunostain. Scale bar, 40 μm. (B) Six-month-old infarct showing an involution of the cytoplasmic hypertrophy. Depicted here are astrocytes with perivascular end-feet (arrows). Scale bar, 40 μm. (C) Four-month-old infarct showing the extent of astrogliosis around an infarct cavity. The amount of astrogliosis is modest and does not form a thick meshwork of astrocytic processes. Scale bar, 120 μm.

which are usually expressed during cell development. GFAP is a sensitive and early marker of reactive astrocytes, and its upregulation may be seen as early as 1 h.[33] Nevertheless, astroglial responses to CNS injury cannot be reduced simply to an upregulation of GFAP. One needs to be cautious about equating the overexpression of GFAP with reactive astrocytosis without examining for other phenotypic characteristics of reactive astrocytes. Unfortunately, that has not always been the case in the literature.

5.3 GENERAL ASPECTS OF ASTROGLIOSIS

5.3.1 Timing

Little is seen before 4 days following an injury on H&E. At this time, slight nuclear pallor and the early development of visible cytoplasm may be detected. The cytoplasm becomes better defined by 7 to 10 days and reaches a maximum degree of hypertrophy at 2–3 weeks. The changes may or may not completely regress. With severe injuries the regression will usually be incomplete, while in milder lesion the reactive astrocyte may return to its normal state.

5.3.2 Age of Subject

The astrocytic changes tend to be much more attenuated in immature animals and humans.[34] The basis for this differential response is not clear. Immature brains do not undergo reactive astrocytosis unless cytokines are administered.[35] The integrity of the immune system in association with intact microglia and macrophages appears to be critical.[36]

5.3.3 Location

The extent of astrogliosis varies inversely with the distance from the lesion. Thus astrocytes adjacent to the lesion show the greatest degree of response, which fades as one moves away from the lesion. It should be emphasized that astrogliosis can extend for great distances from the primary lesion.[37,38]

5.3.4 Proliferation

The extent of proliferation (i.e., mitotic division) is controversial. It does not appear to be a major component of the astrogliotic process.[39–41] Certainly, the identification of mitotic figures in humans is very rare. Nevertheless, unequivocal evidence of astrocyte division has been described.[42,43] In conditions where the blood–brain barrier (BBB) is intact (e.g., in Wallerian degeneration),[44] there is no evidence of proliferation. With a breach of the BBB, some degree of proliferation may be identified. However, in many instances, what is often described as "proliferation" merely reflects a greater recognition of these cells because of their cytoplasmic hypertrophy and increased GFAP immunoreactivity rather than true proliferation. At times, the presence of GFAP in cells possessing proliferation markers has been used as evidence for proliferation. These studies have to be interpreted with caution, as macrophages in injured areas may ingest GFAP from necrotic astrocytes; macrophages/microglia are well known to be capable of proliferation.

5.3.5 Migration

There is no direct evidence for migration of reactive astrocytes *in vivo*, although some may occur *in vitro*.[45–47] The relevance of the *in vitro* to the *in vivo* state is unclear as the environments of these two conditions are very different.

5.3.6 Heterogeneity of Reactive Astrocytes

It is far from clear that astrogliosis represents a stereotypic process. It is quite possible that different types of injuries, distance from the main lesion, and regional and temporal factors may bring about different astroglial responses. For instance, the response may vary in duration, degree of hyperplasia, and the time course of GFAP expression depending on the type of lesion.[48] Tani et al.[49]

note that reactive astrocytes far from lesions of experimental allergic encephalomyelitis (EAE) did not have the IP-10 chemokine, while those close to lesion did. In multiple sclerosis, astrocytes express nitric oxide synthase activity,[50] which may have deleterious consequences, while in cuprizone-induced demyelination, they make IGF-1, which may be involved in tissue repair.[51] An astrocyte-specific antigen recognized by the monoclonal antibody J1-31 is a more intense marker for proximal reactive astrocytes, whereas reactive astrocytes at a distance from the lesion are not detected by this marker.[52]

The phenomena of isomorphic and anisomorphic astrogliosis is relevant to the issue of heterogeneity.[53–55] Destructive lesions such as trauma or ischemia, which are associated with a breakdown of the BBB (along with an emigration of inflammatory cells), elicit anisomorphic gliosis characterized by a disorganized arrangement of reactive astrocytes. On the other hand, lesions not associated with BBB breakdown such as following various toxin/chemical exposure (e.g., kainic acid) or Wallerian degeneration are associated with isomorphic gliosis, characterized by well-organized fascicular arrangement of reactive astrocytes. Iso- and anisomorphic reactive astrocytes have a different profile of proteoglycans, some of which can either promote or inhibit neurite outgrowth.[56] For review, see Reference 57.

The issue of heterogeneity among reactive astrocytes is potentially important and is an area that requires more study. At this time, one needs to be cautious about extrapolating characteristics of reactive astrocytes from one condition or situation to another.

5.3.7   Reversion to an Immature Phenotype

It is of interest that reactive astrocytes tend to revert to an immature phenotype as found during development of the CNS. Thus, reactive astrocytes share features of immature fetal astrocytes such as the expression of the intermediate filaments nestin and vimentin and the neuronal markers MAP-2,[58] GABA,[59] neuron-specific enolase,[58] NCAM,[60] and calbindin-D28K.[61]

5.3.8   Origin/Sources of Reactive Astrocytes

Astrogliosis is generally thought to occur by a phenotypic change in resident astrocytes, and it is the author's bias that this is indeed the case. To what extent astrocytes from adjacent regions migrate and contribute to the population of reactive astrocytes is not known; compelling evidence for such a possibility is absent.

Recently, a number of studies have suggested that stem cells may be a source of reactive astrocytes. It has also been proposed that ependymal cells of the spinal central canal can divide and proliferate depending on the severity of the injury and differentiate into reactive astrocytes within the ependyma.[62,63] These authors were able to identify GFAP+/nestin+ cells. As nestin is a marker of an immature cells, this seems like a plausible view. Holmin et al.[64] and Johansson et al.[65] found that adult, nestin-expressing, subependymal cells could differentiate into astrocytes in response to brain injury. These cells were believed to be capable of migrating towards the lesion and thereby provide a source of reactive astrocytes.

While there is evidence that nestin-positive progenitor cells can differentiate into astrocytes,[65,66] it is uncertain to what extent these cells contribute to the population of reactive astrocytes in association with CNS lesions. It is difficult to be certain that nestin positivity in reactive astrocytes represent the *de novo* development of reactive astrocytes (i.e., from stem cells). It may simply be that resident astrocytes in the process of undergoing reactive astrocytosis simply express primitive cellular markers. While glial progenitors are increased following injury, it remains to be established how quantitatively important this process is relative to astrogliosis.

5.4 SPECIAL FEATURES OF REACTIVE ASTROCYTES

5.4.1 Gap Junctions

Gap junctions were early on identified in reactive astrocytes by ultrastructural methods.[67,68] Such junctions have been observed in reactive astrocytes following trauma,[69] ischemia,[70] and mesial temporal sclerosis[71] and in facial nerve nuclei following facial nerve axotomy.[72] Gap junctions are believed to allow the transfer of small molecules from one cell to another and to contribute to calcium waves following stimulation of astrocytes.[73] The role of gap junctions in injury is uncertain. Some have speculated that gap junctions may contribute to the increase in size of some types of brain lesions[74] and may exacerbate seizures in mesial temporal lobe epilepsy.[71] Additionally, signaling molecules crossing gap junctions may explain the presence of reactive astrocytes at considerable distance from the primary lesion.

5.4.2 Ion Channels

Upregulation of various ion channels (K^+, Ca^{2+}) have been described in reactive astrocytes.[75] The significance of their upregulation is unclear, but they may contribute to the maintenance of ionic homeostasis in injured brain regions. They may also contribute, or represent an adaptive response, to epilepsy.[76–78]

5.4.3 Glutamate Transport

Glutamate transport, a critical function of astrocytes, is carried out by the GLT-1 (EAAT-2) and GLAST (EAAT-1) transporters.[79] Relatively little experimental work has been carried out regarding the state of glutamate transport in astrogliosis. Reactive astrocytes adjacent to CNS grafts showed significant upregulation of GLT-1 by 3 days.[80] Reactive astrocytes in Huntington's disease showed increased GLT-1 mRNA,[81] and increased GLAST mRNA was identified in reactive astrocytes in the periventricular region of rats with kaolin-induced hydrocephalus.[82] These findings support the view that one of the functions of reactive astrocytes is to regulate extracellular levels of glutamate in areas of brain injury, thereby potentially mitigating excitotoxic injury.

Indirect evidence for increased glutamate uptake in reactive astrocytes was provided by Buck et al.,[83] who showed increase in glial level of Ant1, a mitochondrial ATP/ADP exchanger that facilitates the flux of ATP out of the mitochondria, thereby maintaining adequate cellular levels of this nucleotide. Of interest is that glutamate uptake, which requires ATP, is significantly decreased in astrocytes from Ant1-null mutant mice.

5.4.4 Glucose Transport

Reactive astrocytes express the BBB glucose transporter GLUT1,[60] while quiescent astrocytes do not. This is in keeping with ultrastructural and enzyme histochemical evidence of increased metabolic activity in reactive astrocytes.[20,84,85]

5.5 FACTORS PRODUCED BY REACTIVE ASTROCYTES

A plethora of factors have been identified in reactive astrocytes, including enzymes, growth factors, cytokines, extracellular matrix molecules, proteases, protease inhibitors, proto-oncogenes, and heat-shock proteins, among many others, which can only be cursorily treated in this chapter. This extraordinary array of factors only serves to highlight the high metabolic activity and the multitude of functions being carried out by reactive astrocytes. For comprehensive reviews on this subject, the reader is referred to References 20, 48, 55, 86, and 87.

The enzymes that show increased activity or are upregulated include many of the oxidative enzymes (e.g., glutamate dehydrogenase[88]), monoamine oxidase-B, "α1-antichymotrypsin, glutamine synthetase,[89,90] ornithine transcarbamylase, superoxide dismutase, catalase, calcium-ATPase, and glutathione-S-transferase (an enzyme involved in the detoxification of various xenobiotics). Enzymes involved in the biosynthesis and degradation of kynurenic acid and quinolinic acid (an excitotoxic agent), including kynurenine aminotransferase, 3-hydroxyanthranilic acid oxygenase, quinolinic acid, and phosphoribosyltransferase, are also increased.

The inducible form of nitric oxide synthase is elevated in reactive astrocytes following various types of CNS injury.[91,92] The significance of this increase is uncertain. Nitric oxide (NO) possesses cytotoxic properties that could contribute to neuronal death.[93] However, because of its vasodilating effect, NO may improve cerebral blood flow.[94]

An increase in various growth factors including bFGF and NGF has been described in reactive astrocytes.[95] Elevations in ciliary neurotrophic factor and in glia-derived nexin (protease nexin-1), a protease inhibitor with neurite-promoting activity, have been demonstrated. In addition, various growth factor receptors are increased, including epidermal growth factor receptor and truncated forms of trkB receptors (receptors for brain-derived neurotrophic factor and neurotrophins 3, 4, 5), which may be involved in sequestration of growth factors. The levels of transforming growth factor-β (TGFβ) and S-100 protein, which may act as a growth factor for serotonergic neurons,[96] are likewise elevated. As discussed below, these astroglial-derived growth factors may be involved in repair and regeneration of the CNS. For reviews, see References 19, 20, 86, and 97.

Parathyroid hormone-related protein (PTHrP) has been identified in reactive astrocytes.[98] It has neuroprotective actions and its appearance may be in response to inflammation.[99,100] PTHrP apparently can also induce IL-6 in glial cells, a cytokine that has neuroprotective properties.[101,102]

The increased production of extracellular matrix (ECM) proteins, adhesion molecules, and proteoglycans may also have potential implications to CNS regeneration. Such molecules include laminin, chondroitin sulfate proteoglycan, fibronectin (in proliferating but not in nonproliferating reactive astrocytes), and neural cell adhesion molecule (NCAM). Their role in astrogliosis will be discussed below.

Remodeling of the ECM following tissue injury is a consequence of the activation of zinc-dependent metalloproteinases (MMPs). Such activity reflects a balance of MMPs and their inhibitors (tissue inhibitors to metalloproteinases [TIMPs]) (see Reference 103 for review). MMPs are commonly activated in inflammatory processes and can be upregulated by TNF MMPs are produced by reactive astrocytes.[104] When released into the extracellular space, MMPs break down the ECM to allow for cell growth and to facilitate remodeling. An imbalance in proteolytic activity either during the acute injury or during the recovery may aggravate the underlying disease process. MMPs may be involved in the destruction of such neurite inhibitory factors as chondroitin sulfate proteoglycans and tenascin. The importance of these inhibitory factors is discussed below.

Class I[105] and II[106] MHC molecules are synthesized by reactive astrocytes, thereby implicating these cells in inflammatory/immune phenomena. The ability to generate class II molecules enables astrocytes to present antigens to immunocompetent lymphocytes. Astrocytes are capable of synthesizing various cytokines[107,107] and show an increase in IL-6 after injury.[108] On the other hand, factors that may dampen the inflammatory response have also been identified. Lipocortin-1 (annexin-I), a steroid-induced, calcium-dependent membrane-binding protein, is produced in reactive astrocytes.[109,110] Lipocortin has anti–phospholipase A_2 activity, which may diminish the inflammatory response and limit the extent of tissue damage. Substance P (a mediator of vasodilation and local immune response as well as a trophic factor) and apolipoprotein E (which downregulates the immune response and may be involved in the removal of lipids that accumulate after injury) are also increased.[111,112]

Other factors are upregulated in reactive astrocytes, including transcription factors, peripheral benzodiazepine receptors, ganglioside GD_3, endothelin, transferrin, and clusterin (sulfated glycoprotein-2; SGP-2), a glycoprotein associated with responses to cell injury. An increase in "2(α)-macroglobulin receptor/low-density lipoprotein receptor-related protein has been identified, which may be related to trafficking and modulation of TGF-β and PDGF. Levels of metallothionein, a protein involved in metal sequestration and protection against metal toxicity, are increased.[113–115] For reviews, see References 19 and 20.

5.6 REGULATORS OF REACTIVE ASTROCYTES

Cellular sources and factors responsible for the triggering of astrogliosis remain to be more fully defined. Based on the best current evidence, microglia are critically involved in the induction of this process.[116] Their release of proinflammatory cytokines (IL-1β, IL-6, TNFα, IFNγ) is likely the most important factor. Many cytokines are increased following CNS injury,[117] and cytokine receptors have been identified on astrocytes.[118,119] The direct injection of various cytokines in brain clearly results in the formation of reactive astrocytes.[35,120–123] Conversely, application of the anti-inflammatory cytokine IL-10 diminishes the extent of astrogliosis.[124] For review of the role of cytokines in the development of astrogliosis, see Reference 125.

Various growth factors have also been implicated in the induction of astrogliosis. These include NGF,[126] and TrkA receptors which have been identified in reactive astrocytes.[127] Other growth factors implicated in the stimulation of astrogliosis include bFGF,[128–130] TGFβ1,[131,132] and ciliary neurotrophic factor (CNTF).[133–138] The source of these factors is not clear. Astrocytes are capable of producing these factors,[9] and thus it is possible it may represent an autocrine process.

In addition to cytokines and growth factors, other agents have been proposed to play a role in astrogliosis. Brain thrombin levels are increased following CNS injuries that are associated with a breakdown of the BBB. Injection of thrombin into brain produces astrogliosis,[139] while thrombin inhibitors are able to attenuate this response. There is also support for endothelin as a stimulator of astrogliosis.[140–143]

Extracellular adenine nucleotides (including ATP) have been proposed as triggering agents.[144,145] Much of the evidence has been generated in cell culture studies. Nevertheless, the intracerebral injection of the nonhydrolyzable ATP analog, 2-methylthio ATP (2-MeSATP), has resulted in astrogliosis,[146] probably through the involvement P2Y receptor subtypes.[147] The role of ATP as a stimulator of astrogliosis, however, remains highly questionable, as *in situ* astrocytes reportedly do not possess P2 purinergic receptors.[148] Further, treatment of injury-conditioned media with apyrase, which cleaves ATP, caused no reduction in ERK/MAPK-stimulating activity,[149] a signaling kinase that has been strongly implicated in the induction of astrogliosis (see below). As suggested by Franke et al.,[146] it is possible that astrogliosis may be due to activation of microglia by ATP and that microglial-derived signals are the ones that are ultimately responsible for the activation of astrocytes.

5.7 SIGNAL TRANSDUCTION PATHWAYS

The mechanisms by which various triggering factors bring about the transformation of resident astrocytes to the reactive phenotype are only now being actively investigated. Tyrosine phosphorylation, which is regulated by protein tyrosine kinases and phosphatases, appears to play an important role in the activation of astrocytes. The intracellular protein tyrosine phosphatase SHP1 was upregulated following axotomy of peripheral nerves, and a direct cortical lesion led to a massive upregulation of SHP1 in activated microglia and astrocytes, whereas the neuronal expression of SHP1 was not affected.[41] Similar upregulation was observed in reactive astrocytes in response to ischemic brain injury.[150,151]

JAK/STAT is one of the pathways bearing signals from the cell membrane to the nucleus in response to extracellular growth factors and particularly to cytokines. Binding of cytokine leads to activation of receptor-associated tyrosine kinases (JAKs) and to the subsequent phosphorylation, dimerization, and nuclear translocation of STAT3 and STAT1 transcription factors.[152] Activation of JAK/STAT pathway was detected in astrocytes following ischemia in rats.[153] Activation of the STAT3 signaling pathway was identified in reactive astrocytes following entorhinal cortical lesion leading to denervation of the fascia dentata in the hippocampus.[154] Similarly, STAT activation was reported in the facial nucleus following peripheral nerve transection[155] and in ischemic brain.[153]

The mitogen-activated protein kinase (MAPK) signaling pathway is a major signal transduction system.[156,157] Activation (i.e., phosphorylation) of the ERK/MAPK pathway was identified in reactive astrogliosis in various human conditions (trauma, chronic epilepsy, progressive multifocal leukoencephalopathy).[158] The ERK/MAPK pathway can be activated in cultured astrocytes by CNTF, EGF, FGF, PDGF, IGF-1, BDNF, thrombin, and TGF-α.[159,160] Many of these factors have been implicated in the induction of astrogliosis. ERK/MAPK activation has also been detected in animals following stab wound,[161] as well as following mechanical trauma of cultured astrocytes, where this activation was blocked by the MEK inhibitors, PD98059 and U0126.[149] ERK/MAPK activation has been detected in the hippocampus in reperfusion injury after forebrain ischemia in the gerbil.[162]

Activation of the JNK/SAPK-c-Jun pathway has been described in reactive astrocytes of the spinal cord from patients with amyotrophic lateral sclerosis (ALS).[163] This increase was accompanied by NF-kappaB activation, suggesting that this was in response to oxidative stress.

5.8 SIGNIFICANCE OF ASTROGLIOSIS

While the significance of astrogliosis remains unclear, there is a general consensus that in the early postinjury period, reactive astrocytes carry out vital functions aimed at restoring the integrity of the extracellular microenvironment, removing excitotoxins such as glutamate, assisting in the elimination of free radicals, and in restoring the blood–brain barrier.[20,55,120,164–167]

Reactive astrocytes can also be conducive to axon growth[168] and support neurite outgrowth *in vitro*.[169–171] Following injury, reactive astrocytes have been shown to synthesize various growth factors, including NGF,[126,172–174] CNTF,[136,175] and bFGF[176] and to provide various extracellular matrix molecules, including laminin,[177,178] fibronectin,[178–180] NCAM,[178,181] L1,[178] heparan sulfate proteoglycan,[182] and DSD-1 proteoglycan,[178,183] all of which are capable of facilitating the process of repair and regeneration.

In this regard, the studies of Sofroniew's research group are particularly instructive.[165] Reactive astrocytes adjacent to a forebrain stab injury in mice were selectively ablated in adult transgenic mice expressing the herpes simplex thymidine kinase gene under control of the GFAP promoter by treatment with gancyclovir. These mice exhibited a significant increase in inflammatory cell infiltrates, were unable to repair the BBB, and developed an extensive glutamate-mediated neuronal degeneration. These mice did, however, demonstrate an increase in local neurite outgrowth. On balance, therefore, the selective ablation of reactive astrocytes had a more harmful than beneficial effect following injury. A similar conclusion was drawn by Iseda et al.[184] using a different experimental paradigm in which they observed that regeneration failed when astrocytes were absent near the lesion.

Despite these obvious beneficial properties of reactive astrocytes, it has been proposed that these cells actually play a sinister role, particularly in the process of CNS regeneration.[185–189] This detrimental view appears to be a dominant theme in regeneration research. Indeed, many contemporary studies are geared at modifying, ablating, or otherwise destroying astrocytes as a means of improving CNS regeneration. Other adverse effects of reactive astrocytes have been proposed — that they represent an obstacle to remyelination[166,190] and that they possibly contribute to epilepsy/seizures.[191]

Various regeneration inhibitory molecules indeed have been identified in reactive astrocytes, including tenascin-C,[192,193] chondroitin sulfate proteglycan,[166,179,194,195] and NG2-proteoglycan.[178] However, most of these molecules are made by oligodendroglial precursor cells.[196] See Reference 166 for a review of inhibitory factors presumably generated by reactive astrocytes.

This detrimental view requires a more careful analysis as there are a number of issues that blur a true understanding of the role of astrogliosis in neuroregeneration as well as in other neurological conditions. First, a major problem has been the unfortunate tendency of using the term *glial scar* as a synonym for astrogliosis. The scar is often believed to represent nothing more than a meshwork of reactive astrocytes. Yet it must be emphasized that the "scar" contains many other cellular constituents, including oligodendrocytes and their precursors, which are capable of inhibiting axonal growth,[197] possibly through the synthesis of such inhibitory molecules as the proteoglycan NG2, NI250, myelin-associated glycoprotein, and DSD-1/phosphacan, as well as other myelin-derived inhibitory molecules.[198] The scar also contains microglia/macrophages that are capable of generating toxic cytokines as well as releasing glutamate, arachidonic acid metabolites, and reactive oxygen and nitrogen species.[199,200] There are also nonglial elements such as meningeal cells (arachnoidal cells) that express the inhibitory molecule keratan sulfate proteoglycan,[178] fibroblasts, endothelial cells, and Schwann cells (usually in the form of schwannosis).[48,201–203] Particularly in spinal cord lesions, astrocytes and mesenchymal elements cooperate in the laying down of a basal lamina. As discussed by Stichel and Müller,[204] axons can penetrate astrogliotic tissue rich in CSPGs and tenascin-C but stop growing when they meet the basal lamina. It is therefore difficult, if not impossible, to attribute a specific event to one particular cell. Nevertheless, the literature is replete with statements regarding the nefarious properties of reactive astrocytes when, in reality, the changes described are consequences of the glial scar and not the reactive astrocyte *per se*.

Second, astrocytes do not operate in a vacuum, and key interactions with other neural cells occur.[205,206] For example, Ness and David[193] have shown that meningeal cells are capable of inducing a greater expression of CSPGs and tenascin-C in astrocytes.

Third, the extent of astrogliosis is rarely such that it creates an impenetrable physical barrier for the passage of new neurites (Figure 5.1C and Figure 5.2C). Further, axons are capable of entering astrogliotic tissue.[69,207–210] While there is little evidence that reactive astrocytes present a mechanical obstacle to regeneration, it is quite possible that reactive astrocytes may secrete factors that inhibit regeneration, including chondroitin sulfate proteoglycans[166,179,194,195] and the glycoprotein tenascin-C.[192,193]

Fourth, what appears to be detrimental may not necessarily be so. The walling of the lesion by reactive astrocytes may isolate the injured area from the rest of the intact brain, thereby protecting the latter from untoward secondary effects of the injury. Astrogliosis may also prevent the growth of axons into inappropriate areas, thus blocking the development of potentially deleterious neuronal connections, and thereby avoid such untoward consequences as pain and spasticity.[211–214] Further, some of the so-called inhibitory molecules may actually have beneficial effects. Tenascin can both promote and inhibit neurite outgrowth depending on the isoforms,[179,192,215] and some sulfated proteoglycans, especially the dermatan and heparan varieties, have been shown to exhibit neurotrophic activity.[216]

5.9 PERSPECTIVES

Astrogliosis is arguably the most common and ubiquitous response to CNS injury. Significant progress has been made in recent years with regard to factors initiating this response and the possible signal transduction pathways mediating this reaction. Yet much information is still critically needed to define their precise role in neurologic disease. Current opposing views regarding astrogliosis are almost schizophrenic, with some groups regarding reactive astrocytes as vital and beneficial and others viewing them almost as objects of derision that should be eliminated because of their alleged interference with axonal regeneration.

Some caveats regarding astrogliosis should be heeded:

1. Reactive astrocytes are heterogeneous and it is not clear that particular astroglial responses identified in one condition will also be found in others.
2. Cell culture data may not be relevant to *in vivo* states. Critical interactions with other cells absent in cell culture studies may significantly influence the astroglial response *in vivo*.
3. While meaningful *in vitro* models would be highly desirable, the available ones are too limited to make them mechanistically useful; i.e., they either fail to faithfully mimic the *in vivo* condition, or they do not retain the reactive phenotype when transferred to an artificial culture system. Until such time when these limitations can be addressed, we will have to rely largely on *in vivo* studies to continue progress in this field.
4. While GFAP remains a useful marker, its upregulation should not be equated with astrogliosis. While it is unlikely that a reactive astrocyte does not show upregulation of GFAP, its upregulation does not necessarily mean that the astrocyte has undergone a reactive change. Cellular hypertrophy is the hallmark of astrogliosis, and numerous publications have merely documented elevations in GFAP without other concomitant morphologic changes.

In summary, there is no convincing or compelling evidence *in vivo* that reactive astrocytes produce detrimental factors or present an impediment to regeneration. There are many other cellular constituents associated with the "glial scar" that are capable of inhibiting the regenerative response. It would appear that many of the harmful consequences attributed to astrogliosis are simply guilt by association. While there is some evidence that inhibitory molecules are produced in cultured astrocytes, evidence for this occurring *in vivo* is woefully lacking. In view of the many beneficial actions of reactive astrocytes, efforts aimed at eradicating astrogliosis for the purpose of enhancing neuronal and axonal regeneration appear to be misguided and unwise.

ACKNOWLEDGMENTS

This work was supported by grants from the NIH NS38665 and the Department of Veterans Affairs.

REFERENCES

1. Kimelberg, H.K., Jalonen, T., and Walz, W., Regulation of the brain microenvironment: transmitters and ions, in *Astrocytes: Pharmacology and Function*, Murphy, S., Ed., Academic Press, San Diego, CA, 1993, p. 193.
2. Dringen, R., Gebhardt, R., and Hamprecht, B., Glycogen in astrocytes: possible function as lactate supply for neighboring cells, *Brain Res.*, 623, 208, 1993.
3. Westergaard, N., Sonnewald, U., and Schousboe, A., Metabolic trafficking between neurons and astrocytes: the glutamate glutamine cycle revisited, *Dev. Neurosci.*, 17, 203, 1995.
4. Dringen, R. and Hamprecht, B., Glucose, insulin, and insulin-like growth factor I regulate the glycogen content of astroglia-rich primary cultures, *J. Neurochem.*, 58, 511, 1992.
5. Magistretti, P.J. et al., Energy on demand, *Science*, 283, 496, 1999.
6. Keyser, D.O. and Pellmar, T.C., Synaptic transmission in the hippocampus: critical role for glial cells, *Glia*, 10, 237, 1994.
7. Ullian, E.M. et al., Control of synapse number by glia, *Science*, 291, 657, 2001.
8. Pfrieger, F.W. and Barres, B.A., Synaptic efficacy enhanced by glia cells *in vitro*, *Science*, 277, 1684, 1997.
9. Müller, H.W., Junghans, U., and Kappler, J., Astroglial neurotrophic and neurite promoting factors, *Pharmacol. Ther.*, 65, 1, 1995.
10. Norenberg, M.D., The role of astrocytes in hepatic encephalopathy, *Neurochem. Pathol.*, 6, 13, 1987.

11. Abramovitz, M. et al., Characterization and localization of glutathione-S-transferases in rat brain and binding of hormones, neurotransmitters, and drugs, *J. Neurochem.*, 50, 50, 1988.

12. Makar, T.K. et al., Vitamin E, ascorbate, glutathione, glutathione disulfide, and enzymes of glutathione metabolism in cultures of chick astrocytes and neurons: Evidence that astrocytes play an important role in antioxidative processes in the brain, *J. Neurochem.*, 62, 45, 1994.

13. Sawada, J. et al., Induction of metallothionein in astrocytes by cytokines and heavy metals, *Biol. Signals*, 3, 157, 1994.

14. Janzer, R.C. and Raff, M.C., Astrocytes induce blood-brain barrier properties in endothelial cells, *Nature*, 325, 253, 1987.

15. Stitt, T.N., Gasser, U.E., and Hatten, M.E., Molecular mechanisms of glial-guided neuronal migration, *Ann. N.Y. Acad. Sci.*, 633, 113, 1991.

16. Frei, K. and Fontana, A., Immune regulatory functions of astrocytes and microglial cells within the central nervous system, in *Neuroimmune Networks: Physiology and Diseases*, Goetzl, E.J. and Spector, N.H., Eds. Alan R. Liss, New York, 1989, p. 127.

17. O'Callaghan, J.P., Quantitative features of reactive gliosis following toxicant-induced damage of the CNS, *Ann. N.Y. Acad. Sci.*, 679, 195, 1993.

18. O'Callaghan, J.P. and Miller, D.B., Quantification of reactive gliosis as an approach to neurotoxicity assessment, *NIDA Res. Monogr.*, 136, 188, 1993.

19. Kimelberg, H.K. and Norenberg, M.D., Astroglial responses to CNS trauma, in *The Neurobiology of Central Nervous System Trauma*, Salzman, S.K. and Faden, A.I., Eds., Oxford University Press, New York, 1994, p. 193.

20. Norenberg, M.D., Astrocyte responses to CNS injury, *J. Neuropathol. Exp. Neurol.*, 53, 213, 1994.

21. Steward, O. et al., Neuronal activity up-regulates astroglial gene expression, *Proc. Natl. Acad. Sci. U.S.A.*, 88, 6819, 1991.

22. Canady, K.S. and Rubel, E.W., Rapid and reversible astrocytic reaction to afferent activity blockade in chick cochlear nucleus, *J. Neurosci.*, 12, 1001, 1992.

23. Kraig, R.P., Dong, L., Thisted, R., and Jaeger, C.B., Spreading depression increases immunohistochemical staining of glial fibrillary acidic protein, *J. Neurosci.*, 11, 2187, 1991.

24. Anderson, B.J. et al., Glial hypertrophy is associated with synaptogenesis following motor-skill learning, but not with angiogenesis following exercise, *Glia*, 11, 73, 1994.

25. Goldman, J.E. and Corbin, E., Rosenthal fibers contain ubiquitinated αB-crystallin, *Am. J. Pathol.*, 139, 933, 1991.

26. Klemenz, R. et al., AlphaB-crystallin is a small heat shock protein, *Proc. Natl. Acad. Sci. U.S.A.*, 88, 3652, 1991.

27. Nathaniel, E.J.H. and Nathaniel, D.R., The reactive astrocyte, *Adv. Cell. Neurobiol.*, 2, 249, 1981.

28. Eng, L.E. et al., An acidic protein isolated from fibrous astrocytes, *Brain Res.*, 28, 351, 1971.

29. Bignami, A. et al., Localization of the glial fibrillary acidic protein in astrocytes by immunofluorescence, *Brain Res.*, 43, 429, 1972.

30. Bignami, A. and Dahl, D., The astroglial response to stabbing. Immunofluorescence studies with antibodies to astrocyte-specific protein (GFA) in mammalian and submammalian vertebrates, *Neuropathol. Appl. Neurobiol.*, 2, 99, 1976.

31. Dahl, D. et al., Filament proteins in rat optic nerves undergoing wallerian degeneration: localization of vimentin, the fibroblast 100-A filament protein, in normal and reactive astrocytes, *Exp. Neurol.*, 73, 496, 1981.

32. Clarke, S.R. et al., Reactive astrocytes express the embryonic intermediate neurofilament nestin, *NeuroReport*, 5, 1885, 1994.

33. Mucke, L. et al., Rapid activation of astrocyte specific expression of GFAP-lacZ transgene by focal injury, *New Biol.*, 3, 465, 1991.

34. Maxwell, W.L. et al., The response of the cerebral hemisphere of the rat to injury. II. The neonatal rat, *Phil. Trans. Royal Society of London — Series B: Biol. Sci.*, 328, 501, 1990.

35. Balasingam, V. et al., Reactive astrogliosis in the neonatal mouse brain and its modulation by cytokines, *J. Neurosci.*, 14, 846, 1994.

36. Balasingam, V. et al., Astrocyte reactivity in neonatal mice: apparent dependence on the presence of reactive microglia/macrophages, *Glia*, 18, 11, 1996.

37. Mathewson, A.J. and Berry, M., Observations on the astrocyte response to a cerebral stab wound in adult rats, *Brain Res.*, 327, 61, 1985.

38. Ludwin, S.K., Reaction of oligodendrocytes and astrocytes to trauma and implantation, *Lab. Invest.*, 52, 20, 1985.

39. Miyake, T. et al., Quantitative studies on proliferative changes of reactive astrocytes in mouse cerebral cortex, *Brain Res.*, 451, 133, 1988.

40. Takamiya, Y. et al., Immunohistochemical studies on the proliferation of reactive astrocytes and the expression of cytoskeletal proteins following brain injury in rats, *Dev. Brain Res.*, 38, 201, 1988.

41. Horvat, A. et al., A novel role for protein tyrosine phosphatase shp1 in controlling glial activation in the normal and injured nervous system, *J. Neurosci.*, 21, 865, 2001.

42. Latov, N. et al., Fibrillary astrocytes proliferate in response to brain injury. A study combining immunoperoxidase technique for glial fibrillary acid protein and radiography of tritiated thymidine, *Dev. Biol.*, 72, 381, 1979.

43. Villain, M. et al., Macroglial alterations after isolated optic nerve sheath fenestration in rabbit, *Invest. Ophthalmol. Visual Sci.*, 43, 120, 2002.

44. Murray, M. et al., Modification of astrocytes in the spinal cord following dorsal root or peripheral nerve lesions, *Exp. Neurol.*, 110, 248, 1990.

45. Tezel, G., Hernandez, M.R., and Wax, M.B., *In vitro* evaluation of reactive astrocyte migration, a component of tissue remodeling in glaucomatous optic nerve head, *Glia*, 34, 178, 2001.

46. Fitch, M.T. et al., Cellular and molecular mechanisms of glial scarring and progressive cavitation: *In vivo* and *in vitro* analysis of inflammation-induced secondary injury after CNS trauma, *J. Neurosci.*, 19, 8182, 1999.

47. Hernandez, M.R. et al., Differential gene expression in astrocytes from human normal and glaucomatous optic nerve head analyzed by cDNA microarray, *Glia*, 38, 45, 2002.

48. Norton, W.T. et al., Quantitative aspects of reactive gliosis: a review, *Neurochem. Res.*, 17, 877, 1992.

49. Tani, M. et al., *In situ* hybridization analysis of glial fibrillary acidic protein mRNA reveals evidence of biphasic astrocyte activation during acute experimental autoimmune encephalomyelitis, *Am. J. Pathol.*, 148, 889, 1996.

50. Bridges, R.J., Nieto-Sampedro, M., and Cotman, C.W., Stereospecific binding of L-glutamate to astrocyte membranes, *Soc. Neurosci. Abstr.*, 11, 110, 1985.

51. Komoly, S. et al., Insulin-like growth factor I gene expression is induced in astrocytes during experimental demyelination, *Proc. Natl. Acad. Sci. U.S.A.*, 89, 1894, 1992.

52. Malhotra, S.K. et al., Diversity among reactive astrocytes: Proximal reactive astrocytes in lacerated spinal cord preferentially react with monoclonal antibody J1-31, *Brain Res. Bull.*, 30, 395, 1993.

53. Fernaud-Espinosa, I., Nieto-Sampedro, M., and Bovolenta, P., Differential activation of microglia and astrocytes in aniso- and isomorphic gliotic tissue, *Glia*, 8, 277, 1993.

54. Bovolenta, P., Wandosell, F., and Nieto-Sampedro, M., CNS glial scar tissue: a source of molecules which inhibit central neurite outgrowth, in *Neuronal-Astrocytic Interactions: Pathological Implications*, Yu, A.C.H., Hertz, L., Norenberg, M.D., Sykova, E., and Waxman, S.G., Eds., Elsevier, Amsterdam, 1992, p. 367.

55. Ridet, J.L. et al., Reactive astrocytes: cellular and molecular cues to biologic function, *Trends Neurosci.*, 20, 570, 1997.

56. Bovolenta, P., Wandosell, F., and Nieto-Sampedro, M., CNS glial scar tissue: a source of molecules which inhibit central neurite outgrowth, in *Neuronal-Astrocytic Interactions: Implications for Normal and Pathological CNS Function*, Yu, A.C.H. et al., Eds., Elsevier, Amsterdam, 1992, p. 367.

57. Nieto-Sampedro, M., Neurite outgrowth inhibitors in gliotic tissue, *Adv. Exp. Med. Biol.*, 468, 207, 1999.

58. Lin, R.C.S. and Matesic, D.F., Immunohistochemical demonstration of neuron-specific enolase and microtubule-associated protein 2 in reactive astrocytes after injury in the adult forebrain, *Neuroscience*, 60, 11, 1994.

59. Lin, R.C.S., Polsky, K., and Matesic, D.F., Expression of gamma-aminobutyric acid immunoreactivity in reactive astrocytes after ischemia-induced injury in the adult forebrain, *Brain Res.*, 600, 1, 1993.

60. Le Gal La Salle, G., Rougon, G., and Valin, A., The embryonic form of neural cell surface molecule (E-NCAM) in the rat hippocampus and its reexpression on glial cells following kainic acid-induced status epilepticus, *J. Neurosci.*, 12, 872, 1992.

61. Freund, T.F. et al., Relationship of neuronal activity and calcium binding protein immunoreactivity in ischemia, *Exp. Brain Res.*, 83, 55, 1990.

62. Takahashi, M. et al., Ependymal cell reactions in spinal cord segments after compression injury in adult rat, *J. Neuropathol. Exp. Neurol.*, 62, 185, 2003.

63. Yamamoto, S. et al., Proliferation of parenchymal neural progenitors in response to injury in the adult rat spinal cord, *Exp. Neurol.*, 172, 115, 2001.

64. Holmin, S. et al., Adult nestin-expressing subependymal cells differentiate to astrocytes in response to brain injury, *Eur. J. Neurosci.*, 9, 65, 1997.

65. Johansson, C.B. et al., Identification of a neural stem cell in the adult mammalian central nervous system, *Cell*, 96, 25, 1999.

66. Frisén, J. et al., Rapid, widespread, and longlasting induction of nestin contributes to the generation of glial scar tissue after CNS injury, *J. Cell Biol.*, 131, 453, 1995.

67. Anders, J.J. and Brightman, M.W., Assemblies of particles in the cell membrane of developing, mature and reactive astrocytes, *J. Neurocytol.*, 8, 777, 1979.

68. Landis, D.M.D. and Reese, T.S., Membrane structure in mammalian astrocytes: a review of freeze-fracture studies on adult, developing, reactive and cultured astrocytes, *J. Exp. Biol.*, 95, 35, 1981.

69. Alonso, G. and Privat, A., Reactive astrocytes involved in the formation of lesional scars differ in the mediobasal hypothalamus and in other forebrain regions, *J. Neurosci. Res.*, 34, 523, 1993.

70. Hossain, M.Z. et al., Ischemia-induced cellular redistribution of the astrocytic gap junctional protein connexin43 in rat brain, *Brain Res.*, 652, 311, 1994.

71. Fonseca, C.G., Green, C.R., and Nicholson, L.F., Upregulation in astrocytic connexin 43 gap junction levels may exacerbate generalized seizures in mesial temporal lobe epilepsy, *Brain Res.*, 929, 105, 2002.

72. Rohlmann, A. et al., Facial nerve lesions lead to increased immunostaining of the astrocytic gap junction protein (connexin 43) in the corresponding facial nucleus of rats, *Neurosci. Lett.*, 154, 206, 1993.

73. Charles, A., Intercellular calcium waves in glia, *Glia*, 24, 39, 1998.

74. Lin, J.H. et al., Gap-junction-mediated propagation and amplification of cell injury, *Nature Neurosci.*, 1, 494, 1998.

75. Westenbroek, R.E. et al., Upregulation of L-type Ca2+ channels in reactive astrocytes after brain injury, hypomyelination, and ischemia, *J. Neurosci.*, 18, 2321, 1998.

76. O'Connor, E.R. et al., Astrocytes from human hippocampal epileptogenic foci exhibit action potential-like responses, *Epilepsia*, 39, 347, 1998.

77. Gorter, J.A. et al., Sodium channel beta1-subunit expression is increased in reactive astrocytes in a rat model for mesial temporal lobe epilepsy, *Eur. J. Neurosci.*, 16, 360, 2002.

78. Bordey, A. and Sontheimer, H., Properties of human glial cells associated with epileptic seizure foci, *Epilep. Res.*, 32, 286, 1998.

79. Anderson, C.M. and Swanson, R.A., Astrocyte glutamate transport: Review of properties, regulation, and physiological functions, *Glia*, 32, 1, 2000.

80. Krum, J.M., Phillips, T.M., and Rosenstein, J.M., Changes in astroglial GLT-1 expression after neural transplantation or stab wounds, *Exp. Neurol.*, 174, 137, 2002.

81. Arzberger, T. et al., Changes of NMDA receptor subunit (NR1, NR2B) and glutamate transporter (GLT1) mRNA expression in Huntington's disease — an *in situ* hybridization study, *J. Neuropathol. Exp. Neurol.*, 56, 440, 1997.

82. Masago, A. et al., GLAST mRNA expression in the periventricular area of experimental hydrocephalus, *NeuroReport*, 7, 2565, 1996.

83. Buck, C.R. et al., Increased adenine nucleotide translocator 1 in reactive astrocytes facilitates glutamate transport, *Exp. Neurol.*, 181, 149, 2003.

84. Theodosis, D.T. and Poulain, D.A., Contribution of astrocytes to activity-dependent structural plasticity in the adult brain, *Adv. Exp. Med. Biol.*, 468, 175, 1999.

85. McGraw, J., Hiebert, G.W., and Steeves, J.D., Modulating astrogliosis after neurotrauma, *J. Neurosci. Res.*, 63, 109, 2001.

86. Eddleston, M. and Mucke, L., Molecular profile of reactive astrocytes — Implications for their role in neurologic disease, *Neuroscience*, 54, 15, 1993.

87. Montgomery, D.L., Astrocytes: form, functions, and roles in disease, *Vet. Pathol.*, 31, 145, 1994.

88. Hardin, H. et al., Modifications of glial metabolism of glutamate after serotonergic neuron degeneration in the hippocampus of the rat, *Brain Res., Mol. Brain Res.*, 26, 1, 1994.

89. Waniewski, R.A. and McFarland, D., Intrahippocampal kainic acid reduces glutamine synthetase, *Neuroscience*, 34, 305, 1990.

90. Petito, C.K. et al., Brain glutamine synthetase increases following cerebral ischemia in the rat, *Brain Res.*, 569, 275, 1992.

91. Endoh, M. et al., Reactive astrocytes express NADPH diaphorase *in vivo* after transient ischemia, *Neurosci. Lett.*, 154, 125, 1993.

92. Wallace, M.N. and Bisland, S.K., NADPH-diaphorase activity in activated astrocytes represents inducible nitric oxide synthase, *Neuroscience*, 59, 905, 1994.

93. Hewett, S.J., Csernansky, C.A., and Choi, D.W., Selective potentiation of NMDA-induced neuronal injury following induction of astrocytic iNOS, *Neuron*, 13, 487, 1994.

94. Goadsby, P., Kaube, H., and Hoskin, K.L., Nitric oxide synthesis couples cerebral blood flow and metabolism, *Brain Res.*, 595, 167, 1992.

95. Goss, J.R. et al., Astrocytes are the major source of nerve growth factor upregulation following traumatic brain injury in the rat, *Exp. Neurol.*, 149, 301, 1998.

96. Azmitia, E.C., Dolan, K., and Whitaker-Azmitia, P.M., S-100 but not NGF, EGF, insulin or calmodulin is a CNS serotonergic growth factor, *Brain Res.*, 516, 354, 1990.

97. Rudge, J.S., Astrocyte-derived neurotrophic factors, in *Astrocytes: Pharmacology and Function*, Murphy, S., Ed., Academic Press, San Diego, 1993, p. 267.

98. Funk, J.L. et al., Parathyroid hormone-related protein (PTHrP) induction in reactive astrocytes following brain injury: a possible mediator of CNS inflammation, *Brain Res.*, 915, 195, 2001.

99. Brines, M.L., Ling, Z., and Broadus, A.E., Parathyroid hormone-related protein protects against kainic acid excitotoxicity in rat cerebellar granule cells by regulating L-type channel calcium flux, *Neurosci. Lett.*, 274, 13, 1999.

100. Ono, T. et al., Activity-dependent expression of parathyroid hormone-related protein (PTHrP) in rat cerebellar granule neurons. Requirement of PTHrP for the activity-dependent survival of granule neurons, *J. Biol. Chem.*, 272, 14404, 1997.

101. Gadient, R.A. and Otten, U.H., Interleukin-6 (IL-6) — A molecule with both beneficial and destructive potentials, *Prog. Neurobiol.*, 52, 379, 1997.

102. Loddick, S.A., Turnbull, A.V., and Rothwell, N.J., Cerebral interleukin-6 is neuroprotective during permanent focal cerebral ischemia in the rat, *J. Cereb. Blood Flow Metab.*, 18, 176, 1998.

103. Lukes, A. et al., Extracellular matrix degradation by metalloproteinases and central nervous system diseases, *Mol. Neurobiol.*, 19, 267, 1999.

104. Rathke-Hartlieb, S. et al., Elevated expression of membrane type 1 metalloproteinase (MT1-MMP) in reactive astrocytes following neurodegeneration in mouse central nervous system, *FEBS Lett.*, 481, 227, 2000.

105. Suzumura, A., Lavi, E., and Weiss, R., Coronavirus infection induces H-2 antigen expression on oligodendrocytes and astrocytes, *Science*, 232, 231, 1986.

106. Frank, E., Pulver, M., and De Tribolet, N., Expression of class II major histocompatibility antigens on reactive astrocytes and endothelial cells within the gliosis surrounding metastases and abscesses, *J. Neuroimmunol.*, 12, 29, 1986.

107. Merrill, J.E. and Benveniste, E.N., Cytokines in inflammatory brain lesions: helpful and harmful, *Trends Neurosci.*, 19, 331, 1996.

108. Hariri, R.J. et al., Traumatic injury induces interleukin-6 production by human astrocytes, *Brain Res.*, 636, 139, 1994.

109. Johnson, M.D. et al., Lipocortin-1 immunoreactivity in the normal human central nervous system and lesions with astrocytosis, *Am. J. Clin. Pathol.*, 92, 424, 1989.

110. Eberhard, D.A., Brown, M.D., and VandenBerg, S.R., Alterations of annexin expression in pathological neuronal and glial reactions: Immunohistochemical localization of annexins I, II (p36 and p11 subunits), IV, and VI in the human hippocampus, *Am. J. Pathol.*, 145, 640, 1994.

111. Kostyk, S.K., Kowall, N.W., and Hause, S.L., Substance P immunoreactive astrocytes are present in multiple sclerosis plaques, *Brain Res.*, 504, 284, 1989.

112. Boyles, J.K. et al., Apolipoprotein E associated with astrocytic glia of the central nervous system and with nonmyelinating glia of the peripheral nervous system, *J. Clin. Invest.*, 76, 1501, 1985.

113. Carrasco, J. et al., Role of metallothionein-III following central nervous system damage, *Neurobiol. Dis.*, 13, 22, 2003.

114. Espejo, C. et al., Differential expression of metallothioneins in the CNS of mice with experimental autoimmune encephalomyelitis, *Neuroscience*, 105, 1055, 2001.

115. Neal, J.W. et al., Immunocytochemically detectable metallothionein is expressed by astrocytes in the ischaemic human brain, *Neuropathol. Appl. Neurobiol.*, 22, 243, 1996.

116. Giulian, D. et al., The impact of microglia-derived cytokines upon gliosis in the CNS, *Dev. Neurosci.*, 16, 128, 1994.

117. Rostworowski, M. et al., Astrogliosis in the neonatal and adult murine brain post-trauma: elevation of inflammatory cytokines and the lack of requirement for endogenous interferon, *J. Neurosci.*, 17, 3664, 1997.

118. Rubio, N. and De Felipe, C., Demonstration of the presence of a specific interferon-γ receptor on murine astrocyte cell surface, *J. Neuroimmunol.*, 35, 111, 1991.

119. Ban, E.M., Sarlièvre, L.L., and Haour, F.G., Interleukin-1 binding sites on astrocytes, *Neuroscience*, 52, 725, 1993.

120. Herx, L.M. and Yong, V.W., Interleukin-1 beta is required for the early evolution of reactive astrogliosis following CNS lesion, *J. Neuropathol. Exp. Neurol.*, 60, 961, 2001.

121. Da Cunha, A. et al., Control of astrocytosis by interleukin-1 and transforming growth factor-β1 in human brain, *Brain Res.*, 631, 39, 1993.

122. Giulian, D. and Lachman, L.B., Interleukin-1 stimulation of astroglial proliferation after brain injury, *Science*, 228, 497, 1985.

123. Giulian, D. et al., Interleukin-1 injected into mammalian brain stimulates astrogliosis and neovascularization, *J. Neurosci.*, 8, 2484, 1988.

124. Balasingam, V. and Yong, V.W., Attenuation of astroglial reactivity by interleukin-10, *J. Neurosci.*, 16, 2945, 1996.

125. Yong, V.W., Cytokines, astrogliosis, and neurotrophism following CNS trauma, in *Cytokines and the CNS*, Ransohoff, J. and Benveniste, E.N., Eds., CRC Press, Boca Raton, 1996, p. 309.

126. Schwartz, J.P. and Nishiyama, N., Neurotrophic factor gene expression in astrocytes during development and following injury, *Brain Res. Bull.*, 35, 403, 1994.

127. Althaus, H.H. and Richter-Landsberg, C., Glial cells as targets and producers of neurotrophins, *Int. Rev. Cytol.*, 197, 203, 2000.

128. Eclancher, F. et al., Basic fibroblast growth factor (bFGF) injection activates the glial reaction in the injured adult rat brain, *Brain Res.*, 737, 201, 1996.

129. Gomez-Pinilla, F., Vu, L., and Cotman, C.W., Regulation of astrocyte proliferation by FGF-2 and heparan sulfate *in vivo*, *J. Neurosci.*, 15, 2021, 1995.

130. Leme, R.J. and Chadi, G., Distant microglial and astroglial activation secondary to experimental spinal cord lesion, *Arquiv. Neuro-Psiquiat.*, 59, 483, 2001.

131. Reilly, J.F., Maher, P.A., and Kumari, V.G., Regulation of astrocyte GFAP expression by TGF-beta1 and FGF-2, *Glia*, 22, 202, 1998.

132. Krohn, K. et al., Glial fibrillary acidic protein transcription responses to transforming growth factor-beta1 and interleukin-1beta are mediated by a nuclear factor-1-like site in the near-upstream promoter, *J. Neurochem.*, 72, 1353, 1999.

133. Hudgins, S.N. and Levison, S.W., Ciliary neurotrophic factor stimulates astroglial hypertrophy *in vivo* and *in vitro*, *Exp. Neurol.*, 150, 171, 1998.

134. Lisovoski, F. et al., Phenotypic alteration of astrocytes induced by ciliary neurotrophic factor in the intact adult brain, as revealed by adenovirus-mediated gene transfer, *J. Neurosci.*, 17, 7228, 1997.

135. Lee, M.-Y. et al., Transient upregulation of ciliary neurotrophic factor receptor α mRNA in axotomixed rat septal neurons, *Eur. J. Neurosci.*, 9, 622, 1997.

136. Guthrie, K.M. et al., Astroglial ciliary neurotrophic factor mRNA expression is increased in fields of axonal sprouting in deafferented hippocampus, *J. Comp. Neurol.*, 386, 137, 1997.

137. Kahn, M.A. et al., CNTF regulation of astrogliosis and the activation of microglia in the developing rat central nervous system, *Brain Res.*, 685, 55, 1995.

138. Winter, C.G. et al., A role for ciliary neurotrophic factor as an inducer of reactive gliosis, the glial response to central nervous system injury, *Proc. Natl. Acad. Sci. U.S.A.*, 92, 5865, 1995.

139. Nishino, A. et al., Thrombin may contribute to the pathophysiology of central nervous system injury, *J. Neurotrauma*, 10, 167, 1993.

140. Baba, A., Role of endothelin B receptor signals in reactive astrocytes, *Life Sci.*, 62, 1711, 1998.

141. Ishikawa, N. et al., Endothelins promote the activation of astrocytes in rat neostriatum through ET_B receptors, *Eur. J. Neurosci.*, 9, 895, 1997.

142. Hasselblatt et al., Effect of endothelin-1 on astrocytic protein content, *Glia*, 42, 390, 2003.

143. Peters, C.M. et al., Endothelin receptor expression in the normal and injured spinal cord: potential involvement in injury-induced ischemia and gliosis, *Exp. Neurol.*, 180, 1, 2003.

144. Hindley, S., Herman, M.A.R., and Rathbone, M.P., Stimulation of reactive astrogliosis *in vivo* by extracellular adenosine diphosphate or an adenosine A2 receptor agonist, *J. Neurosci. Res.*, 38, 399, 1994.

145. Neary, J.T. et al., Trophic actions of extracellular nucleotides and nucleosides on glial and neuronal cells, *Trends Neurosci.*, 19, 13, 1996.

146. Franke, H., Krügel, U., and Illes, P., P2 receptor-mediated proliferative effects on astrocytes *in vivo*, *Glia*, 28, 190, 1999.

147. Franke, H. et al., P2 receptor-types involved in astrogliosis *in vivo*, *Br. J. Pharmacol.*, 134, 1180, 2001.

148. Jabs, R., Paterson, I.A., and Walz, W., Qualitative analysis of membrane currents in glial cells from normal and gliotic tissue *in situ*: down-regulation of Na+ current and lack of P2 purinergic responses, *Neuroscience*, 81, 847, 1997.

149. Mandell, J.W., Gocan, N.C., and VandenBerg, S.R., Mechanical trauma induces rapid astroglial activation of ERK/MAP kinase: Evidence for a paracrine signal, *Glia*, 34, 283, 2001.

150. Servidei, T. et al., The protein tyrosine phosphatase SHP-2 is expressed in glial and neuronal progenitor cells, postmitotic neurons and reactive astrocytes, *Neuroscience*, 82, 529, 1998.

151. Wishcamper, C.A. et al., Focal cerebral ischemia upregulates SHP-1 in reactive astrocytes in juvenile mice, *Brain Res.*, 974, 88, 2003.

152. Heinrich, P.C. et al., Interleukin-6-type cytokine signaling through the gp130/Jak/STAT pathway, *Biochem. J.*, 334, 297, 1998.

153. Justicia, C., Gabriel, C., and Planas, A.M., Activation of the JAK/STAT pathway following transient focal cerebral ischemia: Signaling through JAK1 and Stat3 in astrocytes, *Glia*, 30, 253, 2000.

154. Xia, X.G. et al., Induction of STAT3 signaling in activated astrocytes and sprouting septal neurons following entorhinal cortex lesion in adult rats, *Mol. Cell. Neurosci.*, 21, 379, 2002.

155. Schwaiger, F.-W. et al., Peripheral but not central axotomy induces changes in Janus kinase (JAK) and signal transducers and activators of transcription (STAT), *Eur. J. Neurosci.*, 12, 1165, 2000.

156. Fukunaga, K. and Miyamoto, E., Role of MAP kinase in neurons, *Mol. Neurobiol.*, 16, 79, 1998.

157. Derkinderen, P., Enslen, H., and Girault, J.A., The ERK/MAP-kinases cascade in the nervous system, *NeuroReport*, 10, R24-R34, 1999.

158. Mandell, J.W. and VandenBerg, S.R., ERK/MAP kinase is chronically activated in human reactive astrocytes, *NeuroReport*, 10, 3567, 1999.

159. Roback, J.D. et al., BDNF-activated signal transduction in rat cortical glial cells, *Eur. J. Neurosci.*, 7, 849, 1995.

160. Tournier, C. et al., MAP kinase cascade in astrocytes, *Glia*, 10, 81, 1994.

161. Carbonell, W.S. and Mandell, J.W., Transient neuronal but persistent astroglial activation of ERK/MAP kinase after focal brain injury in mice, *J. Neurotrauma*, 20, 327, 2003.

162. Namura, S. et al., Intravenous administration of MEK inhibitor U0126 affords brain protection against forebrain ischemia and focal cerebral ischemia, *Proc. Natl. Acad. Sci. U.S.A.*, 98, 11569, 2001.

163. Migheli, A. et al., c-Jun, JNK/SAPK kinases and transcription factor NF-kappa B are selectively activated in astrocytes, but not motor neurons, in amyotrophic lateral sclerosis, *J. Neuropathol. Exp. Neurol.*, 56, 1314, 1997.

164. Yong, V.W., Response of astrocytes and oligodendrocytes to injury, *Ment. Retard. Dev. Disabil. Res. Rev.*, 4, 193, 1998.

165. Bush, T.G. et al., Leukocyte infiltration, neuronal degeneration, and neurite outgrowth after ablation of scar-forming, reactive astrocytes in adult transgenic mice, *Neuron*, 23, 297, 1999.

166. Fawcett, J.W. and Asher, R.A., The glial scar and central nervous system repair, *Brain Res. Bull.*, 49, 377, 1999.

167. Kakinuma, Y. et al., Impaired blood-brain barrier function in angiotensinogen-deficient mice, *Nature Med.*, 4, 1078, 1998.

168. Kawaja, M.D. and Gage, F.H., Reactive astrocytes are substrates for the growth of adult CNS axons in the presence of elevated levels of nerve growth factor, *Neuron*, 7, 1019, 1991.

169. Noble, M., Fok-Seang, J., and Cohen, J., Glia are a unique substrate for the *in vitro* growth of central nervous system neurons, *J. Neurosci.*, 4, 1892, 1984.

170. Fallon, J., Preferential outgrowth of CNS neurites on astrocytes and Schwann cells as compared with non-glial cell *in vitro*, *J. Cell Biol.*, 100, 198, 1985.

171. Neugebauer, K.M. et al., N-cadherin, NCAM, and integrins promote retinal neurite outgrowth on astrocytes *in vitro*, *J. Cell Biol.*, 107, 1177, 1988.

172. Gage, F.H., Olejniczak, P., and Armstrong, D.M., Astrocytes are important for sprouting in the septohippocampal circuit, *Exp. Neurol.*, 102, 2, 1988.

173. Schwartz, J.P. et al., Trophic factor production by reactive astrocytes in injured brain, *Ann. N.Y. Acad. Sci.*, 679, 226, 1993.

174. Gottlieb, M. and Matute, C., Expression of nerve growth factor in astrocytes of the hippocampal CA1 area following transient forebrain ischemia, *Neuroscience*, 91, 1027, 1999.

175. Ip, N.Y. et al., Injury-induced regulation of ciliary neurotrophic factor mRNA in the adult rat brain, *Eur. J. Neurosci.*, 5, 25, 1993.

176. Gómez-Pinilla, F., Lee, J.W., and Cotman, C.W., Basic FGF in adult rat brain: cellular distribution and response to entorhinal lesion and fimbria-fornix transection, *J. Neurosci.*, 12, 345, 1992.

177. Liesi, P., Dahl, D., and Vaheri, A., Laminin is produced by early rat astrocytes in primary culture, *J. Cell Biol.*, 96, 920, 1983.

178. Hirsch, S. and Bähr, M., Immunocytochemical characterization of reactive optic nerve astrocytes and meningeal cells, *Glia*, 26, 36, 1999.

179. McKeon, R.J. et al., Reduction of neurite outgrowth in a model of glial scarring following CNS injury is correlated with the expression of inhibitory molecules on reactive astrocytes, *J. Neurosci.*, 11, 3398, 1991.

180. Niquet, J. et al., Proliferative astrocytes may express fibronectin-like protein in the hippocampus of epileptic rats, *Neurosci. Lett.*, 180, 13, 1994.

181. Smith, G.M. et al., Maturation of astrocytes *in vitro* alters the extent and molecular basis of neurite outgrowth, *Dev. Biol.*, 138, 377, 1990.

182. Ard, M.D. and Bunge, R.P., Heparan sulfate proteoglycan and laminin immunoreactivity on cultured astrocytes: relationship to differentiation and neurite growth, *J. Neurosci.*, 8, 2844, 1988.

183. Butterfield, D.A. et al., β amyloid peptide free radical fragments initiate synaptosomal lipoperoxidation in a sequence-specific fashion: implication to Alzheimer's disease, *Biochem. Biophys. Res. Commun.*, 200, 710, 1994.

184. Iseda, T. et al., Spontaneous regeneration of the corticospinal tract after transection in young rats: collagen type IV deposition and astrocytic scar in the lesion site are not the cause but the effect of failure of regeneration, *J. Comp. Neurol.*, 464, 343, 2003.

185. Ramon y Cajal, S., *Degeneration and Regeneration of the Nervous System*, Oxford University Press, Oxford, 1928.

186. Windle, W.F., Regeneration of axons in the vertebrate nervous system, *Physiol. Rev.*, 36, 427, 1956.

187. Fawcett, J.W., Astrocytic and neuronal factors affecting axon regeneration in the damaged central nervous system, *Cell Tissue Res.*, 290, 371, 1997.

188. Reier, P.J., Stensaas, L.J., and Guth, L., The astrocytic scar as an impediment to regeneration in the central nervous system, in *Spinal Cord Reconstruction*, Kao, C.C., Bunge, R.P., and Reier, P.J., Eds., Raven Press, New York, 1983, p. 163.

189. Fitch, M.T. and Silver, J., Beyond the glial scar: cellular and molecular mechanisms by which glial cells contribute to CNS regenerative failure, in *CNS Regeneration: Basic Science and Clinical Advances*, Tuszynski, M.H. and Kordower, J., Eds., Academic Press, San Diego, 1999, p. 55.

190. Prineas, J.W. et al., Multiple sclerosis. Oligodendrocyte proliferation and differentiation in fresh lesions, *Lab. Invest.*, 61, 489, 1989.

191. Pollen, D.A. and Trachtenberg, M.C., Neuroglia: gliosis and focal epilepsy, *Science*, 167, 1252, 1970.

192. Ajemian, A., Ness, R., and David, S., Tenascin in the injured rat optic nerve and in non-neuronal cells *in vitro*: potential role in neural repair, *J. Comp. Neurol.*, 340, 233, 1994.

193. Ness, R. and David, S., Leptomeningeal cells modulate the neurite growth promoting properties of astrocytes *in vitro*, *Glia*, 19, 47, 1997.

194. Asher, R.A. et al., Chondroitin sulphate proteoglycans: inhibitory components of the glial scar, *Prog. Brain Res.*, 132, 611, 2001.

195. Höke, A. and Silver, J., Proteoglycans and other repulsive molecules in glial boundaries during development and regeneration of the nervous system, *Prog. Brain Res.*, 108, 149, 1996.

196. Levine, J.M., Increased expression of the NG2 chondroitin-sulfate proteoglycan after brain injury, *J. Neurosci.*, 14, 4716, 1994.

197. Fawcett, J.W., Rokos, J., and Bakst, I., Oligodendrocytes repel axons and cause axonal growth cone collapse, *J. Cell Sci.*, 92, 93, 1989.

198. Schwab, M.E. and Bartholdi, D., Degeneration and regeneration of axons in the lesioned spinal cord, *Physiol. Rev.*, 76, 319, 1996.

199. Banati, R.B. et al., Cytotoxicity of microglia, *Glia*, 7, 111, 1993.

200. Gonzalez-Scarano, F. and Baltuch, G., Microglia as mediators of inflammatory and degenerative diseases, *Annu. Rev. Neurosci.*, 22, 219, 1999.

201. Li, M.S. and David, S., Topical glucocorticoids modulate the lesion interface after cerebral cortical stab wounds in adult rats, *Glia*, 18, 306, 1996.

202. Abnet, K., Fawcett, J.W., and Dunnett, S.B., Interactions between meningeal cells and astrocytes *in vivo* and *in vitro*, *Dev. Brain Res.*, 53, 187, 1991.

203. Bruce, J.H. et al., Schwannosis: Role of gliosis and proteoglycan in human spinal cord injury, *J. Neurotrauma*, 17, 781, 2000.

204. Stichel, C.C. and Müller, H.W., The CNS lesion scar: new vistas on an old regeneration barrier, *Cell Tissue Res.*, 294, 1, 1998.

205. Murphy, S., Ed., *Astrocytes: Pharmacology and Function*, Academic Press, San Diego, 1993.

206. Norenberg, M.D., Astrocyte pathophysiology in disorders of the central nervous system, in *Astrocytes in Brain Aging and Neurodegeneration*, Schipper, H.M., Ed., R.G. Landes, Georgetown, TX, 1998, p. 41.

207. Lips, K., Stichel, C.C., and Müller, H.-W., Restricted appearance of tenascin and chondroitin sulphate proteoglycans after transection and sprouting of adult rat postcommissural fornix, *J. Neurocytol.*, 24, 449, 1995.

208. Stichel, C.C. and Müller, H.-W., Relationship between injury-induced astrogliosis, laminin expression and axonal sprouting in the adult rat brain, *J. Neurocytol.*, 23, 615, 1994.

209. Frisén, J. et al., Growth of ascending spinal axons in CNS scar tissue, *Int. J. Dev. Neurosci.*, 11, 461, 1993.

210. Li, Y. and Raisman, G., Sprouts from cut corticospinal axons persist in the presence of astrocytic scarring in long-term lesions of the adult rat spinal cord, *Exp. Neurol.*, 134, 102, 1995.

211. Snow, D., Steindler, D.A., and Silver, J., Molecular and cellular characterization of the glial roof plate of the spinal cord and optic tectum: a possible role for a proteoglycan in the development of an axon barrier, *Dev. Biol.*, 138, 359, 1990.

212. Letourneau, P.C., Condic, M.L., and Snow, D.M., Extracellular matrix and neurite outgrowth, *Curr. Opin. Genet. Dev.*, 2, 625, 1992.

213. Pindzola, R.R., Doller, C., and Silver, J., Putative inhibitory extracellular matrix molecules at the dorsal root entry zone of the spinal cord during development and after root and sciatic nerve lesions, *Dev. Biol.*, 156, 34, 1993.

214. Matsui, F. et al., Transient expression of juvenile-type neurocan by reactive astrocytes in adult rat brains injured by kainate-induced seizures as well as surgical incision, *Neuroscience*, 112, 773, 2002.

215. Fruttiger, M., Schachner, M., and Martini, R., Tenascin-C expression during wallerian degeneration in C55BL/Wlds mice: possible implications for axonal regeneration, *J. Neurocytol.*, 24, 1, 1995.

216. Kappler, J. et al., Chondroitin/dermatan sulphate promotes the survival of neurons from rat embryonic neocortex, *Eur. J. Neurosci.*, 9, 306, 1997.

CHAPTER 6

Molecular Mechanisms of Glutamate and Glutamine Transport in Astrocytes

Stefan Bröer

CONTENTS

6.1 THE CONCEPT OF THE GLUTAMATE/GLUTAMINE CYCLE

Labeling studies performed in the 1960s and 1970s using intact animals demonstrated that label from glucose, glycerol, and lactate was primarily incorporated into glutamate, whereas label from acetate and leucine was primarily incorporated into glutamine.[1] These labeling patterns were interpreted in terms of the presence of two citric acid cycles in the brain, which are now recognized as being located in astrocytes and neurons, respectively. Although glutamine is always synthesized via glutamate as an intermediate, the two labeling patterns were observed because the glial glutamate pool is very small, whereas the neuronal pool is rather big. As a result, glutamate labeled from acetate and leucine is not recognized. The glial glutamate pool is small because it is rapidly converted into glutamine by the activity of glutamine synthetase, which is only expressed in astrocytes.[2] Other studies showed that GABA is rapidly labeled from glucose, but only slowly from glutamate, aspartate, and leucine, suggesting that the two pools are connected and exchange metabolites.[3]

More recent work has confirmed and reinforced this view by more direct evidence:

1. Most of the glutamate is localized in neurons.[4]
2. The vast majority of glutamate transporters, however, are localized on astrocytes.[5]
3. Glutamate transporter currents can be recorded on hippocampal astrocytes but not on neurons.[6]
4. Addition of glutamate to cultured astrocytes induces glutamine release.[7]
5. Inhibition of glutamine synthetase rapidly depletes glutamate pools.[8,9]
6. Inhibition of neuronal glutamine transport depletes GABA pools and glutamate pools.[10,11]
7. NMR studies suggest that the speed of the glutamate–glutamine cycle is tightly coupled to the activity of the brain.[12]

Taken together, it is now recognized that the glutamate–glutamine cycle is an essential pathway in the metabolism of glutamate and GABA. However, it still remains to be established what proportion of neurotransmitters are recycled by this pathway and whether it is used to regulate neuronal activity. Glutamate uptake and glutamine release are two essential steps in this cycle. In the following, transporters will be discussed that mediate the transport of these two metabolites in astrocytes and their physiological role will be evaluated. Reference to neuronal transporters will be given where appropriate.

6.2 GLUTAMATE UPTAKE

6.2.1   EAAT1 (GLAST), EAAT2 (GLT-1) (System X_{AG})

6.2.1.1   Molecular Characteristics

Human EAAT1 and human EAAT2 are proteins of 542 amino acids and 574 amino acids, respectively. Hydropathy plotting indicates the presence of six amino-terminal transmembrane helices followed by a large hydrophobic domain. The accessibility of residues in the large hydrophobic domain has been probed extensively in two studies to elucidate its secondary structure. Both studies developed similar topological models from these data. The study by Grunewald and Kanner[13] proposes two reentry loops: one after transmembrane helix 6 and another after transmembrane helix 7. This array is followed by a helix, which runs in parallel to the membrane followed by transmembrane helix 8 and the cytosolic carboxy terminus. The study by Seal and Amara[14] proposes only one reentrant loop after transmembrane helix 7 but two helices that lie parallel to the plane of the membrane before traversing it by helix 8 to form the cytosolic carboxy terminus. Substrate binding, Na^+ binding, K^+ binding, and H^+ binding are thought to occur in this area.[15] Glutamate transporters also display an anion conductance, which is thought to reside in the amino-terminal half of the protein.[16]

6.2.1.2 Localization

The immunohistochemistry of glial glutamate transporters has been addressed in some detail. From these studies it appears that the glial glutamate transporters EAAT1 (GLAST) and EAAT2 (GLT-1) are responsible for most of the glutamate uptake in the brain.[17] Broadly speaking, EAAT2 is the predominant transporter in the forebrain, whereas EAAT1 is the predominant transporter in the cerebellum. Both EAAT1 and 2 are only found on astroglial membranes in the brain. Screening of large amounts of sections suggests that every astrocyte in the forebrain expresses EAAT2. Expression is particularly high in the hippocampus. Concentration in astrocyte processes in this area reaches 8500/1 μm^2.[18]

6.2.1.3 Mechanism

Studies using synaptosomes established that glutamate uptake is accompanied by the cotransport of at least 2 Na^+ and the antiport of 1 K^+. A binding order was established in which Na^+ binds before the substrate. Return of the substrate-free carrier is facilitated by binding of K^+.[19] Cloning and expression of glutamate transporters combined with electrophysiological analysis of substrate-induced currents revealed modifications to this transport mechanism. First, the Na^+ cotransport stoichiometry was revised from 2 to 3 Na^+ and furthermore a cotransport of 1 H^+ was detected.[20,21] Analysis of substrate-induced currents also revealed the presence of an anion conductance that requires substrate binding but is not coupled to substrate translocation.[22] The complex mechanism of glutamate transport provides a huge driving force for uptake of glutamate into astrocytes, which can be calculated by Equation 6.1:

$$[S]_i/[S]_o = ([Na^+]_o/[Na^+]_i)^3 \times ([H^+]_o/[H^+]_i) \times ([K^+]_i/[K^+]_o) \times 10^{(-2F./2.3RT)} \qquad (6.1)$$

Under physiological conditions this will provide the capacity to accumulate glutamate about 10^6-fold. At an intracellular concentration of 5 mM, this may cause a reduction of extracellular glutamate to a concentration of 5 nM. However, extracellular concentrations in the intact brain are significantly higher, in the lower micromolar range.[17] As a result, glutamate transporters only need to sustain a 10^4-fold gradient, which is well below capacity. This discrepancy suggests the presence of glutamate "leaks" pathways in neurons, astrocytes, or both, which are unrelated to glutamate transporters. Release by reversed transport only occurs in severely deenergized cells, which no longer sustain a Na^+ gradient and a membrane potential.[23]

The most important physiological function of glutamate transporters is to ensure rapid removal of glutamate from the synaptic cleft. Although the mechanism of EAAT1 and 2 is equivalent from a thermodynamical point of view, their kinetic behavior is quite different. Fast application of glutamate to membrane patches containing glutamate transporters elicits a biphasic response: a transient peak current is followed by a smaller steady-state current.[24] The ratio of peak current to steady state is high for EAAT2 but small for EAAT1.[25] This observation has been interpreted in terms of a two-state kinetic model.[24] In the absence of glutamate, the binding site of almost all transporters faces the extracellular space. The prevalent membrane potential ensures this orientation. As a result, glutamate binding occurs at the same time on all transporters in the patch after glutamate is applied. In the case of EAAT2, the next step is a rapid synchronized conformational change that relocates the substrate to the intracellular side, which generates the peak current. After this initial synchronized response, individual transporters may unload glutamate at different times, bind intracellular K^+ at different times, and subsequently reorient the substrate binding site back to the extracellular face of the membrane. As a result, return of the K^+-loaded transporter occurs unsynchronized and more slowly. The whole turnover resulting in net transport, hence, is slower than the initial half-cycle of the glutamate-bound transporter and constitutes the smaller steady-state current. This is in agreement with the view held for most transporters that substrate translocation steps are usually faster than conformational transitions in the absence of substrate. EAAT1, by contrast, has a much slower initial translocation step, and the peak current as a result is small. In fact, determination of the unbinding rate of glutamate from EAAT1 indicates that it is more likely for glutamate to dissociate again after binding to the transporter than to be transported.[25]

Another feature of glutamate transporters is the presence of an anion conductance that is detected in the presence and absence of transporter substrates.[26] However, binding of glutamate to the transporter strongly increases the open probability of the anion conductance. Two arguments suggest that the anion channel is an inherent part of the transporter. First, the on-rate of channel activation is very close to the on-rate of glutamate binding to the transporter.[24] Second, inhibition of the channel conductance by zinc ions can be prevented by site-directed mutagenesis of two histidine residues in the amino-terminal half of the transporter.[27] All substrate-induced currents observed in glutamate transporters are thus the sum of transport currents generated by the coupled movement of Na^+ ions and an uncoupled anion conductance. The ratio of both currents varies strongly with the transporter subtype. Glutamate-induced currents of the neuronal EAAT4 appear to be largely carried by anions, whereas glutamate-induced currents of EAAT1 and EAAT2 largely reflect coupled transporter currents.[28] The ratio can also be varied by exchanging the permeant anion. The anion conductance is small in the presence of chloride but becomes much more conspicuous in the presence of SCN^-. The physiological significance of these currents, at least in the case of EAAT1 and EAAT2, remains to be established.

6.2.1.4 *Physiological Role*

It is well established that glutamate transporters are crucial for the maintenance of low extracellular glutamate concentrations. However, whether glutamate transporters are involved in the shaping of excitatory postsynaptic potentials (EPSPs) is still disputed.[29] Kinetic analysis of glial transporters as outlined above indicates that EAAT2 is the "fast glutamate vacuum cleaner" in the cortex, whereas EAAT1 is the "slow and steady glutamate remover" in the cerebellum. Thus, in principle, EAAT2 would be the candidate to shape EPSPs. However, EAAT2 is usually found in astroglial processes that are at a distance to glutamatergic synapses.[17,30] In the cerebellum, where most glutamatergic synapses are closely ensheathed by astrocyte processes, the slower glutamate transporter EAAT1 is present, which may even extend the time that glutamate is present in the synaptic cleft, because unbinding is more likely than transport. Nevertheless it appears the EPSPs in Purkinje cells can be extended by glutamate transport inhibitors.[29]

In summary, astroglial glutamate transporters rather appear to be important to sustain very low levels of extracellular glutamate but may not shape EPSPs on a millisecond timescale. The overall importance of EAAT2 as the major forebrain glutamate transporter is reflected by the phenotype of knockout mice that develop spontaneous lethal seizures.[31] Knockout of EAAT1 or of the neuronal EAAT3, by contrast, results in less conspicuous phenotypes.[32,33]

6.2.2 4F2hc/xCT (System x^-_c)

6.2.2.1 Molecular Characteristics and Localization

In addition to Na$^+$-dependent glutamate transport, Na$^+$-independent glutamate transport has also been described in mammalian cells.[34] The transport activity is carried out by a heteromeric amino acid transporter, which is composed of the 4F2 heavy chain and the xCT light chain.[35] The human 4F2 heavy chain is a type II membrane protein with a single transmembrane helix and a large extracellular domain. It consists of 529 amino acids and is required for the translocation of the complex to the plasma membrane. The transport itself is thought to be catalyzed by the xCT light chain. Human xCT is a protein of 501 amino acids and has a molecular mass of about 55 kDa. Expression of the xCT light chain is confined to specialized areas in the brain, such as circumventricular organs, plexus choroideus, and the meninges.[36] In culture, cystine uptake mediated via xCT is observed in astrocytes but not in neurons.[37] The presence of cystine/glutamate exchanger in astrocytes was confirmed *in situ* by using aminoadipic acid as a specific marker for the activity of the cystine/glutamate exchanger.[38] Uptake of aminoadipic acid was detected in astrocytes throughout the brain, indicating that the immunofluorescence studies mentioned above may have detected only high concentrations of the xCT light chain.

6.2.2.2 Mechanism

xCT is an obligatory antiporter and mediates the antiport of negatively charged cystine against glutamate. The overall transport process as a result is electroneutral.

6.2.2.3 Physiological Function

Given that the glutamate concentration is high inside the cell and cystine is found only in the oxidizing extracellular milieu, this antiporter mediates the efflux of glutamate in exchange for the uptake of cystine. The transporter is strongly upregulated under oxidative stress and may constitute one of the glutamate "leak" pathways described in Section 6.2.1.3. A role in glutamate release during oxidative stress has been proposed.[38]

6.3 TRANSPORT OF GLUTAMINE

6.3.1 Principles of Glutamine Transport

Amino acid transport in mammalian cells is mediated by a variety of amino acid transporters of overlapping substrate specificity.[39,40] Given the important role of glutamine as a central metabolite in many pathways, it is not surprising that glutamine is transported not only by a single transporter but in fact is a substrate of most amino acid transporters. Characterization of amino acid transport activities in mammalian cells has established the presence of amino acid transport "systems."[41] The acronym given to any amino acid transport system indicates its substrate specificity. Cloning of these transporters revealed that the activities described as amino acid transport systems indeed correspond to different transport proteins. In many cases, however, more than one transport protein corresponds to a particular system.[40,42] A list of astroglial glutamine transporters is presented in Table 6.1. The name of the amino acid transport system and the corresponding cDNAs are indicated in the heading of each paragraph.

In the following, astroglial glutamine transporters and their proposed physiological role will be discussed in alphabetical order.

Table 6.1 Glutamine Transporters in Astrocytes

Transport System	cDNA	Mechanism	Physiological Role	Comment	K_m (Gln) (mM)
A	SNAT1	1Na$^+$/Gln cotransport	Gln uptake into neurons	Neuronal in adult brain	0.2–0.5
	SNAT2	1Na$^+$/Gln cotransport	Net AA uptake when AA are lacking	Widely distributed	1.65
ASC	ASCT2	Antiporter	Cell growth astrogliosis	Na$^+$ dependent	0.07
L	4F2hc/ LAT1	Antiporter	Equilibration of AA levels	Na$^+$ independent	1.6–2.2
	4F2hc/ LAT2	Antiporter	Equilibration of AA levels; Ala and Leu transfer	Na$^+$ independent	0.15–0.30
N	SNAT3	1Na$^+$/Gln cotransport-H$^+$ antiport	Gln release from astrocytes	Mostly close to GABAergic neurons in some areas	1–2
y$^+$L	4F2hc/ y$^+$LAT2	Antiporter	Arginine release	Not expressed in adult brain	0.3

6.3.2 ASCT2 (System ASC)

6.3.2.1 Molecular Characteristics

In cultured astrocytes, system ASC is the prominent transport activity. This transport activity is mediated by ASCT2,[43] a transporter belonging to the family of glutamate transporters.[44] The transporter is composed of 539 amino acids. Its topology is likely to be the same as that of glutamate transporters. ASCT2 transports small neutral amino acids with high affinity, including alanine, serine, cysteine, threonine, glutamine, and asparagine. Leucine and glycine are low-affinity substrates of the transporter. Glutamine is transported with a K_m value of 70 μM.[43]

6.3.2.2 Localization

Although ASCT2 is the prominent glutamine transporter in cultured astrocytes, its mRNA is found only in small amounts in the intact brain.[43] ASCT2 immunoreactivity is mostly found in the thalamus and in the cerebellum (author's unpublished results). The mRNA of ASCT2 is strongly upregulated in reactive astrocytes. Thus, strong expression of ASCT2 is a hallmark of dividing cells and is an indicator of astrogliosis. In agreement with this notion, expression of ASCT2 mRNA is considerably higher in embryonic brain tissue.[43]

6.3.2.3 Mechanism

ASCT2 is a Na$^+$-dependent amino acid exchanger.[45] The Na$^+$ dependence is caused by a high-affinity Na$^+$-binding site, which is saturated both on the intracellular and extracellular face of the transporter. As a result, the transporter carries out a Na$^+$/Na$^+$ exchange and does not use the electrochemical potential of the Na$^+$ gradient. Similar to the glutamate transporters of this family, a substrate-induced anion conductance has been observed in ASCT2.[45] The conductance is almost undetectable in the presence of chloride, i.e., under physiological conditions, but becomes more obvious when Cl$^-$ is replaced by SCN$^-$ or NO$_3^-$.

6.3.2.4 Physiological Role

The physiological role of ASCT2 is still ill-defined. ASCT2 expression might not only be upregulated in dividing cells but may even be essential for cell growth and division. Silencing of ASCT2 mRNA has been shown to induce apoptosis.[46] This feature explains why ASCT2 is the prominent transport activity in cultured astrocytes, whereas in the adult brain it is only weakly expressed. The importance of ASCT2 for cell growth is at variance with its antiport mechanism, which does not allow net uptake of amino acids. However, in conjunction with system A transporters (see Section 6.3.4), it is able to provide entry of small neutral amino acids for cell growth. In summary, ASCT2 is unlikely to be involved in glutamine release from astrocytes but is more likely to be essential during development and for the onset of astrogliosis.

6.3.3 LAT1, LAT2 (System L)

6.3.3.1 Molecular Characteristics and Localization

LAT1 and LAT2 are two molecular isoforms of a Na^+-independent neutral amino acid transporter that has been described as system L activity in mammalian cells (see References 47 and 48 for a review). Both proteins (also called light chains) require association with the type II membrane protein 4F2 heavy chain to be transported to the plasma membrane. In the absence of 4F2, both LAT1 and LAT2 remain in the endoplasmic reticulum because they cannot pass quality-control checkpoints in this compartment.[49] LAT1 is a protein of 506 amino acids and has 12 proposed transmembrane spanning domains. LAT2 has a similar topology and is composed of 535 amino acids. A disulfide bridge is formed between a cysteine of 4F2 that is located just outside the single transmembrane helix and a cysteine that is located in the loop between helix 3 and 4 of the LAT1/2 light chain.[50] The major difference between both transporters is the substrate specificity; LAT1 has a narrow substrate specificity, whereas LAT2 is able to transport almost any neutral amino acid with the exception of proline.[51,52] Glutamine affinity for LAT1 is 2.2 mM, whereas other amino acids such as leucine are transported with K_m values of < 50 µM. LAT2, by contrast, transports glutamine and other neutral amino acids with similar, slightly lower, affinity.[52] The difference between LAT1 and LAT2 is most notable when efflux is investigated, where glutamine has to compete with other intracellular substrates of the transporter.[53] Under those conditions, glutamine is released mainly by LAT2. Both LAT1 and LAT2 are expressed in cultured astrocytes. The combined transport activity amounts to about 40% of the total glutamine transport activity.[54] Glutamine transport in cultured astrocytes is largely mediated by LAT2 with small contributions by LAT1. Although LAT2 is one of the prominent glutamine transport activities in cultured astrocytes, immunohistochemical studies have mainly demonstrated the presence of LAT2 in endothelial cells of the blood–brain barrier.[55,56] However, LAT1 and LAT2 are expressed in both cultured astrocytes and cultured neurons.[54] This might result in ubiquitous immunohistochemical staining being misinterpreted as a nonspecific background. Furthermore, 4F2 heavy chain, which is required for surface expression of LAT1 and LAT2, is widely distributed throughout the brain.[36]

6.3.3.2 Mechanism

LAT1 and LAT2 are obligatory antiporters.[51,57] They mediate the Na^+-independent exchange of neutral amino acids. Intracellular K_m values are severalfold higher than extracellular K_m values.[58] As pointed out previously, this prevents LAT1 from mediating release of glutamine from astrocytes.[53] The obligatory exchange mechanism predicts that efflux of glutamine from astrocytes via LAT2 should increase significantly in the presence of extracellular amino acids. This is indeed the case and can increase total glutamine efflux almost twofold.[53]

6.3.3.3 Physiological Role

The physiological role of LAT2 in glutamine transport remains to be defined. Measurement of metabolite turnover in guinea pig slices demonstrates that BCH, a specific inhibitor of LAT1 and LAT2 in astrocytes, influences the incorporation of [^{13}C]glucose into aspartate (C. Rae and S. Bröer, unpublished observations). This suggests an involvement of LAT1 or LAT2 in the metabolic traffic between neurons and astrocytes. Theoretically, LAT2 would be ideally suited to mediate the transfer of amino acids between astrocytes and neurons. Ammonia released during the glutaminase reaction in neurons is thought to be incorporated into alanine or leucine.[59,60] Both amino acids are then thought to be transferred back to astrocytes to aid in the synthesis of glutamine. As a result, a net exchange of glutamine for alanine or leucine occurs between astrocytes and neurons. This task could be very well supported by LAT2; however, functional and immunohistochemical evidence is still limited.

6.3.4 SNAT1, SNAT2 (ATA1, ATA2)* (System A)

6.3.4.1 Molecular Characteristics

Both SNAT1 and SNAT2 belong to the AAAP family of amino acid and auxin permeases.[61] Human SNAT1 has 485 amino acids; human SNAT2 is a protein of 506 amino acids. Hydropathy analysis indicates the presence of 11 transmembrane helices. SNAT1 and SNAT2 have similar substrate specificity. Both transport most small neutral amino acids including glutamine, asparagine, and histidine. However, SNAT1 shows a preference for glutamine over other amino acids. Interestingly, both SNAT1 and SNAT2 transport serine but not threonine. SNAT1 transports glutamine with a K_m of 230 μM,[62,63] SNAT2 with a K_m of 1.65 mM.[64]

6.3.4.2 Localization

SNAT1 is a fairly brain-specific transporter,[65] whereas SNAT2 appears to be the molecular correlate of the "classical" system A amino acid transporter and is expressed in a wide variety of tissues.[64,66] SNAT1 is found in the membrane of both glutamatergic and GABAergic neurons.[63,67] It appears not to be expressed in astrocytes in situ. SNAT2, in agreement with its more widespread distribution, is found in astrocytes and neurons.[68] Cultured astrocytes express both SNAT1 and SNAT2, but the corresponding transport activity cannot be detected.[54] In agreement with other cell types, system A activity can be induced by amino acid depletion of astrocytes or by hyperosmotic shock.[54,69]

6.3.4.3 Mechanism

Both SNAT1 and SNAT2 cotransport neutral amino acids together with 1 Na$^+$ ion.[62,64,70,71] Oocytes expressing SNAT1 show increased uptake of Na$^+$ even in the absence of transporter substrates. The origin of this Na$^+$ conductance has not been established. A decrease of the extracellular pH strongly increases the K_m for Na$^+$ binding to the transporter.[62] Changes of the extracellular Na$^+$ concentration strongly influence the K_m of glutamine. This has been interpreted in terms of an ordered binding mechanism in which Na$^+$ binds before the substrate to the transporter.[62,70] SNAT1 and SNAT2 display strong pH dependence, being more active at alkaline pH. The pH dependence is caused by a pH-sensitive modifier site on the transporter.[72]

---

* Nomenclature of the SLC38 family has recently been revised.[61] System A isoforms ATA1 and ATA2 are now designated as SNAT1 and SNAT2. System N isoforms SN1 and SN2 are now referred to as SNAT3 and SNAT5.

6.3.4.4 Physiological Role

In contrast to most amino acid transporters, which are antiporters, SNAT1 and SNAT2 are able to mediate net transport of amino acids. It appears likely that antiporters are used to balance the intracellular concentration of all amino acids for protein synthesis.[42] However, net uptake of amino acids requires at least one transporter that is not an antiporter and that has overlapping substrate specificity with antiporters. Alanine and glutamine appear to play the role of universal exchange currency in amino acid transport.[42] Glutamine is a substrate of almost all amino acid antiporters, and for those where this is not the case, alanine is a substrate. Both alanine and glutamine can be synthesized by any cell from metabolic intermediates. In agreement with this notion, system A activity cannot be detected in cultured astrocytes unless being depleted of amino acids.[69] Although both transporters have the same mechanism, SNAT1 and SNAT2 may play quite different roles in the brain. It is likely that SNAT2 represents the transport activity that is upregulated whenever amino acids are required for cell growth to act in concert with amino acid antiporters. SNAT1, on the contrary, is expressed at significant levels even in nondividing fully differentiated GABAergic and glutamatergic neurons. It is thus likely to play an active role in neurotransmitter metabolism. MeAIB, a specific inhibitor of SNAT1 and SNAT2, has been used to address this function. The results of several studies indicate that saturating concentrations of MeAIB deplete neurotransmitter pools in the hippocampus and in brain slices.[10,11] As a result, SNAT1 is thought to mediate uptake of glutamine into neurons after being released from astrocytes. However, it should be kept in mind that two other Na^+-dependent glutamine transport activities have been described in cultured neurons that are different from SNAT1: (1) System N^b, a glutamate-sensitive transporter, which is characterized as a transport system active in the presence of MeAIB, threonine, and histidine,[73] and (2) another glutamate-sensitive glutamine transporter, which is also resistant to MeAIB but is blocked by threonine and histidine.[54] The latter transporter has properties similar to SNAT5 (see Section 6.3.5), but the corresponding transcript has not been detected in cultured neurons.[54]

6.3.5 SNAT3, SNAT5 (SN1, SN2)* (System N)

6.3.5.1 Molecular Characteristics

SNAT3 was initially cloned from liver and brain tissue.[74,75] Both tissues are known to express system N–like transport activity.[76] The transporter belongs to the AAAP family of auxin and amino acid permeases.[77] The mammalian members of this family characterized so far all transport amino acids. They either reside on the plasma membrane, lysosomal membranes, or synaptic vesicles. The human SNAT3 cDNA encodes a protein of 504 amino acids.[78] Hydropathy plots indicate the presence of 11 transmembrane helices, the N-terminus being located in the cytosol.

SNAT3 accepts glutamine, asparagine, and histidine as substrates. Glutamine is transported with an apparent K_m value of 2.4 mM at pH 7.0.[72,74] Histidine has a similar affinity to the transporter as glutamine. For asparagine, a much higher K_m value of 16 mM was reported.[79]

SNAT5 is sequence-related to SNAT3 but has a wider substrate specificity than the latter.[80,81] SNAT5 transports glycine, asparagine, alanine, serine, glutamine, methionine, and histidine. A K_m value of 4.1 mM for glutamine has been reported for the rat isoform of the transporter.[80]

6.3.5.2 Localization

SNAT3 is mainly expressed in liver, kidney, and the brain; its distribution in the brain has been studied in some detail.[82,83] These studie suggest that it is confined to astrocyte processes. Prominent expression was detected in processes adjacent to glutamatergic, GABAergic, and glycinergic synapses.

* Nomenclature of the SLC38 family has recently been revised.[61] System A isoforms ATA1 and ATA2 are now designated as SNAT1 and SNAT2. System N isoforms SN1 and SN2 are now referred to as SNAT3 and SNAT5.

The protein is found in all areas of the brain and shows a relatively even distribution (e.g., less than twofold differences between brain areas). White matter was stained weakly compared to gray matter. In some areas of the brain such as the cerebellum and the hippocampus, SNAT3 was closely positioned to GABAergic synapses, but glutamatergic synapses were not surrounded by SNAT3 containing astrocyte processes.

SNAT5 is found in a variety of tissues including the brain. Its cellular distribution in the brain has not been investigated. However, RT-PCR experiments with cultured astrocytes suggest that it is not expressed in astrocytes.[54]

6.3.5.3 *Mechanism*

Initial characterization of SNAT3 established and confirmed earlier reports that this transporter is Na$^+$-dependent.[74] However, depolarization of cells by addition of KCl had little effect on the transport activity of SNAT3, suggesting that the transporter might not be a simple Na$^+$-glutamine cotransporter. Expression of the transporter in Na$^+$/H$^+$ exchanger-deficient cells revealed that uptake of glutamine caused an alkalinization of the cells. It was concluded that SNAT3 works as a Na$^+$/H$^+$ exchanger in which glutamine is coupled to the transport of Na$^+$.[74] Expression of the transporter in oocytes, however, revealed that the transport of glutamine was accompanied by inward currents, suggesting that more than one Na$^+$ was cotransported together with glutamine.[78] These conflicting data were reconciled by experiments demonstrating that the transporter mediates substrate-induced currents, which are not coupled to substrate translocation.[72,79] Transport of glutamine is energetically coupled to the cotransport of 1 Na$^+$ and the antiport of 1 H$^+$. Further experiments revealed that in the transport cycle, substrate binds prior to Na$^+$ followed by substrate translocation. Return of the substrate-unloaded carrier is facilitated by proton binding. According to Equation 6.2 the transporter will accumulate glutamine inside cells about 20-fold at physiological substrate and ion concentrations.

$$[S]_i/[S]_o = ([Na^+]_o/[Na^+]_i) \times ([H^+]_i/[H^+]_o) \tag{6.2}$$

Small changes of the pH or Na$^+$ gradient, as a result, can change the prevalent transport direction from influx to efflux.[72] Accordingly, it has been demonstrated that release of glutamine from astrocytes is reduced when cells are depleted of intracellular Na$^+$; conversely, efflux increased when the intracellular concentration of Na$^+$ was raised.[53]

The nature of the glutamine-induced currents observed in SNAT3 expressing *Xenopus laevis* oocytes has not been resolved. Several observations suggest that the current is mediated by the transporter itself and not by oocyte-endogenous ion channels. At 1 mM, for example, currents induced by asparagine and glutamine are similar, although asparagine is transported at a much lower rate than glutamine.[79] If oocyte endogenous channels were to be activated by alkalinization of the oocyte, smaller currents would be expected for asparagine than for glutamine. It has been suggested that the currents are generated by a H$^+$ conductance; however, the reversal potential of substrate-induced currents were found to shift with pH in one study but not in another study.[72,79] The amplitude of the currents is strongly Na$^+$ dependent as is the transport activity; however, their reversal potential is unaffected by changes of the Na$^+$ concentration, suggesting that they are also not mediated by Na$^+$ ions.[72,79] The substrate-induced currents are rather small, particularly in the case of glutamine, and therefore may not be of physiological relevance.

The transporter has a strong pH dependence, being more active at alkaline pH than at acidic pH. The pH dependence is not caused by the proton antiport but is caused by a modifier site.[72] The pH dependence of the proton antiport function is observed only at pH values > 9. Further evidence that the pH sensitivity is conferred by a pH modifier site is provided by the similar pH dependence of the related transporters SNAT1 and SNAT2, both of which do not antiport protons[62] (see Section 6.3.4).

The mechanism of SNAT5 appears to be similar to that of SNAT3. The transporter is Na^+ dependent, and substrate-induced inward currents are observed when expressed in oocytes.[80] Moreover, substrate uptake is accompanied by intracellular alkalinization, suggesting that it mediates Na^+/H^+ antiport, with Na^+ transport coupled to substrate translocation.[80] The Na^+ cotransport stoichiometry has not been established. In analogy to SNAT3, it appears likely that the transport mechanism is electroneutral and substrate-induced currents are generated by a cation conductance.

6.3.5.4 *Physiological Role*

Functional studies in cultured astrocytes indicate that SNAT3 activity does not contribute significantly to glutamine uptake (< 10% of the uptake activity),[54] which is in agreement with the low affinity of the transporter for its substrates. By contrast, 50% of the spontaneous glutamine efflux from astrocytes is mediated by SNAT3.[53] In the efflux direction, the low affinity is matched by intracellular glutamine concentrations. Moreover, SNAT3 is specific for glutamine and histidine, which prevents interference by other amino acids. The transporter is likely to be important for efflux also *in situ*.[84] Glutamine efflux via SNAT3 is regulated by the extracellular pH and the intracellular Na^+ concentration.[72] Uptake of glutamate or GABA by astrocytes increases intracellular Na^+ concentration, which will foster glutamine efflux. It has to be pointed out, however, that expressions of glutamate transporters and glutamine transporters do not always coincide. Bergmann glia cells in the cerebellum, for example, express only small amounts of SNAT3 but large amounts of glutamate transporters.[87] This result suggests that other routes of glutamine release exist, which is supported by functional studies.[53] It appears likely that release of glutamine via SNAT3 is more relevant for GABA recycling than for glutamate recycling as suggested by the predominant localization close to GABAergic synapses.

The physiological role of SNAT5 remains to be defined. The transcript is present in brain[80] and a transport activity similar to SNAT5 has been detected in cultured neurons,[54] but the mRNA was notably absent in this preparation.

6.3.6   4F2hc/y⁺LAT2 (System y⁺L)

6.3.6.1 *Molecular Characteristics*

Similar to xCT, LAT1, and LAT2, y⁺LAT2 is a light chain associated with 4F2 (see References 47 and 48 for a review). Human y⁺LAT2 is a membrane protein of 515 amino acids with 12 transmembrane helices. Surface expression of y⁺LAT2 requires coexpression of the 4F2 heavy chain.

6.3.6.2 *Mechanism*

The 4F2/y⁺LAT2 heterodimer, like the other members of the family, is an obligatory antiporter of neutral and cationic amino acids.[85] To balance the charge of the different substrates, neutral amino acids are cotransported together with Na^+. As a result, the overall exchange mechanism is electroneutral. The participation of Na^+ in the transport mechanism influences the directionality of the transporter. The low intracellular Na^+ concentration disables efflux of neutral amino acids. y⁺LAT2 is thought to mediate the release of cationic amino acids in exchange against neutral amino acids. Thus glutamine is a good substrate for uptake via y⁺LAT2 but is not released by this transporter.[85]

6.3.6.3 *Physiological Role*

Severalfold evidence suggests that y⁺LAT2 plays only a limited role in glutamine transport. First, due to the coupling of neutral amino acid transport to Na^+ cotransport, the preferred direction of the transporter is uptake of neutral amino acids and release of cationic amino acids.[85] Second, cultured astrocytes express reasonable amounts of y⁺LAT2 mRNA, but in adult brain the transcript is barely detectable. The transporter may play a role in the release of arginine from selected cell populations or perhaps during embryonic development. The transporter appears not to be important for the glutamate–glutamine cycle.

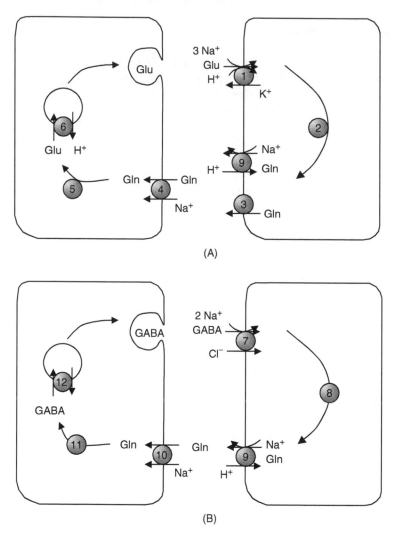

Figure 6.1 Neurotransmitter metabolism in glutamatergic and GABAergic cells. Schematic presentation of major steps in neurotransmitter recycling and intercellular exchange of metabolites. (A) Glutamate metabolism: Glutamate is released from glutamatergic synapses during exocytosis and is taken up largely by glial glutamate transporters EAAT1 and EAAT2 (1). Inside astrocytes, glutamate is converted into glutamine by glutamine synthetase (2). Glutamine is released from astrocytes either by SNAT3 or by an unknown transporter, perhaps system "n" (3). Subsequently, glutamine is taken up by neuronal transporters such as SNAT1, system Nᵇ, or a transporter related to SNAT5 (4). Inside neurons, glutamine is hydrolyzed to glutamate by glutaminase (5), which is then loaded into synaptic vesicles by vesicular glutamate transporters (6). (B) GABA metabolism: GABA is released from GABAergic synapses during exocytosis and is taken up by glial GABA transporters (7). Inside astrocytes, GABA is converted into glutamine by a series of reactions involving part of the TCA cycle and glutamine synthetase (8). Glutamine is released from astrocytes by SNAT3 (9). Subsequently, glutamine is taken up by neuronal transporters such as SNAT1 (10). Inside neurons, glutamine is converted into GABA by glutaminase and glutamate decarboxylase (11). GABA is then loaded into synaptic vesicles by vesicular GABA transporters (12).

6.4 TRANSPORTERS NOT YET DEFINED AT THE MOLECULAR LEVEL

Investigation of glutamine efflux in cultured astrocytes showed that depletion of intracellular Na^+ abolished glutamine efflux via SNAT3 but left 50% of glutamine release activity uninhibited.[53] The remaining Na^+-independent activity appears to be glutamine-specific. This efflux pathway has characteristics similar to "system n," which has been described in liver cells[86] and the blood–brain barrier.[87] Further evidence for additional glutamine release pathways has been provided by immunohistochemical studies of SNAT3 distribution. These data indicate that SNAT3 is predominantly located in the vicinity of GABAergic and even glycinergic neurons. Many glutamatergic synapses, particularly in the cerebellum, by contrast, are surrounded by astrocyte processes devoid of SNAT3 immunoreactivity. Provided that both glutamate and GABA are recycled with the help of astrocytes, it is tempting to speculate that other transporters are involved in recycling.

6.5 SUMMARY

Elucidation of glutamine transport activities in cultured astrocytes combined with immunohistochemical distribution of cloned transporters suggests that the GABA–glutamine cycle makes use of different transporters than the glutamate–glutamine cycle (Figure 6.1). In the GABA–glutamine cycle, GABA is taken up into astrocytes by members of the GAT family. It is then converted into glutamine via the TCA cycle and glutamine is subsequently released via SNAT3. Glutamine is taken up into GABAergic neurons by SNAT1. The glutamate–glutamine cycle is less clear. In the cerebellum, glutamate is taken up into astrocytes by EAAT1 (in the forebrain by EAAT2) and subsequently converted into glutamine. In Bergmann-glia cells of the cerebellum, SNAT3 is lacking. Glutamine release may be mediated by system n, which has not yet been identified on a molecular basis. SNAT1 appears to be involved in glutamine uptake in glutamatergic neurons; however, other systems such as N^B or SNAT5-related transporters are likely to contribute to neuronal glutamine uptake.

LIST OF ABBREVIATIONS

GABA — γ-Aminobutyric acid
NMR — Nuclear magnetic resonance
cDNA — Complementary DNA
MeAIB — N-Methyl-aminoisobutyric acid
BCH — 2-Aminobicyclo[2,2,1]heptane-2-carboxylic acid
RT-PCR — Reverse-transcriptase reaction followed by polymerase chain reaction

REFERENCES

1. Van den Berg, C.J., Krzalic, L., Mela, P., and Waelsch, H., Compartmentation of glutamate metabolism in brain. Evidence for the existence of two different tricarboxylic acid cycles in brain, *Biochem. J.*, 113, 281–290, 1969.
2. Martinez-Hernandez, A., Bell, K.P., and Norenberg, M.D., Glutamine synthetase: glial localization in brain, *Science*, 195, 1356–1358, 1977.
3. van den Berg, C.J. and Garfinkel, D., A stimulation study of brain compartments. Metabolism of glutamate and related substances in mouse brain, *Biochem. J.*, 123, 211–218, 1971.
4. Ottersen, O.P., Zhang, N., and Walberg, F., Metabolic compartmentation of glutamate and glutamine: morphological evidence obtained by quantitative immunocytochemistry in rat cerebellum, *Neuroscience*, 46, 519–534, 1992.

5. Chaudhry, F.A., Lehre, K.P., van Lookeren Campagne, M., Ottersen, O.P., Danbolt, N.C., and Storm-Mathisen, J., Glutamate transporters in glial plasma membranes: highly differentiated localizations revealed by quantitative ultrastructural immunocytochemistry, *Neuron*, 15, 711–720, 1995.

6. Bergles, D.E. and Jahr, C.E., Glial contribution to glutamate uptake at Schaffer collateral-commissural synapses in the hippocampus, *J. Neurosci.*, 18, 7709–7716, 1998.

7. Albrecht, J., L-Glutamate stimulates the efflux of newly taken up glutamine from astroglia but not from synaptosomes of the rat, *Neuropharmacology*, 28, 885–887, 1989.

8. Laake, J.H., Slyngstad, T.A., Haug, F.M., and Ottersen, O.P., Glutamine from glial cells is essential for the maintenance of the nerve terminal pool of glutamate: immunogold evidence from hippocampal slice cultures, *J. Neurochem.*, 65, 871–881, 1995.

9. Pow, D.V. and Robinson, S.R., Glutamate in some retinal neurons is derived solely from glia, *Neuroscience*, 60, 355–366, 1994.

10. Bacci, A., Sancini, G., Verderio, C., Armano, S., Pravettoni, E., Fesce, R., Franceschetti, S., and Matteoli, M., Block of glutamate-glutamine cycle between astrocytes and neurons inhibits epileptiform activity in hippocampus, *J. Neurophysiol.*, 88, 2302–2310, 2002.

11. Rae, C., Hare, N., Bubb, W.A., McEwan, S.R., Broer, A., McQuillan, J.A., Balcar, V.J., Conigrave, A.D., and Broer, S., Inhibition of glutamine transport depletes glutamate and GABA neurotransmitter pools: further evidence for metabolic compartmentation, *J. Neurochem.*, 85, 503–514, 2003.

12. Sibson, N.R., Dhankhar, A., Mason, G.F., Rothman, D.L., Behar, K.L., and Shulman, R.G., Stoichiometric coupling of brain glucose metabolism and glutamatergic neuronal activity, *Proc. Natl. Acad. Sci. U.S.A.*, 95, 316–321, 1998.

13. Grunewald, M. and Kanner, B.I., The accessibility of a novel reentrant loop of the glutamate transporter GLT-1 is restricted by its substrate, *J. Biol. Chem.*, 275, 9684–9689, 2000.

14. Seal, R.P. and Amara, S.G., A reentrant loop domain in the glutamate carrier EAAT1 participates in substrate binding and translocation, *Neuron*, 21, 1487–1498, 1998.

15. Kanner, B.I. and Borre, L., The dual-function glutamate transporters: structure and molecular characterisation of the substrate-binding sites, *Biochim. Biophys. Acta*, 1555, 92–95, 2002.

16. Vandenberg, R.J., Molecular pharmacology and physiology of glutamate transporters in the central nervous system, *Clin. Exp. Pharmacol. Physiol.*, 25, 393–400, 1998.

17. Danbolt, N.C., Glutamate uptake, *Prog. Neurobiol.*, 65, 1–105, 2001.

18. Lehre, K.P. and Danbolt, N.C., The number of glutamate transporter subtype molecules at glutamatergic synapses: chemical and stereological quantification in young adult rat brain, *J. Neurosci.*, 18, 8751–8757, 1998.

19. Kanner, B.I. and Schuldiner, S., Mechanism of transport and storage of neurotransmitters, *CRC Crit. Rev. Biochem.*, 22, 1–38, 1987.

20. Zerangue, N. and Kavanaugh, M.P., Flux coupling in a neuronal glutamate transporter, *Nature*, 383, 634–637, 1996.

21. Levy, L.M., Warr, O., and Attwell, D., Stoichiometry of the glial glutamate transporter GLT-1 expressed inducibly in a Chinese hamster ovary cell line selected for low endogenous Na+-dependent glutamate uptake, *J. Neurosci.*, 18, 9620–9628, 1998.

22. Kavanaugh, M.P., Neurotransmitter transport: models in flux, *Proc. Natl. Acad. Sci. U.S.A.*, 95, 12737–12738, 1998.

23. Rossi, D.J., Oshima, T., and Attwell, D., Glutamate release in severe brain ischaemia is mainly by reversed uptake, *Nature*, 403, 316–321, 2000.

24. Otis, T.S. and Kavanaugh, M.P., Isolation of current components and partial reaction cycles in the glial glutamate transporter EAAT2, *J. Neurosci.*, 20, 2749–2757, 2000.

25. Wadiche, J.I. and Kavanaugh, M.P., Macroscopic and microscopic properties of a cloned glutamate transporter/chloride channel, *J. Neurosci.*, 18, 7650–7661, 1998.

26. Fairman, W.A. and Amara, S.G., Functional diversity of excitatory amino acid transporters: ion channel and transport modes, *Am. J. Physiol.*, 277, F481–F486, 1999.

27. Vandenberg, R.J., Mitrovic, A.D., and Johnston, G.A., Molecular basis for differential inhibition of glutamate transporter subtypes by zinc ions, *Mol. Pharmacol.*, 54, 189–196, 1998.

28. Wadiche, J.I., Amara, S.G., and Kavanaugh, M.P., Ion fluxes associated with excitatory amino acid transport, *Neuron*, 15, 721–728, 1995.

29. Bergles, D.E., Diamond, J.S., and Jahr, C.E., Clearance of glutamate inside the synapse and beyond, *Curr. Opin. Neurobiol.*, 9, 293–298, 1999.

30. Ventura, R. and Harris, K.M., Three-dimensional relationships between hippocampal synapses and astrocytes, *J. Neurosci.*, 19, 6897–6906, 1999.

31. Tanaka, K., Watase, K., Manabe, T., Yamada, K., Watanabe, M., Takahashi, K., Iwama, H., Nishikawa, T., Ichihara, N., Kikuchi, T., Okuyama, S., Kawashima, N., Hori, S., Takimoto, M., and Wada, K., Epilepsy and exacerbation of brain injury in mice lacking the glutamate transporter GLT-1, *Science*, 276, 1699–1702, 1997.

32. Watase, K., Hashimoto, K., Kano, M., Yamada, K., Watanabe, M., Inoue, Y., Okuyama, S., Sakagawa, T., Ogawa, S., Kawashima, N., Hori, S., Takimoto, M., Wada, K., and Tanaka, K., Motor discoordination and increased susceptibility to cerebellar injury in GLAST mutant mice, *Eur. J. Neurosci.*, 10, 976–988, 1998.

33. Peghini, P., Janzen, J., and Stoffel, W., Glutamate transporter EAAC-1-deficient mice develop dicarboxylic aminoaciduria and behavioral abnormalities but no neurodegeneration, *EMBO J.*, 16, 3822–3832, 1997.

34. Bannai, S., Transport of cystine and cysteine in mammalian cells, *Biochim. Biophys. Acta*, 779, 289–306, 1984.

35. Sato, H., Tamba, M., Ishii, T., and Bannai, S., Cloning and expression of a plasma membrane cystine/glutamate exchange transporter composed of two distinct proteins, *J. Biol. Chem.*, 274, 11455–11458, 1999.

36. Sato, H., Tamba, M., Okuno, S., Sato, K., Keino-Masu, K., Masu, M., and Bannai, S., Distribution of cystine/glutamate exchange transporter, system x(c)-, in the mouse brain, *J. Neurosci.*, 22, 8028–8033, 2002.

37. Sagara, J., Miura, K., and Bannai, S., Cystine uptake and glutathione level in fetal brain cells in primary culture and in suspension, *J. Neurochem.*, 61, 1667–1671, 1993.

38. Pow, D.V., Visualising the activity of the cystine-glutamate antiporter in glial cells using antibodies to aminoadipic acid, a selectively transported substrate, *Glia*, 34, 27–38, 2001.

39. Christensen, H.N., Role of amino acid transport and countertransport in nutrition and metabolism, *Physiol. Rev.*, 70, 43–77, 1990.

40. Palacin, M., Estevez, R., Bertran, J., and Zorzano, A., Molecular biology of mammalian plasma membrane amino acid transporters, *Physiol. Rev.*, 78, 969–1054, 1998.

41. Christensen, H.N., Methods for distinguishing amino acid transport systems of a given cell or tissue, *Fed. Proc.*, 25, 850–853, 1966.

42. Broer, S., Adaptation of plasma membrane amino acid transport mechanisms to physiological demands, *Pflugers Arch.*, 444, 457–466, 2002.

43. Broer, A., Brookes, N., Ganapathy, V., Dimmer, K.S., Wagner, C.A., Lang, F., and Broer, S., The astroglial ASCT2 amino acid transporter as a mediator of glutamine efflux, *J. Neurochem.*, 73, 2184–2194, 1999.

44. Utsunomiya-Tate, N., Endou, H., and Kanai, Y., Cloning and functional characterization of a system ASC-like Na+-dependent neutral amino acid transporter, *J. Biol. Chem.*, 271, 14883–14890, 1996.

45. Broer, A., Wagner, C., Lang, F., and Broer, S., Neutral amino acid transporter ASCT2 displays substrate-induced Na+ exchange and a substrate-gated anion conductance, *Biochem. J.*, 346, 705–710, 2000.

46. Fuchs, B.C., Perez, J.C., Suetterlin, J.E., Chaudhry, S.B., and Bode, B.P., Inducible antisense RNA targeting amino acid transporter ATB0/ASCT2 elicits apoptosis in human hepatoma cells, *Am. J. Physiol. Gastrointest. Liver Physiol.*, 2003.

47. Wagner, C.A., Lang, F., and Broer, S., Function and structure of heterodimeric amino acid transporters, *Am. J. Physiol. Cell Physiol.*, 281, C1077–C1093, 2001.

48. Verrey, F., Meier, C., Rossier, G., and Kuhn, L.C., Glycoprotein-associated amino acid exchangers: broadening the range of transport specificity, *Pflugers Arch.*, 440, 503–512, 2000.

49. Nakamura, E., Sato, M., Yang, H., Miyagawa, F., Harasaki, M., Tomita, K., Matsuoka, S., Noma, A., Iwai, K., and Minato, N., 4F2 (CD98) heavy chain is associated covalently with an amino acid transporter and controls intracellular trafficking and membrane topology of 4F2 heterodimer, *J. Biol. Chem.*, 274, 3009–3016, 1999.

50. Pfeiffer, R., Spindler, B., Loffing, J., Skelly, P.J., Shoemaker, C.B., and Verrey, F., Functional heterodimeric amino acid transporters lacking cysteine residues involved in disulfide bond, *FEBS Lett.*, 439, 157–162, 1998.

51. Kanai, Y., Segawa, H., Miyamoto, K., Uchino, H., Takeda, E., and Endou, H., Expression cloning and characterization of a transporter for large neutral amino acids activated by the heavy chain of 4F2 antigen (CD98), *J. Biol. Chem.*, 273, 23629–23632, 1998.

52. Segawa, H., Fukasawa, Y., Miyamoto, K., Takeda, E., Endou, H., and Kanai, Y., Identification and functional characterization of a Na+-independent neutral amino acid transporter with broad substrate selectivity, *J. Biol. Chem.*, 274, 19745–19751, 1999.

53. Deitmer, J.W., Broer, A., and Broer, S., Glutamine efflux from astrocytes is mediated by multiple pathways, *J. Neurochem.*, 87, 127–135, 2003.

54. Heckel, T., Broer, A., Wiesinger, H., Lang, F., and Broer, S., Asymmetry of glutamine transporters in cultured neural cells, *Neurochem. Int.*, 43, 289–298, 2003.

55. Kido, Y., Tamai, I., Uchino, H., Suzuki, F., Sai, Y., and Tsuji, A., Molecular and functional identification of large neutral amino acid transporters LAT1 and LAT2 and their pharmacological relevance at the blood-brain barrier, *J. Pharm. Pharmacol.*, 53, 497–503, 2001.

56. Matsuo, H., Tsukada, S., Nakata, T., Chairoungdua, A., Kim, D.K., Cha, S.H., Inatomi, J., Yorifuji, H., Fukuda, J., Endou, H., and Kanai, Y., Expression of a system L neutral amino acid transporter at the blood-brain barrier, *Neuroreport*, 11, 3507–3511, 2000.

57. Rossier, G., Meier, C., Bauch, C., Summa, V., Sordat, B., Verrey, F., and Kuhn, L.C., LAT2, a new basolateral 4F2hc/CD98-associated amino acid transporter of kidney and intestine, *J. Biol. Chem.*, 274, 34948–34954, 1999.

58. Meier, C., Ristic, Z., Klauser, S., and Verrey, F., Activation of system L heterodimeric amino acid exchangers by intracellular substrates, *EMBO J.*, 21, 580–589, 2002.

59. Yudkoff, M., Daikhin, Y., Grunstein, L., Nissim, I., Stern, J., Pleasure, D., and Nissim, I., Astrocyte leucine metabolism: significance of branched-chain amino acid transamination, *J. Neurochem.*, 66, 378–385, 1996.

60. Zwingmann, C., Richter-Landsberg, C., Brand, A., and Leibfritz, D., NMR spectroscopic study on the metabolic fate of [3-(13)C]alanine in astrocytes, neurons, and cocultures: implications for glia-neuron interactions in neurotransmitter metabolism, *Glia*, 32, 286–303, 2000.

61. Mackenzie, B. and Erickson, J.D., Sodium-coupled neutral amino acid (system N/A) transporters of the SLC38 gene family, *Pflugers Arch.*, 2003.

62. Albers, A., Broer, A., Wagner, C.A., Setiawan, I., Lang, P.A., Kranz, E.U., Lang, F., and Broer, S., Na+ transport by the neural glutamine transporter ATA1, *Pflugers Arch.*, 443, 92–101, 2001.

63. Mackenzie, B., Schafer, M.K., Erickson, J.D., Hediger, M.A., Weihe, E., and Varoqui, H., Functional properties and cellular distribution of the system A glutamine transporter SNAT1 support specialized roles in central neurons, *J. Biol. Chem.*, 278, 23720–23730, 2003.

64. Yao, D., Mackenzie, B., Ming, H., Varoqui, H., Zhu, H., Hediger, M.A., and Erickson, J.D., A novel system A isoform mediating Na+/neutral amino acid cotransport, *J. Biol. Chem.*, 275, 22790–22797, 2000.

65. Varoqui, H., Zhu, H., Yao, D., Ming, H., and Erickson, J.D., Cloning and functional identification of a neuronal glutamine transporter, *J. Biol. Chem.*, 275, 4049–4054, 2000.

66. Sugawara, M., Nakanishi, T., Fei, Y.J., Huang, W., Ganapathy, M.E., Leibach, F.H., and Ganapathy, V., Cloning of an amino acid transporter with functional characteristics and tissue expression pattern identical to that of system A, *J. Biol. Chem.*, 275, 16473–16477, 2000.

67. Weiss, M.D., Derazi, S., Rossignol, C., Varoqui, H., Erickson, J.D., Kilberg, M.S., and Anderson, K.J., Ontogeny of the neutral amino acid transporter SAT1/ATA1 in rat brain, *Brain Res. Dev. Brain Res.*, 143, 151–159, 2003.

68. Reimer, R.J., Chaudhry, F.A., Gray, A.T., and Edwards, R.H., Amino acid transport system A resembles system N in sequence but differs in mechanism, *Proc. Natl. Acad. Sci. U.S.A.*, 97, 7715–7720, 2000.

69. Brookes, N., Neutral amino acid transport in astrocytes: characterization of Na+-dependent and Na+-independent components of alpha-aminoisobutyric acid uptake, *J. Neurochem.*, 51, 1913–1918, 1988.

70. Chaudhry, F.A., Schmitz, D., Reimer, R.J., Larsson, P., Gray, A.T., Nicoll, R., Kavanaugh, M., and Edwards, R.H., Glutamine uptake by neurons: interaction of protons with system a transporters, *J. Neurosci.*, 22, 62–72, 2002.

71. Hatanaka, T., Huang, W., Wang, H., Sugawara, M., Prasad, P.D., Leibach, F.H., and Ganapathy, V., Primary structure, functional characteristics and tissue expression pattern of human ATA2, a subtype of amino acid transport system A, *Biochim. Biophys. Acta*, 1467, 1–6, 2000.

72. Broer, A., Albers, A., Setiawan, I., Edwards, R.H., Chaudhry, F.A., Lang, F., Wagner, C.A., and Broer, S., Regulation of the glutamine transporter SN1 by extracellular pH and intracellular sodium ions, *J. Physiol.*, 539, 3–14, 2002.

73. Tamarappoo, B.K., Raizada, M.K., and Kilberg, M.S., Identification of a system N-like Na(+)-dependent glutamine transport activity in rat brain neurons, *J. Neurochem.*, 68, 954–960, 1997.

74. Chaudhry, F.A., Reimer, R.J., Krizaj, D., Barber, D., Storm-Mathisen, J., Copenhagen, D.R., and Edwards, R.H., Molecular analysis of system N suggests novel physiological roles in nitrogen metabolism and synaptic transmission, *Cell*, 99, 769–780, 1999.

75. Gu, S., Roderick, H.L., Camacho, P., and Jiang, J.X., Identification and characterization of an amino acid transporter expressed differentially in liver, *Proc. Natl. Acad. Sci. U.S.A.*, 97, 3230–3235, 2000.

76. Nagaraja, T.N. and Brookes, N., Glutamine transport in mouse cerebral astrocytes, *J. Neurochem.*, 66, 1665–1674, 1996.

77. Broer, S. and Brookes, N., Transfer of glutamine between astrocytes and neurons, *J. Neurochem.*, 77, 705–719, 2001.

78. Fei, Y.J., Sugawara, M., Nakanishi, T., Huang, W., Wang, H., Prasad, P.D., Leibach, F.H., and Ganapathy, V., Primary structure, genomic organization, and functional and electrogenic characteristics of human system N 1, a Na+- and H+-coupled glutamine transporter, *J. Biol. Chem.*, 275, 23707–23717, 2000.

79. Chaudhry, F.A., Krizaj, D., Larsson, P., Reimer, R.J., Wreden, C., Storm-Mathisen, J., Copenhagen, D., Kavanaugh, M., and Edwards, R.H., Coupled and uncoupled proton movement by amino acid transport system N, *EMBO J.*, 20, 7041–7051, 2001.

80. Nakanishi, T., Kekuda, R., Fei, Y.J., Hatanaka, T., Sugawara, M., Martindale, R.G., Leibach, F.H., Prasad, P.D., and Ganapathy, V., Cloning and functional characterization of a new subtype of the amino acid transport system N, *Am. J. Physiol. Cell Physiol.*, 281, C1757–C1768, 2001.

81. Nakanishi, T., Sugawara, M., Huang, W., Martindale, R.G., Leibach, F.H., Ganapathy, M.E., Prasad, P.D., and Ganapathy, V., Structure, function, and tissue expression pattern of human SN2, a subtype of the amino acid transport system N, *Biochem. Biophys. Res. Commun.*, 281, 1343–1348, 2001.

82. Boulland, J.L., Osen, K.K., Levy, L.M., Danbolt, N.C., Edwards, R.H., Storm-Mathisen, J., and Chaudhry, F.A., Cell-specific expression of the glutamine transporter SN1 suggests differences in dependence on the glutamine cycle, *Eur. J. Neurosci.*, 15, 1615–1631, 2002.

83. Boulland, J.L., Rafiki, A., Levy, L.M., Storm-Mathisen, J., and Chaudhry, F.A., Highly differential expression of SN1, a bidirectional glutamine transporter, in astroglia and endothelium in the developing rat brain, *Glia*, 41, 260–275, 2003.

84. Chaudhry, F.A., Reimer, R.J., and Edwards, R.H., The glutamine commute: take the N line and transfer to the A, *J. Cell Biol.*, 157, 349–355, 2002.

85. Broer, A., Wagner, C.A., Lang, F., and Broer, S., The heterodimeric amino acid transporter 4F2hc/y+LAT2 mediates arginine efflux in exchange with glutamine, *Biochem. J.*, 349, 787–795, 2000.

86. Pacitti, A.J., Inoue, Y., and Souba, W.W., Characterization of Na(+)-independent glutamine transport in rat liver, *Am. J. Physiol.*, 265, G90–G98, 1993.

87. Lee, W.J., Hawkins, R.A., Vina, J.R., and Peterson, D.R., Glutamine transport by the blood-brain barrier: a possible mechanism for nitrogen removal, *Am. J. Physiol.*, 274, C1101–C1107, 1998.

CHAPTER **7**

Neuronal–Astrocytic Interactions (TCA Cycling)

Ursula Sonnewald, Arne Schousboe, and Helle S. Waagepetersen

CONTENTS

7.1 COMPARTMENTATION OF BRAIN METABOLISM

The brains of all vertebrates contain both neurons and glial cells, and the more evolved the organism the higher is the ratio of astrocytes to neurons (see Hertz and Schousboe[1]). It is well known that there are different types of neurons, classified according to their neurotransmitter content. Furthermore, there are different types of glial cells: astrocytes, oligodendrocytes, and microglia. It is commonly believed that astrocytes are a homogeneous cell type, but increasing evidence suggests that heterogeneity also exists among these cells, particularly in different brain regions.[2] The classic concept of astrocytes acting as a scaffold, keeping neurons in their proper location, has during the last decades been replaced with the more dynamic view that astrocytes actively maintain and support neuronal activity.[3,4] The present chapter will focus on metabolic interactions between astrocytes and glutamatergic or GABAergic neurons from cerebellum and cerebral cortex, respectively.

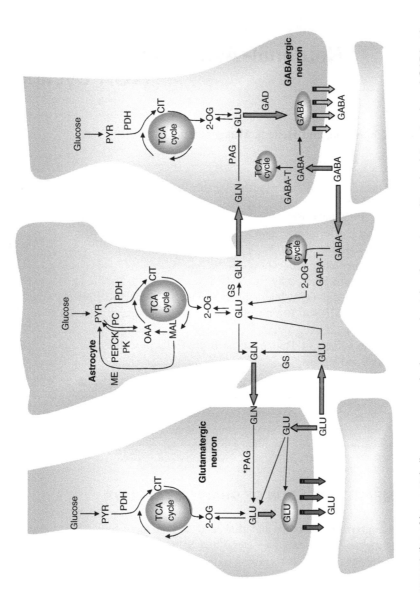

Figure 7.1   Schematic representation of key metabolic processes and release and uptake of neurotransmitters in glutamatergic and GABAergic synapses interacting with a surrounding astrocyte. The vesicular pools of glutamate and GABA are highlighted by ellipses. In the neuronal compartments, glucose and TCA cycle metabolism is indicated. Moreover, the glutamate–glutamine cycle including the glutamine synthetase (GS) reaction is indicated in the glutamatergic neuron–astrocyte interaction. Analogously, the GABA–glutamate–glutamine cycle including the GABA transaminase (GABA-T) and glutamate decarboxylase (GAD) reactions is indicated in the GABAergic neuron–astrocyte interaction. In the astrocytic compartment, pyruvate carboxylation to oxaloacetate via pyruvate carboxylase (PC) is indicated. Additionally, pyruvate recycling by conversion of malate to pyruvate catalyzed by malic enzyme (ME) is shown. One arrow does not imply one reaction only. Abbreviations: CIT, citrate; GAD, glutamate decarboxylase; GABA-T, GABA transaminase; GLN, glutamine; GLU, glutamate; GS, glutamine synthetase; MAL, malate; ME, malic enzyme; 2-OG, 2-oxoglutarate; PAG, phosphate-activated glutaminase; PC, pyruvate carboxylase; PDH, pyruvate dehydrogenase; PYR, pyruvate; TCA, tricarboxylic acid.

The original indication of metabolic compartmentation was based on the observation that a specific labeling of a metabolic product exceeds that of its precursor.[5] Early studies of glutamate and glutamine metabolism in the brain, using ^{14}C-labeled precursors such as glucose, acetate, and bicarbonate, clearly suggested that the metabolism of these amino acids had to take place in different compartments, i.e., astrocytes and neurons.[5,6] Subsequently, the concept of exchange of metabolites between these compartments was developed, advancing the proposal of a glutamate–glutamine cycle, metabolically linking glutamatergic neurons and astrocytes (for schematic representation, see Figure 7.1).[7,8]

7.2 THE GLUTAMATE–GLUTAMINE CYCLE

An important function of astrocytes is to supply glutamatergic neurons with metabolic intermediates, since the neuronal metabolite pool is continuously drained by neurotransmitter release. It should be noted that net synthesis of TCA cycle intermediates, and thus compounds such as glutamate, GABA, and glutamine, depends on an anaplerotic pathway. In the brain, this is preferentially or exclusively achieved by the main anaplerotic enzyme, pyruvate carboxylase (PC), which has a glial localization.[9–11] The carboxylation of pyruvate with carbon dioxide permits *de novo* synthesis of oxaloacetate, which may react with acetyl coenzyme A (acetyl CoA) to provide net synthesis of the TCA cycle intermediate 2-oxoglutarate, from which glutamate can be formed via transamination or NADH-requiring reductive amination by glutamate dehydrogenase.[12] However, the level of glutamate in astrocytes is low compared to neurons.[13] Glutamate is converted to glutamine via glutamine synthetase (GS), which, like PC, is exclusively expressed in glial cells.[14,15] Glutamine is released from astrocytes via astrocyte-specific glutamine transporters[16] and subsequently taken up into neurons by high-affinity glutamine transporters characteristic for these cells.[17] In the neuron, glutamine is converted to glutamate by phosphate-activated glutaminase (PAG).[18] Using postembedding immunogold labeling, high intensities for PAG were found in mitochondria of perikarya and dendrites in most glutamatergic neurons of the cerebellum, whereas nonglutamatergic neurons (Purkinje and Golgi cells) exhibited less labeling for PAG. Glial cell mitochondria were devoid of specific PAG labeling and revealed a much lower glutamate–glutamine ratio than did the mitochondria of mossy fibers.[19] Glutamine from astrocytes is used by the glutamatergic neurons for both neurotransmitter synthesis and energy production.[20] In spite of the above findings, it may be noted that PAG has been found in primary cultures of astrocytes, albeit at a level of activity lower than that in neurons.[21,22]

The major part of glutamate is found in glutamatergic neurons from which it is released during depolarization in a Ca^{2+}-dependent manner.[13,23] Synaptically released glutamate is primarily taken up by astrocytes[24–26] due to the predominant location on these cells of high-affinity glutamate transporters.[27–29] Uptake of glutamate is an energy-demanding process requiring one molecule of adenosine triphosphate (ATP) for uptake of one molecule of glutamate.[30] In addition, the direct conversion of glutamate to glutamine by GS is also ATP dependent with a similar stoichiometry. In order to enter the TCA cycle, glutamate must be transported into mitochondria to be converted to 2-oxoglutarate and subsequently to oxaloacetate producing 9–12 molecules of ATP. The activities of glutamate dehydrogenase and amino transferases have been shown to be high in astrocytes. These enzymes are responsible for conversion of glutamate to 2-oxoglutarate, which may be further metabolized in the TCA cycle.[31–34]

7.3 THE GABA–GLUTAMATE–GLUTAMINE CYCLE

In GABAergic neurons, glutamate is converted to GABA by glutamate decarboxylase (GAD), and glutamate is present in a relatively low concentration.[13,35] Synaptically released GABA is taken up primarily by GABAergic neurons, and to a lesser extent by astrocytes.[36–38] This is schematically represented in Figure 7.1. GABA is metabolized by GABA-transaminase (GABA-T) in the so-called GABA shunt[39] (Figure 7.1), which allows four of the five C atoms from 2-oxoglutarate to reenter the TCA cycle as succinate. GABA-T is a ubiquitous enzyme present in both neurons and astrocytes, making astrocytes and neurons equally capable of metabolizing GABA.[40]

It is well known that astrocytes are intimately involved in glutamate neurotransmission. However, the importance of astrocytic glutamine in GABA synthesis is controversial,[41,42] but both *in vitro* and *in vivo* data imply that glutamine is an important precursor for GABA synthesis.[43–46] Thus, the glutamate–glutamine cycle concept is extended to a GABA–glutamate–glutamine cycle, as seen in Figure 7.1. In keeping with the concept that the drain of GABA from neurons to astrocytes is relatively modest compared with the corresponding drain of glutamate from glutamatergic neurons,[36,47,48] glutamine transport has been shown to be more intense in glutamatergic neurons than in neocortical GABAergic neurons.[49]

7.4 THE INTRACELLULAR COMPARTMENTATION CONCEPT

Mitochondria contain the respiratory chain enzyme complexes that carry out oxidative phosphorylation and produce the main part of cellular energy in the form of ATP. It is conceivable that different types of mitochondria might exist within the same cell due to the diversity of functions.[50] The existence of mitochondrial heterogeneity would expand the number of possible intracellular compartments. Mitochondrial heterogeneity is a well-known phenomenon in patients with defects of the mitochondrial genome (for review, see Taylor et al.[51]). Synaptic and nonsynaptic mitochondria isolated from a homogenate of adult rat brain exhibit differences in the activity of a number of TCA cycle enzymes.[52,53] Intramitochondrial compartmentation could also exist. Free diffusion within the mitochondria has been shown to be unlikely due to the high apparent viscosity of the mitochondrial matrix.[54] Using electron microscopy it has also been suggested that the inner membrane proteins might be compartmentalized.[55] This phenomenon would further increase the number of possible intracellular compartments. Results from several studies indicate that both neuronal and astrocytic metabolism are compartmentalized,[56–65] thus constituting further evidence that mitochondrial heterogeneity may exist. This notion is in line with recent studies demonstrating heterogeneity among mitochondrial populations with regard to expression of pyruvate and 2-oxoglutarate dehydrogenases.[66–68] Needless to say, elucidation of the cellular compartmentation of these neuroactive amino acids is fraught with enormous experimental challenges. However, since the biosynthetic machinery for both glutamate and GABA involves neurons as well as astrocytes at the cytoplasmic and mitochondrial levels, it is imperative to fully understand this compartmentation if the regulatory mechanisms for biosynthesis of these neurotransmitters are to be worked out in detail.

7.5 METABOLIC INTERACTIONS IN BRAIN STUDIED BY ^{13}C NMRS

Nuclear magnetic resonance spectroscopy (NMRS) has several appealing features for applications to metabolic studies. The nuclei that are most commonly used in NMRS for metabolic studies are ^{1}H, ^{31}P, and [13]C. [1]H and ^{31}P are naturally abundant isotopes and therefore constitute the most common basis for studies involving examination of differences in the natural abundance spectra. In contrast, ^{13}C has a natural abundance of 1.1%. This disadvantage normally makes its detection difficult, and ^{13}C NMRS is thus of limited use for studies of endogenous metabolites unless they occur in large amounts. However, the low natural abundance can be an advantage in that ^{13}C-enriched precursors can be used for metabolic pathway mapping with little or no background interference from metabolites endogenously labeled due to the natural abundance of [13]C. Thus, ^{13}C NMRS is a powerful tool for analysis of the metabolic trafficking between the heterogeneous cellular entities of the brain.

Information about biochemical pathways can be gained using ^{13}C-labeled substances.[69–72] These labeling experiments are analogous to those utilizing conventional ^{14}C labeling, but additional information about the location of the label within the molecule can be obtained. Moreover, the potential for *in vivo* applicability exists. Using ^{13}C NMR spectra in a qualitative manner, it is possible to detect unexpected compounds. An example of this is the finding of citrate in the culture medium of astrocytes, explicitly showing that astrocytes release citrate.[73] A wealth of additional information is, however, gained by quantitative analysis of ^{13}C spectra. With specifically labeled precursors, it is possible to distinguish between neuronal and glial pathways. Acetate is selectively taken up by astrocytes since they contain a specialized transport system, which is absent or less active in neurons.[74] In contrast, acetyl CoA derived from glucose has been calculated to be metabolized more actively in the neuronal tricarboxylic acid (TCA) cycle in rats.[75] Thus, by simultaneous injection of [1-^{13}C]glucose and [1,2-^{13}C]acetate and NMRS analysis of brain extracts, information about neuronal and astrocytic metabolism can be obtained in the same animal.[76]

7.6 ^{13}C NMRS ANALYSES OF NEURONAL–GLIAL INTERACTIONS IN ANIMAL MODELS OF EPILEPSY

Epilepsy is one of the most common serious neurological disorders and is characterized by an imbalance of excitatory and inhibitory function mainly due to disturbed neurotransmitter metabolism.[77,78] Animal models of epilepsy are often based on inhibition of the synthesis of GABA. Such inhibition has been shown to promote seizures, as does the administration of GABA$_A$ receptor antagonists and glutamate receptor agonists.[79] In accordance with this, pentylenetetrazol (PTZ) is a frequently used chemical convulsant inducing tonic-clonic seizures.[80] It is generally believed that PTZ exerts its effects by binding to the picrotoxin-recognition site of GABA$_A$ receptors.[81] PTZ is known to decrease the effects of GABA and other inhibitory neurotransmitters, thus enhancing the probability of depolarization of neurons.[77] Kindling is another way to induce seizures.[80] The term *kindling* refers in this context to the continuous application of sub-threshold doses of PTZ leading to a behavior nearly identical to secondary generalized seizures,[82] but other kindling paradigms exist.[80] A model for temporal lobe epilepsy is based on injection of kainic acid, which produces complex partial seizures.[83] Results of NMRS studies using these three models are presented in the following.

Injection of PTZ (70 mg/kg body weight) followed 30 min later by [1-^{13}C]glucose plus [1,2-^{13}C]acetate and decapitation 15 min thereafter showed impairment of amino acid metabolism in glutamatergic neurons, whereas astrocyte metabolism was unchanged at this early postictal stage.[84] Kindling in mice was achieved by 20 intraperitoneal injections of PTZ (35 mg/kg) over a period of 40 days. Metabolism in astrocytes seemed to be impaired as indicated by a decreased labeling of glutamine both from [1-^{13}C]glucose and [1,2-^{13}C]acetate.[85] As mentioned above, astrocytes are closely involved in amino acid homeostasis and play a significant role in epileptogenic foci, where their proliferation (astrogliosis) is a well-known phenomenon.[86] This is especially true for limbic seizures or temporal lobe epilepsy and has been confirmed both in humans and in animal models.[87–89] Thus, limbic seizures are an interesting subject for evaluation by ^{13}C NMRS and, consequently, the model of kainate-induced limbic seizures was used to study metabolic interactions between neurons and glia in this type of epilepsy. In agreement with the results from the PTZ-induced seizures, astrocytic metabolism was impaired 1 day after injection of the glutamate receptor agonist kainate. On day 14, however, no changes were observed in astrocytic metabolites, but labeling from [1-^{13}C]glucose, i.e., reflecting mainly neuronal metabolism, was increased in glutamate, GABA, glutamine, aspartate, and succinate. Thus, it was concluded that turnover of metabolites in the model of kainate-induced limbic seizures is time dependent. Early and only temporary enhanced astrocytic metabolic activity was followed by altered metabolism in neurons with an increased turnover of important metabolites such as the neurotransmitters GABA and glutamate.

7.7 METABOLISM STUDIED BY ^{13}C NMRS IN BRAIN CELL CULTURE

The metabolic fate of glutamate has been studied in cultured neocortical astrocytes,[90,91] cerebellar astrocytes,[92] cerebellar granule neurons,[93] and neocortical neurons[94] using [U-^{13}C]glutamate and NMRS. The main routes for glutamate metabolism is conversion into glutamine, GABA, or peptides, such as GSH, or oxidative metabolism via the TCA cycle for energy production and synthesis of aspartate and various other metabolites including lactate (for review, see Sonnewald et al.[95]). In neocortical and cerebellar astrocytes, it could be shown that glutamate not only was converted to glutamine but to a large extent entered the TCA cycle. [13]C labeling was found in aspartate, glutamate, and glutamine after metabolism of the uniformly labeled carbon skeleton in the TCA cycle. Surprisingly, labeled lactate obtained from glutamate metabolism in the TCA cycle was detected in the medium from these cells.[90–92] McKenna et al.[96] have shown that the route of glutamate metabolism is concentration dependent, in the sense that more glutamate is consumed for direct formation of glutamine at low concentrations (0.01–0.1 mM), whereas more is metabolized via the TCA cycle at higher concentrations (0.2–0.5 mM).

In cerebellar granule cells, [U-^{13}C]glutamate was found to be metabolized in the TCA cycle since both aspartate and glutamate derived from precursors generated in the first and second turns of the TCA cycle were observed in cell extracts.[93] Minor amounts of uniformly labeled lactate were also found, but this could be explained by the small astrocytic contamination of the cultures[93] and thus may not reflect production in the granule neurons of pyruvate/lactate from malate, although these cells express malic enzyme.[63,93]

The metabolism of [U-^{13}C]glutamate in GABAergic cerebral cortical neurons has been found to be rather complex.[45,94] Thus, much more label was seen in aspartate than in GABA,[94] and GABA synthesis has been shown to occur from both a cytoplasmic and a TCA cycle–derived pool of glutamate.[45] Moreover, the glutamate pool, which in these neurons is associated with energy metabolism, is at least partly separated from that acting as a precursor for GABA.[45] It may therefore be concluded that neurons exhibit signs of metabolic compartmentation with regard to glutamate.[45,94]

7.8 PYRUVATE RECYCLING

Pyruvate recycling was first shown in the liver, where [2-^{14}C]pyruvate could be converted to [3-^{14}C]pyruvate and [1-^{14}C]pyruvate, a process that can occur only if pyruvate is incorporated into the TCA cycle via acetyl CoA and subsequently regenerated from TCA cycle constituents[97] (see Figure 7.1). Recycling of pyruvate in the brain has been demonstrated by Cerdan et al.,[98] who found that [1,2-^{13}C]acetate, a substrate that is specifically taken up and therefore metabolized in astrocytes (see above), can be converted in brain to monolabeled [1-^{13}C]- and [2-^{13}C]acetyl CoA and to glutamate labeled either in the C-4 or the C-5 position. This requires entry of acetate into the tricarboxylic acid (TCA) cycle (after formation of acetyl CoA) and exit of a TCA cycle intermediate to form pyruvate, which then is reintroduced into the TCA cycle. Based on the observation that this label from acetate was incorporated into glutamate but not into glutamine, it was concluded that pyruvate recycling took place in a compartment without glutamine synthetase activity, i.e., a neuronal but not an astrocytic compartment. Pyruvate recycling in the brain *in vivo* has been confirmed by Hassel et al.,[99] who found formation of labeled lactate from [2-^{13}C]acetate and of [2-^{13}C]lactate from [1-^{13}C]glucose, and based on a more pronounced formation of TCA cycle–derived lactate from labeled acetate than from labeled glucose, it was concluded that pyruvate recycling was likely to occur in the astrocytic rather than the neuronal compartment. Cell culture work has confirmed this conclusion. Using [U-^{13}C]glutamate or [3-^{13}C]glutamate, it could be shown that neocortical and cerebellar astrocytes are able to generate pyruvate from a TCA cycle intermediate (malate or oxaloacetate) and subsequently reintroduce the carbon skeleton of this pyruvate into the cycle as acetyl CoA.[100,101] In contrast, neocortical and cerebellar neurons did not use this pathway. Incubation of neocortical cocultures (neurons and astrocytes) with [3-^{13}C]glutamate did not show any upregulation of pyruvate recycling.[102] Lack of pyruvate, i.e., hypoglycemia, could presumably lead to activation of pyruvate recycling. However, this was observed neither in cerebellar granule neurons nor in neocortical astrocytes.[103,104]

7.9 ^{13}C NMRS ANALYSIS OF THE EFFECTS OF THIOPENTAL ON BRAIN CELLS IN CULTURE

NMRS can also be used to examine the effects of pharmacological agents on the metabolic pathways in brain cells in culture. In the following discussion, thiopental is used as an example. Barbiturates such as thiopental are intravenously injected anesthetics interacting with the $GABA_A$ receptor at the barbiturate-binding site, prolonging the opening of the Cl[-] channel and thus extending the hyperpolarizing effect of GABA (for review, see Ito et al.[105]).

7.9.1 Glutamate Uptake and Release

In neocortical astrocytes the amount of glutamate taken up was decreased during incubation with 1 mM thiopental.[106] In contrast, thiopental did not affect glutamate uptake in cultured cerebellar astrocytes and granule neurons.[92,106] Glutamate release can occur either via a vesicular mechanism or reversal of the plasma membrane transporters involving the cytoplasmic pool of glutamate.[106–109] In cerebellar granule neurons, thiopental did indeed decrease glutamate and aspartate release,[106] and in neocortical astrocytes, aspartate release was reduced.[91] These findings may be compatible with the previous demonstration that thiopental inhibits evoked glutamate release from synaptosomes.[110] However, since the paradigm used in the cerebellar granule neurons involved repetitive exposure to 200 μM glutamate in addition to a chronic exposure to 100 μM glutamate, it is likely to represent reversal of transport in an outward direction.[106] Thus, it appears that thiopental preferentially inhibits the glutamate transporters operating in the outward direction in cerebellar neurons and neocortical astrocytes, but not in cerebellar astrocytes.[91,92,106]

7.9.2 Glutamate Metabolism

[U-^{13}C]Glutamate was found to be metabolized via the same pathways in cerebellar and neocortical astrocytes.[91,92] However, the amounts of lactate formed from [U-^{13}C]glutamate, especially in the presence of 1 mM thiopental, were higher in cerebellar than in neocortical astrocytes, and alanine formation was only observed in cerebellar astrocytes. Such regional differences were also reported in earlier studies.[99,111,112] Furthermore, the distribution of glutamate into alternative metabolic pathways was different in the astrocyte cultures from these two regions. This underscores the importance of analyzing astrocytes from different parts of the brain. Regional differences in astrocyte function may reflect the fact that astrocytic properties with regard to expression of neurotransmitter transporters, for example, is influenced by the neuronal environment.[47] In cultured astrocytes, it may, however, be considered that differences in cell maturational stages could play a role.[113] In keeping with the observation that in cerebellar granule neurons and astrocytes thiopental had no effect on glutamate uptake, it was found that the total amount consumed was unaffected by thiopental. However, the amounts of most metabolites synthesized from [U-^{13}C]glutamate were increased in the presence of thiopental. This indicates that processes such as CO_2 production that are not detected by the present NMRS method may be decreased.[92,106] Khazanov and Saratikov[114] have shown that phenobarbital and benzonal depress energy production by rat brain mitochondria. Thus, it is not unlikely that the aforementioned results indicate a depression of the energy production by thiopental in mouse cerebellar cells in culture.

7.10 CONCLUDING REMARKS

Based on the original observation that the metabolism of glutamate in the brain must take place in separated pools (compartments) and that this most likely represents distinct metabolic patterns in neurons and glial cells (for references, see Waagepetersen et al.[68]), it is clear that exchange of metabolites involved in this metabolism must occur between these two cell types. As exemplified in this chapter, toxic drugs may be utilized as tools to obtain information about such exchange reactions, as well as the metabolic processes in the individual cell types. Moreover, the use of such drugs combined with ^{13}C labeling to follow the metabolic fate of individual metabolites and even individual C atoms has been instrumental in attempts to delineate a possible metabolic compartmentation at the cellular level occurring in neurons as well as glial cells. During the decade that has passed since the first indications of such cellular compartmentation,[57] it has become increasingly clear that such compartmentation does indeed exist. Future experimentation may lead to a better understanding of the functional implications of this phenomenon. One of the important challenges will be to elucidate the functional significance of regional diversity of astrocytes and the possibility that such diversity may be influenced by the neuronal environment. Any given individual synapse represents a unique microenvironment in which the functional status reflects the integration of such cross talk between the presynaptic, postsynaptic, and astrocytic entities.

ACKNOWLEDGMENT

This research was supported by the Danish MRC (22-00-0503 & 22-00-1011) and the Norwegian Epilepsy, Blix, Lundbeck, and NOVO Nordisk Foundations. The excellent technical assistance of Bente Urfjell is greatly appreciated.

LIST OF ABBREVIATIONS

GABA-T — γ-Aminobutyric acid transaminase
GAD — Glutamate decarboxylase
GS — Glutamine synthetase
NMR(S) — Nuclear magnetic resonance (spectroscopy)
PAG — Phosphate-activated glutaminase
PC — Pyruvate carboxylase
PDH — Pyruvate dehydrogenase
TCA — Tricarboxylic acid

REFERENCES

1. Hertz, L. and Schousboe, A., Ion and energy metabolism of the brain at the cellular level, *Int. Rev. Neurobiol.*, 18, 141, 1975.
2. Schousboe, A. and Divac, I., Difference in glutamate uptake in astrocytes cultured from different brain regions, *Brain Res.*, 177, 407, 1979.
3. Schousboe, A. and Waagepetersen, H., Role of astrocytes in homeostasis of glutamate and GABA during physiological and pathophysiological conditions, *Adv. Mol. Cell Biol.*, 31, 461, 2004.
4. Hansson, E. and Rönnbäck, L., Astrocytic receptors and second messenger systems, *Adv. Mol. Cell Biol.*, 31, 475, 2004.
5. Berl, S. and Clarke, D.D., The metabolic compartmentation concept, in *Glutamine, Glutamate and GABA in the Central Nervous System*, Hertz, L., Kvamme, E., McGeer, E.G., and Schousboe, A., Eds., Alan R. Liss, Inc., New York, 1983, p. 205.

6. Berl, S. and Clarke, D.D., Compartmentation of amino acid metabolism, in *Handbook of Neurochemistry*, 1st ed., Vol. 2, Lajtha, A., Ed., Plenum Press, New York, 1969, p. 447.

7. Van den Berg, C.J. and Garfinkel, D., A simulation study of brain compartments: Metabolism of glutamate and related substances in mouse brain, *Biochem. J.*, 23, 211, 1971.

8. Balázs, R., Patel, A.J., and Richter, D., Metabolic compartments in the brain: Their properties and relation to morphological structures, in *Metabolic Compartmentation in the Brain*, Balázs, R. and Cremer, J.E., Eds., McMillan, London, 1973, p. 167.

9. Yu, A.C.H. et al., Pyruvate carboxylase activity in primary cultures of astrocytes and neurons, *J. Neurochem.*, 41, 1484, 1983.

10. Shank, R.P. et al., Pyruvate carboxylase: An astrocytspecific enzyme implicated in the replenishment of amino acid neurotransmitter pools, *Brain Res.*, 329, 364, 1985.

11. Cesar, M. and Hamprecht, B., Immunocytochemical examination of neural rat and mouse primary cultures using monoclonal antibodies raised against pyruvate carboxylase, *J. Neurochem.*, 64, 2312, 1995.

12. Westergaard, N. et al., Evaluation of the importance of transamination versus deamination in astrocytic metabolism of [U-^{13}C]glutamate, *Glia*, 17, 160, 1996.

13. Ottersen, O.P., Zhang, N., and Walberg, F., Metabolic compartmentation of glutamate and glutamine: morphological evidence obtained by quantitative immunocytochemistry in rat cerebellum, *Neuroscience*, 46, 519, 1992.

14. Norenberg, M.D. and Martinez-Hernandez, A., Fine structural localization of glutamine synthetase in astrocytes of rat brain, *Brain Res.*, 161, 303, 1979.

15. Tansey, F.A., Farooq, M., and Cammer, W., Glutamine synthetase in oligodendrocytes and astrocytes: new biochemical and immunocytochemical evidence, *J. Neurochem.*, 56, 266, 1991.

16. Chaudhry, F.A. et al., Molecular analysis of system N suggests novel physiological roles in nitrogen metabolism and synaptic transmission, *Cell*, 99, 769, 1999.

17. Varoqui, H. et al., Cloning and functional identification of a neuronal glutamine transporter, *J. Biol. Chem.*, 275, 4049, 2000.

18. Kvamme, E., Roberg, B., and Torgner, I.A., Phosphate-activated glutaminase and mitochondrial glutamine transport in the brain, *Neurochem. Res.*, 25, 1407, 2000.

19. Laake, J.H. et al., Postembedding immunogold labelling reveals subcellular localization and pathway-specific enrichment of phosphate activated glutaminase in rat cerebellum, *Neuroscience*, 4, 1137, 1999.

20. Hertz, L. et al., Astrocytes: glutamate producers for neurons, *J. Neurosci. Res.*, 57, 417, 1999.

21. Schousboe, A. et al., Phosphate activated glutaminase activity and glutamine uptake in primary cultures of astrocytes, *J. Neurochem.*, 32, 943, 1979.

22. Kvamme, E. et al., Properties of phosphate activated glutaminase in astrocytes cultured from mouse brain, *Neurochem. Res.*, 7, 761, 1982.

23. Cousin, M.A., Hurst, H., and Nicholls, D.G., Presynaptic calcium channels and field-evoked transmitter exocytosis from cultured cerebellar granule cells, *Neuroscience*, 81, 151, 1997.

24. Hertz, L., Functional interactions between neurons and astrocytes I. Turnover and metabolism of putative amino acid transmitters, *Prog. Neurobiol.*, 13, 277, 1979.

25. Schousboe, A., Transport and metabolism of glutamate and GABA in neurons and glial cells, *Int. Rev. Neurobiol.*, 22, 1, 1981.

26. Erecinska, M., The neurotransmitter amino acid transport systems. A fresh outlook on an old problem, *Biochem. Pharmacol.*, 36, 3547, 1987.

27. Gegelashvili, G. and Schousboe, A., High affinity glutamate transporters: regulation of expression and activity, *Mol. Pharmacol.*, 52, 6, 1997.

28. Gegelashvili, G. and Schousboe, A., Cellular distribution and kinetic properties of high-affinity glutamate transporters, *Brain. Res. Bull.*, 45, 233, 1998.

29. Tanaka, K., Functions of glutamate transporters in the brain, *Neurosci. Res.*, 37, 15, 2000.

30. Magistretti, P.J. et al., Energy on demand, *Science*, 283, 496, 1999.

31. Schousboe, A., Svenneby, G., and Hertz, L., Uptake and metabolism of glutamate in astrocytes cultured from dissociated mouse brain hemispheres, *J. Neurochem.*, 29, 999, 1977.

32. Hertz, L. et al., Astrocytes in primary cultures, in *Neuroscience Approached through Cell Culture*, Vol. 1, Pfeiffer, S.E., Ed., CRC Press, Boca Raton, FL, 1982, p. 175.

33. Yu, A.C.H., Schousboe, A., and Hertz, L., Metabolic fate of 14C-labeled glutamate in astrocytes in primary cultures, *J. Neurochem.*, 39, 954, 1982.

34. Hertz, L., Drejer, J., and Schousboe, A., Energy metabolism in glutamatergic neurons, GABAergic neurons and astrocytes in primary cultures, *Neurochem. Res.*, 13, 605, 1988.

35. Martin, D.L. and Rimvall, K., Regulation of gamma-aminobutyric acid synthesis in the brain, *J. Neurochem.*, 60, 395, 1993.

36. Hertz, L. and Schousboe, A., Primary cultures of GABAergic and glutamatergic neurons as model systems to study neurotransmitter functions. I. Differentiated cells, in *Model Systems of Development and Aging of the Nervous System*, Vernadakis, A., Privat, A., Lauder, J.M., Timiras, P.S., and Giacobini, E., Eds., Martinus Nijhoff Publishing, Boston, 1987, p. 19.

37. Borden, L.A., GABA transporter heterogeneity: pharmacology and cellular localization, *Neurochem. Int.*, 29, 335, 1996.

38. Schousboe, A., Pharmacological and functional characterization of astrocytic GABA transport: a short review, *Neurochem. Res.*, 25, 1241, 2000.

39. Balazs, R. et al., The operation of the γ-aminobutyrate bypath of the tricarboxylic acid cycle in brain tissue in vitro, *Biochem. J.*, 116, 445, 1970.

40. Baxter, C.F., Some recent advances in studies of GABA metabolism and compartmentation, in *GABA in Nervous System Function*, Roberts, E., Chase, T.N., and Tower, D.B., Eds., Raven Press, New York, 1976, 61.

41. Fonnum, F. and Paulsen, R.E., Comparison of transmitter amino acid levels in rat globus pallidus and neostriatum during hypoglycemia or after treatment with methionine sulfoximine or gamma-vinyl gamma-aminobutyric acid, *J. Neurochem.*, 54, 1253, 1990.

42. Preece, N.E. and Cerdan, S., Metabolic precursors and compartmentation of cerebral GABA in vigabatrin-treated rats, *J. Neurochem.*, 67, 1718, 1996.

43. Reubi, J-C., Van der Berg, C., and Cuénod, M., Glutamine as precursor for the GABA and glutamate transmitter pools, *Neurosci. Lett.*, 10, 171, 1978.

44. Sonnewald, U. et al., Direct demonstration by [^{13}C]NMR spectroscopy that glutamine from astrocytes is a precursor for GABA synthesis in neurons, *Neurochem. Int.*, 22, 19, 1993.

45. Waagepetersen, H.S. et al., Synthesis of vesicular GABA from glutamine involves TCA cycle metabolism in neocortical neurons, *J. Neurosci. Res.*, 57, 342, 1999.

46. Rothman, D.L. et al., In vivo nuclear magnetic resonance spectroscopy studies of the relationship between the glutamate-glutamine neurotransmitter cycle and functional neuroenergetics, *Phil. Trans. R. Soc. Lond.*, 354, 1165, 1999.

47. Schousboe, A. et al., Role of astrocytic transport processes in glutamatergic and GABAergic neurotransmission, *Neurochem. Int.*, 45, 52, 2004.

48. Peng, L., et al., Utilization of glutamine and of TCA cycle constituents as precursors for transmitter glutamate and GABA, *Dev. Neurosci.*, 15, 367, 1993.

49. Su, T.Z., Campbell, G.W., and Oxender, D.L., Glutamine transport in cerebellar granule cells in culture, *Brain Res.*, 757, 69, 1997.

50. Sonnewald, U., Hertz, L., and Schousboe, A., Mitochondrial heterogeneity in the brain at the cellular level, *J. Cereb. Blood Flow Metab.*, 18, 231, 1998.

51. Taylor, R.W. et al., Treatment of mitochondrial disease, *J. Bioenerg. Biomembr.*, 29, 195, 1997.

52. Leong, S.F. et al., The activities of some energy-metabolising enzymes in nonsynaptic (free) and synaptic mitochondria derived from selected brain regions, *J. Neurochem.*, 42, 1306, 1984.

53. Lai, J.C., Leung, T.K., and Lim, L., Heterogeneity of monoamine oxidase activities in synaptic and non-synaptic mitochondria derived from three brain regions: some functional implications, *Metab. Brain Dis.*, 9, 53, 1994.

54. Lopez-Beltran, E.A., Mate, M.J., and Cerdan, S., Dynamics and environment of mitochondrial water as detected by ^{1}H NMR, *J. Biol. Chem.*, 271, 10648, 1996.

55. Perkins, G.A. and Frey, T.G., Recent structural insight into mitochondria gained by microscopy, *Micron*, 31, 97, 2000.

56. Schousboe, A. et al., Glutamate and glutamine metabolism and compartmentation in astrocytes, *Dev. Neurosci.*, 15, 359, 1993.

57. Sonnewald, U. et al., NMR spectroscopic studies of ^{13}C acetate and ^{13}C glucose metabolism in neocortical astrocytes: evidence for mitochondrial heterogeneity, *Dev. Neurosci.*, 15, 351, 1993.

58. Waagepetersen, H.S. et al., Comparison of lactate and glucose metabolism in cultured neocortical neurons and astrocytes using ^{13}C NMR spectroscopy, *Dev. Neurosci.*, 20, 310, 1998.

59. Waagepetersen, H.S. et al., Multiple compartments with different metabolic characteristics are involved in biosynthesis of intracellular and released glutamine and citrate in astrocytes, *Glia*, 35, 246, 2001.

60. Bouzier, A.K. et al., Compartmentation of lactate and glucose metabolism in C6 glioma cells. A ^{13}C and ^{1}H NMR study, *J. Biol. Chem.*, 273, 27162, 1998.

61. McKenna, M.C. et al., The metabolism of malate by cultured rat brain astrocytes, *Neurochem. Res.*, 15, 1211, 1990.

62. McKenna, M.C. et al., New insight into the compartmentation of glutamate and glutamine in cultured rat brain astrocytes, *Dev. Neurosci.*, 18, 380, 1996.

63. McKenna, M.C. et al., Mitochondrial malic enzyme activity is much higher in mitochondria from cortical synaptic terminals compared with mitochondria from primary cultures of cortical neurons or cerebellar granule cells, *Neurochem. Int.*, 36, 451, 2000.

64. Cruz, F. et al., Intracellular compartmentation of pyruvate in primary cultures of cortical neurons as detected by (13)C NMR spectroscopy with multiple (13)C labels, *J. Neurosci. Res.*, 66, 771, 2001.

65. Zwingmann, C., Richter-Landsberg, C., and Leibfritz, D., ^{13}C isotopomer analysis of glucose and alanine metabolism reveals cytosolic pyruvate compartmentation as part of energy metabolism in astrocytes, *Glia*, 34, 200, 2001.

66. Margineantu, D.H. et al., Heterogenous distribution of pyruvate dehydrogenase in the matrix of mitochondria, *Mitochondrion*, 1, 327, 2002.

67. Waagepetersen, H.S. et al., Mitochondrial heterogeneity in brain cells, *J. Neurochem.*, 85 (Suppl. 1), 56, 2003.

68. Waagepetersen, H.S., Sonnewald, U., and Schousboe, A., Compartmentation of glutamine, glutamate and GABA metabolism in neurons and astrocytes: Functional implications, *Neuroscientist*, 9, 398, 2003.

69. Badar-Goffer, R.S., Bachelard, H.S., and Morris, P.G., Cerebral metabolism of acetate and glucose studied by 13C-n.m.r. spectroscopy. A technique for investigating metabolic compartmentation in the brain, *Biochem. J.*, 266, 133, 1990.

70. Hassel, B. et al., Trafficking of amino acids between neurons and glia *in vivo*. Effects of inhibition of glial metabolism by fluoroacetate, *J. Cereb. Blood Flow Metab.*, 17, 1230, 1997.

71. Håberg, A. et al., *In vivo* injection of [1-13C]glucose and [1,2-13C]acetate combined with *ex vivo* 13C nuclear magnetic resonance spectroscopy: a novel approach to the study of middle cerebral artery occlusion in the rat, *J. Cereb. Blood Flow Metab.*, 18, 1223, 1998.

72. Chapa, F. et al., Metabolism of (1-(13)C) glucose and (2-(13)C, 2-(2)H(3)) acetate in the neuronal and glial compartments of the adult rat brain as detected by [(13)C, (2)H] NMR spectroscopy, *Neurochem. Int.*, 37, 217, 2000.

73. Sonnewald U. et al., First direct demonstration of preferential release of citrate from astrocytes using [13C]NMR spectroscopy of cultured neurons and astrocytes, *Neurosci. Lett.*, 128, 235, 1991.

74. Waniewski, R.A. and Martin, D.L., Exogenous glutamate is metabolized to glutamine and exported by rat primary astrocyte cultures, *J. Neurochem.*, 47, 304, 1986.

75. Qu, H. et al., (13)C MR spectroscopy study of lactate as substrate for rat brain, *Dev. Neurosci.*, 22, 429, 2000.

76. Taylor, A. et al., Approaches to studies on neuronal/glial relationships by 13C-MRS analysis. *Dev. Neurosci.*, 18, 434, 1996.

77. Bradford, H.F., Glutamate, GABA and epilepsy, *Epilepsia*, 30, 17, 1989.

78. Sander, J.W. and Shorvon, S.D., Epidemiology of the epilepsies, *J. Neurol. Neurosurg. Psychiatry*, 61, 433, 1996.

79. Hosford, D.A., Models of primary generalized epilepsy, *Curr. Opin. Neurol.*, 8, 121, 1995.

80. Löscher, W., Animal models of epilepsy for the development of antiepileptogenic and disease-modifying drugs. A comparison of the pharmacology of kindling and post-status epilepticus models of temporal lobe epilepsy, *Epilepsy Res.*, 50, 105, 2002.

81. Macdonald, R.L. and Barker, J.L., Phenobarbital enhances GABA-mediated postsynaptic inhibition in cultured mammalian neurons, *Trans. Am. Neurol. Assoc.*, 102, 139, 1977.

82. Sonnewald, U. and Kondziell D., Neuronal glial interaction in different neurological diseases studied by ex vivo ^{13}C NMR spectroscopy, *NMR Biomed.*, 16, 424, 2003.

83. Ben-Ari, Y., Limbic seizure and brain damage produced by kainic acid: mechanisms and relevance to human temporal lobe epilepsy, *Neuroscience*, 14, 375, 1985.

84. Eloqayli, H. et al., Pentylenetetrazole decreases metabolic glutamate turnover in rat brain, *J. Neurochem.*, 85, 1200, 2003.

85. Kondziella, D. et al., The pentylenetetrazole-kindling model of epilepsy in SAMP8 mice: glial-neuronal metabolic interactions, *Neurochem. Int.*, 43, 629, 2003.

86. Khurgel, M. and Ivy, G.O., Astrocytes in kindling: relevance in epileptogenesis, *Epilepsy Res.*, 26, 163, 1996.

87. Hudson, L.P. et al., Amygdaloid sclerosis in temporal lobe epilepsy, *Ann. Neurol.*, 33, 622, 1993.

88. Amano, S. et al., Development of a novel rat mutant with spontaneous limbic-like seizures, *Am. J. Patol.*, 149, 329, 1996.

89. Garzillo, C.L. and Mello, L.E., Characterization of reactive astrocytes in the chronic phase of the pilocarpine model of epilepsy, *Epilepsia*, 43, 107, 2002.

90. Sonnewald, U. et al., Metabolism of [U-^{13}C]glutamate in astrocytes studied by ^{13}C NMR spectroscopy: incorporation of more label into lactate than into glutamine demonstrates the importance of the tricarboxylic acid cycle, *J. Neurochem.*, 61, 1179, 1993.

91. Qu, H. et al., Decreased glutamate metabolism in cultured astrocytes in the presence of thiopental, *Biochem. Pharmacol.*, 58, 1075, 1999.

92. Qu, H. et al., The effect of thiopental on glutamate metabolism in cerebellar astrocytes, *Neurosci. Lett.*, 304, 141, 2001.

93. Sonnewald, U. et al., MRS study of glutamate metabolism in cultured neurons/glia, *Neurochem. Res.*, 21, 987, 1996.

94. Westergaard, N. et al., Glutamate and glutamine metabolism in cultured GABAergic neurons studied by ^{13}C NMR spectroscopy may indicate compartmentation and mitochondrial heterogeneity, *Neurosci. Lett.*, 185, 24, 1995.

95. Sonnewald, U., Westergaard, N., and Schousboe, A., Glutamate transport and metabolism in astrocytes, *Glia*, 21, 56, 1997.

96. McKenna, M.C. et al., Exogenous glutamate concentration regulates the metabolic fate of glutamate in astrocytes, *J. Neurochem.*, 66, 386, 1996.

97. Freidmann, B. et al., An estimation of pyruvate recycling during gluconeogenesis in the perfused rat liver, *Arch. Biochem. Biophys.*, 143, 566, 1971.

98. Cerdan, S., Künnecke, B., and Seelig, J., Cerebral metabolism of [1,2-^{13}C2]acetate as detected by *in vivo* and *in vitro* ^{13}C NMR, *J. Biol. Chem.*, 365, 12916, 1990.

99. Hassel, B., Sonnewald, U., and Fonnum, F., Glial-neuronal interactions as studied by cerebral metabolism of [2-^{13}C]acetate and [1-^{13}C]glucose: an *ex vivo* ^{13}C NMR spectroscopic study, *J. Neurochem.*, 64, 2773, 1995.

100. Sonnewald, U. et al., Metabolism of [U-^{13}C$_5$] glutamine in cultured astrocytes studied by NMR spectroscopy: first evidence of astrocytic pyruvate recycling, *J. Neurochem.*, 67, 2566, 1996.

101. Håberg, A. et al., *In vitro* and *ex vivo* ^{13}C-NMR spectroscopy studies of pyruvate recycling in brain, *Dev. Neurosci.*, 20, 389, 1998.

102. Waagepetersen, H.S. et al., Demonstration of pyruvate recycling in primary cultures of neocortical astrocytes but not in neurons, *Neurochem. Res.*, 11, 1431, 2002.

103. Bakken, I.J. et al., [U-13C]Glutamate metabolism in astrocytes during hypoglycemia and hypoxia, *J. Neurosci. Res.*, 51, 636, 1998.

104. Bakken, I.J. et al., [U-13C] Aspartate metabolism in cultured cortical astrocytes and cerebellar granule neurons studied by NMR spectroscopy, *Glia*, 23, 271, 1998.

105. Ito, T. et al., Pharmacology of barbiturate tolerance/dependence: GABAA receptors and molecular aspects, *Life Sci.*, 59, 169, 1996.

106. Qu, H. et al., Effects of thiopental on transport and metabolism of glutamate in cultured cerebellar granule neurons, *Neurochem. Int.*, 37, 207, 2000.

107. Nicholls, D. and Attwell, D., The release and uptake of excitatory amino acids, *Trends Pharmacol. Sci., 11, 477*, 1990.

108. Belhage B. et al., ^{3}H-D-Aspartate release from cerebellar granule neurons is differentially regulated by glutamate- and K$^+$-stimulation, *J. Neurosci. Res.*, 33, 436, 1992.

109. Bak, L.K., Schousboe, A., and Waagepetersen, H.S., Characterization of depolarization-coupled release of glutamate from cultured mouse cerebellar granule cells using DL-threo-β-benzyloxyaspartate (DL-TBOA) to distinguish between the vesicular and the cytoplasmic pools, *Neurochem. Int.*, 43, 417, 2003.

110. Bak, L.K., Waagepetersen, H.S., and Schousboe, A., Role of astrocytes in depolarization-coupled release of glutamate in cerebellar cultures, *Neurochem. Res.*, 29, 257, 2004.

111. Pastuszko, A., Wilson, D.F., and Erecinska, M., Amino acid neurotransmitters in the CNS: effect of thiopental, *FEBS Lett.*, 177, 249, 1984.

112. Merle, M. et al., [1-13C]Glucose metabolism in brain cells: isotopomer analysis of glutamine from cerebellar astrocytes and glutamate from granule cells, *Dev. Neurosci.*, 18, 460, 1996.

113. Martin, M., Canioni, P., and Merle, M., Analysis of carbon metabolism in cultured cerebellar and cortical astrocytes, *Cell Mol. Biol. (Noisy-le-grand)*, 43, 631, 1997.

114. Khazanov, V.A. and Saratikov, A.S., Effect of phenobarbital and benzonal on succinate and alpha-ketoglutarate oxidation by rat brain mitochondria, *Biull. Eksp. Biol. Med.*, 100, 692, 1985.

CHAPTER **8**

Cytokines in Neuronal–Glial Interaction

Barbara Viviani and Marina Marinovich

CONTENTS

8.1 INTRODUCTION

Cytokines are multifunctional proteins that act as humoral regulators at femto- to nanomolar concentrations, which can modulate under physiological, pathological, and toxicological conditions the functional activities of individual cells and tissues.

Cytokines are grouped into several different families whose names derive from the function they were originally thought to exert (Table 8.1). As it appears from this table, the term *cytokines* defines a broad range of molecules, some of which (i.e., interleukins, chemokines, tumor necrosis factors, interferons) were described as immune cell mediators in the periphery. Research during

Table 8.1   Cytokines

Family	Major Activities and Features
Interleukins	Cell activation; differentiation; cell-to-cell interaction
Chemokines (CXC-, CC-)	Chemotaxis; activation of inflammatory cells; control of viral infection and replication
Tumor necrosis factors	Multiple tissue and immunoregulatory activities; tumor cytotoxicity
Interferons	Inhibition of viral DNA replication; antiproliferative activity; immunomodulation
Colony stimulating factors	Proliferation of hematopoietic progenitor cells; synthesis of growth factors; maintenance of cell viability
Growth factors	Cell growth and differentiation

the past decade has clearly demonstrated that this group of cytokines is also produced within the central nervous system (CNS) and modulates several neurological functions and dysfunctions, thus disproving the idea of the CNS as a site devoid of any immunological response.

In general, the CNS responds to cytokines produced either in the periphery or within the CNS itself. Cytokines secreted by peripheral macrophages during bacterial infections, such as tumor necrosis factor-α (TNF-α) and interleukin-1 β (IL-1β), induce sickness behavior with fever.[1,2] IL-1β is also involved in the control of the hypothalamic-pituitary-adrenal axis.[3]

Within the CNS, a large number of cytokines and cytokine receptors are expressed in astrocytes, microglia, neurons, and oligodendrocytes, either constitutively or they can be induced following brain damage. IL-1 has been implicated in the mediation of physiological sleep,[4] synaptic plasticity, and maintenance of long-term potentiation,[5] while chemokines such as CXCL12 or CXCL8, respectively, enhance spontaneous synaptic activity[6] and suppress the induction of long-term depression.[7]

Whether in the healthy brain cytokines are absent or expressed at very low concentrations (femtomolar to low picomolar), they dramatically increase following a pathological or toxic insult. CNS ischemia, human immunodeficiency virus (HIV)–associated dementia, Alzheimer's disease, Parkinson's disease, and exposure to neurotoxicants such as trimethyltin, 1-methyl 1-4-phenyl-1,2,3,4 tetrahydropyridine (MPTP), and sarin increase mRNA and protein levels of several different pro- and anti-inflammatory cytokines (Table 8.2).[8–12] Such an increased production within the CNS is sustained by two possible cellular sources, immune cells infiltrating the CNS from the periphery and resident cells within the CNS. The former event occurs when the permeability of the blood–brain barrier is compromised, a condition usually associated with severe brain damage. While the involvement

Table 8.2   Cytokine Synthesis in the Brain

Cytokines	Stimulus
IL-1β, IL-6, TNF-α	Peripheral infection; endotoxin
IL-1β, IL-6, TNF-α, TGF-β, IFN-γ, MIP-I, MIP-2	CNS infection (malaria, HIV, meningitis, cytomegalovirus)
IL-1β, IL-2, IL-6, TNFα, IL-8, LIF, NGF, FGF, PDGF, EGF, TGF-β	Brain injury
TNF-α, MIP-1α, IL-1β, IL-1α, IL-6, MCP-1	Exposure to neurotoxicants:
TNF-α, IL-1α, IL-1β, IL-6, IFN-γ, IL-10	TMT
TNF-α, IL-1β, IL-6, E-selectin	MPTP
	Soman
IL-1β, IL-6, TNF-α, LIF	Convulsants
IL-1β, IL-6, FGF, TGF-β	Ischemia
IL-1β, IL-2, IL-6, TNF-α, TNF-β, INF-γ	Multiple sclerosis
TNF-α, IL-1α, IL-1β, IL-6, IL-10, IL-12, MCP-1	Alzheimer's disease
TNF-α, IL-1α, IL-1β, IL-6, IL-2, IL-4, TGF-1α, TGF-1β	Parkinson's disease

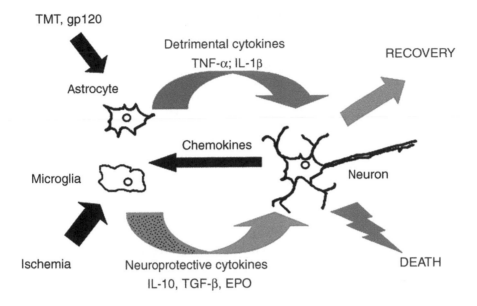

Figure 8.1 *(A color version of this figure follows page 236.)* The cytokine cycle in glial–neuronal interaction. Exposure of glial cells to a pathological (i.e., ischemia) or a toxic (i.e., TMT, gp120) insult triggers a reactive response, leading to the release of proinflammatory or anti-inflammatory cytokines capable of affecting neuronal function. In general, it is assumed that proinflammatory cytokines (i.e., TNF-α and IL-1β) exacerbate and sustain neurodegeneration, while anti-inflammatory cytokines (i.e., IL-10 and TGF-β) promote neuronal survival. In turn, cytokines derived from neurons can modulate glial response.

of neurons in cytokine production is controversial, there is no doubt about the ability of both astrocytes and microglia to express, produce, and release a wide range of pro- and anti-inflammatory cytokines. It seems that CNS injury generates signals that instruct glial cell transformation and activation. This process might be ruled (1) by a direct effect of neurotoxicants on glial cells, proteins with a disease-related production (i.e., amyloid β), or bacterial toxins, or (2) by the neuronal response to injury. For example, glia might be activated by altered levels of molecules associated with neurotransmission, tightly controlled during a physiological activity. This is the case of high ATP levels or extracellular K^+ exceeding normal ranges, which would signal damage and facilitate microglia reactions.[13,14] A primary neurodegenerative event may also induce glial response by (1) releasing neurospecific protein factors through the activation of caspases,[15] (2) inducing the expression of neurospecific chemokines,[16] or (3) interrupting a downregulating input that constitutively provides an inhibitory effect on glial cells (i.e., fractalkine, the neuronal glycoprotein CD200).[17,18] In turn, glial-derived cytokines have a significant impact on neuronal function and survival, underlying the relevance of cytokines in the bidirectional communication between glial cells and neurons (Figure 8.1).

The observation that glial cells can be activated to produce cytokines that are able to modulate neuronal functions has revolutionized the concept of neurotoxicity. Classically, neurotoxicity has always been regarded as the consequence of a direct interaction of toxicants with neurons. However, it is now appreciated that glia may be a direct target for neurotoxicants and that the resulting activation may influence the response of neurons to injury. Glia can either exacerbate neuronal damage or favor recovery that is dependent upon the types of cytokines produced.[10,19,20]

In this chapter, we will review the cytokines involved in glia–neuron cross talk. Due to the vastness of the topic, we will focus on inflammatory cytokines that were initially characterized in the context of the immune system and that are relevant for the development of a neurotoxic insult.

8.2 CYTOKINES THAT CONTRIBUTE TO NEURODEGENERATION AND NEUROTOXICITY

The process of neurodegeneration is likely to involve a shifting in the levels of the anti-inflammatory cytokines toward the production of proinflammatory ones. Conventionally, TNF-α together with IL-1β belong to the proinflammatory cytokines class. The action of these compounds should be counterbalanced by anti-inflammatory cytokines, such as transforming growth factor-β (TGF-β) and interleukin 10 (IL-10). Actually, both groups of cytokines indirectly suppress the synthesis of each other, and, more recently, TNF-α and IL-1β have been shown to inhibit directly the signaling of anti-inflammatory cytokine receptors on neurons.

8.2.1   IL-1β and "Its Family"

IL-1β belongs to the IL-1 family of three closely related proteins: the agonists IL-1α and IL-1β and the antagonist IL-1ra. All IL-1 members are formed as precursors, pro- IL-1α, pro-IL-1β, and pro-IL-1ra. While IL-1α and IL-1ra precursors are biologically active, pro-IL-1β is inactive and in order to assume reactivity it must be cleaved by caspase 1 (IL-1 converting enzyme, ICE).[21]

Several new members of the IL-1 family have been recently identified. Of these, IL-18 was the first IL-1 homolog to be found. IL-18 shares many common features with IL-1; however, despite these similarities, the two cytokines seems to exert different effects.

Among the different IL-1 family members, IL-1β appears to have a most critical role in sustaining a neuropathological cascade. Four major pieces of evidence support this contention[22–27]:

1. Expression of IL-1β rapidly increases following ischemia, hypoxia, excitotoxicity, and exposure to neurotoxicants (Table 8.2), which all lead to neurodegeneration. In addition, elevated levels of IL-1β characterize several different human neurodegenerative diseases (Table 8.2).
2. Intracerebral injection of recombinant IL-1β markedly exacerbates ischemic and excitotoxic injury *in vivo*.
3. *In vivo* administration of either IL-1 receptor antagonist (IL-1ra) or IL-1β antibody reduces ischemic, excitotoxic, and traumatic brain injury.
4. IL-1β (and α) gene polymorphism has been associated with multiple sclerosis[28] and temporal lobe epilepsy and appears to confer a higher risk for the development of Alzheimer's disease.[29,30]

8.2.1.1   *Molecular Mechanisms Involved in IL-1β–Induced Neuronal Damage*

In general, the "classic IL-1 family" (IL-1α, IL-1β, and IL-1ra) is a potent activator of host defense responses to infection, injury, inflammation, and disease. Within the CNS, this results in sustained glial activation and further release of cytokines. Neuronal degeneration most likely occurs when this effect is prolonged due to the persistence of the noxa or defect in mechanisms that may restrict or terminate the response. Under inflammatory conditions, glial subtypes, particularly microglia and partly astrocytes, can express neurotoxic factors. From *in vivo* and *in vitro* data we can deduce that the contribution of IL-1 to neurodegeneration might depend on the balance between neuroprotective and neurotoxic factors released from glia. It has been reported that IL-1 stimulates the release of neurotoxins from glia[31] such as nitric oxide[32] or α_1-antichymotrypsin, a serine proteinase inhibitor that is able to influence both formation and destabilization of β-amyloid fibrils.[33] On the other hand, IL-1 induces NGF expression and release in astrocytes, which exerts neuroprotective action *in vitro*.[34,35]

Direct action of glial-derived IL-1β can contribute to neuronal cell death. Primary glial cells exposed to the HIV-glycoprotein Gp120 show an upregulation of IL-1β, which is directly and specifically involved in neuronal death since this effect is completely prevented by either a neutralizing antibody to IL-1β or reducing glial IL-1β overexpression by means of antioxidants.[19]

It is interesting to note that IL-1β does not cause neuronal death in healthy brain tissue or normal neurons.[36] Nevertheless, it enhances ischemic and excitotoxic brain damage, thus suggesting that its neurotoxic effect has to be unraveled by the interaction with one or more factors. Substantial evidence suggests the existence of a reciprocal functional interaction between IL-1β and NMDA receptors, whose overstimulation is involved in several pathological conditions (excitotoxicity, ischemia, convulsions). We recently observed that IL-1β potentiates the NMDA receptor function, enhancing NMDA-induced intracellular Ca^{2+} increase and neuronal death.[37] Potentiation of intracellular Ca^{2+} increase persists for several minutes after removal of the cytokine, suggesting that even a transient exposure of neurons to IL-1β may favor long-lasting activation of the NMDA receptor. Thus, these results suggest that hippocampal neurons exposed to IL-1β are more susceptible to glutamatergic excitation through the NMDA receptor component. This may have functional relevance for either neuronal excitability or excitotoxicity.

IL-1β also induces the expression of multiple genes through the activation of the transcription factors nuclear factor-kappa B (NF-κB) and activator protein-1 (AP-1).[38] Consistent with the activation of transcription factors, IL-1β upregulates potentially neurotoxic proteins such as the β-amyloid precursor protein (APP), the Aβ-associated proteins (e.g., $α_1$-antichymotrypsin), and the hypothalamic neuropeptide corticotrophin-releasing factor, which has been implicated in ischemic and traumatic brain injury.[39,40]

8.2.1.2 Regulation of IL-1 Bioavailability and Bioactivity

IL-1 is believed to signal through a single receptor (IL-1RI) that requires association with an accessory protein (IL-1 receptor accessory protein [IL-1RAcP]) to increase the affinity of IL-1 binding and initiate signal transduction.[1] Formation of the IL-1RI/IL-1RAcP complex together with the adaptor molecule MyD88 leads to recruitment of the IL-1 receptor kinases (IRAKs) and to the sequential activation of a cascade of kinases. IL-1RAcP is not recruited following the binding of IL-1RI to IL-1ra. IL-1ra is thus considered an antagonist, since it fails to nucleate a functional receptor complex and prevents the binding of IL-1α and β.

A second receptor exists, named IL-1RII or decoy receptor, since IL-1 binding fails to induce cell signaling. IL-1RII binds both pro-IL-1β and IL-1β, thus preventing IL-1β processing to a mature form and blocking IL-1β binding with the functional receptor. It thus appears that IL-1 actions are finely tuned by the upregulation or downregulation of any one component of the family and of the accessory protein necessary to mediate their function.

8.2.2 Tumor Necrosis Factor-α

TNF-α is clearly involved in neurodegeneration. TNF-α initiates a sequence of events associated with neuronal apoptosis and neurological damage and, like IL-1β, increases activity of JUN amino terminal kinase/stress-activated protein kinases (JNK/SAPK), sphyngomyelinases, caspases, and nitric oxide.[41–44] As for IL-1β, the direct parenchymal injection of TNF-α does not cause neurotoxicity in healthy neurons, suggesting additional mechanisms for its function other than a direct effect.

All the cells present in the brain synthesize and release TNF-α,[45,46] particularly as a consequence of an insult to brain tissues.[47]

The pleiotropic action exerted by TNF-α is due to its interaction with two different receptors, TNFR1 (p55) and TNFR2 (p75), present both on glia and neurons.[48,49] The activation of p55 receptor has been associated with neuronal apoptosis in vitro.[50] An intracellular protein termed TRADD (TNF receptor-associated death domain) is associated with the death domain (DD) of the p55 receptor.[51] A second protein, FADD (Fas-associated death domain), binds to the TRADD-DD complex and initiates the apoptotic signal cascade that includes activation of acid sphingomyelinase through caspase 8. This causes the cleavage of sphingomyelin to ceramide and phosphocholine.

Ceramide, in turn, is a second messenger capable of activating ceramide-activated protein kinase, ceramide-activated protein phosphatase, and protein kinase Cζ. In TNFR2-deficient conditions, TNF aggravates cell death in different retinal layers after ischemia.[52] This not only indicates that TNFR1 signaling is sufficient for TNF-mediated retinal damage, but also suggests that an unbalanced activation of TNFR1 enhances this damage. Western blot analyses of retina revealed that the reduced neuronal cell loss in TNFR1 −/− animals correlated with the presence of activated Akt/protein kinase B (PKB). Inhibition of phosphatidylinositol 3-kinase signaling pathway reverted neuroprotection in TNFR1-deficient mice, indicating an instrumental role of Akt/PKB in neuroprotection and TNFR2 dependence of this pathway.

TNF-α has been reported to be neuroprotective, particularly in conditions of excitotoxic death.[48] The exact factors responsible for shifting the effect of TNF-α from neurotoxicity to neuroprotection is yet unknown. These opposite effects could depend on parameters such as the site, degree, and duration of the insult, the amount of TNF produced, the expression level of the two receptors, and the cellular environment of affected neurons.[54]

As already mentioned, signals released by damaged neurons can amplify the synthesis and the release of TNF-α by surrounding glial cells. In turn, such a release exacerbates the injury.[10] The nature of these signals is largely unknown, but it is well established that protease processing seems necessary for the activity,[15] as well as the withdrawal of a neuronal signal associated with the recruitment of glial cells.[46] For example, TNF-α release from microglia is normally inhibited by the chemokine fractalkine, which is predominantly expressed in the CNS.[18]

The p55 receptor-associated sphingomyelinase cleavage is not the only way in which TNFR1 receptor can kill neurons. Recently, it has been proposed that TNF-α induces neuronal death through the silencing of survival signals (SOSS)[55] activated by insulin-like growth factor (IGF-1) receptor. Actually, both TNF-α and IGF are produced in the brain during cerebral ischemia.[47,56]

Since this event occurs at concentrations lower than those necessary to cause neurotoxicity, it is more likely to represent a pathophysiological event. The effect is reached by blocking the IGF-1− mediated phosphorylation of tyrosine residues in the major insulin-receptor substrate protein in neurons. Thus, IGF-1 will be unable to activate phosphatidylinositol 3′ kinase, an enzyme involved in cell growth, survival, and differentiation. Since IL-1 inhibits insulin and IGF-1–induced cell growth,[57] this cytokine could act as silencer of survival signals as well.

8.2.3 Chemokines

The chemokine family includes greater than 50 members interacting with at least 20 receptors.[46,58] In addition to their activity in immunosurveillance and inflammations, they have been implicated to play important roles in angiogenesis, tumor progression, and neurodegeneration.[59–61]

Chemokines are classified on the basis of the relative position of their first N-terminal cysteine residues (Table 8.3), and consequently the receptors' classification parallels the nomenclature of their ligands (XCR, CCR, CXCR, and CX3CR). They activate G-proteins (Gαi), are sensitive to the *Bordetella pertussis* toxin, and cause activation of phospholipase C, which leads to inositol-1,4-5-triphosphate generation and increase of intracellular calcium.[62–64] Consequently, protein kinase C is also activated.

Other transduction pathways that mediate chemokine effects are the mitogen-activated protein kinase (MAPK) cascade, phosphoinositide 3′ kinase (PI3-K), and RAC, RhoA, and CDC2H (small GTP-binding proteins).[63,65–69]

Although the chemokines field is expanding very rapidly, the information concerning the physiological and pathological role of these molecules in the brain remains scarce. Astrocytes, microglial cells, oligodendrocytes, neurons, and brain endothelial cells can express both chemokines and their receptors,[46,69,70] either in physiological (chemokines drive cell migration during the development of the CNS) or pathological conditions. Expression of various chemokines, such as RANTES, MIP-1α, MCP-1, C10, and IP-10, has been implicated in neuroimmune response associated with cerebral

Table 8.3 Chemokines

Class	Structure	Members (Old Terminology)
C (γ chemokines)	Two conserved cysteines	XCL1, XCL2 (lymphotactin α, lymphotactin β)
CC (β chemokines)	Four conserved cysteines; two adjacent cysteines	CCL 1, 2, 3, 4, 5, 7, 8, 11, 13, 14, 15,16, 17, 19, 20, 21, 22, 22, 23, 24, 25, 26, 27, 28 (MCP-1, MIP-1α, RANTES, eotaxin)
CXC (α chemokines)	Four conserved cysteines; two cysteines separated by an amino acid	CXCL 1, 2, 3, 5, 6, 7, 8, 9, 10, 11, 12, 13, 16 (Groα, β, γ, IL-8, IP-10, SDF-1)
CX$_3$C (δ chemokines)	The first two cysteines are separated by three amino acids	CX$_3$CL1 (fractalkine)

ischemia, multiple sclerosis, and Alzheimer's disease. In almost all neurodegenerative diseases, chemokines have been associated with the activation and attraction of glial cells and with the recruitment of leukocytes. Interestingly, the neuronal expression of chemokines during the damage always precedes glial chemokine expression,[16,71] suggesting a role for the chemokine in the communication between dying neurons and glia in the early phases of the damage.[72,73]

In Alzheimer's disease, chemokines expressed in the proximity of amyloid plaques initially attract and activate glial cells,[74] and CCR2 knockout mice develop no signs of multiple sclerosis pathology and no macrophage infiltration.[75] CCR1 immunoreactivity was found in dystrophic neurites that were associated with senile plaques containing amyloid beta peptides of the 1–42 species.[76]

Another neuropathology where chemokines seem to play a decisive role is the AIDS-dementia complex. In this case neurons die, but there are glial cells to be infected, mainly through the copresence of CXCR4 and CD4 receptors. In fact, as in T-lymphocytes, CXCR4 is an obligatory coreceptor along with the lineage marker CD4 for HIV entry. Recently, it has been demonstrated that SDF-1 (stromal cell-derived factor-1), a chemokine highly expressed in HIV-infected astrocytes, is converted to a truncated form selectively toxic to neurons.[77] The cleavage is operated by prometalloproteinase-2, induced and secreted from HIV-infected macrophages, and activated on contact with neurons. SDF-1 could be involved also in another neurological complication of AIDS, the sensory neuropathy. CXCR4 binding on Schwann cells by SDF-1 or gp120 results in the release of RANTES, which induces TNF-α-production by dorsal root ganglion neurons, and subsequent TNFR-1–mediated neurotoxicity in an autocrine/paracrine fashion.[78]

A completely different function has been invoked for CX3C chemokine fractalkine and its receptor CX3CR1, both of which are present on neurons and microglia. Their interaction contributes to maintain microglia in a resting state. Whenever neurons are damaged, fractalkine message is reduced, resulting in microglia recruitment and activation.[79]

8.3 CYTOKINES WITH AMBIGUOUS BEHAVIOR

8.3.1 Interleukin-6 (IL-6)

IL-6 is a cytokine with major regulating effects on the inflammatory response, playing a particular role in the initiation and regulation of acute phase responses. Among the resident cells within the CNS, astrocytes seems to be the predominant source of this cytokine.[80] Analogous to other cytokines, IL-6 levels are extremely low under physiological conditions but dramatically increase during brain injury.[81] Unlike IL-1, for which specific inhibitors are available, the absence of specific inhibitors for IL-6 has hampered progress in defining its role and whether endogenous

IL-6 exerts any physiological action and neuroprotective or neurotoxic effects. IL-6 displays neurotrophic effects[82] and protects neurons from excitotoxicity and hypoxia.[20,83–85] IL-6–induced neuroprotection seems to be due to the induction of factors such as nerve growth factor[86,87] and antioxidants such as metallothionein I+II.[88] There are a number of studies to substantiate the neurodegenerative role of IL-6. Mice overexpressing IL-6 show marked neurodegeneration[89,90] unless exposed to brain injury.[91] These data indicate that chronic IL-6 expression leads to neurotoxic effects. Nevertheless, during an acute neuropathological insult, a neuroprotective effect is evident.

8.3.1.1 Regulation of IL-6 Bioactivity and Intracellular Signaling

IL-6 act on target cells through the interaction with a ligand-specific receptor, which associates with a gp130 homodimer to transduce the cellular signaling and activate STAT-1α and STAT-3 transcription factors.[92] IL-6 ligand-specific receptor (IL-6R) is both soluble (sIL-6R) and membrane bound.[93] It has been observed that sIL-6R greatly enhances IL-6–mediated responses both in astrocytes and neurons,[93] indicating that sIL-6R may offer an additional level of regulation that depends on its expression and availability. sIL-6R levels increase in the CSF of patients with cerebral trauma,[94] suggesting that a pathological response might increase sIL-6R production.

8.4 CYTOKINES THAT SUPPORT NEUROPROTECTION

8.4.1 Interleukin-10 (IL-10)[95,96]

IL-10 displays features of anti-inflammatory, immunosuppressive, and neuroprotective actions reducing the clinical symptoms associated with stroke, multiple sclerosis, Alzheimer's disease, and meningitis. While IL-10 expression is low in normal brain, it increases following CNS injury or disease. This cytokine is produced and acts on both neurons and glial cells by promoting the expression of survival signals and counteracting the actions of elevated IL-1 and tumor necrosis factor-α (TNF-α). The anti-inflammatory effect of IL-10 is exerted through (1) the reduction of proinflammatory cytokines synthesis and (2) the suppression of the expression and the inhibition of their receptors. JAK/STAT3, PI3-kinase, MAPK, and NF-kB pathways seem to be involved in the IL-10–induced effects.

8.4.2 Transforming Growth Factor-β (TGF-β)

TGF-β belongs to a superfamily of growth factors of approximately 30 members. In mammals, the TGF-β subfamily includes three isoforms: TGF-β1, -2, and -3. Immunohistochemistry shows a widespread expression of TGF-β2 and -3 in both developing and adult CNS, while TGF-β1 displays a scant staining.[97] In general, TGF-β1 expression is activated by ischemia, physical wounds, degenerative diseases such as Alzheimer's and Parkinson's diseases, and cellular stress.[97,98] The principal sources of TGF-β1 in the brain appear to be astrocytes and microglia, though neurons can produce it as well.[97] Since the increased expression of TGF-β in activated glial cells is common to several different neurodegenerative diseases, it is conceivable that the neurodegenerative process itself may stimulate TGF-β expression and production.

Within the nervous system, TGF-β plays an important role in maintaining neuronal integrity and survival, in controlling neuronal differentiation, and in regulating glial activation.[99] *In vivo*, intracerebral administration of TGF-β1 markedly reduces the extension of ischemic damage,[97] and lack of TGF-β1 expression results in an increase in degenerating neurons and microgliosis.[100] Lack of TGF-β1 also reduces survival of primary neurons,[100] whereas TGF-β1–treated neurons are protected against the aberrant effects of different neurotoxic agents (i.e., MTPT, β-amyloid, gp120).[97]

Because both neurons and glial cells express TGF-β receptors,[97] neuronal protection could be direct or the result of a complex cross talk between these two cell populations.

TGF-β may favor neuronal survival by interaction with several neurotrophins. This cytokine can synergize with or potentiate the effects of low concentrations of NGF, BDNF, NT-3, and NT-4.[101] TGF-β may also directly inhibit neuronal apoptosis by increasing the expression of Bcl-2 and Bc-lx$_L$ or by reducing the intracellular Ca^{2+} increase due to excitotoxic injury or exposure to neurotoxins.[97]

Another feature that characterizes TGF-β1 knockout (ko) mice is a reduction in the level of extracellular laminin.[100] Degradation of laminin by tissue plasminogen activator has been proposed to be responsible for neuronal death following kainate cytotoxicity.[102] Since TGF-β1 has been reported to be an inducer of plasminogen activator inhibitor,[103] Brionne et al.[100] suggested that neuronal survival in TGF-β1 ko mice may be reduced due to accelerated degradation of laminin.

8.4.2.1 Control of TGF-β Bioactivity and Intracellular Signaling

Astroglia, microglia, oligodendrocytes, and neurons have been shown to be targets for TGF-β. Biological actions of TGF-β are triggered by the binding to type I and II heterodimeric transmembrane serine/threonine kinase receptors.[104] The function of type II receptors is to activate the type I receptors, which transduce the signal by phosphorylating the Smad transcription factors. These translocate to the nucleus and regulate gene transcription only following association with other transcription factors.[104]

As previously described for other cytokines, TGF-β bioavailability is also tightly controlled. TGF-β access to the receptors may be prevented by soluble proteins such as the latency-associated protein (LAP, the propeptide from the TGF-β precursor). In contrast, other cell surface receptors (type III receptor and endoglin) facilitate the binding of TGF-β to the type II receptors.[104]

8.4.3 Erythropoietin (EPO)

EPO is a glycoprotein originally identified as the principal regulator of erythroid progenitor cells. During the prenatal period, EPO is produced in the liver and after birth in the kidney. EPO and its receptor (EPOR) are functionally expressed in the nervous system of rodents, primates, and humans, both in astrocytes and in neurons. Oxygen deficiency is the stimulus that triggers their expression in brain cells.[105–107]

EPO seems to play a critical role in neuronal survival. Several independent groups have reported that EPO prevents neuronal injury following exposure to neurotoxicants such as glutamate[108] and trimethyltin[109] or during hypoxic conditions,[106] hypoglycemia,[107] and free-radical injury.[110] *In vivo*, systemic administration of EPO is neuroprotective in animal models of cerebral ischemia, mechanical trauma, excitotoxicity, and neuroinflammation,[111] and neutralization of endogenous EPO with soluble EPOR augments ischemic brain damage.[112] EPO-induced neuroprotection is also coupled to a significant inhibition of leukocyte infiltration in the ischemic brain and to a reduction by more than 50% of the levels of proinflammatory cytokines such as TNF-α, IL-6, and MCP-1. This suggests that its neuroprotective effect could be the result of a reduced proinflammatory response sustained by both glia and infiltrating immune cells. However, *in vitro* EPO did not inhibit lipopolysaccharide (LPS)-induced TNF-α production by human peripheral blood mononuclear cells or rat glial cells.[109] Interestingly, EPO reduced TNF-α production in mixed glial–neuronal cultures treated with selective neurotoxicants, in which glial cells are activated by dying neurons.[15,109] This suggests that EPO acts indirectly as an anti-inflammatory cytokine in neurodegenerative diseases, by promoting neuronal survival.

8.4.3.1 Molecular Mechanisms Involved in EPO Effect

EPO binds to a surface receptor, which when activated dimerizes and induces JAK2 tyrosine kinases, triggering a cascade of phosphorylating events leading to the intracellular activation of the Ras/mitogen-activated kinase pathway, phosphatidylinositol 3 kinase, and STATs transcriptional factors.[113] It has been hypothesized that the activation of these pathways can rescue neuronal cells by affecting caspase activation and Bcl-2/Bcl-x_L expression.

8.5 CONCLUSIONS

The past several years have consecrated cytokines as neuromodulators that are involved in the bidirectional cross talk between neuronal and glial cells. Contemporary dogmas derived from the recent literature suggest that cytokines are mainly produced by activated glial cells, a condition that occurs following neurotoxic or pathologic damage. Glial-derived cytokines thus modulate neuronal functions, exacerbating the damage or favoring the recovery. To the contrary, neurons seem to induce or enhance glial activation through their death by either releasing neurospecific cytokines or interrupting a downregulating input that is sustained by cytokines (e.g., fractalkine).

The overall picture on the role of cytokines in injury and regeneration remains extremely complex and controversial. Complex, because the same cytokine can be produced by and act on several different cell types within the CNS (subpopulation of glial cells as well as neurons) and because of the concomitant production of different kinds of cytokines. Controversial, because the same cytokine can exert opposite effects. All these factors are determinants in the evolution of a specific pathology or of a neurotoxic insult. Thus, to better understand the role of cytokines in glial–neuronal cross talk in the progression of the damage, it will be essential to discriminate between the functions a cytokine exerts and the specific role it plays in a pathological or toxic context.

LIST OF ABBREVIATIONS

AP-1 — Activator protein-1
APP — β-Amyloid precursor protein
BDNF — Brain-derived growth factor
CNS — Central nervous system
EPO — Erythropoietin
EPOR — Erythropoietin receptor
FADD — Fas-associated death domain
IL-1β — Interleukin-1β
IL-6 — Interleukin-6
IL-6R — IL-6 receptor
sIL-6R — IL-6 soluble receptor
IL-10 — Interleukin-10
ICE — IL-1 converting enzyme
IL-1RacP — IL-1 receptor accessory protein
IL-1ra — IL-1 receptor antagonist
IRAKs — IL-1 receptor kinases
IGF-1 — Insulin-like growth factor I
JNK/SAPK — JUN amino terminal kinase/stress-activated protein kinases
LAP — Latency-associated protein
LPS — Lipopolysaccharide

MPTP — 1-Methyl-1-4-phenyl-1,2,3,4 tetrahydropyridine
MAPK — Mitogen-activated protein kinase
NGF — Nerve growth factor
NF-κB — Nuclear factor-kappa B
PI3-K — Phosphoinositide 3′ kinase
PKB — Protein kinase B
SOSS — Silencing of survival signals
SDF-1 — Stromal cell-derived factor-1
TNF-α — Tumor necrosis factor-α
TRADD — TNF receptor-associated death domain
TGF-β — Transforming growth factor-β

REFERENCES

1. Dinarello, C., Cannon, J., and Wolff, S., New concepts on the pathogenesis of fever, *Rev. Infect. Dis.*, 10, 168, 1988.
2. Kent, S., Bret-Dibat, J., and Kelley, K., Mechanism of sickness-induced decreases in food-motivated behaviour. *Neurosci. Biobehav. Rev.*, 20, 171, 1996.
3. Del Rey, A., Besedovsky, H., and Sorkin, E., Interleukin-1 and glucocorticoid hormones integrate an immunoregulatory feedback circuit, *Ann. N.Y. Acad. Sci.*, 496, 85, 1987.
4. Takahashi, S., Kapas, L., Fang, J., Seyer, J.M., Wang, Y., and Krueger, J.M., An interleukin-1 receptor fragment inhibits spontaneous sleep and muramyl dipeptide-induced sleep in rabbits, *Am. J. Physiol.*, 271, R101, 1996.
5. Scheider, H., Pitossi, F., Balschun, D., Wagner, A., Del Rey, W., and Besedovsky, H.O., A neuromodulatory role of interleukin 1-β in the hippocampus, *Proc. Natl. Acad. Sci. U.S.A.*, 95, 7778, 1998.
6. Limatola, C., Giovannelli, A., Maggi, L., Ragozzino, D., Castellani, L., Ciotti, M.T., Vacca, F., Mercanti, D., Santoni, A., and Eusebi, F., SDF-1 alpha-mediated modulation of synaptic transmission in rat cerebellum, *Eur. J. Neurosci.*, 12, 2497, 2000.
7. Giovannelli, A., Limatola, C., Ragozzino, D., Mileo, A.M., Ruggieri, A., Ciotti, M.T., Mercanti, D., Santoni, A., and Eusebi F., CXC chemokines interleukin-8 (IL-8) and growth-related gene product alpha (GROalpha) modulate Putkinje neuron activity in mouse cerebellum, *J. Neuroimmunol.*, 92, 122, 1998.
8. Allan, S.M. and Rothwell, N.J., Cytokines and acute neurodegeneration, *Nature*, 2, 734, 2001.
9. Sriram, K., Matheson, J.M., Benkovich, S.A., Miller, D.B., Lister, M.I., and O'Callaghan, J.P., Mice deficient in TNF-a receptors are protected against dopaminergic neurotoxicity: implications for Parkinson's disease, *Faseb J.*, 16, 1474, 2002.
10. Viviani, B., Corsini, E., Galli, C.L., and Marinovich, M., Glia increase neurodegeneration of hippocampal neurons through release of tumor necrosis factor-alpha, *Toxicol. Appl. Pharmacol.*, 150, 271, 1998.
11. Harry, J.G., Bruccoleri, A., and Lefebvre d'Hellencourt, C., Differential modulation of hippocampal chemical-induced injury response by ebselen, pentoxifylline, and TNFalpha-, IL-1alpha- and IL-6-neutralizing antibodies, *J. Neurosci. Res.*, 73, 526, 2003.
12. Henderson, R.F., Barr, E.B., Blackwell, W.B., Clark, C.R., Conn, C.A., Kalra, R., March, T.H., Sopori, M.L., Tesfaigzi, Y., Menache, M.G., and Mash, D.C., Response of rats to low levels of sarin, *Toxicol. Appl. Pharmacol.*, 184, 67, 2002.
13. Hide, I., Tanaka, M., Inoue, A., Nakajima, K., Kohsaka, S., Inoue, K., and Nakata, Y., Extracellular ATP triggers tumor necrosis factor-alpha release from rat microglia, *J. Neurochem.*, 75, 965, 2000.
14. Sanz, J.M. and DiVirgilio, F., Kinetics and mechanism of ATP-dependent IL-1β release from microglial cells, *J. Immunol.*, 164, 4893, 2000.
15. Viviani, B., Corsini, E., Galli, C.L., Padovan,i A., Ciusani, E., and Marinovich, M., Dying neural cells activate glia through the release of a protease product, *Glia*, 32, 84, 2000.
16. Biber, K., Sauter, A., Brouwer, N., Copray, J.C.V.M., and Boddeke, H.V.G.M., Ischemia-induced neuronal expression of the microglia attracting chemokine secondary lymphoid-tissue chemokine (SLC), *Glia*, 31, 121, 2001.

17. Neumann, H., Control of glial immune function by neurons, *Glia*, 36, 191, 2001.

18. Zujovic, V., Benavides, J., Vige, X., Carter, C., and Taupin, V., Fractalkine modulates TNF-alpha secretion and neurotoxicity induced by microglial activation, *Glia*, 29, 305, 2000.

19. Viviani, B., Corsini, E., Binaglia, M., Galli, C.L., and Marinovich, M., Reactive oxygen species generated by glia are responsible for neuron death induced by human immunodeficiency virus-glycoprotein 120 *in vitro*, *Neuroscience*, 107, 51, 2001.

20. Maeda, Y., Matsumoto, M., Hori, O., Kuwabara, K., Ogawa, S., Yan, S.D., Ohtsuki, T., Kinoshita, T., Kamada, T., and Stern, D.M., Hypoxia/reoxygenation-mediated induction of astrocyte interleukin 6: a paracrine mechanism potentially enhancing neuron survival, *J. Exp. Med.*, 180, 2297, 1994.

21. Thornberry, N.A., Bull, H.G., Calaycay, J.R., Chapman, K.T., Howard, A.D., Kostura, M.J., Miller, D.K., Molineaux, S.M., Weidner, J.R., Aunins, J. et al., A novel heterodimeric cysteine protease is required for IL-1beta processing in monocytes, *Nature*, 356, 768, 1992.

22. Griffin, W.S., Stanley, L.C., Ling, C., White, L., MacLeod, V., Perrot, L.J., White, C.L., and Araoz, C., Brain interleukin 1 and S-100 immunoreactivity are elevated in Down syndrome and Alzheimer disease, *Proc. Natl. Acad. Sci. U.S.A.*, 86, 7611, 1989.

23. Loddick, S.A. and Rothwell, N.J., Neuroprotective effects of human recombinant interleukin-1 receptor antagonist in focal cerebral ischaemia in the rat, *J. Cereb. Blood Flow Metab.*, 16, 932, 1996.

24. Allan, S.M., Parker, L.C., Collins, B., Davies, R., Luheshi, G.N., and Rothwell, N.J., Cortical cell death induced by IL-1 is mediated via actions in the hypothalamus of the rat, *Proc. Natl. Acad. Sci. U.S.A.*, 97, 5580, 2000.

25. Lawrence, L.C., Allan, S.M., Rothwell, N.J., Interleukin-1 beta and the interleukin-1 receptor antagonist act in the striatum to modify excitotoxic brain damage in the rat. *Eur. J. Neurosci.*, 10, 1188, 1998.

26. Touzani, O., Boutin, H., Chuquet, J., and Rothwell, N.J., Potential mechanisms of interleukin-1 involvement in cerebral ischaemia, *J. Neuroimmunol.*, 100, 203, 1999.

27. Yamasaki, Y., Matsuura, N., Shozuhara, H., Onodera, H., Itoyama, Y., and Kogure, K., Interleukin-1 as a pathogenetic mediator of ischemic brain damage in rats, *Stroke*, 26, 676, 1995.

28. Sciacca, F.L., Ferri, C., Vandenbroeck, K., Veglia, F., Gobbi, C., Martinelli, F., Franciotta, D., Zaffaroni, M., Marrosu, M., Martino, G., Martinelli, V., Comi, G., Canal, N., and Grimaldi, L.M., Relevance of Interleukin-1 receptor antagonist intron 2 polymorphism in Italian MS patients, *Neurology*, 52, 1896, 1999.

29. Grimaldi, L.M., Casadei, V.M., Ferri, C., Veglia, F., Licastro, F., Annoni, G., Biunno, I., De Bellis, G., Sorbi, S., Mariani, C., Canal, N., Griffin, W.S., and Franceschi, M., Association of early-onset Alzheimer's disease with an interleukin-1a gene polymorphism, *Ann. Neurol.*, 47, 361, 2000.

30. Nicoll, J.A., Mrak, R.E., Graham, D.I., Stewart, J., Wilcock, G., MacGowan, S., Esiri, M.M., Murray, L.S., Dewar, D., Love, S., Moss, T., and Griffin, W.S., Association of Interleukin-1 gene polymorphisms with Alzheimer's disease, *Ann. Neurol.*, 47, 365, 2000.

31. Giulian, D., Vaca, K., and Corpuz, M., Brain glia release factors with opposing actions upon neuronal survival, *J. Neurosci.*, 13, 29, 1993.

32. Boje, K.M. and Arora, P.K., Microglial produced nitric oxide and reactive nitrogen oxides mediate neuronal cell death, *Brain Res.*, 587, 250, 1992.

33. Kordula, T., Bugno, M., Rydel, R.E., and Travis, J., Mechanism of Interleukin-1- and tumor necrosis factor α-dependent regulation of the α_1-antichimotrypsin gene in human astrocytes, *J. Neurosci.*, 20, 7510, 2000.

34. Carman-Krzan, M., Vigé, X., and Wise, B.C., Regulation by interleukin-1 of nerve growth factor secretion and nerve growth factor mRNA expression in rat primary astroglia cultures, *J. Neurochem.*, 56, 636, 1991.

35. Strijbos, P.J. and Rothwell, N.J., Interleukin-1 beta attenuates excitatory amino acid-induced neuro-degeneration *in vitro*: involvement of nerve growth factor, *J. Neurosci.*, 15, 3468, 1995.

36. Rothwell, N.J., Sixteenth Gaddum Memorial Lecture December 1996. Neuroimmune interactions: the role of cytokines. *Br. J. Pharmacol.*, 121, 841, 1997.

37. Viviani, B., Bartesaghi, S., Gardoni, F., Vezzani, A., Behrens, M.M., Bartfai, T., Binaglia, M., Corsini, E., Di Luca, M., Galli, C.L., and Marinovich, M., Interleukin-1β enhances NMDA receptor-mediated intracellular calcium increase through activation of the Src family of kinases, *J. Neurosci.*, 23, 8692, 2003.

38. Baeuerle, P.A. and Henkel, T., Function and activation of NF-kappa B in the immune system, *Annu. Rev. Immunol.*, 12, 141, 1994.

39. Strijbos, P.J., Relton, J.K., and Rothwell, N.J., Corticotrophin-releasing factor antagonist inhibits neuronal damage induced by focal cerebral ischemia or activation of NMDA receptors in the rat brain, *Brain Res.*, 656, 405, 1994.

40. Roe, S.Y., McGowan, E.M., and Rothwell, N.J., Evidence for the involvement of corticotrophin releasing hormone in the pathogenesis of traumatic brain injury, *Eur. J. Neurosci.*, 10, 553, 1998.

41. Zhang, P., Miller, B.S., Rosenzweig, S.A., and Bhat, R., Activation of JNK/SAPK in primary glial cultures: II. Differential activation of kinase isoforms corresponds to their differential expression, *J. Neurosci. Res.*, 46, 114, 1996.

42. Singh, I., Pahan, K., Khan, M., and Singh, A.K., Citokine-mediated induction of ceramide production is redox-sensitive. Implications to proinflammatory citokine-mediated apoptosis in demyelinating diseases, *J. Biol. Chem.*, 273, 20354, 1998.

43. Wright, K., Kolios, G., Westwick, J., and Ward, S.G., Cytokine-induced apoptosis in epithelial HT-29 cells is independent of nitric oxide formation. Evidence for an interleukin-13-driven phosphatidylinositol 3-kinase-dependent survival mechanism, *J. Biol. Chem.*, 274, 17193, 1999.

44. Liu, J., Zhao, M.L., Brosnan, C.F., and Lee, S.C., Expression of type II nitric oxide synthase in primary human astrocytes and microglia: role of IL-1beta and IL-1 receptor antagonist, *J. Immunol.*, 157, 3569, 1996.

45. Hopkins, S.J. and Rothwell, N.J., Cytokines and the nervous system, I: expression and recognition, *Trends Neurosci.*, 18, 83, 1995.

46. Hanisch, U.-K., Microglia as a source and target of cytokines, *Glia*, 40, 140, 2002.

47. Botchkina, G.I., Meistrell, M.E., Botchkina, I.L., and Tracey, K.J., Expression of TNF and TNF receptors (p55 and p75) in the rat brain after focal cerebral ischemia, *Mol. Med.*, 3, 765, 1997.

48. Cheng, B., Tumor necrosis factors protect neurons against metabolic-excitotoxic insults and promote maintenance of calcium homeostasis, *Neuron*, 12, 139, 1994.

49. Courtney, M.J., Neurotrophins protect cultured cerebellar granule neurons against the early phase of cell death by a two-component mechanism, *J. Neurosci.*, 17, 4201, 1997.

50. Haviv, R. and Stein, R., The intracellular domain of p55 tumor necrosis factor receptor induces apoptosis which requires different caspases in naïve and neuronal large PC12 cells, *J. Neurosci. Res.*, 52, 380, 1998.

51. Weiss, T., Grell, M., Siemienski, K., Muhlenbeck, F., Durkop, H., Pfizenmaier, K., Scheurich, P., and Wajant, H., TNFR80-dependent enhancement of TNFR60-induced cell death is mediated by TNFR-associated factor 2 and is specific for TNFR60, *J. Immunol.*, 161, 3136, 1998.

52. Fontaine, V., Mohand-Said, S., Hanoteau, N., Fuchs, C., Pfizenmaier, K., and Eisel, U., Neurodegenerative and neuroprotective effects of tumor necrosis factor (TNF) in retinal ischemia: opposite roles of TNF receptor 1 and TNF receptor 2, *J. Neurosci.*, 22, 1, 2002.

54. Loddick, S.A. and Rothwell, N.J., Mechanisms of tumor necrosis factor a action on neurodegeneration: interaction with insulin-like growth factor-1, *Proc. Natl. Acad. Sci.*, 96, 9449, 1999.

55. Venters, H.D., Dantzer, R., and Kelley, K.W., A new concept in neurodegeneration: TNFα is a silencer of survival signals, U.S.A., *Trends Neurosci.*, 23, 175, 2000.

56. Gluckman, P., A role for IGF-1 in the rescue of CNS neurons following hypoxic-ischemic injury, *Biochem. Biophys. Res. Commun.*, 182, 593, 1992.

57. Costantino, A., Vinci, C., Mineo, R., Frasca, F., Pandini, G., Milazzo, G., Vigneri, R., and Belfiore, A., Interleukin-1 β locks insulin and insulin-like growth factor-stimulated growth in MCF-7 human breast cancer cells by inhibiting receptor tyrosine kinase activity, *Endocrinology*, 137, 4100, 1996.

58. Bajetto, A., Bonavia, R., Barbero, S., and Schettini, G., Characterization of chemokines and their receptors in the central nervous system: physiopathological implications, *J. Neurochem.*, 82, 1311, 2002.

59. Bajetto, A., Bonavia, R., Barbero, S., Florio, T., and Schettini, G., Chemokines and their receptors in the central nervous system, *Front. Neuroendocrinol.*, 22, 147, 2001.

60. Belperio, J.A., Keane, M.P., Arenberg, D.A., Addison, C.L., Ehlert, J.E., Burdick, M.D., and Strieter, R.M., CXC chemokines in angiogenesis, *J. Leukoc. Biol.*, 68, 1, 2000.

61. Rossi, D. and Zlotnik, A., The biology of chemokines and their receptors, *Annu. Rev. Immunol.*, 18, 217, 2000.

62. Bokoch, G.M., Chemoattractant signaling and leukocyte activation, *Blood*, 86, 1649, 1995.
63. Thelen, M., Dancing to the tune of chemokines, *Nat. Immunol.*, 2, 129, 2001.
64. Premack, B.A. and Schall, T.J., Chemokine receptors: gateways to inflammation and infection, *Nat. Med.*, 2, 1174, 1996.
65. Sanchez-Madrid, F., and del Pozo, M.A., Leukocyte polarization in cell migration and immune interactions, *EMBO J.*, 18, 501, 1999.
66. Sotsios, Y., Whittaker, G.C., Westwick, J., and Ward, S.G., The CXC chemokine stromal cell-derived factor activates a Gi-coupled phosphoinositide 3-kinase in T-lymphocytes, *J.Immunol.*, 163, 5954, 1999.
67. Ganju, R.K., Brubaker, S.A., Meyer, J., Dutt, P., Yang, Y., Qin, S., Newman, W., and Groopman, J.E., The alpha-chemokine, stromal cell derived factor-1alpha, binds to the transmembrane G-protein-coupled CXCR-4 receptor and activates multiple signal transduction pathways, *J. Biol. Chem.*, 273, 23169, 1998.
68. Tilton, B., Ho, L., Oberlin, E., Loetscher, P., Baleux, F., Clark-Lewis, I., and Thelen, M., Signal transduction by CXC chemokine receptor 4. Stromal cell-derived factor 1 stimulates prolonged protein kinase B and extracellular signal-regulated kinase 2 activation in T lymphocytes, *J. Exp. Med.*, 192, 313, 2000.
69. Bajetto, A., Barbero, S., Bonavia, R., Piccioli, P., Pirani, P., Florio, T., and Schettini, G., Stromal cell-derived factor 1alpha induces astrocyte proliferation through the activation of extracellular signal-regulated kinases 1/2 pathway, *J. Neurochem.*, 77, 1126, 2001.
70. Dorf, M.E., Berman, M.A., Tanabe, S., Heesen, M., and Luo, Y., Astrocytes express functional chemokine receptors, *J. Neuroimmunol.*, 111, 109, 2000.
71. Che, X., Ye, W., Panga, L., Wu, D.C., and Yang, G.Y., Monocyte chemoattractant protein-1 expressed in neurons and astrocytes during focal ischemia in mice, *Brain Res.*, 902, 171, 2001.
72. Streit, W.J., Walter, S.A., and Pennell, N.A., Reactive microgliosis, *Prog. Neurobiol.*, 57, 563,1999.
73. Biber, K., Zuurman, M.W., Dijkstra, I.M., and Boddeke, H.V.G.M., Chemokines in the brain: neuroimmunology and beyond, *Curr. Opin. Pharmacol.*, 2, 63, 2002.
74. Xia, M.Q. and Hyman, B.T., Chemokines/chemokine receptors in the central nervous system and Alzheimer's disease, *J. Neurovirol.*, 5, 32, 1999.
75. Fife, B.T., Huffnagle, G.B., Kuziel, W.A., and Karpus, W.J., CC chemokine receptor 2 is critical for induction of experimental autoimmune encephalomyelitis, *J. Exp. Med.*, 192, 899, 2000.
76. Halks-Miller, M., Schroeder, M.L., Haroutunian, V., Moenning, U., Rossi, M., Achim, C., Purohit, D., Mahmoudi, M., and Horuk, R., CCR1 is an early and specific marker of Alzheimer's disease, *Ann. Neurol.*, 54, 638, 2003.
77. Zhang, K., McQuibban, G.A., Silva, C., Butler, G.S., Johnston, J.B., Holden, J., Clark-Lewis, I., Overall, C.M., and Power, C., HIV-induced metalloproteinase processing of the chemokine stromal cell derived factor-1 causes neurodegeneration, *Nat. Neurosci.*, 6, 1064, 2003.
78. Keswani, S.C., Polley, M., Pardo, C.A., Griffin, J.W., McArthur, J.C., and Hoke, A., Schwann cell chemokine receptor mediate HIV-1 gp120 toxicity to sensory neurons, *Ann. Neurol.*, 54, 287, 2003.
79. Harrison, J.K., Jiang, Y., Chen, S., Xia, Y., Maciejewski, D., McNamara, R.K., Streit, W.J., Salafranca, M.N., Adhikari, S., Thompson, D.A., Botti, P., Bacon, K.B., and Feng, L., Role of neuronally-derived fractalkine in mediating interactions between neurons and CX3CR1-expressing microglia, *Proc. Natl. Acad. Sci. U.S.A.*, 95, 10896, 1998.
80. Gruol, D.L. and Nelson, T.E., Physiological and pathological roles of interleukin 6 in the central nervous system, *Mol. Neurobiol.*, 15, 307, 1997.
81. Van Wagoner, N.J. and Benveniste, E.N., Interleukin-6 expression and regulation in astrocytes, *J. Neuroimmunol.*, 100, 124, 1999.
82. Marz, P., Heese, K., Dimitriades-Schmutz, B., Rose-Jhon, S., and Otten, U., Role of interleukin 6 and soluble IL-6 receptor in region specific induction of astrocytic differentiation and neurotrophin expression, *Glia*, 26, 191, 1999.
83. Toulmond, S., Vige, X., Fage, D., and Benavides, J., Local infusion of interleukin-6 attenuates the neurotoxic effects of NMDA on rat striatal cholinergic neurons, *Neurosci. Lett.*, 144, 49, 1992.
84. Yamada, M. and Hatanaka, H., Interleukin-6 protects cultured rat hippocampal neurons against glutamate-induced cell death, *Brain. Res.*, 643, 173, 1994.

85. Loddick, S.A., Turnbull, A.V., and Rothwell, N.J., Cerebral interleukin-6 is neuroprotective during permanent focal cerebral ischemia in the rat, *J. Cereb. Blood Flow Metab.*, 18, 176, 1998.

86. Frei, K., Malipiero, U.V., Leist, T.P., Zinkernagel, R.M., Schwab, M.E., and Fontana, A., On the cellular source and function of interleukin 6 produced in the central nervous system in viral diseases, *Eur. J. Immunol.*, 19, 689, 1989.

87. Kossmann, T., Hans, V., Imhof, H.G., Trentz, O., and Morganti-Kossmann, M.C., Interleukin-6 released in human cerebrospinal fluid after traumatic brain injury may trigger nerve growth factor production in astrocytes, *Brain. Res.*, 713, 143, 1996.

88. Hidalgo, J., Castellano, B., and Campbell, I.L., Regulations of brain metallothioneins, *Curr. Top Neurochem.*, 1, 1, 1997.

89. Campbell, I.L., Abraham, C.R., Masliah, E., Kemper, P., Inglis, J.D., Oldstone, M.B., and Mucke, L., Neurologic disease induced in transgenic mice by cerebral overexpression of interleukin 6, *Proc. Natl. Acad. Sci. U.S.A.*, 90, 10061, 1993.

90. Hernandez, J., Molinero, A., Campbell, I.L., and Hidalgo, J., Transgenic expression of interleukin 6 in the central nervous system regulates brain metallothionein-I and -III expression in mice, *Mol. Brain Res.*, 48, 125, 1997.

91. Penkowa, M., Giralt, M., Lago, N., Camats, J., Carrasco, J., Hernandez, J., Molinero, A., Campbell, I.L., and Hidalgo, J., Astrocyte-targeted expression of IL-6 protects the CNS against a focal brain injury, *Exp. Neurol.*, 181, 130, 2003.

92. Hibi, M., Nakajima, K., and Hirano, T., IL-6 cytokine family and signal transduction: a model of the cytokine system, *J. Mol. Med.*, 74, 1, 1996.

93. Van Wagoner, N., Oh, J.-W., Repovic, P., and Benveniste, E.N., IL6 production by astrocytes: autocrine regulation by IL6 and the soluble IL-6 receptor, *J. Neurosci.*, 19, 5236, 1999.

94. Hans, V.H.J., Kossmann, T., Joller, H., Otto, V., and Morganti-Kossmann, M.C., Interleukin-6 and its soluble receptor in serum and cerebrospinal fluid after cerebral trauma., *NeuroReport*, 10, 409, 1999.

95. Vitkovich, L., Maeda, S., and Sternberg, E., Anti-inflammatory cytokines: expression and action in the brain, *Neuroimmunomodulation*, 9, 295, 2001.

96. Strle, K., Zhou, J.H., Shen, W.H., Broussard, S.R., Johnson, R.W., Freund, G.G., Dantzer, R., and Kelley, K.W., Interleukin-10 in the brain, *Crit. Rev. Immunol.*, 21, 427, 2001.

97. Flanders, K.C., Fen, R.F., and Lippa, C.F., Transforming growth factor-βs in neurodegenerative disease, *Prog. Neurobiol.*, 54, 71, 1998.

98. Roberts, A.B. and Sporn, M.B., Transforming growth factor-b, in *The Molecular and Cellular Biology of Wound Repair*, Clark, R.A.F., Ed., Plenum Press, New York, 1996, p. 275.

99. Unsicker, K. and Strelau, J., Functions of transforming growth factor-b isoforms in the nervous system. Cues based on localization and experimental *in vitro* and *in vivo* evidence, *Eur. J. Biochem.*, 267, 6972, 2000.

100. Brionne, T.C., Tesseur, I., Masliah, E., and Wyss-Coray, T., Loss of TGF-b1 leads to increased neuronal cell death and microgliosis in mouse brain, *Neuron*, 40, 1133, 2003.

101. Unsicker, K. and Krieglstein, K., Co-activation of TGF-b and cytokine signaling pathways are required for neurotrophic functions, *Cytokine Growth Factor Rev.*, 11, 97, 2000.

102. Chan, Z.L. and Strickland, S., Neuronal death in the hippocampus is promoted by plasmin-catalyzed degradation of laminin, *Cell*, 91, 917, 1997.

103. Loskutoff, D.J., Sawdey, M., and Mimuro, J., Type 1 plasminogen activator inhibitor, *Prog. Hemost. Thromb.*, 9, 87, 1989.

104. Massagué, J., Blain, S.W., and Lo, R.S., TGFb signaling in growth control, cancer, and heritable disorders, *Cell*, 103, 295, 2000.

105. Bernaudin, M., Marti, H.H., Roussel, S., Divoux, D., Nouvelot, A., MacKemzie, E.T., and Petit, E., A potential role for erythropoietin in focal permanent cerebral ischemia in mice, *J. Cereb. Blood Flow Metab.*, 19, 643, 1999.

106. Marti, H.H., Wenger, R.H., Rivas, L.A., Straumenn, U., Digicaylioglu, M., Henn, V., Yonekawa, Y., Bauer, C., and Gassamann, M., Erythropoietin gene expression in human, monkey and murine brain, *Eur. J. Neurosci.*, 8, 666, 1996.

107. Lewczuc, P., Hasselblatt, M., Kamrowski-Kruch, H., Heyer, A., Unzicker, C., Siren, A.L., and Ehrenreich, H., Survival of hippocampal neurons in culture upon hypoxia: effect of erithropoietin, *Neuroreport*, 11, 3485, 2000.

108. Morishita, E., Masuda, S., Nagao, M., Yasuda, Y., and Sasaki, R., Erythropoietin receptor is expressed in rat hippocampal and cerebral cortical neurons and erythropoietin prevents *in vitro* glutamate-induced neuronal death, *Neuroscience*, 76, 105, 1997.

109. Villa, P., Bigini, P., Mennini, T., Agnello, D., Laragione, T., Cagnotto, A., Viviani, B., Marinovich, M., Cerami, A., Coleman, T.R., Brines, M., and Ghezzi, P., Erythropoietin selectively attenuates cytokine production and inflammation in cerebral ischemia by targeting neuronal apoptosis, *J. Exp. Med.*, 198, 971, 2003.

110. Chong, Z.Z., Kang, J.Q., and Maiese, K., Erythropoietin prevents ischemic neuronal apoptosis through the serine-threonine kinase Akt1 and cytochrome *c* release, *Soc. Neurosci. Abstr.*, 31, 394, 2001.

111. Brines, M.L., Ghezzi, P., Keenan, S., Agnello, D., de Lanerolle, N.C., Cerami, C., Itri, L.M., and Cerami, A., Erythropoietin crosses the blood-brain barrier to protect against experimental brain injury, *Proc. Natl. Acad. Sci. U.S.A.*, 97, 10526, 2000.

112. Sakanaka, M., Wen, T.C., Matsuda, S., Masuda, S., Morishita, E., Nagao, M., and Sasaki, R., *In vivo* evidence that erythropoietin protects neurons from ischemic damage, *Proc. Natl. Acad. Sci. U.S.A.*, 95, 4635, 1998.

113. Chong, Z.Z., Kang, J.Q., and Maiese, K., Hematopoietic factor erythropoietin fosters neuroprotection through novel signal transduction cascades. *J. Cereb. Blood Flow Metab.*, 22, 503, 2002.

CHAPTER 9

Signaling Mechanisms Underlying Toxicant-Induced Gliosis

Krishnan Sriram and James P. O'Callaghan

CONTENTS

9.1 INTRODUCTION

Damage to diverse regions of the central nervous system (CNS) by a broad array of known neurotoxic agents engenders activation of microglia and astroglia at the sites of injury. These cellular responses are collectively known as glial activation or, more commonly, as gliosis. The generality of the glial reaction to injury, despite the regional- and cell-type specific targets of individual neurotoxic insults, implies that there are common "signals" underlying this cellular response.

Discovery and characterization of the signaling events that lead to this common response to brain damage would enhance our ability to detect early "biomarkers" of the neurotoxic condition and would thereby offer the potential for therapeutic intervention, prior to the onset of potentially irreversible effects. These biomarkers may serve as the basis for an early warning system to detect neurotoxic hazards.

Signal transduction events, especially in the CNS, rely on protein phosphorylation, a reversible process that can alter the function of target proteins and control diverse biological responses. Indeed, protein phosphorylation has long been recognized as the major posttranslational modification through which numerous physiological processes are regulated. Protein kinases and protein phosphatases, the key controlling enzymes in this process, are themselves regulated by a myriad of extracellular and intracellular signals. The brain expresses the greatest diversity and highest concentration of protein kinases and protein phosphatases (Nestler and Greengard, 1984; da Cruz e Silva and O'Callaghan, 1997; Greengard, 2001). Many extracellular messengers exert their effects in the CNS by regulating the intracellular concentration of specific second messengers, which in turn activate specific kinases or phosphatases (Nestler and Greengard, 1984). Many of the complex functions of the mammalian CNS, from receptor activation to gene expression, are known to be regulated by protein phosphorylation. With respect to neurotoxicity, in general, and glial activation, in particular, little data exist as to the role of protein phosphorylation in these processes (O'Callaghan, 1994a; da Cruz e Silva and O'Callaghan, 1997; Hebert and O'Callaghan, 2000; O'Callaghan and Sriram, 2004; Sriram et al., 2004). Given the central role of protein phosphorylation in neural function, it is not a question of whether protein phosphorylation is affected by neurotoxic insults but rather a question of which phosphorylation pathways are affected and how they relate to specific cellular changes *in vivo*.

In this chapter we will update our previous review (Martin and O'Callaghan, 1996) with respect to documenting the conditions, diseases, and injury models that result in gliosis. We will reiterate the key role of GFAP expression as a means of defining the astrogliotic condition. Finally, we will review recent evidence for the role of specific signal transduction modules in the induction of gliosis and examine novel technologies for identifying the earliest signaling mechanisms responsible for induction of gliosis. Clearly, the latest data lead to the conclusion that gliosis is a rapid and dynamic response to neural injury, a cellular reaction that no longer can be viewed simply as an impediment to regeneration or recovery from an injured state.

9.2 REACTIVE GLIOSIS IS ASSOCIATED WITH SEVERAL FORMS OF NEURAL INJURY AND NEURODEGENERATION: ASTROGLIAL AND MICROGLIAL RESPONSES

Both microglia and astroglia become "activated" or "reactive" in response to diverse insults of the CNS (Kreutzberg, 1996; Ridet et al., 1997; Streit, 1999). Indeed, as noted by Kreutzberg (1996), the most accurate way to view these cellular responses is in the context of microsensors, i.e., cells poised to sense the subtlest damage at a cellular level.

Of the major glial subtypes, microglia have been relatively neglected until recently. Although this lack of attention can be attributed to the historical absence of selective stains, the fact remains that microglia constitute a major glial cell component of the CNS (Del Rio-Hortega, 1932), likely representing as much as 10–12% of the total glial cell population (Vaughan and Peters, 1974). Microglia are localized in the vicinity of neurons (Lawson et al., 1990) and play an important role in host defense, sharing many phenotypic features with hematogenic macrophages. As the primary immune cell in the CNS, microglia play a dual role in cellular responses to neuronal injury: a pathogenic role that initiates inflammation and exacerbates degeneration, and a neuroprotective role (Gonzalez-Scarano and Baltuch, 1999; Stoll and Jander, 1999; Streit et al., 1999). Microglial activation occurs in several neurological disease states and is associated with neuronal damage in brain injury and neurotoxic injury models (Table 9.1). The advent of sensitive lectin stains for detecting activated microglia has led to the observation that, in general, microglia are the first responders to injury. As such, factors released from activated microglia, such as cytokines and growth factors, may serve as ligands to initiate the activation of astrocytes, glial cells known to respond to chemokines, cytokines, and growth factors. However, such molecular signaling mechanisms underlying microglial–astroglial interaction remain to be characterized.

Astroglia contribute approximately 40% of the total CNS cell population and participate in several important brain functions that include regulation of neuronal growth, extracellular ion and neurotransmitter concentrations, and immune response. Like microglia, astrocytes respond to subtle changes in their environment and are widely recognized to undergo hypertrophy in response to local neural damage. The hallmark of this response is enhanced expression of the intermediate filament protein, GFAP. As is thought to be the case for microglia, astrocytes also may play a dual role in CNS injury: a neuroprotective role, by producing neurotrophic/growth factors, and a pathogenic/detrimental role, by forming glial scars and preventing neuronal function and regeneration. Reactive astrocytes produce several structural proteins (GFAP, vimentin, nestin, microtubule associated protein-2), enzymes (superoxide dismutase, catalase, glutathione-S-transferase), growth factors (basic fibroblast growth factor, nerve growth factor, ciliary neurotrophic factor, transforming growth factor, glial-derived neurotrophic factor), matrix molecules (laminin, proteoglycans, fibronectin); adhesion molecules (neural cell adhesion molecule-1, intracellular cell adhesion molecule-1), cytokines (IL-1, IL-6, IL-10, IFN-α, IFN-β), chemokines (MCP-1, RANTES), and immunological factors (class I and II MHC molecules, CD40) associated with astrocyte hypertrophy and

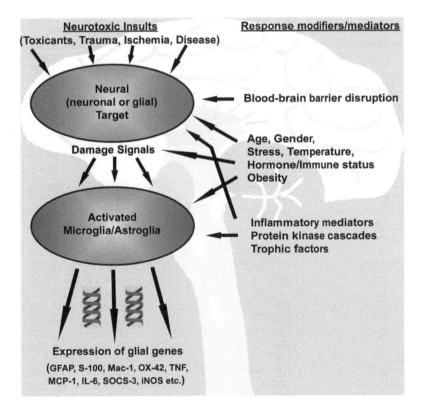

Figure 9.1   *(A color version of this figure follows page 236.)* Cell-signaling events associated with injury-induced glial activation. Neurotoxic insults elicit activation of a glial response that leads to expression of glial-specific genes. These insults can be mediated through or modulated by several external (environmental) or internal factors to exaggerate the injury and subsequent glial response.

Table 9.1   Disease, Injury, Toxic or Chemical Agents that Induce Microglial Activation *In Vivo*

Disease/Injury/Neurotoxicant	Brain Region Examined	Ref.
Neurologic Conditions		
Alzheimer's disease	CTX-FR; TE; OC; CER; BS	Wierzba-Bobrowicz et al., 2002
Alzheimer's disease	CTX-FR	Versijpt et al. 2003
Alzheimer-type dementia	CTX-ER; CN; TE	Cagnin et al., 2001
Amyotrophic lateral sclerosis	CTX; HIP	Wilson et al. 2001
Amyotrophic lateral sclerosis	—	Strong and Rosenfeld, 2003
Amyotrophic lateral sclerosis	—	Wilms et al., 2003b
Corticobasal degeneration	Supratentorial structures	Ishizawa and Dickson, 2001
Creutzfeldt-Jackob disease	CTX-FR; TE; STR; TH; CER	Gray et al., 1999
Creutzfeldt-Jackob disease	—	Rezaie and Lantos, 2001
Kuru	—	Rezaie and Lantos, 2001
Multiple sclerosis	—	De Simone et al., 1995
Multiple sclerosis	—	Gveric et al., 1998
Multiple system atrophy	STR	Schwarz et al., 1998
Parkinsonism—MPTP induced	SN	Langston et al., 1999
Parkinson's disease	SN	Banati et al., 1998
Parkinson's disease	SN	Hirsch et al., 2003
Progressive supranuclear palsy	Infratentorial structures	Ishizawa and Dickson, 2001
Viral encephalitis	CTX-FR; TE; OC; CER; BS	Wierzba-Bobrowicz et al., 2002
Wilson's disease	CTX-FR; TE; OC; CER; BS	Wierzba-Bobrowicz et al., 2002
Closed head injury	—	Engel et al., 1996
Brain Injury		
Brain lesions	Perforant path axons	Jensen et al., 2000
Global cerebral ischemia	—	Kato et al., 2000
Mild focal brain ischemia	STR	Katchanov et al., 2003
Stab injury	CTX	Isono et al., 2003
Stroke	—	Gerhard et al., 2000
Transient focal ischemia	SN	Huh et al., 2003
Transient global ischemia	STR; HIP	Lin et al., 1998
Transient MCAO	SN	Dihne and Block, 2001
Transient MCAO	—	Schilling et al., 2003
Traumatic brain injury	CTX	DeKosky et al., 1996
Toxic/Chemical Agents		
β-Amyloid peptide	Nucleus basalis	Scali et al., 1999
3-nitropropionic acid	STR	Ryu et al., 2003
6-hydroxydopamine	CER	Podkletnova et al., 2001
6-hydroxydopamine	STR	Rodrigues et al., 2001
6-hydroxydopamine	SN	Depino et al., 2002
Aluminium	STR; HIP; CTX	Platt et al., 2001
Diethyldithiocarbamate	CTX-ER; HIP; HYP	Zucconi et al., 2002
Domoic acid	HIP	Ananth et al., 2003
Ibotenic acid	—	Dommergues et al., 2003
Kainic acid	HIP	Jorgensen et al., 1993
Kainic acid	HIP	Araujo and Wandosell, 1996
Kainic acid	HIP	Tanikawa et al., 1996
Kainic acid	HIP	Chuang and Han, 2003
Lipopolysaccharide	SN	Arimoto and Bing, 2003
Methamphetamine	STR; HIP; CER	Escubedo et al., 1998
Methamphetamine	STR	Asanuma et al., 2003
Methamphetamine	STR	Guilarte et al., 2003
Methamphetamine	STR	Thomas et al., 2004
Methylazoxymethanol	STR; TH	Ashwell, 1992
Methylazoxymethanol	CER	Lafarga et al., 1998
Monosodium glutamate	CTX-FP	Martinez-Contreras et al., 2002

(*Continued*)

Table 9.1   Disease, Injury, Toxic or Chemical Agents that Induce Microglial Activation *In Vivo* (*Continued*)

Disease/Injury/Neurotoxicant	Brain Region Examined	Ref.
Toxic/Chemical Agents		
MPTP	STR	Francis et al., 1995
MPTP	STR; SN	Czlonkowska et al., 1996
MPTP	STR; SN	Kohutnicka et al., 1998
MPTP	SN	Liberatore et al., 1999
MPTP	STR; SN	Kurkowska-Jastrzebska et al., 1999
MPTP	—	Wu et al., 2002
MPTP	SN	Sugama et al., 2003
MPTP	SN	McGeer et al., 2003
MPTP	SN	Cardenas and Bolin, 2003
Phencyclidine	CTX-CN	Rajdev et al., 1998
Quinolinic acid	STR	Lindenau et al., 1998
Quinolinic acid	STR; SN	Dihne et al., 2001
Rotenone	STR; SN	Sherer et al., 2003
S-AMPA	Basal forebrain	Oliveira et al., 2003
Tetanus toxin	HIP	Shaw et al., 1990
Thrombin	SN	Carreno-Muller et al., 2003
Trimethyltin	HIP	Koczyk & Oderfeld-Nowak, 2000
Trimethyltin	HIP	Fiedorowicz et al., 2001
Volkensin	CER	Cevolani et al., 2001

Note: CTX = cortex; CTX-FR = frontal cortex; CTX-TE = temporal cortex; CTX-OC = occipital cortex; CTX-ER = entorhinal cortex; CTX-CN = cingulate cortex; CTX-FP = frontoparietal cortex; CER = cerebellum; BS = brainstem; STR = striatum; TH = thalamus; SN = substantia nigra; HIP = hippocampus; HYP = hypothalamus; PUT = putamen; AMG = amygdala.

response to neuronal injury (Norenberg, 1994; Dong and Benveniste, 2001). Astrogliosis is associated with neurological disease states, traumatic brain injuries, and neural damage due to chemical neurotoxicity (Table 9.2).

9.3 GFAP EXPRESSION AS AN INDEX OF ASTROGLIOSIS

GFAP is a classical marker of mature astrocytes. GFAP is a Class III intermediate filament protein, sharing a common tripartite structure with other intermediate filament proteins, comprising a highly conserved central α-helical rod domain, flanked by nonhelical N-terminal head and C-terminal tail domains (Steinert and Roop, 1988; Chen and Liem, 1994). Increased expression of GFAP in astrocytes is one of the most widely recognized responses to injuries of the CNS. It occurs following a diverse array of insults and is observed across a wide variety of animal species, including humans. Although less well documented, enhanced expression of GFAP also occurs in response to damage of the developing CNS. Clearly, neural injury is the major stimulus for the induction of GFAP expression. Nevertheless, it is important to keep in mind that GFAP also is developmentally regulated and it is under hormonal regulation, primarily mediated through the adrenal gland (e.g., O'Callaghan et al., 1991).

9.3.1   The GFAP Gene

The murine *Gfap* gene (Figure 9.2) is located on chromosome 11 and maps to position 11D,11 62.0 cM; the rat Gfap gene is located on chromosome 10 and maps to position 10q31.1; and the human *GFAP* gene is located on chromosome 17 and maps to position 17q21 (Bongcam-Rudloff et al., 1991; Schuler et al., 1996). The GFAP gene has been cloned in mouse (Lewis et al., 1984; Balcarek and Cowan, 1985; Miura et al., 1990; Sarid, 1991), human (Reeves et al., 1989; Brenner et al., 1990; Bongcam-Rudloff et al., 1991; Issacs et al., 1998), and rat (Feinstein et al., 1992;

Table 9.2   Disease, Injury, Toxic, or Chemical Agents that Induce Astrogliosis *In Vivo*

Disease/Injury/Neurotoxicant	Brain Region Examined	Ref.
Neurologic Conditions		
Alzheimer's disease	CTX-TE	Panter et al., 1985
Alzheimer's disease	HIP	Yamaguchi et al., 1987
Alzheimer's disease	CN; TH; CER; BS	Delacourte, 1990
Alzheimer's disease	HIP	Vijayan et al., 1991
Alzheimer's disease	CTX; PUT; AMG	Cullen, 1997
Alzheimer's-type dementia	—	Duffy et al., 1980
Alzheimer's-type dementia	CTX-TE	Harpin et al., 1990
Amyotrophic lateral sclerosis	CTX	Troost et al., 1992
Creutzfeldt-Jackob disease	CTX	Akoi et al., 1999
Multiple sclerosis	CTX	Petzold et al., 2002
Multiple sclerosis	—	Malmestrom et al., 2003
Schizophrenia with dementia	CTX-FR, TE	Arnold et al., 1996
Brain Injury		
Cerebellar stab injury	CER	Ajtai and Kalman, 1998
Cortical stab injury	CTX	Krum et al., 2002
Cortical stab injury	CTX	Isono et al., 2003
Facial nerve lesion	CTX	Laskawi et al., 1997
Forebrain stab lesion	HIP	Carbonell and Mandell, 2003
Hippocampal stab wound	HIP	Zhu et al., 2003
Ischemic brain injury	—	Gabryel and Trzeciak, 2001
Severe focal brain ischemia	CTX	Cheung et al., 1999
Stab injury (needle-insertion lesion)	—	Lee et al., 2003
Stroke	—	Badan et al. 2003
Transient global ischemia	HIP	Endoh et al., 1994
Transient global ischemia	HIP	Soltys et al., 2003
Transient MCAO	STR; HIP; CTX	Butler et al., 2002
Toxic/Chemical Agents		
2'-NH2-MPTP	STR; CTX; HIP; BS	Luellen et al., 2003
3-Nitropropionic acid	STR	Page et al., 1998
3-Nitropropionic acid	STR	Vis et al., 1999
3-Nitropropionic acid	STR	Teunissen et al., 2001
5,7-Dihydroxytryptamine	HIP	Dugar et al., 1998
6-Hydroxydopamine	STR	Rodrigues et al., 2001
Chlorpyrifos	STR	Garcia et al., 2002
Kainic acid	HIP	Represa et al., 1993
Kainic acid	HIP	Lenz et al., 1997
Kainic acid	HIP	Ferraguti et al., 2001
Kainic acid	HIP	Penkowa et al., 2001
Kainic acid	HIP	Sriram et al., 2002b
MDMA	STR	Johnson et al., 2002
Methamphetamine	Caudate nucleus	Hess et al., 1990
Methamphetamine	STR	Fukumura et al., 1998
Methamphetamine	Caudate-putamen	Cappon et al., 2000
Methamphetamine	STR	Miller et al., 2000
Methamphetamine	STR	Sriram et al., 2002b
Methamphetamine	STR	Miller and O'Callaghan, 2003
Monosodium glutamate	CTX-FP	Martinez-Contreras et al., 2002
MPTP	STR	Stromberg et al., 1986
MPTP	STR	Schneider and Denaro, 1988
MPTP	STR	O'Callaghan et al., 1990
MPTP	STR	O'Callaghan and Seidler, 1992
MPTP	STR	Francis et al., 1995
MPTP	STR	O'Callaghan et al., 1998

(Continued)

Table 9.2   Disease, Injury, Toxic or Chemical Agents that Induce Astrogliosis *In Vivo* (*Continued*)

Disease/Injury/Neurotoxicant	Brain Region Examined	Ref.
Toxic/Chemical Agents		
MPTP	STR	Sriram et al., 2002a
Quinolinic acid	STR	Schiefer et al., 1998
Quinolinic acid	STR; SN	Dihne et al., 2001
Trimethyltin	HIP	Kuhlmamm and Guilarte, 2000
Trimethyltin	HIP	Fiedorowicz et al., 2001
Volkensin	CER	Cevolani et al., 2001

Note: CTX = cortex; CTX-FR = frontal cortex; CTX-TE = temporal cortex; CTX-OC = occipital cortex; CTX-ER = entorhinal cortex; CTX-CN = cingulate cortex; CTX-FP = frontoparietal cortex; CER = cerebellum; BS = brainstem; STR = striatum; TH = thalamus; SN = substantia nigra; HIP = hippocampus; HYP = hypothalamus; PUT = putamen; AMG = amygdala.

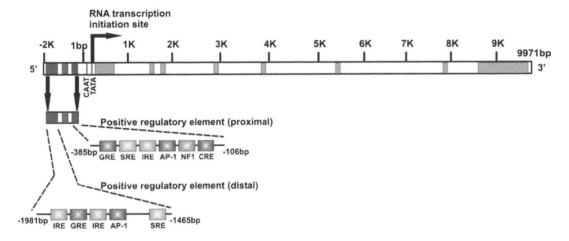

Figure 9.2 *(A color version of this figure follows page 236.)* The murine *Gfap* gene. The expression of the gene is controlled by positive (enhancers) and negative (silencers) regulatory elements situated 2 kb upstream of transcription start site. Depicted here are the positive regulatory elements in the promoter region that are essential for activation of the gene. This region contains binding sites for specific transcription factors that are involved in transcriptional activation of GFAP.

Kaneko and Sueoka, 1993; Condorelli et al., 1994; Condorelli et al., 1999). The gene extends to nearly 10 kb and comprises nine exons. Across species, the GFAP gene shows a high degree of homology in the coding region, but less so in the 3′ untranslated region (Brenner, 1994). The GFAP mRNA of 2.9 kb is the predominant isoform in the CNS, known as GFAPα. In addition, minor isoforms, GFAPβ (Feinstein et al., 1992; Galea et al., 1995), GFAPγ (Zelenika et al., 1995), GFAPδ (Condorelli et al., 1999), and GFAPε (Nielsen et al., 2002) occurring as a result of alternative splicing have been described in rodents and humans.

9.3.2   Regulation of GFAP Gene Expression

The GFAP gene, like most mammalian genes, is transcribed by RNA polymerase II (*Pol* II) and contains a basal promoter region within the first 100 bp from the transcription start site. The basal promoter elements in this region regulate the initiation and rate of transcription. The most common transcriptional initiator is a repeated DNA nucleotide sequence of thymidine-adenosine called the "TATA" sequence. In the *Gfap* gene basal promoter region, these sequences are found 29 to 33 bp upstream of the transcription start site. Basal promoter elements that regulate/modulate the rate of transcription include GC box, CAAT, etc. Apart from regulation by these basal promoter elements, transcriptional regulation of the *Gfap* gene also is mediated by DNA sequences typically found within 2 kb upstream of the transcription initiation site (Brenner and Messing, 1996). The human *GFAP* gene is activated by positive regulatory elements (enhancers) located between −250 and −80 bp and between −1980 and −1500 bp (Brenner, 1994). The murine and rat *Gfap* genes are controlled by regulatory elements located between −385 and −106 bp and between −1981 and −1465 bp. Similarly, negative regulatory elements (silencers) located between −650 and −360 bp are thought to inhibit gene transcription. The 5′ upstream promoter of the *Gfap* gene contains several *cis*-acting elements (Miura et al., 1990; Brenner et al., 1994; Kahn et al., 1997; Krohn et al., 1999) and binding sites for transcription factors such as CREB, Sp-1, NF-1, AP-1, AP-2, etc., that activate gene expression. In addition, potential STAT3 binding sites have been reported in the promoter region of the *Gfap* gene (Nakashima et al., 1999; Takizawa et al., 2001), which regulate its expression. Since STAT3 is activated by several gp130-related cytokines, such as IL-6, IL-11, LIF, OSM, CNTF, and CT-1 (Taga and Kishimoto, 1997; Heinrich et al., 1998; Catteneo et al., 1999), and the GFAP promoter has been shown to contain potential STAT3 binding sites, it seems likely that STAT3 may play an important role in transcriptional activation of GFAP.

9.4 SIGNAL TRANSDUCTION PATHWAYS ASSOCIATED WITH GLIOSIS

9.4.1 Protein Phosphorylation Cascades and Gliosis

As we noted previously, all cellular responses to extracellular stimuli are mediated by protein kinase/phosphatase signaling cascades. Given that the CNS contains the highest concentration of these signaling effectors, and that microglia and astroglia constitute a far greater percentage of the nervous system content than do neurons, cell-signaling events underlying reactive gliosis also must be regulated by protein phosphorylation.

While a variety of signal transduction pathways have been shown to be involved in the "activation" of astrocytes *in vitro* (Guillemin et al., 1996; Rajan and McKay, 1998; Bajetto et al., 2001; Dunn et al., 2002), very few signaling modules have been linked to induction of reactive gliosis *in vivo*. The limited evidence obtained to date with traumatic and ischemic models of brain damage points to an involvement of proinflammatory cytokine/chemokine signaling pathways in astroglial responses to injury. These include JNK, ERK, and JAK/STAT pathways as downstream effectors potentially relevant to reactive gliosis.

9.4.2 Use of Focused Microwave Irradiation to Preserve *In Vivo* Steady-State Protein Phosphorylation

Protein phosphorylation is achieved through protein kinase catalyzed transfer of phosphate to serine, threonine, or tyrosine residues of a given protein substrate. These protein- and residue-specific signaling events can now be studied using widely available phospho-state specific antibodies. Due to the rapidly reversible nature of protein phosphorylation, any given phosphorylation event often can be quite transient. Therefore, net phosphorylation represents the contribution of protein kinase and protein phosphatase activities affecting a specific site on a given substrate. Thus, preserving phosphorylation state becomes a critical issue, especially when analyzed *in vivo*, because variations in postmortem activities of kinases and phosphatases are likely to affect net protein phosphorylation, unless these enzymes are rapidly inactivated. These variations can be significantly minimized by focused microwave irradiation sacrifice (O'Callaghan and Sriram, 2004). Moreover, we recently demonstrated (O'Callaghan and Sriram, 2004) that this procedure may be required to achieve biologically relevant data for the *in vivo* protein phosphorylation state of many phosphoproteins (Figure 9.3). The work from our laboratory described in the following sections relied heavily on the preservation of steady-state phosphorylation, *in vivo*, that is preserved in microwaved tissue. Whether

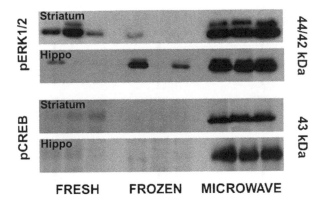

Figure 9.3 Microwave irradiation preserves phosphorylation of protein kinases and transcription factors in various brain regions. Mice ($n = 3$ per group) were sacrificed by decapitation (FRESH), decapitation into liquid nitrogen (FROZEN), or focused microwave irradiation (MICROWAVE). The striatum and hippocampus were rapidly dissected and homogenized in hot 1% SDS. An aliquot of the total protein (20 µg) was separated by SDS-PAGE and transferred onto nitrocellulose, and immunoblots with antibodies to phospho-ERK1/2 (Thr202/Tyr204) and phospho-CREB (Ser133) were performed. Following appropriate secondary antibody incubation, the signals were developed using an ECL chemiluminescent substrate and exposed to film.

this technology will have to be used for preservation of a given phosphorylation site will likely have to be evaluated on a phospho-site and specific substrate basis (O'Callaghan and Sriram, 2004).

9.4.3 The MAPK Pathway: Its Involvement in Neuronal and Glial Response to Injury

The mitogen-activated protein kinases (MAPKs) are a family of intracellular signaling proteins that are strongly conserved through evolution and regulate diverse cellular functions following sequential activation of a three-kinase signal module (Figure 9.4). A canonical MAPK pathway thus consists of three protein kinases: a MAPK kinase kinase (MAPKKK or MEKK) that activates a MAPK kinase (MAPKK or MEK), which, in turn, activates the MAPK enzyme. In mammalian cells, the MAPK family comprises four separate cascades: (1) Extracellular signal-regulated kinases (ERK1/ERK2, p44/42 MAPK); (2) the c-Jun-N-terminal kinases/stress-activated protein kinases (JNKs/SAPKs); (3) p38 MAPK; and (4) ERK5, also known as the big MAP kinase 1 (BMK1).

The MAPK superfamily of enzymes coordinates the extracellular and intracellular signals generated by a variety of mediators/ligands by phosphorylation and sequential activation of the MAPK module and transmits the signal down the cascade, leading to phosphorylation of proteins that are associated with regulation of cellular function, such as protein kinases, transcription factors, cytoskeletal proteins, enzymes, etc. The ERK1/2 cascade is stimulated by both G protein–coupled receptors as well as receptor tyrosine kinases, which classically involves Shc tyrosine phosphorylation, recruitment of the Grb2-Sos complex, and the subsequent sequential activation of Ras, Raf kinase, MEK, and ERK (Marshall, 1994; Fukunaga and Miyamoto, 1998; Wilsbacher et al., 1999). Phosphorylated ERK1/2 (Thr202/Tyr204) translocates to the nucleus where it activates (phosphorylates) transcription factors such as c-Myc, N-Myc, c-fos, c-jun, Elk-1, Stat1, Stat3, CREB, etc. (Davis, 1993, Ginty et al., 1994; Seger and Krebs, 1995; Xing et al., 1996; Vossler et al., 1997; Sgamboto et al., 1998). Other ERK1/2 substrates include phospholipase A2, microtubule-associated protein-2, neurofilament proteins, tau, epidermal growth factor receptor, and nerve growth factor receptor (Davis, 1993; Seger and Krebs, 1995; Jovanovic et al., 1996, Matsubara et al., 1996).

The ERK1/2 signaling pathway plays a predominant role in regulation of cell growth and differentiation following activation by growth/neurotrophic factors. In the brain, ERK1/2 pathway is associated with neural plasticity and memory (Grewal et al., 1999; Sweatt, 2001), cellular responses

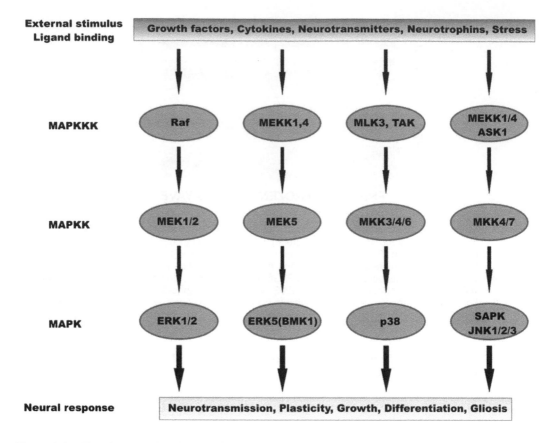

Figure 9.4   *(A color version of this figure follows page 236.)* Schematic representation of the canonical
MAPK pathway indicating activation of the pathways by upstream ligands and the concomitant
downstream neural responses.

to stress, immune response, apoptosis, neurotransmission, cytoskeletal dynamics, ion channel function, behavioral responses to addictive drugs, and regulation of neural cell growth and differentiation
(Hunter and Cooper, 1984; Blenis et al., 1993; Cowley et al., 1994; Alessi et al., 1995; Hill and
Treisman, 1995, Vanhoutte et al., 1999; Yan et al., 1999; Valjent et al., 2001). Activation of the MAPK
pathway has also been reported to be associated with neural injury. Phosphorylation of ERK1/2 has
been observed following exposure to occupational, environmental, chemical, or neurotoxic agents
(Kim et al., 1994, O'Callaghan et al., 1998; Fuller et al., 2001; Caughlan et al., 2004), electroconvulsive shocks (Kang et al., 1994), and cerebral ischemia (Hu and Wieloch, 1994; Irving et al.,
2000). Pharmacological modulation of ERK1/2 using the ERK1/2 inhibitor PD098059 has been
shown to reduce neuronal death (Murray et al., 1998; Alessandrini et al., 1999).

ERK1/2 phosphorylation has also been reported to occur in glial cells following disease or
injury-induced damage to neurons. Phosphorylation of ERK1/2 has been reported in reactive
astrocytes from patients with Alzheimer's disease (Hyman et al., 1994; Perry et al., 1999; Russo
et al., 2002) and in experimental models following ischemia (Jiang et al., 2003), endotoxin treatment
(Xie et al., 2004), infarct, mechanical trauma, chronic epilepsy, and leukoencephalopathy (Mandell
and Vandenberg, 1999). The ERK1/2 pathway also plays a critical role in microglial activation *in
vivo* in chronic neurodegenerative diseases such as Alzheimer's disease and following acute brain
injury, as in stroke and ischemia (Koistinaho and Koistinaho, 2002).

The stress-activated protein kinases/Jun-N-terminal kinases (SAPK/JNKs) are activated by a
variety of growth factors, cytokines, and environmental stresses including ultraviolet and ionizing
radiation, heat shock, etc. (Kyriakis et al., 1994; Kyriakis and Avruch, 1996; Paul et al., 1997).

Following activation of this pathway, SAPK/JNK translocates to the nucleus where it regulates the activity of several transcription factors such as c-Jun, ATF-2, Elk-1, Smad4, p53, NFAT4, NFATc1, etc. SAPK/JNK stimulation is associated with induction of apoptosis (Yang et al., 1997), cellular proliferation (Chen et al., 1998), and differentiation (Yao et al., 1997). In the CNS, JNK/SAPK has been shown to be involved in the neurodegeneration seen in Alzheimer's disease (Mielke and Herdegen, 2000; Okazawa and Estus, 2002), in ischemic/reperfusion injury (Hayashi et al., 2000), and following neurotoxic injury (Glicksman et al., 1998; Saporito et al., 1999).

The p38 MAPK, another member of the MAPK superfamily, is activated by a variety of inflammatory cytokines and environmental stress factors. Following activation, p38 MAPK regulates the activity of transcription factors such as Stat1, Myc, ATF-2, MEF2, Elk-1, and CREB. Activation of p38 MAPK is predominantly associated with apoptosis. The role of p38 in the nervous system is poorly understood; nevertheless, activation of this pathway during neuronal response to stress is evident. Activation of p38 has been reported to occur following cerebral ischemia and stroke (Barone et al., 2001; Nozaki et al., 2001) and has been shown to play a critical role in transducing microglial responses to activation stimuli (Giovannini et al., 2002; Piao et al., 2002; Choi et al., 2003; Wilms et al., 2003a).

Activation of the MAPK pathway by specific ligands or following neural injury occurs in glial cells and in many instances is associated with glial activation. Recent studies addressing the activation of MAPK signaling in glial cells and glial activation are listed in Table 9.3.

Table 9.3   Activation of MAPK Pathway in Glial Cells and Its Association with Glial Activation

Mode of Neural Injury/Activation	Glial Cell Type	MAPK Pathway Activated	Ref.
α,β-Methylene ATP	Astrocytes	ERK1/2	Brambilla et al., 2002
ATP	Astrocytes	ERK	Neary and Zhu, 1994
Benzoyl-benzoyl ATP (Bz-ATP)	Astrocytes	ERK1/2; p38	Panenka et al., 2001
Beta-amyloid (Aβ1–42)	Microglia	p38	Giovannini et al., 2002
Dorsal root transection	Microglia	ERK1/2	Cheng et al., 2003
Endotoxin (LPS)	Astrocytes and microglia	ERK1/2, p38, JNK	Xie et al., 2004
Fibroblast growth factor-2	Astrocytes	ERK	Bayatti and Engele, 2001
Forebrain stab lesion	Astrocytes	ERK	Carbonell and Mandell, 2003
Hypoxia	Microglia	p38	Kim et al., 2003
Interleukin-1	Astrocytes	p38	Dunn et al., 2002
Isoproterenol (β-adrenergic agonist)	Astrocytes	ERK	Gharami and Das, 2004
Mechanical trauma	Astrocytes	ERK	Mandell and VandenBerg, 1999
Methamphetamine	?	ERK1/2; JNK	Hebert and O'Callaghan, 2000
MPTP	?	ERK1/2	O'Callaghan et al., 1998
MPTP	?	ERK1/2	Sriram et al., 2004
Postischemia (normoxic incubation)	Astrocytes	ERK1/2	Jiang et al., 2003
SOD1G93A (ALS mouse model)	Astrocytes and microglia	p38	Tortarolo et al., 2003
Spinal cord ligation	Microglia	p38	Jin et al., 2003
Subacute and chronic lesions	Astrocytes	ERK	Mandell and VandenBerg, 1999
Thrombin	Microglia	ERK1/2	Suo et al., 2003
Thrombin	Microglia	ERK1/2, p38	Choi et al., 2003
TNF-α	Astrocytes	JNK	Zhang et al., 1996
TNF-α	Astrocytes	ERK2	Zhang et al., 2000
Transient focal ischemia	Astrocytes and microglia	p38	Piao et al., 2002
Transient focal ischemia	Astrocytes	p38	Piao et al., 2003
Traumatic brain injury	Astrocytes	ERK1/2	Otani et al., 2002

Note: ERK = extracellular signal-regulated kinase; JNK = jun-N-terminal kinase.

9.4.4 The JAK/STAT Pathway: Its Involvement in Neuronal and Glial Response to Injury

Signal transducers and activators of transcription (STATs) are tyrosine phosphorylated transcription factors activated by tyrosine protein kinases. Various ligands, such as cytokines, growth factors, and hormones, induce activation of STATs to modulate a variety of cellular events such as acute-phase response, differentiation, growth, and proliferation (Darnell et al., 1994; Ihle et al., 1995; Hirano et al., 2000; Takeda and Akira, 2000). STAT3 was initially identified as a DNA-binding protein stimulated by the cytokine IL-6, with distinct ability to interact with response elements in the promoter region of acute-phase genes (Akira et al., 1994; Zhong et al., 1994). Activation of STAT3 is regulated mainly by posttranslational modification, particularly phosphorylation at tyrosine and serine residues. Several tyrosine kinases such as Janus kinases, receptor-tyrosine kinases, and certain Src family kinases can mediate tyrosine phosphorylation of STAT3 (Guschin et al., 1995). Serine phosphorylation occurs at a MAPK consensus site and is required for maximal transcriptional activity. Upon tyrosine phosphorylation, STAT3 is released from the receptor, undergoes dimerization through SH2 domain interaction, translocates to the nucleus, and binds to response elements in the promoter region of target genes, leading to transcriptional activation of those genes. Target genes of STAT3 include apoptosis-related genes (Bcl-2, Bcl-xL), cell cycle–related genes (p21, Cyclin D,

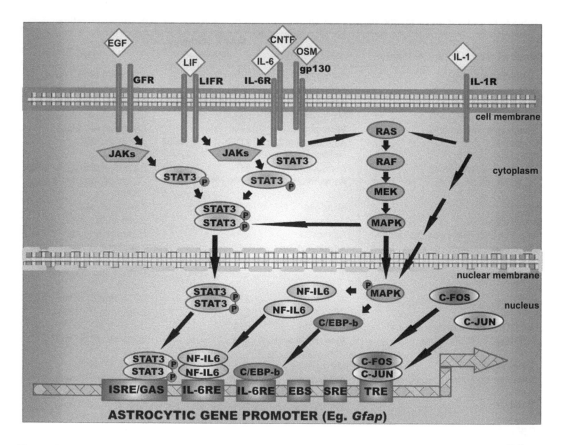

Figure 9.5 *(A color version of this figure follows page 236.)* Schematic diagram depicting the probable association of the gp130-JAK/STAT3 and MAPK pathway in regulating astroglial activation and GFAP expression. Proinflammatory cytokines and growth factors, including IL-6–type cytokines, are potential effectors (ligands) for activation of glial genes following disease- or injury-induced neurodegeneration. A myriad of transcription factors activated by these pathways may converge and bind to specific sites on the astrocytic gene promoter (e.g., GFAP) to regulate its expression.

Table 9.4 Activation of JAK/STAT Pathway in Glial Cells and Its Association with Glial Activation/Differentiation

Mode of Neural Injury/Activation	Glial Cell Type	JAK/STAT Pathway Activated	Ref.
Axokine (CNTF analog)	Retinal (Muller) glia	STAT1, STAT3	Peterson et al., 2000
BMP-2/LIF	Astrocyte differentiation	STAT3	Nakashima et al., 1999
Bone morphogenetic protein-7	Astrocyte differentiation	STAT3	Yanagisawa et al., 2001
Brain inflammation	Microglia	JAK1/JAK2-STAT1/STAT3	Park et al., 2003
Cardiotrophin-1	Astrocyte differentiation	STAT3	Ochiai et al., 2001
CNTF	Astrocyte differentiation	JAK-STAT	Bonni et al., 1997
CNTF	Astrocyte differentiation	STAT1α	Kahn et al., 1997
CNTF	Astrocyte differentiation	JAK-STAT	Rajan and McKay, 1998
CNTF	Astrocyte differentiation	STAT3	Aberg et al., 2001
CNTF	Astrocytes	JAK-STAT	Monville et al., 2001
CNTF/EGF cotreatment	Astrocytes	STATs	Levison et al., 2000
EAE	Astrocytes and microglia	STAT1	Maier et al., 2002
Entorhinal cortex lesion	Astrocytes	STAT3	Xia et al., 2002
Focal cerebral ischemia	Astrocytes	JAK1-STAT3	Justica et al., 2000
FP6 (sIL-6R/Il-6 fusion protein)	Astrocyte differentiation	STAT3	Takizawa et al., 2001
Ganglioside-induced inflammation	Microglia	JAK1/JAK2-STAT1/STAT3	Kim et al., 2002
GM-CSF	Microglia	JAK2-STAT5α/β	Liva et al., 1999
LPS	Astrocytes	STAT1α/β	Dell'Albani et al., 2001
Methamphetamine	?	STAT3	Hebert and O'Callaghan, 2000
MPTP	Astrocytes	STAT3	Sriram et al., 2004
Prolactin	Astrocytes	JAK2-STAT1/STAT3	Mangoura et al., 2000

Note: JAK = Janus kinase; STAT = signal transducer and activator of transcription.

Cyclin E), acute phase response factors (C-reactive protein, α2-macroglobulin), cytokines and growth factors (interleukin-10, gp130, Oncostatin M), and certain transcription factors (c-Fos, c-Jun, Stat3). In addition, potential STAT3 binding sites have been identified on the promoter of the astroglial marker, GFAP (Kahn et al., 1997; Nakashima et al., 1999), and earlier studies have linked activation of STAT3 to gliogenesis (Bonni et al., 1997). These findings suggest that STAT3 signaling pathways may be associated with glial activation (Figure 9.5). The JAK/STAT pathway is also known to be associated with astrocyte differentiation, glial activation, and astroglial and microglial gene responses following neuronal injury (Rajan and McKay, 1998; Liva et al., 1999; Levison et al., 2000; Dell'Albani et al., 2001; Kim et al., 2002; Park et al., 2003; Sriram et al., 2004). Table 9.4 provides a broad comparison of different ligands/injury models that activate the JAK/STAT signaling pathway in glial cells.

9.4.5 Phosphorylation of JAK2/STAT3 as a Key Signaling Pathway for Astrogliosis

Induction of gliosis may be influenced by a variety of factors emanating from multiple neural cell types. Moreover, nonneural, blood-borne cell types also may be involved in neuronal and glial responses to injury due to their entry via a breached blood–brain barrier. Thus, the discovery of signaling mechanisms that lead to gliosis would best be achieved through the use of injury models

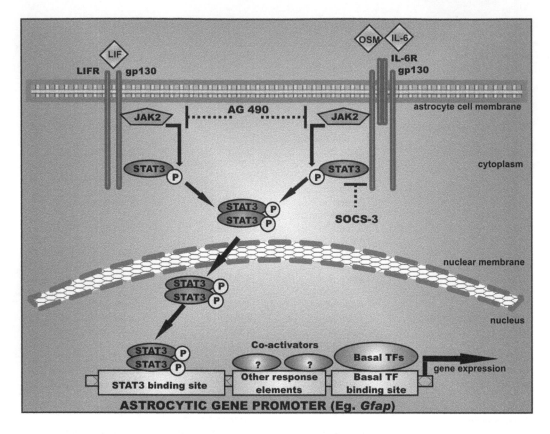

Figure 9.6   *(A color version of this figure follows page 236.)* Schematic diagram depicting the involvement of gp130-mediated phosphorylation of JAK2/STAT3 pathway prior to induction of GFAP in the MPTP model of dopaminergic neurodegeneration. Upon putative ligand binding (e.g., IL-6, LIF, OSM), JAK2 and STAT3 are recruited to the gp130 signal transducer. Following docking of STAT3 to gp130 via its SH2 domain, JAK2 phosphorylates the Tyr705 residue on STAT3. The phosphorylated STAT3 dimerizes, translocates to the nucleus, and in association with other transcription factors (TFs) mediates transcriptional activation of astrocytic genes such as *Gfap*. MPTP-induced activation of STAT3 is attenuated by pharmacological intervention with the JAK2 inhibitor, AG490. Similarly, MPTP treatment causes upregulation of SOCS-3 mRNA, resulting in negative feedback regulation of STAT3 phosphorylation. (Adapted from Sriram, K. et al. *J. Biol. Chem.*, 279, 19936, 2004. With permission.)

that target defined cell type without disrupting the blood–brain barrier. Ideally, the affected cell type also should be well characterized in terms of its morphology, biochemistry, and pharmacology. Known neurotoxicants serve as selective denervation tools that meet these criteria and, therefore, make ideal agents to discover and characterize the signals that lead to gliosis. Using this approach we have obtained strong evidence for involvement of the gp130-mediated activation of JAK2/STAT3 pathway, *in vivo*, in the subsequent astroglial response (Sriram et al., 2004; Sriram and O'Callaghan, unpublished observations). A schematic representation of this involvement is illustrated in Figure 9.6. Following 1-methyl-4-phenyl-1,2,3,6-tetrahydropyridine (MPTP)–mediated and methamphetamine (METH)-mediated striatal dopaminergic neurodegeneration or kainic acid (KA)–mediated hippocampal neuronal injury, we have demonstrated that phosphorylation of STAT3 at Tyr705 displays all of the characteristics of an astroglial activation response linked to selective chemical denervation. Thus, phosphorylation of STAT3 after neuronal injury with MPTP, METH, or KA (1) followed the onset of damage to the dopaminergic nerve terminal (↓Dopamine and tyrosine hydroxylase) or hippocampal neurons (↓microtubule-associated protein 2, ↑Fluoro Jade B); (2) preceded the induction of astrogliosis (↑GFAP mRNA and protein); (3) was terminated prior to peak astrogliosis; (4) was only found at the site of damage; (5) was completely abolished by pharmacological neuroprotection of the DA terminals in the MPTP model; and (6) resulted in its translocation to the nucleus of astrocytes prior to the induction of GFAP. Since the *Gfap* gene contains several *cis*-acting elements (Miura et al., 1990; Brenner et al., 1994; Kahn et al., 1997; Krohn et al., 1999) that include potential STAT3 binding sites (Nakashima et al., 1999; Takizawa et al., 2001), it seems likely that STAT3 may play an important role in transcriptional activation of GFAP.

Dimerization and translocation of pSTAT3 is achieved by phosphorylation at Tyr705 residue catalyzed by the Janus family of tyrosine kinases, JAK1, JAK2, JAK3, and TYK2, while phosphorylation of Ser727 by MAPK or mTOR is thought to maximize transcriptional ability of STAT3, in a variety of *in vitro* models (Darnell, 1997; Leonard and O'Shea, 1998). Among the several tyrosine kinases known to be associated with activation of STAT3, we observed activation of only JAK2 in a manner temporally consistent with the subsequent phosphorylation of STAT3 at Tyr705 in all the chemical-injury models that were evaluated in our laboratory (Sriram et al., 2004). While JAK1 has been implicated in the activation of STAT3 in neurons and glia in ischemic-injury models (Justica et al., 2000), this kinase was not activated after chemical injury. Further, activated STAT3 was confined to astrocytes in the MPTP damage model, suggesting that JAK2 is linked to activation of STAT3 in reactive astrocytes, whereas JAK1 may play a role in other neural cell–type responses in other injury models. As further evidence of the cellular specificity of the observed JAK2/STAT3 activation, pharmacological modulation of JAK/STAT signaling with the JAK2 inhibitor, AG490, significantly attenuated MPTP-induced phosphorylation of STAT3 and astrogliosis, but not dopaminergic nerve terminal damage. These *in vivo* findings suggest that it should be possible to distinctly modulate astroglial responses from neuronal damage by altering JAK2/STAT3 signaling. Nevertheless, these observations remain to be established across multiple injury models and neurologic disease states. The results of such experiments should further our understanding of signaling events associated with glial activation, in general, and neuron–glia interactions during injury and repair, in particular.

9.5 APPLYING GENOMIC, PROTEOMIC, AND PROTEIN PHOSPHORYLATION ANALYSES TO IDENTIFY POTENTIAL LIGANDS ACTIVATING SIGNALING PATHWAYS ASSOCIATED WITH REACTIVE ASTROGLIOSIS

Astrocytes and microglia are thought to play a major role in brain inflammatory responses (Raivich et al., 1996; Ransohoff et al., 1996). Both of these cell types exhibit a reactive phenotype in association with neurodegenerative diseases (Norton et al., 1992; Boka et al., 1994; Lippa et al., 1995; Unger, 1998; Giasson et al., 2000; Masliah and LiCastro, 2000; Vila et al., 2001) as well as in response to neurotoxic insults (Balban et al., 1988; O'Callaghan, 1993; Gramsbergen and van den Berg, 1994; O'Callaghan and Miller, 1994; O'Callaghan et al., 1995; Miller et al., 1998; O'Callaghan et al., 1998; Sriram et al., 2002a; Sriram et al., 2002b; Sriram et al., 2004). Recent evidence indicates that brain injury is associated with enhanced expression of cytokines, chemokines, and growth factors, such as tumor necrosis factor (TNF)-α, interleukin (IL)-1β, IL-2, IL-4, and IL-6, MCP-1, and transforming growth factor (TGF)-α and TGF-β1 (Merrill and Chen, 1991; De Bock et al., 1996; Botchkina et al., 1997; Little and O'Callaghan, 2002; Sriram et al., 2002a). Indeed, an elevation in these factors has been linked to neurodegenerative disorders such as Parkinson's disease (Boka et al., 1994; Mogi et al., 1996), Alzheimer's disease (Bauer et al., 1991; Fillit et al., 1991), multiple sclerosis (Merrill, 1992), and stroke (Sairanen et al., 2001), conditions also associated with activation of microglia and astroglia. Thus, ample evidence exists to implicate cytokines/chemokines as participants in the pathological processes underlying both neuronal and glial responses associated with disease-, injury-, or chemical-induced neurodegeneration in the CNS. Such glial activation signals may emanate from the neurons or their terminals and likely only involves signaling among the two activated cell types, microglia and astrocytes. Since the astroglial response is predominantly characterized by the upregulation of GFAP, analysis of the signaling mechanisms associated with the induction of this gene may provide meaningful insights into the role of this gene, in particular, and astroglia, in general, following neural injury.

One way to achieve this is through a detailed study of the changes in genomic and proteomic expression induced by a neurotoxicant, which would help characterize early effectors and mediators of neuronal degeneration and associated gliosis. Microarray analysis and real-time PCR analyses are widely used tools for analysis of genomewide changes in gene expression. Large-scale gene expression profiling studies help identify novel gene expression patterns and provide key insight into gene function and interactions within and across cellular pathways that can then be verified by real-time PCR. However, since not all changes in gene expression may result in a functional protein, and as many cellular events do not require *de novo* synthesis, profiling changes in protein expression or protein levels, including posttranslational modifications such as protein phosphorylation, are critical to envisage a complete picture of the cellular and molecular events. Antibody microarray and ProteinChip® analysis may provide additional data to corroborate genomic analysis. Further, immunoblot analyses of posttranslational modifications, particularly protein phosphorylation using phospho-state specific antibodies, are critical for elucidating the signal transduction pathways. Thus, comprehensive analyses of genomic, proteomic, and posttranslational changes may be the key to understanding the molecular mechanisms underlying glial activation in neurodegenerative disorders, neuronal injury, or neurotoxicity.

In response to the striatal nerve terminal injury mediated by MPTP or METH, or hippocampal neuronal damage caused by KA, the expression of the astrocytic marker, GFAP, increases in a time-dependent manner. GFAP mRNA levels increase by 12 h, while GFAP protein is detectable by 48 h after dosing with MPTP, METH, or KA (Figure 9.7). Having established the time course of neurodegeneration and associated astrogliosis caused by these chemical agents, we have analyzed gene expression changes that occurred within the time frame of neurotoxicant administration and GFAP upregulation (1–72 h after exposure) to identify candidate genes that may contribute to the neuronal degeneration and reactive gliosis. cDNA and antibody microarrays used in these gene expression studies identified several genes whose expression were altered in a time-dependent manner following MPTP. Cytokines, chemokines, trophic factors, transcription factors, cell-adhesion

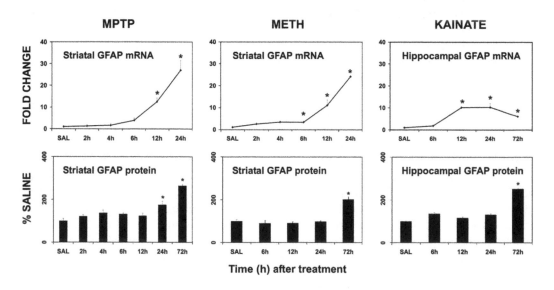

Figure 9.7   Time-dependent expression of GFAP mRNA and protein in multiple models of chemically-induced neurodegeneration. Following systemic administration of MPTP (12.5 mg/kg, *s.c.*), METH (20 mg/kg, *s.c.*), or KA (20 mg/kg, *s.c.*), GFAP mRNA and protein levels were measured at various time points. GFAP mRNA was quantified by real-time PCR and the protein was quantified by ELISA in the target regions of these neurotoxicants.

Table 9.5   Genomic and Proteomic Analysis of MPTP-Induced Striatal Dopaminergic Neurodegeneration

Gene	cDNA Microarray Analysis Fold Change (1–24 h)	Antibody Microarray Analysis Fold Change (48 h)
Ataxin-2	—	2.4
Cyclin E	1.9 to 2.4	1.7
DAT-1	1.7 to 2.1	—
E2f-1	2.0	1.5
Egfr	2.5	1.8
Gfap	2.2 to 2.7	2.0
HO-1	1.7 to 2.7	1.9
Mdm2	2.0 to 3.0	1.8
MEF-1	1.7 to 3.0	—
NeuroD	2.2 to 2.9	—
Neurogenin-3	—	2.6
Neuronal death protein DP-5	1.9	—
Nfat-1	1.6 to 2.8	2.0
NPY receptor	1.7 to 2.3	—
p35	−2.0 to −14.0	—
p53	3.8 to −2.1	1.8
p73	1.6 to 2.0	1.7
Pkc-a	1.9 to 2.3	1.6
Pkc-b	1.7 to 2.6	1.5
Rad50	1.8	1.7
Synuclein-α	—	2.4
Vimentin	1.7 to 2.7	—
Zfp-37	2.7 to 3.7	2.0

Note: — = Not present on the microarray.

Note: Genomic analysis was performed on custom glass mouse microarrays (Clontech) containing 732 genes spotted in duplicate. Total RNA (3 µg) was reverse-transcribed, labeled with Cy3 or Cy5 using the 3DNA Submicro Oligo expression array detection kit (Genisphere) and hybridized (16 h) to the microarray. Antibody microarray analysis was performed using BD Clontech Ab microarray. Total protein (400 µg) was labeled with Cy3 and Cy5 dyes, and 50-µg aliquot of each of the labeled protein was incubated with Ab microarrays. Following appropriate washes, the cDNA or antibody microarrays were scanned for Cy3 (532 nm) and Cy5 (635 nm) Fluor in a GenePix 4000B microarray scanner (Molecular Dynamics) and analyzed. The data are expressed as fold change over saline-treated controls.

molecules, cell-surface antigens, and cytoskeletal proteins were among the genes that were differentially expressed. A medley of genes that are differentially expressed (up- or downregulated) and are consistent across genomic and proteomic analysis are listed in Table 9.5. Many of these findings were validated by real-time PCR and immunoblot analyses. Further, by using the Ciphergen® ProteinChip platform, we have detected, following MPTP treatment, the striatal expression of a ~ 17-kDa protein, which we have putatively identified as the microglial response factor-1 (Figure 9.8). The expression of this factor is consistent with the activation of microglia in the striatum following MPTP (Francis et al., 1995). Thus, genomic and proteomic studies helped identify early effectors/mediators of chemically induced neurotoxicity and have provided a foundation for (1) further characterization of candidate genes to understand their role in the neurotoxicity or the neurodegenerative process, (2) identifying potential signaling pathways associated with glial response to injury, and (3) identifying potential targets for therapeutic intervention.

Using such approaches, we have also identified several JAK/STAT-related ligands that are potential activators of STAT3 in multiple models of neurodegeneration (Sriram and O'Callaghan, unpublished observations). Following MPTP, METH, or KA, the *in vivo* expression of CNTF, IL-6, LIF, and OSM occurred prior to the activation of JAK2/STAT3 pathway and the upregulation of

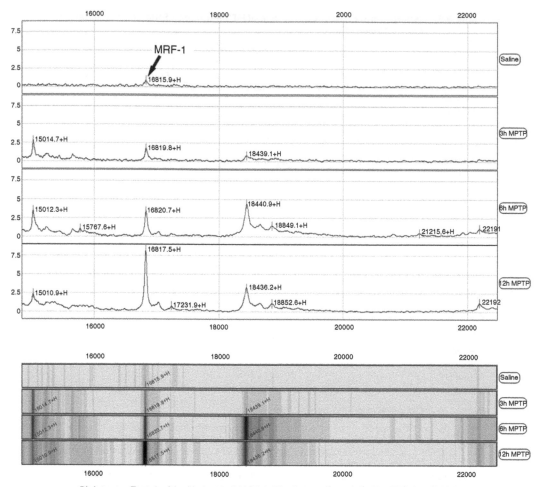

Ciphergen Proteinchip: Hydrophobic (H4); Binding and wash buffer-PBS/0.5M NaCl

Figure 9.8 Proteomic profiling of proteins was performed using Ciphergen® ProteinChip® arrays. Protein samples for proteomic analysis were prepared in 0.5% β-octylglucopyranoside at a concentration of 5 mg/ml. Aliquots of the protein were diluted in appropriate binding buffers and 50 µl of these aliquots were loaded onto each spot using a bioprocessor. Using ProteinChip arrays with hydrophobic (H4) surface, proteins were captured and subject to surface-enhanced laser desorption and ionization time-of-flight mass spectrophotometry (SELDI-TOF). The precise mass was derived based on the recorded time-of-flight of the analyte, and the putative identities of the proteins were determined from the SwissProt protein database.

GFAP, suggesting that these gp130-related cytokines may activate astrocytes following neuronal damage (Figure 9.9). Further, the MPTP-induced expression of the mRNAs for these ligands was blocked by prior neuroprotection with nomifensine, thereby linking their induction to nerve terminal damage. Among these effectors, LIF and OSM serve as likely candidates for "astrocyte activation" because, *in vitro*, they induce the differentiation of astrocytes (Yanigasawa et al., 2001). Further, transgenic mice lacking LIF or LIFR genes exhibit impaired astrocytic differentiation (Bugga et al., 1998; Koblar et al., 1998), suggesting the involvement of this factor in astroglial activation. Although enhanced expression of IL-6 has also been associated with neural injury responses (Raivich et al., 1999; Van Wagoner and Benveniste, 1999), mice lacking IL-6 exhibit reactive gliosis and activation of STAT3 in response to MPTP (Sriram and O'Callaghan, unpublished observations). Thus, contrary to earlier reports, predominantly *in vitro* observations, that associate the induction

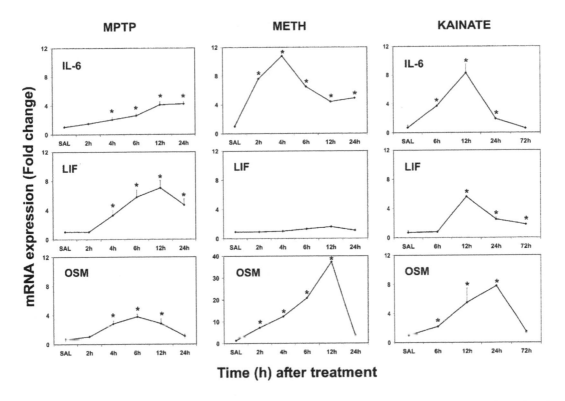

Figure 9.9 Time-dependent expression of gp130–related cytokines in multiple models of chemically induced neurodegeneration. Following systemic administration of MPTP (12.5 mg/kg, *s.c.*), METH (20 mg/kg, *s.c.*), or KA (20 mg/kg, *s.c.*), IL-6, LIF, and OSM mRNAs were quantified at various time points by real-time PCR in the target regions of these neurotoxicants. In all the cases, the expression of these STAT3 ligands preceded the expression of the astrocytic marker, GFAP, as seen in Figure 9.7.

of IL-6 to astroglial activation (Klein et al., 1997), our findings in the MPTP model fail to verify such a link *in vivo*. This observation was further validated using another chemical-injury model, trimethyltin, wherein massive neuronal loss in the hippocampus and the subsequent astroglial activation were not associated with enhanced expression of IL-6 (Little et al., 2002), but resulted in activation of STAT3 (Little and O'Callaghan, unpublished observations). These findings suggest that effectors other than IL-6 activate STAT3 pathway and regulate the expression of GFAP. Indeed, cytokines and chemokines that do not belong to the gp130 class of ligands have been shown to activate STAT3 in nonneuronal cells (Guo et al., 1998; Vila-Coro et al., 1999; Rizk et al., 2001). Thus, it is possible that cytokines and chemokines that are not upstream effectors for the gp130-JAK/STAT3 pathway but that are linked to early injury responses following chemical denervation (e.g., TNF-α and MCP-1) (Little et al., 2002; Sriram et al., 2002a) can play an important role in the induction of astrogliosis. Such additional effectors may emanate from the damaged neural element itself or from activated microglia, prior to contributing to astroglial activation. Thus, in an *in vivo* situation, several cellular signaling pathways may independently lead to, crosstalk with, and/or may converge to activate astrocytic STAT3.

Based on the activation of JAK2/STAT3 in astrocytes across several chemical-injury models (Sriram and O'Callaghan, unpublished observations), it seems likely that the gp130-mediated activation of JAK2/STAT3 signaling pathway may play a predominant role in the regulation of GFAP and astrogliosis in response to chemically induced neurodegeneration. Whether this pathway is key to astrocytic activation, in general, regardless of the injury stimulus or its location will require experimental confirmation using a large variety of brain injury models.

9.6 CONCLUSION

The existence of the glial response to injury of the CNS has been known for over a century. Greatly improved molecular and cellular techniques now allow us to examine gliosis in detail. Studies using these approaches document the role of microglia and astroglia as finely tuned and sensitive monitors of the local neural microenvironment. The rapid response of microglia and astroglia to selective chemical denervation suggests that these cells evolved as rapid responders to the disruptions in local cellular homeostasis. Clearly, finely regulated cell-signaling events must underlie glial activation. Identification and characterization of these glial activation signaling cascades will permit modulation of gliosis to promote neuroprotection and regeneration in the face of toxic, traumatic, and disease-induced assaults of the developing and adult CNS.

REFERENCES

Aberg, M.A. et al. (2001). Selective introduction of antisense oligonucleotides into single adult CNS progenitor cells using electroporation demonstrates the requirement of STAT3 activation for CNTF-induced gliogenesis, *Mol. Cell Neurosci.*, 17, 426.

Ajtai, B.M. and Kalman, M. (1998). Glial fibrillary acidic protein expression but no glial demarcation follows the lesion in the molecular layer of cerebellum, *Brain Res.*, 802, 285.

Akira, S. et al. (1994). Molecular cloning of APRF, a novel IFN-stimulated gene factor 3 p91-related transcription factor involved in the gp130-mediated signaling pathway, *Cell*, 77, 63.

Alessandrini, A. et al. (1999). MEK1 protein kinase inhibition protects against damage resulting from focal cerebral ischemia, *Proc. Natl. Acad. Sci. U.S.A.*, 96, 12866.

Alessi, D.R. et al. (1995). PD 098059 is a specific inhibitor of the activation of mitogen-activated protein kinase kinase *in vitro* and *in vivo*, *J. Biol. Chem.*, 270, 27489.

Ananth, C., Gopalakrishnakone, P., and Kaur, C. (2003). Induction of inducible nitric oxide synthase expression in activated microglia following domoic acid (DA)-induced neurotoxicity in the rat hippocampus, *Neurosci. Lett.*, 338, 49.

Aoki, T., Kobayashi, K., and Isaki, K. (1999). Microglial and astrocytic change in brains of Creutzfeldt-Jakob disease: an immunocytochemical and quantitative study, *Clin. Neuropathol.*, 18, 51.

Araujo, M. and Wandosell, F. (1996). Differential cellular response after glutamate analog hippocampal damage, *J. Neurosci. Res.*, 44, 397.

Arimoto, T. and Bing, G. (2003). Up-regulation of inducible nitric oxide synthase in the substantia nigra by lipopolysaccharide causes microglial activation and neurodegeneration, *Neurobiol. Dis.*, 12, 35.

Arnold, S.E. et al. (1996). Glial fibrillary acidic protein-immunoreactive astrocytosis in elderly patients with schizophrenia and dementia, *Acta Neuropathol. (Berl.)*, 91, 269.

Asanuma, M. et al. (2003). Methamphetamine-induced neurotoxicity in mouse brain is attenuated by keto-profen, a non-steroidal anti-inflammatory drug, *Neurosci. Lett.*, 352, 13.

Ashwell, K.W. (1992). The effects of pre-natal exposure to methylazoxymethanol acetate on microglia, *Neuropathol. Appl. Neurobiol.*, 18, 610.

Badan, I. et al. (2003). Accelerated glial reactivity to stroke in aged rats correlates with reduced functional recovery, *J. Cereb. Blood Flow Metab.*, 23, 845.

Bajetto, A. et al. (2001). Stromal cell-derived factor-1alpha induces astrocyte proliferation through the activation of extracellular signal-regulated kinases 1/2 pathway, *J. Neurochem.*, 77, 1226.

Balaban, C.D., O'Callaghan, J.P., and Billingsley, M.L. (1988). Trimethyltin-induced neuronal damage in the rat brain: comparative studies using silver degeneration stains, immunocytochemistry and immunoassay for neurotypic and gliotypic proteins, *Neuroscience*, 26, 337.

Balcarek, J.M. and Cowan, N.J. (1985). Structure of the mouse glial fibrillary acidic protein gene: implications for the evolution of the intermediate filament multigene family, *Nucleic Acids Res.*, 13, 5527.

Banati, R.B., Daniel, S.F., and Blunt, S.B. (1998). Glial pathology but absence of apoptotic nigral neurons in long-standing Parkinson's disease, *Mov. Disord.*, 13, 221.

Barone, F.C. et al. (2001). Inhibition of p38 mitogen-activated protein kinase provides neuroprotection in cerebral focal ischemia, *Med. Res. Rev.*, 21, 129.

Bauer, J. et al. (1991). Interleukin-6 and alpha-2-macroglobulin indicate an acute-phase state in Alzheimer's disease cortices, *FEBS Lett.*, 285, 111.

Bayatti, N. and Engele, J. (2001). Cyclic AMP modulates the response of central nervous system glia to fibroblast growth factor-2 by redirecting signaling pathways, *J. Neurochem.*, 78, 972.

Blenis, J. (1993). Signal transduction via the MAP kinases: proceed at your own RSK, *Proc. Natl. Acad. Sci. U.S.A.*, 90, 5889.

Boka, G. et al. (1994). Immunocytochemical analysis of tumor necrosis factor and its receptors in Parkinson's disease, *Neurosci. Lett.*, 172, 151.

Bongcam-Rudloff, E. et al. (1991). Human glial fibrillary acidic protein: complementary DNA cloning, chromosome localization, and messenger RNA expression in human glioma cell lines of various phenotypes, *Cancer Res.*, 51, 1553.

Bonni, A. et al. (1997). Regulation of gliogenesis in the central nervous system by the JAK-STAT signaling pathway, *Science*, 278, 477.

Botchkina, G.I. et al. (1997). Expression of TNF and TNF receptors (p55 and p75) in the rat brain after focal cerebral ischemia, *Mol. Med.*, 3, 765.

Brambilla, R. et al. (2002). Induction of COX-2 and reactive gliosis by P2Y receptors in rat cortical astrocytes is dependent on ERK1/2 but independent of calcium signaling, *J. Neurochem.*, 83, 1285.

Brenner, M. (1994). Structure and transcriptional regulation of GFAP gene, *Brain Pathol.*, 4, 245.

Brenner, M. et al. (1990). Characterization of human cDNA and genomic clones for glial fibrillary acidic protein, *Brain Res. Mol Brain Res.*, 7, 277.

Brenner, M. et al. (1994). GFAP promoter directs astrocyte-specific expression in transgenic mice, *J. Neurosci.*, 14, 1030.

Brenner, M. and Messing, A. (1996). GFAP transgenic mice, *Methods*, 10, 351.

Bugga, L. et al. (1998). Analysis of neuronal and glial phenotypes in brains of mice deficient in leukemia inhibitory factor, *J. Neurobiol.*, 36, 509.

Butler, T.L. et al. (2002). Neurodegeneration in the rat hippocampus and striatum after middle cerebral artery occlusion, *Brain Res.*, 929, 252.

Cagnin, A. et al. (2001). *In-vivo* measurement of activated microglia in dementia, *Lancet*, 358, 461.

Cappon, G.D., Pu C., and Vorhees, C.V. (2000). Time-course of methamphetamine-induced neurotoxicity in rat caudate-putamen after single-dose treatment, *Brain Res.*, 863, 106.

Carbonell, W.S. and Mandell, J.W. (2003). Transient neuronal but persistent astroglial activation of ERK/MAP kinase after focal brain injury in mice, *J. Neurotrauma*, 20, 327.

Cardenas, H. and Bolin, L.M. (2003). Compromised reactive microgliosis in MPTP-lesioned IL-6 KO mice, *Brain Res.*, 985, 89.

Carreno-Muller, E. et al. (2003). Thrombin induces *in vivo* degeneration of nigral dopaminergic neurons along with the activation of microglia, *J. Neurochem.*, 84, 1201.

Cattaneo, E., Conti, L., and De-Fraja, C. (1999). Signalling through the JAK-STAT pathway in the developing brain, *Trends Neurosci.*, 22, 365.

Caughlan, A. et al. (2004). Chlorpyrifos induces apoptosis in rat cortical neurons that is regulated by a balance between p38 and ERK/JNK MAP kinases, *Toxicol Sci.*, 78, 125.

Cevolani, D., Bentivoglio, M., and Strocchi P. (2001). Glial reaction to volkensin-induced selective degeneration of central neurons, *Brain Res. Bull.*, 54, 353.

Chen, J. et al. (1998). Effects of ethanol on mitogen-activated protein kinase and stress-activated protein kinase cascades in normal and regenerating liver, *Biochem. J.*, 334, 669.

Chen, W.J. and Liem, R.K. (1994). The endless story of the glial fibrillary acidic protein, *J. Cell Sci.*, 107, 2299.

Cheng, X.P. et al. (2003). Phosphorylation of extracellular signal-regulated kinases 1/2 is predominantly enhanced in the microglia of the rat spinal cord following dorsal root transaction, *Neuroscience*, 119, 701.

Cheung, W.M. et al. (1999). Changes in the level of glial fibrillary acidic protein (GFAP) after mild and severe focal cerebral ischemia, *Chin. J. Physiol.*, 42, 227.

Choi, J.S. et al. (2003). Upregulation of gp130 and STAT3 activation in the rat hippocampus following transient forebrain ischemia, *Glia*, 41, 237.

Choi, S.H. et al. (2003). Thrombin-induced microglial activation produces degeneration of nigral dopaminergic neurons *in vivo*, *J. Neurosci.*, 23, 5877.

Chung, S.Y. and Han, S.H. (2003). Melatonin attenuates kainic acid-induced hippocampal neurodegeneration and oxidative stress through microglial inhibition, *J. Pineal Res.*, 34, 95.

Condorelli, D.F. et al. (1994). Tissue-specific DNA methylation patterns of the rat glial fibrillary acidic protein gene, *J. Neurosci. Res.*, 39, 694.

Condorelli, D.F. et al. (1999). Structural features of the rat GFAP gene and identification of a novel alternative transcript, *J. Neurosci. Res.*, 56, 219.

Cowley, S. et al. (1994). Activation of MAP kinase kinase is necessary and sufficient for PC12 differentiation and for transformation of NIH 3T3 cells, *Cell*, 77, 841.

Cullen, K.M. (1997). Perivascular astrocytes within Alzheimer's disease plaques, *Neuroreport*, 8, 1961.

Czlonkowska, A. et al. (1996). Microglial reaction in MPTP (1-methyl-4-phenyl-1,2,3,6-tetrahydropyridine) induced Parkinson's disease mice model, *Neurodegeneration*, 5, 137.

Da Cruz e Silva, E.F. and O'Callaghan, J.P. (1997). Phosphoprotein phosphatases as potential mediators of neurotoxicity, in *Comprehensive Toxicology, Vol. 11, Nervous System and Behavioral Toxicology*, Sipes, G.I., McQueen, C.A., and Gandolfi, A.J., Eds., Elsevier Science, London, p. 181.

Darnell, J.E., Jr., Kerr, I.M., and Stark, G.R. (1994). Jak-STAT pathways and transcriptional activation in response to IFNs and other extracellular signaling proteins, *Science*, 264, 1415.

Darnell, J.E., Jr. (1997). STATs and gene regulation, *Science*, 277, 1630.

Davis, R.J. (1993). The mitogen-activated protein kinase signal transduction pathway, *J. Biol. Chem.*, 268, 14553.

De Bock, F., Dornand, J., and Rondouin, G. (1996). Release of TNF alpha in the rat hippocampus following epileptic seizures and excitotoxic neuronal damage, *Neuroreport*, 7, 1125.

De Simone, R. et al. (1995). The costimulatory molecule B7 is expressed on human microglia in culture and in multiple sclerosis acute lesions, *J. Neuropathol. Exp. Neurol.*, 54, 175.

DeKosky, S.T. et al. (1996). Interleukin-1 receptor antagonist suppresses neurotrophin response in injured rat brain, *Ann. Neurol.*, 39, 123.

Delacourte, A. (1990). General and dramatic glial reaction in Alzheimer brains, *Neurology*, 40, 33.

Dell'Albani, P. et al. (2001). JAK/STAT signaling pathway mediates cytokine-induced iNOS expression in primary astroglial cell cultures, *J. Neurosci. Res.*, 65, 417.

Del Rio-Hortega, P. (1932). Microglia, in *Cytology and Cellular Pathology of the Nervous System*, Penfield, W., Ed., Hoeber, New York, p. 481.

Depino, A.M, et al. (2003). Microglial activation with atypical proinflammatory cytokine expression in a rat model of Parkinson's disease, *Eur. J. Neurosci.*, 18, 2731.

Dihne, M. and Block, F. (2001). Focal ischemia induces transient expression of IL-6 in the substantia nigra pars reticulata, *Brain Res.*, 889, 165.

Dihne, M. et al. (2001). Time course of glial proliferation and glial apoptosis following excitotoxic CNS injury, *Brain Res.*, 902, 178.

Dommergues, M.A. et al. (2003). Early microglial activation following neonatal excitotoxic brain damage in mice: a potential target for neuroprotection, *Neuroscience*, 121, 619.

Dong, Y. and Benveniste, E.N. (2001). Immune function of astrocytes, *Glia*, 36, 180.

Duffy, P.E., Rapport, M., and Graf, L. (1980). Glial fibrillary acidic protein and Alzheimer-type senile dementia, *Neurology*, 30, 778.

Dugar, A. et al. (1998). Immunohistochemical localization and quantification of glial fibrillary acidic protein and synaptosomal-associated protein (mol. wt 25000) in the ageing hippocampus following administration of 5,7-dihydroxytryptamine, *Neuroscience*, 85, 123.

Dunn, S.L. et al. (2002). Activation of astrocyte intracellular signaling pathways by interleukin-1 in rat primary striatal cultures, *Glia*, 37, 31.

Endoh, M., Maiese, K., and Wagner, J. (1994). Expression of the inducible form of nitric oxide synthase by reactive astrocytes after transient global ischemia, *Brain Res.*, 65, 92.

Engel, S., Wehner, H.D., and Meyermann, R. (1996). Expression of microglial markers in the human CNS after closed head injury, *Acta Neurochir. Suppl. (Wien)*, 66, 89.

Escubedo, E. et al. (1998). Microgliosis and down-regulation of adenosine transporter induced by methamphetamine in rats, *Brain Res.*, 814, 120.

Feinstein, D.L., Weinmaster, G.A., and Milner, R.J. (1992). Isolation of cDNA clones encoding rat glial fibrillary acidic protein: expression in astrocytes and in Schwann cells, *J. Neurosci. Res.*, 32, 1.

Ferraguti, F. et al. (2001). Activated astrocytes in areas of kainite-induced neuronal injury upregulate the expression of the metabotropic glutamate receptors 2/3 and 5, *Exp. Brain Res.*, 137, 1.

Fiedorowicz, A. et al. (2001). Dentate granule neuron apoptosis and glia activation in murine hippocampus induced by trimethyltin exposure, *Brain Res.*, 912, 116.

Fillit, H. et al. (1991). Elevated circulating tumor necrosis factor levels in Alzheimer's disease, *Neurosci. Lett.*, 129, 318.

Francis, J.W. et al. (1995). Neuroglial responses to the dopaminergic neurotoxicant 1-methyl-4-phenyl-1,2,3,6-tetrahydropyridine in mouse striatum, *Neurotoxicol Teratol.*, 17, 7.

Fukumura, M. et al. (1998). A single dose model of methamphetamine-induced neurotoxicity in rats: effects on neostriatal monoamines and glial fibrillary acidic protein, *Brain Res.*, 806, 1.

Fukunaga, K. and Miyamoto, E. (1998). Role of MAP kinase in neurons, *Mol. Neurobiol.*, 16, 79.

Fuller, G. et al. (2001). Activation of p44/p42 MAP kinase in striatal neurons via kainate receptors and PI3 kinase, *Brain Res. Mol. Brain Res.*, 89, 126.

Gabryel, B. and Trzeciak, H.I. (2001). Role of astrocytes in pathogenesis of ischemic brain injury, *Neurotox. Res.*, 3, 205.

Galea, E., Dupouey, P., and Feinstein, D.L. (1995). Glial fibrillary acidic protein mRNA isotypes: expression *in vitro* and *in vivo*, *J. Neurosci. Res.*, 41, 452.

Garcia, S.J. et al. (2002). Chlorpyrifos targets developing glia: effects on glial fibrillary acidic protein, *Brain Res. Dev. Brain Res.*, 133, 151.

Gerhard, A. et al. (2000). *In vivo* imaging of activated microglia using [11C]PK1195 and positron emission tomography in patients after ischemic stroke, *Neuroreport*, 11, 2957.

Geveric, D. et al. (1998). Transcription factor NF-kappaB and inhibitor I kappaBalpha are localized in macrophages in active multiple sclerosis lesions, *J. Neuropathol. Exp. Neurol.*, 57, 168.

Gharami, K. and Das, S. (2004). Delayed but sustained induction of mitogen activated protein kinase activity is associated with beta-adrenergic receptor-mediated morphological differentiation of astrocytes, *J. Neurochem.*, 88, 12.

Giasson, B.I. et al. (2000). The cellular and molecular pathology of Parkinson's disease, in *Neurodegenerative Dementias*, Clark, C.M. and Trojanowski, J.Q., Eds., McGraw-Hill, New York, pp. 219–228.

Ginty, D.D., Bonni, A., and Greenberg, M.E. (1994). Nerve growth factor activates a Ras-dependent protein kinase that stimulates c-fos transcription via phosphorylation of CREB, *Cell*, 77, 713.

Giovannini, M.G. et al. (2002). Beta-amyloid-induced inflammation and cholinergic hypofunction in the rat brain *in vivo*: involvement of the p38MAPK pathway, *Neurobiol. Dis.*, 11, 257.

Glicksman, M.A. et al. (1998). CEP-1347/KT7515 prevents motor neuronal programmed cell death and injury-induced dedifferentiation *in vivo*, *J. Neurobiol.*, 35, 361.

Gonzalez-Scarano, F. and Baltuch, G. (1999). Microglia as mediators of inflammatory and degenerative diseases, *Annu. Rev. Neurosci.*, 22, 219.

Gramsbergen, J.B. and van den Berg, K.J. (1994). Regional and temporal profiles of calcium accumulation and glial fibrillary acidic protein levels in rat brain after systemic injection of kainic acid, *Brain Res.*, 667, 216.

Gray, F. et al. (1999). Neuronal apoptosis in Creutzfeldt-Jakob disease, *J. Neuropathol. Exp. Neurol.*, 58, 321.

Greengard, P. (2001). The neurobiology of slow synaptic transmission, *Science*, 294, 1024.

Grewal, S.S., York, R.D., and Stork, P.J. (1999). Extracellular-signal-regulated kinase signalling in neurons, *Curr. Opin. Neurobiol.*, 9, 544.

Guilarte, T.R. et al. (2003). Methamphetamine-induced deficits of brain monoaminergic neuronal markers: distal axotomy or neuronal plasticity, *Neuroscience*, 122, 499.

Guillemin, G. et al. (1996). Granulocyte macrophage colony stimulating factor stimulates *in vitro* proliferation of astrocytes derived from simian mature brains, *Glia*, 16, 71.

Guo, D. et al. (1998). Induction of Jak/STAT signaling by activation of the type 1 TNF receptor, *J. Immunol.*, 160, 2742.

Guschin, D. et al. (1995). A major role for the protein tyrosine kinase JAK1 in the JAK/STAT signal transduction pathway in response to interleukin-6, *EMBO J.*, 14, 1421.

Harpin, M.L. et al. (1990). Glial fibrillary acidic protein and beta A4 protein deposits in temporal lobe of aging brain and senile dementia of the Alzheimer type: relation with the cognitive state and with quantitative studies of senile plaques and neurofibrillary tangles, *J. Neurosci. Res.*, 27, 587.

Hayashi, T. et al. (2000). C-Jun N-terminal kinase (JNK) and JNK interacting protein response in rat brain after transient middle cerebral artery occlusion, *Neurosci. Lett.*, 284, 195.

Hebert, M.A. and O'Callaghan, J.P. (2000). Protein phosphorylation cascades associated with methamphetamine-induced glial activation, *Ann. N.Y. Acad. Sci.*, 914, 238.

Heinrich, P.C. et al. (1998). Interleukin-6-type cytokine signalling through the gp130/Jak/STAT pathway, *Biochem. J.*, 334, 297.

Hess, A., Desiderio, C., and McAuliffe, W.G. (1990). Acute neuropathological changes in the caudate nucleus caused by MPTP and methamphetamine: immunohistochemical studies, *J. Neurocytol.*, 19, 338.

Hill, C.S. and Treisman, R. (1995). Transcriptional regulation by extracellular signals: mechanisms and specificity, *Cell*, 80, 199.

Hirano, T., Ishihara, K., and Hibi, M. (2000). Roles of STAT3 in mediating the cell growth, differentiation and survival signals relayed through the IL-6 family of cytokine receptors, *Oncogene*, 19, 2548.

Hirsch, E.C. et al. (2003). The role of glial reaction and inflammation in Parkinson's disease, *Ann. N.Y. Acad. Sci.*, 991, 214.

Hu, B.R. and Wieloch, T. (1994). Tyrosine phosphorylation and activation of mitogen-activated protein kinase in the rat brain following transient cerebral ischemia, *J. Neurochem.*, 62, 1357.

Huh, Y. et al. (2003). Microglial activation and tyrosine hydroxylase immunoreactivity in the substantia nigral region following transient focal ischemia in rats, *Neurosci. Lett.*, 349, 63.

Hunter, T. and Cooper, J.A. (1984). Tyrosine protein kinases and their substrates: an overview, *Adv. Cyclic. Nucleotide Protein Phosphorylation Res.*, 17, 443.

Hyman, B.T., Elvhage, T.E., and Reiter, J. (1994). Extracellular signal regulated kinases. Localization of protein and mRNA in the human hippocampal formation in Alzheimer's disease, *Am. J. Pathol.*, 144, 565.

Ihle, J.N. et al. (1995). Signaling through the hematopoietic cytokine receptors, *Annu. Rev. Immunol.*, 13, 369.

Irving, E.A. et al. (2000). Differential activation of MAPK/ERK and p38/SAPK in neurones and glia following focal cerebral ischemia in the rat, *Brain Res. Mol. Brain Res.*, 77, 65.

Ishizawa, K. and Dickson, D.W. (2001). Microglial activation parallels system degeneration in progressive supranuclear palsy and corticobasal degeneration, *J. Neuropathol. Exp. Neurol.*, 60, 647.

Isono, M. et al. (2003). TGF-alpha over-expression induces astrocytic hypertrophy after cortical stab wound injury, *Neurol. Res.*, 25, 546.

Issacs, A. et al. (1998). Determination of the gene structure of human GFAP and absence of coding region mutations associated with fronto-temporal dementia with parkinsonism linked to chromosome 17, *Genomics*, 51, 152.

Jensen, M.G. et al. (2000). IFNgamma enhances microglial reactions to hippocampal axonal degeneration, *J. Neurosci.*, 20, 3612.

Jiang, Z. et al. (2003). Apoptosis and activation of Erk1/2 and Akt in astrocytes post-ischemia, *Neurochem. Res.*, 28, 831.

Jin, S.X. et al. (2003). p38 mitogen-activated protein kinase is activated after a spinal nerve ligation in spinal cord microglia and dorsal root ganglion neurons and contributes to the generation of neuropathic pain, *J. Neurosci.*, 23, 4017.

Johnson, E.A., O'Callaghan, J.P., and Miller, D.B. (2002). Chronic treatment with supraphysiological levels of corticosterone enhances D-MDMA-induced dopaminergic neurotoxicity in the C57BL/6J female mouse, *Brain Res.*, 933, 130.

Jorgensen, M.B. et al. (1993). Microglial and astroglial reactions to ischemic and kainic acid–induced lesions of the adult rat hippocampus, *Exp. Neurol.*, 120, 70.

Jovanovic, J.N. et al. (1996). Neurotrophins stimulate phosphorylation of synapsin I by MAP kinase and regulate synapsin I-actin interactions, *Proc. Natl. Acad. Sci. U.S.A.*, 93, 3679.

Justica, C., Gabriel, C., and Planas, A.M. (2000). Activation of the JAK/STAT pathway following transient focal cerebral ischemia: signaling through Jak1 and Stat3 in astrocytes, *Glia*, 30, 253.

Kahn, M.A. et al. (1997). Ciliary neurotrophic factor activates JAK/Stat signal transduction cascade and induces transcriptional expression of glial fibrillary acidic protein in glial cells, *J. Neurochem.*, 68, 1413.

Kaneko, R. and Sueoka, N. (1993). Tissue-specific versus cell type-specific expression of the glial fibrillary acidic protein, *Proc. Natl. Acad. Sci. U.S.A.*, 90, 4698.

Kang, U.G. et al. (1994). Activation and tyrosine phosphorylation of 44-kDa mitogen-activated protein kinase (MAPK) induced by electroconvulsive shock in rat hippocampus, *J. Neurochem.*, 63, 1979.

Katchanov, J. et al. (2003). Selective neuronal vulnerability following mild focal brain ischemia in the mouse, *Brain Pathol.*, 13, 452.

Kato, H. et al. (2000). Expression of microglial response factor-1 in microglia and macrophages following cerebral ischemia in the rat, *Brain Res.*, 882, 206.

Kim, N.G. et al. (2003). Hypoxia induction of caspase-11/caspase-1/interleukin-1beta in brain microglia, *Brain Res. Mol. Brain Res.*, 114, 107.

Kim, O.S. et al. (2002). JAK-STAT signaling mediates gangliosides-induced inflammatory responses in brain microglial cells, *J. Biol. Chem.*, 277, 40594.

Kim, Y.S. et al. (1994). Phosphorylation and activation of mitogen-activated protein kinase by kainic acid-induced seizure in rat hippocampus, *Biochem. Biophys. Res. Commun.*, 202, 1163.

Klein, M.A. et al. (1997). Impaired neuroglial activation in interleukin-6 deficient mice, *Glia*, 19, 227.

Koblar, S.A. et al. (1998). Neural precursor differentiation into astrocytes requires signaling through the leukemia inhibitory factor receptor, *Proc. Natl. Acad. Sci. U.S.A.*, 95, 3178.

Koczyk, D. and Oderfeld-Nowak, B. (2000). Long-term microglial and astroglial activation in the hippocampus of trimethyltin-intoxicated rat: stimulation of NGF and TrkA immunoreactivities in astroglia but not in microglia, *Int. J. Dev. Neurosci.*, 18, 591.

Kohutnicka, M. et al. (1998). Microglial and astrocytic involvement in a murine model of Parkinson's disease induced by 1-methyl-4phenyl-1,2,3,6-tetrahydropyridine (MPTP), *Immunopharmacology*, 39, 167.

Koistinaho, M. and Koistinaho, J. (2002). Role of p38 and p44/42 mitogen-activated protein kinases in microglia, *Glia*, 40, 175.

Kreutzberg, G.W. (1996). Microglia: A sensor for pathological events in the CNS, *Trends Neurosci.*, 19, 312.

Krohn, K. et al. (1999). Glial fibrillary acidic protein transcription responses to transforming growth factor-β1 and interleukin-1β are mediated by a nuclear factor-1 like site in the near-upstream promoter, *J. Neurochem.*, 72, 1353.

Krum, J.M., Phillips, T.M., and Rosenstein, J.M. (2002). Changes in astroglial GLT-1 expression after neural transplantation or stab wounds, *Exp. Neurol.*, 174, 137.

Kuhlmann, A.C. and Guilarte, T.R. (2000). Cellular and subcellular localization of peripheral benzodiazepine receptors after trimethyltin neurotoxicity, *J. Neurochem.*, 74, 1694.

Kurkowaska-Jastrzebska, I. et al. (1999). The inflammatory reaction following 1-methyl-4-phenyl-1,2,3,6-tetrahydropyridine intoxication in mouse, *Exp. Neurol.*, 156, 50.

Kyriakis, J.M. and Avruch, J. (1996). Sounding the alarm: protein kinase cascades activated by stress and inflammation, *J. Biol. Chem.*, 271, 24313.

Kyriakis, J.M. et al. (1994). The stress-activated protein kinase subfamily of c-Jun kinases, *Nature*, 369, 156.

Lafarga, M. et al. (1998). Reactive gliosis of immature Bergmann glia and microglial cell activation in response to cell death of granule cell precursors induced by methylazoxymethanol treatment in developing rat cerebellum, *Anat. Embryol. (Berl.)*, 198, 111.

Langston, J.W. et al. (1999). Evidence of active nerve cell degeneration in the substantia nigra of humans years after 1-methyl-4-phenyl-1,2,3,6-tetrahydropyridine exposure, *Ann. Neurol.*, 46, 598.

Laskawi, R., et al. (1997). Rapid astroglial reactions in the motor cortex of adult rats following peripheral facial nerve lesions, *Eur. Arch. Otorhinolaryngol.*, 254, 81.

Lawson, L.J. et al. (1990). Heterogeneity in the distribution and morphology of microglia in the normal adult mouse brain, *Neuroscience*, 39, 151.

Lee, C.Y. et al. (2003). Proliferation of a subpopulation of reactive astrocytes following needle-insertion lesion in rat, *Neurol. Res.*, 25, 767.

Lenz, G. et al. (1997). Temporal profiles of the *in vitro* phosphorylation rate and immuno-content of glial fibrillary acidic protein (GFAP) after kainic acid-induced lesions in area CA1 of the rat hippocampus: demonstration of a novel phosphoprotein associated with gliosis, *Brain Res.*, 764, 188.

Leonard, W.J. and O'Shea, J.J. (1998). Jaks and STATs: biological implications, *Annu. Rev. Immunol.*, 16, 293.

Levison, S.W. et al. (2000). IL-6-type cytokines enhance epidermal growth factor-stimulated astrocyte proliferation, *Glia*, 32, 328.

Lewis, S.A. et al. (1984). Sequence of a cDNA clone encoding mouse glial fibrillary acidic protein: structural conservation of intermediate filaments, *Proc. Natl. Acad. Sci. U.S.A.*, 81, 2743.

Liberatore, G.T et al. (1999). Inducible nitric oxide synthase stimulates dopaminergic neurodegeneration in the MPTP model of Parkinson disease, *Nat. Med.*, 5, 1403.

Lin, B. et al. (1998). Sequential analysis of subacute and chronic neuronal, astrocytic and microglial alterations after transient global ischemia in rats, *Acta Neuropathol. (Berl.)*, 95, 511.

Lindenau, J. et al. (1998). Enhanced cellular glutathione peroxidase immunoreactivity in activated astrocytes and in microglia during excitotoxin induced neurodegeneration, *Glia*, 39, 167.

Lippa, C.F., Smith, T.W., and Flanders, K.C. (1995). Transforming growth factor-beta: neuronal and glial expression in the CNS degenerative diseases, *Neurodegeneration*, 4, 425.

Little, A.R., Jr. and O'Callaghan, J.P. (2002). The astrocyte response to neural injury: a review and reconsideration of key features, in *Site-Selective Neurotoxicity*, Lester, D., Slikker, W., and Lazarovici, P., Eds., Taylor and Francis Publishers, London, pp. 233–265.

Liva, S.M. et al. (1999). Signal transduction pathways induced by GM-CSF in microglia: significance in the control of proliferation, *Glia*, 26, 344.

Luellen, B.A. et al. (2003). Neuronal and astroglial responses to the serotonin and norepinephrine neurotoxin: 1-methyl-4-(2'-aminophenyl)-1,2,3,6-tetrahydropyridine, *J. Pharmacol. Exp. Ther.*, 307, 923.

Maier, J. et al. (2002). Regulation of signal transducer and activator of transcription and suppressor of cytokine-signaling gene expression in the brain of mice with astrocyte-targeted production of interleukin-12 or experimental autoimmune encephalomyelitis, *Am. J. Pathol.*, 160, 271.

Malmestrom, C. et al. (2003). Neurofilament light protein and glial fibrillary acidic protein as biological markers in MS, *Neurology*, 61, 1720.

Mandell, J.W. and VandenBerg, S.R. (1999). ERK/MAP kinase is chronically activated in human reactive astrocytes, *Neuroreport*, 10, 3567.

Mangoura, D. et al. (2000). Prolactin concurrently activates src-PLD and JAK/Stat signaling pathways to induce proliferation while promoting differentiation in embryonic astrocytes, *Int. J. Dev. Neurosci.*, 18, 693.

Marshall, C.J. (1994). MAP kinase kinase kinase, MAP kinase kinase and MAP kinase, *Curr. Opin. Genet. Dev.*, 4, 82.

Martin, P.M. and O'Callaghan, J.P. (1996). Gene expression in astrocytes after neural injury, in *The Role of Glia in Neurotoxicity*, Aschner, M. and Kimbelberg, H.K., Eds., CRC Press, Boca Raton, p. 285.

Martinez-Contreras, A. et al. (2002). Astrocytic and microglia cells reactivity induced by neonatal administration of glutamate in cerebral cortex of the adult rats, *J. Neurosci. Res.*, 67, 200.

Masliah, E. and LiCastro, F. (2000). Neuronal and synaptic loss, reactive gliosis, microglial response, and induction of the complement cascade in Alzheimer's disease, in *Neurodegenerative Dementias*, Clark, C.M. and Trojanowski, J.Q., Eds., McGraw-Hill, New York, pp. 131–146.

Matsubara, M. et al. (1996). Site-specific phosphorylation of synapsin I by mitogen-activated protein kinase and Cdk5 and its effects on physiological functions, *J. Biol. Chem.*, 271, 21108.

McGeer, P.L. et al. (2003). Presence of reactive microglia in monkey substantia nigra years after 1-methyl-4-phenyl-1,2,3,6-tetrahydropyridine administration, *Ann. Neurol.*, 54, 599.

Merrill, J.E. (1992). Proinflammatory and anti-inflammatory cytokines in multiple sclerosis and central nervous system acquired immunodeficiency syndrome, *J. Immunother.*, 12, 167.

Merrill, J.E. and Chen, I.S. (1991). HIV-1, macrophages, glial cells, and cytokines in AIDS nervous system disease, *FASEB J.*, 5, 2391.

Mielke, K. and Herdegen, T. (2000). JNK and p38 stress kinases-degenerative effectors of signal-transduction-cascades in the nervous system, *Prog. Neurobiol.*, 61, 45.

Miller, D.B. and O'Callaghan, J.P. (2003). Elevated environmental temperature and methamphetamine neurotoxicity, *Environ. Res.*, 92, 48.

Miller, D.B. et al. (1998). The impact of gender and estrogen on striatal dopaminergic neurotoxicity, *Ann. N.Y. Acad. Sci.*, 844, 153.

Miller, D.B., O'Callaghan, J.P., and Ali, S.F. (2000). Age as a susceptibility factor in the striatal dopaminergic neurotoxicity observed in the mouse flowing substituted amphetamine exposure, *Ann. N.Y. Acad. Sci.*, 914, 194.

Miura, M., Tamura, T., and Mikoshiba, K. (1990). Cell-specific expression of the mouse glial fibrillary acidic protein gene: identification of the cis-and trans-acting promoter elements for astrocyte-specific expression, *J. Neurochem.*, 55, 1180.

Mogi, M. et al. (1996). Interleukin (IL)-1 beta, IL-2, IL-4, IL-6 and transforming growth factor-alpha levels are elevated in ventricular cerebrospinal fluid in juvenile parkinsonism and Parkinson's disease, *Neurosci. Lett.*, 211, 13.

Monville, C. et al. (2001). Ciliary neurotrophic factor may activate mature astrocytes via binding with the leukemia inhibitory factor receptor, *Mol. Cell Neurosci.*, 17, 373.

Murray, B. et al. (1998). Inhibition of the p44/42 MAP kinase pathway protects hippocampal neurons in a cell-culture model of seizure activity, *Proc. Natl. Acad. Sci. U.S.A.*, 95, 11975.

Nakashima, K. et al. (1999). Synergistic signaling in fetal brain by STAT3-Smad1 complex bridged by p3000, *Science*, 284, 479.

Neary, J.T. and Zhu, Q. (1994). Signaling by ATP receptors in astrocytes, *Neuroreport*, 5, 1617.

Nestler, E.J. and Greengard, P. (1984). Neuron-specific phosphoproteins in mammalian brain, *Adv. Cyclic Nucleotide Protein Phosphorylation Res.*, 17, 483.

Nielsen, A.L. et al. (2002). A new splice variant of glial fibrillary acidic protein, GFAP epsilon, interacts with the presenilin proteins, *J. Biol. Chem.*, 277, 29983.

Norenberg, M.D. (1994). Astrocyte responses to CNS injury, *J. Neuropathol. Exp. Neurol.*, 53, 213.

Norton, W.T. et al. (1992). Quantitative aspects of reactive gliosis: a review, *Neurochem. Res.*, 17, 877.

Nozaki, K., Nishimura, M., and Hashimoto, N. (2001). Mitogen-activated protein kinases and cerebral ischemia, *Mol. Neurobiol.*, 23,1.

O'Callaghan, J.P. (1993). Quantitative features of reactive gliosis following toxicant-induced damage of the CNS, *Ann. N.Y. Acad. Sci.*, 679, 195.

O'Callaghan, J.P. (1994). A potential role for altered protein phosphorylation in the mediation of developmental neurotoxicity, *Neurotoxicology*, 15, 29.

O'Callaghan, J.P. and Miller, D.B. (1994). Neurotoxicity profiles of substituted amphetamines in the C57BL/6J mouse, *J. Pharmacol. Exp. Ther.*, 270, 741.

O'Callaghan, J.P., Miller, D.B., and Reinhard, J.F., Jr. (1990). Characterization of the origins of astrocyte response to injury using the dopaminergic neurotoxicant, 1-methyl-4-phenyl-1,2,3,6-tetrahydropyridine, *Brain Res.*, 521, 73.

O'Callaghan, J.P. and Sriram, K. (2004). Focused microwave irradiation of the brain preserves *in vivo* protein phosphorylation: Comparison with other methods of sacrifice and analysis of multiple phosphoproteins, *J. Neurosci. Methods*, 135, 159.

O'Callaghan, J.P., Brinton, R.E., and McEwen, B.S. (1991). Glucocorticoids regulate the synthesis of glial fibrillary acidic protein in intact and adrenalectomized rats but do not affect its expression following brain injury, *J. Neurochem.*, 57, 860.

O'Callaghan, J.P., Jensen, K.F., and Miller, D.B. (1995). Quantitative aspects of drug and toxicant-induced astrogliosis, *Neurochem. Int.*, 26, 115.

O'Callaghan, J.P., Martin, P.M., and Mass, M.J. (1998). The MAP kinase cascade is activated prior to the induction of gliosis in the 1-methyl-4-phenyl-1,2,3,6-tetrahydropyridine (MPTP) model of dopaminergic neurotoxicity, *Ann. N.Y. Acad. Sci.*, 844, 40.

O'Callaghan, J.P. and Seidler, F.J. (1992). 1-Methyl-4-phenyl-1,2,3,6-tetrahydropyridine (MPTP)-induced astrogliosis does not require activation of ornithine decarboxylase, *Neurosci. Lett.*, 148, 105.

Ochiai, W. et al. (2001). Astrocyte differentiation of fetal neuroepithelial cells involving cardiotrophin-1-induced activation of STAT3, *Cytokine*, 14, 264.

Okazawa, H. and Estus, S. (2002). The JNK/c-Jun cascade and Alzheimer's disease, *Am. J. Alzheimers Dis. Other Demen.*, 17, 79.

Oliveira, A., Hodges, H., and Rezaie, P. (2003). Excitotoxic lesioning of the rat basal forebrain with S-AMPA: consequent mineralization and associated glial response, *Exp. Neurol.*, 179, 127.

Otani, N. et al. (2002). Temporal and spatial profile of phosphorylated mitogen activated protein kinase pathways after lateral fluid percussion injury in the cortex of the rat brain, *J. Neurotrauma*, 19, 1587.

Page, K.J., Dunnett, S.B., and Everitt, B.J. (1998). 3-Nitropropionic acid-induced changes in the expression of metabolic and astrocyte mRNAs, *Neuroreport*, 9, 2881.

Panenka, W. et al. (2001). P2X7-like receptor activation in astrocytes increases chemokine monocyte chemoattractant protein-1 expression via mitogen-activated protein kinase, *J. Neurosci.*, 21, 7135.

Panter, S.S. et al. (1985). Glial fibrillary acidic protein and Alzheimer's disease, *Neurochem. Res.*, 10, 1567.

Park, E.J. et al. (2003). 15d-PGJ2 and rosiglitazone suppress Janus kinase-STAT inflammatory signaling through induction of suppressor of cytokine signaling 1 (SOCS1) and SOCS3 in glia, *J. Biol. Chem.*, 278, 14747.

Paul, A. et al. (1997). Stress-activated protein kinases: activation, regulation and function, *Cell Signal.*, 9, 403.

Penkowa, M. et al. (2001). Interleukin-6 deficiency reduces the brain inflammatory response and increases oxidative stress and neurodegeneration after kainic acid-induced seizures, *Neuroscience*, 102, 805.

Perry, G. et al. (1999). Activation of neuronal extracellular receptor kinase (ERK) in Alzheimer disease links oxidative stress to abnormal phosphorylation, *Neuroreport*, 10, 2411.

Peterson, W.M. et al. (2000). Ciliary neurotrophic factor and stress stimuli activate the Jak-STAT pathway in retinal neurons and glia, *J. Neurosci.*, 20, 4081.

Petzold, A. et al. (2002). Markers for different glial cell responses in multiple sclerosis: clinical and pathological correlations, *Brain*, 125, 1462.

Piao, C.S. et al. (2002). Delayed and differential induction of p38 MAPK isoforms in microglia and astrocytes in the brain after transient global ischemia, *Brain Res. Mol. Brain Res.*, 107, 137.

Piao, C.S. et al. (2003). Dynamic expression of p38beta MAPK in neurons and astrocytes after transient focal ischemia, *Brain Res.*, 976, 120.

Platt, B. et al. (2001). Aluminium toxicity in the rat brain: histochemical and immunocytochemical evidence, *Brain Res. Bull.*, 55, 257.

Podkletnova, I. et al. (2001). Microglial response to the neurotoxicity of 6-hydroxydopamine in neonatal rat cerebellum, *Int. J. Dev. Neurosci.*, 19, 47.

Raivich, G., Bluethmann, H., and Kreutzberg, G.W. (1996). Signaling molecules and neuroglial activation in the injured central nervous system, *Keio J. Med.*, 45, 239.

Raivich, G. et al. (1999). Molecular signals for glial activation: pro- and anti-inflammatory cytokines in the injured brain, *Acta Neurochir. Suppl. (Wien)*, 73, 21.

Rajan, P. and McKay, R.D.G. (1998). Multiple routes to astrocytic differentiation in the CNS, *J. Neurosci.*, 18, 3620.

Rajdev, S., Fix, A.S., and Sharp, F.R. (1998). Acute phencyclidine neurotoxicity in rat forebrain: induction of haem oxygenase-1 and attenuation by the antioxidant dimethylthiourea, *Eur. J. Neurosci.*, 10, 3840.

Ransohoff, R.M., Glabinski, A., and Tani, M. (1996). Chemokines in immune-mediated inflammation of the central nervous system, *Cytokine Growth Factor Rev.*, 7, 35.

Reeves, S.A. et al. (1989). Molecular cloning and primary structure of human glial fibrillary acidic protein, *Proc. Natl. Acad. Sci. U.S.A.*, 86, 5178.

Represa, A. et al. (1993). Reactive astrocytes in the kainic acid-damage hippocampus have the phenotypic features of type-2 astrocytes, *J. Neurocytol.*, 22, 299.

Rezaie, P. and Lantos, P.L. (2001). Microglia and the pathogenesis of spongiform encephalopathies, *Brain Res. Brain Res. Rev.*, 35, 55.

Ridet, J.L. et al. (1997). Reactive astrocytes: cellular and molecular cues to biological function, *Trends Neurosci.*, 20, 570.

Rizk, N.M. et al. (2001). Leptin and tumor necrosis factor-alpha induce the tyrosine phosphorylation of signal transducer and activator of transcription proteins in the hypothalamus of normal rats *in vivo*, *Endocrinology*, 142, 3027.

Rodrigues, R.W., Gomide, V.C., and Chadi, G. (2001). Astroglial and microglial reaction after a partial nigrostriatal degeneration induced by the striatal injection of different doses of 6-hydroxydopamine, *Int. J. Neurosci.*, 109, 91.

Russo, C. et al. (2002). Signal transduction through tyrosine-phosphorylated carboxy-terminal fragments of APP via an enhanced interaction with Shc/Grb2 adaptor proteins in reactive astrocytes of Alzheimer's disease brain, *Ann. N.Y. Acad. Sci.*, 973, 323.

Ryu, J.K. et al. (2003). Microglial activation and cell death induced by the mitochondrial toxin 3-nitropropionic acid: *in vitro* and *in vivo* studies, *Neurobiol. Dis.*, 12, 121.

Sairanen, T. et al. (2001). Evolution of cerebral tumor necrosis factor-alpha production during human ischemic stroke, *Stroke*, 32, 1750.

Saporito, M.S. et al. (1999). CEP-1347/KT-7515, an inhibitor of c-jun N-terminal kinase activation, attenuates the 1-methyl-4-phenyl tetrahydropyridine-mediated loss of nigrostriatal dopaminergic neurons *in vivo*, *J. Pharmacol. Exp. Ther.*, 288, 421.

Sarid, J. (1991). Identification of a cis-acting positive regulatory element of the glial fibrillary acidic protein gene, *J. Neurosci. Res.*, 28, 217.

Scali, C. et al. (1999). Beta (1-40) amyloid peptide injection into the nucleus basalis of rats induces microglia reaction and enhances cortical gamma-amino butyric acid release *in vivo*, *Brain Res.*, 831, 319.

Schiefer, J. et al. (1998). Expression of interleukin 6 in the rat striatum following stereotaxic injection of quinolinic acid, *J. Neuroimmunol.*, 89, 168.

Schilling, M. et al. (2003). Microglial activation precedes and predominates over macrophage infiltration in transient focal cerebral ischemia: a study in green fluorescent protein transgenic bone marrow chimeric mice, *Exp. Neurol.*, 183, 25.

Schneider, J.S. and Denaro, F.J. (1988). Astrocytic responses to the dopaminergic neurotoxin 1-methyl-4-phenyl-1,2,3,6-tetrahydropyridine (MPTP) in cat and mouse brain, *J. Neuropathol. Exp. Neurol.*, 47, 452.

Schuler, G.D. et al. (1996). A gene map to the human genome, *Science*, 274, 540.

Schwartz, S.C. et al. (1998). Microglial activation in multiple system atrophy: a potential role for NF-kappaB/rel proteins, *Neuroreport*, 9, 3029.

Seger, R. and Krebs, E.G. (1995). The MAPK signaling cascade, *FASEB J.*, 9, 726.

Sgambato, V. et al. (1998). *In vivo* expression and regulation of elk-1, a target of the extracellular-regulated kinase signaling pathway, in the adult rat brain, *J. Neurosci.*, 18, 214.

Shaw, J.A., Perry, V.H., and Mellanby, J. (1990). Tetanus toxin-induced seizures cause microglial activation in rat hippocampus, *Neurosci. Lett.*, 120, 66.

Sherer, T.B. et al. (2003). Selective microglial activation in the rat rotenone model of Parkinson's disease, *Neurosci. Lett.*, 341, 87.

Soltys, Z. et al. (2003). Morphological transformations of cells immunopositive for GFAP, TrkA or p75 in the CA1 hippocampal area following transient global ischemia in the rat. A quantitative study, *Brain Res.*, 987, 186.

Sriram, K. et al. (2002a). Mice deficient in TNF receptors are protected against dopaminergic neurotoxicity: implications for Parkinson's disease, *FASEB J.*, 16, 1474.

Sriram, K. et al. (2002b). Obesity exacerbates chemically induced neurodegeneration, *Neuroscience*, 115, 1335.

Sriram, K. et al. (2004). Induction of gp130-related cytokines and activation of JAK2/STAT3 pathway in astrocytes precedes upregulation of GFAP in the MPTP model of neurodegeneration: Key signaling pathway for astrogliosis *in vivo*?, *J. Biol. Chem.*, 279, 19936.

Steinert, P.M. and Roop, D.R. (1988). Molecular and cellular biology of intermediate filaments, *Annu. Rev. Biochem.*, 57, 593.

Stoll, G. and Jander, S. (1999). The role of microglia and macrophages in the pathophysiology of the CNS, *Prog. Neurobiol.*, 58, 233.

Streit, W.J., Walter, S.A., and Pennell, N.A. (1999). Reactive microgliosis, *Prog. Neurobiol.*, 57, 563.

Stromberg, I. et al. (1986). Astrocyte responses to dopaminergic denervations by 6-hydroxydopamine and 1-methyl-4-phenyl-1,2,3,6-tetrahydropyridine as evidenced by glial fibrillary acidic protein immunohistochemistry, *Brain Res. Bull.*, 17, 225.

Strong, M. and Rosenfeld, J. (2003). Amyotrophic lateral sclerosis: a review of current concepts, *Amyotroph. Lateral Scler. Other Motor Neuron Disord.*, 4, 136.

Sugama, S. et al. (2003). Age-related microglial activation in 1-methyl-4-phenyl-1,2,3,6-tetrahydropyridine (MPTP)-induced dopaminergic neurodegeneration in C57BL/6 mice, *Brain Res.*, 964, 288.

Suo, Z. et al. (2003). Persistent protease-activated receptor 4 signaling mediates thrombin-induced microglial activation, *J. Biol. Chem.*, 278, 31177.

Sweatt, J.D. (2001). The neuronal MAP kinase cascade: a biochemical signal integration system subserving synaptic plasticity and memory, *J. Neurochem.*, 76, 1.

Taga, T. and Kishimoto, T. (1997). gp130 and the interleukin-6 family of cytokines, *Annu. Rev. Immunol.*, 15, 797.

Takeda, K. and Akira, S. (2000). STAT family of transcription factors in cytokine-mediated biological responses, *Cytokine Growth Factor Rev.*, 11, 199.

Takizawa, T. et al. (2001). Directly linked soluble IL-6 receptor-IL-6 fusion protein induces astrocyte differentiation from neuroepithelial cells via activation of STAT3, *Cytokine*, 13, 272.

Taniwaki, Y. et al. (1996). Microglial activation by epileptic activities through the propagation pathway of kainic acid-induced hippocampal seizures in the rat, *Neurosci. Lett.*, 217, 29.

Teunissen, C.E. et al. (2001). Behavioural correlates of striatal glial fibrillary acidic protein in the 3-nitropropionic acid rat model: disturbed walking pattern and spatial orientation, *Neuroscience*, 105, 153.

Thomas, D.M. et al. (2004). Identification of differentially regulated transcripts in mouse striatum following methamphetamine treatment—an oligonucleotide microarray approach, *J. Neurochem.*, 88, 380.

Tortarolo, M. et al. (2003). Persistent activation of p38 mitogen-activated protein kinase in a mouse model of familial amyotrophic lateral sclerosis correlates with disease progression, *Mol. Cell Neurosci.*, 23, 180.

Troost, D. et al. (1992). Neurofilament and glial alterations in the cerebral cortex in amyotrophic lateral sclerosis, *Acta Neuropathol.*, 84, 664.

Unger, J.W. (1998). Glial reaction in aging and Alzheimer's disease, *Microsc. Res. Tech.*, 43, 24.

Valjent, E., Caboche, J., and Vanhoutte, P. (2001). Mitogen-activated protein kinase/extracellular signal-regulated kinase induced gene regulation in brain: a molecular substrate for learning and memory?, *Mol. Neurobiol.*, 23, 83.

Van Wagoner, N.J. and Benveniste, E.N. (1999). Interleukin-6 expression and regulation in astrocytes, *J. Neuroimmunol.*, 100, 124.

Vanhoutte, P. et al. (1999). Glutamate induces phosphorylation of Elk-1 and CREB, along with c-fos activation, via an extracellular signal-regulated kinase-dependent pathway in brain slices, *Mol. Cell Biol.*, 19, 136.

Vaughan, D.W. and Peters, A. (1974). Neuroglial cells in the cerebral cortex of rats from young adulthood to old age: an electron microscope study, *J. Neurocytol.*, 3, 405.

Versijpt, J.J. et al. (2003). Assessment of neuroinflammation and microglial activation in Alzheimer's disease with radiolabelled PK11195 and single photon emission computed tomography. A pilot study, *Eur. Neurol.*, 50, 39.

Vijayan, V.K. et al. (1991). Astrocyte hypertrophy in the Alzheimer's disease hippocampal formation, *Exp. Neurol.*, 112, 72.

Vila, M. et al. (2001). The role of glial cells in Parkinson's disease, *Curr. Opin. Neurol.*, 14, 483.

Vila-Coro, A.J. et al. (1999). The chemokine SDF-1alpha triggers CXCR4 receptor dimerization and activates the JAK/STAT pathway, *FASEB J.*, 13, 1699.

Vis, J.C. et al. (1999). 3-Nitropropionic acid induces a spectrum of Huntington's disease-like neuropathology in rat striatum, *Neuropathol. Appl. Neurobiol.*, 25, 513.

Vossler, M.R. et al. (1997). cAMP activates MAP kinase and Elk-1 through a B-Raf- and Rap1-dependent pathway, *Cell*, 89, 73.

Wierzba-Bobrowicz, T. et al. (2002). Morphological analysis of active microglia-rod and ramified microglia in human brains affected by some neurological diseases (SSPE, Alzheimer's disease and Wilson's disease), *Folia Neuropathol.*, 40, 125.

Wilms, H. et al. (2003a). Activation of microglia by human neuromelanin is NF-kappaB dependent and involves p38 mitogen-activated protein kinase: implications for Parkinson's disease, *FASEB J.*, 17, 500.

Wilms, H. et al. (2003b). Intrathecal synthesis of monocyte chemoattractant protein-1 (MCP-1) in amyotrophic lateral sclerosis: further evidence for microglial activation in neurodegeneration, *J. Neuroimmunol.*, 144, 139.

Wilsbacher, J.L., Goldsmith, E.J., and Cobb, M.H. (1999). Phosphorylation of MAP kinases by MAP/Erk involves multiple regions of MAP kinases, *J. Biol. Chem.*, 274, 16988.

Wilson, C.M. et al. (2001). Cognitive impairment in sporadic ALS: a pathologic continuum underlying a multisystem disorder, *Neurology*, 57, 651.

Wu, D.C. et al. (2002). Blockade of microglial activation is neuroprotective in the 1-methyl-4-phenyl-1,2,3,6-tetrahydropyridine mouse model of Parkinson disease, *J. Neurosci.*, 22, 1763.

Xia, X.G. et al. (2002). Induction of STAT3 signaling in activated astrocytes and sprouting septal neurons following entorhinal cortex lesion in adult rats, *Mol. Cell Neurosci.*, 21, 379.

Xie, Z., Smith, C.J., and Van Eldik, L.J. (2004). Activated glia induce neuron death via MAP kinase signaling pathways involving JNK and p38, *Glia*, 45, 170.

Xing, J., Ginty, D.D., and Greenberg, M.E. (1996). Coupling of the Ras-MAPK pathway to gene activation by rsk2, a growth factor-regulated CREB kinase, *Science*, 273, 959.

Yamaguchi, H. et al. (1987). Alzheimer's neurofibrillary tangles are penetrated by astroglial processes and appear eosinophilic in their final stages, *Acta Neuropathol. (Berl.)*, 72, 214.

Yan, Z. et al. (1999). D(2) dopamine receptors induce mitogen-activated protein kinase and cAMP response element-binding protein phosphorylation in neurons, *Proc. Natl. Acad. Sci. U.S.A.*, 96, 11607.

Yang, D.D. et al. (1997). Absence of excitotoxicity-induced apoptosis in the hippocampus of mice lacking the Jnk3 gene, *Nature*, 389, 865.

Yanigisawa, M. et al. (2001). Signaling crosstalk underlying synergistic induction of astrocyte differentiation by BMPs and IL-6 family of cytokines, *FEBS Lett.*, 489, 139.

Yao, G.L. et al. (1997). Selective upregulation of cytokine receptor subchain and their intracellular signalling molecules after peripheral nerve injury, *Eur. J. Neurosci.*, 9, 1047.

Zelenika, D. et al. (1995). A novel glial fibrillary acidic protein mRNA lacking exon 1, *Brain Res. Mol Brain Res.*, 30, 251.

Zhang, L. et al. (2000). TNF-alpha induced over-expression of GFAP is associated with MAPKs, *Neuroreport*, 11, 409.

Zhang, P. et al. (1996). Activation of C-jun N-terminal kinase/stress-activated protein kinase in primary glial cultures, *J. Neurosci. Res.*, 46, 114.

Zhong, Z., Wen, Z., and Darnell, J.E., Jr. (1994). Stat3: a STAT family member activated by tyrosine phosphorylation in response to epidermal growth factor and interleukin-6, *Science*, 264, 95.

Zhu, W. et al. (2003). Different glial reactions to hippocampal stab wounds in young adult and aged rats, *J. Gerontol. A. Biol. Sci. Med. Sci.*, 58, 117.

Zucconi, G.G. et al. (2002). Microglia activation and cell death in response to diethyl-dithiocarbamate acute administration, *J. Comp. Neurol.*, 446, 135.

CHAPTER 10

Astrocyte Cellular Swelling: Mechanisms and Relevance to Brain Edema

Herminia Pasantes-Morales and Rodrigo Franco

CONTENTS

10.1 INTRODUCTION

Astrocytes have an essential role in supporting neuronal function by maintaining the normal composition of the extracellular milieu.[1] By this clearance activity, astrocytes protect neurons from the potential injuring or the disturbing effects of excessive concentration of molecules such as glutamate, K^+, ammonium, or lactate. Specific mechanisms of uptake or metabolism operate in astrocytes in optimal conditions to accomplish this homeostatic function. During the progress of pathologies such as ischemia, trauma, epilepsies, and hepatic encephalopathy, the clearance or metabolic capacities of astrocytes initially fulfill their protective role on neurons, but at the cost of altering their own homeostasis, including the increase in cell volume. If the pathological conditions persist or progress, the mechanisms of protection are overwhelmed and astrocytes not only fail to restore the brain homeostasis, but also may trigger responses that exacerbate and spread the original damage. Astrocyte swelling may occur as consequence of a decrease in external osmolarity or in isosmotic conditions, by changes in ion redistribution, or accumulation of ammonia or lactate. This is called cytotoxic or cellular swelling.[2] These two types of swelling are characteristically different from the vasogenic edema, in which the hallmark is the brain–blood barrier disruption. However, pathologies leading to brain edema are not, in general, entirely associated with one type of edema, as during the progress of the pathology the initially pure vasogenic edema gradually results in the development of cellular edema, just as much as cellular edema can induce blood–brain barrier disruption and enable the development of vasogenic edema. In the present review, the emphasis will be put recent findings about the molecular aspects of mechanisms leading to osmotic or cellular astrocyte swelling. Other issues addressed in this chapter include the reasons why astrocytes fail to efficiently operate mechanisms of cell volume correction during cellular swelling and the consequences of astrocyte swelling in the pathology evolution. The consistent observation of a preferential swelling of astrocytes over neurons is discussed in light of the mechanisms operating in each type of cell in response to the swelling-generating factors.

10.2 ASTROCYTE SWELLING IN ISCHEMIA, TRAUMA, AND EPILEPSIES

Ischemia, trauma, and epilepsy have in common a number of causal factors of astrocyte swelling. In epilepsy, the abnormally high firing of neurons elevates K^+ levels, and the resulting depolarization activates the release of glutamate from the widespread glutamatergic synapses. The astrocyte capacity of K^+ clearance by spatial buffering is exceeded, K^+ intracellularly accumulates, and astrocytes swell by the subsequent influx of Cl^- and water. The ineffective clearance of K^+ sustains neuronal depolarization, and glutamate release continues. Excessive glutamate is accumulated into the intracellular compartment by the highly effective glial carriers, which by their intrinsic mechanism, coupled to Na^+, Cl^-, and water, may be, *per se*, swelling inductors.

In ischemia, several conditions or factors successively generate at the various steps of an ischemic episode, leading to astrocyte swelling. At the early phases, the interruption of glucose and oxygen supply leads to energy failure, ATP depletion, and the progressive collapse of transmembrane ionic gradients, resulting in glial and neuronal depolarization, and the outcome of extracellular K^+ and glutamate rise. Additional elements for astrocyte swelling then generate by the activation of anaerobic glycolysis, a pathway that remains operating in astrocytes under hypoxic conditions, as long as there is a remnant delivery of glucose. This alternate route contributes to swelling by generation of protons and lactate. In the late ischemic phases, the mechanism of glutamate removal is inefficiently operating, as the disturbed ionic gradients make the carriers work in reverse, further increasing the concentration of extracellular glutamate.[3,4] At this moment, the cascade of excitotoxicity is triggered, with its inevitable sequel of intracellular Ca^{2+} rise, phospholipid degradation, free fatty acid release, and formation of reactive oxygen species. All these factors substantially contribute, by different mechanisms, to further aggravate astrocyte swelling.

Traumatic brain injury is a multifactorial situation, in which a large number of mechanical and biochemical events are triggered, with different time course and strong interplaying. Among factors potentially inducing swelling, the most directly related to trauma is the postraumatic widespread depolarization, with its associated increase in intracellular Na^+ and Cl^- and water influx. Additional swelling factors derive from the transient sheer forces that mechanically deform membranes, resulting in ion permeability changes. When restriction of cerebral blood flow occurs associated with trauma, the cascade of reactions characteristic of ischemia is also triggered. An indirect mechanism of swelling after brain trauma is the hyponatremia evolved as consequence of a disturbed secretion of the antidiuretic hormone.

10.2.1 Extracellular K+ Clearance and Astrocyte Swelling

The extracellular K+ concentration in brain in physiological conditions is tightly kept within a range of 3–5 mM. This is regulated mainly by the net K+ accumulation into astrocytes, accomplished by the joint operation of the Na+/K+ ATPase and the NKCC cotransporter.[5] The accumulated K+ inside the cell is then redistributed into astrocytes nearby through the gap junction network, to be finally released in the extracellular space at sites with low K+ concentration. This is known as the K+ spatial buffering and is one of the most important functions of the astrocytic compartment in maintaining normal brain excitability and homeostasis.[5] In ischemia and epilepsy, the extracellular K+ levels markedly increase, attaining concentrations of about 15 mM during intense neuronal activity as in epilepsy, and up to 30–80 mM in ischemia.[5] The cellular astrocyte swelling observed in these conditions results in good proportion from the overfunctioning of the K+ uptake mechanisms, exceeding the capacity of spatial buffering. This occurs even when the gap junction operation is not restricted, but rather upregulated by high K+ (see Section 10.3).

Both electroneutral cotransporters and channel-operated mechanisms appear involved in the K+-induced cytotoxic swelling, although the relative contribution of each of them is as yet undefined. The K+-dependent astrocyte swelling can be examined in astrocytes from primary cultures, which show high coupling via gap junctions. The evidence currently available points to the NKCC cotransporter as a main contributor to the K-induced astrocyte swelling, as supported by the following results:

1. In rat cortical cultured astrocytes, K+-evoked swelling is reduced by 10 μM bumetanide, a potent and specific blocker of the carrier.[6]
2. Astrocytes from NKCC-deficient mice show less K+-induced cell swelling.[7]
3. Increased extracellular K+ activates the NKCC transporter in mice astrocytes.[7]
4. In a preparation of astrocytes from the enucleated optic nerve, K+-evoked swelling is blocked by furosemide and reduced by bumetanide.[8]

The KCC cotransporter, found recently in astrocytes, may also contribute to K+ accumulation and astrocyte swelling. Normally this carrier extrudes K+, but under conditions of high extracellular K+, its driving force direction is changed, being then capable of an efficient uptake. The passive permeation of K+ (and Cl−) through ion channels may be an additional mechanism for K+ sequestering. The driving force operating those channels would come from the Donnan forces.[5] This mechanism requires a significant Cl− permeability of the astrocyte membrane, a point that is not fully demonstrated. However, if cells swell, Cl− permeability may increase through volume-activated Cl− channels, allowing Cl− entry until the electrochemical equilibrium is reached. The role for K+ and Cl− channels in accumulating extracellular K+ is supported by a study in rat cortical astrocytes, showing a marked inhibitory effect of KCl-induced swelling by the Cl− channel blockers NPPB and L-644711.[9] The external K+ concentration of 100 mM tested in this study is higher than that used in the cotransporter studies. This may suggest that both carrier- and channel-operating mechanisms may coexist and complement each other, with their relative contribution depending on the extent of the K+ levels reached at the extracellular space. The mechanisms of K+-induced astrocyte swelling are summarized in Figure 10.1.

10.2.2 Glutamate-Induced Astrocyte Swelling

Glutamate clearance from the extracellular space is an important function of astrocytes for maintaining normal excitability and preventing neuronal injury by excitotoxicity. In physiological conditions, glutamate released from neurons and astrocytes is removed by a carrier-mediated uptake. Glutamate transporters are expressed in neurons, astrocytes, oligodendrocytes, and microglia, but uptake by astrocytes is quantitatively the most important, in line with their critical role in glutamate

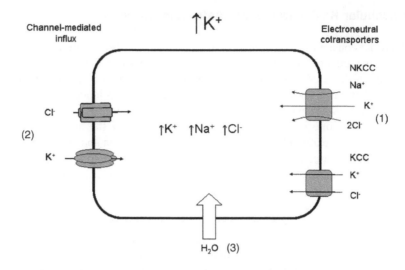

Figure 10.1 K+-induced astrocyte swelling is likely mediated by (1) uptake by the electroneutral cotransporters, NKCC and KCC (KCC3 isoform); and (2) passive influx through K+ channels, driving Cl- influx and osmotically obligated water.

clearance.[10] The glutamate transporter subtypes EAAT1 and EAAT2 (also called GLAST and GLT1) are those expressed in glia. These are Na+-dependent transporters, working by coupling the inward movement of 1 glutamate with 3 Na+ and 1 H+ and the outward movement of 1 K+. The transport is therefore electrogenic and potentially an inductor of astrocyte swelling. In addition to the ion movements directly associated with the transporter, glutamate uptake activates an anion current, which does not appear to be thermodynamically coupled to glutamate transport.[10,11] Increased extracellular glutamate also activates an inwardly rectifying K+ channel. All this ion redistribution leads to osmotic water influx and swelling.[4,12] Moreover, water transport may also occur directly by the EAAT1 operation.[13] Water is cotransported within the protein transporter molecule, together with glutamate and Na+. A fixed number of water molecules is coupled to the influx of each unit charge, and for the stoichiometry of the glutamate transporter, the calculated water influx is of 436 water molecules per one glutamate molecule. The cotransport of water occurs against the water chemical potential difference, and seems energized by coupling of the electrochemical driving force of the downhill Na+ influx.[13] The water pore in the EAAT1 transporter also allows passive fluxes of urea and glycerol.

Extracellular glutamate concentrations rise as a consequence of the increased neuronal activity in epilepsies and contribute, together with K+, to cytotoxic astrocyte swelling. In ischemia, swelling by glutamate uptake occurs mainly at the early steps of the pathology[4] when the transporters in astrocytes are still operating for uptake. The reduced ability to accumulate glutamate occurs earlier in neurons than in astrocytes, a difference based on the different competence of the two cell types to generate the necessary ATP to drive the ion gradients required for the transporter operation. Thus, when neuronal glutamate uptake is interrupted, astrocytes continue to accumulate it. This is illustrated in a study by Torp and coworkers[14] showing a redistribution of glutamate from neurons to astrocytes in the ischemic condition. The normal pattern characterized by higher glutamate levels in neurons than in astrocytes is reversed in ischemia. Glutamate accumulated in astrocytes is neither properly metabolized due to the energy failure nor adequately redistributed due to the impaired astrocyte coupling (Section 10.3), all this resulting in astrocyte swelling. With the energy depletion persistence, the glial glutamate uptake is also finally interrupted, and the transporter operates in reversal, resulting in glutamate efflux.[3,4] At this moment, glutamate release may also occur by activation of the swelling-sensitive glutamate pathways.[15] At this point, excitotoxicity is inevitable.

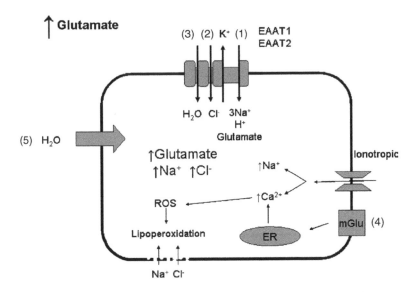

Figure 10.2 Mechanisms of astrocyte swelling induced by high extracellular glutamate. (1) Glutamate, Na^+, and H^+ transport by the glial carriers EAAT1 and EAAT2. (2) Cl^- influx associated with the carrier operation. (3) Water transport by the carrier. (4) Glutamate receptor activation, Na^+ and Cl^- influx, intracellular Ca^{2+} accumulation and outcome of the cascade of ROS generation, mitochondrial permeability transition, and membrane lipoperoxidation, leading to further ion overload. Increased glutamate accumulation and intra cellular ion overload leads to further influx of osmotically obligated water (5) ER, endoplasmic reticulum; mGlu, metabotropic glutamate receptor; ROS, reactive oxygen species.

The excitotoxic condition activates additional mechanisms for astrocyte swelling, now mediated by activation of glutamate receptors present in astrocytes, as suggested by the effects of agonists and antagonists of these receptors on astrocyte volume.[4,16] Excessive Na^+ and Ca^{2+} influx by the receptor hyperactivation triggers a cascade of injuring elements such as arachidonic acid and reactive oxygen species (ROS) (Section 10.2.4). Swelling then results from Na^+ and Cl^- overload through the membranes deteriorated by lipid peroxidation, and the additional Ca^{2+} entry refuels the chain of swelling-inducing elements. All these mechanisms are summarized in Figure 10.2.

10.2.3 Lactacidosis

Lactacidosis is a condition known to accompany ischemia and traumatic brain injury and is a main component of the glial edema associated with these pathological situations. During cerebral ischemia, lactate concentration may increase up to 20–30 mM, and the interstitial pH drops to 6.8–5.8. Lactacidosis results from the ability of astrocytes to continue synthesizing ATP from glycolysis to meet the energy demands, as long as the glucose supply is not fully interrupted and the glycogen pool is not exhausted. Glycolysis generates lactate in proportion of two molecules of lactate for one glucose molecule. The immediate hydrolysis of the ATP formed in this reaction, and the associated increase in H^+ concentration, produce astrocyte swelling by overactivation of the H^+ buffering mechanisms, i.e., the Na^+/H^+ antiporter coupled to the Cl^-/HCO_3^- antiporter. This chain of reactions for lactacidosis-induced swelling is supported by studies in C6 glioma cells, showing the Na^+-dependence of swelling and its reduction in Cl^--free or bicarbonate-free medium, as well as by the protective effect of Na^+/H^+ antiporter or Cl^-/HCO_3^- exchanger blockers.[17] In addition, acidosis-induced swelling is partly prevented by $ZnCl_2$, an inhibitor of selective proton channels.[18] Besides the role of H^+ in generating astrocyte swelling by the glycolytic pathway, lactate is, *per se*, an osmotically active solute and a potential swelling inductor. Lactate redistribution by either active transport or permeation through the gap junctions may spread swelling into other cells in the vicinity. Lactate transport occurs by a monocarboxylate transporter, whose subtype MCT1 is

Lactacidosis

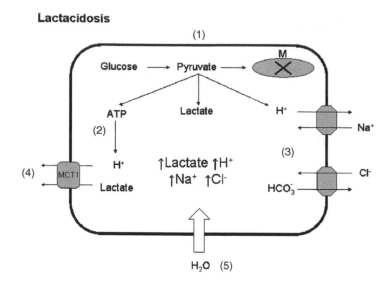

Figure 10.3 Elements of the lactacidosis-induced astrocyte swelling. (1) Excessive lactate formation by increased glycolytic pathway and impaired pyruvate oxidation in the defective mitochondria. (2) Increased H^+ formation coupled to ATP synthesis. (3) Overactivation of the H^+ buffering mechanisms; coupled operation of Na^+/H^+ and Cl^-/HCO_3^- antiporters. (4) Lactate redistribution by the monocarboxylate transporters. Increased lactate accumulation and intracellular ion overload leads to osmotically obligated water influx (5). MCT1, monocarboxylate transporter; M, mitochondria.

almost exclusively expressed in cultured astrocytes, while MCT2 subtype is found predominantly in neurons but is also expressed in astrocytic foot processes around capillaries.[19] Elements of the lactacidosis-induced edema are illustrated in Figure 10.3.

10.2.4 Arachidonic Acid and Generation of Reactive Oxygen Species

Increased free fatty acid levels and generation of reactive oxygen species (ROS) occur during ischemia and brain trauma and are potent swelling inductors in both neurons and astrocytes. Phospholipid hydrolysis and the consequent increase of free fatty acids result from the action of phospholipases. Phospholipase A_2 (PLA$_2$) seems particularly involved in arachidonic acid release.[20] The Ca^{2+}-dependent PLA$_2$ isoform (cPLA$_2$) may be the subtype related to phospholipid degradation in ischemia, as several factors concurrent with the ischemic condition, such as a rise in extracellular K^+ and increased glutamate receptor activity, enhance intracellular Ca^{2+} levels, thus activating cPLA$_2$. Astrocytes express cPLA$_2$ both *in vivo* and *in vitro* and are the cells preferentially showing cPLA$_2$ in human brain.[20] The free fatty acids released by phospholipase activity, by their detergent properties, have, *per se*, a deteriorating influence in membrane fluidity and permeability. In addition, arachidonic acid causes intracellular acidosis and generation of ROS, causing further membrane injury by lipid peroxidation and oxidative damage to membrane proteins. The consequence of this series of reactions is a cell membrane with a highly disturbed permeability and an unregulated influx of Na^+, Cl^-, and Ca^{2+}. This ion overload not only causes swelling but also contributes to autopropagation of the cell membrane insult.

Cellular edema associated with arachidonic, linoleic, and docosahexanoic acids in different brain preparations has been known since the early work by Chan and coworkers.[21] In primary cultured astrocytes and in C6 glioma cells, exposure to arachidonic acid results in a rapid, concentration-dependent swelling, as well as a decrease in intracellular pH and elevated lactate production, two conditions leading to further swelling.[22,23] Glial swelling by arachidonic acid is prevented by antagonists of lipid peroxidation and by replacement of Na^+ or Cl^- in the external medium, in support to the chain of events previously discussed as causative of astrocyte swelling. An effect of arachidonic acid also likely contributing to astrocyte swelling is the stimulation of glycogenolysis, with its concurrent generation of lactate.[24] Astrocytic coupling (Section 10.3) is also affected by elements of the arachidonic acid metabolic cascade and ROS.[25] At this point, the spatial buffering of extracellular glutamate and K+ is disturbed, and astrocyte swelling necessarily occurs.

A consequence of ROS generation that may be relevant for astrocyte swelling in some pathologies is the development of the membrane permeability transition (MPT). MPT is characterized by opening of the permeability transition pore in the inner mitochondrial membrane, allowing the passage of protons, other ions, and some small solutes. This increased permeability collapses the mitochondrial inner membrane potential, which results in colloidosmotic swelling of the mitochondrial matrix, defective oxidative phosphorylation, and interruption of ATP synthesis. Further generation of ROS aggravates the MPT condition. MPT is found in ischemia, trauma, and hepatic encephalopathy. There is evidence showing that MPT blockade by cyclosporine A increases neuronal survival in ischemic conditions, preventing apoptosis, but a connection with astrocyte swelling has been so far described only in hepatic encephalopathy, as will be discussed in the corresponding section.

10.2.5 Mechanical Stress

Activation by mechanical injury of stretch-activated, nonselective cation currents has been recently reported in astrocytes.[26] Cation passage through these channels, driving Cl− and water, may result in cell swelling.

10.3 THE ROLE OF GAP JUNCTIONS IN ASTROCYTE SWELLING

Clearance of K+ and glutamate by astrocytes is determinant for protecting neurons in physiological and pathological conditions. This essential function, accomplished by the mechanisms of spatial buffering, is based on astrocyte coupling through the gap junctions. The adequate operation of this network of intercellular communication is then crucial in preventing astrocyte swelling by excessive accumulation of K+ and glutamate and its known sequel of neuronal injury. However, a number of elements in ischemia and trauma may affect astrocyte coupling by way of gap junctions. The gap junctions are aqueous channels formed by the apposition of two hemichannels, or connexons, each one located in the two interconnected cells. Ions and small molecules diffuse through this channel. The molecular constituents of gap junction hemichannels are the connexins, a family of homolog protein subunits. Connexin 43 (Cx43) is the predominant form in astrocytes, although low levels of other connexins have also been detected. Gap junctions and connexin expression are particularly abundant in astrocytic processes around chemical synapses, as well as at the astrocytic endfeet surrounding blood vessels.[27] Gap junction operation is regulated by long-term and short-term mechanisms and is influenced by a large number of effectors, including high levels of K+ and glutamate. Increased extracellular K+ levels induce a concentration-dependent upregulation of astrocytic coupling, achieved by increased hemichannel opening. Similarly, glutamate-evoked depolarization increases astrocyte gap junction communication.[28,29] Thus, in epilepsies and in early ischemic phases, upregulated, functional gap junctions contribute to K+ and glutamate clearance. However, as the ischemic episode progresses, the efficacy of the astrocyte coupling is reduced. Although increased Cx43 immunoreactivity has been observed in experimental models of ischemia, the overexpressed Cx43 appears to correspond mainly to internalized connexin, a situation rather reducing the efficiency of astrocyte coupling.[30] In fact, studies in brain slices and in cultured astrocytes treated with metabolic inhibitors have shown that gap junctions remain open, but the astrocytic coupling is markedly reduced.[31–33] Dephosphorylation of Cx43 is observed in conjunction with the uncoupling, but it is so far unclear whether these two processes have a causal relationship, as enhancement as well as closure or disruption of gap junctions has been found associated with

Cx43 dephosphorylation.[25] Arachidonic acid, generated during ischemia and trauma, exerts a concentration- and time-dependent reduction in gap junction–mediated astrocyte coupling. This effect is greatly reduced by inhibitors of arachidonic acid metabolism, by external Ca^{2+} removal, or by antioxidants. Astrocyte uncoupling by arachidonic acid occurs without changes in Cx43 phosphorylation, but the connexin expression is markedly decreased.[25] This effect of arachidonic acid on astrocyte coupling, by reducing the efficacy of spatial buffering of extracellular glutamate and K^+, is also instrumental in astrocyte swelling.

10.4 ASTROCYTE SWELLING IN HEPATIC ENCEPHALOPATHY

Hepatic encephalopathy (HE) refers to a complex neurological and neuropsychiatric syndrome occurring as a consequence of acute liver failure or chronic liver disease. The neuropathology in both the acute and chronic conditions involves a marked neuronal dysfunction, resulting from numerous alterations in ion channels and in neurotransmitter transporters and receptors.[34] In acute, but not in chronic, liver failure, brain edema is the most characteristic neuropathological feature, and it is the major cause of death. The fatal outcome is a consequence of raised intracranial pressure and herniation. Brain edema in acute liver failure is essentially of the cellular type, with low contribution of vasogenic edema. The blood–brain barrier appears anatomically normal, with intact intercellular tight junctions, although marked swelling is observed at the perivascular astroglial foot processes. Then, brain edema in acute liver failure has an important component of cellular edema, which is, in essence, restricted to astrocytes. However, the increases in intracranial pressure and brain herniation are likely due to cerebral vasodilatation and higher cerebral blood flow rather than to cellular swelling.[35]

The astrocytic cellular swelling in acute liver failure seems to involve, as initial and causal factor, an increase in brain ammonia levels, subsequent to its rise in blood. Ammonia concentration in arterial blood of patients with liver failure may be up to 45-fold higher than normal, and the physiological brain/blood ammonia ratio of 1.5-3.0 increases up to 8 in hyperammonemia. The rise in brain ammonia, present in all conditions of liver dysfunction, seems due in part to its enhanced permeability through the blood–brain barrier. The ammonia metabolic rate is also elevated in the brains of patients with HE, as detected by ^{15}N-ammonia turnover.[36] Ammonia detoxification in brain in both normal and hyperammonemic conditions is essentially carried out by astrocytes. Since the brain lacks the key enzymes to remove ammonia in the form of urea, its metabolism occurs basically via the synthesis of glutamine, through the amidation of glutamate by the glutamine synthetase, an enzyme localized almost exclusively in astrocytes.[37] Consequently, brain glutamine production in astrocytes is dramatically increased following hyperammonemia caused by acute liver failure.[36] Astrocyte swelling, and the resulting brain edema in acute liver failure, seems to result from the interplay of several mechanisms initiated by intracellular glutamine accumulation. Blockade of key enzymes in the oxidative metabolism and lactate production, free-radical generation, and induction of mitochondrial permeability transition all occur in association with the ammonium/glutamine rise in astrocytes. How these different elements in the pathogenic chain are interconnected and their precise hierarchy are still unresolved.

Consistent with the critical role of increased cellular glutamine levels in astrocyte swelling, treatment with methionine sulfoximine, an inhibitor of glutamine synthesis, reduces glutamine increase and brain water content in models of experimental hyperammonemia *in vivo*, as well as swelling in ammonia-exposed cultured astrocytes.[38,39] A suggested possibility is that swelling results solely from the osmotic imbalance produced by this intracellular glutamine accumulation. This explanation supposes that the mechanisms activated to adjust cell volume, i.e., the extrusion of intracellular osmolytes, are impaired or that their capacity is exceeded by the glutamine accumulation. Another potential mechanism for astrocyte swelling is that derived from the mitochondrial dysfunction resulting from the ammonia or glutamine induction of the MPT.[40] The MPT is a Ca^{2+}-dependent mechanism and is induced by oxidative stress, nitric oxide, and peripheral benzodiazepine receptor activation, conditions all associated with the ammonia increase. Induction of MPT by ammonia has been shown in cultured astrocytes, but interestingly, this effect is not observed in neurons, suggesting that glutamine, and not directly ammonia, is the critical factor inducing MPT.[40] In line with this interpretation is the effect of methionine sulfoximine preventing MPT.[40] The mechanisms by which glutamine induces MPT are so far unclear. Glutamine mitochondrial accumulation together with osmotically obligated water may be an initial step in the MPT occurrence in astrocytes.[41] The implication of an MPT-mediated effect of ammonia/glutamine on astrocyte swelling is demonstrated by its suppression by cyclosporin A, a well-known MPT blocker.[42] MPT may contribute to swelling by mechanisms linked to the ammonia-induced upregulation of AQP4 in astrocytes, which is also inhibited by cyclosporin A.[43]

A crucial element in astrocyte swelling in astrocytes appears to be the free-radical generation. Markers of a condition of oxidative/nitrosative stress in brain have been found to be associated with acute ammonia toxicity. In cultured astrocytes, ammonia and glutamine induce a rapid increase of free-radical production, and conversely, antioxidants reduce ammonia-induced astrocyte swelling.[44] Oxidative/nitrosative stress may contribute to astrocyte swelling by several mechanisms, one of the most plausible being the development of the MPT, which as previously discussed has a key role in ammonia-induced astrocyte swelling. Once MPT has developed, the consequent mitochondrial dysfunction leads, in turn, to further increased free-radical production, in an autogenerating chain of oxidative stress and astrocyte swelling.

Extracellular glutamate levels are enhanced in hyperammonemia, possibly due to the ammonia effect decreasing the expression of EAAT1, the glial glutamate transporter.[36] Hyperactivation of glutamate receptors by both glutamate and depolarization, triggers the cascade of ROS generation and MPT induction, all of which contribute to brain swelling. This is supported by the effects of the NMDA receptor blockers memantine and MK-801, which reduce intracranial pressure and water content in experimental HE, as well as the ammonia-induced swelling in brain cortical slices.[45,46] However, a contribution of the astrocytic compartment to glutamate-induced swelling in HE is not documented.

Brain lactate concentration is enhanced in both acute and chronic liver failure, and lactate levels correlate with the severity of the encephalopathy.[47] Lactate accumulation is found to precede the rise in intracranial pressure, stressing the lactate role in the generation of brain edema.[48] Raised lactate levels may result from the effects of ammonia (1) reducing the activity of the α-ketoglutarate dehydrogenase, (2) stimulating phosphofructokinase, and (3) preventing pyruvate oxidation by the functionally impaired mitochondria. Although the demonstration of astrocyte lactate formation is missing, these are the cells predominantly showing the ability to form lactate, and they are therefore the most plausible candidates for its synthesis in HE. The basic mechanisms leading to astrocyte swelling in acute liver failure are summarized in Figure 10.4.

In chronic liver disease, brain edema seems not to develop in the dramatic way occurring in the acute hepatic failure, and fatal outcome from brain edema rarely occurs in chronic liver dysfunction. Marked and generalized astrocytic cellular swelling is not characteristic of chronic liver failure, although a mild form of swelling occurs in the Alzheimer type II astrocytes, the morphological degenerative change characteristic of this pathology. However, recent studies using magnetization transfer imaging to directly assess brain water content have detected mild edema in patients with chronic liver disease.[49] The reason for the absence of pronounced brain edema in chronic liver failure compared with the acute condition is unclear, since the cerebral glutamine enhancement, a crucial element in generating edema, is similar in the two conditions. A possible explanation, suggested in Rovira et al.[49] and in Haussinger et al.,[50] is that the more rapid evolution of the pathology in the acute condition prevents the volume compensatory changes presumably operating in the chronic condition. Another explanation for the marked difference between acute and chronic liver failure regarding intracranial pressure, edema, and brain herniation points to cerebral vasodilatation as the crucial event. Patients with signs of cerebral edema and intracranial pressure have a higher cerebral blood flow compared to patients without brain swelling.[35]

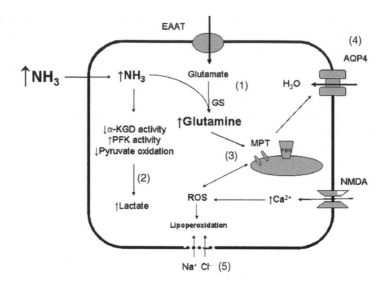

Figure 10.4 Mechanisms of astrocyte swelling in acute liver failure. (1) Ammonia-induced glutamine accumulation. (2) Lactate accumulation by ammonia-induced increase in α-KGD and PFK activity and impaired pyruvate oxidation. (3) MPT induction by glutamine, Ca^{2+} increase, ROS formation, and PBR activation. (4) MPT upregulation of AQP4. (5) Membrane lipoperoxidation and ion overload. α-KGD, α-ketoglutarate dehydrogenase; PFK, phosphofructokinase; MPT, mitochondrial permeability transition; EAAT, glial excitatory amino acid transporter; ROS, reactive oxygen species; PBR, peripheral benzodiazepine receptor; AQP4, glial aquaporin subtype; NMDA, glutamate ionotropic receptor.

10.5 HYPOSMOTIC ASTROCYTE SWELLING AND REGULATORY VOLUME DECREASE

Hyponatremia, a common cause of hyposmotic brain cell swelling, results from an imbalance between intake and excretion of water and electrolytes. Water imbalance occurs by excessive water intake, as in psychotic polydipsia or more commonly from impaired renal elimination as in renal or hepatic failure or by the inappropriate secretion of the antidiuretic hormone. Na^+ loss results from mineralocorticoid deficiency, nephrotic syndrome, osmotic diuresis, vomiting, diarrhea, or excessive sweating. Brain edema occurs in conditions of acute severe hyponatremia but not in chronic hyponatremia. This essential difference likely reflects the ability of brain cells to adjust properly to the external osmolarity decrease and its consequent cell water accumulation, when the osmotic imbalance is not too rapidly imposed.

The first response of brain facing hyponatremia is a compensatory displacement of liquid from the interstitial space to the cerebrospinal fluid and thereafter to the systemic circulation. The next adaptive brain reaction is the activation of cell volume control mechanisms to counteract the intracellular water influx resulting from hyponatremia. This mechanism, termed *regulatory volume decrease*, is accomplished by the extrusion of intracellular solutes and osmotically obligated water. The osmolytes involved are essentially the most concentrated intracellular ions, K^+ and Cl^-, and a number of small organic molecules including myo-inositol, phosphocreatine/creatine, glycerophosphoryl choline, phosphoethanolamine, and the most abundant amino acids, glutamate, glutamine, taurine, and glycine.[51] While decreases of electrolytes are rapid and transitory, those of organic osmolytes are sustained as long as the hyponatremic conditions persist.[51] This difference suggests that the efflux of K^+ and Cl^- is an emergency action for rapidly reducing brain swelling, but because this ion redistribution at the long term alters neuronal function and excitability, the required osmotic equilibrium is achieved by the extrusion of relatively innocuous organic osmolytes. Taurine in particular may be a perfect osmolyte because it is metabolically inert and exhibits only weak synaptic interactions.

Most studies about the mechanisms of swelling and volume regulation have been carried out in brain cells from primary cultures. In these preparations, no substantial differences between neurons and astrocytes have been observed, neither in the hyposmotic-induced swelling nor in the subsequent volume correction, but it is unclear whether this similarity is also present *in vivo*. Water movements across cell membranes are in large part driven by hydrostatic and osmotic pressure, the latter depending on the differences in concentration of osmotically active solutes. Therefore, hyposmotic cell swelling should not be different, in principle, in neurons and astrocytes. However, predominant astrocyte swelling may occur either by differences in osmolyte handling during cell volume regulation or by intrinsic cell properties, such as the distribution or regulation of aquaporins. A preferential astrocyte swelling may be linked to the predominance in astrocytes of AQP4, the most common aquaporin in brain that exhibits exceptionally high water permeability, which upregulates in hyponatremia.[52] This most important aspect about possible differences in swelling and volume adjustment between brain cell types *in vivo* has not been investigated in detail.

Cultured astrocytes suddenly exposed to solutions of decreased osmolarity activate the volume regulatory decrease by the extrusion of K^+, Cl^-, and organic osmolytes. By this adaptive response, cells tend to recover their original volume, despite the persistence of the hyposmotic condition (Figure 10.5A, B). Cultured astrocytes as well as the C6 glioma cell line, which exhibits many astrocyte features, have been widely used to characterize the mechanisms of the corrective osmolyte fluxes. Swelling-evoked K^+ and Cl^- efflux occurs through separate channels with marginal contribution

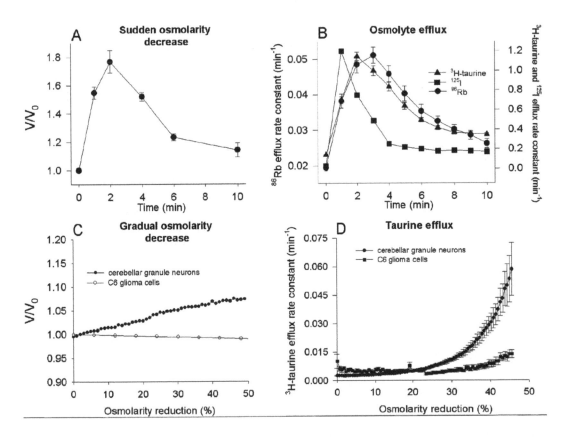

Figure 10.5   Changes in cell volume and osmolyte efflux in hyposmotic conditions. (A) Regulatory volume decrease in cultured cerebellar astrocytes exposed to a 50% hyposmotic solution. (B) The swelling-activated efflux of taurine, K^+, and Cl^-, followed by labeled tracers. (C) Volume changes in cultured cerebellar granule neurons and in C6 glioma cells exposed to gradual osmolarity reductions of 1.8 mOsm/min. (D) Taurine efflux in response to the gradual osmolarity change in cerebellar granule neurons and in C6 cells.

of electroneutral cotransporters. The swelling-activated Cl⁻ channel, examined in detail in C6 glioma cells by Jackson and Strange,[53] is basically similar to that found in other cell types. It is typically an outward rectifier, with unitary conductance of 40–78 pS, requires nonhydrolyzable ATP for activation, and is inhibited by general Cl⁻ channel blockers. The osmosensitive K^+ efflux, also characterized in C6 cells, occurs by activation, at small osmolarity changes, of a Ca^{2+}-dependent, charybdotoxin-sensitive K^+ channel and, at more pronounced osmolarity reductions, of a Ca^{2+}- and voltage-independent channel, blocked by clofilium.[54] The efflux of organic osmolytes occurs, in general, through leak pathways, with the net solute movement following the concentration gradient direction and with essentially no contribution of the Na^+-dependent transporters.[55,56]

The sudden and stringent osmolarity reductions used in most of these experiments may approach the conditions of acute severe hyponatremia, while in chronic hyponatremia the decrease in osmolarity may be small and gradual, allowing brain cells to more efficiently adapt, thus preventing the development of edema. In fact, studies in cerebellar granule neurons and in C6 glioma cells, reproducing more closely the condition of chronic hyponatremia, have shown essentially no swelling in neurons and some swelling in glial cells (Figure 10.5C). This difference might be attributed, at least in part, to an earlier and more important efflux of taurine (and possibly other organic osmolytes) in neurons compared with C6 cells (Figure 10.5D).[57,58] In the same line, a study by Nagelhus et al.[59] in experimental hyponatremia *in vivo* showed cerebellar astrocyte swelling due to transfer of the large pool of taurine from the Purkinje cells to astrocytes. By this mechanism, neurons are spared from swelling. This is another interesting situation in which astrocyte swelling results from their protective role on neuronal homeostasis.

10.6 VOLUME REGULATORY MECHANISMS DO NOT OPERATE IN CYTOTOXIC EDEMA

The persistence of cytotoxic astrocyte swelling in experimental models of ischemia, trauma, and hepatic encephalopathy or in brain edema in patients with acute liver failure reveals the absence of efficient mechanisms for cell volume adjustment. Insufficient volume regulation may be caused by a number of factors. As discussed in the precedent section, volume regulation relies on the extrusion of intracellular osmolytes, mainly K^+, Cl^-, and organic molecules, together with osmotically obligated water. The mechanism of extrusion occurs, in general, via diffusion pathways through which osmolytes move following their concentration gradient. In ischemic conditions, the change in balance of K^+, Na^+, and Cl^- gradients across the membrane, caused by reduced energy production, excludes ion extrusion as an effective mechanism for volume correction, since gradients may favor influx instead of efflux. Besides this, the swelling-sensitive Cl^- channel, which is crucial for an adequate process of volume regulation, requires ATP for activation and will thus be inoperative in conditions of ATP depletion. Furthermore, a recent study[60] has shown impaired activity of the volume-sensitive Cl^- channel during lactacidosis, a condition associated with various situations generating cellular astrocyte swelling.

The release of organic osmolytes, amino acids, myo-inositol, and polyamines could occur, in principle, if the extrusion pathways are not impaired, and this may then help reducing astrocyte volume or at least limit the extent of cell swelling. *In vivo* studies in experimental ischemic models have shown rapid increases in the extracellular concentration of aspartate, glutamate, taurine, and GABA, and the release continues during reoxygenation. A more delayed efflux of glycine, alanine, serine, and phosphoethanolamine occurs only upon reperfusion.[61] Since the efflux is measured in brain *in vivo*, from areas containing neurons and astrocytes, the source of the amino acid efflux cannot be properly identified. The efflux of GABA and glutamate, which are neurotransmitters as well as osmolytes, may respond to events concurrent with ischemia, such as depolarization and cytosolic Ca^{2+} increase, activating neuronal, and not astrocytic, release. In a pure astrocyte preparation, as the primary cultured astrocytes, taurine and glutamate efflux is observed in ischemic-like conditions.[62] The release seems to occur by the carrier reversal operation, driven by the change in ion gradients in the ischemic condition. It is then possible that upon swelling, astrocytes respond by amino acid release, which may be helpful to limit the extent of swelling but is clearly insufficient to correct it. In the case of glutamate, an increased efflux, although potentially useful to relieve swelling, is an additional element of neurotoxicity.

In the section corresponding to liver failure, we referred to the differences regarding brain edema between the acute and the chronic conditions. This difference could be due, as in the case of acute and chronic hyponatremia, to the impossibility of an efficient volume correction in the acute condition due to its rapid evolution, exceeding the capacity of mechanisms for volume control.[49] In the case of ammonia-induced swelling, there is no sufficient information about the reasons for the inefficacy of volume regulatory mechanisms, as those activated in hyponatremia. Depletion of myo-inositol, an osmolyte in astrocytes, is reported in association with increased glutamine levels during hyperammonemia, both *in vivo* and *in vitro*,[63] and this result has been taken as a demonstration of the activation of volume regulatory mechanisms. However, a similar response for taurine, another important osmolyte, is unclear. Ammonia-induced taurine efflux is reported in cultured astrocytes but seems not to be the typical swelling-evoked response,[64] and *in vivo*, no clear correlation between taurine efflux and cell swelling has been established in experimental models of HE or in patients.[34] There is no information about a potential effect of ammonia on the swelling-activated Cl^- and K^+ channels.

10.7 AQUAPORINS AND ASTROCYTE SWELLING

Aquaporins (AQP) are small integral membrane pore proteins that serve as selective pathways for water transport. At least 10 aquaporins have been cloned from mammals, with specific distribution in different organs. Functionally, aquaporins are classified in two groups, those permeable only to water (AQP0, AQP1, AQP2, AQP4, and AQP5) and those transporting water as well as a number of small solutes. AQP3, AQP7, and AQP8 transport glycerol and urea, AQP9 permeates carbamides, polyols, purines, pyrimidines, urea, and monocarboxylates, and AQP6 conducts anions.[65,66] AQP1, AQP3, AQP4, AQP5, AQP8, and AQP9 are expressed in brain. AQP1 expression is restricted to the ventricular-facing surface of the choroid plexus. It is also found in human astrocytoma. AQP3, AQP5, and AQP8 are localized in neurons as well as in glial cells, while AQP4 and AQP9 are present almost exclusively in astrocytes.[65,66] These two astrocytic aquaporins have high intrinsic water permeability. AQP4 is the predominant form in the central nervous system. It has a highly polarized distribution, with selective expression in astrocyte endfeet processes around blood vessels, as well as in astrocytes adjacent to the ependymal and pial surfaces in the ventricular system. The polarized distribution of AQP4 requires the anchorage to an adapter protein related to dystrophin. AQP4 is also heavily expressed in the osmosensory areas of the brain, including hypothalamic supraoptic and paraventricular nuclei. AQP4 expression in glial cells in these areas shows little or no polarization.[65,67] Cultured astrocytes obtained from rat brain also express nonpolarized AQP4.

The specific localization of aquaporins, particularly of AQP4, in glial cells at blood–brain and ventricular interfaces suggests their role in brain water homeostasis in physiological conditions and during the regulation of water fluxes associated with K^+ clearance. Aquaporins may also function as sensors of plasma osmolarity in the brain areas with high vascular density.[68] The expression of AQP4 in astrocytes is sensitive to injuring conditions, including hyponatremia, ischemia, trauma, and hyperammonemia, suggesting that this aquaporin may play a role in the development of brain edema.[66] Supporting this possibility are studies showing the following:

1. An increase in AQP4 expression in astrocytes, which parallels the progress of ischemic brain edema, as monitored by magnetic resonance imaging[69]
2. Significantly less brain edema, better neurological outcome, and less mortality in AQP4–/– transgenic mice compared with AQP4+/+ with induced experimental ischemia and hyponatremia[70]

3. A marked reduction in osmotic water permeability in primary astrocyte cultures from AQP4-deficient mice[71]
4. Delayed onset of brain edema following the AQP4 mislocalization in dystrophin-null transgenic mice
5. AQP4 overexpression in ammonia-treated astrocytes, which appears to precede the onset of astrocyte swelling[43]

The inductors of AQP4 overexpression in pathologies associated with brain edema are not well known. In hyperammonemia, the mitochondrial permeability transition (MPT) induced by glutamine in astrocytes may be an inductive factor, since cyclosporine A blocks both MPT and AQP4 overexpression.[43] MPT induction may also occur in some other conditions associated with brain edema, but a direct correlation between MPT and AQP4 expression has not been established in these pathologies.

AQP9, another predominantly astrocytic aquaporin, may also participate in the development of brain edema in ischemic conditions. AQP9 is found on astrocytic processes and cell bodies, and its distribution is coincident with AQP4 in white matter tracts.[66] AQP9 has an intrinsically high water permeability. It is also permeable to lactate, and this permeability is highly increased by acidosis, which may result in the uptake of excess lactate by astrocytes. By this mechanism, AQP9 may also contribute to lactacidosis-evoked astrocyte swelling.

Besides their possible role in generating edema in pathological conditions, aquaporins may have an important function in the control of cell water content and its influence on cell growth and plasticity. An interesting study in primary cultured astrocytes revealed that suppression of AQP4 determined a much lower water content in astrocytes, accompanied by a new morphological cell phenotype and a strong reduction in cell growth.[72]

10.8 CONSEQUENCES OF ASTROCYTE SWELLING

Astrocyte swelling in the pathologies here examined can be considered as a two-facet process. In the first phase, swelling results from setting in motion a number of reactions to preserve neurons from a risky condition, including swelling itself. Brain edema at this point is mild and transient and has no major consequences. In the second phase, the conditions generating swelling exceed the astrocytes capacities for buffering or active volume regulation, resulting in uncontrolled and extensive cellular edema. At this point, astrocytes are no longer protective and may even exacerbate and spread the damage to sites distant from the initial lesion. Permeation of excessive glutamate, lactate, or Ca^{2+}, through the gap junction network, propagates hyperexcitability, excitotoxicity, and further swelling by Ca^{2+}-associated ROS generation.

Astrocyte swelling influences brain excitability. Epileptiform activity and increased susceptibility to seizures often occurs in acute hyponatremia, as well as in small proportion in hepatic encephalopathy. Narrowing of the extracellular space due to astrocyte swelling generates hyperexcitability by reducing diffusion of neurotransmitters and by enhancing ephaptic interactions. Studies showing that reducing astrocyte swelling with furosemide decreases epileptiform activity and prevents synchronized burst discharges stress the importance of maintaining the normal astrocyte volume and extracellular space size.[73]

Another risk of brain damage related to astrocyte swelling derives from the rapid correction of a hyponatremic condition without considering the changes in the intracellular pool of organic osmolytes resulting from the cell adaptive changes to a hyposmotic environment. Restoring too fast the normal plasma Na^+ levels, without enough time for a new adaptation of intracellular osmolyte levels, results in brain cell dehydration, damage to the blood–brain barrier, and demyelination.

Astrocyte swelling, initially restricted to the cellular type, may evolve to blood–brain barrier disruption and development of vasogenic edema. In such an event, the increased intracranial pressure and the resulting microvessel rupture initiates or aggravates an ischemic condition. Brain herniation, the extreme consequence of brain edema, is life threatening. Brain edema occurs in association with trauma (more vasogenic than cellular edema), stroke (cellular edema in the first stage), acute hyponatremia (vasogenic and cellular edema), and acute liver failure (cellular edema and possibly brain vasodilatation and increased cerebral blood flow). Brain edema associated with severe, acute hyponatremia occurs in psychogenic polydipsia, rapid correction of uremia by excessive hemodialysis, infusion of hypotonic solutions in the preoperative period, or diuretic (thiazide) overuse. Fatal hyponatremia-induced cerebral edema has been recently associated with "ecstasy" use as well as with the noncardiogenic pulmonary edema during strenuous exercise. As previously discussed, fatal brain edema occurs during acute liver failure. In severe head trauma, brain edema is initially only vasogenic, but astrocyte cellular swelling may develop with time and trigger the injuring cascade of increased intracranial pressure, vasculature disruption, ischemia, neuronal death, and extreme brain herniation. In ischemia and hyponatremia, premenopausal women are more at risk of fatal brain edema because of the combined action of estrogens and antidiuretic hormone to constrict brain blood vessels.

10.9 WHY DO ASTROCYTES SWELL MORE THAN NEURONS?

A consistent observation in all pathologies associated with brain edema is that astrocytes, rather than neurons, are the predominant swollen cells. As shown in the present review, this is essentially a consequence of intrinsic properties or mechanisms in the astrocyte directed to protect neuronal homeostasis. They can be summarized as follows:

1. Removal of the highly neurotoxic glutamate occurs preferentially in astrocytes, by the higher efficiency of the glial glutamate transporters.
2. Clearance of excessive extracellular K^+ concentrations is essentially in charge of astrocytes.
3. Activation in astrocytes of the anaerobic glycolysis, an alternate metabolic route facing energy failure, eventually results in swelling-generating factors.
4. The neuron-to-astrocyte transfer of osmolytes such as taurine, glutamate, and possibly others, sparing neurons from swelling.
5. Localization restricted to astrocytes, of the ammonia-detoxification mechanisms.
6. The astroglial predominance of aquaporins with exceptionally high intrinsic permeability to water.

Thus, swelling in astrocytes can be basically considered as a consequence of their quality as homeostatic neuron protectors.

ACKNOWLEDGMENTS

We deeply acknowledge Dr. Michael D. Norenberg for critical review of the hepatic encephalopathy section.

REFERENCES

1. Landis, D.M., The early reactions of non-neuronal cells to brain injury, *Annu. Rev. Neurosci.*, 17, 133, 1994.
2. Kimelberg, H.K., Current concepts of brain edema. Review of laboratory investigations, *J. Neurosurg.*, 83, 1051, 1995.
3. Rossi, D.J., Oshima, T., and Attwell, D., Glutamate release in severe brain ischaemia is mainly by reversed uptake, *Nature*, 403, 316, 2000.
4. Hansson, E. et al., Astroglia and glutamate in physiology and pathology: aspects on glutamate transport, glutamate-induced cell swelling and gap-junction communication, *Neurochem. Int.*, 37, 317, 2000.

5. Walz, W., Role of astrocytes in the clearance of excess extracellular potassium, *Neurochem. Int.*, 36, 291, 2000.

6. Su, G. et al., Astrocytes from Na(+)-K(+)-Cl(-) cotransporter-null mice exhibit absence of swelling and decrease in EAA release, *Am. J. Physiol. Cell Physiol.*, 282, C1147, 2002.

7. Su, G., Kintner, D.B., and Sun, D., Contribution of Na(+)-K(+)-Cl(-) cotransporter to high-[K(+)](o)-induced swelling and EAA release in astrocytes, *Am. J. Physiol. Cell Physiol.*, 282, C1136, 2002.

8. MacVicar, B.A. et al., Intrinsic optical signals in the rat optic nerve: role for K(+) uptake via NKCC1 and swelling of astrocytes, *Glia*, 37, 114, 2002.

9. Rutledge, E.M., Aschner, M., and Kimelberg, H.K., Pharmacological characterization of swelling-induced D-[3H]aspartate release from primary astrocyte cultures, *Am. J. Physiol.*, 274, C1511, 1998.

10. Anderson, C.M. and Swanson, R.A., Astrocyte glutamate transport: review of properties, regulation, and physiological functions, *Glia*, 32, 1, 2000.

11. Fairman, W.A. and Amara, S.G., Functional diversity of excitatory amino acid transporters: ion channel and transport modes, *Am. J. Physiol.*, 277, 481, 1999.

12. Bender, A.S. et al., Ionic mechanisms in glutamate-induced astrocyte swelling: role of K+ influx, *J. Neurosci. Res.*, 52, 307, 1998.

13. MacAulay, N. et al., Passive water and urea permeability of a human Na(+)-glutamate cotransporter expressed in Xenopus oocytes, *J. Physiol.*, 542, 817, 2002.

14. Torp, R. et al., Cellular and subcellular redistribution of glutamate-, glutamine- and taurine-like immunoreactivities during forebrain ischemia: a semiquantitative electron microscopic study in rat hippocampus, *Neuroscience*, 41, 433, 1991.

15. Kimelberg, H.K. et al., Swelling-induced release of glutamate, aspartate, and taurine from astrocyte cultures, *J. Neurosci.*, 10, 1583, 1990.

16. Koyama, Y. et al., Transient treatments with L-glutamate and threo-beta-hydroxyaspartate induce swelling of rat cultured astrocytes, *Neurochem. Int.*, 36, 167, 2000.

17. Staub, F. et al., Effects of lactacidosis on glial cell volume and viability, *J. Cereb. Blood Flow Metab.*, 10, 866, 1990.

18. Plesnila, N. et al., Effect of lactacidosis on cell volume and intracellular pH of astrocytes, *J. Neurotrauma*, 16, 831, 1999.

19. Pellerin, L., Lactate as a pivotal element in neuron-glia metabolic cooperation, *Neurochem. Int.*, 43, 331, 2003.

20. Farooqui, A.A. et al., Phospholipase A2 and its role in brain tissue, *J. Neurochem.*, 69, 889, 1997.

21. Chan, P.H. et al., Induction of brain edema following intracerebral injection of arachidonic acid, *Ann. Neurol.*, 13, 625, 1983.

22. Staub, F. et al., Swelling, acidosis, and irreversible damage of glial cells from exposure to arachidonic acid *in vitro*, *J. Cereb. Blood Flow Metab.*, 14, 1030, 1994.

23. Winkler, A.S., Mechanisms of arachidonic acid induced glial swelling, *Mol. Brain Res.*, 76, 419, 2000.

24. Sorg, O. et al., Adenosine triphosphate and arachidonic acid stimulate glycogenolysis in primary cultures of mouse cerebral cortical astrocytes, *Neurosci. Lett.*, 188, 109, 1995.

25. Martinez, A.D. and Saez, J.C., Regulation of astrocyte gap junctions by hypoxia-reoxygenation, *Brain Res. Rev.*, 32, 250, 2000.

26. Di, X. et al., Mechanical injury alters volume activated ion channels in cortical astrocytes, *Acta Neurochir. Suppl.*, 76, 379, 2000.

27. Rouach, N. et al., Gap junctions and connexin expression in the normal and pathological central nervous system, *Biol. Cell.*, 94, 457, 2002.

28. Enkvist, M.O. and McCarthy, K.D., Astroglial gap junction communication is increased by treatment with either glutamate or high K+ concentration, *J. Neurochem.*, 62, 489, 1994.

29. De Pina-Benabou, M.H. et al., Calmodulin kinase pathway mediates the K+-induced increase in Gap junctional communication between mouse spinal cord astrocytes, *J. Neurosci.*, 21, 6635, 2001.

30. Nagy, J.I. and Li, W.E., A brain slice model for *in vitro* analyses of astrocytic gap junction and connexin 43 regulation: actions of ischemia, glutamate and elevated potassium, *Eur. J. Neurosci.*, 12, 4567, 2000.

31. Cotrina, M.L. et al., Astrocytic gap junctions remain open during ischemic conditions, *J. Neurosci.*, 18, 2520, 1998.

32. Lin, J.H. et al., Gap-junction-mediated propagation and amplification of cell injury, *Nat. Neurosci.*, 1, 494, 1998.

33. Contreras, J.E. et al., Metabolic inhibition induces opening of unapposed connexin 43 gap junction hemichannels and reduces gap junctional communication in cortical astrocytes in culture, *Proc. Natl. Acad. Sci. U.S.A.*, 99, 495, 2002.

34. Butterworth, R.F., Molecular neurobiology of acute liver failure, *Semin. Liver Dis.*, 23, 251, 2003.

35. Blei, A.T. and Larsen, F.S., Pathophysiology of cerebral edema in fulminant hepatic failure, *J. Hepatol.*, 31, 771, 1999.

36. Felipo, V. and Butterworth, R.F., Neurobiology of ammonia, *Prog. Neurobiol.*, 67, 259, 2002.

37. Martinez-Hernandez, A., Bell, K.P., and Norenberg, M.D., Glutamine synthetase: glial localization in brain, *Science*, 195, 1356, 1997.

38. Takahashi, H. et al., Restoration of cerebrovascular CO_2 responsivity by glutamine synthesis inhibition in hyperammonemic rats, *Circ. Res.*, 71, 1220, 1992.

39. Willard-Mack, C.L. et al., Inhibition of glutamine synthetase reduces ammonia-induced astrocyte swelling in rat, *Neuroscience*, 71, 589, 1996.

40. Bai, G. et al., Ammonia induces the mitochondrial permeability transition in primary cultures of rat astrocytes, *J. Neurosci. Res.*, 66, 981, 2001.

41. Zieminska, E. et al., Induction of permeability transition and swelling of rat brain mitochondria by glutamine, *Neurotoxicology*, 21, 295, 2000.

42. Rama Rao, K.V. et al., Suppression of ammonia-induced astrocyte swelling by cyclosporin A, *J. Neurosci. Res.*, 74, 891, 2003.

43. Rama Rao, K.V. et al., Increased aquaporin-4 expression in ammonia-treated cultured astrocytes, *Neuroreport*, 14, 2379, 2003.

44. Rama Rao, K.V., Jayakumar, A.R., and Norenberg, M.D., Ammonia neurotoxicity: role of the mitochondrial permeability transition, *Met. Brain Dis.*, 18, 113, 2003.

45. Vogels, B.A. et al., Memantine, a noncompetitive NMDA receptor antagonist improves hyperammonemia-induced encephalopathy and acute hepatic encephalopathy in rats, *Hepatology*, 25, 820, 1997.

46. Zielinska, M., Law, R.O., and Albrecht, J., Excitotoxic mechanism of cell swelling in rat cerebral cortical slices treated acutely with ammonia, *Neurochem. Int.*, 43, 299, 2003.

47. Chatauret, N. et al., Mild hypothermia prevents cerebral edema and CSF lactate accumulation in acute liver failure, *Metab. Brain Dis.*, 16, 95, 2001.

48. Tofteng, F. et al., Cerebral microdialysis in patients with fulminant hepatic failure, *Hepatology*, 36, 1333, 2002.

49. Rovira, A. et al., Magnetic resonance imaging measurement of brain edema in patients with liver disease: resolution after transplantation, *Curr. Opin. Neurol.*, 15, 731, 2002.

50. Haussinger, D. et al., Hepatic encephalopathy in chronic liver disease: a clinical manifestation of astrocyte swelling and low-grade cerebral edema?, *J. Hepatol.*, 32, 1035, 2000.

51. Verbalis, J.G. and Gullans, S.R., Hyponatremia causes large sustained reductions in brain content of multiple organic osmolytes in rats, *Brain Res.*, 567, 274, 1991.

52. Vajda, Z. et al., Increased aquaporin-4 immunoreactivity in rat brain in response to systemic hyponatremia, *Biochem. Biophys. Res. Commun.*, 270, 495, 2000.

53. Jackson, P.S. and Strange, K., Characterization of the voltage-dependent properties of a volume-sensitive anion conductance, *J. Gen. Physiol.*, 105, 661, 1995.

54. Ordaz, B. et al., Volume changes and whole cell membrane currents activated during gradual osmolarity decrease in C6 glioma cells. Contribution of two types of K+ channels, *Am. J. Physiol. Cell Physiol.*, 2004.

55. Sanchez-Olea, R. et al., Hyposmolarity-activated fluxes of taurine in astrocytes are mediated by diffusion, *Neurosci. Lett.*, 130, 233, 1991.

56. Isaacks, R.E. et al., Effect of osmolality and anion channel inhibitors on myo-inositol efflux in cultured astrocytes, *J. Neurosci. Res.*, 57, 866, 1999.

57. Tuz, K. et al., Isovolumetric regulation mechanisms in cultured cerebellar granule neurons, *J. Neurochem.*, 79, 143, 2001.

58. Ordaz, B. et al., Osmolytes and mechanisms involved in regulatory volume decrease under conditions of sudden or gradual osmolarity decrease, *Neurochem. Res.*, 29, 65, 2004.

59. Nagelhus, E.A., Lehmann, A., and Ottersen, O.P., Neuronal-glial exchange of taurine during hypo-osmotic stress: a combined immunocytochemical and biochemical analysis in rat cerebellar cortex, *Neuroscience*, 54, 615, 1993.

60. Nabekura, T. et al., Recovery from lactacidosis-induced glial cell swelling with the aid of exogenous anion channels, *Glia*, 41, 247, 2003.

61. Phillis, J.W. and O'Regan, M.H., Characterization of modes of release of amino acids in the ischemic/reperfused rat cerebral cortex, *Neurochem. Int.*, 43, 461, 2003.
62. Saransaari, P. and Oja, S.S., Taurine release is enhanced in cell-damaging conditions in cultured cerebral cortical astrocytes, *Neurochem. Res.*, 24, 1523, 1999.
63. Haussinger, D. et al., Proton magnetic resonance spectroscopy studies on human brain myo-inositol in hypo-osmolarity and hepatic encephalopathy, *Gastroenterology*, 107, 1475, 1994.
64. Zielinska, M., Effects of ammonia *in vitro* on endogenous taurine efflux and cell volume in rat cerebrocortical minislices: influence of inhibitors of volume-sensitive amino acid transport, *Neuroscience*, 91, 631, 1999.
65. Venero, J.L. et al., Aquaporins in the central nervous system, *Prog. Neurobiol.*, 63, 321, 2001.
66. Badaut, J. et al., Aquaporins in brain: distribution, physiology, and pathophysiology, *J. Cereb. Blood Flow Metab.*, 22, 367, 2002.
67. Hasegawa, H. et al., Molecular cloning of a mercurial-insensitive water channel expressed in selected water-transporting tissues, *J. Biol. Chem.*, 269, 5497, 1994.
68. Wells, T., Vesicular osmometers, vasopressin secretion and aquaporin-4: a new mechanism for osmoreception?, *Mol. Cell Endocrinol.*, 136, 103, 1998.
69. Taniguchi, M. et al., Induction of aquaporin-4 water channel mRNA after focal cerebral ischemia in rat, *Brain Res. Mol. Brain Res.*, 78, 131, 2000.
70. Manley, G.T. et al., Aquaporin-4 deletion in mice reduces brain edema after acute water intoxication and ischemic stroke, *Nat. Med.*, 6, 159, 2002.
71. Solenov, E. et al., Sevenfold-reduced osmotic water permeability in primary astrocyte cultures from AQP-4-deficient mice, measured by a fluorescence quenching method, *Am. J. Physiol. Cell Physiol.*, 286, C426, 2004.
72. Nicchia, G.P. et al., Inhibition of aquaporin-4 expression in astrocytes by RNAi determines alteration in cell morphology, growth, and water transport and induces changes in ischemia-related genes, *FASEB J.*, 17, 1508, 2003.
73. Hochman, D.W. et al., Dissociation of synchronization and excitability in furosemide blockade of epileptiform activity, *Science*, 270, 99, 1995.

CHAPTER 11

Astrocyte Metallothioneins and Physiological and Pathological Consequences to Brain Injury

Juan Hidalgo

CONTENTS

11.1 INTRODUCTION

Almost half a century ago, an unusual cadmium binding protein was isolated from horse kidney.[1] Due to its high content of metals and cysteine residues, this protein was named metallothionein (MT).[2,3] Over the years, an exponential number of reports have been published concerning structural, biochemical, regulatory, and physiological aspects of MT.[4–9] It is generally agreed that metallothioneins play major roles in the body, but their physiological functions remain elusive to some extent. It is precisely in the brain where MTs appear to accomplish important functions against injury.

MTs occur throughout the animal kingdom, but also in eukaryotic and prokaryotic microorganisms and higher plants. On the basis of structural relationships, MTs have been subdivided into classes or families.[10] Four, closely linked *Mt* genes (*Mt1–4*) are present in rodents,[11,12] but multiple *Mt1* gene variants are present in ungulates and primates.[4,13,14] *Mt1* and *Mt2* (*Mt1&2*) are expressed coordinately in most tissues including the central nervous system (CNS),[15–17] while *Mt3* and *Mt4* show a much more restricted tissue expression (basically CNS and stratified squamous epithelia, respectively).

All CNS MTs are composed of a single polypeptide chain of 61–68 amino acids, 20 of which are highly conserved cysteine residues, and remarkably no aromatic amino acids or histidine are present. A major feature of their amino acid sequences is the occurrence of Cys-Xaa-Cys and Cys-Cys repeats, where Xaa stands for an amino acid residue other than Cys.[18] These proteins usually bind seven divalent metal ions (Zn(II) in physiological conditions, Cd(II)) and up to 12 monovalent copper ions through thiolate bonds, partitioned into two metal-thiolate clusters.[19–21] Each cluster is located in a separate protein domain designated α (residues 32–61) and β (residues 1–31). When compared with the classical MT-I&II sequences (61–62 amino acids), the sequence of MT-III (68 amino acids) shows two inserts: a single Thr in the N-terminal region and an acidic hexapeptide in the C-terminal region.[22] Such inserts may underlie their apparent different functions in the CNS.

11.2 CENTRAL NERVOUS SYSTEM METALLOTHIONEINS ARE DIFFERENTIALLY EXPRESSED

11.2.1 Metallothionein-I&II Are Preferentially Expressed in Reactive Astrocytes

The interest in investigating MTs in the brain was launched by the finding of a possible involvement of MT-III in Alzheimer's disease.[22] Nevertheless, because most of what is known of this family of proteins has been generated with MT-I&II, I will discuss first the expression of these MT isoforms.

By using conventional chromatographic procedures and cadmium binding to MTs, Chen and Ganther showed in the 1970s that MTs were readily present in the rat brain.[23] More specific analyses revealed MT-I transcripts and upregulation following cadmium administration in the mouse brain a few years later.[24] MT-I&II expression in the brain has been reported in many other species including human,[25] monkey,[26] dog,[27] sheep,[28] and cattle.[29]

By Northern blot,[30–37] in situ hybridization,[17,27,32,35,38–49] microarray,[50,51] RT-PCR,[52] immunohistochemistry,[25,27,28,35,44–46,52–80] Western blot,[36,37,79,80] and radioimmunoassay,[52,58,59,81–86] it has been demonstrated that MT-I&II occur throughout the brain and spinal cord, and that the main cell expressing these MT isoforms is the astrocyte, especially the reactive astrocyte. Nevertheless, MT-I&II expression is also found in ependymal cells, epithelial cells of choroid plexus, meningeal cells of the pia mater, and endothelial cells of blood vessels. Neurons appear to express MT-I&II to a much lower extent than astrocytes,[59,87–89] and thus only a few reports show positive immunostaining of these cells.[25,46,59,60,64,71,90] In the normal brain, microglia and oligodendrocytes are essentially devoid of MT-I&II, but the former cells do upregulate MT-I&II expression in response to injury.[44,45,48,68,70,73,75,91–96]

11.2.2 Expression of Metallothionein-III in the Central Nervous System

The identification of MT-III by Uchida and coworkers[22] was a major breakthrough in the MT field. This protein was unexpectedly discovered in human brain while pursuing the rationale for the decreased neurotrophic activity of brain extracts from Alzheimer's disease (AD) patients in a bioassay where the survival of rat neonatal cortical neurons is examined.[97,98] Because of its activity in such bioassay, MT-III was originally named growth inhibitory factor (GIF), a protein of 68 amino acids whose amino acid sequence was 70% identical to that of human MT-II and that had a single amino acid insert and a unique six amino acid insert in the NH_2- and the COOH-terminal portions, respectively.[22] The biological activity identified with the bioassay was unique to MT-III/GIF, since MT-I+II did not show such inhibitory effects on neuronal survival. A significant decrease of MT-III/GIF was observed in AD brains, and thus these authors proposed that this protein could be underlying the neuropathology of this devastating disease. The molecular cloning of human MT-III/GIF cDNA confirmed the previous results.[11,99] These molecular studies showed that human and mouse MT-III genes had similar size and shared intron/exon boundaries with the MT-I+II genes, and, furthermore, that they were closely linked to the latter in the chromosomes of the respective species (16 in humans, eight in mice).

In contrast to MT-I&II, which are widely expressed in all tissues, the expression of MT-III is much more restricted. It was originally proposed that MT-III was expressed solely in the CNS,[11,22,99,100] but it is now agreed that this protein is expressed at various degrees in other tissues, in some cases in a development-dependent manner.[101–107] Besides the human and mouse brain, MT-III has been reported in other species such as horse and cow,[108] rat,[31] pig,[109] dog,[110] and sheep.[111]

In contrast to MT-I&II, there is still significant uncertainty regarding the cellular source of MT-III/GIF in the CNS. Using antibodies raised against the unique six amino acid insert of GIF, Uchida et al.[22] observed that GIF immunoreactivity was prominent in the gray matter of normal human brains, and in astrocytes but not in neurons or other brain cells. Further confirmation by the same group of these results, by *in situ* hybridization experiments in human brains, and by Northern blot analysis of cultured rat astrocytes, neurons, or microglia, was soon reported.[100,112] However, a flurry of studies have appeared that do not support an astrocyte-only site of MT-III expression, and indeed major discrepancies still exist in the literature. By *in situ* hybridization experiments and with transgenic mice generated using 11.5 kb of the mouse MT-3 5′ flanking region fused to the *E. coli lacZ* gene, Masters et al.[40] clearly demonstrated that in mice MT-III is expressed predominantly in neurons, with little or no signal of MT-III in glial cells; MT-III mRNA was also reported in ependymal cells and in endothelial cells of the choroid plexus. Although in most cases no thorough cell identification has been established by robust methods such as double labeling, the data for MT-III mRNA signal are more consistent with neuronal than with glial cells through the brain parenchyma in normal conditions, with astrocyte MT-III upregulation eventually occurring following injury.[32,35,41,42,45,49,73,110,113–120] In contrast to the MT-III messenger, the MT-III protein has been suggested to be present almost exclusively in astrocytes by a number of studies.[22,45,70,73,79,91,119,121–123] Other reports, however, show convincing nearly neuron-only MT-III protein presence.[80,117,118,124,125] A report demonstrates MT-III protein not only in neurons, astrocyte, and microglia but also in oligodendrocytes following LPS injection.[126] All in all, the results suggest that the cellular source identified depends heavily on the antibody used. Whether this reflects a problem of specificity or a more biological problem is currently unknown and deserves further work.

11.3 TRANSGENIC MICE DEMONSTRATE IMPORTANT ROLES OF METALLOTHIONEINS IN THE CENTRAL NERVOUS SYSTEM

11.3.1 Metallothionein-I&II Are Essential Proteins for Coping with Brain Damage

There is increasing experimental evidence that oxidative stress contributes significantly to the neuropathology of several adult neurodegenerative disorders as well as in stroke, trauma, seizures, and neuronal degeneration caused by persistent activation of glutamate-gated ion channels.[127] MT-I&II are potent antioxidant proteins and are induced by oxidative stress.[7,128,129] Not surprisingly, brain MT-I&II levels have been shown to be increased in Alzheimer's disease,[78,112,130–132] Pick's disease,[130] short-course Creutzfeld-Jakob disease,[79] amyotrophic lateral sclerosis,[54,56,133] multiple sclerosis,[51,96] and aging.[134]

In accordance with the above human data, experiments carried out in animal models clearly demonstrate dramatic upregulations of MT-I&II in the brain by the inflammatory factor endotoxin,[11,15,38,135] stress,[42,82,136,137] glutamate analogs,[31,32,48,50,75,138,139] cryogenic injury,[44,45,62,70] stroke/ischemia,[17,37,46,50,74] familial amyotrophic lateral sclerosis models,[52,86,120,140] multiple sclerosis models,[141–143] and gliotoxins.[66,76,144] In all these models, cytokines and/or oxidative stress are likely to be involved.[145–148] Transgenic mice expressing either interleukin-6 (IL-6), tumor necrosis factor-α (TNF-α), interleukin-3 (IL-3), or interferon-α (IFN-α) under the control of the glial fibrillary acidic protein (GFAP) gene promoter, thereby causing a targeted expression of the astrocyte, show evident signs of gliosis and neuronal damage.[149,150] As could be expected, a dramatic upregulation of MT-I&II was observed in clinically symptomatic GFAP-IL6,[35,151] GFAP-TNF,[73] GFAP-IL3, and GFAP-IFNα[92] mice. A major role of IL-6 in brain MT-I&II regulation was also concluded by studies with IL-6 null mice.[42,45,144,152,153]

Taken together, the above studies strongly suggest a significant role of the MT-I&II during neurodegenerative diseases and in response to brain injury. In order to establish their putative role(s), the generation of genetically modified mice[154–156] has greatly potentiated this research. Mice overexpressing MT-I were partially protected against mild focal cerebral ischemia and reperfusion, since the volume of affected tissue was smaller and the motor performance (3 weeks after the lesion) better.[46] Conversely, MT-I&II null mice developed approximately threefold larger infarcts than wild-type mice and a significantly worse neurological outcome.[37] These results highlight an essential role of MT-I&II to cope with ischemic damage of the brain. Other studies have clearly demonstrated a similar essential role to cope with damage elicited by kainic acid–induced seizures,[48] the gliotoxin 6-aminonicotinamide,[76,157] 6-hydroxydopamine,[158] mutated Cu,Zn-superoxide dismutase,[52,140] multiple sclerosis models,[143,159] traumatic brain injury,[44,93,160] and transgenic IL-6–induced neuropathology.[94,95,161] Over all the results obtained in these studies are compatible with a role of MT-I&II decreasing oxidative stress, inflammation, and apoptosis in the CNS, which is in accordance with results in other tissues.[162–172] The main effects caused by MT-I&II deficiency and MT-I overexpression following injury to the CNS are shown in Figure 11.1. Interestingly enough, exogenously applied MT-II protein mimics MT-I transgenic overexpression, causing a significant clinical improvement in the animal model of multiple sclerosis experimental autoimmune encephalomyelitis,[141,173] and protecting against traumatic brain injury,[93,174] which opens exciting perspectives about the use of the MT family as therapeutic agents.

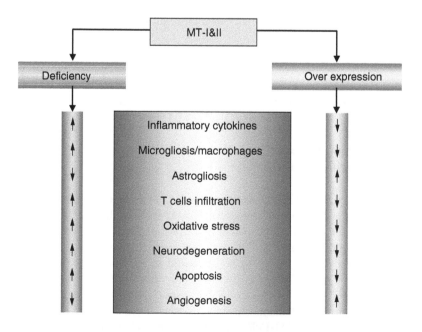

Figure 11.1   Roles of metallothionein-I&II isoforms in the brain following injury.

11.3.2 Metallothionein-III Shows Different Functions

MT-III was discovered unexpectedly while pursuing putative mechanisms underlying the neuropathology of Alzheimer's disease.[22] It was originally suggested that MT-III is downregulated during AD, but unfortunately this is not a consistent finding.[22,70,80,99,175-177] Whether this is due to methodological inconsistencies or to different AD ethiologies is currently unclear. MT-III expression has also been analyzed in other human diseases such as Down syndrome,[178] Creutzfeld-Jakob disease,[79] pontosubicular necrosis,[123] Parkinson's disease, meningitis, or amyotrophic lateral sclerosis.[177] In some of these situations MT-III levels appear to be decreased, while the opposite is observed in others. This is also the case in animal models. Brain damage induced by stab wound injury increased MT-III expression in reactive astrocytes,[113,122,179] as did the administration of kainic acid into the ventricles in reactive astrocytes around the degenerated neurons in the CA3 field of the Ammon's horn 3 days after injury.[113] In contrast, cortical ablation of the somatosensory cortex decreased MT-III expression in the cortex ipsilateral to the injury 1 day after the ablation, but increased it transiently at 4 days.[114] Facial nerve transection decreased MT-III mRNA levels significantly in the ipsilateral facial nucleus for at least 5 weeks,[115] while middle cerebral artery occlusion decreased progressively MT-III levels for 7 days and thereafter returned steadily to normal.[180] More recent studies have also shown a biphasic response of MT-III to CNS injury, with initial downregulation followed by upregulation in response to NMDA[119] or to a cryolesion.[45,160] Taken together, the results indicate that brain damage is associated with significant alterations in MT-III expression, and that the type and temporal pattern of the MT-III response depends on the nature of the insult used to inflict brain damage.

As for MT-I&II, the generation of genetically modified mice[181,182] will help significantly to understand the potential biological roles of MT-III. In normal conditions, MT-III null mice do not show any appreciable effects in brain weight, morphology, histology, or behavior even after aging. Nevertheless, old mice showed a higher expression of GFAP, which might be related to an increased astrocytic reactivity associated with senescent changes in neuronal viability not detectable by conventional histology, thereby suggesting a role of MT-III in the aging process.[182] When these mice were challenged with the seizures elicited with the glutamate analog kainic acid, they showed enhanced sensitivity, convulsing longer and having greater mortality than littermate controls, and showing increased neuronal death in the CA3 pyramidal cell layer of hippocampus; quite importantly, transgenic mice overexpressing MT-III showed the opposite trends.[182] Consistent with these results, MT-III was found to prevent glutamate neurotoxicity in primary cultures of cerebellar neurons.[183] The increased neuronal death following kainic acid of the CA3 area has been confirmed recently, but these authors have also shown a decreased neuronal death in the CA1 area.[125] The results thus indicate a protective or detrimental role of MT-III depending on the brain area, which likely will also happen depending on the type on injury. In transgenic G93A SOD1 mice crossed with MT-III null mice,[140] the deficiency of MT-III potentiated motor neuron impairment as revealed by stride length and grip strength, likely because of increased motor neuron death. In line with this study, a recent report shows that an adenoviral vector encoding rat MT-III prevents the degeneration of injured motoneurons.[184] In the cryolesion model, MT-III deficiency did not affect the inflammatory response, oxidative stress, or apoptotic death,[185] in sharp contrast to what is observed in MT-I&II deficient mice. However, MT-III deficiency did increase the expression of some neurotrophins and other factors that may significantly affect neuronal survival and/or growth, such as GAP43,[186,187] which would support an inhibitory role of MT-III in line with some *in vitro* bioassays.[22,174,175,188-190] A recent report analyzing the response of MT-III deficient mice in a peripheral nerve model supports a role of MT-III as an inhibitory factor of neuronal sprouting,[191] since axonal regeneration was faster as substantiated electrophysiologically and histologically. Thus, it might be envisaged that MT-III could serve different functions in the CNS, promoting neuronal survival (perhaps only in specific neuronal populations such as motoneurons or hippocampus CA3 neurons) or death (hippocampus CA1 neurons) while inhibiting neuronal sprouting

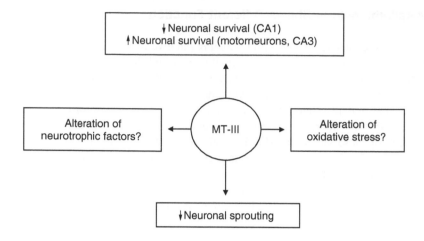

Figure 11.2 Roles of metallothionein-III isoforms in the brain following injury.

(Figure 11.2). The mechanisms underlying these effects remain to be established despite much effort with *in vitro* studies since again significant discrepancies are observed in the literature (i.e., Ref. 192 vs. 193 and 194, for instance).

ACKNOWLEDGMENTS

Support by Ministerio de Ciencia y Tecnología and Feder (SAF2002-01268) and Direcció General de Recerca (2001SGR 00203) is fully acknowledged. Thanks are given to the members of my laboratory and to all of you who have contributed so much to my work.

LIST OF ABBREVIATIONS

AD — Alzheimer's disease
CNS — Central nervous system
GAP-43 — Growth-associated protein-43
GFAP — Glial fibrillary acidic protein
GFAP-IL3 — Transgenic mice expressing IL-3 under the GFAP gene promoter
GFAP-IL6 — Transgenic mice expressing IL-6 under the GFAP gene promoter
GFAP-IFN — Transgenic mice expressing IFN-α under the GFAP gene promoter
GIF — Growth inhibitory factor
IFN-α — Interferon-α
IL-3 — Interleukin-3
IL-6 — Interleukin-6
NMDA — N-Methyl-D-aspartate
MT — Metallothionein
RT-PCR — Real time–polymerase chain reaction
SOD1 — Superoxide dismutase 1
TNF-α — Tumor necrosis factor-α

REFERENCES

1. Margoshes, M. and Vallee, B.L., A cadmium protein from equine kidney cortex, *J. Am. Chem. Soc.*, 79, 4813, 1957.
2. Kägi, J.H.R. and Vallee, B.L., Metallothionein: A cadmium- and zinc-containing protein from equine renal cortex, *J. Biol. Chem.*, 235, 3460, 1960.
3. Kägi, J.H.R. and Valle, B.L., Metallothionein: A cadmium- and zinc-containing protein from equine renal cortex. II. Physicochemical properties, *J. Biol. Chem.*, 236, 2435, 1961.
4. Hamer, D.H., Metallothionein, *Annu. Rev. Biochem.*, 55, 913, 1986.
5. Bremner, I., Interactions between metallothionein and trace elements, *Prog. Food Nutr. Sci.*, 11, 1, 1987.
6. Vasák, M. and Hasler, D.W., Metallothioneins: New functional and structural insights, *Curr. Opin. Chem. Biol.*, 4, 177, 2000.
7. Andrews, G.K., Regulation of metallothionein gene expression by oxidative stress and metal ions, *Biochem. Pharmacol.*, 59, 95, 2000.
8. Ghoshal, K. and Jacob, S.T., Regulation of metallothionein gene expression., *Prog. Nucleic Acid Res. Mol. Biol.*, 66, 357, 2001.
9. Coyle, P., Philcox, J.C., Carey, L.C., and Rofe, A.M., Metallothionein: The multipurpose protein, *Cell Mol. Life Sci.*, 59, 627, 2002.
10. Binz, P.-A. and Kägi, J.H.R., Metallothionein: Molecular evolution and classification, in *Metallothionein IV*, Klaassen, C. D., Ed., Birkhäuser Verlag, Basel, 1999, p. 7.
11. Palmiter, R.D., Findley, S.D., Whitmore, T.E., and Durnam, D.M., MT-III, a brain-specific member of the metallothionein gene family, *Proc. Natl. Acad. Sci. U.S.A.*, 89, 6333, 1992.
12. Quaife, C.J., Findley, S.D., Erickson, J.C., Froelick, G.J., Kelly, E.J., Zambrowicz, B.P., and Palmiter, R.D., Induction of a new metallothionein isoform (MT-IV) occurs during differentiation of stratified squamous epithelia, *Biochemistry*, 33, 7250, 1994.
13. West, A.K., Stallings, R., Hildebrand, C.E., Chiu, R., Karin, M., and Richards R.I., Human metallothionein genes: Structure of the functional locus at 16q13, *Genomics*, 8, 513, 1990.
14. Samson, S.L. and Gedamu, L., Molecular analyses of metallothionein gene regulation, *Prog. Nucleic Acid Res. Mol. Biol.*, 59, 257, 1998.
15. Searle, P.F., Davison, B.L., Stuart, G.W., Wilkie, T.M., Norstedt, G., and Palmiter, R.D., Regulation, linkage, and sequence of mouse metallothionein I and II genes, *Mol. Cell Biol.*, 4, 1221, 1984.
16. Yagle, M.K. and Palmiter, R.D., Coordinate regulation of mouse metallothionein I and II genes by heavy metals and glucocorticoids, *Mol. Cell Biol.*, 5, 291, 1985.
17. van Lookeren Campagne, M., Thiobodeaux, H., van Bruggen, N., Cairns, B., and Lowe, D.G., Increased binding activity at an antioxidant-responsive element in the metallothionein-1 promoter and rapid induction of metallothionein-1 and -2 in response to cerebral ischemia and reperfusion, *J. Neurosci.*, 20, 5200, 2000.
18. Kägi, J.H. and Schäffer, A., Biochemistry of metallothionein, *Biochem.*, 27, 8509, 1988.
19. Vasák, M. and Kägi, J.H.R., Metallothionein, in *Encyclopedia of Inorganic Chemistry,* King, R. B., Ed., J. Wiley & Sons Ltd., New York, 1994, p. 2229.
20. Bogumil, R., Faller, P., Binz, P.A., Vasák, M., Charnock, J.M., and Garner, C.D., Structural characterization of Cu(I) and Zn(II) sites in neuronal-growth-inhibitory factor by extended X-ray absorption fine structure (EXAFS), *Eur. J. Biochem.*, 255, 172, 1998.
21. Faller, P., Hasler, D.W., Zerbe, O., Klauser, S., Winge, D.R., and Vasák, M., Evidence for a dynamic structure of human neuronal growth inhibitory factor and for major rearrangements of its metal-thiolate clusters, *Biochemistry*, 38, 10158, 1999.
22. Uchida, Y., Takio, K., Titani, K., Ihara, Y., and Tomonaga, M., The growth inhibitory factor that is deficient in the Alzheimer's disease brain is a 68 amino acid metallothionein-like protein, *Neuron*, 7, 337, 1991.
23. Chen, R.W. and Ganther, H.E., Relative cadmium-binding capacity of metallothionein and other cytosolic fractions in various tissues of the rat, *Environ. Physiol. Biochem.*, 5, 378, 1975.
24. Durnam, D.M. and Palmiter, R.D., Transcriptional regulation of the mouse metallothionein-I gene by heavy metals, *J. Biol. Chem.*, 256, 5712, 1981.

25. Blaauwgeers, H.G., Sillevis Smitt, P.A., De Jong, J.M., and Troost D., Distribution of metallothionein in the human central nervous system, *Glia*, 8, 62, 1993.

26. Gulati, S., Paliwal, V.K., Sharma, M., Gill, K.D., and Nath, R., Isolation and characterization of a metallothionein-like protein from monkey brain, *Toxicology*, 45, 53, 1987.

27. Kojima, S., Shimada, A., Morita, T., Yamano, Y., and Umemura, T., Localization of metallothioneins-I & -II and -III in the brain of aged dog, *J. Vet. Med. Sci.*, 61, 343, 1999.

28. Holloway, A.F., Stennard, F.A., Dziegielewska, K.M., Weller, L., and West, A.K., Localisation and expression of metallothionein immunoreactivity in the developing sheep brain, *Int. J. Dev. Neurosci.*, 15, 195, 1997.

29. Hanlon, J., Monks, E., Hughes, C., Weavers, E., and Rogers, M., Metallothionein in bovine spongiform encephalopathy, *J. Comp. Pathol.*, 127, 280, 2002.

30. Saijoh, K., Kuno, T., Shuntoh, H., Tanaka, C., and Sumino, K., Molecular cloning of cDNA for rat brain metallothionein-II and regulation of its gene expression, *Pharmacol. Toxicol.*, 64, 464, 1989.

31. Dalton, T., Pazdernik, T.L., Wagner, J., Samson, F., and Andrews, G.K., Temporalspatial patterns of expression of metallothionein-I and -III and other stress related genes in rat brain after kainic acid-induced seizures, *Neurochem. Int.*, 27, 59, 1995.

32. Zheng, H., Berman, N.E., and Klaassen, C.D., Chemical modulation of metallothionein I and III mRNA in mouse brain, *Neurochem. Int.*, 27, 43, 1995.

33. Belloso, E., Hernández, J., Giralt, M., Kille, P., and Hidalgo, J., Effect of stress on mouse and rat brain metallothionein I and III mRNA levels, *Neuroendocrinology*, 64, 430, 1996.

34. Aschner, M., Lorscheider, F.L., Cowan, K.S., Conklin, D.R., Vimy, M.J., and Lash L.H., Metallothionein induction in fetal rat brain and neonatal primary astrocyte cultures by *in utero* exposure to elemental mercury vapor (Hg0). *Brain Res.*, 778, 222, 1997.

35. Carrasco, J., Hernández, J., González, B., Campbell, I.L., and Hidalgo, J., Localization of metallothionein-I and -III expression in the CNS of transgenic mice with astrocyte-targeted expression of interleukin 6, *Exp. Neurol*, 153, 184, 1998.

36. Pazdernik, T.L., Emerson, M.R., Cross, R., Nelson, S.R., and Samson, F.E., Soman-induced seizures: Limbic activity, oxidative stress and neuroprotective proteins, *J. Appl. Toxicol.*, 21, S87, 2001.

37. Trendelenburg, G., Prass, K., Priller, J., Kapinya, K., Polley, A., Muselmann, C., Ruscher, K., Kannbley, U., Schmitt, A.O., Castell, S., Wiegand, F., Meisel, A., Rosenthal, A., and Dirnagl, U., Serial analysis of gene expression identifies metallothionein-II as major neuroprotective gene in mouse focal cerebral ischemia., *J. Neurosci.*, 22, 5879, 2002.

38. Itano, Y., Noji, S., Koyama, E., Taniguchi, S., Taga, N., Takahashi, T., Ono, K., and Kosaka, F., Bacterial endotoxin-induced expression of metallothionein genes in rat brain, as revealed by in situ hybridization, *Neurosci. Lett.*, 124, 13, 1991.

39. Hao, R., Cerutis, D.R., Blaxall, H.S., Rodriguez-Sierra, J.F., Pfeiffer, R.F., and Ebadi, M., Distribution of zinc metallothionein I mRNA in rat brain using in situ hybridization, *Neurochem. Res.*, 19, 761, 1994.

40. Masters, B.A., Quaife, C.J., Erickson, J.C., Kelly, E.J., Froelick, G.J., Zambrowicz, B.P., Brinster, R.L., and Palmiter R.D., Metallothionein III is expressed in neurons that sequester zinc in synaptic vesicles, *J. Neurosci.*, 14, 5844, 1994.

41. Choudhuri, S., Kramer, K.K., Berman, N.E., Dalton, T.P., Andrews, G.K., and Klaassen, C.D., Constitutive expression of metallothionein genes in mouse brain, *Toxicol. Appl. Pharmacol.*, 131, 144, 1995.

42. Carrasco, J., Hernández, J., Bluethmann, H., and Hidalgo, J., Interleukin-6 and tumor necrosis factor-alpha type 1 receptor deficient mice reveal a role of IL-6 and TNF-alpha on brain metallothionein-I and -III regulation, *Mol. Brain Res.*, 57, 221, 1998.

43. Molinero, A., Carrasco, J., Hernández, J., and Hidalgo, J., Effect of nitric oxide synthesis inhibition on mouse liver and brain metallothionein expression, *Neurochem. Int.*, 33, 559, 1998.

44. Penkowa, M., Carrasco, J., Giralt, M., Moos, T.. and Hidalgo, J., CNS wound healing is severely depressed in metallothionein I- and II-deficient mice, *J. Neurosci.*, 19, 2535, 1999.

45. Penkowa, M., Moos, T., Carrasco, J., Hadberg, H., Molinero, A., Bluethmann, H., and Hidalgo, J., Strongly compromised inflammatory response to brain injury in interleukin-6-deficient mice, *Glia*, 25, 343, 1999.

46. van Lookeren Campagne, M., Thibodeaux, H., van Bruggen, N., Cairns, B., Gerlai, R., Palmer, J.T., Williams, S.P., and Lowe, D.G., Evidence for a protective role of metallothionein-1 in focal cerebral ischemia, *Proc. Natl. Acad. Sci. U.S.A.*, 96, 12870, 1999.

47. Giralt, M., Molinero, A., Carrasco, J., and Hidalgo, J., Effect of dietary zinc deficiency on brain metallothionein-I and -III mRNA levels during stress and inflammation, *Neurochem. Int.*, 36, 555, 2000.

48. Carrasco, J., Penkowa, M., Hadberg, H., Molinero, A., and Hidalgo, J., Enhanced seizures and hippocampal neurodegeneration following kainic acid induced seizures in metallothionein-I+II deficient mice, *Eur. J. Neurosci.*, 12, 2311, 2000.

49. Kim, D., Kim, E.H., Kim, C., Sun, W., Kim, H.J., Uhm, C.S., Park, S.-H., and Kim, H., Differential regulation of metallothionein-I, II, and III mRNA expression in the rat brain following kainic acid treatment, *Neuroreport.*, 14, 679, 2003.

50. Tang, Y., Lu, A., Aronow, B.J., Wagner, K.R., and Sharp, F.R., Genomic responses of the brain to ischemic stroke, intracerebral haemorrhage, kainate seizures, hypoglycemia, and hypoxia., *Eur. J. Neurosci.*, 15, 1937, 2002.

51. Lock, C., Hermans, G., Pedotti, R., Brendolan, A., Schadt, E., Garren, H., Langer-Gould, A., Strober, S., Cannella, B., Allard, J., Klonowski, P., Austin, A., Lad, N., Kaminski, N., Galli, S.J., Oksenberg, J.R., Raine, C.S., Heller, R., and Steinman L., Gene-microarray analysis of multiple sclerosis lesions yields new targets validated in autoimmune encephalomyelitis, *Nat. Med.*, 8, 500, 2002.

52. Nagano, S., Satoh, M., Sumi, H., Fujimura, H., Tohyama, C., Yanagihara, T., and Sakoda, S., Reduction of metallothioneins promotes the disease expression of familial amyotrophic lateral sclerosis mice in a dose-dependent manner, *Eur. J. Neurosci.*, 13, 1363, 2001.

53. Young, J.K., Garvey, J.S., and Huang, P.C., Glial immunoreactivity for metallothionein in the rat brain, *GLIA*, 4, 602, 1991.

54. Sillevis Smitt, P.A., Blaauwgeers, H.G., Troost, D., and de Jong, J.M., Metallothionein immunoreactivity is increased in the spinal cord of patients with amyotrophic lateral sclerosis, *Neurosci. Lett.*, 144, 107, 1992.

55. Blaauwgeers, H.G., Sillevis Smitt, P.A., de Jong, J.M., and Troost, D., Localization of metallothionein in the mammalian central nervous system, *Biol. Signals*, 3, 181, 1994.

56. Sillevis Smitt, P.A., Mulder, T.P., Verspaget, H.W., Blaauwgeers, H.G., Troost, D., and de Jong, J.M., Metallothionein in amyotrophic lateral sclerosis, *Biol. Signals*, 3, 193, 1994.

57. Suzuki, K., Nakajima, K., Otaki, N., and Kimura, M., Metallothionein in developing human brain, *Biol. Signals*, 3, 188, 1994.

58. Suzuki, K., Nakajima, K., Otaki, N., Kimura, M., Kawaharada, U., Uehara, K., Hara, F., Nakazato, Y., and Takatama, M., Localization of metallothionein in aged human brain, *Pathol. Int.*, 44, 20, 1994.

59. Hidalgo, J., García, A., Oliva, A.M., Giralt, M., Gasull, T., González, B., Milnerowicz, H., Wood, A., and Bremner, I., Effect of zinc, copper and glucocorticoids on metallothionein levels of cultured neurons and astrocytes from rat brain, *Chem. Biol. Interact.*, 93, 197, 1994.

60. Leyshon-Sorland, K., Jasani, B., and Morgan, A.J., The localization of mercury and metallothionein in the cerebellum of rats experimentally exposed to methylmercury, *Histochem. J.*, 26, 161, 1994.

61. Young, J.K., Glial metallothionein, *Biol. Signals*, 3, 169, 1994.

62. Penkowa, M. and Moos, T., Disruption of the blood-brain interface in neonatal rat neocortex induces a transient expression of metallothionein in reactive astrocytes, *Glia*, 13, 217, 1995.

63. Nakajima, K. and Suzuki, K., Immunochemical detection of metallothionein in brain, *Neurochem. Int.*, 27, 73, 1995.

64. Kiningham, K., Bi, X., and Kasarskis, E.J., Neuronal localization of metallothioneins in rat and human spinal cord, *Neurochem. Int.*, 27, 105, 1995.

65. Heal, J.W., Singhrao, S.K., Jasan, B., and Newman, G.R., Immunocytochemically detectable metallothionein is expressed by astrocytes in the ischaemic human brain, *Neuropathol. Appl. Neurobiol.*, 22, 243, 1996.

66. Penkowa, M., Hidalgo, J., and Moos, T., Increased astrocytic expression of metallothioneins I + II in brainstem of adult rats treated with 6-aminonicotinamide, *Brain Res.*, 774, 256, 1997.

67. Shimada, A., Yanagida, M., and Umemura, T., An immunohistochemical study on the tissue-specific localization of metallothionein in dogs, *J. Comp. Pathol.*, 116, 1, 1997.

68. Vela, J.M., Hidalgo, J., González, B., and Castellano, B., Induction of metallothionein in astrocytes and microglia in the spinal cord from the myelin-deficient jimpy mouse, *Brain Res.*, 767, 345, 1997.

69. Shimada, A., Uemura, T., Yamamura, Y., Kojima, S., Morita, T., and Umemura, T., Localization of metallothionein-I and -II in hypertrophic astrocytes in brain lesions of dogs, *J. Vet. Med. Sci.*, 60, 351, 1998.

70. Carrasco, J., Giralt, M., Molinero, A., Penkowa, M., Moos, T., and Hidalgo, J., Metallothionein (MT)-III: Generation of polyclonal antibodies, comparison with MT-I+II in the freeze lesioned rat brain and in a bioassay with astrocytes, and analysis of Alzheimer's disease brains, *J. Neurotrauma*, 16, 1115, 1999.

71. Dincer, Z., Haywood, S., and Jasani, B., Immunocytochemical detection of metallothionein (MT1 and MT2) in copper-enhanced sheep brains, *J. Comp. Pathol.*, 120, 29, 1999.

72. Kawashima, T., Adachi, T., Tokunaga, Y., Furuta, A., Suzuki, S.O., Doh ura, K., and Iwaki, T., Immunohistochemical analysis in a case of idiopathic Lennox-Gastaut syndrome, *Clin. Neuropathol.*, 18, 286, 1999.

73. Carrasco, J., Giralt, M., Penkowa, M., Stalder, A.K., Campbell, I.L., and Hidalgo, J., Metallothioneins are upregulated in symptomatic mice with astrocyte-targeted expression of tumor necrosis factor-α, *Exp. Neurol.*, 163, 46, 2000.

74. Neal, J.W., Singhrao, S.K., Jasani, B., and Newman, G.R., Immunocytochemically detectable metallothionein is expressed by astrocytes in the ischaemic human brain, *Neuropathol. Appl. Neurobiol.*, 22, 243, 1996.

75. Acarin, L., González, B., Hidalgo, J., Castro, A.J., and Castellano B., Primary cortical glial reaction versus secondary thalamic glial response in the excitotoxically injured young brain: astroglial response and metallothionein expression, *Neuroscience*, 92, 827, 1999.

76. Penkowa, M., Giralt, M., Moos, T., Thomsen, P.S., Hernández, J., and Hidalgo, J., Impaired inflammatory response to glial cell death in genetically metallothionein-I- and -II-deficient mice, *Exp. Neurol.*, 156, 149, 1999.

77. Penkowa, M., Nielsen, H., Hidalgo, J., Bernth, N., and Moos, T., Distribution of metallothionein I + II and vesicular zinc in the developing central nervous system: Correlative study in the rat, *J. Comp. Neurol.*, 412, 303, 1999.

78. Zambenedetti, P., Giordano, R., and Zatta, P., Metallothioneins are highly expressed in astrocytes and microcapillaries in Alzheimer's disease, *J. Chem. Neuroanat.*, 15, 21, 1998.

79. Kawashima, T., Doh-ura, K., Torisu, M., Uchida, Y., Furuta, A., and Iwaki, T., Differential expression of metallothioneins in human prion diseases, *Dement. Geriatr. Cogn. Disord.*, 11, 251, 2000.

80. Yu, W.H., Lukiw, W.J., Bergeron, C., Niznik, H.B., and Fraser, P.E., Metallothionein III is reduced in Alzheimer's disease, *Brain Res.*, 894, 37, 2001.

81. Nolan, C.V. and Shaikh, Z.A., Determination of metallothionein in tissues by radioimmunoassay and by cadmium saturation method, *Anal. Biochem.*, 154, 213, 1986.

82. Gasull, T., Giralt, M., García, A., and Hidalgo, J., Regulation of metallothionein-I+II levels in specific brain areas and liver in the rat: Role of catecholamines, *Glia*, 12, 135, 1994.

83. Gasull, T., Giralt, M., Hernández, J., Martínez, P., Bremner, I., and Hidalgo, J., Regulation of metallothionein concentrations in rat brain: Effect of glucocorticoids, zinc, copper, and endotoxin, *Am. J. Physiol.*, 266, E760, 1994.

84. Hidalgo, J., Belloso, E., Hernández, J., Gasull, T., and Molinero, A., Role of glucocorticoids on rat brain metallothionein-I and -III response to stress, *Stress*, 1, 231, 1997.

85. Rojas, P., Hidalgo, J., Ebadi, M., and Rios, C., Changes of metallothionein I + II proteins in the brain after 1-methyl-4-phenylpyridinium administration in mice, *Prog. Neuropsychopharmacol. Biol. Psych.*, 24, 143, 2000.

86. Fukada, K., Nagano, S., Satoh, M., Tohyama, C., Nakanishi, T., Shimizu, A., Yanagihara, T., and Sakoda, S., Stabilization of mutant Cu/Zn superoxide dismutase (SOD1) protein by coexpressed wild SOD1 protein accelerates the disease progression in familial amyotrophic lateral sclerosis mice, *Eur. J. Neurosci.*, 14, 2032, 2001.

87. Kramer, K.K., Zoelle, J.T., and Klaassen, C.D., Induction of metallothionein mRNA and protein in primary murine neuron cultures, *Toxicol. Appl. Pharmacol.*, 141, 1, 1996.

88. Kramer, K.K., Liu, J., Choudhuri, S., and Klaassen, C.D., Induction of metallothionein mRNA and protein in murine astrocyte cultures, *Toxicol. Appl. Pharmacol.*, 136, 94, 1996.

89. Suzuki, Y., Apostolova, M.D., and Cherian, M.G., Astrocyte cultures from transgenic mice to study the role of metallothionein in cytotoxicity of tert-butyl hydroperoxide, *Toxicology*, 145, 51, 2000.

90. Skabo, S.J., Holloway, A.F., West, A.K., and Chuah, M.I., Metallothioneins 1 and 2 are expressed in the olfactory mucosa of mice in untreated animals and during the regeneration of the epithelial layer, *Biochem. Biophys. Res. Commun.*, 232, 136, 1997.

91. Penkowa, M., Giralt, M., Thomsen, P., Carrasco, J., and Hidalgo, J., Zinc or copper deficiency-induced impaired inflammatory response to brain trauma may be caused by the concomitant metallothionein changes, *J. Neurotrauma*, 18, 447, 2001.

92. Giralt, M., Carrasco, J., Penkowa, M., Morcillo, M.A., Santamaría, J., Campbell, I.L., and Hidalgo, J., Astrocyte-targeted expression of interleukin-3 and interferon-α causes specific changes in metallothionein expression in the brain, *Exp. Neurol.*, 168, 334, 2001.

93. Giralt, M., Penkowa, M., Lago, N., Molinero, A., and Hidalgo, J., Metallothionein-1+2 protect the CNS after a focal brain injury, *Exp. Neurol.*, 173, 114, 2002.

94. Giralt, M., Penkowa, M., Hernández, J., Molinero, A., Carrasco, J., Lago, N., Camats, J., Campbell, I.L., and Hidalgo, J., Metallothionein-1+2 deficiency increases brain pathology in transgenic mice with astrocyte-targeted expression of interleukin 6, *Neurobiol. Dis.*, 9, 319, 2002.

95. Molinero, A., Penkowa, M., Hernández, J., Camats, J., Giralt, M., Lago, N., Carrasco, J., Campbell, I.L., and Hidalgo, J., Metallothionein-I overexpression decreases brain pathology in transgenic mice with astrocyte-targeted expression of interleukin 6, *J. Neuropathol. Exp. Neurol.*, 62, 315, 2003.

96. Penkowa, M., Espejo, C., Ortega-Aznar, A., Hidalgo, J., Montalban, X., and Martínez-Cáceres, E.M., Metallothionein expression in the central nervous system of multiple sclerosis patients, *Cell. Mol. Life Sci.*, 60, 1258, 2003.

97. Uchida, Y., Ihara, Y., and Tomonaga, M., Alzheimer's disease brain extract stimulates the survival of cerebral cortical neurons from neonatal rats, *Biochem. Biophys. Res. Commun.*, 150, 1263, 1988.

98. Uchida, Y. and Tomonaga, M., Neurotrophic action of Alzheimer's disease brain extract is due to the loss of inhibitory factors for survival and neurite formation of cerebral cortical neurons, *Brain Res.*, 481, 190, 1989.

99. Tsuji, S., Kobayashi, H., Uchida, Y., Ihara, Y., and Miyatake, T., Molecular cloning of human growth inhibitory factor cDNA and its down-regulation in Alzheimer's disease, *EMBO J.*, 11, 4843, 1992.

100. Kobayashi, H., Uchida, Y., Ihara, Y., Nakajima, K., Kohsaka, S., Miyatake, T., and Tsuji, S., Molecular cloning of rat growth inhibitory factor cDNA and the expression in the central nervous system, *Mol. Brain Res.*, 19, 188, 1993.

101. Hoey, J.G., Garrett, S.H., Sens, M.A., Todd, J.H., and Sens, D.A., Expression of MT-3 mRNA in human kidney, proximal tubule cell cultures, and renal cell carcinoma, *Toxicol. Lett.*, 92, 149, 1997.

102. Moffatt, P. and Séguin, C., Expression of the gene encoding metallothionein-3 in organs of the reproductive system, *DNA Cell. Biol.*, 17, 501, 1998.

103. Garrett, S.H., Sens, M.A., Todd, J.H., Somji, S., and Sens, D.A., Expression of MT-3 protein in the human kidney, *Toxicol. Lett.*, 105, 207, 1999.

104. Garrett, S.H., Sens, M.A., Shukla, D., Nestor, S., Somji, S., Todd, J.H., and Sens, D.A., Metallothionein isoform 3 expression in the human prostate and cancer-derived cell lines, *Prostate*, 41, 196, 1999.

105. Sens, M.A., Somji, S., Garrett, S.H., Beall, C.L., and Sens, D.A., Metallothionein isoform 3 overexpression is associated with breast cancers having a poor prognosis., *Am. J. Pathol.*, 159, 21, 2001.

106. Cyr, D.G., Dufresne, J., Pillet, S., Alfieri, T.J., and Hermo, L., Expression and regulation of metallothioneins in the rat epididymis, *J. Androl.*, 22, 124, 2001.

107. Yamashita, M., Glasgow, E., Zhang, B.J., Kusano, K., and Gainer, H., Identification of cell-specific messenger ribonucleic acids in oxytocinergic and vasopressinergic magnocellular neurons in rat supraoptic nucleus by single-cell differential hybridization, *Endocrinology*, 143, 4464, 2002.

108. Pountney, D.L., Fundel, S.M., Faller, P., Birchler, N.E., Hunziker, P., and Vasak, M., Isolation, primary structures and metal binding properties of neuronal growth inhibitory factor (GIF) from bovine and equine brain, *FEBS Lett.*, 345, 193, 1994.

109. Chen, C.F., Wang, S.H., and Lin, L.Y., Identification and characterization of metallothionein III (growth inhibitory factor) from porcine brain, *Comp. Biochem. Physiol. Biochem. Mol. Biol.*, 115, 27, 1996.

110. Kojima, S., Shimada, A., Kodan, A., Kobayashi, K., Morita, T., Yamano, Y., and Umemura, T., Molecular cloning and expression of the canine metallothionein- III gene, *Can. J. Vet. Res.*, 62, 148, 1998.

111. Chung, R.S., Holloway, A.F., Eckhardt, B.L., Harris, J.A., Vickers, J.C., Chuah, M.I., and West, A.K., Sheep have an unusual variant of the brain-specific metallothionein, metallothionein-III, *Biochem. J.*, 365, 323, 2002.

112. Uchida, Y., Growth inhibitory factor in brain, in *Metallothionein III*, Suzuki, K. T., Imura, N., and Kimura, M., Eds., Birkhäuser Verlag, Basel, 1993, p. 315.

113. Anezaki, T., Ishiguro, H., Hozumi, I., Inuzuka, T., Hiraiwa, M., Kobayashi, H., Yuguchi, T., Wanaka, A., Uda, Y., Miyatake, T., Yamada, K., Tohyama, M., and Tsuji, S., Expression of growth inhibitory factor (GIF) in normal and injured rat brains, *Neurochem. Int.*, 27, 89, 1995.

114. Yuguchi, T., Kohmura, E., Yamada, K., Sakaki, T., Yamashita, T., Otsuki, H., Kataoka, K., Tsuji, S., and Hayakawa, T., Expression of growth inhibitory factor mRNA following cortical injury, *J. Neurotrauma*, 12, 299, 1995.

115. Yuguchi, T., Kohmura, E., Yamada, K., Sakaki, T., Yamashita, T., Otsuki, H., Wanaka, A., Tohyama, M., Tsuji, S., and Hayakawa, T., Changes in growth inhibitory factor mRNA expression compared with those in c-jun mRNA expression following facial nerve transection, *Mol. Brain Res.*, 28, 181, 1995.

116. Yuguchi, T., Kohmura, E., Sakaki, T., Nonaka, M., Yamada, K., Yamashita, T., Kishiguchi, T., Sakaguchi, T., and Hayakawa, T., Expression of growth inhibitory factor mRNA after focal ischemia in rat brain, *J. Cereb. Blood Flow Metab.*, 17, 745, 1997.

117. Kojima, S., Shimada, A., Morita, T., Yamano, Y., and Umemura, T., Localization of metallothioneins-I & -II and -III in the brain of aged dog, *J. Vet. Med. Sci.*, 61, 343, 1999.

118. Velázquez, R.A., Cai, Y., Shi, Q., and Larson, A.A., The distribution of zinc selenite and expression of metallothionein- III mRNA in the spinal cord and dorsal root ganglia of the rat suggest a role for zinc in sensory transmission, *J. Neurosci.*, 19, 2288, 1999.

119. Acarin, L., Carrasco, J., González, B., Hidalgo, J., and Castellano, B., Expression of growth inhibitory factor (metallothionein- III) mRNA and protein following excitotoxic immature brain injury, *J. Neuropathol. Exp. Neurol.*, 58, 389, 1999.

120. Gong. Y.H. and Elliott. J.L., Metallothionein expression is altered in a transgenic murine model of familial amyotrophic latyeral sclerosis, *Exp. Neurol.*, 162, 27, 2000.

121. Yamada, M., Hayashi, S., Hozumi, I., Inuzuka, T., Tsuji, S., and Takahashi, H., Subcellular localization of growth inhibitory factor in rat brain: Light and electron microscopic immunohistochemical studies, *Brain Res.*, 735, 257, 1996.

122. Hozumi, I., Inuzuka, T., Ishiguro, H., Hiraiwa, M., Uchida, Y., and Tsuji, S., Immunoreactivity of growth inhibitory factor in normal rat brain and after stab wounds—an immunocytochemical study using confocal laser scan microscope, *Brain Res.*, 741, 197, 1996.

123. Isumi, H., Uchida, Y., Hayashi, T., Furukawa, S., and Takashima, S., Neuron death and glial response in pontosubicular necrosis. The role of the growth inhibition factor, *Clin. Neuropathol.*, 19, 77, 2000.

124. Yanagitani, S., Miyazaki, H., Nakahashi, Y., Kuno, K., Ueno, Y., Matsushita, M., Naitoh, Y., Taketani, S., and Inoue, K., Ischemia induces metallothionein III expression in neurons of rat brain, *Life Sci.*, 64, 707, 1999.

125. Lee, J.Y., Kim, J.H., Palmiter, R.D., and Koh, J.Y., Zinc released from metallothionein-III may contribute to hippocampal CA1 and thalamic neuronal death following acute brain injury, *Exp. Neurol.*, 184, 337, 2003.

126. Miyazaki, I., Asanuma, M., Higashi, Y., Sogawa, C.A., Tanaka, K., and Ogawa, N., Age-related changes in expression of metallothionein-III in rat brain, *Neurosci. Res.*, 43, 323, 2002.

127. Coyle, J. and Puttfarcken, P., Oxidative stress, glutamate, and neurodegenerative disorders, *Science*, 262, 689, 1993.

128. Sato, M. and Bremner, I., Oxygen free radicals and metallothionein, *Free Rad. Biol. Med.*, 14, 325, 1993.

129. Aschner, M., The functional significance of brain metallothioneins, *FASEB J.*, 10, 1129, 1996.

130. Duguid, J.R., Bohmont, C.W., Liu, N.G., and Tourtellotte, W.W., Changes in brain gene expression shared by scrapie and Alzheimer disease, *Proc. Natl. Acad. Sci. U.S.A.*, 86, 7260, 1989.

131. Adlard, P.A., West, A.K., and Vickers, J.C., Increased density of metallothionein I/II-immunopositive cortical glial cells in the early stages of Alzheimer's disease, *Neurobiol. Dis.*, 5, 349, 1998.

132. Chuah, M.I. and Getchell, M.L., Metallothionein in olfactory mucosa of Alzheimer's disease patients and apoE-deficient mice, *Neuroreport*, 10, 1919, 1999.

133. Blaauwgeers, H.G., Anwar Chand, M., van den Berg, F.M., Vianney de Jong, J.M., and Troost, D., Expression of different metallothionein messenger ribonucleic acids in motor cortex, spinal cord and liver from patients with amyotrophic lateral sclerosis, *J. Neurol. Sci.*, 142, 39, 1996.

134. Suzuki, K., Nakajima, K., Kawaharada, U., Uehara, K., Hara, F., Otaki, N., Kimura, M., and Tamura, Y., Metallothionein in the human brain, *Acta. Histochem. Cytochem.*, 25, 617, 1992.

135. De, S.K., McMaster, M.T., and Andrews, G.K., Endotoxin induction of murine metallothionein gene expression, *J. Biol. Chem.*, 265, 15267, 1990.

136. Hidalgo, J., Borras, M., Garvey, J.S., and Armario, A., Liver, brain, and heart metallothionein induction by stress, *J. Neurochem.*, 55, 651, 1990.

137. Jacob, S.T., Ghoshal, K., and Sheridan, J.F., Induction of metallothionein by stress and its molecular mechanisms, *Gene Exp.*, 7, 301, 1999.

138. Hidalgo, J., Castellano, B., and Campbell, I.L., Regulation of brain metallothioneins, *Curr. Top. Neurochem.*, 1, 1, 1997.

139. Montpied, P., de Bock, F., Baldy Moulinier, M., and Rondouin, G., Alterations of metallothionein II and apolipoprotein J mRNA levels in kainate-treated rats, *Neuroreport*, 9, 79, 1998.

140. Puttaparthi, K., Gitomer, W.L., Krishnan, U., Son, M., Rajendran, B., and Elliott, J.L., Disease progression in a transgenic model of familial amyotrophic lateral sclerosis is dependent on both neuronal and non-neuronal zinc binding proteins, *J. Neurosci.*, 22, 8790, 2002.

141. Penkowa, M. and Hidalgo, J., Metallothionein I+II expression and their role in experimental autoimmune encephalomyelitis, *Glia*, 32, 247, 2000.

142. Espejo, C., Carrasco, J., Hidalgo, J., Penkowa, M., García, A., Sáez-Torres, I., and Martínez-Cáceres, E.M., Differential expression of metallothioneins in the CNS of mice with experimental autoimmune encephalomyelitis, *Neuroscience*, 105, 1055, 2001.

143. Penkowa, M., Espejo, C., Martínez-Cáceres, E.M., Poulsen, C.B., Montalban, X., and Hidalgo, J., Altered inflammatory response and increased neurodegeneration in metallothionein I+II deficient mice during experimental autoimmune encephalomyelitis, *J. Neuroimmunol.*, 119, 248, 2001.

144. Penkowa, M. and Hidalgo, J., IL-6 deficiency leads to reduced metallothionein-I+II expression and increased oxidative stress in the brain stem after 6-aminonicotinamide treatment, *Exp. Neurol.*, 163, 72, 2000.

145. Hopkins, S. and Rothwell, N., Cytokines and the nervous system. I: Expression and recognition, *Trends Neurosci.*, 18, 83, 1995.

146. Rothwell, N.J. and Hopkins, S.J., Cytokines and the nervous system II: Actions and mechanisms of action, *Trends Neurosci.*, 18, 130, 1995.

147. Stichel, C. and Verner Müller, H., Experimental strategies to promote axonal regeneration after traumatic central nervous system injury, *Prog. Neurobiol.*, 56, 119, 1998.

148. McIntosh, T., Juhler, M., and Wieloch, T., Novel pharmacologic strategies in the treatment of experimental traumatic brain injury, *J. Neurotrauma*, 15, 731, 1998.

149. Campbell, I.L., Abraham, C.R., Masliah, E., Kemper, P., Inglis, J.D., Oldstone, M.B.A., and Mucke, L., Neurologic disease in transgenic mice by cerebral overexpression of interleukin 6, *Proc. Natl. Acad. Sci. U.S.A.*, 90, 10061, 1993.

150. Stalder, A.K., Carson, M.J., Pagenstecher, A., Asensio, V.C., Kincaid, C., Benedict, M., Powell, H.C., Masliah, E., and Campbell I.L., Late-onset chronic inflammatory encephalopathy in immune-competent and severe combined immune-deficient (SCID) mice with astrocyte-targeted expression of tumor necrosis factor, *Am. J. Pathol.*, 153, 767, 1998.

151. Hernández, J., Molinero, A., Campbell, I.L., and Hidalgo, J., Transgenic expression of interleukin 6 in the central nervous system regulates brain metallothionein-I and -III expression in mice, *Mol. Brain Res.*, 48, 125, 1997.

152. Penkowa, M., Giralt, M., Carrasco, J., Hadberg, H., and Hidalgo, J., Impaired inflammatory response and increased oxidative stress and neurodegeneration after brain injury in interleukin-6-deficient mice, *Glia*, 32, 271, 2000.

153. Penkowa, M., Molinero, A., Carrasco, J., and Hidalgo, J., Interleukin-6 deficiency reduces the brain inflammatory response and increases oxidative stress and neurodegeneration after kainic acid-induced seizures, *Neuroscience*, 102, 805, 2001.

154. Palmiter, R.D., Sandgren, E.P., Koeller, D.M., and Brinster, R.L., Distal regulatory elements from the mouse metallothionein locus stimulate gene expression in transgenic mice, *Mol. Cell. Biol.*, 13, 5266, 1993.

155. Michalska, A.E. and Choo, K.H., Targeting and germ-line transmission of a null mutation at the metallothionein I and II loci in mouse, *Proc. Natl. Acad. Sci. U.S.A.*, 90, 8088, 1993.

156. Masters, B.A., Kelly, E.J., Quaife, C.J., Brinster, R.L., and Palmiter, R.D., Targeted disruption of metallothionein I and II genes increases sensitivity to cadmium, *Proc. Natl. Acad. Sci. U.S.A.*, 91, 584, 1994.

157. Penkowa, M., Giralt, M., Camats, J., and Hidalgo, J., Metallothionein 1+2 protect the CNS during neuroglial degeneration induced by 6-aminonicotinamide, *J. Comp. Neurol.*, 444, 174, 2002.

158. Asanuma, M., Miyazaki, I., Higashi, Y., Tanaka, K., Haque, M.E., Fujita, N., and Ogawa, N., Aggravation of 6-hydroxydopamine-induced dopaminergic lesions in metallothionein-I and -II knock-out mouse brain, *Neurosci. Lett.*, 327, 61, 2002.

159. Penkowa, M., Espejo, C., Martínez-Cáceres, E.M., Montalban, X., and Hidalgo, J., Increased demyelination and axonal damage in metallothionein I+II-deficient mice during experimental autoimmune encephalomyelitis, *Cell. Mol. Life Sci.*, 60, 185, 2003.

160. Penkowa, M., Carrasco, J., Giralt, M., Molinero, A., Hernández, J., Campbell, I.L., and Hidalgo, J., Altered central nervous system cytokine-growth factor expression profiles and angiogenesis in metallothionein-I+II deficient mice, *J. Cereb. Blood Flow Metab.*, 20, 1174, 2000.

161. Penkowa, M., Camats, J., Giralt, M., Molinero, A., Hernández, J., Carrasco, J., Campbell, I.L., and Hidalgo, J., Metallothionein-I overexpression alters brain inflammation and stimulates brain repair in transgenic mice with astrocyte-targeted interleukin-6 expression, *Glia*, 42, 287, 2003.

162. Lazo, J.S., Kondo, Y., Dellapiazza, D., Michalska, A.E., Choo, K.H., and Pitt, B.R., Enhanced sensitivity to oxidative stress in cultured embryonic cells from transgenic mice deficient in metallothionein I and II genes, *J. Biol. Chem.*, 270, 5506, 1995.

163. Kang, Y.J., Chen, Y., Yu, A., Voss McCowan, M., and Epstein, P.N., Overexpression of metallothionein in the heart of transgenic mice suppresses doxorubicin cardiotoxicity, *J. Clin. Invest.*, 100, 1501, 1997.

164. Kondo, Y., Rusnak, J., Hoyt, D., Settineri, C., Pitt, B., and Lazo, J., Enhanced apoptosis in metallothionein null cells, *Mol. Pharmacol.*, 52, 195, 1997.

165. Liu, J., Liu, Y., Habeebu, S.S., and Klaassen, C.D., Metallothionein (MT)-null mice are sensitive to cisplatin-induced hepatotoxicity, *Toxicol. Appl. Pharmacol.*, 149, 24, 1998.

166. Fu, K., Tomita, T., Sarras, M.P., Jr., De Lisle, R.C., and Andrews, G.K., Metallothionein protects against cerulein-induced acute pancreatitis: Analysis using transgenic mice, *Pancreas*, 17, 238, 1998.

167. Hanada, K., Sawamura, D., Tamai, K., Baba, T., Hashimoto, I., Muramatsu, T., Miura, N., and Naganuma, A., Novel function of metallothionein in photoprotection: Metallothionein-null mouse exhibits reduced tolerance against ultraviolet B injury in the skin, *J. Invest. Dermatol.*, 111, 582, 1998.

168. Kang, Y.J., Li, G., and Saari, J.T., Metallothionein inhibits ischemia-reperfusion injury in mouse heart, *Am. J. Physiol.*, 276, H993, 1999.

169. Kang, Y.J., The antioxidant function of metallothionein in the heart, *Proc. Soc. Exp. Biol. Med.*, 222, 263, 1999.

170. Wang, G.W., Schuschke, D.A., and Kang, Y.J., Metallothionein-overexpressing neonatal mouse cardiomyocytes are resistant to H2O2 toxicity, *Am. J. Physiol.*, 276, H167, 1999.

171. Liu, J., Liu, Y., Hartley, D., Klaassen, C.D., Shehin Johnson, S.E., Lucas, A., and Cohen, S.D., Metallothionein-I/II knockout mice are sensitive to acetaminophen-induced hepatotoxicity, *J. Pharmacol. Exp. Ther.*, 289, 580, 1999.

172. Youn, J., Hwang, S.H., Ryoo, Z.Y., Lynes, M.A., Paik, D.J., Chung, H.S., and Kim, H.Y., Metallothionein suppresses collagen-induced arthritis via induction of TGF-beta and down-regulation of proinflammatory mediators, *Clin. Exp. Immunol.*, 129, 232, 2002.

173. Penkowa, M. and Hidalgo, J., Treatment with metallothionein prevents demyelination and axonal damage and increases oliogodendrocyte precursors and tissue reapir during experimental autoimmune encephalomyelitis (EAE). *J. Neurosci. Res.*, 72, 574, 2003.

174. Chung, R.S., Vickers, J.C., Chuah, M.I., and West, A.K., Metallothionein-IIA promotes initial neurite elongation and postinjury reactive neurite growth and facilitates healing after focal cortical brain injury, *J. Neurosci.*, 23, 3336, 2003.

175. Erickson, J.C., Sewell, A.K., Jensen, L.T., Winge, D.R., and Palmiter, R.D., Enhanced neurotrophic activity in Alzheimer's disease cortex is not associated with down-regulation of metallothionein-III (GIF), *Brain Res.*, 649, 297, 1994.

176. Amoureux, M.C., Van Gool, D., Herrero, M.T., Dom, R., Colpaert, F.C., and Pauwels, P.J., Regulation of metallothionein-III (GIF) mRNA in the brain of patients with Alzheimer disease is not impaired, *Mol. Chem. Neuropathol.*, 32, 101, 1997.

177. Uchida, Y., Growth-inhibitory factor, metallothionein-like protein, and neurodegenerative diseases, *Biol. Signals*, 3, 211, 1994.

178. Arai, Y., Uchida, Y., and Takashima, S., Developmental immunohistochemistry of growth inhibitory factor in normal brains and brains of patients with Down syndrome, *Pediatr. Neurol.*, 17, 134, 1997.

179. Hozumi, I., Inuzuka, T., Hiraiwa, M., Uchida, Y., Anezaki, T., Ishiguro, H., Kobayashi, H., Uda, Y., Miyatake, T., and Tsuji, S., Changes of growth inhibitory factor after stab wounds in rat brain, *Brain Res.*, 688, 143, 1995.

180. Inuzuka T., Hozumi, I., Tamura, A., Hiraiwa, M., and Tsuji, S., Patterns of growth inhibitory factor (GIF) and glial fibrillary acidic protein relative level changes differ following left middle cerebral artery occlusion in rats, *Brain Res.*, 709, 151, 1996.

181. Erickson, J.C., Masters, B.A., Kelly, E.J., Brinster, R.L., and Palmiter, R.D., Expression of human metallothionein-III in transgenic mice, *Neurochem. Int.*, 27, 35, 1995.

182. Erickson, J.C., Hollopeter, G., Thomas, S.A., Froelick, G.J., and Palmiter, R.D., Disruption of the metallothionein-III gene in mice: Analysis of brain zinc, behavior, and neuron vulnerability to metals, aging, and seizures, *J. Neurosci.*, 17, 1271, 1997.

183. Montoliu, C., Monfort, P., Carrasco, J., Palacios, O., Capdevila, M., Hidalgo, J., and Felipo, V., Metallothionein-III prevents glutamate and nitric oxide neurotoxicity in primary cultures of cerebellar neurons, *J. Neurochem.*, 75, 266, 2000.

184. Sakamoto, T., Kawazoe, Y., Uchida, Y., Hozumi, I., Inuzuka, T., and Watabe, K., Growth inhibitory factor prevents degeneration of injured adult rat motoneurons, *Neuroreport*, 14, 2147, 2003.

185. Carrasco, J., Penkowa, M., Giralt, M., Camats, J., Molinero, A., Campbell, I.L., Palmiter, R.D., and Hidalgo, J., Role of metallothionein-III following central nervous system damage, *Neurobiol. Dis.*, 13, 22, 2003.

186. Bibel, M. and Barde, Y.-A., Neurotrophins: Key regulators of cell fate and cell shape in the vertebrate nervous system, *Genes Dev.*, 14, 2919, 2000.

187. Benowitz, L.I. and Routtenberg, A., GAP-43: An intrinsic determinant of neuronal development and plasticity, *Trends Neurosci.*, 20, 84, 1997.

188. Sewell, A.K., Jensen, L.T., Erickson, J.C., Palmiter, R.D., and Winge, D.R., Bioactivity of metallothionein-3 correlates with its novel beta domain sequence rather than metal binding properties, *Biochem.*, 34, 4740, 1995.

189. Chung, R.S., Vickers, J.C., Chuah, M.I., Eckhardt, B.L., and West, A.K., Metallothionein-III inhibits initial neurite formation in developing neurons as well as postinjury, regenerative neurite sprouting, *Exp. Neurol.*, 178, 1, 2002.

190. Chung, R.S. and West, A.K., A role for extracellular metallothioneins in CNS injury and repair, *Neuroscience*, in press, 2004.

191. Ceballos, D., Lago, N., Verdu, E., Penkowa, M., Carrasco, J., Navarro, X., Palmiter, R.D., and Hidalgo, J., Role of metallothioneins in peripheral nerve function and regeneration, *Cell. Mol. Life Sci.*, 60, 1209, 2003.

192. Uchida, Y., Gomi, F., Masumizu, T., and Miura, Y., Growth inhibitory factor prevents neurite extension and death of cortical neurons caused by high oxygen exposure through hydroxyl radical scavenging, *J. Biol. Chem.*, 277, 32353, 2002.

193. Chen, Y., Irie, Y., Keung, W.M., and Maret, W., S-Nitrosothiols react preferentially with zinc thiolate clusters of metallothionein III through transnitrosation, *Biochemistry*, 41, 8360, 2002.

194. Shi, Y., Wang, W., Mo, J., Du, L., Yao, S., and Tang, W., Interactions of growth inhibitory factor with hydroxyl and superoxide radicals, *Biometals*, 16, 383, 2003.

CHAPTER **12**

Peripheral Benzodiazepine Receptor (PBR) Imaging in Glial Cells: Molecular Sensors of Brain Pathology

Tomás R. Guilarte

CONTENTS

12.1 INTRODUCTION

Glial cells have been described as having a supporting role to neurons in normal brain function and pathology. However, during the last two decades important new information has emerged on new roles for glial cells in brain development, as neuronal precursors, in the regulation of synaptic function, and in brain injury and repair (Goldman, 2003; Horner and Plamer, 2003; Newman, 2003; Ransom et al., 2003; Slezak and Pfrieger, 2003). Some of these functions are discussed in other chapters of this book. These new biological functions ascribed to glial cells bring them to the forefront of neuroscience research. I should note that although oligodendrocytes are an important glial cell type, in this chapter I will refer to glial cells primarily as microglia and astrocytes.

The peripheral benzodiazepine receptor (PBR) is a protein that is primarily localized to glial cells in the brain and plays an important role in brain injury and repair (Brown and Papadopoulos, 2001). The fact that PBR in the brain is localized to glial cells makes it uniquely suited to monitor glial cell activation under a variety of experimental paradigms. In this chapter, I will discuss the role of the PBR in glial cells as a sensor of brain pathology and the emerging evidence that activation of the PBR may be a potential target for therapeutic intervention of brain inflammation and neuronal injury. Further, I will discuss *in vitro* and *in vivo* methodologies to visualize PBR expression in the living brain in order to assess brain inflammation and pathology.

12.2 CENTRAL AND PERIPHERAL BENZODIAZEPINE RECEPTORS

Benzodiazepines are a class of chemicals used clinically as sedatives, anxiolytics, and anticonvulsants. There are two classes of benzodiazepine receptors. The "central-type" is associated with the GABAa receptor supramolecular complex and mediates the anxiolytic and anti-convulsive properties of benzodiazepines such as diazepam (Valium™) in the central nervous system (CNS) (Costa and Guidotti, 1979; Olsen, 1982). Central-type receptors are known to be present at high concentrations in neuronal membranes but not in peripheral organs. The second type of benzodiazepine receptors is the "peripheral-type," which was first identified as a high-affinity binding site for diazepam in the rat kidney (Braestrup and Squires, 1977a, 1977b). Peripheral benzodiazepine receptors (PBRs) are present at high levels in peripheral organs such as the lungs, adrenal glands, liver, kidneys, testes, ovaries, blood cells, and placenta and at low levels in the brain (Gavish et al., 1992; Woods and Williams, 1996; Zisterer and Williams, 1997). Of peripheral tissues tested, the adrenals express the

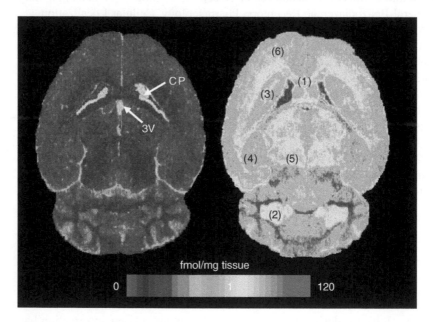

Figure 12.1 *(A color version of this figure follows page 236.)* Representative pseudocolor autoradiograms of [³H]-(*R*)-PK11195 binding to horizontal rat brain sections from a control (left image) and a cuprizone-treated rat (right image). Cuprizone treatment was 0.2% w/w in the diet for 4 weeks. Note the low level of [³H]-(*R*)-PK11195 binding to peripheral benzodiazepine receptors in the brain neuropil of the control rat brain. High levels of binding in the control rat brain are present in the choroid plexus and in the lining of the ventricles. In the cuprizone-treated rat, there is widespread increases in the levels of [³H]-(*R*)-PK11195 binding in the brain consistent with the known neuropathology of cuprizone. (See Chen et al., 2004, for more details.) The horizontal color bar is representative of levels of [³H]-(*R*)-PK11195 binding in fmol/mg tissue. Abbreviations: CP = choroid plexus; 3V = third ventricle; (1) = corpus collosum; (2) = deep cerebellar nuclei; (3) = corpus striatum; (4) = hippocampus; (5) = superior colliculus; (6) = frontal cortex.

highest density of PBR followed by kidney, heart, testes, ovaries, uterus, and liver (Gavish et al., 1999). Since PBR is also found in the brain, albeit at much lower levels than in other organ systems, the term "peripheral benzodiazepine receptor" is a misnomer. However, from a historical perspective the name has been maintained in the literature to avoid confusion. Other names commonly used in the published literature are "mitochondrial benzodiazepine receptor" and "ω3 sites." As indicated above, PBRs are found at low levels in glial cells in the brain parenchyma and at higher concentrations in nonneuronal structures within the brain such as the ependymal cells of the ventricles and in the choroid plexus (Benavides et al., 1983; Anholt et al., 1984; see Figure 12.1). PBR has also been detected in the nerve fiber and glomerular layers of the olfactory bulb (Anholt et al., 1984).

12.3 PERIPHERAL BENZODIAZEPINE RECEPTORS: PHARMACOLOGY, CELLULAR LOCALIZATION, AND MOLECULAR COMPOSITION

The PBR differs from the central-type in its pharmacology, subcellular and anatomical distribution, and importantly in its localization to glial cells in the brain (Gavish et al., 1992; Woods and Williams, 1996; Zisterer and Williams, 1997). The differences in pharmacology between the central and peripheral benzodiazepine receptors was first identified with the use of the benzodiazepine clonazepam, which binds to the central-type receptor with nanomolar affinity but exhibited very low affinity for the PBR (Awad and Gavish, 1987; Bender and Hertz, 1987; Langer and Arbilla, 1988). On the other hand, the benzodiazepine Ro5-4864 (4′-chlorodiazepam) and the nonbenzodiazepine isoquinoline carboxamide derivative PK11195 bind to the PBR with high affinity but have essentially no affinity for the central-type (Brown and Papadopoulos, 2001). Other classes of chemicals that also express affinity for the PBR are indoleacetamide derivatives such as FGIN-1-27 (Farges et al., 1994) and more recently, a new class of high affinity PBR ligands, the pyridazinoindole derivates with SSR180575 as prototype (Vin et al., 2003).

There are naturally occurring substances that have high affinity for the PBR and have been proposed to act as endogenous ligands. These include the porphyrins (Verma et al., 1987) and the endozepines, peptides isolated from rat brain that displace the binding of benzodiazepines from their receptors (Guidotti et al., 1983). In particular, the parent compound of all endozepines, the so called diazepam-binding inhibitor (DBI) is widely distributed in the CNS and is predominantly expressed in glial cells (Tonon et al., 1990; Vidnyanszky et al., 1994).

Studies on the subcellular distribution of the PBR indicate that it is associated with the outer membrane of mitochondria (Anholt et al., 1986; Antkiewcz-Michaluk and Guidotti et al., 1988; Bribes et al., 2004). However, low expression of PBR has also been shown to be localized to the plasma membrane of peripheral tissues (Woods and Williams, 1996) and in the purinuclear/nuclear compartment of cancer cell lines (Hardwick et al., 1999) and glial cells in situ (Kuhlmann and Guilarte, 2000).

The development of photoaffinity probes for PBR was critical for its purification and cloning of its cDNA (Antkiewicz-Michaluk and Mukhin et al., 1988; Riond et al., 1989). The cDNA has been cloned from several species from rat to human and it encodes a reading frame of 169 amino acids with approximately 80% homology among all species (Sprengel et al., 1989; Parola et al., 1991; Riond et al., 1991; Chang et al., 1992; Garnier et al., 1994). PBR is highly hydrophobic and it possesses five transmembrane domains. Purification of PBR from rat kidney mitochondria that preserves the binding of PK11195 and Ro5-4864 resulted in a complex that contained three polypeptides of molecular weights 18, 30, and 32 kDa (McEnery et al., 1992). The heavier molecular weight components of this complex were identified as the adenine nucleotide carrier (ADC) and the voltage-dependent anion channel (VDAC), respectively (McEnery et al., 1992). The 18-kDa subunit was the polypeptide that contained the binding site for the isoquinoline carboxamide PK11195. However, the binding of Ro5-4864 to the 18-kDa polypeptide also required VDAC, suggesting that the binding domains of these two PBR ligands are not identical despite the fact that they can displace each other from binding. VDAC by itself does not possess binding affinity for PBR ligands. In mitochondria PBR appears to be associated with the mito-chondrial permeability transition pore (MPTP) in conjunction with ADC and VDAC as well as other cytosolic PBR-interacting proteins such as PRAX-1 and Star (Papadopoulos et al., 2001). The association of PBR with VDAC and ADC as a larger MPTP complex is supported by recent studies showing that PBR ligands can modulate cellular apoptosis (Bono et al., 1999; Castedo et al., 2002).

Molecular modeling and topography studies have demonstrated that the native PBR is organized in clusters of four to six of the 18-kDa molecules associated with possibly one VDAC subunit (Garnier et al., 1994). On the other hand, upon activation, PBR polymerizes within seconds to form larger clusters of 15–25 particles (Papadopoulos et al., 1994; Boujrad et al., 1996; Delavoie et al., 2003). The formation of these large clusters appears to produce a change in the mitochondrial membrane such that it forms contact sites between the outer and inner surfaces to facilitate molecular transfer of cholesterol (Boujrad et al., 1996; Delavoie et al., 2003).

12.4 FUNCTION OF THE PERIPHERAL BENZODIAZEPINE RECEPTOR

A number of studies have demonstrated the involvement of the PBR in a variety of cellular functions including the regulation of mitochondrial respiration, cellular proliferation, heme biosynthesis, modulation of immune cells, apoptosis and steroidogenesis (for reviews, see Gavish et al., 1999; Brown and Papadopoulos, 2001; Casellas et al., 2002). From these diverse list of functions, the role of PBR in steroidegenesis is the best understood. Following the initial observation that the PBR is highly expressed in steriodogenic tissues, it was demonstrated that PBR ligands are able to stimulate steroid synthesis in a number of organ systems including the brain (Gavish et al., 1999; Brown and Papadopoulos, 2001; Casellas et al., 2002). Elegant studies by Papadopoulos and coworkers demonstrate that the PBR is the rate-limiting step in the conversion of pregnenolone from cholesterol in the synthesis of steroids (Papadopoulos et al., 1997a,b).

12.5 PERIPHERAL BENZODIAZEPINE RECEPTORS AS MOLECULAR SENSORS OF BRAIN PATHOLOGY

Following the discovery that PBR is expressed in brain parenchyma, Gallager et al. (1981) reported for the first time a dramatic increase in the levels of PBR in the brain of rats that had been injected with kainic acid into the striatum. This study used the benzodiazepine Ro5-4864 as radioactive ligand and the authors concluded that these binding sites were of nonneuronal origin and pharma-cologically distinct from the clonazepam-sensitive neuronal binding sites (Gallager et al., 1981). Schoemaker et al. (1982) showed a robust increase in [³H]-Ro5-4864 binding ipsilateral but not in the contralateral striatum of rats injected with kainic acid. They concluded that the increase in [³H]-Ro5-4864 binding was localized to glial cells. Subsequent studies showed that a variety of approaches to produce brain injury resulted in increased [³H]-PK11195 binding to PBR that was selective to the brain regions damaged (Benavides et al., 1987; Miyazawa et al., 1995; Gehlert et al., 1997; Altar and Baudry, 1990; Leong et al., 1994; Guilarte et al., 1995). Our laboratory has examined the degree of damage necessary to produce an increase in PBR levels following exposure to a variety of neurotoxicants. From these studies it is clear that neuronal damage ranging from frank neuronal loss to subtle terminal damage in a number of brain regions results in marked increases in PBR levels at the primary sites of damage (Guilarte et al., 1995; Kuhlmann and Guilarte, 1997, 1999, 2000; Chen et al., 2004). Further, the PBR is a much more sensitive indicator in detecting early damage than histological techniques (Kuhlmann and Guilarte, 1997; Chen et al., 2004).

It is now widely recognized that measurement of PBR levels in brain can be used as a marker of gliosis and thus as a surrogate marker of neuronal injury (Banati, 2002; Guilarte, 2002; Weissman and Raveh, 2003). Further, increased levels of PBR are also measured in secondary areas of brain injury resulting from the primary insult, providing a more extensive assessment of damaged neuronal networks (Kuhlmann and Guilarte, 1999; Banati et al., 2000).

12.6 CELLULAR CONTRIBUTIONS TO INCREASED PBR LEVELS IN BRAIN INJURY

Although brain PBRs are known to be localized in both microglia and astrocytes, it has been unclear until recently what contribution each cell type makes to the increases in PBR levels measured following brain damage. Studies using ischemic models have suggested that activated astrocytes are responsible for increased PBR expression (Benavides et al., 1990). Subsequent studies indicated that elevated PBR levels are only consistently correlated with activated microglia rather than astrocytes using ischemic models (Stephenson et al., 1995), axotomy models (Banati et al., 1997), and multiple sclerosis and experimental autoimmune encephalomyelitis (Vowinckel et al., 1997; Banati et al., 2000). On the other hand, *in situ* colocalization of PBR expression has been demonstrated in activated microglia and/or astrocytes using double-labeling fluorescence immunohistochemistry in neurotoxicant-induced brain injury (Kuhlmann and Guilarte, 2000), in the brain of jimpy and shiverer mice expressing mutations in genes associated with myelination (Le Goascogne et al., 2000), and in temporal lobe epilepsy in humans (Sauvageau et al., 2002). One possible explanation for these different conclusions is the fact that some of the previous studies that only found an association with microglia and not with astrocytes did not examine the temporal profile of increased PBR levels following injury in conjunction with both glial cell types. This is important because the contribution of astrocytes to increased PBR levels occurs later in time relative to microglia following brain injury.

More recently, work performed in our laboratory using [³H]-R-PK11195 emulsion microautoradiography to measure binding to PBR coupled with immunohistochemical labeling of microglia and astrocytes indicates that in fact PBR colocalizes to both glial cell types. Importantly, we show that the contribution of microglia and astrocytes have different temporal profiles. In general, we have found that the early increase in PBR level is associated mostly with microglia (Kuhlmann and Guilarte, 2000; Chen et al., 2004). However, as time progresses, astrocytes become activated and provide a significant contribution to the increases in [³H]-R-PK11195 binding to PBR. Therefore, at later time points after injury the increase in PBR levels is from both microglia and astrocytes (Kuhlmann and Guilarte, 2000; Chen et al., 2004) (see Figure 12.2.).

The functional significance of increased PBR levels in glial cells following neuronal injury is still unclear. Microglia are known to migrate and proliferate in response to many types of brain injury (Streit el al., 1988; Kreutzberg, 1996). It is possible that increased PBR levels are linked to both of these functions. Recent evidence suggests that PBR ligands can influence both the rate of DNA synthesis and the chemotactic potential of tumor cell lines (Hardwick et al., 1999). Another possibility is that glial cells upregulate PBR levels to increased neurosteroid synthesis at the sites of damage. Studies have shown that PBR activation in glial cells promotes the synthesis of pregnenolone and progesterone (Le Goascogne et al., 2000), two neurosteroids that possess neurotrophic and neuroprotective activity (Le Goascogne et al., 2000). It is possible that increased PBR levels in glial cells during the process of brain injury are associated with the induction and secretion of substances such as neurosteroids that promote repair and/or survival of neuronal cells.

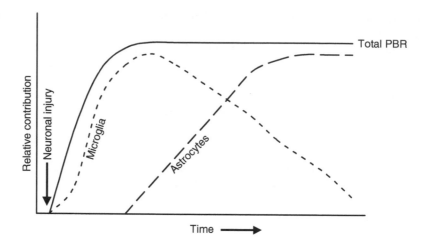

Figure 12.2 Schematic of proposed temporal contribution of peripheral benzodiazepine receptor (PBR) levels by glial cell types following brain injury. The early PBR response to injury seems to be primarily mediated by microglia followed by an astrocytic component. According to the type and degree of brain injury produced, the microglial response may decay or be sustained as a function of time.

12.7 PERIPHERAL BENZODIAZEPINE RECEPTOR IMAGING *IN VITRO* AND *IN VIVO* TO ASSESS BRAIN PATHOLOGY

12.7.1 *In Vitro* Imaging

The use of PBR autoradiography with the PBR specific ligand [³H]-(*R*)-PK11195 has gained a great deal of popularity to assess chemical-induced as well as other types of brain injury. Until recently, most studies used the racemic form of [³H]-PK11195. However, it is known that one of the isomers is less active and during the last 5 years laboratories began to use the more specific *R*-isomer of [³H]-PK11195. The advantage of using the enantiomerically active isomer is that it reduces the level of nonspecific binding and thus it enhances the signal-to-noise ratio making the assay more sensitive to small changes in receptor levels (Guilarte, 2002). Another advance that has occurred in the last few years is the use of [¹²³I]- or [¹²⁵I]-(*R*)-PK11195. The advantage of using the iodinated compound versus the tritiated compound for PBR autoradiography is that with the iodinated compound one obtains autoradiography images in less than 24 hours compared to 4–5 weeks for the tritiated compound. Thus, the use of [¹²⁵I]-(*R*)-PK11195 markedly reduces the time and increases the volume of work that can be accomplished in a given amount of time.

In our laboratory, we have used PBR expression as a molecular biomarker of neurotoxicity using a variety of model neurotoxicants. These include trimethyltin (Guilarte et al., 1995; Kuhlmann and Guilarte, 2000), domoic acid (Kuhlmann and Guilarte, 1997), MPTP (Kuhlmann and Guilarte, 1999), methamphetamine (Guilarte et al., 2003), and most recently cuprizone, a copper chelator that induces demyelination in the central nervous system (Chen et al., 2004). Figure 12.1 shows PBR autoradiography using the enantiomerically active [³H]-(*R*)-PK11195 in animals exposed to cuprizone in the diet for 4 weeks. See the increased levels of [³H]-(*R*)-PK11195 in white matter and gray matter areas undergoing demyelination (Chen et al., 2004).

Our work has shown that the temporal pattern and the degree of PBR expression following chemical-induced brain injury are a direct reflection of the glial response and the type of damage produced by neurotoxicant exposure. For example, PBR expression following a single injection of TMT is not significantly increased until 3 days after exposure, but once elevated at 7 days, it remains increased in affected areas for a long period of time (Guilarte et al., 1995). This is a reflection of the slow but progressive neuronal cell loss that TMT produces in limbic structures such as the hippocampus (Balaban et al., 1988). We have also shown that the early increase in PBR expression is associated with microglia activation, while the later response appears to be primarily related to astrocytes (Kuhlmann and Guilarte, 2000). A similar effect as with TMT was observed with domoic acid, another neurotoxicant that damages limbic structures (Kuhlmann and Guilarte, 1997). In this case, PBR levels are increased prior to any indication of seizure-induced neuronal cell loss. On the other hand, the PBR response is increased much more rapidly in the striatum and substantia nigra of animals exposed to the dopaminergic neurotoxicant MPTP, and the response is dose-dependent (Kuhlmann and Guilarte, 1999). MPTP is a neurotoxicant that produces degeneration of dopaminergic terminal fields in the striatum and cell bodies in the substantia nigra. The PBR response to MPTP neurotoxicity appears to persist in the dorsal aspects of the striatum and in the substantia nigra, but it is transient in the ventral aspects of the striatum, possibly reflecting different degrees of sensitivity of these brain structures to MPTP exposure. These findings are consistent with a dorso-ventral gradient in the astrocytic response to MPTP damage in the corpus striatum (Francis et al., 1995).

We are currently assessing the PBR response resulting from exposure to cuprizone. Cuprizone is a copper chelator that has been used to produce a model of demyelination in the central nervous system (Matsuchima and Morell, 2001). The corpus collosum is a white matter region in which demyelination is readily observable following cuprizone ingestion in mice (Kesterson and Carlton, 1971). Our studies have shown that PBR levels are increased in this and other brain structures in a time-dependent manner consistent with known demyelination patterns produced by cuprizone (Kesterson and Carlton, 1971). It should be noted that in our studies as well as those of others, it has been shown that the increases in [^{3}H]-R-PK11195 binding to PBR is the result of an increase in the maximal number of receptors (Bmax) and not in receptor affinity for the ligand (Kd) (Guilarte et al., 1995; Chen et al., 2004).

12.7.2 *In Vivo* PBR Imaging Using Positron Emission Tomography

Positron emission tomography (PET) has made it possible to noninvasively visualize and quantitatively measure specific biochemical processes in the living brain with exquisite resolution. A detailed technical description of this technique is not suited to this chapter but can be found in published reviews (Myers et al., 1998; Phelps, 2000). Briefly, PET is based on the use of short-lived (minutes), positron-emitting radioisotopes (carbon-11, fluorine-18) for labeling organic molecules or drugs that are part of specific biochemical reactions or their interaction with specific "binding sites" in the brain such as receptors or enzymes. Following the intravenous injection of a radioactively labeled ligand, the distribution of the radioactivity in the brain can be measured by coincidence detection of geometrically opposite gamma rays (0.511 KeV) that are emitted when the positron radiation emitted decays. The gamma rays are detected in coincidence by a ring of detectors in the PET scanner and the distribution of the radioactivity is computer-reconstructed to form an image. The image is a direct reflection of the amount and spatial distribution of the interaction of the radiolabeled ligand with its biological target.

The use of PBR-PET has received increased attention in recent years. Although PBR-PET studies were performed in the late 1980s when PET scanners were becoming available at research institutions, very few studies were performed. The lack of studies was the result of poor image quality due to the use of racemic PK11195, poor spatial resolution of PET scanners, inherently low levels of PBR in the brain, and the lack of appropriate mathematical models to accurately quantify PBR expression. The limited number of PET studies were primarily done for the detection of gliomas, brain tumors that express high levels of PBR (Junck et al., 1989; Pappata et al., 1991), and the damage produced by stroke (Ramsay et al., 1992). More recently, PBR-PET has been used in a number of human neurological diseases, because advances in all of the areas that previously contributed to poor image quality have been made. For example, current PET scanners have better spatial resolution and greater axial sampling, the pharmacologically active (R)-PK11195 enantiomer is available, and advanced mathematical models are currently used to accurately quantify brain radioactivity.

The use of PBR-PET has now been described in human studies of multiple sclerosis (Vowinckel et al., 1997; Banati et al., 2000), Rasmussen's encephalitis (Banati et al., 1999), cerebral vasculitis associated with refractory epilepsy (Goerres et al., 2001), improved imaging of ischemic stroke (Pappata et al., 2000), to assess the neuroinflammation and gliosis in Alzheimer's disease (Cagnin et al., 2001a; Versijpt et al., 2002), multiple system atrophy (Gerhard et al., 2003), and herpes encephalitis (Cagnin et al., 2001b). From the aforementioned studies, it appears that the PBR is a useful marker of inflammation and ongoing gliosis, and thus in an indirect way a marker of brain injury. In summary, PBR-PET has the potential to be a molecular marker of disease activity that can be monitored in the living brain. This approach may be useful to follow disease progression as well as to assess the effectiveness of therapeutic interventions.

12.8 POTENTIAL USE OF PBR LIGANDS FOR THERAPEUTIC INTERVENTIONS

There is emerging evidence that PBR activation may be useful in the treatment of certain inflammatory conditions (Torres et al., 2000), attenuation of seizures, and brain injury (Ferzaz et al., 2002; Veenman et al., 2002). The exact mechanisms by which these protective effects of PBR ligands occur are currently not known. However, at least in the CNS, it is presumed that these effects may be associated with the ability of PBR ligands to stimulate neurosteroid synthesis through PBR activation (Lacapere and Papadopoulos, 2003). It is known that glial cells, in particular microglia, are closely associated with injured neurons presumably to remove and clean up neuronal debris. Microglia are also capable of secreting a number of substances such as cytokines and growth factor that may serve to improve the neuronal milieu in order to facilitate repair and recovery (Bruce-Keller, 1999; Raivich et al., 1999). Similarly, astrocytes also become reactive following injury and demonstrate a rapid induction of growth factors such as insulin growth factor (IGF) and platelet-derived growth factor (PDGF). Microglia and astrocytes both expressed increased PBR levels and PBR activation in both cell types may increase neurosteroid synthesis (Schumacher et al., 2000). This is particularly important because neurosteroids have been shown to exert protective effects on both neurons and glial cells and promote regenerative processes (Schumacher et al., 2000). The mechanism by which neurosteroids produce neurotrophic and neuroprotective effects may be via their ability to regulate the transcription of hormone-sensitive genes or by the modulation of neuronal proteins (Schumacher et al., 2000). Clearly, there is much more to learn about the role of the PBR in brain pathology and in recovery. In this context, current advances in our ability to visualize the PBR under a number of pathological conditions should provide a valuable tool for *in vitro* as well as *in vivo* studies.

ACKNOWLEDGMENTS

This work was supported from a grant from the National Institute of Environmental Health Sciences ES07062.

REFERENCES

Altar, C.A. and Baudry, M. (1990). Systemic injections of kainic acid: Gliosis is olfactory and limbic brain regions quentified with [³H]-PK 11195 binding autoradiography, *Exp. Neurol.,* 109, 333–341.
Anholt, R.R., Murphy, K.M., Mack, G.E., and Snyder, S.H. (1984). Peripheral-type benzodiazepine receptors in the central nervous system: Localization to olfactory nerves, *J. Neurosci.,* 4, 593–603.

Anholt, R.R.H., Pederson, P.L., De Souza, E.B., and Snyder, S.H. (1986). The peripheral benzodiazepine receptor localization to the mitochondrial outer membrane, *J. Biol. Chem.*, 261, 576–583.

Antkiewicz-Michaluk, L., Guidotti, A., and Krueger, K.E. (1988). Molecular characterization and mitochondrial density of a recognition site for peripheral-type benzodiazepine ligands, *Mol. Pharmacol.*, 34, 272–278.

Antkiewicz-Michaluk, L., Mukhin, A.G., Guidotti, A., and Krueger, K.E. (1988). Purification and characterization of a protein associated with peripheral-type benzodiazepine binding sites, *J. Biol. Chem.*, 263, 17317–17321.

Awad, M. and Gavish, M. (1987). Binding of [^{3}H]-RO5-4864 and [^{3}H]-PK11195 to cerebral cortex and peripheral tissue of various species: Species differences and heterogeneity in peripheral benzodiazepine binding sites, *J. Neurochem.*, 49, 1407–1414.

Balaban, C.D., O'Callaghan, J.P., and Billingsley, M.L. (1988). Trimethyltin-induced neuronal damage in the rat brain: Comparative studies using silver degeneration stains, immunocytochemistry and immunoassay for neurotypic and gliotypic proteins, *Neuroscience*, 26, 337–361.

Banati, R.B. (2002). Visualizing microglial activation *in vivo*, *Glia*, 40, 206–217.

Banati, R.B., Goerres, G.W., Myers, R., Gunn, R.N., Turkheimer, F.E., Kreutzberg, G.W., Brooks, D.J., Jones, T., and Duncan, J.S. (1999). [11C](R)-PK11195 positron emission tomography imaging of activated microglia in vivo in Rasmussen's encephalitis, *Neurology*, 53, 2199–2203.

Banati, R.B., Myers, R., and Kreutzberg, G.W. (1997). PK ('peripheral benzodiazepine')—binding sites in the CNS indicate early and discrete brain lesions: Microautoradiographic detection of [^{3}H]PK11195 binding to activated microglia, *J. Neurocytol.* 26, 77–82.

Banati, R.B., Newcombe, J., Gunn, R.N., Cagnin, A., Turkheimer, F., Heppner, F., Price, G., Wegner, F., Giovannoni, G., Miller, D.H., Perkin, G.D., Smith, T., Hewson, A.K., Bydder, G., Kreutzberg, G.W., Jones, T., Cuzner, M.L., and Myers, R. (2000). The peripheral benzodiazepine binding site in the brain in multiple sclerosis: quantitative *in vivo* imaging of microglia as a measure of disease activity, *Brain*, 123, 2321–2337.

Benavides, J., Dubois, A., Gotti, B., Bourdiol, F., and Scatton, B. (1990). Cellular distribution of ω3 (peripheral type benzodiazepine) binding sites in the normal and ischaemic rat brain: An autoradiographic study with the photoaffinity ligand [^{3}H]-PK14105, *Neurosci. Lett.*, 114, 32–38.

Benavides, J., Fage, D., Carter, C., and Scatton, B. (1987). Peripheral type benzodiazepine binding sites are a sensitive indirect index of neuronal damage, *Brain Res.*, 421, 167–172.

Benavides, J., Quarteronet, D., Imbault, F., Malgouris, C., Uzan, A., Renault, C., Dubroeucq, M.C., Gueremy, C., and Le Fur, G. (1983). Labelling of "peripheral-type" benzodiazepine binding sites in the rat brain by using [^{3}H]PK 11195, an isoquinoline carboxamide derivative: Kinetic studies and autoradiographic localization, *J. Neurochem.*, 41, 1744–1750.

Bender, A.S. and Hertz, L. (1987). Pharmacological characteristics of diazepam receptors in neurons and astrocytes in primary cultures, *J. Neurosci Res.*, 18, 366–372.

Bono, F., Lamarche, I., Prabonnaud, V., LeFur, G., and Herbert, J.M. (1999). Peripheral benzodiazepine receptor agonists exhibit potent antiapoptotic activities, *Biochem. Biophys. Res. Commuh.*, 265, 457–461.

Boujrad, N, Vidic, B, and Papadopoulos, V. (1996). Acute action of choriogonadotropin on Leydig tumor cells: Changes in the topography of the mitochondrial peripheral-type benzodiazepine receptor, *Endocrinology*, 137, 5727–5730.

Braestrup, C. and Squires, R.F. (1977a). Benzodiazepine receptors in rat brain, *Nature*, 266, 732–734.

Braestrup, C. and Squires, R.F. (1977b). Specific benzodiazepines receptors in the rat brain characterized by high-affinity [^{3}H]-diazepam binding, *Proc. Natl. Acad. Sci. U.S.A.*, 74, 3805–3809.

Bribes, E., Carriere, D., Goubet, C., Galiegue, S., Casellas, P., and Simony-Lafontaine, J. (2004). Immunohistochemical assessment of the peripheral benzodiazepine receptor in human tissues, *J. Histochem. Cytochem.*, 52, 19–28.

Brown, R.C. and Papadopoulos, V. (2001). Role of the peripheral-type benzodiazepine receptor in adrenal and brain steroidogenesis, *Int. Rev. Neurobiol.*, 46, 117–143.

Bruce-Keller, A. (1999). Microglial-neuronal interactions in synaptic damage and repair, *J. Neurosci. Res.*, 58, 191–201.

Cagnin, A., Brooks, D.J., Kennedy, A.M., Gunn, R.N., Myers, R., Turkheimer, F.E., Jones, T., and Banati, R.B. (2001a). *In vivo* measurement of activated microglia in dementia, *Lancet*, 358, 461–467.

Cagnin, A., Myers, R., Gunn, R.N., Lawrence, A.D., Stevens, T., Kreutzberg, G.W., Jones, T., and Banati, R.B. (2001b). *In vivo* visualization of activated glia by [^{11}C]-(R)-PK11195-PET following herpes encephalitis reveals projected neuronal damage beyond the primary focal lesion, *Brain*, 124, 2014–2027.

Casellas, P., Galiegue, S., and Basile, A.S. (2002). Peripheral benzodiazepine receptors and mitochondrial function, *Neurochem. Int.,* 40, 475–486.

Castedo, M., Perfettini, J.L., and Kroemer, G. (2002). Mitochondrial apoptosis and the peripheral benzodiazepine receptor: A novel target for viral and pharmacological manipulations, *J. Exp. Med.,* 196, 1121–1125.

Chang, Y.J., McCabe, R.T., Rennert, H., Budart, M.L., Sayegh, R., Emanuel, B.S., Skolnick, P., and Strauss, J.F. (1992). The human "peripheral-type" benzodiazepine receptor: Regional mapping of the gene and characterization of the receptor expressed from cDNA, *DNA Cell Biol.,* 11, 471–480.

Costa, E. and Guidotti, A. (1979). Molecular mechanisms in the receptor activation of benzodiazepines, *Ann. Rev. Pharmacol. Toxicol.,* 19, 531–545.

Chen, M.K., Baidoo, K., Verina, T., and Guilarte, T.R. (2004). Peripheral benzodiazepine receptor imaging in CNS demyelination: Functional implications of anatomical and cellular localization, *Brain,* 127, 1379–1392.

Delavoie, F., Li, H., Hardwick, M., Robert, J.C., Giatzakis, C., Peranzi, G., Yao, Z.X., Maccario, J., Lacapere, J.J., and Papadopoulos, V. (2003). *In vivo* and *in vitro* peripheral-type benzodiazepine receptor polymerization: Functional significance in drug ligand binding and cholesterol binding, *Biochemistry,* 42, 4506–4519.

Farges, R., Joseph-Liauzun, E., Shire, D., Caput, D., LeFur, G., and Ferrara, P. (1994). Site-directed mutagenesis of the peripheral benzodiazepine receptor: Identification of amino acids implicated in the binding site of Ro5-4864, *Mol. Pharmacol.,* 46, 1160–1167.

Ferzaz, B., Brault, E., Bourliaud, G., Robert, J.P., Poughon, G., Claustre, Y., Marguet, F., Liere, P., Schumacher, M., Nowicki, J.P., Fournier, J., Marabout, B., Sevrin, M., George, P., Soubrie, P., Benavides, J., and Scatton, B. (2002). SSR180575 (7-chloro-*N,N,*5-trimethyl-4-oxo-3-phenyl-3,5-dihydro-4*H*-pyridazino-[4,5-*b*]indole-1-acetamide), a peripheral benzodiazepine receptor ligand, promotes neuronal survival and repair, *J. Pharmacol. Exp. Ther.,* 301, 1067–1078.

Francis, J.W., Visger, J.V., Markelonis, G.J. and Oh, T.H. (1995). Neuroglial responses to the dopaminergic neurotoxicant 1-methyl-4-phenyl-1,2,3,6-tetrahydropyridine in mouse striatum, *Neurotoxicol. Teratol.,* 17, 7–12.

Gallager, D.W., Mallorga, P., Oertel, W., Henneberry, R., and Tallman, J. (1981). [3H]-Diazepam binding in mammalian central nervous system: A pharmacological characterization, *J. Neurosci.,* 1, 218–225.

Garnier, M., Dimchev, A.B., Boujard, N., Price, J.M., Musto, N.A., and Papadapoulos, V. (1994). *In vitro* reconstitution of a functional peripheral-type benzodiazepine receptor from mouse Leydig tumor cells, *Mol. Pharmacol.,* 45, 201–211.

Gavish, M., Katz, Y., Bar-Ami, S., and Weizman, R. (1992). Biochemical, physiological, and pathological aspects of the peripheral benzodiazepine receptor, *J. Neurochem.,* 58, 1589–1601.

Gavish, M., Bachman, I., Shoukrun, R., Katz, Y., Veenman, L., Weisinger, G., and Weizman, A. (1999). Enigma of the peripheral benzodiazepine receptor, *Pharmacol. Rev.,* 51, 629–650.

Gehlert, D.R., Stephenson, D.T., Schober, D.A., Rash, K., and Clemens, J.A. (1997). Increased expression of peripheral benzodiazepine receptors in the facial nucleus following motor neuron axotomy, *Neurochem. Int.,* 31, 705–713.

Gerhard, A, Banati, R.B., Goerres, G.B., Cagnin, A., Myers, R., Gunn, R.N., Turkheimer, F., Good, C.D., Mathias, C.J., Quinn, N., Schwarz, J., and Brooks, D.J. (2003). [11C]-(R)-PK11195 PET imaging of microglia activation in multiple system atrophy, *Neurology,* 61, 686–689.

Goerres, G.W., Revesz, T., Duncan, J., and Banati, R.B. (2001). Imaging cerebral vasculitis in refractory epilepsy using [11C](*R*)-PK11195 positron emission tomography, *Am. J. Radiol.,* 176, 1016–1018.

Goldman, S. (2003). Glia as neuronal progenitor cells, *Trends Neurosci.,* 26, 590–596.

Guidotti, A., Forchetti, C.M., Corda, M.G., Konkel, D., Bennett, C.D., and Costa, E. (1983). Isolation, characterization and purification to homogeneity of an endogenous polypeptide with agonistic action on benzodiazepine receptors, *Proc. Natl. Acad. Sci. U.S.A.,* 80, 3531–3535.

Guilarte, T.R. (2002). Peripheral benzodiazepine receptors: Molecular biomarkers of neurotoxicity, in *Biomarkers of Environmentally Associated Disease—Technologies, Concepts and Perspectives,* Wilson, S.H. and Suk, W.A.., Eds., Lewis Publishers, Boca Raton, Florida, pp. 411–427.

Guilarte, T.R., Kuhlmann, A.C., O'Callaghan, J.P., and Miceli, R.C. (1995). Enhanced expression of peripheral benzodiazepine receptors in trimethyltin-exposed rat brain: A biomarker of neurotoxicity, *Neurotoxicology,* 16, 441–450.

Guilarte, T.R., Nihei, M.K., McGlothan, J.L., and Howard, A.S. (2003). Methamphetamine-induced deficits of brain monoaminergic neuronal markers: distal axotomy or neuronal plasticity, *Neuroscience,* 122, 499–513.

Hardwick, M., Fertikh, D., Culty, M., Li, H., Vidic, B., and Papadopoulos, V. (1999). Peripheral-type benzo-diazepine receptor (PBR) in human breast cancer: correlation of breast cancer cell aggressive phenotype with PBR expression, nuclear localization, and PBR mediated cell proliferation and nuclear transport of cholesterol, *Cancer Res.*, 59, 831–842.

Horner, P.J. and Plamer, T.D. (2003). New roles for astrocytes: The nightlife of an "astrocyte." La vida loca! *Trends Neurosci.*, 26, 597–603.

Junck, L., Olson, J.M.M., Ciliax, B.J., Koeppe, R.A., Watkins, G.L., Jewett, D.M., McKeever, P.E., Wieland, D.M., Kilbourn, M.R., Starosta-Rubinstein, S., Mancini, W.R., Kuhl, D.E., Greenberg, H.S., and Young, A.B. (1989). PET imaging of human glioma with ligands for the peripheral benzodiazepine site, *Ann. Neurol.*, 26, 752–758.

Kesterson, J.W. and Carlton, W.W. (1971). Histopathologic and enzyme histochemical observations of the cuprizone-induced brain edema, *Exp. Mol. Pathol.*, 15, 82–96.

Kreutzberg, G.W. (1996). Microglia: A sensor for pathological events in the CNS, *Trends Neurosci.*, 19, 312–318.

Kuhlmann, A.C. and Guilarte, T.R. (1997). The peripheral benzodiazepine receptor is a sensitive indicator of domoic acid neurotoxicity, *Brain Res.*, 751, 281–288.

Kuhlmann, A.C. and Guilarte, T.R. (1999). Regional and temporal expression of the peripheral benzodiazepine receptor in MPTP neurotoxicity, *Toxicol. Sci.*, 48, 107–116.

Kuhlmann, A.C. and Guilarte, T.R. (2000). Cellular and subcellular localization of peripheral benzodiazepine receptors after trimethyltin neurotoxicity, *J. Neurochem.*, 74, 1694–1704.

Lacapere, J.J. and Papadopoulos, V. (2003). Peripheral-type benzodiazepine receptors: Structure and function of a cholesterol-binding protein in steroid and bile acid biosynthesis, *Steroids*, 68L, 569 585.

Langer, S.Z. and Arbilla, S. (1988). Imidazopyridines as a tool for the characterization of benzodiazepine receptors: A proposal for the pharmacological classification as Omega receptor subtypes, *Pharmacol. Biochem. Behav.*, 29, 763–766.

Le Goascogne, C., Eychenne, B., Tonon, M.C., Lachapelle, F., Baumann, N., and Robel, P. (2000). Neurosteroid progesterone is up-regulated in the brain of jimpy and shiverer mice, *Glia*, 29, 14–24.

Leong, D.K., Le, O., Oliva, L., and Butterworth, R.F. (1994). Increased densities of binding sites for the "peripheral type" benzodiazepine receptor ligand [3H]-PK11195 in vulnerable regions of the rat brain in thiamine deficiency encephalopathy, *J. Cereb. Blood Flow Metab.*, 14, 100–105.

Matsushima, G.K. and Morell, P. (2001). The neurotoxicant, cuprizone, as a model to study demyelination and remyelination in the central nervous system, *Brain Pathol.*, 11, 107–116.

McEnery, M.W., Snowman, A.M., Trifiletti, R.R., and Snyder, S.H. (1992). Isolation of the mitochondrial benzodiazepine receptor: Association with the voltage-dependent anion channel and the adenine nucleotide carrier, *Proc. Natl. Acad. Sci. U.S.A.*, 89, 3170–3174.

Miyazawa, N., Diksic, M., and Yamamoto, Y. (1995). Chronological study of peripheral benzodiazepine binding sites in the rat brain stab wounds using [3H]-PK-11195 as a marker of gliosis, *Acta Neurochir.*, 137, 207–216.

Myers, R., Banati, R.B., Paulesu, E., Thorpe, J., Miller, D.H., and Jones, T. (1998). Use of two- and three-dimensional PET and [11C](R)-PK11195 to image focal and regional brain pathology, in *Quantitative Functional Brain Imaging with Positron Emission*, Carson, R.E., Daube-Witherspoon, M.E., and Herscovitch, P., Eds., Academic Press, pp. 195–200.

Newman, E.S. (2003). New roles for astrocytes: Regulation of synaptic transmission, *TINS*, 26, 536–542.

Olsen, R.W. (1982). Drug interactions at the GABAa receptor ionophore complex, *Annu. Rev. Pharmacol. Toxicol.*, 22, 242–277.

Papadopoulos, V., Amri, H., Boujrad, N., Cascio, C., Culty, M., Garnier, M., Hardwick, M., Li, H., Vidic, B., Brown, A.S., Reversa, J.L., and Bernassau, J.M. (1997a). Peripheral benzodiazepine receptor in cholesterol transport and steroidogenesis, *Steroids*, 62, 21–28.

Papadopoulos, V., Amri, H., Li, H., Boujrad, N., Vidic, B., and Garnier, M. (1997b). Targeted disruption of the peripheral-type benzodiazepine receptor gene inhibits steroidogenesis in the R2C Leydig tumor cell line, *J. Biol. Chem.*, 272, 32129–32135.

Papadopoulos, V., Amri, H., Li, H., Yao, Z., Brown, R.C., Vidic, B., and Culty, M. (2001). Structure, function and regulation of the mitochondrial peripheral-type benzodiazepine receptor, *Therapie*, 56, 549–556.

Papadopoulos, V., Boujrad, N., Ikonomovic, M.D., Ferrara, P., and Vidic, B. (1994). Topography of the Leydig cell mitochondrial peripheral-type benzodiazepine receptor, *Mol. Cell. Endocrinol.*, 104, R5–R9.

Pappata, S., Cornu, P., Samson, Y., Prenant, C., Benavides, J., Scatton, B., Crouzel, C., Hauw, J.J., and Syrota, A. (1991). PET study of carbon-11-PK11195 binding to peripheral type benzodiazepine sites in glioblastoma: a case report, *J. Nucl. Med.,* 32, 1608–1610.

Pappata, S., Levasseur, M., Gunn, R.N., Myers, R., Crouzel, C., Syrota, A., Jones, T., Kreutzberg, G.W., and Banati, R.B. (2000). Thalamic microglia activation in ischemic stroke detected *in vivo* by PET and [^{11}C]PK11195, *Neurology,* 55, 1052–1054.

Parola, A.L., Stump, D.G., Pepperl, D.J., Krueger, K.E., Regan, J.W., and Laird, H.E. (1991). Cloning and expression of a pharmacologically unique bovine peripheral-type benzodiazepine receptor isoquinoline binding protein, *J. Biol. Chem.,* 266, 14082–14087.

Phelps, M.E. (2000). PET: The merging of biology and imaging into molecular imaging, *J. Nuclear Med.,* 41, 661–681.

Raivich, G., Jones, L.L., Werner, A., Bluthmann, H., Doestschmann, T., and Kreutzberg, G.W. (1999). Molecular signals for glial activation: Pro- and anti-inflammatory cytokines in the injured brain, *Acta Neurochir.,* 73, 21–30.

Ramsay, S.C., Weiller, C., Myers, R., Cremer, J.E., Luthra, S.K., Lammertsa, A.A., and Frackowiak, R.S.J. (1992). Monitoring by PET of macrophage accumulation in brain after ischaemic stroke, *Lancet,* 339, 1054–1055.

Ransom, B., Behar, T., and Nedergaard, M. (2003). New roles for astrocytes (stars at last), *Trends Neurosci.,* 26, 520–522.

Riond, J., Vita, N., LeFur, G., and Ferrara, P. (1989). Characterization of a peripheral-type benzodiazepine binding site in the mitochondria of Chinese hamster ovary cells, *FEBS Lett.,* 245, 238–244.

Riond, J., Mattei, M.G., Kaghad, M., Dumont, X., Guillemot, J.C., LeFur, G., Caput, D., and Ferrara, P. (1991). Molecular cloning and chromosomal localization of a human peripheral-type benzodiazepine receptor, *Eur. J. Biochem.,* 195, 305–311.

Sauvageau, A., Desjardins, P., Lozeva, V., Rose, C., Hazell, A.S., Bouthillier, A., and Boutherwort, R.F. (2002). Increased expression of "peripheral type" benzodiazepine receptors in human temporal lobe epilepsy: Implications for PET imaging of hippocampal sclerosis, *Metab. Brain Dis.,* 17, 3–11.

Schoemaker, H., Morelli, M., Deshmukh, P., and Yamamura, H.I. (1982). [^{3}H]-Ro5-4864 benzodiazepine binding in the kainate lesioned striatum and Huntington's disease basal ganglia, *Brain Res.,* 248, 396–401.

Schumacher, M., Akwa, Y., Guennoun, R., Robert, F., Labombarda, F., Desarnaud, F., Robel, P., DeNicola, A.F., and Baulieu, E.E. (2000). Steroid synthesis and metabolism in the nervous system: Trophic and protective effects, *J. Neurocytol.,* 29, 307–326.

Slezak, M. and Pfrieger, F.W. (2003). New roles for astrocytes: Regulation of CNS synaptogenesis, *Trends Neurosci.,* 26, 531–535.

Sprengel, R., Werner, P., Seeberg, P.H., Mukhin, A.G., Santi, M.R., Grayson, D.R., Guidotti, A., and Krueger, K.E. (1989). Molecular cloning and expression of cDNA encoding a peripheral-type benzodiazepine receptor, *J. Biol. Chem.,* 264, 20415–20421.

Stephenson, D.T., Schober, D.A., Smalstig, E.B., Mincy, R.E., Gehlert, D.R., and Clemens, J.A. (1995). Peripheral benzodiazepine receptors are colocalized with activated microglia following transient global forebrain ischemia in the rat, *J. Neurosci.,* 15, 5263–5274.

Streit, W.J., Graeber, M.B., and Kreutzberg, G.W. (1988). Functional plasticity of microglia: A review, *Glia,* 1, 301–307.

Tonon, M.C., Desy, L., Nicolas, P., Vaudry, H., and Pelletier, G. (1990). Immunocytochemical localization of the endogenous benzodiazepine ligand octadecaneuropeptide (ODN) in the rat brain, *Neuropeptides,* 15, 17–24.

Torres, S.R., Frode, T.S., Nardi, G.M., Vita, N., Reeb, R., Ferrera, P., Ribeiro-do-Valle, R.M., and Farges, R.C. (2000). Anti-inflammatory effects of peripheral benzodiazepine receptor ligands in two mouse models of inflammation, *Eur. J. Pharmacol.,* 408, 199–211.

Veenman, L., Leschiner, S., Spanier, I., Weisinger, G., Weizman, A., and Gavish, M. (2002). PK11195 attenuates kainic acid-induced seizures and alterations in peripheral-type benzodiazepine receptor (PBR) protein components in the rat brain, *J. Neurochem.,* 80, 917–927.

Verma, A., Nye, J.S., and Snyder, S.H. (1987). Porphyrins are endogenous ligands for the mitochondrial (peripheral-type) benzodiazepine receptor, *Proc. Natl. Acad. Sci. U.S.A.,* 84, 2256–2260.

Versijpt, J.J., Dumont, F., Laere, K.J.V., Decoo, D., Santens, P., Audenaert, K., Achten, E., Slegers, G., Dierckx, R.A., and Korf, J. (2002). Assessment of neuroinflammation and microglial activation in Alzheimer's disease with radiolabeled PK11195 and single photon emission computed tomography, *Eur. Neurol.,* 50, 39–47.

Vidnyanszky, Z., Gorcs, T.J., and Hamori, J. (1994). Diazepam binding inhibitor fragments 33-50 (octadeca-neuropeptide) immunoreactivity in the cerebellar cortex is restricted to glial cells, *Glia,* 10, 132–141.

Vin, V., Leducq, N., Bono, F., and Herbert, J.M. (2003). Binding characteristics of SSR180575, a potent and selective peripheral benzodiazepine ligand, *Biochem. Bioophys. Res. Commun.,* 310, 785–790.

Vowinckel, E., Reutens, D., Becher, B., Verge, G., Evans, A., Owens, T., and Antel, J.P. (1997). PK11195 binding to the peripheral benzodiazepine receptor as a marker of microglia activation in multiple sclerosis and experimental autoimmune encephalomyelitis, *J. Neurosci. Res.,* 50, 345–353.

Weissman, B.A. and Raveh, L. (2003). Peripheral benzodiazepine receptors: On mice and human brain imaging, *J. Neurochem.,* 84, 432–437.

Woods, M.J. and Williams, D.C. (1996). Multiple forms and locations for the peripheral-type benzodiazepine receptor, *Biochem. Pharmacol.,* 52, 1805–1814.

Zisterer, D.M. and Williams, D.C. (1997). Peripheral-type benzodiazepine receptors, *Gen. Pharmacol.,* 29, 305–314.

CHAPTER **13**

Astrocyte–Neuronal Interaction and Oxidative Injury: Role of Glia in Neurotoxicity

Valerie Chock and Rona Giffard

CONTENTS

13.1 INTRODUCTION

The pathophysiology of many neurologic disorders involves the generation of pathological amounts and types of reactive oxygen species (ROS). These highly reactive molecules are implicated in normal aging and acute and chronic neurodegeneration. While ROS have normal physiologic signaling roles, they can also damage nearly all essential cellular constituents. Oxidative stress occurs when ROS production exceeds detoxification and the normal antioxidant mechanisms fail. Astrocytes normally protect neuronal function by maintenance of extracellular ion concentrations, release of neurotrophic factors, neurotransmitter uptake, regulation of synaptic function, and anti-oxidant defense. Oxidative damage can result in the loss of these essential functions and accelerate neuronal damage. Glial cells are central to regulation of the inflammatory response in brain. Inflammation is also linked to oxidative stress and contributes to many kinds of neurodegeneration. This chapter will briefly review the sources and scavenging of ROS, with a focus on the role played by astrocytes and data from studies of stroke. Both neuroprotective and neurotoxic effects of glia will be considered.

13.2 GENERATION, INTERCONVERSION, AND SCAVENGING OF REACTIVE OXYGEN SPECIES

An extensive review of free radicals in biology by Halliwell and Gutteridge[1] and several useful reviews are recommended.[2–5]

13.2.1 Superoxide

Superoxide ($O_2 \bullet^-$) is formed by the reduction of oxygen. Mitochondria are the principal source of superoxide, with about 5% of electron flow resulting in superoxide formation.[6] Superoxide is also produced by several prooxidant enzymes including NADPH oxidase, especially in phagocytic cells, xanthine oxidase during purine degradation, and cyclooxygenases in arachidonic acid metabolism. Xanthine oxidase is present at higher levels in astrocytes compared with neurons.[3] In ischemia, increased levels of intracellular calcium activate proteases that convert xanthine dehydrogenase to xanthine oxidase, further exacerbating the problem. Excess superoxide can cause damage by promoting toxic hydroxyl radical formation (OH•) via hydrogen peroxide (by the Haber–Weiss reaction), by combination with nitric oxide to form the highly reactive peroxynitrite radical (ONOO•−),[7] or by dismutation to singlet oxygen, a powerful oxidant, and hydrogen peroxide.[8]

Normally superoxide is metabolized by superoxide dismutase (SOD). There are three forms of SOD: copper/zinc SOD (CuZnSOD or SOD1), manganese SOD (MnSOD or SOD2), and extra-cellular SOD (ECSOD or SOD3). SOD1 is localized to the cytosol, MnSOD to mitochondria, and SOD3 to cerebral spinal fluid and cerebral vessels.[5,9] During oxidative stress when superoxide formation exceeds dismutation, superoxide can exit cells via anion channels and diffuse to neighboring cells. This is especially dangerous if nitric oxide is available for generation of peroxynitrite.

13.2.2 Hydrogen Peroxide

Hydrogen peroxide is produced by mitochondrial respiration and in the cytosol through the action of monoamine oxidase. In the Fenton reaction, ferrous iron (Fe^{2+}) catalyzes formation of the highly reactive hydroxyl radical from hydrogen peroxide. Hydroxyl radical immediately attacks other molecules in the vicinity, generating secondary radicals after a mean diffusion distance of only 0.3 nanometers.[7] In comparison, hydrogen peroxide, nitric oxide, and superoxide easily cross cell membranes and can diffuse for several micrometers before they interact with another molecule. During reperfusion iron homeostasis is altered to favor the more reactive ferrous state (Fe^{2+}). The iron stores in a cell influence its ability to withstand oxidative stress. For example, younger neurons contain more free iron and are more prone to oxidative damage than mature neurons.[10] GSH peroxidase (GPx) and catalase are endogenous antioxidant enzymes that metabolize H_2O_2 to produce water. Astrocytes contain more endogenous GPx and catalase than neurons. While both enzymes metabolize H_2O_2, GPx is significantly more abundant in the brain compared with catalase and can metabolize a range of peroxides.[11]

13.2.3 Nitric Oxide

Nitric oxide is a diffusible free radical and intra- and intercellular messenger.[12] However, when present in excessive amounts, it can react with superoxide ion to form toxic peroxynitrite,[13] which can then inhibit mitochondrial respiration and damage multiple other targets in the cell. Three isoforms of nitric oxide synthase are known in the brain: (1) neuronal NOS (nNOS), a constitutive isoform localized to neurons; (2) eNOS, a constitutive isoform localized to endothelial cells; and (3) iNOS, an inducible isoform that is induced in astrocytes, microglia, and endothelial cells. Of these isoforms, only iNOS is Ca^{2+} independent. Studies with knockout mice have shown that eNOS is protective against cerebral ischemia, while nNOS and iNOS can contribute to brain injury.[5,14]

In stroke there are data demonstrating a role for several prooxidant enzymes including nitric oxide synthase, cyclooxygenases, and monoamine oxidase.[2,5] The high metabolic rate and mitochondrial dysfunction associated with reperfusion contribute to excessive ROS generation at a time when the capacity of antioxidant defenses may be reduced. Nitric oxide contributes to injury after ischemia in a biphasic pattern due to the multiple sources of NOS.[14] Decreased cerebral perfusion first leads to an increase in eNOS and nNOS isoforms, with a subsequent increase in nitric oxide. Although the vasodilation resulting from this increase is protective, there are also damaging effects of nitric oxide. After 24 to 48 hours, release of inflammatory cytokines induces peak iNOS levels in astrocytes, microglia, and infiltrating neutrophils.[14] The role of these glial cells in NO-mediated neuronal injury is discussed below. This nitric oxide production contributes to energy failure and oxidative damage.

In summary, superoxide and hydrogen peroxide are central mediators of oxidative stress in the cell. Superoxide can dismutate to hydrogen peroxide and singlet oxygen and can reduce ferric iron to ferrous iron, which in turn can catalyze free-radical generation. Nitric oxide contributes to the formation of peroxynitrite. Hydrogen peroxide contributes to formation of the strong oxidants hydroxyl radical and peroxynitrite.

13.2.4 Low-Molecular-Weight Antioxidants: Glutathione (GSH), Ascorbic Acid, and α-Tocopherol

GSH plays a pivotal role in the scavenging of ROS. It can reduce ROS directly, serve as a source of electrons in GPx-mediated metabolism of peroxides, and reduce the oxidized form of vitamin C (DHAA) back to ascorbic acid. The availability of GSH is a key regulator of ischemia-induced damage. GSH depletion worsens injury and GSH supplementation can reduce injury.[15,16] Both vitamins E (α-tocopherol) and C (ascorbic acid) can react with free radicals such as hydroxyl radical, which may otherwise escape cellular antioxidant systems. Vitamin E is lipid soluble and can terminate propagation of lipid peroxidation chains. Vitamin C can be taken up by both astrocytes and neurons to directly inactivate oxidizing species and regenerate both GSH and vitamin E. Vitamin E and GSH levels are higher in astrocytes than neurons, while neurons have more vitamin C.[17,18] This reinforces the point that different cell types within the CNS differ in their antioxidant defenses and their vulnerability to oxidative stress.

13.3 OXIDATIVE STRESS, ASTROCYTES, AND MECHANISMS OF ISCHEMIC BRAIN INJURY

Oxidative stress contributes to ischemic brain injury directly and via excitotoxicity, apoptosis, and inflammation (Figure 13.1). Astrocytes protect neurons by multiple mechanisms during ischemia until they are injured and their normal function is disrupted. Astrocyte neuroprotective mechanisms include upregulation of antioxidants, maintenance of extracellular ionic homeostasis, glutamate uptake, and release of neurotrophic factors. We will discuss these mechanisms individually.

13.3.1 Astrocytes Are Central to Brain Antioxidant Defense

Glial cells are more resistant to oxidative stress than neurons largely due to their greater use of anaerobic glycolysis and their higher GSH content.[17,18] Although neurons have higher levels of ascorbate than astrocytes,[17] astrocytes take up oxidized vitamin C (dehydroascorbic acid) through plasma membrane transporters, reduce it to ascorbate, and release it extracellularly where it may contribute to the antioxidant defense of neurons.[19] Astrocytes also express the metal-binding proteins metallothionein I and II, while neurons do not.[20] These proteins suppress metal-catalyzed free-radical production and may directly scavenge ROS.[21]

Astrocytes possess a more efficient GSH antioxidant system and are therefore more resistant to peroxynitrite-mediated mitochondrial damage than neurons.[22] When stressed with oxygen-glucose deprivation, an ischemia-like insult, astrocytes preserved mitochondrial function more effectively than did neurons; in addition, astrocytes maintained higher GSH levels and had lower generation of free radicals.[23] The ability of astrocytes to use glycolysis allows them to continue producing ATP in the face of mitochondrial inhibition. Astrocyte death from substrate deprivation is closely linked to GSH depletion; cell death could be increased or decreased by manipulating GSH levels before removing substrate from the medium.[24]

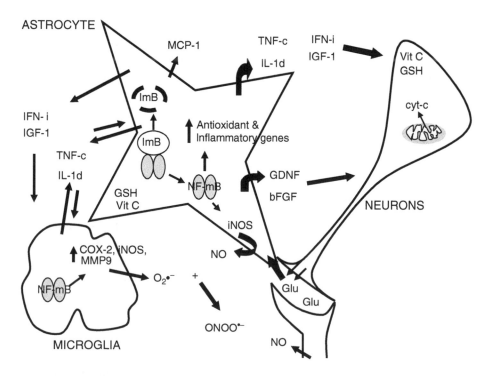

Figure 13.1 Diagram of cellular oxidant and inflammatory mechanisms and cell–cell interactions involved in CNS injury.

13.3.2 Astrocytes Bolster Neuronal Antioxidant Defenses

Cultured astrocytes are more resistant to oxidants including nitric oxide, hydrogen peroxide, and 6-hydroxydopamine compared with neurons and protect neurons from oxidative stress.[11,25,26] In response to oxidative challenge, astrocytes transcriptionally upregulate the rate-limiting enzyme for GSH synthesis, γ-glutamylcysteine synthetase, and increase GSH levels. Only neurons maintained in glial-conditioned medium upregulated GSH synthesis; without glial-conditioned medium, neurons failed to increase GSH in response to oxidative challenge.[25]

Neuronal GSH concentration was reported to double after 24 hours with astrocytes.[27] Co-cultured neurons may be protected due to glial release of cysteine and pyruvate. Neuronal uptake of cysteine bolsters GSH synthesis. Astrocytes also protect neurons via release of vitamin C and other soluble factors. The large surface/volume ratio of astrocytes facilitates the diffusion of ROS into these cells.[11] Thus, more active detoxification of ROS by astrocytes and their ability to upregulate neuronal antioxidant capability comprise important aspects of brain antioxidant defense and neuroprotection.

13.3.3 Oxidative Stress and Apoptosis

A link between oxidative injury and the mitochondrial-signaling pathway of apoptosis has also been demonstrated.[27] Excessive superoxide production after ischemia may result in mitochondrial injury and consequent release of multiple proapoptotic factors into the cytosol, including cytochrome c. The cytochrome c–dependent caspase pathway is then initiated with activation of caspase-9, caspase-3, and ensuing cellular apoptosis. Mitochondrial cytochrome c release induced by ROS has been documented.[28] This mechanism is relevant to brain injury as transgenic rats overexpressing CuZn SOD demonstrated less superoxide production, cytochrome c release, and caspase 3 activation and decreased apoptotic neuronal death after global ischemia.[29] The amount of ROS present may thus influence a cell's entry into the apoptotic pathway.

13.3.4 Oxidative Stress and Gene Expression

Oxidative stress activates transcription factors, notably NF-κB, to regulate gene expression. While NF-κB activation is often protective,[30] in some cases it has been found to contribute to injury.[31] NF-κB is persistently activated in response to cerebral ischemia, proinflammatory stimuli, or ROS, but the resulting changes are cell-type specific. In astrocytes and microglia, activated NF-κB can induce expression of inflammatory genes such as COX-2, iNOS, matrix metalloproteinase-9, leukocyte adhesion molecules, and proinflammatory cytokines, eventually leading to generation of neurotoxic ROS and breakdown of the blood–brain barrier.[5,31] Both induction of antiapoptotic genes and induction of damaging pathways has been demonstrated.[31] For example, NF-κB activation was associated with increased neuronal injury after cerebral ischemia, while reduced NK-κB activation achieved by inhibiting the proteosome was associated with protection.[32] In addition to NF-κB, other transcription factors sensitive to oxidative stress include AP-1, HIF-1, SP-1, and EIK-1. Their regulation of gene expression deserves further study in neurodegeneration.

13.3.5 Astrocytes Protect by Production of Neurotrophic Factors

Several neurotrophic factors produced by astrocytes such as GDNF (glial-derived neurotrophic factor) and bFGF (basic fibroblast growth factor) have been reported to protect neurons against oxidative stress. Neurotrophic factors have been shown to reduce cerebral ischemic injury. GDNF, a member of the TGF-β family, reduced both focal ischemic injury and the associated nitric oxide production.[33] Infarct volume was reduced by bFGF when the trophic factor was provided by grafted cells engineered to express bFGF[34] or when given intravenously.[35] Another report did not find a reduced infarct volume but observed augmentation of changes in progenitor cells induced by ischemia.[36] Astrocyte damage may lead to increased brain injury if levels of neurotrophic factors are reduced.

13.3.6 Astrocyte Glutamate Uptake Is Impaired by Oxidative Stress

Glutamate uptake is an essential physiological function of astrocytes and an important neuroprotective mechanism.[37] Knockout of glial glutamate transporters produces neurodegeneration characteristic of excitotoxicity.[38] After ischemia, failure of energy production causes depolarization and increased transmitter release. Reduced uptake further aggravates the situation. Depolarization, oxidative stress, and inflammatory stimuli all specifically impair glial glutamate uptake.[39–42] Astrocyte exposure to TNF-α also inhibited glutamate uptake,[43] possibly related to nitric oxide production. All of these conditions occur during ischemia/reperfusion, and impaired glutamate uptake has been demonstrated after ischemia *in vivo*.[44] Pathological overactivation of glutamate receptors elevates intracellular Ca^{2+}, which contributes to oxidative stress in several ways. These include activation of nNOS to produce toxic nitric oxide and calcium overload of mitochondria causing increased ROS production,[45] as well as release of proapoptotic factors.[28] There is evidence that ROS lead to further release of excitatory amino acids, creating positive feedback.[46] Increased intracellular Ca^{2+} can also increase oxidative stress by activation of the arachidonic acid pathway.[47] Thus, in ischemia and reperfusion the excitotoxic cascade and oxidative injury interact to potentiate cell death, of which inhibition of glial high-affinity glutamate uptake is a central element.

13.4 DIRECT CONTRIBUTION OF ASTROCYTES TO NEUROTOXICITY

13.4.1 Glia Can Promote Inflammation

Astrocytes and microglia, when stimulated by oxidative stress, initiate an inflammatory response that can ultimately be toxic to neurons. The role of cytokine signaling by glia and their possible contribution to neuronal loss in Alzheimer's disease has been discussed.[48] An important role for inflammation in many neurodegenerative diseases is increasingly appreciated. Activation of NF-κB in astrocytes results in the production of proinflammatory cytokines and nitric oxide. Astrocytes are known to express IGF-1, TNF-α, IL-1β, and IFN-γ and to respond to stimulation by TNF-α, IL-1β, and IFN-γ with release of growth factors and cytokines.[49,50] In addition, activated microglia release ROS, NO, and cytokines that activate astrocytes. In response to the proinflammatory cytokines TNF-α, IFN-γ, or IL-1B, astrocytes produce the chemokine MCP-1, which induces the migration of leukocytes across the blood–brain barrier.[51] While microglia may respond first to an insult, activation of astrocytes may result in the loss of their negative feedback on microglia, and they themselves can contribute to inflammation by release of nitric oxide, ROS, and inflammatory cytokines.[48] Cyclic nucleotide signaling is implicated in this switch from an anti-inflammatory, neuroprotective phenotype to a proinflammatory and potentially neurotoxic phenotype.

The antioxidant enzyme GPx has been shown to modulate the inflammatory response in brain. Overexpression of GPx in mice inhibited neutrophil migration, the expression of various chemokines, TNF-α, IL-6, and FAS ligand after focal ischemia.[52] These inflammatory mediators are predominantly produced by astrocytes and microglia, implicating a protective effect involving glia. GPx overexpression reduced translocation of cytochrome c and increased the proportion of bcl-2–positive cells, linking antioxidant upregulation with attenuation of apoptosis.[53] Thus GPx overexpression is associated with direct reduction of ROS, decreased inflammation, improved synaptic function, and reduced apoptosis in the setting of protection from stroke.[52,54]

13.4.2 Astrocytes as a Source of Toxic NO

Nitric oxide is an important neurotoxin in inflammation.[55] Nitric oxide produced by iNOS has been implicated in ischemic neuronal injury *in vitro* and *in vivo*. Nitric oxide is known to diffuse and injure surrounding cells. Cell culture studies have shown that after oxygen-glucose deprivation, induction of iNOS contributes to cerebral endothelial cell apoptosis.[56] Bacterial endotoxin and various cytokines induce iNOS and increase nitric oxide. Astrocytes stimulated with LPS and interferon-γ express iNOS, produce nitric oxide, and generate increased cytosolic and extracellular ROS.[55] During *in vivo* ischemia, iNOS was induced in reactive astrocytes.[57] In cocultures of neurons and astrocytes, induction of astrocyte iNOS by IL-1β and IFN-γ increased neuronal sensitivity to *N*-methyl-D-aspartate–mediated excitotoxicity.[58]

Neuronal ATP loss[27] and mitochondrial damage[59] have been reported in cocultures of neurons and iNOS-producing astrocytes. Inflammatory stimulation of astrocytes, likely both directly and via activated microglia, leads to production of nitric oxide and inhibition of neuronal mitochondrial respiration. Mitochondrial damage may lead to release of apoptotic factors. In a piglet model of hypoxia/ischemia, 2-iminobiotin, an inhibitor of nNOS and iNOS, reduced vasogenic edema and apoptotic markers (TUNEL, caspase-3 activity), decreased tyrosine nitration, and improved cell survival.[60] Thus, decreasing nitric oxide production by glial cells is another potential neuroprotective strategy (Figure 13.1).

13.5 NEUROPROTECTIVE STRATEGIES THAT INCLUDE REDUCING OXIDATIVE STRESS

13.5.1 Antioxidant Enzymes

Adult mice and rats overexpressing CuZn SOD (SOD1) showed reduced ischemic injury compared to wild-type animals.[61,62] These transgenic mice also had reduced blood–brain barrier disruption and oxidative cellular injury after photothrombotic ischemia.[63] A likely component of reduced infarct volume and improved maintenance of the blood–brain barrier involves glial cell protection by SOD1. SOD1-overexpressing mice deficient in GPx lost most of the protection from SOD1 overexpression after ischemia/reperfusion due to accumulation of hydrogen peroxide.[64] Thus the protective effect of SOD1 overexpression is dependent on adequate detoxification of the increased peroxide production. As noted above, glia play a central role in detoxification of hydrogen peroxide. Unlike adult animals, neonatal mice overexpressing SOD1 had increased ischemic brain damage.[65] GPx levels are low in the neonatal period, likely accounting for the observed increased accumulation of hydrogen peroxide and greater injury since overexpressing GPx resulted in protection.[66] These findings illustrate the changing role of oxidative stress and antioxidant mechanisms during development.

Differential efficacy of SOD1 overexpression was also observed in neurons compared with astrocytes. Cultured neurons overexpressing SOD1 exhibited improved survival after exposure to nitric oxide, but decreased survival after exposure to the superoxide generators menadione and paraquat.[67] Since astrocytes exhibit greater catabolism of hydrogen peroxide compared with neurons,[11] improved astrocytic survival in SOD1 transgenic mice would be predicted after exposure to ROS. Astrocytes overexpressing SOD1 were indeed resistant to superoxide oxidative injury, but the mechanism appeared to be independent of astrocytic GPx and catalase activity.[68] Different astrocytic antioxidant mechanisms may be active depending on the type of oxidative injury incurred.

MnSOD-deficient mice had increased mitochondrial superoxide production and worse neurologic deficits compared to wild-type littermates,[69] while transgenic overexpression of MnSOD reduced infarct volume, decreased lipid peroxidation, and decreased protein nitration after cerebral ischemia.[70] MnSOD is transcriptionally upregulated in response to hydrogen peroxide or paraquat in cultured astrocytes.[71] Mitochondrial preservation by MnSOD in glial cells may therefore contribute to the neuroprotection seen after injury by ROS. Soluble factors released by astrocytes enhanced MnSOD expression in endothelial cells,[72] which could contribute to better maintained blood–brain barrier integrity.

The ability of GPx overexpression to decrease the inflammatory response and provide brain protection was discussed previously.

13.5.2 Antiapoptotic Proteins

Overexpression of the prototypic antiapoptotic gene Bcl-2 protects neurons and glia against cell death. Bcl-2 can heterodimerize with proapoptotic proteins such as Bax, thereby impeding the release of cytochrome c from mitochondria and blocking apoptotic cell death.[73,74] However, Bcl-2 is also known to reduce necrotic cell death, and other mechanisms, including an antioxidant effect, have been suggested.[75,76] Bcl-2 has been shown to increase the level of GSH and some antioxidant enzymes in neurons[77] and astrocytes,[16] though the importance of an antioxidant mechanism depends on the cell type. Bcl-2 overexpression reduced neurotoxicity from an excitotoxic insult and reduced superoxide accumulation, but had no effect on lipid peroxidation.[73]

Astrocytes overexpressing a related antiapoptotic protein, Bcl-x_L, had reduced ROS accumulation after exposure to hydrogen peroxide or glucose deprivation as well as increased GSH levels (Figure 13.2).[78,79] The antioxidant actions of Bcl-2 are not well understood, but may be indirect via upregulation of cellular antioxidant effects in the absence of stress.[75] Bcl-2 may induce an antioxidant response by acting as a pro-oxidant itself. This is consistent with the observation of increased superoxide levels in unstressed astrocytes overexpressing Bcl-2[16] or bcl-x_L[79] and the observation of a small increase in DNA fragmentation in bcl-2 transfected cells in the absence of any external apoptotic stress.[77] Another redox-dependent mechanism of action proposed for Bcl-2 is changes in gene transcription due to a reduced nuclear redox state reflecting increased nuclear levels of GSH.[80] Further investigation into the mechanisms of protection by Bcl-2 from oxidative injury is warranted.

13.5.3 Heat-Shock Proteins

Another protective physiologic response to oxidative stress is the induction of heat-shock proteins (Hsps). Hsps are induced in response to increased levels of denatured proteins as occurs during ischemia, heat shock, and other stresses. These endogenous Hsps are protein chaperones that prevent aggregation of unfolded proteins and facilitate refolding to restore native function. Conditions in cerebral ischemia that are known to increase protein aggregation include oxidative stress and acidosis. Protein aggregates occur in stroke[81] and in many chronic neurological degenerative diseases including Huntington's disease, Alzheimer's disease, Parkinson's disease, and prion diseases.[82]

The Hsp70 family is the best-studied heat-shock or heat-stress protein family. Both constitutively expressed and inducible forms are known. Hsp72 (inducible Hsp70) is synthesized at high levels in response to heat, ischemia, and other stresses that produce denatured proteins. The use of transgenic animals and viral vector–mediated gene transfer has allowed a more detailed investigation of the potential of these proteins to provide neuroprotection. Hsp70 overexpression using a Herpes amplicon vector protected rat neurons when introduced either before or after ischemic injury.[83] Other authors found protection from stroke in transgenic mice that overexpress Hsp70.[84,85] In primary cultures Hsp70 protected astrocytes from heat shock, hydrogen peroxide exposure, glucose deprivation, and oxygen-glucose deprivation.[86,87] Although resting levels of GSH were not changed by Hsp70 overexpression, after injury it was found that GSH was better preserved in astrocytes overexpressing Hsp70.[79] Overexpression of Hsp70 in astrocytes was sufficient to protect wild-type neurons cocultured with the transfected astrocytes against oxygen-glucose deprivation,[88] again emphasizing that interactions between cell types influences outcome.

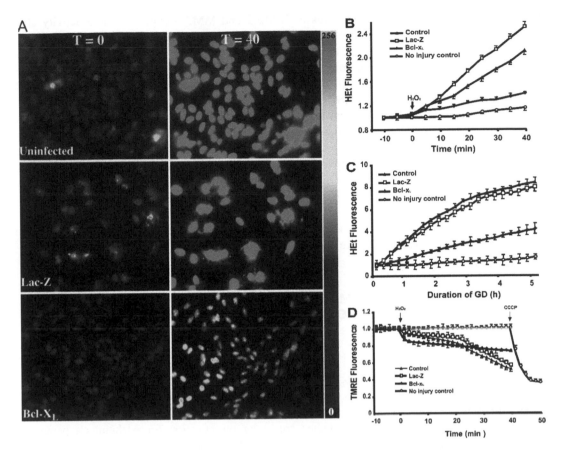

Figure 13.2 *(A color version of this figure follows page 236.)* Overexpression of Bcl-x$_L$ reduces ROS production by stressed astrocytes and inhibits changes in mitochondrial membrane potential. (A) Astrocytes were loaded with the ROS-sensitive dye hydroethidine, and their fluorescence was monitored. Photos shown represent cultures under control condition ($T = 0$) and after 40-min exposure to 400-μM hydrogen peroxide ($T = 40$). LacZ or Bcl-x$_L$ overexpression was induced with retroviral vectors. Increased levels of ROS corresponding to increased fluorescence are indicated on the pseudocolor scale to the right. (B) Quantitation of the fluorescence averaged from at least 50 cells per condition. (C) ROS production by astrocytes stressed by removal of substrate from the medium, assessed with hydroethidine. Data are averaged from at least 50 cells per condition. (D) Mitochondrial membrane potential was assessed with tetramethylrhodamine ethylester in astrocytes exposed to hydrogen peroxide. The conditions associated with increased ROS production show more mitochondrial depolarization with time. CCCP was added at the end of the experiment to show the level of fluorescence associated with full depolarization. (Data from Ouyang, Y.B. et al., *Free Radic. Biol. Med.*, 33, 544, 2002. With permission.)

Cochaperones interact with chaperones such as Hsp70 to promote protein folding and to regulate other chaperone activities. Overexpression of Hdj-2, a member of the Hsp40 family, protected astrocytes from ischemia-like injury and markedly reduced the number of cells with protein aggregates.[89] Work in protein aggregation diseases has suggested that there is a toxic gain of function that may in some cases include generation of oxidative stress. Whether this contributes to ischemic brain injury is not yet clear.

The effect of Hsp70 on transcription of NF-kB and iNOS also links this heat-shock protein with both oxidative injury and the inflammatory process. As discussed above, NF-κB is an important regulator of the inflammatory role of glial cells. Hsp70 has been reported to reduce NF-κB levels and iNOS activation in glia.[90,91] As a result, fewer downstream inflammatory mediators such as iNOS, COX-2, and MMP-9 would be produced and fewer ROS generated. We specifically found that Hsp70 in cultured astrocytes suppressed MMP-2 and MMP-9 expression and activity after ischemic injury.[92] Hsp70 suppressed MMP-9 at the transcriptional level, perhaps by inhibiting NF-κB activation. As MMPs can disrupt the blood–brain barrier and potentiate cerebral edema, hemorrhage, and leukocyte infiltration, downregulation of specific MMPs may be an important protective and anti-inflammatory function of Hsp70. Several genes including TNF-α, IL-1β, and SOD1 also contain heat-shock elements (HSE) in their promoters[93] and would therefore be regulated in response to stresses that activate heat-shock transcription factor, which binds the HSE. More research needs to be conducted regarding the ability of heat-shock proteins to regulate inflammatory and antioxidant gene products.

After ischemia, astrocytes show marked induction of Hsp27, another heat-shock protein with ties to antioxidant status. Expression of this gene protects against hydrogen peroxide–induced injury and TNF-α[94] by increasing GSH levels. Hsp27 may chaperone cytoskeletal proteins like intermediate filaments and play a role in cytoskeletal changes that occur during astrocyte activation.[95] While much remains to be learned of the mechanisms responsible for protection by Hsps, induction of Hsps in astrocytes may be a future strategy for neuroprotection.

13.6 COMMON THEMES IN NEURODEGENERATION: A ROLE FOR ASTROCYTES IN OXIDATIVE INJURY

13.6.1 Alzheimer's Disease

Oxidative stress, caused at least in part by β-amyloid and inflammation, and impaired glutamate uptake by astrocytes are thought to contribute to the pathogenesis of Alzheimer's disease.[96–98] MnSOD overexpression can reduce apoptosis, lipid peroxidation, and protein nitration induced by β-amyloid.[70] Multiple antioxidant strategies are under investigation for Alzheimer's patients.[99] Astrocytic and microglial activation and inflammation are also considered important in the pathogenesis of Alzheimer's disease.[48,100]

13.6.2 Amyotrophic Lateral Sclerosis (ALS)

Oxidative damage to motor neurons and reactive astrocytosis are observed in patients with ALS.[42,101] About 1–2% of cases of ALS are linked to mutations in SOD1.[102] SOD1 is upregulated in astrocytes in ALS, emphasizing an antioxidant role of astrocytes in this disease.[103] Astrocyte glutamate uptake may play a critical role in determining the onset of symptoms in this disease, with oxidative damage reducing glutamate uptake and precipitating irreversible neuronal damage.[39,42] A feedforward cycle in which neuronally produced oxidants decrease astrocytic glutamate uptake leading to increased excitotoxic damage would contribute to disease progression and illustrates the central role of reciprocal interactions between neurons and astrocytes in normal physiology and their disruption in neurodegeneration.

13.6.3 Parkinson's Disease

Oxidative stress is central to neuronal injury in Parkinson's disease.[99] Dopamine oxidation leads to oxidative damage and loss of nigrostriatal dopaminergic neurons, a central feature of Parkinson's disease. Glial-conditioned medium protected neurons *in vitro* against L-DOPA toxicity.[104] GDNF increased neuronal survival in rats given 6-hydroxydopamine infusions to model Parkinson's disease.[105] GDNF promoted survival of fetal dopaminergic neurons and increased dopamine metabolism in adult rats.[105] In Parkinson's-diseased brains, decreased GSH levels were found with mitochondrial damage.[106]

13.7 CONCLUSION

We have described how oxidative injury, due to both increased generation of ROS and decreased scavenging, is a common mechanism involved in many neurodegenerative diseases. Astrocytes are more resistant to oxidative injury due to their enhanced antioxidant capabilities including higher intracellular GSH levels and higher levels of antioxidant enzymes. Astrocytes improve neuronal antioxidant reserves by regenerating cysteine and ascorbate and by releasing neurotrophic factors for use by neurons. The failure of these mechanisms contributes to brain injury. In addition, glia may become activated and directly contribute to neurotoxicity through their release of inflammatory mediators. Therapies to reduce oxidative injury include upregulation of antioxidant genes and induction of protective heat-shock and antiapoptotic proteins. Future research into oxidative injury and the interaction between neurons, astrocytes, and microglia will undoubtedly provide better strategies to reduce neurologic injury.

REFERENCES

1. Halliwell, B. and Gutteridge, G., *Free Radicals in Biology and Medicine*, Clarendon Press, Oxford, 1989, p. 543.
2. Juurlink, B.H., Response of glial cells to ischemia: roles of reactive oxygen species and glutathione, *Neurosci. Biobehav. Rev.*, 21, 151, 1997.
3. Peuchen, S. et al., Interrelationships between astrocyte function, oxidative stress and antioxidant status within the central nervous system, *Prog. Neurobiol.*, 52, 261, 1997.
4. Halliwell, B., Reactive oxygen species and the central nervous system, *J. Neurochem.*, 59, 1609, 1992.
5. Chan, P.H., Reactive oxygen radicals in signaling and damage in the ischemic brain, *J. Cereb. Blood Flow Metab.*, 21, 2, 2001.
6. Boveris, A. and Chance, B., The mitochondrial generation of hydrogen peroxide. General properties and effect of hyperbaric oxygen, *Biochem. J.*, 134, 707, 1973.
7. Beckman, J.S., Peroxynitrite versus hydroxyl radical: the role of nitric oxide in superoxide-dependent cerebral injury, *Ann. N.Y. Acad. Sci.*, 738, 69, 1994.
8. Khan, A.U., Singlet molecular oxygen from superoxide anion and sensitized fluorescence of organic molecules, *Science*, 168, 476, 1970.
9. Fridovich, I., Superoxide radical and superoxide dismutases, *Annu. Rev. Biochem.*, 64, 97, 1995.
10. Palmer, C. et al., Changes in iron histochemistry after hypoxic-ischemic brain injury in the neonatal rat, *J. Neurosci. Res.*, 56, 60, 1999.
11. Desagher, S., Glowinski, J., and Premont, J., Astrocytes protect neurons from hydrogen peroxide toxicity, *J. Neurosci.*, 16, 2553, 1996.
12. Bredt, D.S. and Snyder, S.H., Nitric oxide: a physiologic messenger molecule, *Annu. Rev. Biochem.*, 63, 175, 1994.
13. Beckman, J.S. et al., Apparent hydroxyl radical production by peroxynitrite: implications for endothelial injury from nitric oxide and superoxide, *Proc. Natl. Acad. Sci. U.S.A.*, 87, 1620, 1990.
14. Iadecola, C., Bright and dark sides of nitric oxide in ischemic brain injury, *Trends Neurosci.*, 20, 132, 1997.
15. Mizui, T., Kinouchi, H., and Chan, P.H., Depletion of brain glutathione by buthionine sulfoximine enhances cerebral ischemic injury in rats, *Am. J. Physiol.*, 262, H313, 1992.
16. Papadopoulos, M.C. et al., Potentiation of murine astrocyte antioxidant defence by bcl-2: protection in part reflects elevated glutathione levels, *Eur. J. Neurosci.*, 10, 1252, 1998.
17. Rice, M.E. and Russo-Menna, I., Differential compartmentalization of brain ascorbate and glutathione between neurons and glia, *Neuroscience*, 82, 1213, 1998.
18. Raps, S.P. et al., Glutathione is present in high concentrations in cultured astrocytes but not in cultured neurons, *Brain Res.*, 493, 398, 1989.
19. Wilson, J.X., Antioxidant defense of the brain: a role for astrocytes, *Can. J. Physiol. Pharmacol.*, 75, 1149, 1997.

20. Nakajima, K. and Suzuki, K., Immunochemical detection of metallothionein in brain, *Neurochem. Int.*, 27, 73, 1995.

21. Sato, M. and Bremner, I., Oxygen free radicals and metallothionein, *Free Rad. Biol. Med.*, 14, 325, 1993.

22. Bolanos, J.P. et al., Effect of peroxynitrite on the mitochondrial respiratory chain: differential susceptibility of neurones and astrocytes in primary culture, *J. Neurochem.*, 64, 1965, 1995.

23. Almeida, A. et al., Oxygen and glucose deprivation induces mitochondrial dysfunction and oxidative stress in neurones but not in astrocytes in primary culture, *J. Neurochem.*, 81, 207, 2002.

24. Papadopoulos, M.C. et al., Vulnerability to glucose deprivation injury correlates with glutathione levels in astrocytes, *Brain Res.*, 748, 151, 1997.

25. Iwata-Ichikawa, E. et al., Glial cells protect neurons against oxidative stress via transcriptional upregulation of the glutathione synthesis, *J. Neurochem.*, 72, 2334, 1999.

26. Gegg, M.E. et al., Differential effect of nitric oxide on glutathione metabolism and mitochondrial function in astrocytes and neurones: implications for neuroprotection/neurodegeneration?, *J. Neurochem.*, 86, 228, 2003.

27. Heales, S.J. and Bolanos, J.P., Impairment of brain mitochondrial function by reactive nitrogen species: the role of glutathione in dictating susceptibility, *Neurochem. Int.*, 40, 469, 2002.

28. Petrosillo, G., Ruggiero, F.M., and Paradies, G., Role of reactive oxygen species and cardiolipin in the release of cytochrome c from mitochondria, *FASEB J.*, 17, 2202, 2003.

29. Sugawara, T. et al., Overexpression of copper/zinc superoxide dismutase in transgenic rats protects vulnerable neurons against ischemic damage by blocking the mitochondrial pathway of caspase activation, *J. Neurosci.*, 22, 209, 2002.

30. Lezoualc'h, F. et al., High constitutive NF-kappaB activity mediates resistance to oxidative stress in neuronal cells, *J. Neurosci.*, 18, 3224, 1998.

31. Mattson, M.P. and Camandola, S., NF-kappaB in neuronal plasticity and neurodegenerative disorders, *J. Clin. Invest.*, 107, 247, 2001.

32. Williams, A.J. et al., Delayed treatment with MLN519 reduces infarction and associated neurologic deficit caused by focal ischemic brain injury in rats via antiinflammatory mechanisms involving nuclear factor-kappaB activation, gliosis, and leukocyte infiltration, *J. Cereb. Blood Flow Metab.*, 23, 75, 2003.

33. Wang, Y. et al., Protective effects of glial cell line-derived neurotrophic factor in ischemic brain injury, *Ann. N.Y. Acad. Sci.*, 962, 423, 2002.

34. Fujiwara, K. et al., Reduction of infarct volume and apoptosis by grafting of encapsulated basic fibroblast growth factor-secreting cells in a model of middle cerebral artery occlusion in rats, *J. Neurosurg.*, 99, 1053, 2003.

35. Li, Q. and Stephenson, D., Postischemic administration of basic fibroblast growth factor improves sensorimotor function and reduces infarct size following permanent focal cerebral ischemia in the rat, *Exp. Neurol.*, 177, 531, 2002.

36. Wada, K. et al., Effect of basic fibroblast growth factor treatment on brain progenitor cells after permanent focal ischemia in rats, *Stroke*, 34, 2722, 2003.

37. Anderson, C.M. and Swanson, R.A., Astrocyte glutamate transport: review of properties, regulation, and physiological functions, *Glia*, 32, 1, 2000.

38. Rothstein, J.D. et al., Knockout of glutamate transporters reveals a major role for astroglial transport in excitotoxicity and clearance of glutamate, *Neuron*, 16, 675, 1996.

39. Trotti, D., Danbolt, N.C., and Volterra, A., Glutamate transporters are oxidant-vulnerable: a molecular link between oxidative and excitotoxic neurodegeneration?, *Trends Pharmacol. Sci.*, 19, 328, 1998.

40. Miralles, V.J. et al., Na+ dependent glutamate transporters (EAAT1, EAAT2, and EAAT3) in primary astrocyte cultures: effect of oxidative stress, *Brain Res.*, 922, 21, 2001.

41. Korcok, J. et al., Sepsis inhibits reduction of dehydroascorbic acid and accumulation of ascorbate in astroglial cultures: intracellular ascorbate depletion increases nitric oxide synthase induction and glutamate uptake inhibition, *J. Neurochem.*, 81, 185, 2002.

42. Rao, S.D., Yin, H.Z., and Weiss, J.H., Disruption of glial glutamate transport by reactive oxygen species produced in motor neurons, *J. Neurosci.*, 23, 2627, 2003.

43. Fine, S.M. et al., Tumor necrosis factor alpha inhibits glutamate uptake by primary human astrocytes. Implications for pathogenesis of HIV-1 dementia, *J. Biol. Chem.*, 271, 15303, 1996.

44. Bruhn, T., Christensen, T., and Diemer, N.H., *In vivo* cellular uptake of glutamate is impaired in the rat hippocampus during and after transient cerebral ischemia: a microdialysis extraction study, *J. Neurosci. Res.*, 66, 1118, 2001.

45. Dugan, L.L. et al., Mitochondrial production of reactive oxygen species in cortical neurons following exposure to N-methyl-D-aspartate, *J. Neurosci.*, 15, 6377, 1995.

46. Pellegrini-Giampietro, D.E. et al., Excitatory amino acid release and free radical formation may cooperate in the genesis of ischemia-induced neuronal damage, *J. Neurosci.*, 10, 1035, 1990.

47. Taylor, A.L. and Hewett, S.J., Potassium-evoked glutamate release liberates arachidonic acid from cortical neurons, *J. Biol. Chem.*, 277, 43881, 2002.

48. Schubert, P. et al., Cascading glia reactions: a common pathomechanism and its differentiated control by cyclic nucleotide signaling, *Ann. N.Y. Acad. Sci.*, 903, 24, 2000.

49. Eddleston, M. and Mucke, L., Molecular profile of reactive astrocytes—implications for their role in neurologic disease, *Neuroscience*, 54, 15, 1993.

50. Pearson, V.L., Rothwell, N.J., and Toulmond, S., Excitotoxic brain damage in the rat induces inter-leukin-1beta protein in microglia and astrocytes: correlation with the progression of cell death, *Glia*, 25, 311, 1999.

51. Weiss, J.M. et al., Astrocyte-derived monocyte-chemoattractant protein-1 directs the transmigration of leukocytes across a model of the human blood-brain barrier, *J. Immunol.*, 161, 6896, 1998.

52. Ishibashi, N. et al., Inflammatory response and glutathione peroxidase in a model of stroke, *J. Immunol.*, 168, 1926, 2002.

53. Hoehn, B. et al., Glutathione peroxidase overexpression inhibits cytochrome C release and proapoptotic mediators to protect neurons from experimental stroke, *Stroke*, 34, 2489, 2003.

54. Weisbrot-Lefkowitz, M. et al., Overexpression of human glutathione peroxidase protects transgenic mice against focal cerebral ischemia/reperfusion damage, *Brain Res. Mol. Brain Res.*, 53, 333, 1998.

55. Brown, G.C. and Bal-Price, A., Inflammatory neurodegeneration mediated by nitric oxide, glutamate, and mitochondria, *Mol. Neurobiol.*, 27, 325, 2003.

56. Xu, J. et al., Oxygen-glucose deprivation induces inducible nitric oxide synthase and nitrotyrosine expression in cerebral endothelial cells, *Stroke*, 31, 1744, 2000.

57. Endoh, M., Maiese, K., and Wagner, J., Expression of the inducible form of nitric oxide synthase by reactive astrocytes after transient global ischemia, *Brain Res.*, 651, 92, 1994.

58. Hewett, S.J., Csernansky, C.A., and Choi, D.W., Selective potentiation of NMDA-induced neuronal injury following induction of astrocytic iNOS, *Neuron*, 13, 487, 1994.

59. Bal-Price, A. and Brown, G.C., Inflammatory neurodegeneration mediated by nitric oxide from activated glia-inhibiting neuronal respiration, causing glutamate release and excitotoxicity, *J. Neurosci.*, 21, 6480, 2001.

60. Peeters-Scholte, C. et al., Neuroprotection by selective nitric oxide synthase inhibition at 24 hours after perinatal hypoxia-ischemia, *Stroke*, 33, 2304, 2002.

61. Chan, P.H. et al., Neuroprotective role of CuZn-superoxide dismutase in ischemic brain damage, *Adv. Neurol.*, 71, 271, 1996.

62. Chan, P.H. et al., Overexpression of SOD1 in transgenic rats protects vulnerable neurons against ischemic damage after global cerebral ischemia and reperfusion, *J. Neurosci.*, 18, 8292, 1998.

63. Kim, G.W. et al., The cytosolic antioxidant, copper/zinc superoxide dismutase, attenuates blood-brain barrier disruption and oxidative cellular injury after photothrombotic cortical ischemia in mice, *Neuroscience*, 105, 1007, 2001.

64. Crack, P.J. et al., Glutathione peroxidase-1 contributes to the neuroprotection seen in the superoxide dismutase-1 transgenic mouse in response to ischemia/reperfusion injury, *J. Cereb. Blood Flow Metab.*, 23, 19, 2003.

65. Ditelberg, J.S. et al., Brain injury after perinatal hypoxia-ischemia is exacerbated in copper/zinc superoxide dismutase transgenic mice, *Pediatr. Res.*, 39, 204, 1996.

66. Fullerton, H.J. et al., Copper/zinc superoxide dismutase transgenic brain accumulates hydrogen peroxide after perinatal hypoxia ischemia, *Ann. Neurol.*, 44, 357, 1998.

67. Ying, W. et al., Differing effects of copper,zinc superoxide dismutase overexpression on neurotoxicity elicited by nitric oxide, reactive oxygen species, and excitotoxins, *J. Cereb. Blood Flow Metab.*, 20, 359, 2000.

68. Chen, Y., Chan, P.H., and Swanson, R.A., Astrocytes overexpressing Cu,Zn superoxide dismutase have increased resistance to oxidative injury, *Glia*, 33, 343, 2001.

69. Kim, G.W. et al., Manganese superoxide dismutase deficiency exacerbates cerebral infarction after focal cerebral ischemia/reperfusion in mice: implications for the production and role of superoxide radicals, *Stroke*, 33, 809, 2002.

70. Keller, J.N. et al., Mitochondrial manganese superoxide dismutase prevents neural apoptosis and reduces ischemic brain injury: suppression of peroxynitrite production, lipid peroxidation, and mitochondrial dysfunction, *J. Neurosci.*, 18, 687, 1998.

71. Rohrdanz, E. et al., The influence of oxidative stress on catalase and MnSOD gene transcription in astrocytes, *Brain Res.*, 900, 128, 2001.

72. Schroeter, M.L. et al., Astrocytes induce manganese superoxide dismutase in brain capillary endothelial cells, *Neuroreport*, 12, 2513, 2001.

73. Howard, S. et al., Neuroprotective effects of bcl-2 overexpression in hippocampal cultures: interactions with pathways of oxidative damage, *J. Neurochem.*, 83, 914, 2002.

74. Cory, S., Huang, D.C., and Adams, J.M., The Bcl-2 family: roles in cell survival and oncogenesis, *Oncogene*, 22, 8590, 2003.

75. Steinman, H.M., The Bcl-2 oncoprotein functions as a pro-oxidant, *J. Biol. Chem.*, 270, 3487, 1995.

76. Jang, J.H. and Surh, Y.J., Potentiation of cellular antioxidant capacity by Bcl-2: implications for its antiapoptotic function, *Biochem. Pharmacol.*, 66, 1371, 2003.

77. Kane, D.J. et al., Bcl-2 inhibition of neural death: decreased generation of reactive oxygen species, *Science*, 262, 1274, 1993.

78. Ouyang, Y.B., Carriedo, S.G., and Giffard, R.G., Effect of Bcl-x(L) overexpression on reactive oxygen species, intracellular calcium, and mitochondrial membrane potential following injury in astrocytes, *Free Rad. Biol. Med.*, 33, 544, 2002.

79. Xu, L. et al., Overexpression of bcl-xl protects astrocytes from glucose deprivation and is associated with higher glutathione, ferritin, and iron levels, *Anesthesiology*, 91, 1036, 1999.

80. Voehringer, D.W. et al., Bcl-2 expression causes redistribution of glutathione to the nucleus, *Proc. Natl. Acad. Sci. U.S.A.*, 95, 2956, 1998.

81. Hu, B.R. et al., Protein aggregation after focal brain ischemia and reperfusion, *J. Cereb. Blood Flow Metab.*, 21, 865, 2001.

82. Beissinger, M. and Buchner, J., How chaperones fold proteins, *Biol. Chem.*, 379, 245, 1998.

83. Hoehn, B. et al., Overexpression of HSP72 after induction of experimental stroke protects neurons from ischemic damage, *J. Cereb. Blood Flow Metab.*, 21, 1303, 2001.

84. Plumier, J.C. et al., Transgenic mice expressing the human inducible Hsp70 have hippocampal neurons resistant to ischemic injury, *Cell Stress Chaperones*, 2, 162, 1997.

85. Rajdev, S. et al., Mice overexpressing rat heat shock protein 70 are protected against cerebral infarction, *Ann. Neurol.*, 47, 782, 2000.

86. Xu, L. and Giffard, R.G., HSP70 protects murine astrocytes from glucose deprivation injury, *Neurosci. Lett.*, 224, 9, 1997.

87. Papadopoulos, M.C. et al., Over-expression of HSP-70 protects astrocytes from combined oxygen-glucose deprivation, *Neuroreport*, 7, 429, 1996.

88. Xu, L., Lee, J.E., and Giffard, R.G., Overexpression of bcl-2, bcl-XL or hsp70 in murine cortical astrocytes reduces injury of co-cultured neurons, *Neurosci. Lett.*, 277, 193, 1999.

89. Qiao, Y., Ouyang, Y.B., and Giffard, R.G., Overexpression of HDJ-2 protects astrocytes from ischemia-like injury and reduces redistribution of ubiquitin staining *in vitro*, *J. Cereb. Blood Flow Metab.*, 23, 1113, 2003.

90. Feinstein, D.L. et al., Heat shock protein 70 suppresses astroglial-inducible nitric-oxide synthase expression by decreasing NFkappaB activation, *J. Biol. Chem.*, 271, 17724, 1996.

91. Heneka, M.T. et al., The heat shock response inhibits NF-kappaB activation, nitric oxide synthase type 2 expression, and macrophage/microglial activation in brain, *J. Cereb. Blood Flow Metab.*, 20, 800, 2000.

92. Lee, J.E. et al., The 70kDa heat shock protein suppresses matrix metalloproteinases in astrocytes, *Neuroreport*, in press, 2004.

93. Yoo, H.Y., Chang, M.S., and Rho, H.M., The activation of the rat copper/zinc superoxide dismutase gene by hydrogen peroxide through the hydrogen peroxide-responsive element and by paraquat and heat shock through the same heat shock element, *J. Biol. Chem.*, 274, 23887, 1999.

94. Mehlen, P. et al., Human hsp27, Drosophila hsp27 and human alphaB-crystallin expression-mediated increase in glutathione is essential for the protective activity of these proteins against TNFalpha-induced cell death, *EMBO J.*, 15, 2695, 1996.

95. Wisniewski, T. and Goldman, J.E., Alpha B-crystallin is associated with intermediate filaments in astrocytoma cells, *Neurochem. Res.*, 23, 385, 1998.

96. Mattson, M.P., Oxidative stress, perturbed calcium homeostasis, and immune dysfunction in Alzheimer's disease, *J. Neurovirol.*, 8, 539, 2002.

97. Butterfield, D.A., Amyloid beta-peptide (1-42)-induced oxidative stress and neurotoxicity: implications for neurodegeneration in Alzheimer's disease brain. A review, *Free Rad. Res.*, 36, 1307, 2002.

98. Harris, M.E. et al., Amyloid beta peptide (25-35) inhibits Na^+-dependent glutamate uptake in rat hippocampal astrocyte cultures, *J. Neurochem.*, 67, 277, 1996.

99. Di Matteo, V. and Esposito, E., Biochemical and therapeutic effects of antioxidants in the treatment of Alzheimer's disease, Parkinson's disease, and amyotrophic lateral sclerosis, *Curr. Drug Target CNS Neurol. Disord.*, 2, 95, 2003.

100. Mrak, R.E. and Griffin, W.S., The role of activated astrocytes and of the neurotrophic cytokine S100B in the pathogenesis of Alzheimer's disease, *Neurobiol. Aging*, 22, 915, 2001.

101. Cleveland, D.W. and Rothstein, J.D., From Charcot to Lou Gehrig: deciphering selective motor neuron death in ALS, *Nat. Rev. Neurosci.*, 2, 806, 2001.

102. Rosen, D.R. et al., Mutations in Cu/Zn superoxide dismutase gene are associated with familial amyotrophic lateral sclerosis, *Nature*, 362, 59, 1993.

103. Blaauwgeers, H.G. et al., Enhanced superoxide dismutase-2 immunoreactivity of astrocytes and occasional neurons in amyotrophic lateral sclerosis, *J. Neurol. Sci.*, 140, 21, 1996.

104. Mena, M.A. et al., Glia conditioned medium protects fetal rat midbrain neurones in culture from L-DOPA toxicity, *Neuroreport*, 7, 441, 1996.

105. Kearns, C.M. and Gash, D.M., GDNF protects nigral dopamine neurons against 6-hydroxydopamine *in vivo*, *Brain Res.*, 672, 104, 1995.

106. Jenner, P. et al., Oxidative stress as a cause of nigral cell death in Parkinson's disease and incidental Lewy body disease. The Royal Kings and Queens Parkinson's Disease Research Group, *Ann. Neurol.*, 32, S82, 1992.

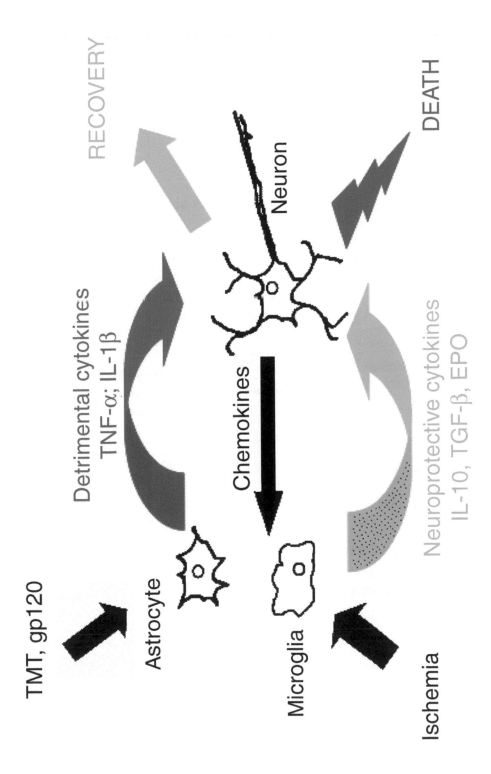

Figure 8.1 The cytokine cycle in glial–neuronal interaction. Exposure of glial cells to a pathological (i.e., ischemia) or a toxic (i.e., TMT, gp120) insult triggers a reactive response, leading to the release of proinflammatory or anti-inflammatory cytokines capable of affecting neuronal function. In general, it is assumed that proinflammatory cytokines (i.e., TNF-α and IL-1β) exacerbate and sustain neurodegeneration, while anti-inflammatory cytokines (i.e., IL-10 and TGF-β) promote neuronal survival. In turn, cytokines derived from neurons can modulate glial response.

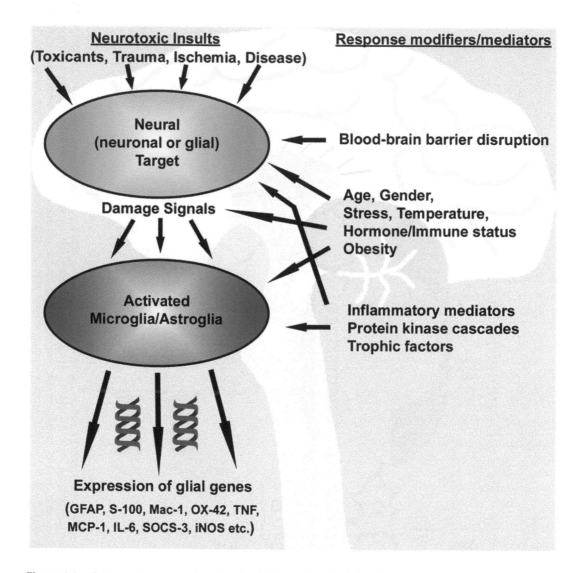

Figure 9.1 Cell-signaling events associated with injury-induced glial activation. Neurotoxic insults elicit activation of a glial response that leads to expression of glial-specific genes. These insults can be mediated through or modulated by several external (environmental) or internal factors to exaggerate the injury and subsequent glial response.

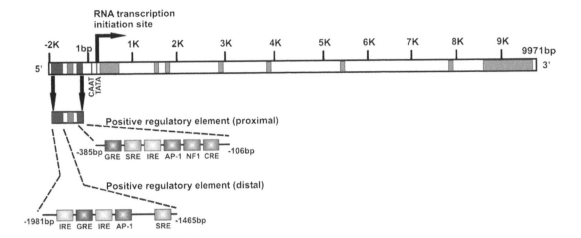

Figure 9.2 The murine *Gfap* gene. The expression of the gene is controlled by positive (enhancers) and negative (silencers) regulatory elements situated 2 kb upstream of transcription start site. Depicted here are the positive regulatory elements in the promoter region that are essential for activation of the gene. This region contains binding sites for specific transcription factors that are involved in transcriptional activation of GFAP.

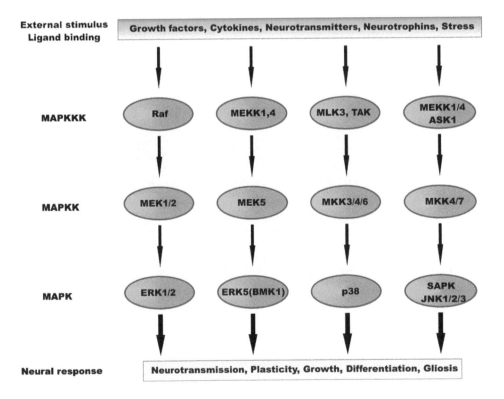

Figure 9.4 Schematic representation of the canonical MAPK pathway indicating activation of the pathways by upstream ligands and the concomitant downstream neural responses.

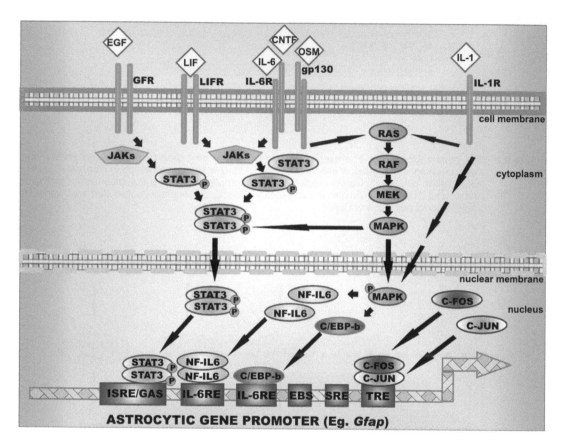

Figure 9.5 Schematic diagram depicting the probable association of the gp130-JAK/STAT3 and MAPK pathway in regulating astroglial activation and GFAP expression. Proinflammatory cytokines and growth factors, including IL6-type cytokines, are potential effectors (ligands) for activation of glial genes following disease- or injury-induced neurodegeneration. A myriad of transcription factors activated by these pathways may converge and bind to specific sites on the astrocytic gene promoter (e.g., GFAP) to regulate its expression.

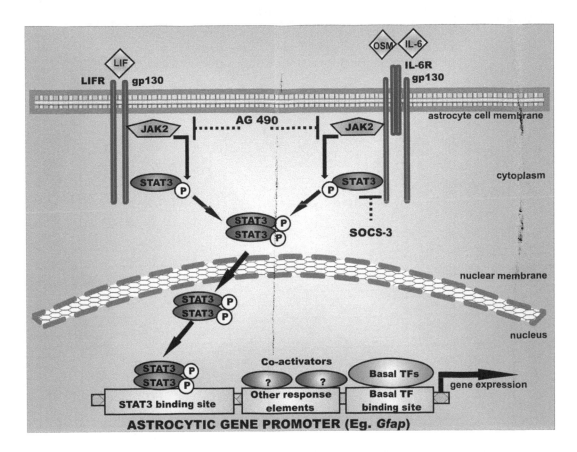

Figure 9.6 Schematic diagram depicting the involvement of gp130-mediated phosphorylation of JAK2/STAT3 pathway prior to induction of GFAP in the MPTP model of dopaminergic neurodegeneration. Upon putative ligand binding (e.g., IL-6, LIF, OSM), JAK2 and STAT3 are recruited to the gp130 signal transducer. Following docking of STAT3 to gp130 via its SH2 domain, JAK2 phosphorylates the Tyr705 residue on STAT3. The phosphorylated STAT3 dimerizes, translocates to the nucleus, and in association with other transcription factors (TFs) mediates transcriptional activation of astrocytic genes such as *Gfap*. MPTP-induced activation of STAT3 is attenuated by pharmacological intervention with the JAK2 inhibitor, AG490. Similarly, MPTP treatment causes upregulation of SOCS-3 mRNA, resulting in negative feedback regulation of STAT3 phosphorylation. (Adapted from Sriram, K. et al. *J. Biol. Chem.*, 279, 19936, 2004. With permission.)

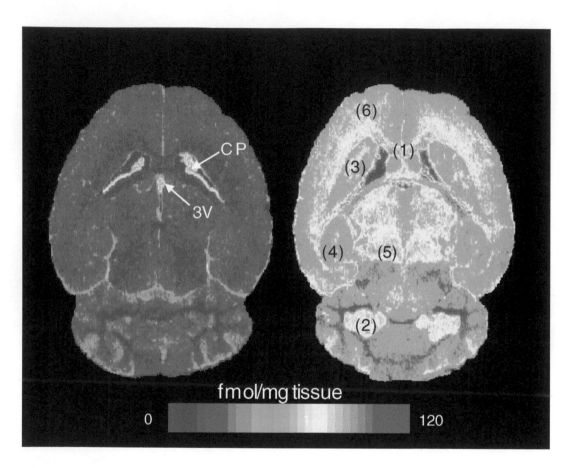

Figure 12.1 Representative pseudocolor autoradiograms of [³H]-(*R*)-PK11195 binding to horizontal rat brain sections from a control (left image) and a cuprizone-treated rat (right image). Cuprizone treatment was 0.2% w/w in the diet for 4 weeks. Note the low level of [³H]-(*R*)-PK11195 binding to peripheral benzodiazepine receptors in the brain neuropil of the control rat brain. High levels of binding in the control rat brain are present in the choroid plexus and in the lining of the ventricles. In the cuprizone-treated rat, there are widespread increases in the levels of [³H]-(*R*)-PK11195 binding in the brain consistent with the known neuropathology of cuprizone. (See Chen et al., 2004, for more details.) The horizontal color bar is representative of levels of [³H]-(*R*)-PK11195 binding in fmol/mg tissue. Abbreviations: CP = choroid plexus; 3V = third ventricle; (1) = corpus collosum; (2) = deep cerebellar nuclei; (3) = corpus striatum; (4) = hippocampus; (5) = superior colliculus; (6) = frontal cortex.

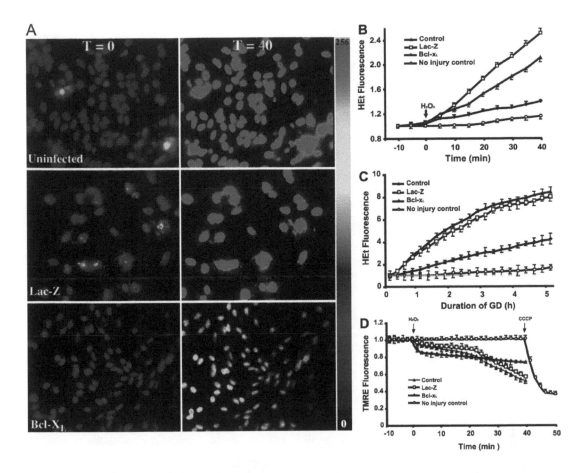

Figure 13.2 Overexpression of Bcl-x$_L$ reduces ROS production by stressed astrocytes and inhibits changes in mitochondrial membrane potential. (A) Astrocytes were loaded with the ROS-sensitive dye hydroethidine, and their fluorescence was monitored. Photos shown represent cultures under control condition ($T = 0$) and after 40-min exposure to 400-μM hydrogen peroxide ($T = 40$). LacZ or Bcl-x$_L$ overexpression was induced with retroviral vectors. Increased levels of ROS corresponding to increased fluorescence are indicated on the pseudocolor scale to the right. (B) Quantitation of the fluorescence averaged from at least 50 cells per condition. (C) ROS production by astrocytes stressed by removal of substrate from the medium, assessed with hydroethidine. Data are averaged from at least 50 cells per condition. (D) Mitochondrial membrane potential was assessed with tetramethylrhodamine ethylester in astrocytes exposed to hydrogen peroxide. The conditions associated with increased ROS production show more mitochondrial depolarization with time. CCCP was added at the end of the experiment to show the level of fluorescence associated with full depolarization. (Data from Ouyang, Y.B. et al., *Free Radic. Biol. Med.*, 33, 544, 2002. With permission.)

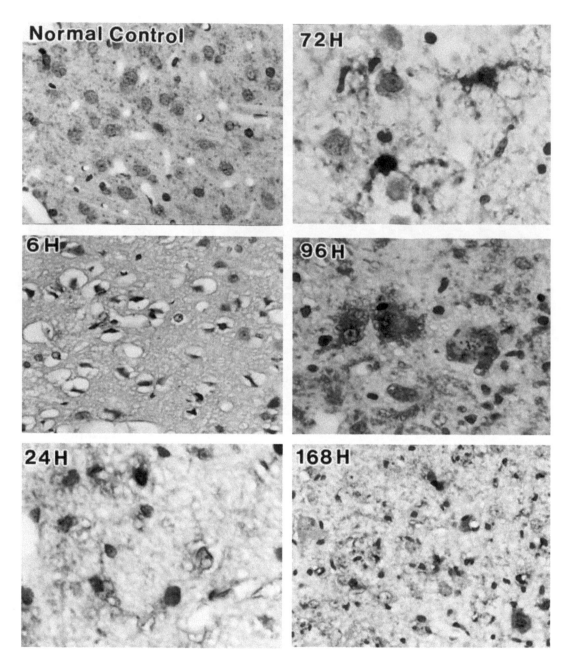

Figure 17.4   Immunohistochemical detection of S100 in reactive astrocytes around the infarct border in the rat brain subjected to permanent middle cerebral artery occlusion (pMCAO). (Courtesy of Dr. Fusahiro Ikuta, Niigata Brain Research Institute.)

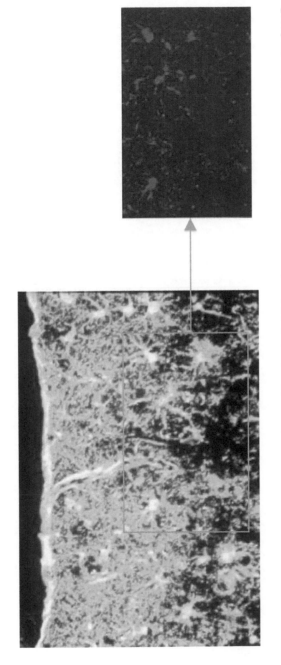

Figure 17.5 The left photo shows double immunohistochemical staining of reactive astrocytes in the rat cortex region for S100B (green) and inducible nitric oxide synthesis (iNOS) (red) as examined by f uorescence microscopy. The right photo shows the immunohistochemical staining for iNOS alone. (Courtesy of Dr. Setsuya Fujita, Louis Pasteur Center for Medical Research, Kyoto, Japan.)

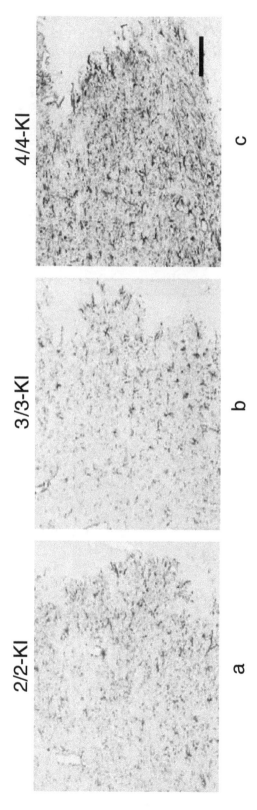

Figure 17.9 Expression of glial fibrillary acidic protein in the peri-infarct area at 5 days (a, b, c) after permanent cerebral artery occlusion (pMCAO) in homozygous human apolipoprotein E (apoE)2 (2/2)–, apoE3 (3/3)–, or apoE4 (4/4)-knock-in mice. Note the pronounced astrocytic activation in 4/4–KI mice compared to 2/2– or 3/3–KI mice at 5 days after pMCAO. The infarct border lies in the lowermost part of each figure. Scale bar denotes 100 μm.

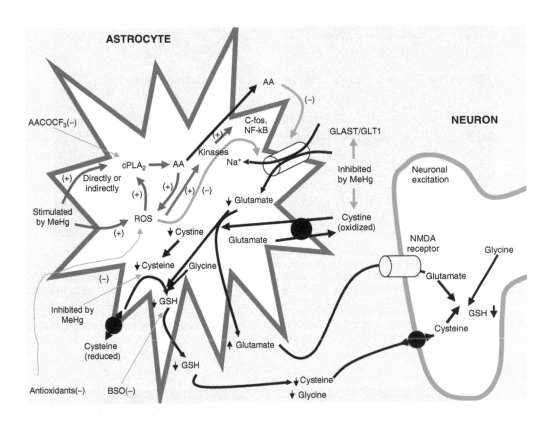

Figure 24.1 A schematic astrocyte/neuronal model depicting various MeHg-initiated processes resulting in neurotoxicity. Please refer to text for details. Abbreviations: AA, arachidonic acid; GSH, glutathione; BSO, buthionine sulfoximine; ROS, reactive oxygen species; cPLA₂, cytosolic phospholipase A₂; AACOCF3, arachidonyl trifluoromethyl ketone; GLAST/GLT-1, astrocytic glutamate transporters in rodents; NF-kB, nuclear factor kappa B.

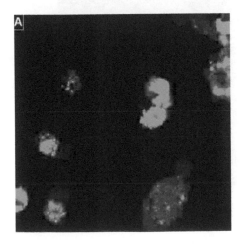 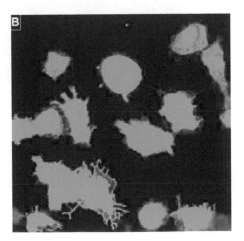

Figure 24.2 ROS formation in rat primary cerebral astrocytes exposed to 10 μM MeHg as assessed by changes in DCF fluorescence. After growing the neonatal rat cerebral astrocytes (4–6 weeks) in poly-d-lysine–coated Corning 35-mm dishes (MatTek Co., Ashland, MA), they were treated with 10 μM MeHg for 30 min at 37°C. Subsequent to the addition of 10 μM CM-H₂DCFDA dye, fluorescence images were recorded at 1-min intervals using a laser-scanning (Zeiss LSM 510, inverted Axiovert 100M) microscope, at excitation wavelength of 488 nm (argon laser) and emission wavelength of 515 nm. The above pictures (A = control and B = MeHg treated) were taken at 25 min after dye addition.

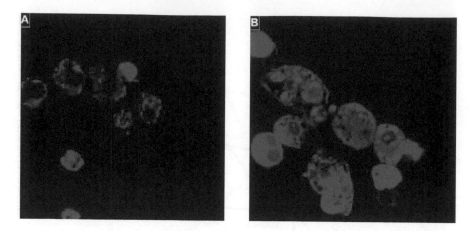

Figure 24.4 Mitochondrial ROS formation in cerebral astrocytes exposed to 10 μM MeHg as assessed by changes in fluorescence of the oxidized form of X-rosamine mitotracker dye. Neonatal rat cerebral astrocytes, grown in poly-d-lysine-coated Corning 35-mm dishes, were treated for 30 min at 37°C with 10 μM MeHg. After the addition of 250 nM CM-H₂XRos, the fluorescence images were taken at 1-min intervals using a laser-scanning confocal microscope equipped with a rhodamine laser (543-nm excitation and 570–600-nm emission). The indicated pictures (A = control and B = MeHg treated) were taken 25 min after dye addition.

GFP

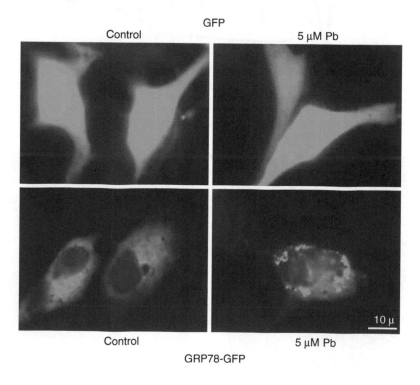

Figure 26.3 Lead-induced GRP78 aggregation in human CCF-STTG1 astrocytoma cells. Fluorescence was detected by digital microscopic imaging. Representative micrographs of transfected control cells and cells that had been exposed to 5 μM Pb in medium for 10 h. The clumping of the fluorescent chimeric protein is observed only in the Pb-treated cells transfected with GRP78-GFP, suggesting that Pb causes GRP78 aggregation. The clumping does not appear in Pb-treated cells transfected with GFP only.

CHAPTER 14

Astrocytes in Acute Energy Deprivation Syndromes

E. Spencer Williams, Amanda D. Phelka, David E. Ray, and Martin A. Philbert

CONTENTS

14.1 INTRODUCTION

14.1.1 Acute Energy Deprivation Syndromes

The term *acute energy deprivation syndromes* was coined by Cavanagh to describe a set of neuropathological disorders of energy metabolism. These disorders, which disrupt mitochondrial bioenergetics, lead to a remarkable array of neuropathological disorders. Several reviews have focused on clinical and biochemical aspects of mitochondrial encephalopathies in recent years.[1,2] The precise mechanisms underlying the susceptibility of these regions have yet to be determined. However, it is generally agreed that disruption of energy metabolism through genetic deficiencies, dietary cofactor insufficiencies, or toxicant-induced dysfunction can cause coincident injury in the auditory and vestibular centers of the mammalian brain.

Experimental evidence suggests glucose and oxygen requirement alone do not determine susceptibility of these brainstem regions to toxicants. Forebrain and mesencephalic structures such as the occipital cortex, hippocampal formation, amygdala, occipital cortex, and substantia nigra, like vulnerable brainstem nuclei, have a high requirement for glucose, oxygen, and oxidative phosphorylation[3–5] but are unaffected by AEDS and assorted chemical stressors. Energy deprivation alone therefore does not sufficiently explain the selective vulnerability of brainstem nuclei to attack by these chemicals.

It is fairly clear that levels of neuronal activity in the affected regions correlate well with the development of lesions. Neuropathologies associated with experimental models of AEDS were ameliorated by reducing input to the auditory and vestibular nuclei through tympanic puncture and anaesthesia.[6] These toxic effects of chemicals that induce AEDS could be exacerbated by tremorigenic pyrethroids (Bifenthrin) or increased noise,[7] both of which increase metabolic load on affected regions. Despite the apparent importance of neuronal activity, it is important to note that the astrocyte provides the earliest indication of injury and is the most readily rescued by tympanic rupture.

14.1.2 Pathology

While the neuropathological picture of AEDS varies, distribution of gliovascular edematous lesions frequently spans the periaqaductal structure of the IVth ventricle and includes the vestibular, superior olivary and deep cerebellar nuclei, inferior colliculi, and invariably the ventral cochlear nuclei.

Astrocytes in susceptible brainstem regions are the primary target with secondary neuronal and endothelial involvement. Indeed, it is not until the astrocytic component of the neuropil is significantly affected that other components of the central nervous system display evidence of morphological disturbance.[8] Early lesions are typified by expansion of the Virchow-Robin spaces, swelling of perivascular astrocyte foot processes, and retraction of perineuronal branches. As the lesion progresses, vascular elements can be observed in the expanded perivascular spaces with marked increase in extracellular edema in the adjacent neuropil. As injury progresses, plasma cells may infiltrate affected regions with concomitant loss of neuronal elements. With greater involvement of oligodendrocytes, there is moderate to severe injury to myelinated axons. Intramyelinic edema is frequently seen at all stages of intoxication.

14.1.3 Astrocytes

Astrocytes are the predominant nonneuronal cell type in the brain, outnumbering neurons approximately 1.4 to 1 in the human brain.[9] Though once considered part of a passive supporting matrix for neurons, astrocytes perform several critical functions that include controlling ion homeostasis and pH, protecting neurons from glutamate excitotoxicity, and providing metabolic substrates and glutathione for neurons.[10,11] These cells are closely associated with axon terminals and nodes of Ranvier, allowing them to regulate synapse formation and synaptic strength.[12,13] *In vitro*, neurons form more numerous and more effective synaptic junctions when cocultured with astrocytes.[14,15] Bidirectional signaling between neurons and astrocytes using glutamate and Ca^{2+} waves reinforces physiological mechanisms of synaptic signaling, though the molecular mechanism has yet to be elucidated.[16–18]

Astrocytes are typically less sensitive to injury by most xenobiotics than are neurons. This lack of vulnerability may reflect the vital role astrocytes play in protecting neurons from cellular stresses by providing the cells with antioxidants and metabolic intermediates and subsequently removing harmful substrates such as excitotoxic neurotransmitters.[19–21] However, the supportive functions that astrocytes provide to neurons and the rest of the CNS are highly dependent on energy, and damage to astrocytic mitochondria has been shown to precede neuronal injury in a variety of neurotoxic states.[20]

The pattern of damage seen with chemicals that produce AEDS is remarkably similar in topology and progression to that seen in a variety of clinically relevant entities. Many of these disorders affect juveniles and are associated with single or multiple defects in mitochondrial function.

14.2 CLINICAL SYNDROMES

14.2.1 mtDNA Disorders

Mitochondrial DNA (mtDNA) encodes 37 genes involved in protein synthesis and bioenergetics within the mitochondrial compartment, including tRNAs and subunits of complex I, III, and IV of the electron transport chain.[22] Several defects in mtDNA have been associated with clinical diseases, including missense mutations, protein synthesis mutations, and rearrangements.[1] Missense mutations in mtDNA commonly result in Leber's hereditary optic neuropathy (LHON), a condition characterized by degeneration of the optic nerve and retinal ganglia. Disruption of tRNA[Lys] sequences by point mutation causes two similar conditions, myoclonus epilepsy with ragged red fibers (MERRF) and mitochondrial encephalopathy with lactic acidosis and stroke-like episodes (MELAS). Diseases that result from rearrangements of mtDNA include chronic progressive external ophthalmoplegia (CPEO) and Kearns-Sayre syndrome.[23] Several of these syndromes preferentially affect bioenergetics within the mitochondrial compartment.

Each of these disorders produce clinical manifestations that may include seizures, lactic acidosis, vomiting, stroke-like episodes, myoclonus, ataxia, and signs of brainstem dysfunction such as abnormal respiration, nystagmus, and ophthalmoparesis (reviewed in Reference 24). In general, these diseases are progressive after delayed onset.[1]

Mitochondria in patients with these diseases generally possess mutated and wild-type mtDNA in varying proportions, a phenomenon known as heteroplasmy. As the proportion of mutated DNA rises in regard to wild type, symptoms of these syndromes begin to manifest.[23] For example, in MELAS and MERRF, a 20-year-old patient with only 15% wild-type mtDNA may express little or no symptomology.[25]

14.2.1.1 Leigh's Syndrome

Mutations or rearrangements of mitochondrial DNA or nuclear DNA cause Leigh's syndrome. Deficiencies in cytochrome oxidase, pyruvate dehydrogenase, complex I, complex II, and ATP synthase have all been implicated in the pathogenesis of Leigh's syndrome.[26] The disruptions in energy metabolism lead to elevations of lactic acid in blood and cerebrospinal fluid.[2] In terms of frequency, 18% of causes involved mtDNA defects, 10% PDHC deficiencies, 19% complex I, 14% cytochrome c oxidase, and the remaining 39% are not attributable to any specific cause, though complex II deficiencies and mtDNA depletion are possibilities.[27] These defects result in bilaterally symmetrical lesions in the brainstem (specifically, the substantia nigra, inferior colliculus, cerebellar nuclei), basal ganglia, and the thalamus. Clinically, Leigh's syndrome is manifested as signs of brainstem dysfunction, including abnormal breathing, ataxia, dystonia, and nystagmus. Onset of this syndrome is usually within the first year of life, and death often follows within 2 years.[27]

The most common mtDNA defect associated with onset of Leigh's syndrome is a T-to-G mutation in the ATP6 gene, resulting in a leucine-to-arginine change.[28] This mutation causes blockade of the F_0 ATP synthase proton channel.[29]

14.2.1.2 MELAS/MERRF

Related to Leigh's syndrome are two neurodegenerative conditions termed myoclonus epilepsy with ragged red fibers (MERRF) and mitochondrial myopathy, encephalopathy, lactic acidosis, and stroke-like episodes (MELAS). MERRF is caused by an mtDNA point mutation in the tRNA[Lys] coding sequence, resulting in deficiencies in complex I and IV.[2] MELAS has a similar deficiency; a mutation in tRNA[Leu] results in reduced complex I and cytochrome oxidase (COX) activity.

The pattern of neuropathology in these diseases differs significantly from Leigh's syndrome. Lesions in MELAS span the same regions as Leigh's (basal ganglia and cerebellum) but also involve spongiform degeneration in the cerebral cortex. These findings are of particular interest considering that there is considerable overlap between deficiencies in electron transport chain components between Leigh's and MELAS. However, the pathology of MELAS may be attributable to aberrant transcriptional termination of 16S rRNA.[30] Most of the clinical manifestations of these diseases are described by the acronym assigned to them.

14.2.2 Wernicke's/Thiamine Deficiency

Wernicke's encephalopathy is an acute energy deprivation syndrome brought on by thiamine deficiency. The causes of thiamine deficiency include chronic alcoholism, voluntary starvation, and anorexia nervosa. Paralysis of ocular motor muscles secondary to nerve dysfunction, ataxia, disorientation, and generally altered consciousness, possibly reaching as far as Korsakoff syndrome, an irreversible amnesia are frequently observed. Pathology includes lesions in periaquaductal gray matter, specifically, degeneration of glia in brainstem mammillary bodies and deep cerebellar roof nuclei, followed by neuronal loss (reviewed in Reference 31). Acute thiamine deficiency caused by poor diet is known by another name, beriberi. These terms have become interchangeable in the literature, which is appropriate, as each condition conveys the same pattern of neuropathology and symptomatology. In many cases, the neuropathology resulting from deficiency is reversible by treatment with thiamine.[32–34]

14.3 EXPERIMENTAL MODELS OF AEDS

14.3.1 Chlorosugars

6-Chloro-6-deoxyglucose and similar chlorosugars produce glio-vascular lesions in prone regions, including inferior colliculi and superior olivary nuclei.[35,36] Metabolism of these sugars leads to production of 3-chlorolactaldehyde, inhibiting glycolysis by interfering with glucose-3-phosphate dehydrogenase and triose isomerase reactions.

The earliest neuropathological changes are detected in astrocytes, including swelling and retraction of astroglial processes.[37] In response to 6-chloro-6-deoxyglucose treatment, mice developed vacuolated lesions caused by swelling of astrocytes, which appeared to contain enlarged, disorganized mitochondria.[35]

14.3.2 6-Aminonicotinamide

6-Aminonicotinamide forms an ineffective product through NAD(P)-glycohydrolase, leading to a pattern of edematous glial lesions reminiscent of those observed with thiamine deficiency, though neurons are more profoundly affected than in other models.[38–40] Recent experiments by Tyson and coworkers demonstrate that 6-AN inhibits the pentose phosphate pathway as a result of NADPH loss. This inhibition results in a buildup of 6-phosphogluconate.[41]

14.3.3 Nitroaromatic Chemicals (Misonidazol/Metronidazol/m-Dinitrobenzene)

Several nitroaromatic compounds cause neuropathological profiles consistent with acute energy deprivation syndromes. These include the industrial reagent 1,3-dinitrobenzene (m-DNB) and the drugs misonidazol and metronidazol.

The stepwise metabolic reduction of aromatic nitrocompounds found in the environment is as follows:

Equation 1: $R\text{-}NO_2 + \varepsilon^- \rightarrow R\text{-}NO2^{\bullet-}$ (Nitro-anion radical)

Equation 2: $R\text{-}NO_2^{\bullet-} + 3\varepsilon^- \rightarrow R\text{-}NO$ (Nitroso intermediate)

Equation 3: $R\text{-}NO + 2\varepsilon^- \rightarrow R\text{-}NHOH$ (Hydroxylamino derivative)

Equation 4: $R\text{-}NHOH + 2\varepsilon^- \rightarrow R\text{-}NH_2$ (Amine derivative)

Equation 5: $R\text{-}NH_2 + CH_3\text{-}CO_2\text{-}H \rightarrow R\text{-}NH\text{-}O\text{-}CO\text{-}CH_3$ (*N*-acetylated derivative)

The neurotoxicological effects of m-DNB are the best studied among the chemical inducers of AEDS. Studies have suggested that the mechanism of m-DNB–induced neurodegeneration involves oxidative stress, possibly as a consequence of redox cycling between the parent compound and a nitro-anion metabolite.[42] Inhibition of dehydrogenase enzymes intimately associated with energy metabolism, including GAPDH, SDH, and isocitrate dehydrogenase, may be another critical effect of m-DNB.

NMR tomography reveals that m-DNB induces lesions in a number of brainstem regions, including the inferior colliculus, cerebellar roof nuclei, and vestibular complex (Figure 14.1).

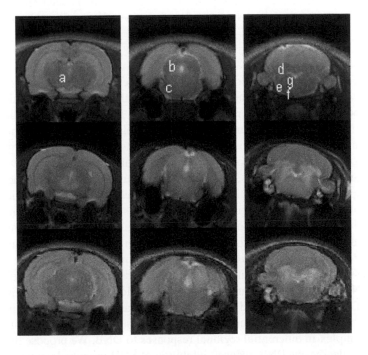

Figure 14.1 Summary of the principal lesions caused by m-DNB in the rat brain as shown in RARE NMR tomographs with anatomical structures from roughly the same level of the brain aligned in each column. The principal lesion sites are labeled on the first row: (a) red and pararubral nuclei; (b) inferior colliculus; (c) nuclei of the lateral lemniscus; (d) cerebellar roof nuclei; (e) spinal trigeminal nucleus; (f) facial nucleus; (g) vestibular complex. Images in the first row were taken from a rat imaged 5 hours after dosing with DMSO solvent. The lower two rows are NMR images from affected rat brains, the middle row 26 hours after the last of three doses of 10 mg/kg m-DNB (given at 0, 4, and 24 h) and the bottom row 25 hours after the last of four daily doses of 7.5 mg/kg m-DNB.

14.4 MOLECULAR MECHANISMS OF ASTROCYTE CELL DEATH

14.4.1 Mitochondrial Dysfunction

14.4.1.1 Lactic Acidosis

Acidosis and cellular damage are induced by the production and accumulation of lactate resulting from the inhibition of aerobic metabolism under hypoxic conditions.[43] Disruption of mitochondrial metabolism results in the buildup of lactic acid. Friede and Van Houton postulated that lactic acidosis was the major source of neuropathy in these conditions.[44] Lactic acidosis, as a result of complex I deficiency, in Leigh's syndrome is a common cause of neonatal mortality.[45] After m-DNB intoxication, lactate levels are increased in regions that are not susceptible to lesions as well as those that are preferentially affected. However, the magnitude of increase was much more pronounced in susceptible regions including the inferior colliculi and vestibular nuclei,[46] suggesting that DNB disrupts mitochondrial function in these regions by interfering with aerobic energy production. ^{1}H NMR spectroscopy demonstrates higher basal lactate concentrations, which lead to considerably higher levels after m-DNB intoxication (Figure 14.2). This may suggest a lower threshold for disruption of oxidative energy metabolism in regions susceptible to m-DNB–induced neuropathology.

Alternatively, elevated levels of lactate in sensitive regions might suggest a compensatory, protective mechanism by astrocytes to provide neurons with an alternate carbon source, which has been shown by others to be sufficient for maintaining neuronal energy metabolism.[47,48] Our observation of elevated lactate concentrations prior to onset of DNB-induced encephalopathy might provide an explanation for the initial sparing of neurons in DNB-sensitive brain regions.

14.4.1.2 SDH Inhibition

m-DNB induces region-dependent alterations in succinate dehydrogenase activity *in vivo*. Moderate decreases in activity are observed in the white matter, granular layer, and molecular layer of DNB-exposed rat cerebellum at 24 h, while relative SDH activity in the 3-h rat cerebellum and in the rat hippocampus remain relatively constant between the control and DNB-treated animals. These data provide additional evidence that exposure to DNB disrupts mitochondrial energy metabolism in neurons and glia of sensitive brain regions. DNB-induced inhibition of SDH is, however, comparable in rat brainstem and cortical astrocytes, suggesting that disruption of oxidative phosphorylation and the electron transport chain are not likely to be principal mechanisms defining regional vulnerability to DNB. These observations are in agreement with data resulting from 3-nitropropionate, an irreversible SDH inhibitor. The pattern of neuropathy that results from 3-NPA treatment does not correlate well with SDH inhibition, which is similar in both sensitive and resistant areas.[49]

Although mitochondrial permeability transition pore inhibitors including cyclosporin A and bongkrekic acid significantly reduced inhibition of SDH in brainstem astrocytes following exposure to 100 μ*M* m-DNB, these inhibitors are ineffective in cortical astrocytes. Significant decreases in DNB-induced SDH inhibition limited to only brainstem astrocytes following pretreatment with known mtPTP inhibitors provides additional evidence of potential involvement of the permeability transition pore complex in differential regional responses to DNB. We propose that CsA may have been ineffective in decreasing DNB-induced inhibition of SDH in rat and mouse cortical astrocytes because the dose of DNB used (100 μ*M*) has been shown by our laboratory to be incapable of inducing mtPTP opening in cortex-derived astrocytes.[50] Therefore, we suggest that CsA is not so much ineffective as it is unnecessary to prevent onset of the MPT, as the pore likely remains in the closed state at this dose in cortical astrocytes. However, the possibility exists that the observed regional differences in the ability of CsA (and bongkrekic acid) to significantly decrease DNB-induced inhibition of SDH may be a reflection of regional mitochondrial heterogeneity.

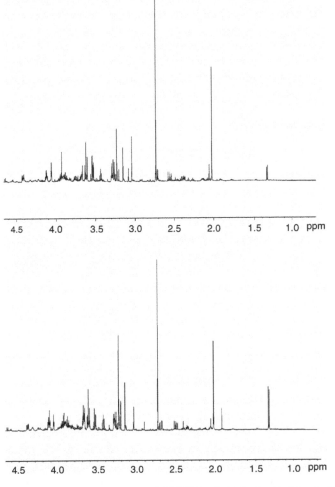

ANOVA: Effect of Dinitrobenzene $p < 0.001$, Tukey's Pairwise Test for Individual Regions ($p = 0.0014$ in midbrain)

Lactate (μmol/g)	Cerebral Cortex	Midbrain	Brainstem	Cerebellum
Control	2.96 ± 0.78	3.81 ± 0.70	6.18 ± 0.56	4.52 ± 0.77
m-Dinitrobenzene	5.28 ± 0.59	7.96 ± 0.62	8.55 ± 0.62	6.15 ± 0.62

Figure 14.2 Representative ^{1}H NMR spectra from TCA extracts of the brainstem of individual rats treated with four daily i.p. injections of DMSO (top) and four daily i.p injections of 8.0 mg/kg m-DNB (bottom) and subjected to microwave inactivation 4 hours after the last dose. Note the pronounced intensification of the lactate doublet resonance at 1.33 ppm in the latter spectrum.

14.4.1.3 ATP Loss

Any disruption in mitochondrial bioenergetics leads to reduced ATP synthesis. Rapid disruption of mitochondrial function is frequently accompanied by a dramatic decrease in cellular ATP. Necrotic cell death is initiated under these conditions since induction and continuance of the apoptotic signaling cascade are dependent on maintenance of ATP levels.[51] However, if mitochondrial function is more slowly disrupted and ATP levels are either preserved or other sources of ATP generation remain intact, apoptotic signaling may proceed. Secondary necrosis is observed in the event that ATP levels eventually collapse.

Experiments in our laboratory have verified that cellular ATP levels are greatly reduced in response to m-DNB intoxication.[52] In Leigh's syndrome, defects in the ATP6 gene lead to decreased ATP synthesis, though the exact nature of the defect is still unclear.[53] Also, loss of mitochondrial integrity can cause rapid hydrolysis of ATP via reverse action of the F_1F_0 ATPase. The activity of caspase enzymes, which are responsible for cleavage of chromatin in the final stages of apoptosis, is ATP-dependent. Thus, the loss of ATP due to disrupted bioenergetics may cause secondary necrosis in susceptible astrocytes, though apoptotic pathways are initiated.[51]

14.4.1.4   ROS/Oxidative Stress

The intuitive connection of perturbation of energy generation with increased generation of reactive oxygen species (ROS) has been verified experimentally in a number of organs.[54,55] The brain consumes one-fourth of the total oxygen intake, generates more oxygen radicals than most organs, and is particularly susceptible to oxidative damage because it contains relatively high concentrations of polyunsaturated fatty acids that are easily peroxidized and relatively low levels of antioxidant enzymes.[56] Complex I is the primary source of ROS in the brain.[57]

These findings are consistent with recent reports indicating that oxidative damage to astrocytic mitochondria precedes neuronal injury.[20] GSH depletion in astrocytes follows exposure to m-DNB,[42] followed by a rebound increase in total GSH levels. Further experiments demonstrated that depletion of glutathione exacerbated brainstem lesions after m-DNB intoxication.[58] Also, recent experiments by Pannunzio et al. suggest that supplementation with antioxidants alleviate eventual neuronal death precipitated by thiamine deficiency.[59]

One of the important physiological roles of astrocytes is providing neurons with glutathione precursors.[11] In response to insult, glutathione synthesis machinery is upregulated in astrocytes, which then plays a role in protecting neurons from oxidative stress.[60] The importance of astrocytic glutathione was underlined by Papadopoulos and colleagues, who noted that reductive capacity in these cells was linked to survival of metabolic disruptions.[61]

The ability to enzymatically conjugate glutathione to classes of substrate is unequally distributed by region and cell type in the central and peripheral nervous system.[62–66] In general, α-isoforms of glutathione S-transferase are expressed in the nuclei of selected neurons throughout the nervous system. Expression of μ- and π-isoforms are confined to astrocytes and oligodendrocytes respectively.

The distribution of glutathione within cellular compartments may be critical to understanding oxidative mechanisms of m-DNB injury. Mitochondrial pools of glutathione are particularly susceptible to attack either by *in situ* generation of free-radical species or by release of mitochondrial glutathione through the transition pore following oxidative stress elsewhere in the cell.[67]

14.4.1.5   Mitochondrial Permeability Transition

Cell death itself appears to involve activation of the mitochondrial permeability transition (MPT). However, the absence of apoptotic bodies suggests a necrotic event. Possibly, this phenomenon occurs as a result of ATP loss due to glutathione regeneration or disruption of energy metabolism, resulting in necrapoptosis as suggested by Lemasters.[51]

Disruption of the cellular redox potential places the cell in a state of chemically induced hypoxia, which inhibits greatly the ability of the mitochondria to function adequately by limiting the oxygen necessary for mitochondrial oxidations.[68] Oxidative phosphorylation is severely lessened as a result and intracellular creatinine phosphate is rapidly depleted, leading to a concomitant rise in inorganic phosphate (P_i). Decreased oxidative phosphorylation and increased P_i both stimulate heightened glycolysis. The hydrolytic activity of the F_0F_1 ATPase (ATP synthase) is also induced by hypoxic conditions in an attempt to restore physiological ATP concentrations. However, the ATPase inadvertently consumes the negligible ATP generated by glycolysis under hypoxic conditions, making futile these attempts to restore ATP to physiological levels.

Mitochondria can accumulate calcium as long as adenine nucleotides are available and intramitochondrial pyridine nucleotides are maintained in a reduced state.[69] Oxidative stress brings about a slow, progressive increase in cytosolic free calcium, which in turn gives rise to a proportionally greater increase in mitochondrial calcium until mitochondrial calcium overload occurs.[69–71] Mitochondrial calcium overload can induce a sudden increase in inner mitochondrial membrane permeability.[69,72]

Decreased cellular ATP, increased inorganic phosphate, and oxidative stress are common to mitochondrial dysfunction and are also among the known potent inducers of the mitochondrial permeability transition. The MPT has been characterized as an abrupt increase in mitochondrial membrane permeability to solutes smaller than 1500 Da, and is attributed to opening of the mitochondrial permeability transition pore (mtPTP[70,71,73]).

The mtPTP participates in matrix calcium, pH, mitochondrial membrane potential, and volume regulation and serves as a calcium, voltage, pH, and redox-gated channel with limited ion selectivity.[74] Conductance of the mtPTP is so great that opening of only a small percentage of the pores in a single mitochondrion is sufficient to cause membrane depolarization, large amplitude swelling, uncoupling of oxidative phosphorylation, and release of intramitochondrial ions and metabolic intermediates.[75,76] Loss of the electrochemical gradient that drives ATP production may be a functional molecular endpoint of AEDS. Our laboratory has previously demonstrated region- and cell-specific susceptibility to DNB-induced onset of the MPT.[50,52] Primary rat brainstem astrocytes display an order of magnitude greater sensitivity to DNB-induced onset of the MPT than do rat cortical astrocytes, and neuronal-derived SY5Y neuroblastoma cells are approximately 10-fold more sensitive to DNB-induced mitochondrial depolarization than their glial-derived C6 glioma counterparts. These data imply that the mtPTP may be regulated differentially in DNB-sensitive versus non-DNB-sensitive brain regions and cell types. That the known mtPTP inhibitor, cyclosporin A, effectively increases the dose and duration of DNB exposure required to induce MPT onset in brainstem and cortical astrocytes suggests that the DNB-induced loss in mitochondrial membrane potential is likely the result of stabilization of the mtPTP in the open conformation.

14.4.2 Bcl-2 Proteins

The role of Bcl-2 family proteins in regulation of MPT assembly has been extensively investigated.[77] Relative expression of Bcl-2 family proteins may play a role in patterns of regional sensitivity. Bcl-2 protein family members are generally regarded as either proapoptotic (Bax, Bad, Bak) or antiapoptotic (Bcl-2, Bcl-X$_L$, Bcl-w). The levels at which these critical proteins are expressed in relation to each other, where they are located in the cell, and whether or not they are posttranslationally modified all influence greatly the balance of active agonistic to active antagonistic Bcl-2 proteins within a cell, and this balance is a critical determinant to how a cell manages a potentially lethal insult.

Preliminary research has provided evidence in primary rat astrocytes that Bcl-XL is expressed at constitutively lower levels in brainstem astrocytes than in their cortical counterparts, while BCL-2 and Bax are comparably expressed in astrocytes acquired from DNB-sensitive and non-DNB-sensitive brain regions.[50] Western blot analyses of brainstem and cortical tissue homogenates confirm that Bcl-XL is more highly expressed in the non-DNB-sensitive cortex than in the DNB-sensitive brainstem. It has been concluded, based on these results, that Bcl-XL may play a central role in defining regional astrocytic sensitivity to DNB, and while likely less critical, Bcl-2 and Bax may have functional relevance in DNB-induced neurotoxicity.

14.5 SUMMARY

The causes of disrupted mitochondrial bioenergetics in acute energy deprivation syndromes are varied. In the clinical setting, AEDS is caused by deficiencies in critical metabolic enzymes including those involved in glycolysis, oxidative phosphorylation, and ATP synthesis. Disruption of energy metabolism itself by inhibitory substrates or nitroaromatic compounds causes a similar array of symptoms and pathologies. The underlying mechanisms of regional susceptibility remain unclear, though data from the m-DNB model indicate that maintenance of mitochondrial polarity is a critical determinant of eventual cell death. Though apoptosis is initiated through activation of the MPT, depletion of ATP causes diversion of cellular fate from apoptosis to necrosis. The activity and relative expression of Bcl-2 family proteins as regulators of the MPT may be critical in determining susceptibility of these separate populations of astrocytes. The involvement of redox capacity is also a subject for future study, as the differential ability of these cellular populations to detoxify mitochondrially generated ROS is an attractive hypothesis to explain regional susceptibility.

REFERENCES

1. Wallace, D.C., Mitochondrial defects in neurodegenerative disease, *Ment. Retard. Dev. Disabil. Res. Rev.*, 7, 158–166, 2001.
2. Schmiedel, J. et al., Mitochondrial cytopathies, *J. Neurol.*, 250, 267–277, 2003.
3. Bagley, P.R. et al., Anatomical mapping of glucose transporter protein and pyruvate dehydrogenase in rat brain: an immunogold study, *Brain Res.*, 499, 214–224, 1989.
4. Mastrogiacomo, F., Bergeron, C., and Kish, S.J., Brain alpha-ketoglutarate dehydrogenase complex activity in Alzheimer's disease, *J. Neurochem.*, 61, 2007–2014, 1993.
5. Calingasan, N.Y. et al., Distribution of the alpha-ketoglutarate dehydrogenase complex in rat brain, *J. Comp. Neurol.*, 346, 461–479, 1994.
6. Holton, J.L. et al., Increasing or decreasing nervous activity modulates the severity of the glio-vascular lesions of 1,3-dinitrobenzene in the rat: effects of the tremorgenic pyrethroid, Bifenthrin, and of anaesthesia, *Acta Neuropathol. (Berl.)*, 93, 159–165, 1997.
7. Ray, D.E. et al., Functional/metabolic modulation of the brain stem lesions caused by 1,3-dinitrobenzene in the rat, *Neurotoxicology*, 13, 379–388, 1992.
8. Philbert, M.A. et al., 1,3-Dinitrobenzene-induced encephalopathy in rats, *Neuropathol. Appl. Neurobiol.*, 13, 371–389, 1987.
9. Bass, N.H. et al., Quantitative cytoarchitectonic distribution of neurons, glia, and DNa in rat cerebral cortex, *J. Comp. Neurol.*, 143, 481–490, 1971.
10. Nedergaard, M., Ransom, B., and Goldman, S.A., New roles for astrocytes: redefining the functional architecture of the brain, *Trends Neurosci.*, 26, 523–530, 2003.
11. Wang, X.F. and Cynader, M.S., Astrocytes provide cysteine to neurons by releasing glutathione, *J. Neurochem.*, 74, 1434–1442, 2000.
12. Fields, R.D. and Stevens-Graham, B., New insights into neuron-glia communication, *Science*, 298, 556–562, 2002.
13. Hatton, G.I., Glial-neuronal interactions in the mammalian brain, *Adv. Physiol. Educ.*, 26, 225–237, 2002.
14. Pfrieger, F.W. and Barres, B.A., Synaptic efficacy enhanced by glial cells *in vitro*, *Science*, 277, 1684–1687, 1997.
15. Ullian, E.M. et al., Control of synapse number by glia, *Science*, 291, 657–661, 2001.
16. Araque, A. et al., Calcium elevation in astrocytes causes an NMDA receptor-dependent increase in the frequency of miniature synaptic currents in cultured hippocampal neurons, *J. Neurosci.*, 18, 6822–6829, 1998.
17. Parpura, V. and Haydon, P.G., Physiological astrocytic calcium levels stimulate glutamate release to modulate adjacent neurons, *Proc. Natl. Acad. Sci. U.S.A.*, 97, 8629–8634, 2000.
18. Hansson, E. and Ronnback, L., Glial neuronal signaling in the central nervous system, *FASEB J.*, 17, 341–348, 2003.
19. Xu, L. et al., Overexpression of bcl-xL protects astrocytes from glucose deprivation and is associated with higher glutathione, ferritin, and iron levels, *Anesthesiology*, 91, 1036–1046, 1999.
20. Robb, S.J. and Connor, J.R., An *in vitro* model for analysis of oxidative death in primary mouse astrocytes, *Brain Res.*, 788, 125–132, 1998.

21. Peuchen, S. et al., Interrelationships between astrocyte function, oxidative stress and antioxidant status within the central nervous system, *Prog. Neurobiol.*, 52, 261–281, 1997.

22. Anderson, S. et al., Sequence and organization of the human mitochondrial genome, *Nature*, 290, 457–465, 1981.

23. Chinnery, P.F. and Turnbull, D.M., Mitochondrial DNA mutations in the pathogenesis of human disease, *Mol. Med. Today*, 6, 425–432, 2000.

24. Tanji, K. et al., Neuropathological features of mitochondrial disorders, *Semin. Cell Dev. Biol.*, 12, 429–439, 2001.

25. Wallace, D.C. et al., Familial mitochondrial encephalomyopathy (MERRF): genetic, pathophysiological, and biochemical characterization of a mitochondrial DNA disease, *Cell*, 55, 601–610, 1988.

26. Brown, G.K. and Squier, M.V., Neuropathology and pathogenesis of mitochondrial diseases, *J. Inherit. Metab. Dis.*, 19, 553–572, 1996.

27. Dahl, H.H., Getting to the nucleus of mitochondrial disorders: identification of respiratory chain-enzyme genes causing Leigh syndrome, *Am. J. Hum. Genet.*, 63, 1594–1597, 1998.

28. Holt, I.J. et al., A new mitochondrial disease associated with mitochondrial DNA heteroplasmy, *Am. J. Hum. Genet.*, 46, 428–433, 1990.

29. Trounce, I., Neill, S., and Wallace, D.C., Cytoplasmic transfer of the mtDNA nt 8993 T—>G (ATP6) point mutation associated with Leigh syndrome into mtDNA-less cells demonstrates cosegregation with a decrease in state III respiration and ADP/O ratio, *Proc. Natl. Acad. Sci. U.S.A.*, 91, 8334–8338, 1994.

30. Hess, J.F. et al., Impairment of mitochondrial transcription termination by a point mutation associated with the MELAS subgroup of mitochondrial encephalomyopathies, *Nature*, 351, 236–239, 1991.

31. Cavanagh, J.B., Selective vulnerability in acute energy deprivation syndromes, *Neuropathol. Appl. Neurobiol.*, 19, 461–470, 1993.

32. Bergui, M. et al., Diffusion-weighted MR in reversible wernicke encephalopathy, *Neuroradiology*, 43, 969–972, 2001.

33. Oka, M. et al., Diffusion-weighted MR findings in a reversible case of acute Wernicke encephalopathy, *Acta Neurol. Scand.*, 104, 178–181, 2001.

34. Sparacia, G., Banco, A., and Lagalla, R., Reversible MRI abnormalities in an unusual paediatric presentation of Wernicke's encephalopathy, *Pediatr. Radiol.*, 29, 581–584, 1999.

35. Jacobs, J.M. and Ford, W.C., The neurotoxicity and antifertility properties of 6-chloro-6-deoxyglucose in the mouse, *Neurotoxicology*, 2, 405–417, 1981.

36. Cavanagh, J.B. and Nolan, C.C., The neurotoxicity of alpha-chlorohydrin in rats and mice: II. Lesion topography and factors in selective vulnerability in acute energy deprivation syndromes, *Neuropathol. Appl. Neurobiol.*, 19, 471–479, 1993.

37. Cavanagh, J.B., Nolan, C.C., and Seville, M.P., The neurotoxicity of alpha-chlorohydrin in rats and mice: I. Evolution of the cellular changes, *Neuropathol. Appl. Neurobiol.*, 19, 240–252, 1993.

38. Schneider, H. and Cervos-Navarro, J., Acute gliopathy in spinal cord and brain stem induced by 6-aminonicotinamide, *Acta Neuropathol. (Berl.)*, 27, 11–23, 1974.

39. Watanabe, I., Pyrithiamine-induced acute thiamine-deficient encephalopathy in the mouse, *Exp. Mol. Pathol.*, 28, 381–394, 1978.

40. Watanabe, I. and Kanabe, S., Early edematous lesion of pyrithiamine induced acute thiamine deficient encephalopathy in the mouse, *J. Neuropathol. Exp. Neurol.*, 37, 401–413, 1978.

41. Tyson, R.L., Perron, J., and Sutherland, G.R., 6-Aminonicotinamide inhibition of the pentose phosphate pathway in rat neocortex, *Neuroreport*, 11, 1845–1848, 2000.

42. Romero, I.A. et al., Early metabolic changes during m-Dinitrobenzene neurotoxicity and the possible role of oxidative stress, *Free Radi. Biol. Med.*, 18, 311–319, 1995.

43. Ishii, H. et al., Effects of propofol on lactate accumulation and oedema formation in focal cerebral ischaemia in hyperglycaemic rats, *Br. J. Anaesth.*, 88, 412–417, 2002.

44. Friede, R.L. and Van Houten, W.H., Relations between postmortem alterations and glycolytic metabolism in the brain, *Exp. Neurol.*, 4, 197–204, 1961.

45. Sue, C.M. and Schon, E.A., Mitochondrial respiratory chain diseases and mutations in nuclear DNA: a promising start?, *Brain Pathol.*, 10, 442–450, 2000.

46. Phelka, A.D., Beck, M.J., and Philbert, M.A., 1,3-Dinitrobenzene inhibits mitochondrial complex II in rat and mouse brainstem and cortical astrocytes, *Neurotoxicology*, 24, 403–415, 2003.

47. Abi-Saab, W.M. et al., Striking differences in glucose and lactate levels between brain extracellular fluid and plasma in conscious human subjects: effects of hyperglycemia and hypoglycemia, *J. Cereb. Blood Flow Metab.*, 22, 271–279, 2002.

48. Rice, A.C. et al., Lactate administration attenuates cognitive deficits following traumatic brain injury, *Brain Res.*, 928, 156–159, 2002.

49. Nony, P.A. et al., 3-Nitropropionic acid (3-NPA) produces hypothermia and inhibits histochemical labeling of succinate dehydrogenase (SDH) in rat brain, *Metab. Brain Dis.*, 14, 83–94, 1999.

50. Tjalkens, R.B., Phelka, A.D., and Philbert, M.A., Regional variation in the activation threshold for 1,3-DNB-induced mitochondrial permeability transition in brainstem and cortical astrocytes, *Neurotoxicology*, 24, 391–401, 2003.

51. Lemasters, J.J., V. *Necrapoptosis* and the mitochondrial permeability transition: shared pathways to necrosis and apoptosis, *Am. J. Physiol.*, 276, G1–G6, 1999.

52. Tjalkens, R.B., Ewing, M.M., and Philbert, M.A., Differential cellular regulation of the mitochondrial permeability transition in an *in vitro* model of 1,3-dinitrobenzene-induced encephalopathy, *Brain Res.*, 874, 165–177, 2000.

53. Mattiazzi, M. et al., The mtDNA T8993G (NARP) mutation results in an impairment of oxidative phosphorylation that can be improved by antioxidants, *Hum. Mol. Genet.*, 13, 869–879, 2004.

54. Ganey, P.E. et al., Ethanol potentiates oxygen uptake and toxicity due to menadione bisulfite in perfused rat liver, *Mol. Pharmacol.*, 38, 959–964, 1990.

55. Keller, B.J. et al., O2-dependent hepatotoxicity due to ethylhexanol in the perfused rat liver: mitochondria as a site of action, *J. Pharmacol. Exp. Ther.*, 252, 1355–1360, 1990.

56. Sanberg, P.R., Willing, A.E., and Cahill, D.W., Neurosurgery for the 21st century. What does neurobiology justify for repairing neurodegenerative disorders?, *Clin. Neurosurg.*, 48, 113–126, 2001.

57. Turrens, J.F., Mitochondrial formation of reactive oxygen species, *J. Physiol.*, 552, 335–344, 2003.

58. Hu, H.L. et al., Glutathione depletion increases brain susceptibility to m-dinitrobenzene neurotoxicity, *Neurotoxicology*, 20, 83–90, 1999.

59. Pannunzio, P. et al., Thiamine deficiency results in metabolic acidosis and energy failure in cerebellar granule cells: an *in vitro* model for the study of cell death mechanisms in Wernicke's encephalopathy, *J. Neurosci. Res.*, 62, 286–292, 2000.

60. Iwata-Ichikawa, E. et al., Glial cells protect neurons against oxidative stress via transcriptional upregulation of the glutathione synthesis, *J. Neurochem.*, 72, 2334–2344, 1999.

61. Papadopoulos, M.C. et al., Vulnerability to glucose deprivation injury correlates with glutathione levels in astrocytes, *Brain Res.*, 748, 151–156, 1997.

62. Abramovitz, M. et al., Expression of an enzymatically active Yb3 glutathione S-transferase in Escherichia coli and identification of its natural form in rat brain, *J. Biol. Chem.*, 263, 17627–17631, 1988.

63. Abramovitz, M. and Listowsky, I., Developmental regulation of glutathione S-transferases, *Xenobiotica*, 18, 1249–1254, 1988.

64. Cammer, W. et al., Differential localization of glutathione-S-transferase Yp and Yb subunits in oligodendrocytes and astrocytes of rat brain, *J. Neurochem.*, 52, 876–883, 1989.

65. Johnson, J.A. et al., Glutathione S-transferase isoenzymes in rat brain neurons and glia, *J. Neurosci.*, 13, 2013–2023, 1993.

66. Philbert, M.A. et al., Glutathione S-transferases and gamma-glutamyl transpeptidase in the rat nervous systems: a basis for differential susceptibility to neurotoxicants, *Neurotoxicology*, 16, 349–362, 1995.

67. Huang, J. and Philbert, M.A., Distribution of glutathione and glutathione-related enzyme systems in mitochondria and cytosol of cultured cerebellar astrocytes and granule cells, *Brain Res.*, 680, 16–22, 1995.

68. Di Lisa, F. and Bernardi, P., Mitochondrial function as a determinant of recovery or death in cell response to injury, *Mol. Cell. Biochem.*, 184, 379–391, 1998.

69. Crompton, M., The mitochondrial permeability transition pore and its role in cell death, *Biochem. J.*, 341, 233–249, 1999.

70. Hunter, D.R. and Haworth, R.A., The Ca2+-induced membrane transition in mitochondria. I. The protective mechanisms, *Arch. Biochem. Biophys.*, 195, 453–459, 1979.

71. Haworth, R.A. and Hunter, D.R., The Ca2+-induced membrane transition in mitochondria. II. Nature of the Ca2+ trigger site, *Arch. Biochem. Biophys.*, 195, 460–467, 1979.

72. Huser, J. and Blatter, L.A., Fluctuations in mitochondrial membrane potential caused by repetitive gating of the permeability transition pore, *Biochem. J.*, 343, 311–317, 1999.

73. Lemasters, J.J. et al., The mitochondrial permeability transition in toxic, hypoxic and reperfusion injury, *Mol. Cell. Biochem.*, 174, 159–165, 1997.

74. Marzo, I. et al., The permeability transition pore complex: a target for apoptosis regulation by caspases and bcl-2-related proteins, *J. Exp. Med.*, 187, 1261–1271, 1998.

75. Lemasters, J.J. et al., The mitochondrial permeability transition in cell death: a common mechanism in necrosis, apoptosis and autophagy, *Biochim. Biophys. Acta*, 1366, 177–196, 1998.

76. Lemasters, J.J. et al., Mitochondrial dysfunction in the pathogenesis of necrotic and apoptotic cell death, *J. Bioenerg. Biomembr.*, 31, 305–319, 1999.

77. Tsujimoto, Y., Cell death regulation by the Bcl-2 protein family in the mitochondria, *J. Cell Physiol.*, 195, 158–167, 2003.

CHAPTER 15

The Role of Multiple Drug Resistance in Neuroglial Survival: Lessons from Human Epileptic Brain

Kerri L. Hallene, Kelly M. Kight, Gabriele Dini, and Damir Janigro

CONTENTS

15.1 BLOOD–BRAIN BARRIER

At the cellular level, astrocytes and endothelial cells (EC) form a selectively permeable barrier known as the blood–brain barrier (BBB). The role of the BBB is to serve as a controlled, functioning gateway protecting the brain from systemic influences, while maintaining homeostasis and allowing transport of nourishment to neurons in the parenchyma.[1] Tight junctions formed by microvascular endothelial cells, lack of fenestrations, and minimal pinocytic vesicles restrict the diffusion of proteins and other molecules into the intracellular space of the brain. However, under certain conditions, the BBB may have an increased permeability, allowing the potential entrance of toxins

For reasons of clarity, several cellular mechanisms involved in the toxicity of epileptic brain will be introduced as separate entities. Figure 15.1 summarizes possible mechanisms relating various central nervous system (CNS) contributors in neuroglial survival.

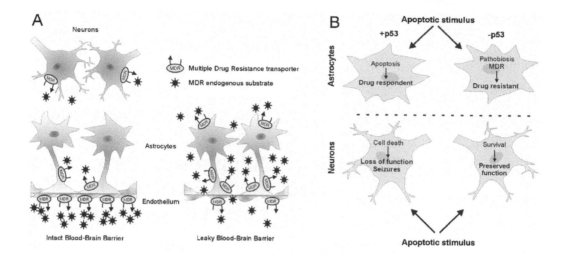

Figure 15.1 Possible mechanisms involved in neuroglial toxicity and survival. (A) An intact blood–brain barrier prevents potential toxic agents from entering the CNS compartment (left), while a leaky barrier can lead to overexpression of multidrug-resistant proteins by astrocytes, in response to a potential toxic stimulus (right). (B) Cells lacking functional p53 have a different response to apoptotic stimuli (glutamate, toxic agents, seizures, traumas) compared with their normal counterpart, possibly leading to multidrug resistance.

into the brain. For compounds capable of this, such as glutamate, efflux mechanisms exist. Therefore, it is generally accepted that the BBB plays a crucial role in the determination of neurotoxicity and its prevention.[1,2]

There has been considerable progress made towards the understanding of the pathophysiology and mechanisms involved in blood–brain barrier permeability. In neurological disorders affecting the brain, the cerebral endothelium plays a crucial etiologic role when the BBB becomes disrupted or modified in a way that increases the vascular permeability.[1] Various ways exist in which several molecules are able to pass the endothelium, including intercellular routes or direct transcellular penetration through a damaged endothelium. Dysfunction of the BBB may be caused by or a consequence of a particular disease process including but not limited to, neoplasia, ischemia, hypertension, dementia, epilepsy, infection, and trauma. It is questionable whether or not the BBB disturbance constitutes the main pathogenic factor that could elicit a sequence of events forming the final pathological state.[1]

15.2 ASTROGLIA AND ITS ROLE IN NEUROTOXICITY

Glial cells comprise a heterogeneous population of cells and are the predominant cell type found in the adult human brain. However, unlike the majority of other cell types, many astrocytes never fully exit the cell cycle in the adult CNS; consequently, cell contact may negatively regulate their proliferation. The classic view of glia was that they had a passive role in brain activity, serving as metabolic and support cells for neurons. In the past several years, it has been established that they take part in a more dynamic role in central nervous system function than previously believed by having a significant role in blood–brain barrier formation by releasing soluble factors to induce the BBB phenotype and other cellular properties in endothelial cells. Glial cells provide extensive structural and physiological support to neurons by phagocytizing neuronal debris, maintaining the extracellular levels of ions, regulating neurotransmitter uptake, and controlling synaptogenesis, synapse number, function, and plasticity.[3–8] Additionally, it has been found that glial cells orchestrate responses to cerebral inflammatory diseases and other CNS injuries.[9] It may be noted that glia are also involved in almost all aspects of brain function and development.[10]

One of the functions attributed to astrocytes is the buffering of extracellular [K+], which appears to be of critical importance to ensure normal neuronal excitability.[11] It has been demonstrated that astrocytes are liberally endowed with voltage-activated K channels,[12] which mediate the diffusional uptake of K+. Inwardly rectifying K (K_{IR}) channels in particular play an important role in K+ buffering by astrocytes.[11] While the biochemical and antigenic changes that occur in conjunction with reactive gliosis have been well studied,[13] the changes in biophysical properties of reactive astrocytes are just being revealed. A decrease in K_{IR} current amplitudes has been reported after injury of cultured astrocytes[14] and after a posttraumatic injury *in vivo*,[11] leading to an impaired K+ homeostasis. Intercellular coupling by gap junctions, another important aspect of potassium buffering,[15] is also affected after injury.[16,17] Finally, it has been repeatedly shown that reactive gliotic changes include attempted reentry into cell cycle.[14,18,19]

In a model of reactive gliosis, MacFarlane and Sontheimer found that proliferative astrocytes were concentrated in specific regions surrounding the lesion, where these glia lacked expression of K_{IR} channels and were not coupled by gap junctions as judged from dye injections.[14] By contrast, astrocytes in the hyperexcitable zone were not proliferating and displayed increased intercellular coupling associated with expression of K_{IR}. These findings emphasize the need for accurate localization of electrical activity in tissue resections from human brain and underscore the fact that astrocytes form a pathologically heterogeneous population of cells.

These results demonstrate the complexity of toxic events in the CNS. In fact, as in other organs, xenobiotics may be chiefly involved, with or without the intervention of the immune system. In the CNS, local endogenous signals such as K+ or glutamate may be sufficient to warrant specific mechanisms of clearance or buffering to prevent neurotoxic events. At the cellular level, the neurovascular unit composed of astrocytes and the BBB is clearly relevant. However, it is at the

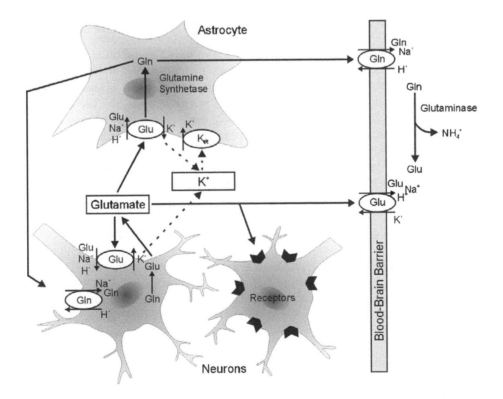

Figure 15.2 Molecular mechanisms involved in glutamate and K+ regulation and clearance at the CNS and blood–brain barrier level. Environments such as the epileptic brain can present alterations of these mechanisms in one or both compartments.

molecular level that these contributors assist. As shown in Figure 15.2, some of these molecular mechanisms are expressed by both cell types, while others are not. Alterations in timing of these mechanisms are probably important, but they become critical in situations where cell cycle and apoptosis are more significant than in normal adult CNS, such as epilepsy.

15.3 p53 AND p21

p53 is a nuclear phosphoprotein and is the most frequently mutated gene in human cancers. This protein likely exists as a tetramer, and when associated with a mutant form, the tetramer takes on the mutant confirmation. Wild-type p53 is present at low levels in all normal cells to restrain cell growth; however, it was originally classified as an oncogene because its mutations took an oncogenic phenotype.

Wild type p53 is now known as a tumor suppressor protein because the missense mutations serve an inhibitory function.[20] The absence of p53 or a nonfunctional p53 is common in human tumors and may contribute to the progression of other tumor types with a lack of cell specificity. p53 may also be silent, primarily due to deletions of both wild-type alleles or by a dominant mutation in one allele.[20] Wild-type p53 may also be silent in many malignant and premalignant tissues as a result of abnormal sequestration in the cytoplasm where it is functionally muted. Although present in constitutive levels in all cells, the activity level of p53 increases at the site of DNA damage and therefore activates one of two known pathways. Early in the cell cycle, p53 activates a checkpoint that prevents further progression into the cell cycle until damage has been repaired. On the other hand, if the cell has entered the S-phase of the cell cycle and is prepared to divide, programmed cell death occurs, preventing proliferation of damaged cells. In the brain, loss of p53 influences the cell cycle and therefore allows the survival of potentially damaged glial cells to proliferate.[21–25]

p21 is a 21-kDa protein and is a found downstream of p53. The discovery of p21 came about as a senescent cell-derived inhibitor of DNA replication in diploid human fibroblasts.[26] The promoter region includes recognition sequences for the TP53 protein, and thus, p21 is regulated by the levels of WT p53 present in the cell. Mutant p53, however, is unable to do so. The p21 protein is a universal inhibitor of cyclin dependent kinases (cdk) and binds to an extensive array of cyclin/CDK complexes, including but not limited to, cdk4,6-cyclin D and cdk2-cyclin E, and G_1 cyclins that are complexed to CDK2, which are most pertinent to G_1/S-phase arrest mediated by TP53.[20,27] In cells that have been perpetuated successfully in culture, cdk-cylin complexes tend to lack p21, suggesting that p21 may be implicated in the control of G_1/S-phase.[20]

15.4 MULTIPLE DRUG RESISTANCE PROTEIN

The multiple drug resistance protein (MDR1) is a 170-kDa protein, encoding a plasma membrane glycoprotein, P-glycoprotein (P-gp). Both are members of the ATP-binding cassette (ABC) superfamily of membrane transporters and are energy-dependent multidrug efflux pumps conferring resistance to chemotherapeutic drugs. As a result of P-gp activity, a reduced accumulation of intracellular drug levels ensues.[28,29] Consequently, a poor response to chemotherapeutic drugs arises. However, MDR is commonly expressed in tissues with excretory or absorbing functions, such as kidney, liver, and intestine.[28,30] MDR1 is also involved in blood–brain barrier function, but its expression levels are increased in a variety of neurological diseases (Figure 15.1A).

Our results suggested that the presence of MDR1 serves as a neuroglial protectant. If drugs produce stress in the cell as demonstrated by Aghdasi et al.,[31,32] in the presence of MDR1, drugs would be actively extruded, preventing any type of stress on the glial cells, therefore exhibiting healthy phenotypes. However, with a lack of MDR1, drugs are able to penetrate the cell, causing stress and death to the cells, possibly by mechanisms involving p53. In the event that there is no MDR1 and no p53, or if p53 is present in a mutated form, the cell suffers stress from the drug and is unable to undergo apoptosis leading to abnormal growth and tumorigenesis. The main aspect of the yet poorly understood interactions between drugs, intrinsic toxicity, MDR1, and apoptosis is summarized in the next section.

15.5 MULTIPLE DRUG RESISTANCE GENE AND PROTEIN EXPRESSION IN HUMAN EPILEPTIC BRAIN

Drug resistance in neurological disorders may be due to an alteration of the pharmacokinetic or pharmacodynamic mechanisms of central nervous system drug penetration.[29] The multiple drug resistance protein is often overexpressed in glia and in blood–brain barrier endothelium in patients who fail CNS drug treatments. The contribution of each of these cell types to multiple drug resistance is still unclear.

One of the limitations of basic science studies on mechanisms of human epileptogenesis is the fact that almost invariably these are performed on tissue resected from multiple drug resistant epileptics. The possibility that intrinsic molecular differences exist between MDR1-expressing epileptic brain and tissue from drug-respondent patients has only recently been explored.[29] Studies have shown that glial MDR1 expression is limited to gliotic regions, while endothelial cells overexpress MDR1 throughout the epileptic brain.[29,33] The preferential expression for astrocytic MDR1 in dysplastic or otherwise mainly abnormal regions of the resected tissue further calls for intraoperative or radiologic characterization of the tissue for accurate data analysis.

Several lines of evidence support the hypothesis that astrocytes play a role in epileptogenesis, and in particular that cortical astrocytes are in part responsible for development of multiple drug resistance in epilepsy.[34–42] Abnormal levels of GFAP and S100β are expressed in epileptic brain,[29,43,44] but abnormal glial gene expression and morphology involve only a subpopulation of glia, usually localized in island-like structures at the border of gray and white matter.[29,37] Interestingly, these regions colocalize with intense staining for multiple drug resistance protein.[29,37] It has recently been discovered that expression of mutated p53 or absence of p53 and p21 proteins is also a hallmark of multiple-drug-resistant epileptic brain.[29,37] Incidentally, abnormal levels of p53 are linked to expression of MDR1 in a variety of tumors.[21–25] This suggests a significant overlap between abnormal expression of p53/MDR1 in tumors and multiple-drug-resistant epileptic glia. A complex scenario can thus be outlined where yet unknown noxious events, perhaps triggered by seizures, lead to apoptotic cell death. In brain regions where p53 function is lost, survival of potentially damaged glial cells is possible. These abnormal astrocytes express MDR1 and may lead to a drug-resistant phenotype. Loss of p53 influence on cell cycle and apoptosis will result in astrocytic proliferation. The relationship between MDR1 and p53 is currently being investigated.

Although much data have been acquired regarding the p53-MDR relationship in human cancers, their correlation in human brain remains unclear. Future experiments using molecular technology in cell lines will elucidate some of these mechanisms. These findings may then be applied to specific CNS disease models in an attempt to unveil a possible role in pathogenesis. Numerous studies indicate that both multidrug transporters and p53 play critical roles in chemoresistance.[45] Initial reports demonstrated a transcriptional dependence of the MDR promoter by p53.[45] Early studies suggested that the ABCB1 promoter was downregulated by WT p53, while MT p53 upregulated the expression of ABCB1.[45,46]

15.6 p53, MDR, AND EPILEPSY

The molecular pathways that regulate cell death after seizure remain somewhat unclear compared with other types of injury in the brain, yet no systematic studies on p53 expression in human epileptic brain are available. The tumor suppressor p53 was among the first apoptosis-regulating elements to be identified as affected by seizure activity[47,48] and continues to be among the most consistently activated cell death pathways in the setting of seizure-induced neuronal death.[49,50] Activation of the p53 stress response pathway has been implicated in excitotoxic neuronal cell death, and induction of both p53 mRNA and protein in the rat brain following lithium-pilocarpine-mediated status epilepticus has been demonstrated.[49] Other proteins that have been shown to participate in the p53 cascade are also induced by seizures, including Mdm2, BAX, CD95/Fas/APO-1, ATM, Ref-1, and ubiquitin. Although nuclear Mdm2 was found to be increased, free ubiquitin expression was lower in cells with p53 and Mdm2 accumulation. Increased immunoreactivity for CD95/Fas/APO-1 and Bax was also detected in the same p53-positive cells. Moreover, expression of Ref-1 and ATM, which are involved in the response to oxidative stress-induced DNA damage and regulation of p53 function, were similarly increased. These results indicate that p53 induction following seizures in an animal model may activate downstream proapoptotic genes, leading to neurodegeneration.

Seizures *per se* do not elicit a consistently homogenous, model-independent cell death phenotype. Animal studies generally support necrosis as the principal morphological phenotype of dying cells after seizures.[51-54] True apoptotic morphology may compose only a small percentage of seizure-induced cell death. The exception may be within the dentate gyrus after seizures and the setting of kindling in which apoptosis may contribute substantially.[55] In animal models, seizure-damaged neurons upregulate p53 at both mRNA and protein level, while functional evidence for p53 comes from the findings that p53 DNA binding is promoted after seizures[56]; Bax, a key transcriptional target, is upregulated by seizures[44,57,58]; a synthetic p53 inhibitor protects against kainic acid excitotoxicity[59]; and p53-deficient mouse neurons are resistant to seizure and excitotoxin-induced apoptosis.[60-62] The targets and consequences of the engagement of p53 in seizure-induced neuronal death remain unclear, however, not least because of the multitude of pathways in which p53 may function.[60,61] Nevertheless, animal models suggest that p53 remains a key candidate mediator of programmed cell death after seizures. However, apoptotic cell death is not a common event in human epileptic brain, perhaps due to the fact that in human epileptic brain p53 is either mutated or absent in the astrocytic lineage.

15.7 p53 AND BRAIN TUMORS

The loss of the tumor suppressor gene p53 is the earliest and most common genetic lesion in astrocytic tumors.[63] Thus, 40–50% of all gliomas have compromised p53 function. The functional significance of p53 loss in nontumor glia is less understood. The relevance of p53 for astrocytic pathophysiology is probably best documented by the studies performed by Cavenee and colleagues.[21,22,64-66] In brief, their results can be summarized as follows:

1. Astrocytes isolated from p53$^{-/-}$ mice grow rapidly, lack contact inhibition, and are immortal.
2. After several passages in culture, p53$^{-/-}$ astrocytes form colonies in soft agarose.
3. Loss of p53 *per se* is not sufficient to promote malignant transformation of these glia, and other factors are needed for malignant progression.
4. Environmental cues present *in vitro* cause malignant transformation of p53 null astrocytes. These include EGF and BFGF.

Historically, it has always been assumed that brain tumors are a frequent cause of epileptic seizures. Seizures occur in 50% of patients with intracranial brain tumors (BT) and AED therapy is prophylactically administered to most brain tumor patients. In addition, chronic epilepsy can be the only symptom of low-grade brain tumors.[67-69] There is a significant overlap between genes that are associated with genetic forms of epilepsy and alterations in tumor suppressor genes. For example, the tuberous sclerosis complex–associated genes TSC1-TSC2 are linked to a phenotype characterized by an increased number of astrocytes and a propensity for tumor growth. Clinically, the majority of these patients are affected by seizures.[70] Similar considerations can be made for neurofibromatosis, associated with mutations of NF1 and NF2, two genes also involved in cell cycle control and tumor suppression. Finally, the benign brain tumors known as gangliogliomas are almost always (90% of cases) associated with epileptic seizures. It is not clear, however, whether a cause-effect relationship exists between epileptogenesis and tumorigenesis.

While it is widely believed that brain tumors lead to seizures, the possibility that brain tumors can be caused by epilepsy has not been investigated, nor is it known whether seizures have any effect on tumor growth. It has been suggested that chronic epilepsy prolongs survival of BT patients. In fact, several patients with low-grade gliomas that present with seizures are sometimes placed on a "wait-and-see" surgical list, as these tumors may remain stable for prolonged periods without treatment.[71] Although this practice exists, the risk of further progression of a low-grade tumor into a malignant one warrants surgical resection. However, a link between preexisting drug-resistant seizures and reduced rate of growth of WHO grade I–III tumors suggested that epileptic brain is an unfavorable environment for tumor development.[72] This study demonstrated that patients with brain tumors and intractable epilepsy had a 10-year survival rate of 90% compared with other seizure-free brain tumors, which had a 5-year survival rate.

15.8 ANTIEPILEPTIC DRUGS AND APOPTOSIS

Antiepileptic drugs (AED) are used for the prevention and treatment of seizures. As described by many, they act through three distinct mechanisms. Neuronal firing is limited by the blockade of voltage-dependent sodium channels, by a blockade of glutamatergic excitatory neurotransmission, or through the enhancement of γ-aminobutyric acid (GABA)–mediated transmission. There are numerous neurotoxic effects of these types of drugs, yet the underlying mechanisms are still unclear.[73] It has been demonstrated by many that the activation of GABA A receptors or the suppression of synaptic neurotransmission may lead to apoptotic neurodegeneration.[73–75] In a recent study, it was shown that sensitive neurons underwent apoptotic cell death in the developing forebrain of a rat and that this death increased when introduced to a combination of AEDs at plasma concentrations used for seizure control in humans. Although this was reported in neurons of the rodent, it has been suggested that AEDs induce functional modifications and may affect the biochemical pathways of cortical rat astrocytes in a dose-dependent fashion.[32]

15.9 CURRENT QUESTIONS

1. Based on our current knowledge of the role of glia in epileptogenesis and drug resistance, the question arises whether these are beneficial or detrimental for the patient. For example, is multiple drug resistance gene expression in glial cells a mechanism of neuroglial survival, or does it participate in the multiple-drug-resistant phenotype?
2. At what stage of development or epileptogenesis does glial MDR1 expression become important?
3. What are the interrelationships between p53 and MDR1 in the mammalian brain? How do they relate to specific disease phenotypes?
4. Is p53 expression altered in epileptic brain? If so, are these changes somatic or constitutive? Does altered expression of p53 imply a deletion or mutation? What cell types are involved?
5. Do these defects (p53 and MDR1) correlate topographically with the active region of the epileptic brain? Do these regions correlate with hypometabolism or other focal abnormalities?
6. Do antiepileptic drugs play a role in all these phenomena? Are MDR1 properties bestowed to glial cells by exposure to antiepileptic drugs? Do antiepileptic drugs participate to the apoptotic process in the epileptic brain?

15.10 CONCLUSIONS AND PERSPECTIVES

Given these premises, it appears that future studies will have to use appropriate models to develop further knowledge on the mechanisms of gliogenesis and maturation of glial cells in a variety of conditions. These experiments have been so far performed on tissue sections from human epileptic brain. This was due to the availability of these samples and to the fact that surgical interventions on epileptics encompass a broad range of ages (from neonatal to aging). Translation of this knowledge into animal models will be difficult, but it is nevertheless necessary to establish a clear, cause-effect relationship between, for example, p53 deficiency and MDR1 expression. Parallel studies using genetically modified cells or cell lines will also be of importance.

ACKNOWLEDGMENTS

This work was supported by NIH-2RO1 HL51614 and NIH-RO1 NS38195 to Damir Janigro.

REFERENCES

1. Grant, G.A. and Janigro, D., The blood-brain barrier, in *Youmans Neurological Surgery*, Vol. 1, Winn, H.R., Ed., Saunders, Philadelphia, PA, 2004, pp. 153–174.
2. Schinkel, A.H., P-Glycoprotein, a gatekeeper in the blood-brain barrier, *Adv. Drug Deliv. Rev.*, 36, 179–194, 1999.
3. Araque, A., Parpura, V., Sanzgiri, R.P., and Haydon, P.G., Tripartite synapses: glia, the unacknowledged partner, *Trends Neurosci.*, 22, 208–215, 1999.
4. Bezzi, P., Carmignoto, G., Pasti, L., Vesce, S., Rossi, D., Rizzini, B.L., Pozzan, T., and Volterra, A., Prostaglandins stimulate calcium-dependent glutamate release in astrocytes, *Nature*, 391, 281–285, 1998.
5. Parpura, V., Basarsky, T.A., Liu, F., Jeftinija, K., Jeftinija, S., and Haydon, P.G., Glutamate-mediated astrocyte-neuron signaling, *Nature*, 369, 744–747, 1994.
6. Chen, Z.L. and Strickland, S., Neuronal death in the hippocampus is promoted by plasmin-catalyzed degradation of laminin, *Cell*, 91, 917–925, 1997.
7. del Zoppo, G.J. and Mabuchi, T., Cerebral microvessel responses to focal ischemia, *J. Cereb. Blood Flow Metab.*, 23, 879–894, 2003.
8. Privat, A., Astrocytes as support for axonal regeneration in the central nervous system of mammals, *Glia*, 43, 91–93, 2003.
9. Descamps, L., Coisne, C., Dehouck, B., Cecchelli, R., and Torpier, G., Protective effect of glial cells against lipopolysaccharide-mediated blood-brain barrier injury, *Glia*, 42, 46–58, 2003.
10. Rakic, P., Adult neurogenesis in mammals: an identity crisis, *J. Neurosci.*, 22, 614–618, 2002.
11. D'Ambrosio, R., Maris, D.O., Grady, M.S., Winn, H.R., and Janigro, D., Impaired K homeostasis and altered electrophysiological properties of post-traumatic hippocampal glia, *J. Neurosci.*, 19, 8152–8162, 1999.
12. Barres, B.A., Glial ion channels, *Curr. Opin. Neurobiol.*, 1, 354–359, 1991.
13. Roitbak, T. and Syková, E., Diffusion barriers evoked in the rat cortex by reactive astrogliosis, *Glia*, 28, 40–48, 1999.
14. MacFarlane, S.N. and Sontheimer, H., Electrophysiological changes that accompany reactive gliosis *in vitro*, *J. Neurosci.*, 17, 7316–7329, 1997.
15. McKhann, G.M., D'Ambrosio, R., and Janigro, D., Heterogeneity of astrocyte resting membrane potentials and intercellular coupling revealed by whole-cell and gramicidin-perforated patch recordings from cultured neocortical and hippocampal slice astrocytes, *J. Neurosci.*, 17, 6850–6863, 1997.
16. Lee, S.H., Kim, W.T., Cornell Bell, A.H., and Sontheimer, H., Astrocytes exhibit regional specificity in gap-junction coupling, *Glia*, 11, 315–325, 1994.
17. Lee, S.H., Magge, S., Spencer, D.D., Sontheimer, H., and Cornell-Bell, A.H., Human epileptic astrocytes exhibit increased gap junction coupling, *Glia*, 15, 195–202, 1995.

18. Bordey, A. and Sontheimer, H., Postnatal development of ionic currents in rat hippocampal astrocytes *in situ*, *J. Neurophysiol.*, 78, 461–477, 1997.

19. MacFarlane, S.N. and Sontheimer, H., Changes in ion channel expression accompany cell cycle progression of spinal cord astrocytes, *Glia*, 30, 39–48, 2000.

20. Lewin, B., Oncogenes and cancer, in *Genes*, Vol. VII, Oxford University Press, Oxford, 2000, pp. 875–912.

21. Chevillard, S., Lebeau, J., Pouillart, P., de Toma, C., Beldjord, C., Asselain, B., Klijanienko, J., Fourquet, A., Magdelenat, H., and Vielh, P., Biological and clinical significance of concurrent p53 gene alterations, MDR1 gene expression, and S-phase fraction analyses in breast cancer patients treated with primary chemotherapy or radiotherapy, *Clin. Cancer Res.*, 3, 2471–2478, 1997.

22. Chin, K.V., Ueda, K., Pastan, I., and Gottesman, M.M., Modulation of activity of the promoter of the human MDR1 gene by Ras and p53, *Science*, 255, 459–462, 1992.

23. de Kant, E., Heide, I., Thiede, C., Herrmann, R., and Rochlitz, C.F., MDR1 expression correlates with mutant p53 expression in colorectal cancer metastases, *J. Cancer Res. Clin. Oncol.*, 122, 671–675, 1996.

24. Nguyen, K.T., Liu, B., Ueda, K., Gottesman, M.M., Pastan, I., and Chin, K.V., Transactivation of the human multidrug resistance (MDR1) gene promoter by p53 mutants, *Oncol. Res.*, 6, 71–77, 1994.

25. Thottassery, J.V., Zambetti, G.P., Arimori, K., Schuetz, E.G., and Schuetz, J.D., p53-dependent regulation of MDR1 gene expression causes selective resistance to chemotherapeutic agents, *Proc. Natl. Acad. Sci. U.S.A.*, 94, 11037–11042, 1997.

26. el Deiry, W.S., Tokino, T., Velculescu, V.E., Levy, D.B., Parsons, R., Trent, J.M., Lin, D., Mercer, W.E., Kinzler, K.W., and Vogelstein, B., WAF1, a potential mediator of p53 tumor suppression, *Cell*, 75, 817–825, 1993.

27. Kleihues, P., Louis, D.N., Scheithauer, B.W., Rorke, L.B., Reifenberger, G., Burger, P.C., and Cavenee, W.K., The WHO classification of tumors of the nervous system, *J. Neuropathol. Expo. Neuro.*, 61, 215–2, 2002.

28. Farrell, R.J., Menconi, M.J., Keates, A.C., and Kelly, C.P., P-glycoprotein-170 inhibition significantly reduces cortisol and ciclosporin efflux from human intestinal epithelial cells and T lymphocytes.

29. Marroni, M., Agarwal, M., Kight, K., Hallene, K., Hossain, M., Cucullo, L., Signorelli, K., Namura, S., and Janigro, D., Relationship between expression of multiple drug resistance proteins and p53 tumor suppressor gene proteins in human brain astrocytes, *Neuroscience*, 121, 605–617, 2003.

30. Fromm, M.F., P-glycoprotein: a defense mechanism limiting oral bioavailability and CNS accumulation of drugs, *Int. J. Clin. Pharmacol. Ther.*, 38, 69–74, 2000.

31. Aghdasi, B., Reid, M.B., and Hamilton, S.L., Nitric oxide protects the skeletal muscle Ca2+ release channel from oxidation induced activation, *J. Biol. Chem.*, 272, 25462–25467, 1997.

32. Pavone, A. and Cardile, V., An *in vitro* study of new antiepileptic drugs and astrocytes, *Epilepsia*, 44 suppl 10, 34–39, 2000.

33. Dombrowski, S., Desai, S., Marroni, M., Cucullo, L., Bingaman, W., Mayberg, M.R., Bengez, L., and Janigro, D., Overexpression of multiple drug resistance genes in endothelial cells from patients with refractory epilepsy, *Epilepsia*, 42, 1504–1507, 2001.

34. Aronica, E., Gorter, J.A., Jansen, G.H., Leenstra, S., Yankaya, B., and Troost, D., Expression of connexin 43 and connexin 32 gap-junction proteins in epilepsy-associated brain tumors and in the perilesional epileptic cortex, *Acta Neuropathol. (Berl.)*, 101, 449–459, 2001.

35. Emmi, A., Wenzel, H.J., Schwartzkroin, P.A., Taglialatela, M., Castaldo, P., Bianchi, L., Nerbonne, J., Robertson, G.A., and Janigro, D., Do glia have heart? Expression and functional role for ether-a-go-go currents in hippocampal astrocytes, *J. Neurosci.*, 20, 3915–3925, 2000.

36. Janigro, D., Gasparini, S., D'Ambrosio, R., McKhann, G., and DiFrancesco, D., Reduction of K+ uptake in glia prevents long-term depression maintenance and causes epileptiform activity, *J. Neurosci.*, 17, 2813–2824, 1997.

37. Marroni, M., Marchi, N., Cucullo, L., Abbott, N.J., Signorelli, K., and Janigro, D., Vascular and parenchymal mechanisms in multiple drug resistance: a lesson from human epilepsy, *Curr. Drug Targets*, 4, 297–304, 2003.

38. Tishler, D.M., Weinberg, K.I., Hinton, D.R., Barbaro, N., Annett, G.M., and Raffel, C., MDR1 gene expression in brain of patients with medically intractable epilepsy, *Epilepsia*, 36, 1–6, 1995.

39. Wolf, H.K., Wellmer, J., Muller, M.B., Wiestler, O.D., Hufnagel, A., and Pietsch, T., Glioneuronal malformative lesions and dysembryoplastic neuroepithelial tumors in patients with chronic pharma-coresistant epilepsies, *J. Neuropathol. Exp. Neurol.*, 54, 245–254, 1995.

40. Bordey, A. and Sontheimer, H., Properties of human glial cells associated with epileptic seizure foci, *Epilepsy Res.*, 32, 286–303, 1998.

41. Bordey, A. and Sontheimer, H., Ion channel expression by astrocytes *in situ*: comparison of different CNS regions, *Glia*, 30, 27–38, 2000.

42. Bordey, A., Lyons, S.A., Hablitz, J.J., and Sontheimer, H., Electrophysiological characteristics of reactive astrocytes in experimental cortical dysplasia, *J. Neurophysiol.*, 85, 1719–1731, 2001.

43. Griffin, W.S., Yeralan, O., Sheng, J.G., Boop, F.A., Mrak, R.E., Rovnaghi, C.R., Burnett, B.A., Feoktistova, A., and Van Eldik, L.J., Overexpression of the neurotrophic cytokine S100 beta in human temporal lobe epilepsy, *J. Neurochem.*, 65, 228–233, 1995.

44. Proper, E.A., Oestreicher, A.B., Jansen, G.H., Veelen, C.W., van Rijen, P.C., Gispen, W.H., and de Graan, P.N., Immunohistochemical characterization of mossy fibre sprouting in the hippocampus of patients with pharmaco-resistant temporal lobe epilepsy, *Brain*, 123, 19–30, 2000.

45. Bush, J.A. and Li, G., Cancer chemoresistance: the relationship between p53 and multidrug trans-porters, *Int. J. Cancer*, 98, 323–330, 2002.

46. Wang, Q. and Beck, W.T., Transcriptional suppression of multidrug resistance-associated protein (MRP) gene expression by wild-type p53, *Cancer Res.*, 58, 5762–5769, 1998.

47. Sakhi, S., Bruce, A., Sun, N., Tocco, G., Baudry, M., and Schreiber, S.S., p53 induction is associated with neuronal damage in the central nervous system, *Proc. Natl. Acad. Sci. U.S.A.*, 91, 7525–7529, 1994.

48. Sakhi, S., Sun, N., Wing, L.L., Mehta, P., and Schreiber, S.S., Nuclear accumulation of p53 protein following kainic acid-induced seizures, *Neuroreport*, 7, 493–496, 1996.

49. Tan, Z., Sankar, R., Tu, W., Shin, D., Liu, H., Wasterlain, C.G., and Schreiber, S.S., Immunohis-tochemical study of p53-associated proteins in rat brain following lithium-pilocarpine status epilep-ticus, *Brain Res.*, 929, 129–138, 2002.

50. Tan, Z., Sankar, R., Shin, D., Sun, N., Liu, H., Wasterlain, C.G., and Schreiber, S.S., Differential induction of p53 in immature and adult rat brain following lithium-pilocarpine status epilepticus, *Brain Res.*, 928, 187–193, 2002.

51. Puig, B. and Ferrer, I., Caspase-3-associated apoptotic cell death in excitotoxic necrosis of the entorhinal cortex following intraperitoneal injection of kainic acid in the rat, *Neurosci. Lett.*, 321, 182–186, 2002.

52. Kubova, H., Druga, R., Lukasiuk, K., Suchomelova, L., Haugvicova, R., Jirmanova, I., and Pitkanen, A., Status epilepticus causes necrotic damage in the mediodorsal nucleus of the thalamus in immature rats, *J. Neurosci.*, 21, 3593–3599, 2001.

53. Fujikawa, D.G., Shinmei, S.S., and Cai, B., Seizure-induced neuronal necrosis: implications for programmed cell death mechanisms, *Epilepsia*, 41 (Suppl. 6), S9–S13, 2000.

54. Ebert, U., Brandt, C., and Loscher, W., Delayed sclerosis, neuroprotection, and limbic epileptogenesis after status epilepticus in the rat, *Epilepsia*, 43 (Suppl. 5), 86–95, 2002.

55. Sloviter, R.S., Status epilepticus-induced neuronal injury and network reorganization, *Epilepsia*, 40 (Suppl. 1), S34–S39, 1999.

56. Liu, W., Rong, Y., Baudry, M., and Schreiber, S.S., Status epilepticus induces p53 sequence-specific DNA binding in mature rat brain, *Brain Res. Mol. Brain Res.*, 63, 248–253, 1999.

57. Lopez, E., Pozas, E., Rivera, R., and Ferrer, I., Bcl-2 and Bax expression following methylazoxymeth-anol acetate-induced apoptosis in the external granule cell layer of the developing rat cerebellum, *Brain Res. Dev. Brain Res.*, 112, 149–153, 1999.

58. Gillardon, F., Wickert, H., and Zimmermann, M., Up-regulation of bax and down-regulation of bcl-2 is associated with kainate-induced apoptosis in mouse brain, *Neurosci. Lett.*, 192, 85–88, 1995.

59. Culmsee, C., Zhu, X., Yu, Q.S., Chan, S.L., Camandola, S., Guo, Z., Greig, N.H., and Mattson, M.P., A synthetic inhibitor of p53 protects neurons against death induced by ischemic and excitotoxic insults, and amyloid beta-peptide, *J. Neurochem.*, 77, 220–228, 2001.

60. Morrison, R.S., Kinoshita, Y., Johnson, M.D., Guo, W., and Garden, G.A., p53-dependent cell death signaling in neurons, *Neurochem. Res.*, 28, 15–27, 2003.

61. Morrison, R.S. and Kinoshita, Y., The role of p53 in neuronal cell death, *Cell Death Differ.*, 7, 868–879, 2000.

62. Xiang, H., Hochman, D.W., Saya, H., Fujiwara, T., Schwartzkroin, P.A., and Morrison, R.S., Evidence for p53-mediated modulation of neuronal viability, *J. Neurosci.*, 16, 6753–6765, 1996.

63. Vogelstein, B. and Kinzler, K.W., p53 function and dysfunction, *Cell*, 70, 523–526, 1992.

64. Bogler, O., Nagane, M., Gillis, J., Huang, H.J., and Cavenee, W.K., Malignant transformation of p53-deficient astrocytes is modulated by environmental cues *in vitro*, *Cell Growth Differ.*, 10, 73–86, 1999.

65. Bogler, O., Huang, H.J., Kleihues, P., and Cavenee, W.K., The p53 gene and its role in human brain tumors, *Glia*, 15, 308–327, 1995.

66. Bogler, O., Huang, H.J., and Cavenee, W.K., Loss of wild-type p53 bestows a growth advantage on primary cortical astrocytes and facilitates their *in vitro* transformation, *Cancer Res.*, 55, 2746–2751, 1995.

67. Furnari, F.B., Huang, H.J., and Cavenee, W.K., Genetics and malignant progression of human brain tumours, *Cancer Surv.*, 25, 233–275, 1995.

68. Kleihues, P., Cavence, W. K., eds., *Pathology and Genetics of the Nervous System*, IARC Press, Lyon, France, 2000.

69. Maher, E.A., Furnari, F.B., Bachoo, R.M., Rowitch, D.H., Louis, D.N., Cavenee, W.K., and DePinho, R.A., Malignant glioma: genetics and biology of a grave matter, *Genes Dev.*, 15, 1311–1333, 2001.

70. Munari, C., Talairach, J., Bancaud, J., Rovei, V., Sanjuan, E., Peschanski, M., and Morselli, P.L., Brain levels of antiepileptic drugs in man, *Monogr. Neural Sci.*, 5, 213–220, 1980.

71. Recht, L.D., Lew, R., and Smith, T.W., Suspected low-grade glioma: is deferring treatment safe?, *Ann. Neurol.*, 31, 431–436, 1992.

72. Luyken, C., Blumcke, I., Fimmers, R., Urbach, H., Elger, C.E., Wiestler, O.D., and Schramm, J., The spectrum of long-term epilepsy-associated tumors: long-term seizure and tumor outcome and neurosurgical aspects, *Epilepsia*, 44, 822–830, 2003.

73. Bittigau, P., Sifringer, M., Genz, K., Reith, E., Pospischil, D., Govindarajalu, S., Dzietko, M., Pesditschek, S., Mai, I., Dikranian, K., Olney, J.W., and Ikonomidou, C., Antiepileptic drugs and apoptotic neurodegeneration in the developing brain, *Proc. Natl. Acad. Sci. U.S.A.*, 99, 15089–15094, 2002.

74. Ikonomidou, C., Stefovska, V., and Turski, L., Neuronal death enhanced by N-methyl-D-aspartate antagonists, *Proc. Natl. Acad. Sci. U.S.A.*, 97, 128855–12890, 2000.

75. Ikonomidou, C., Bosch, F., Miksa, M., Bittigau, P., Vockler, J., Dikranian, K., Tenkova, T.I., Stefovska, V., Turski, L., and Olney, J.W., Blockade of NMDA receptors and apoptotic neurodegeneration in the developing brain, *Science*, 283, 70–74, 1999.

CHAPTER **16**

Pathogenesis of HIV-Associated Dementia and Multiple Sclerosis: Role of Microglia and Astrocytes

Alireza Minagar, Paul Shapshak, and J. Steven Alexander

CONTENTS

16.1 INTRODUCTION

The central nervous system (CNS) has long been viewed as an immunologically privileged and isolated organ beyond the reach of immune system and its components. This immunoseparation is established and maintained through the presence of the unique structure of the blood–brain barrier and CNS immunosuppressive milieu. However, during pathogenesis of HIV-1–associated dementia and multiple sclerosis, both immune-mediated diseases, macrophage/microglia (MØ), and astrocytes participate in destruction, protection, and attempted repair of the CNS. Complicated pathogenesis of these two dementing and disabling diseases involves activation and interaction of MØ, astrocytes, and Th1- and Th2-T lymphocytes. Expression of MHC and adhesion molecules and release of various reactive oxygen/nitrogen intermediates, quinolinic acid, chemokines, cytokines, and other components of inflammation are the consequences of these inflammatory cascades. The role of MØ and astrocytes in cellular/molecular mechanisms of pathogenesis HAD and MS are examined and reviewed. HIV-1 infection is the original insult in HAD, and macrophages are involved in CNS invasion by HIV-1. The etiology of MS remains unknown, and MØ are involved in loss of myelin/oligodendrocyte complex. Astrocytes are also activated in pathogenesis of HAD and MS. In both diseases, cytokine/chemokine communication between microglia, astrocytes, and Th1- and Th2-T lymphocytes occurs and leads to both destruction and attempted repair of the CNS.

16.2 MACROPHAGE/MICROGLIA (MØ)

Macrophage/microglial (MØ) cells are the resident macrophage cell population in the nervous system and represent the primary immunocompetent cells. These cells are involved in mounting an immune reaction against invading infectious agents and tumors and are involved in phagocytosis of cellular debris and antigen presentation, as well as generation and secretion of cytokines, eicosanoids, complement components, and excitatory amino acids including glutamate, oxidative radicals, and nitric oxide (NO).[1–3] MØ represent 10–20% of the glial cell population in the brain. MØ are particularly sensitive to alterations in their microenvironment and readily become activated in response to infection or injury. In their "resting" state, MØ show a ramified configuration with extended pseudopodia with reduced genomic activity, whereas upon activation, MØ cells down-regulate surface-bound keratin sulfate proteoglycans and transform into an amoeboid shape. Activated phagocytic MØ show upregulation of gene expression and generate a large array of potentially neurotoxic mediators. *In vivo*, MØ can express class I and II major histocompatibility (MHC) antigens, costimulatory B7-1 (CD80) and B7-2 (CD86) molecules, Fc receptors (I-III), intercellular adhesion molecule-1 (ICAM-1/CD54), and other surface markers including CD68, CD45, CD14, CD11c (CR4), and CD11b (CR3/MAC1)[4–13] as well as *in vitro* culture.[14]

Astrocytes, the most numerous glial cell type, outnumber neurons 10 to 1 and have significant roles in the development and support of neurons, repair of damaged neurons, formation and maintenance of the blood–brain barrier, neurotransmitter uptake, and maintenance of ion and metabolite homeostasis within the brain. Morphologically, astrocytes are divided into two large groups, protoplasmic and fibrous. Protoplasmic astrocytes are found primarily in gray matter, while fibrous astrocytes are located primarily in the white matter. Astrocytes express glial fibrillary acidic protein (GFAP), which makes these cells unique within the CNS. Cultured astrocytes also express various neurotrophic molecules including nerve growth factor (NGF), glial-derived growth factor (GDFD), and ciliary neurotrophic protein (CNTF)[15] as well as class I and II MHC molecules in response to activation by TNF-α or IFN-.[16] Activated astrocytes also generate NO, which can damage oligodendrocytes.

Interaction of MØ and astrocytes, both of which play dual destructive and protective roles in pathogenesis of HIV-1 associated dementia (HAD) and multiple sclerosis (MS), has major implications in the development of these neurological diseases.[17] The authors review roles of MØ and astrocytes, their interactions, and their effects on the balance of Th1- and Th2-cytokines in the context of pathogenesis of HIV-1–associated dementia.

16.3 HIV-1-ASSOCIATED DEMENTIA

Direct infection of the brain by human immunodeficiency virus type 1 (HIV-1) occurs in the early stages of HIV-1 infection,[18,19] and the virus evolves separately in different brain regions.[20,21] It has been hypothesized that HIV-1 mainly enters the CNS from peripheral circulation by ingress of infected macrophages and that this occurs throughout the course of disease.[22,23]

The spectrum of clinical manifestation of HIV-1 infection of brain consists of HAD, HIV-1–associated minor cognitive/motor disorder, and other neuropsychiatric syndromes.[20,24–26] The term HIV-1–associated dementia applies to a characteristic progressive encephalopathy, which results from HIV-1–related impairment of normal brain function in the absence of detectable primary influences from other pathogens or conditions. HAD is the leading cause of dementia in individuals aged 20 to 59 years.[27] HAD is a subcortical dementia in which the usual cortical features such as aphasia, alexia, and agraphia are absent. These cortical features occur when HAD has advanced with more global damage to the brain. HAD is characterized by progressive motor abnormalities (tremor, gait ataxia, and loss of fine motor movements), cognitive disorders (mental slowing, forgetfulness, and poor concentration), and behavioral abnormalities (including mania at the outset, apathy, and emotional lability).[25,28–33]

Direct infection of the brain by HIV-1 eventually leads to HAD, which prior to introduction of antiretroviral therapy affected 20% of patients.[34] However, with introduction of highly active antiretroviral therapy (HAART), the incidence of HAD has been reduced, whereas prevalence is increasing since patients with HIV live longer.[35–37] Prior to introduction of HAART, up to half of virus-infected individuals showed neuropathological changes at autopsy, and one-fourth had clinical presentations such as behavioral, cognitive, and motor abnormalities, ranging from mild motor/cognitive deficits to overt HAD.[38,39] The incidence rates of HAD in a cohort of 2734 HIV-1–seropositive homosexual men declined significantly by 53% from 21.3 per 1000 person-years from 1990 to 1992, to 10.0 per 1000 person-years from 1996 to 1998, reflecting the impact of HAART.[40,41]

Cerebral atrophy is a common finding in patients with HAD and often occurs with a fronto-temporal distribution. Neuropathological features of HIV infection of brain consist of HIV encephalitis (HIVE) consisting of HIV infection of the brain and demonstrated as multiple small nodules containing macrophages, lymphocytes, and microglia (also known as microglial nodules), which are often disseminated in the brain. Frequently, pathology is present more in white matter than in subcortical gray matter in regions such as the basal ganglia, thalamus, and brainstem.[30,31,42] HIV encephalitis results from fusion of the infected microglial cells, macrophages, or microglial nodules. Other significant neuropathological changes of HAD include dendritic loss, vacuolization, and neuronal loss.[43]

16.4 MØ AND ASTROCYTE ACTIVATION IN HAD

Application of techniques including immunocytochemistry, *in situ* hybridization, and *in situ* PCR combined with *in situ* hybridization to HIV-1–infected brains has shown that MØ are the primary cell types infected with HIV-1 and capable of HIV-1 production; MØ are thus an HIV-1 reservoir[30,31,42,44–46] (reviewed by Minagar et al.[17]). The susceptibility of MØ to infection with HIV-1 is related to expression of specific external surface proteins. HIV-1 in brain environment recognizes CD4 as its primary receptor and chemokine receptors including CCR-5 as the coreceptor (reviewed by Shapshak et al.[47]). These surface antigens are expressed profusely by macrophage/microglia. MØ are typical targets of HIV-1 infection, and it has been hypothesized that HIV-1 gains access to the CNS mainly through infected MØ destined to become brain-resident cells or perivascular macrophages.[48] The significance of MØ in pathogenesis of HAD is further understood by emergence of specific monocyte subpopulations in the peripheral circulation of patients with HAD. These specific macrophage/microglia cells express CD14/CD16 and CD14/CD69 surface markers and show increase capability to migrate and secrete neurotoxins.[49–52] The development of such subpopulations of macrophage cells precedes development of HAD. Activation of the immune system in HIV patients is also associated with increased transendothelial infiltration of activated macrophages through a failing blood–brain barrier (BBB) into brain and secretion of a number of viral and immune factors that alter neuronal function.[23,53] Once activated macrophages are in the brain, they become perivascular macrophages, which explains why HIV-1 expression and multinucleated giant cells are often perivascular in location.[22,23,45] MØ move within the ventricular system, within the white matter pathways, and across the corpus callosum and cause massive neuronal destruction. A subset of macrophages/monocytes (CD14+/CD16+) antigen positive emerge during advanced HIV-1 infection and correlates with cognitive impairment.[54] In addition, Luo et al.,[55] in a proteomic study of macrophages extracted from HIV-1–infected patients with cognitive impairment, identified

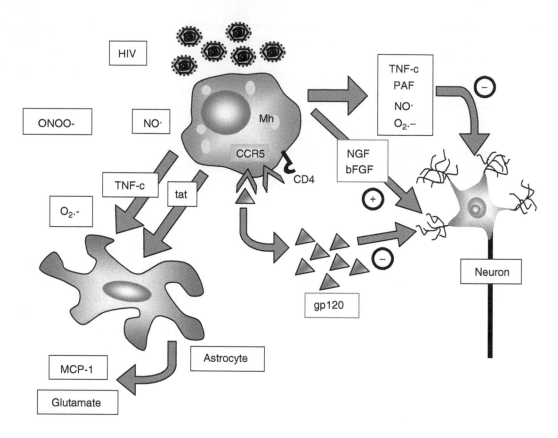

Figure 16.1 HIV-1–infected MØ release tat and TNF-α, which will affect astrocytes and cause release of monocytes chemoattractant protein (MCP-1) and glutamate by these cells. Activated astrocytes also release superoxide anions, which will react with NO released from MØ and generate peroxynitrite.

unique phenomic macrophages in these patients, and their findings supported the concept that alteration in monocyte function parallels the development of HAD. Prior studies demonstrated that virus-infected MØ secrete or induce other neural cells to generate neurotoxic factors that lead to neuronal loss in the context of HIV encephalitis. Some of these neurotoxins include TNF-α, platelet-activating factor (PAF),[56] nitric oxide (NO),[57] quinolinic acid (QUIN),[58–62] and glutamate.[63] In addition, HIV proteins such as gp120,[64] gp41,[62] and Tat,[65,66] secreted by infected brain MØ, can cause neuronal loss or dysfunction (Figure 16.1).

Astrocytes, another major participant in the pathogenesis of HIV-1 infection of brain, are not infected productively with HIV-1. In AIDS patients, astrocytes also reveal evidence of HIV multispliced transcripts related to rev, nef, and tat control proteins rather than unspliced mRNAs related to structural proteins such as envelope and gag.[67–70] HIV-1 infection in these cells is defective due to predominance of multispliced mRNA and insufficient production of HIV-1 structural protein.[70–72] Another possible obstacle to HIV-1 infection of astrocytes may be at virus entry in that astrocytes do not possess intrinsic intracellular support for efficient HIV-1 replication.[73] Infection of astrocytes in cell cultures with HIV-1 results in an initially productive and noncytopathic infection that diminishes to a restricted viral latent phase.[74] In addition, intracellular HIV tat expression in astrocytes promotes astrocyte survival and ensures its generation but induces potent neurotoxicity at distant sites via axonal transport.[75]

TNF-α upregulates the generation of HIV-1 in infected cells[76,77] and induces astrocytes to secrete other inflammatory cytokines.[7] TNF-α is toxic to cultured oligodendrocytes and myelin[78,79] and alters the function of cultured neurons.[80,81] In the pathogenesis of HIV-1 infection of the brain, TNF-α is produced by both MØ and astrocytes; however, on the brains of AIDS patients dying of AIDS, MØ are the dominant cell type producing TNF-α. Elevated levels of mRNA for TNF-α in brains of adult demented AIDS patients compared with those AIDS patients without dementia have been shown, which emphasizes the role of this proinflammatory cytokine in pathogenesis of HAD.[82–84]

PAF is produced in HIV-infected monocyte and astrocyte cultures.[56] Elevated levels of PAF have been reported in patients with AIDS and HAD, as well as in the patients with neoplastic and metabolic diseases.[56] This molecule, through an NMDA-mediated process, is neurotoxic to human fetal cortical cultures and rat retinal cultures.[85]

NO and its intermediates have been implicated in excitotoxic neuronal injury in patients with HIV-1 infection.[86,87] Elevated levels of iNOS along with elevated levels of HIV gp24 have been shown in brains of patients dying of AIDS.[88] In addition, levels of iNOS have been correlated with severity of dementia.[89] A role for HIV-1 tat in inducing the expression of iNOS in human astrocytes and pathogenesis of HAD has been reported.[90] Interestingly, a protective role for NO—directly, indirectly, or both—decreasing or blocking HIV-1 replication through inhibition of viral enzymes such as reverse transcriptase, protease, or cellular nuclear transcription factor (NFkappa B) and long-terminal repeat-driven transcription-in pathogenesis of HAD has been reported.[91]

Quinolinic acid (QUIN) is a metabolite of L-tryptophan and is highly neurotoxic. QUIN may gain access to the brain by crossing the BBB or could be generated locally by activated MØ or astrocytes within brain in response to IFN-γ or other cytokines.[60,92] QUIN is produced by macrophages stimulated by platelet-activating factor, nef, and tat, and it is believed that some of the neurotoxicity of PAF, nef, and tat may be mediated by QUIN.[62] QUIN acts on the NMDA receptors and causes elevation of the intracellular calcium and possibly cell death. QUIN is a major player in the pathogenesis of neuronal injury in AIDS patients. QUIN induces chemokine and chemokine receptor expression in astrocytes and is at least as potent as classical mediators such as inflammatory cytokines (TNF-α). QUIN is critical in the amplification of brain inflammation, particularly in dementia associated with HIV-1 infection.[93]

Glutamate, which is released by HIV-1–infected MØ, causes neuronal apoptosis.[65,94] Glutamate toxicity acts via two distinct pathways: an excitotoxic one in which glutamate receptors are hyperactivated and an oxidative one in which cystine uptake is inhibited, resulting in glutathione depletion and oxidative stress. Normally, astrocytes take up glutamate, keeping extracellular glutamate concentration low in the brain and preventing excitotoxicity. This action is inhibited in HIV infection, probably due to the effects of inflammatory mediators and viral proteins. In patients with HAD, glutamate production by macrophages is enhanced by mitochondrial glutaminase.[95]

Interactions of MØ and astrocytes is a significant component in HAD pathogenesis. Studies on cocultures of MØ and astrocytes have shown the concentrations of neurotoxins are affected by the interactions between these two cell groups. For example, expression of nef, TNF-α, and IL-6 is enhanced in coculture of astrocytes with MØ.[96] HIV-1–infected MØ, after interacting with astrocytes, secrete eicosanoids, i.e., arachidonic acid and its metabolites, including platelet-activating factor as well as other proinflammatory cytokines such as IL-1β and TNF-α.[44,97] Such interactions may also decrease the level of neurotoxins that are generated within the CNS during development of HAD. These protective effects are independent of TNF-α, whose production by astrocytes suppresses MØ activation.

Interestingly, both MØ and astrocytes exert neuroprotective effects in the context of HAD pathogenesis.[17] HIV-1–infected MØ express and secrete monocyte chemoattractant protein-1 (MCP-1),[98] which protects human neurons and astrocytes from NMDA or HIV-tat–induced apoptosis.[99] Human astrocytes express a novel chemokine, known as fractalkine, while MØ express the receptor for fractalkine (CX3CR1).[100] Fractalkine has been found to function as a survival factor for hippocampal neurons by decreasing neuronal apoptosis induced by HIV-1 protein gp120[101] and is neuroprotective in rat hippocampal neuron cultures against neuronal injury.[102] Astrocytes also are neuroprotective by minimizing QUIN production (a significant player in development of HAD) and maximizing synthesis of kynurenic acid.[93] Astrocytes secrete neurotrophic factors such as bFGF and NGF[103–105] and play a role in neuroregeneration.

16.5 MULTIPLE SCLEROSIS

Multiple sclerosis (MS) is an immune-mediated degenerative disease of the central nervous system (CNS) that is probably triggered by an environmental factor in genetically susceptible individuals.[106–108] MS manifests with a relapsing-remitting or progressive course and clinically manifests with overwhelming fatigue, cognitive decline, motor and sensory loss, sphincter dysfunction, and cerebellar, brainstem, and optic nerve involvement. Neuropathologically, MS is characterized by demyelinating plaques that contain many CD4+ T lymphocytes as well as lipid-laden macrophages and activated microglial cells.[109] Pathogenesis of MS involves a complicated inflammatory cascade that involves white matter more than gray matter and uncommonly leads to dementia. Activated macrophages and CD4 T lymphocytes, the two major cell types, move in from peripheral circulation into CNS and destroy myelin/oligodendrocyte complex.[110] MØ and astrocytes are significantly involved in the inflammatory cascade and immune responses in the CNS of MS patients. It has been proposed that net balance between MØ and astrocytes and their interactions play a major role in the pathogenesis of myelin/oligodendrocyte complex loss in MS patients.[111]

MØ constitutively express MHC class II molecules, and this expression is enhanced in response to dynamic inflammatory cascade of MS. Certain cytokines such as IFN-γ and IL-12 as well as lipopolysaccharide (LPS) and adhesion molecules stimulate MØ to further express MHC class II molecules. Reactive MØ within MS lesions express the receptor for the constant region of immunoglobulin (FcR), B7-1 (CD80), B7-2 (CD86), and ICAM-1, which indicates that these cells possibly function as antigen-presenting cells (APC) in T cell activation.[12,13,112,113] MØ also express the complement receptors CR1 and CR2. The binding of complement results in activation of MØ and gene ration of IL-1, IL-6, and TNF-α.[114] Activated MØ secrete TNF-α, NO, and proteases, which in turn damage myelin/oligodendrocyte complex and promote demyelinating process.

Astrocytes are the other major players in pathogenesis of MS. Astrocytes do not constitutively express MHC class II molecules, and many investigators do not consider them as the major APC in MS patients. Certain neurotransmitters including norepinephrine, glutamate, and vasoactive intestinal polypeptide are involved in suppression of the induction of MHC class II molecules and adhesion of costimulatory molecules by the astrocytes,[115–119] and loss of the receptors for these neurotransmitters, particularly β2-adrenergic receptors in astrocytes of MS patients, has been reported.[120,121] Astrocytes, under stimulation with cytokines such as IFN-γ or TNF-α, are capable of expressing a number of immunologically relevant molecules such as MHC class I and II, complement components, and adhesion molecules ICAM-1, VCAM-1, and E-selectin.[122,123] Astrocytes are capable of secreting or responding to a variety of immunoregulatory cytokines/chemokines, such as IL-1, IL-4, IL-6, IL-10, IL-12, leukemia inhibitory factor (LIF), CNTF, granulocyte macrophage colony-stimulating factor, IFN-γ, IFN-β, transforming growth factor-β (TGF-β), TNF-α, RANTES, IP-10, IL-8, and neurotrophins.

Based on our current level of understanding of the function of MØ and astrocytes in the context of MS, any scientific model of MS must explain these issues in genetically susceptible individuals and following exposure to as yet unidentified environmental antigens:

1. Clonal expansion of myelin-specific CD4+ T lymphocytes (specific for putative MS antigens such as myelin basic protein, myelin oligodendrocyte glycoprotein, or proteolipid protein) and their transendothelial migration into the CNS.
2. Myelin-specific CD4+ T lymphocytes within the CNS do not directly cause demyelination and loss of myelin/oligodendrocyte complex since these elements do not express MHC class II molecules, which are necessary for the antigen presentation to the CD4+ T lymphocytes.
3. Activated MØ products directly damage the myelin/oligodendrocyte complex, participate in demyelination, and possibly cause neuronal loss.

This proposed model places more emphasis on MØ rather than astrocytes as APC and major effectors in the demyelination process.

16.6 MØ VERSUS ASTROCYTES: THEIR EFFECTS ON THE TH1/TH2 CELLS AND THEIR INTERACTIONS IN PATHOGENESIS OF MS

The central theme of T helper 1 (Th1) versus T helper 2 (Th2) cells focuses on the fact that cytokines generated by either type of response are critical for both the regulatory and effector aspects of immune function. The balance between generation and action of these Th1 versus Th2 cytokines plays a significant role in pathogenesis of MS and is affected also by the interactions between MØ and astrocytes. For example, IL-12 that is generated by MØ, regulates the development of Th1-type lymphocyte populations producing IFN-γ. A surge of serum levels of IFN-γ precedes clinical relapses of MS.[124] On the other hand, IL-4, which is an anti-inflammatory cytokine and has been located within chronic active and chronic inactive MS lesions,[125] promotes expansion of Th2-type lymphocyte population and assists in recovery from relapses in MS. In the context of MS, there is a balance between MØ and astrocytes in adjusting local immune reactions, particularly Th1/Th2 responses (Figure 16.2).

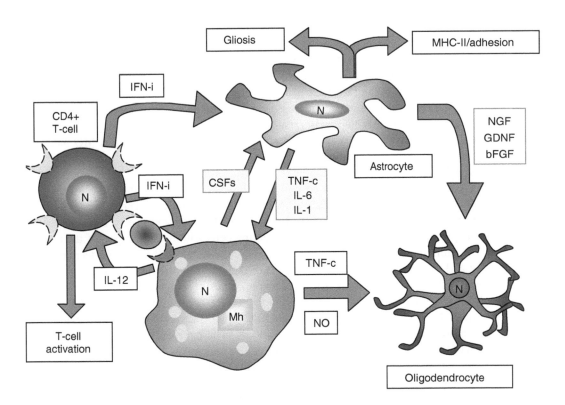

Figure 16.2 Schematic figure showing the complex interactions between MØ, astrocytes, and oligodendrocytes in the inflammatory cascade of MS. Following exposure of genetically susceptible individual to certain unknown environmental antigens, macrophages are activated and migrate into the CNS. Activated MØ present putative MS antigens in the context of MHC class II molecules to CD4+ T lymphocytes (trimolecular complex), which in turn secrete proinflammatory cytokines such as IFN-γ, which further activates MØ. The activated MØ generate other inflammatory mediators such as TNF-α and NO, which participate in demyelination and loss of myelin/oligodendrocyte complex. Both MØ and astrocytes play dual roles in pathogenesis of MS. Astrocytes also generate colony-stimulating factors (CSFs) that are significant for maintaining the growth of MØ.

The capability of MØ and astrocytes to activate Th1 and Th2 lymphocyte populations depends on the presence of certain surface molecules such as MHC class II, B7, and CD40. Myelin basic protein-reactive CD4+ T lymphocytes, which are the major cells in perpetuating the inflammatory cascade of MS, are induced by MØ that express MHC class II on their surface. The activated myelin basic protein-reactive CD4+ T lymphocytes in turn generate IFN-γ and TNF-α. These proinflammatory cytokines induce expression of CD40 molecules by the MØ, which again leads to enhancement of the Th1-type inflammatory response.[126] Th1-type lymphocytes stimulate MØ to generate prostaglandin E2 (PGE2), which inhibits Th1 cell–type activity through a negative feedback mechanism. Upon presentation of putative MS antigens to activated MØ and Th1 T lymphocytes within the CNS milieu, these cells produce IFN-γ, which induces astrocytes to produce PGE2, which in turn adjusts the activity of MØ and Th1 cell T lymphocyte activity.[127]

MØ, as the main effector cells within CNS that are activated in response to active inflammation and infection, are directly involved in loss of myelin/oligodendrocyte complex through a number of mechanisms such as generation of proinflammatory cytokines (particularly TNF-α and IFN-γ), action of matrix metalloproteinases, nitrative and oxidative stress, and phagocytosis of myelin debris. Interestingly, MØ as a source of TGF-β1 (which is significant for development of oligodendrocytes) may play a potential role in remyelination by modulating the growth factor repertoire in demyelinating disease.[128] In MS patients, MØ may exert deleterious effects on the neural stem cell population, which may contribute to the depletion of the stem cell population.[129] On the other hand, IFN-γ-activated astrocytes are major cellular sources of nerve growth factor (NGF), which prevents development of experimental autoimmune encephalomyelitis (EAE),[111] expresses functional receptors for IL-4, and generates more NGF upon exposure to IL-4.[130] Enhancement of IL-4 production by astrocytes is one of the proposed mechanisms of action of beta-interferons in treatment of MS. It has also been shown that astrocytes synthesize the growth factors basic fibroblast growth factor (bFGF) and platelet-derived growth factor (PDGF) in MS lesions and that these growth factors act as oligodendrocyte progenitor mitogens and chemoattractant signals. Examination of the CNS lesions in EAE in Lewis rats, using terminal transferase deoxyribonucleotidyl dUTP nuclear end labeling (TUNEL) assay, has shown that astrocytes rather than MØ induce the programmed cell death of infiltrating inflammatory cells.[131]

Activation of MØ during CNS pathologies is associated with a number of alterations including expression of leukocyte function-associated molecule 1 (LFA-1), intercellular adhesion molecule 1 (ICAM-1), and MHC class II molecules, phagocytosis, and cytotoxicity. The expression of these immune reactions and adhesion molecules are downregulated by astrocyte-derived factors, emphasizing the concept that deactivation of antigen-presenting cells within CNS is part of the astrocyte function in MS. Human astrocytes produce transforming growth factor-β (TGF-β), decrease generation of IFN-γ, and decrease T cell proliferation.[132] TGF-β and IL-4 both suppress phagocytic activity of the reactive MØ. In addition, increased expression of nestin, embryonic neural cell adhesion molecule (eNCAM), epidermal growth factor receptor (EGFR), nerve growth factor (NGF), connective tissue growth factor (CTGF), and basic fibroblast growth factor (bFGF) in astrocytes of MS patients has been reported, which further implicates these cells in repair process in MS.[133]

Generation of IL-12 p75 and p40 by IFN-γ/endotoxin-stimulated MØ decreases after coculturing these cells with astrocytes, indicating the regulatory role of astrocytes in generation of IL-12 by reactive MØ.[126] On the contrary, stimulation of astrocytes by endotoxin increases their secretion of IL-10.[126] IL-10 as an anti-inflammatory cytokine suppresses inflammation in the brain by (1) reducing synthesis of proinflammatory cytokines, (2) suppressing cytokine receptor expression, and (3) inhibiting receptor activation. In addition, IL-10 induces anergy in brain-infiltrating T cells by inhibiting cell signaling through the costimulatory CD28-CD80/86 pathway.[134] Dysregulation of IL-10 and IL-12 p40 in patients with MS has been reported.[135] Another protective role of astrocytes in the pathogenesis of MS includes suppression of endotoxin-induced NO production by MØ.[136,137] These findings indicate that both MØ and astrocytes play dual protective and destructive roles, depending on the balance between Th1 and Th2 cytokines, in pathogenesis of MS.

16.7 CONCLUSIONS

In normal individuals, the CNS environment contains and restricts local immune responses of MØ and astrocytes. However, under pathological conditions such as HAD and MS, which represent two different disease mechanisms, immune suppression versus aberrant immune activation, the interaction between microglia, astrocytes, and invading activated immune cells from the periphery is augmented by the dynamic inflammatory response and leads to generation of a large number of pro- and anti-inflammatory cytokines/chemokines as well as other inflammatory mediators such NO and its derivatives. An imbalance between such interactions leads to deviation of the whole system including Th1 toward Th2 responses, which in turn has significant implications in further destruction of the CNS or its attempted repair during the pathogenesis of HAD and MS.

The literature review presented in this chapter outlines physiological and molecular processes for both HAD and MS that result not only from abnormal immune response but also from the failure of the protective mechanisms within the CNS to contain inflammatory cascades during the course of these disorders. In addition, we emphasize that any effective immune therapy must be designed to address the reduction of MØ destructive effects as well as to enhance astrocytes' protective effects.

ACKNOWLEDGMENTS

This work was supported in part by NIH grants DA 14533, DA 12580, and GM056529 (to PS) and a grant from Berlex (to AM and JSA).

LIST OF ABBREVIATIONS

MØ — Macrophage/microglia
CNS — Central nervous system
HAD — HIV-1–associated dementia
NO — Nitric oxide
MHC — Major histocompatibility
ICAM-1/CD54 — Intercellular adhesion molecule-1
GFAP — Glial fibrillary acidic protein
NGF — Nerve growth factor
GDFD — Glial-derived growth factor
CNTF — Ciliary neurotrophic protein
HIV-1 — Human immunodeficiency virus type 1
HAART — Highly active antiretroviral therapy
HIVE — HIV encephalitis
BBB — Blood–brain barrier
PAF — Platelet-activating factor
QUIN — Quinolinic acid
MCP-1 — Monocyte chemoattractant protein-1
CX3CR1 — Fractalkine
LPS — Lipopolysaccharide
APC — Antigen-presenting cells
Th1 — T helper 1 cells
Th2 — T helper 2 cells
PGE2 — Prostaglandin E2
EAE — Experimental autoimmune encephalomyelitis

bFGF — Basic fibroblast growth factor
PDGF — Platelet-derived growth factor
TUNEL — Terminal transferase deoxyribonucleotidyl dUTP nuclear end labeling
LFA-1 — Leukocyte function-associated molecule 1
ICAM-1 — Intercellular adhesion molecule 1
TGF-β — Transforming growth factor-β
eNCAM — Embryonic neural cell adhesion molecule
EGFR — Epidermal growth factor receptor
NGF — Nerve growth factor
CTGF — Connective tissue growth factor

REFERENCES

1. Banati, R.B. et al., Cytotoxicity of microglia, *Glia*, 7, 111, 1993.
2. Gordon, S., The macrophage, *Bioassays*, 17, 977, 1995.
3. Gehrmann, J., Matsumoto, Y., and Kreutzenberg, G.W., Microglia: intrinsic immunoeffector cell of the brain, *Brain Res. Rev.*, 20, 269, 1995.
4. Delgado, S. et al., Heterogeneity of macrophages in the peripheral nervous system in HIV infected individuals, *J. Neuro. AIDS*, 5, 79, 1998.
5. Shapshak, P. et al., Brain macrophage surface marker expression with HIV-1 infection & drug abuse; a preliminary study, *J. Neuro. AIDS*, 2, 37, 2002.
6. Walker, D.G., Kim, S.U., and McGeer, P.L., Complement and cytokine gene expression in cultured microglia derived from postmortem human brains, *J. Neurosci. Res.*, 40, 478, 1995.
7. Shrikant, P. et al., Intercellular adhesion molecule-1 gene expression by glial cells. Differential mechanisms of inhibition by IL-10 and IL-6, *J. Immunol.*, 155, 1489, 1995.
8. Panek, R.B. and Benveniste, E.N., Class II MHC gene expression in microglia. Regulation by the cytokines IFN-gamma, TNF-alpha, and TGF-beta, *J. Immunol.*, 154, 2846, 1995.
9. Frei, K. et al., Antigen presentation and tumor cytotoxicity by interferon-gamma-treated microglial cells, *Eur. J. Immunol.*, 17, 1271, 1987.
10. Suzumura, A. et al., MHC antigen expression on bulk isolated macrophage-microglia from newborn mouse brain: induction of Ia antigen expression by gamma-interferon, *J. Neuroimmunol.*, 15, 263, 1987.
11. Williams, K., Ulvestad, E., and Antel, J.P., B7/BB-1 antigen expression on adult human microglia studied *in vitro* and *in situ*, *Eur. J. Immunol.*, 24, 3031, 1994.
12. Williams, K. et al., Biology of adult human microglia in culture: comparisons with peripheral blood monocytes and astrocytes, *J. Neuropathol. Exp. Neurol.*, 51, 538, 1992.
13. De Simone, R. et al., The costimulatory molecule B7 is expressed on human microglia in culture and in multiple sclerosis acute lesions, *J. Neuropathol. Exp. Neurol.*, 54, 175, 1995.
14. Stewart, R.V. et al., Effects of cocaethylene on surface marker expression of HIV-1 infected macrophages isolated from human brain using antibody-coated magnetic beads: a preliminary study, *Vision Res.*, 4, 66, 1997.
15. Merrill, J.E. and Jonakait, G.M., Interactions of the nervous and immune systems in development, normal brain homeostasis, and disease, *FASEB J.*, 9, 611, 1995.
16. Kraus, E. et al., Augmentation of major histocompatibility complex class I and ICAM-1 expression on glial cells following measles virus infection: evidence for the role of type-1 interferon, *Eur. J. Immunol.*, 22, 175, 1992.
17. Minagar, A. et al., Microglia and astrocytes in neuro-AIDS, Alzheimer's disease and multiple sclerosis, *J. Neurol. Sci.*, 202, 13, 2002.
18. Resnick, L. et al., Early penetration of the blood-brain-barrier by HTLV-III/LAV, *Neurology*, 38, 9, 1988.
19. Singer, E.J. et al., Cerebrospinal fluid p24 antigen levels and intrathecal immunoglobulin G synthesis are associated with cognitive disease severity in HIV-1, *AIDS*, 8, 197, 1994.
20. Goodkin, K. et al., HIV-1 infection of brain: A region-specific approach to its neuropathophysiology and therapeutic prospects, *Psychiatr. Ann.*, 31, 182, 2001.

21. Shapshak, P. et al., Independent evolution of HIV-1 in different brain regions, *AIDS Res. Hum. Retroviruses*, 15, 811, 1999.

22. Gartner, S. and Liu, Y., Insights into the role of immune activation in HIV neuropathogenesis, *J. Neurovirol.*, 8, 69, 2002.

23. Gartner, S., HIV infection and dementia, *Science*, 287, 602, 2000.

24. Ellis, R.J. et al., Progression to neuropsychological impairment in human immunodeficiency virus infection predicted by elevated cerebrospinal fluid levels of human immunodeficiency virus RNA, *Arch. Neurol.*, 59, 923, 2002.

25. Goodkin, K., Shapshak, P. et al., Immune function, brain, and HIV-1 infection, in *Psychoneuroimmunology: Stress, Mental Disorders & Health*, Goodkin, K. and Visser, A., Eds., American Psychiatric Assoc. Press, Inc., Washington, D.C., 1999.

26. Goodkin, K. et al., Subtle neuropsychologic impairment and minor cognitive-motor disorder in HIV-1 infection, *Neuroimag. Clin. N. Am. Neuroimag. AIDS II*, 7, 561, 1997.

27. Janssen, R.S. et al., Epidemiology of human immunodeficiency virus encephalopathy in the United States, *Neurology*, 42, 1472, 1992.

28. Mindt, M.R. et al., The functional impact of HIV-associated neuropsychological impairment in Spanish-speaking adults: a pilot study, *J. Clin. Exp. Neuropsychol.*, 2003.

29. Deutsch, R. et al., AIDS-associated mild neurocognitive impairment is delayed in the era of highly active antiretroviral therapy, *AIDS*, 15, 1898, 2001.

30. Navia, B.A., Jordan, B.D., and Price, R.W., The AIDS dementia complex: I. Clinical features, *Ann. Neurol.*, 19, 517, 1986.

31. Navia, B.A. et al., The AIDS dementia complex: II. Neuropathology, *Ann. Neurol.*, 19, 525, 1986.

32. Maher, J. et al., AIDS dementia complex with generalized myoclonus, *Mov. Disord.*, 12, 593, 1997.

33. Mirsattari, S.M. et al., Paroxysmal dyskinesias in patients with HIV infection, *Neurology*, 52, 109, 1999.

34. McArthur, J.C. et al., Dementia in AIDS patients: incidence and risk factors: Multicenter AIDS Cohort Study, *Neurology*, 43, 2245, 1993.

35. Dore, G.J. et al., Changes to AIDS dementia complex in the era of HAART, *AIDS*, 13, 1249, 1999.

36. Kandanearatchi, A., Williams, B., and Everall, I.P., Assessing the efficacy of highly active antiretroviral therapy in the brain, *Brain Pathol.*, 13, 10, 2003.

37. Sperber, K. and Shao, L., Neurologic consequences of HIV infection in the ERA of HAART, *AIDS Patient Care STDs*, 17, 509, 2003.

38. Gendelman, H.E. et al., The neuropathogenesis of the AIDS dementia complex, *AIDS*, 11 (Suppl. A), S35, 1997.

39. Masliah, E. et al., Changes in pathological findings at autopsy in AIDS cases for the last 15 years, *AIDS*, 14, 69, 2000.

40. Sacktor, N. et al., HIV-associated neurologic disease incidence changes: Multicenter AIDS Cohort Study, 1990–1998, *Neurology*, 56, 257, 2001.

41. McArthur, J.C. et al., Human immunodeficiency virus-associated dementia: an evolving disease. *J. Neurovirol.*, 9, 205, 2003.

42. de la Monte, S.M. et al., Subacute encephalomyelitis of AIDS and its relation to HTLV-III infection, *Neurology*, 37, 562, 1987.

43. Gray, F. et al., The changing pattern of HIV neuropathology in the HAART era, *J. Neuropathol. Exp. Neurol.*, 62, 429, 2003.

44. Genis, P. et al., Cytokines and arachidonic metabolites produced during human immunodeficiency virus (HIV)-infected macrophage-astroglia interactions: implications for the neuropathogenesis of HIV disease, *J. Exp. Med.*, 176, 1703, 1992.

45. Vazeux, R., AIDS encephalopathy and tropism of HIV for brain monocytes/macrophages and microglial cells, *Pathobiology*, 59, 214, 1991.

46. Grant, I. et al., Evidence for early central nervous system involvement in the acquired immunodeficiency syndrome (AIDS) and other human immunodeficiency virus (HIV) infections. Studies with neuropsychologic testing and magnetic resonance imaging, *Ann. Intern. Med.*, 107, 828, 1987.

47. Shapshak, P. et al., Dementia and the neurovirulence of HIV-1, CNS spectrums, *Int. Psych. J.*, 5, 31, 2000.

48. Hafler, D.A. et al., TCR usage in human and experimental demyelinating disease, *Trends Immunol. Today*, 17, 152, 1996.

49. Raine, C.S., The Dale E. McFarlin Memorial Lecture: the immunology of the multiple sclerosis lesion, *Ann. Neurol.*, 36, S61, 1994.

50. Raine, C.S., Multiple sclerosis: a pivotal role for the T cell in lesion development, *Neuropathol. Appl. Neurobiol.*, 17, 265, 1991.

51. Zamvil, S.S. and Steinman, L., The T lymphocyte in experimental allergic encephalomyelitis. *Annu. Rev. Immunol.*, 8, 579, 1990.

52. Sriram, S. et al., *In vivo* immunomodulation by monoclonal anti-CD4+ Ab, effect on T cell response to MBP and EAE, *J. Immunol.*, 141, 464, 1988.

53. Traugott, U., Characterization and distribution of lymphocyte subpopulations in MS plaques vs autoimmune demyelinating lesions, *Semin. Immunopathol.*, 8, 71, 1985.

54. Pulliam, L. et al., Unique monocyte subset in patients with AIDS dementia, *Lancet*, 349, 692, 1997.

55. Luo, X. et al., Macrophage proteomic fingerprinting predicts HIV-1-associated cognitive impairment, *Neurology*, 60, 1931, 2003.

56. Gelbard, H.A. et al., Platelet-activating factor: a candidate human immunodeficiency virus type 1-induced neurotoxin, *J. Virol.*, 68, 4628, 1994.

57. Morgan, M.J., Kimes, A.S., and London, E.D., Possible roles for nitric oxide in AIDS and associated pathology, *Med. Hypoth.*, 40, 142, 1993.

58. Heyes, M.P. et al., Cerebrospinal fluid quinolinic acid concentrations are increased in acquired immune deficiency syndrome, *Ann. Neurol.*, 26, 275, 1989.

59. Heyes, M.P. et al., Increased ratio of quinolinic acid to kynurenic acid in cerebrospinal fluid of D retrovirus-infected Rhesus macaques: relationship to clinical and viral status, *Ann. Neurol.*, 27, 666, 1990.

60. Heyes, M.P. et al., Quinolinic acid in cerebrospinal fluid and serum in HIV-1 infection: relationship to clinical and neurologic status, *Ann. Neurol.*, 29, 202, 1991.

61. Heyes, M.P. et al., Inter-relationships between quinolinic acid, neuroactive kynurenines, neopterin and beta 2-microglobulin in cerebrospinal fluid and serum of HIV-1-infected patients, *J. Neuroimmunol.*, 40, 71, 1992.

62. Nath, A. and Geiger, J., Neurobiological aspects of human immunodeficiency virus infection: neurotoxic mechanisms, *Prog. Neurobiol.*, 54, 19, 1998.

63. Jiang, Z.G. et al., Glutamate is a mediator of neurotoxicity in secretions of activated HIV-1-infected macrophages, *J. Neuroimmunol.*, 117, 97, 2001.

64. Corasaniti, M.T. et al., Neurobiological mediators of neuronal apoptosis in experimental neuroAIDS, *Toxicol. Lett.*, 139, 199, 2003.

65. Self, R.L. et al., The human immunodeficiency virus type-1 transcription factor Tat produces elevations in intracellular Ca(2+) that require function of an N-methyl-D-aspartate receptor polyamine-sensitive site, *Brain Res.*, 995, 39, 2004.

66. Pu, H. et al., HIV-1 Tat protein upregulates inflammatory mediators and induces monocyte invasion into the brain, *Mol. Cell. Neurosci.*, 24, 224, 2003.

67. Furtado, M.R. et al., Analysis of alternatively spliced human immunodeficiency virus type-1 mRNA species, one of which encodes a novel tat-env fusion protein, *Virology*, 185, 258, 1991.

68. Saito, Y. et al., Overexpression of nef as a marker for restricted HIV-1 infection of astrocytes in postmortem pediatric central nervous tissues, *Neurology*, 44, 474, 1994.

69. Takahashi, K. et al., Localization of HIV-1 in human brain using polymerase chain reaction/*in situ* hybridization and immunocytochemistry, *Ann. Neurol.*, 39, 705, 1996.

70. Tornatore, C. et al., HIV-1 infection of subcortical astrocytes in the pediatric central nervous system, *Neurology*, 44, 481, 1994.

71. Ranki, A. et al., Abundant expression of HIV Nef and Rev proteins in brain astrocytes *in vivo* is associated with dementia, *AIDS*, 9, 1001, 1995.

72. Gorry, P. et al., Restricted HIV-1 infection of human astrocytes: potential role of nef in the regulation of virus replication, *J. Neurovirol.*, 4, 377, 1998.

73. Canki M. et al., Highly productive infection with pseudotyped human immunodeficiency virus type 1 (HIV-1) indicates no intracellular restrictions to HIV-1 replication in primary human astrocytes, *J. Virol.*, 75, 7925, 2001.

74. Messam, C.A. and Major, E.O., Stages of restricted HIV-1 infection in astrocyte cultures derived from human fetal brain tissue, *J. Neurovirol.*, 6 (Suppl. 1), S90, 2000.

75. Chauhan, A. et al., Intracellular human immunodeficiency virus Tat expression in astrocytes promotes astrocyte survival but induces potent neurotoxicity at distant sites via axonal transport, *J. Biol. Chem.*, 278, 13512, 2003.

76. Griffin, G.E. et al., Induction of NF-kappa B during monocyte differentiation is associated with activation of HIV-gene expression, *Res. Virol.*, 142, 233, 1991.

77. Mellors, J.W. et al., Tumor necrosis factor-alpha/cachectin enhances human immunodeficiency virus type 1 replication in primary macrophages, *J. Infect. Dis.*, 78, 163, 1991.

78. Robbins, D.S. et al., Production of cytotoxic factor for oligodendrocytes by stimulated astrocytes, *J. Immunol.*, 139, 2593, 1987.

79. Selmaj, K. et al., Non-specific oligodendrocyte cytotoxicity mediated by soluble products of activated T cell lines, *J. Neuroimmunol.*, 35, 261, 1991.

80. Shibata, M. and Blatteis, C.M., Human recombinant tumor necrosis factor and interferon affect the activity of neurons in the organum vasculosum laminae terminalis, *Brain Res.*, 562, 323, 1991.

81. Soliven, B. and Albert, J., Tumor necrosis factor modulates the inactivation of catecholamine secretion in cultured sympathetic neurons, *J. Neurochem.*, 58, 1073, 1992.

82. Wesselingh, S.L. et al., Cellular localization of tumor necrosis factor mRNA in neurological tissue from HIV-infected patients by combined reverse transcriptase/polymerase chain reaction *in situ* hybridization and immunohistochemistry, *J. Neuroimmunol.*, 74, 1, 1997.

83. Saha, R.N. and Pahan, K., Tumor necrosis factor-alpha at the crossroads of neuronal life and death during HIV-associated dementia, *J. Neurochem.*, 86, 1057, 2003.

84. Miura, Y. et al., Tumor necrosis factor-related apoptosis-inducing ligand induces neuronal death in a murine model of HIV central nervous system infection, *Proc. Natl. Acad. Sci. U.S.A.*, 100, 2777, 2003.

85. Lipton, S.A., Yeh, M., and Dreyer, E.B., Update on current models of HIV-related neuronal injury: platelet-activating factor, arachidonic acid and nitric oxide, *Adv. Neuroimmunol.*, 4, 181, 1994.

86. Dawson, V.L. et al., Nitric oxide mediates glutamate neurotoxicity in primary cortical structures, *Proc. Natl. Acad. Sci. U.S.A.*, 88, 6368, 1991.

87. Lee, J. et al., Translational control of inducible nitric oxide synthase expression by arginine can explain the arginine paradox, *Proc. Natl. Acad. Sci. U.S.A.*, 100, 4843, 2003.

88. Adamson, D.C. et al., Immunologic NO synthase: elevation in severe AIDS dementia and induction by HIV-1 gp41, *Science*, 274, 1917, 1996.

89. Adamson, D.C. et al., Rate and severity of HIV-associated dementia (HAD): correlations with Gp41 and iNOS, *Mol. Med.*, 5, 98, 1999.

90. Liu, X. et al., Human immunodeficiency virus type 1 (HIV-1) tat induces nitric-oxide synthase in human astroglia, *J. Biol. Chem.*, 277, 39312, 2002.

91. Torre, D., Pugliese, A., and Speranza, F., Role of nitric oxide in HIV-1 infection: friend or foe?, *Lancet Infect. Dis.*, 2, 273, 2002.

92. Heyes, M.P. et al., Elevated cerebrospinal fluid quinolinic acid levels are associated with region-specific cerebral volume loss in HIV infection, *Brain*, 124, 1033, 2001.

93. Guillemin, G.J. et al., Quinolinic acid upregulates chemokine production and chemokine receptor expression in astrocytes, *Glia*, 41, 371, 2003.

94. Chen, W. et al., Development of a human neuronal cell model for human immunodeficiency virus (HIV)-infected macrophage-induced neurotoxicity: apoptosis induced by HIV type 1 primary isolates and evidence for involvement of the Bcl-2/Bcl-xL-sensitive intrinsic apoptosis pathway, *J. Virol.*, 76, 9407, 2002.

95. Zhao, J. et al., Mitochondrial glutaminase enhances extracellular glutamate production in HIV-1-infected macrophages: Linkage to HIV-1 associated dementia, *J. Neurochem.*, 88, 169, 2004.

96. Fiala, M. et al., Regulation of HIV-1 infection in astrocytes: expression of Nef, TNF-alpha and IL-6 is enhanced in coculture of astrocytes with macrophages, *J. Neurovirol.*, 2, 158, 1996.

97. Hopkins, S.J. and Rothwell, N.J., Cytokines and the nervous system. I: Expression and recognition, *Trends Neurosci.*, 18, 83, 1995.

98. Mengozzi, M. et al., Human immunodeficiency virus replication induces monocyte chemotactic protein-1 in human macrophages and U937 promonocytic cells, *Blood*, 93, 1851, 1999.

99. Eugenin, E.A. et al., MCP-1 (CCL2) protects human neurons and astrocytes from NMDA or HIV-tat-induced apoptosis, *J. Neurochem.*, 85, 1299, 2003.

100. Hatori, K. et al., Fractalkine and fractalkine receptors in human neurons and glial cells, *J. Neurosci. Res.*, 69, 418, 2002.

101. Meucci, O. et al., Chemokines regulate hippocampal neuronal signaling and gp120 neurotoxicity, *Proc. Natl. Acad. Sci. U.S.A.*, 95, 14500, 1998.

102. Meucci, O. et al., Expression of CX3CR1 chemokine receptors on neurons and their role in neuronal survival, *Proc. Natl. Acad. Sci. U.S.A.*, 97, 8075, 2000.

103. Boven, L.A. et al., Overexpression of nerve growth factor and basic fibroblast factor in AIDS dementia complex, *J. Neuroimmunol.*, 3690, 154, 1999.

104. Pechan, P.A., Chowdhury, K., and Siefert, W., Free radicals induce gene expression of NGF and bFGF in rat astrocyte culture, *Neuroreport*, 3, 469, 1992.

105. Ballabriga, J. et al., BFGF and FGFR-3 in reactive astrocytes, PGFR-3 in reactive microglia, *Brain Res.*, 752, 315, 1997.

106. Tourtellotte, W.W. et al., Cerebrospinal fluid (CSF) immunoglobulin-G(IgC) of extravascular origin in normals and patients with multiple sclerosis (MS): clinical correlation, *Trans. Am. Neurol. Assoc.*, 100, 250, 1975.

107. Noseworthy, J.H. et al., *Multiple sclerosis*, N. Engl. J. Med., 343, 938, 2000.

108. Tourtellotte, W.W., Baumheffner, R.W., Shapshak, P., and Osborne, M., The status of intra-blood brain barrier IgG synthesis in multiple sclerosis, *Rev. Neurol.*, 57, 236, 1987.

109. Kornek, B. and Lassmann, H., Neuropathology of multiple sclerosis-new concepts, *Brain Res. Bull.*, 61, 321, 2003.

110. Minagar, A. and Alexander J.S., Blood-brain barrier disruption in multiple sclerosis, *Multiple Sclerosis*, 9, 540, 2003.

111. Xiao, B.G. and Link, H., Is there a balance between microglia and astrocytes in regulating Th1/Th2-cell responses and neuropathologies?, *Trends Immunol. Today*, 20, 477, 1999.

112. Minagar, A. et al., Serum from multiple sclerosis patients down-regulates occludin and VE-cadherin expression in cultured endothelial cells, *Multiple Sclerosis*, 9, 235, 2003.

113. Gerritse, K. et al., CD40-CD40 ligand interactions in experimental allergic encephalomyelitis and multiple sclerosis, *Proc. Natl. Acad. Sci. U.S.A.*, 93, 2499, 1996.

114. Rajan, A.J. et al., A pathogenic role for gamma delta T cells in relapsing-remitting experimental allergic encephalomyelitis in the SJL mouse, *J. Immunol.*, 157, 941, 1996.

115. Frohman, E.M. et al., Norepinephrine inhibits gamma-interferon-induced major histocompatibility class II (Ia) antigen expression on cultured astrocytes via beta-2-adrenergic signal transduction mechanisms, *Proc. Natl. Acad. Sci. U.S.A.*, 85, 1292, 1988.

116. Frohman, E.M. et al., Norepinephrine inhibits gamma-interferon-induced MHC class II (Ia) antigen expression on cultured brain astrocytes, *J. Neuroimmunol.*, 17, 89, 1988.

117. Frohman, E.M. et al., Vasoactive intestinal polypeptide inhibits the expression of the MHC class II antigens on astrocytes, *J. Neurol. Sci.*, 88, 339, 1988.

118. Lee, S.C. et al., Glutamate differentially inhibits the expression of class I MHC antigens on astrocytes and microglia, *J. Immunol.*, 148, 3391, 1992.

119. Cross, A.H. and Ku, G., Astrocytes and central nervous system endothelial cells do not express B7-1 (CD80) or B7-2 (CD86) immunoreactivity during experimental autoimmune encephalomyelitis, *J. Neuroimmunol.*, 110, 76, 2000.

120. De Keyser, J. et al., Astrocytes in multiple sclerosis lack beta-2 adrenergic receptors, *Neurology*, 53, 1628, 1999.

121. Zeinstra, E., Wilczak, N., and De Keyser, J., [³H]Dihydroalprenolol binding to beta adrenergic receptors in multiple sclerosis brain, *Neurosci. Lett.*, 289, 75, 2000.

122. Fontana, A., Fierz, W., and Wekerle, H., Astrocytes represent myelin basic protein to encephalitogenic T-cell lines, *Nature*, 307, 273, 1984.

123. Wong, G.H. et al., Inducible expression of H-2 and Ia antigens in multiple sclerosis, *Brain*, 117, 59, 1994.

124. Martino, G. et al., Interferon-gamma induced increases in intracellular calcium in T lymphocytes from patients with multiple sclerosis precede clinical exacerbations and detection of active lesions on MRI, *J. Neurol. Neurosurg. Psychiatry*, 63, 339, 1997.

125. Hulshof, S. et al., Cellular localization and expression patterns of interleukin-10, interleukin-4, and their receptors in multiple sclerosis lesions, *Glia*, 38, 24, 2002.

126. Aloisi, F. et al., *J. Immunol.*, 159, 1604, 1997.
127. Oh, L.Y. and Yong, V.W., Astrocytes promote process outgrowth by adult human oligodendrocytes *in vitro* through interaction between bFGF and astrocyte extracellular matrix, *Glia*, 17, 237, 1996.
128. Diemel, L.T., Jackson, S.J., and Cuzner, M.L., Role for TGF-beta1, FGF-2 and PDGF-AA in a myelination of CNS aggregate cultures enriched with macrophages, *J. Neurosci. Res.*, 74, 858, 2003.
129. Brundin, L. et al., Neural stem cells: a potential source for remyelination in neuroinflammatory disease, *Brain Pathol.*, 13, 322, 2003.
130. Brodie, C. et al., Functional IL-4 receptors on mouse astrocytes: IL-4 inhibits astrocyte activation and induces NGF secretion, *J. Neuroimmunol.*, 81, 20, 1998.
131. Kohji, T., Tanuma, N., Aikawa, Y., Kawazoe, Y., Suzuki, Y., Kohyama, K., and Matsumoto, Y., Interaction between apoptotic cells and reactive brain cells in the central nervous system of rats with autoimmune encephalomyelitis, *J. Neuroimmunol.*, 82, 168, 1998.
132. Meinl, E. et al., Multiple sclerosis. Immunomodulatory effects of human astrocytes on T cells, *Brain*, 117, 1323, 1994.
133. Holley, J.E. et al., Astrocyte characterization in the multiple sclerosis glial scar, *Neuropathol. Appl. Neurobiol.*, 29, 434, 2003.
134. Strle, K. et al., Interleukin-10 in the brain, *Crit. Rev. Immunol.*, 21, 427, 2001.
135. Soldan, S.S. et al., Dysregulation of IL-10 and IL-12p40 in secondary progressive multiple sclerosis, *J. Neuroimmunol.*, 146, 209, 2004.
136. Tran, E.H. et al., Astrocytes and microglia express inducible nitric oxide synthase in mice with experimental allergic encephalomyelitis, *J. Neuroimmunol.*, 74, 121, 1997.
137. Hua, L.L. et al., Modulation of astrocyte inducible nitric oxide synthase and cytokine expression by interferon beta is associated with induction and inhibition of interferon gamma-activated sequence binding activity, *J. Neurochem.*, 83, 1120, 2002.

CHAPTER 17

Possible Detrimental Role of Astrocytic Activation during the Subacute Phase of Permanent Focal Cerebral Ischemia in the Rat

Takao Asano, Narito Tateishi, and Takashi Mori

CONTENTS

17.1 INTRODUCTION

Despite the time-honored concept that astrocytes are "the silent partners of the working brain,"[1] recent evidence suggests that they might be obtrusive neighbors of threatened neurons in certain central nervous system (CNS) disorders such as Alzheimer's disease.[2,3] With regard to cerebral ischemia, it has been common to assign protective and harmful roles to astrocytes and microglia, respectively.[4,5] However, such a clear dichotomy of the roles of glial cells seems rather unlikely in the face of the long list of substances that astrocytes produce on activation.[6,7] After permanent focal ischemia, astrocytes in the peri-infarct area are activated, most markedly along the outer border of the infarct. Reactive astrocytes, which are hallmarked by augmented expression of glial fibrillary acidic protein (GFAP), appear around 24 h after the onset of ischemia and increase in number thereafter for weeks, accompanied by considerable changes in their morphological appearance.[8–13] In the chronic phase of cerebral ischemia, the roles of astrocytes that have so far been suspected include the formation of glial limitans,[14] angiogenesis,[7,15] and possibly neurogenesis.[5,16] During the subacute phase (24–168 h after the onset of ischemia), on the other hand, the long-held belief that the infarct expansion after permanent focal ischemia comes to a complete halt by 12 h, well before astrocytic activation becomes manifest,[17,18] has made it difficult to assign any detrimental role to reactive astrocytes. Therefore, it was not until the phenomenon of "delayed infarct expansion" was discovered that the possible detrimental role of astrocytic activation in the subacute phase of cerebral infarction came to be recognized as a definite target of research.[19,20]

Using the permanent middle cerebral artery occlusion (pMCAO) model in the rat, Garcia et al. (1993) for the first time reported a significant increase in the infarct area during the time interval between 6 and 72 h after the onset of ischemia.[12] Du et al. (1996) showed that the infarction after mild focal (transient) ischemia could develop in a surprisingly delayed fashion.[21] Subsequently, using the rat pMCAO model, we showed that after the rapid expansion during the initial 24 h, the infarct volume continues to increase slowly but steadily until it reaches a peak at 168 h, with a significant increase of about 41% compared with the infarct volume at 24 h.[19,20] Now that magnetic resonance imaging studies have established that stroke patients exhibit delayed infarct expansion, the time course and magnitude of which are comparable to those in the rat pMCAO model,[22–24] delayed infarct expansion is considered to be a viable therapeutic target. Subsequently, the association between the appearance of numerous reactive astrocytes and delayed infarct expansion has emerged as an issue of utmost clinical importance. In this chapter, findings pertinent to this issue are reviewed and discussed.

17.2 ALTERATIONS IN THE INFARCT VOLUME AFTER pMCAO IN THE RAT

In the aforementioned study of Garcia et al. (1993),[12] the surface area of the infarct was measured at a predetermined coronal plane, employing the histological criteria that included vacuolation (sponginess) of the neuropil, diffuse pallor of the eosinophilic background, and alterations in the shape and stainability of both neuronal perikarya and astrocytic nuclei. The above study demonstrated a significant increase in the infarct area only between the 6-h and the 72-h groups. Using an animal model and a histological criteria similar to Garcia's, we carried out a study to clarify the temporal profile of infarct expansion after pMCAO.[19,20] Under strict randomization, rats were subjected to pMCAO and sacrificed at 1, 3, 6, 9, 14, 24, 48, 72, 120, or 168 h thereafter. The results are shown in Figure 17.1. The increase in the infarct volume, which was precipitous during the initial 12 h, markedly slowed down around 24 h and continued thereafter at a slow pace until 168 h. The difference between the infarct volume at 24 and 168 h was significant, the latter being larger than the former by 41%. This result indicates that infarct expansion after pMCAO can crudely be divided into the acute (0–24 h) and the delayed (24–168 h) phases, which appear to be mediated by different pathogenic mechanisms.

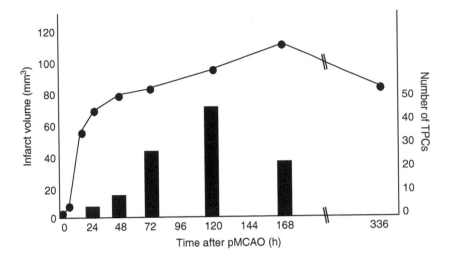

Figure 17.1 Alternations in the infarct volume (line) and the number of terminal deoxynucleotidyl transferase-mediated 2′-deoxyuridine 5′-triphosphate-biotin nick end labeling (TUNEL)–positive cells (TPCs) in the peri-infarct area (column bars) after permanent middle cerebral artery occlusion (pMCAO).

With regard to acute infarct expansion, there is little question that the occurrence of the infarct as discerned by histological evaluation, computed tomography, or magnetic resonance imaging primarily represents astrocytic death.[12,25] Immediately after ischemia, energy failure within the ischemic core induces astrocytic swelling, which reflects the operation of various astrocytic mechanisms to compensate for the altered ionic composition of the extracellular fluid.[26] Spreading depression is triggered by cellular depolarization, and its propagation to the peri-infarct area is at least partly mediated by the astrocytic gap-junctional communication.[27,28] Spreading depression accelerates the acute infarct expansion through aggravation of energy failure.[29] Astrocytes are particularly vulnerable to the combination of hypoxia and acidosis, both of which are conspicuous in the ischemic penumbra.[25] Subsequent astrocytic demise precedes neuronal death,[30] and the consequent creation of the pseudoextracellular space leads to the decreased stainability by eosin, termed *pallor*, which is the earmark of infarct. These and other lines of evidence clearly show that the functional derangements of astrocytes accompanied by concomitant morphological alterations within the ischemic core and in its immediate vicinity are primarily induced by energy depletion, imparting the classical histological features to the infarct and playing the major role in its acute expansion.

In contrast, the mechanisms relevant to delayed infarct expansion have remained elusive. Around 24 h after pMCAO, GFAP-positive reactive astrocytes begin to appear along the infarct border, and thereafter they increase in number accompanied by morphological changes such as hypertrophy and hyperplasia.[8,9,31] It is known that reactive astrocytes exhibit functional activation in terms of antigen presentation and augmented expression of a wide variety of bioactive substances such as adhesion molecules, cytokines, growth factors, eicosanoids, receptors, enzymes, and proteins related to the cytoskeleton as well as the extracellular matrix.[6,7] It has been established that microglial activation precedes the activation of astrocytes and that the production of cytokines and neurotoxins by activated microglia plays a major pathogenic role in almost every type of CNS disorder, including stroke.[4] Nonetheless, the third wave of gene expression peaking around 24–48 h after the onset of ischemia comprises the expression of various cytokines, adhesion molecules, and growth factors that are mainly produced by microglia/neurons, endothelial cells, and astrocytes, respectively.[32] While the production of growth factors rather than cytokines by astrocytes has tended to be emphasized, recent evidence indicates that the inflammatory process leading to the progression of the infarct involves the activation of both microglia and astrocytes.[33,34] It has been suggested that interleukin-1β produced by microglia as well as astrocytes may be the driving force of the inflammatory process in ischemic brain damage.[2,35] There is also a strong line of evidence that an astrocytic protein S100B possesses prominent cytotoxicity, as is discussed in the following sections.

17.3 PROPERTIES AND CYTOTOXICITY OF S100B

The S100 family of calcium-binding proteins comprises 19 members that are differentially expressed in a large number of cell types.[36] Members of this protein family have been implicated in the Ca^{2+}-dependent regulation of a variety of intracellular activities such as protein phosphorylation, enzyme activities, cell proliferation and differentiation, the dynamics of cytoskeleton constituents, the structural organization of membranes, intracellular Ca^{2+} homeostasis, inflammation, and protection from oxidative cell damage.[36–39] In the mammalian brain, S100B (formerly termed S100β) and S100A1 (S100α) are primarily expressed by astrocytes and neurons, respectively.[40,41] Brain S100 is actually a mixture of S100B and S100A1, which form the homodimers $S100B_2$ (S100b) and $S100A1_2$ (S100ao) as well as the heterodimers S100A1/S100B (S100a).[42] Whereas S100 proteins have been suggested to have a role in intracellular signal transduction by linking elevation of the cytosolic Ca^{2+} concentration to the phosphorylation state of a variety of target proteins, S100B secreted by astrocytes has also been shown to exert diverse extracellular actions.

In the normal brain, S100B is present in nanomolar concentrations, and it has been shown to promote neurite extension[43] and enhance survival of neurons during development[44] and after injury.[45,46] Although the nature of S100B receptor has not been firmly established,[36] the binding of S100B to receptive cells results in an increase in the intracellular free Ca^{2+} concentration from internal Ca^{2+} stores via activation of phospholipase C and the ensuing formation of inositol triphosphate.[47] How elevation of the cytosolic Ca^{2+} transduces the S100B binding into trophic effects on responsive neurons and stimulation of glial proliferation has not been firmly established.

On the other hand, high levels of S100B in the micromolar range have been detected in brains from patients with Down's syndrome and Alzheimer's disease.[48] In Alzheimer's disease, the levels of S100B are elevated in the activated astrocytes associated with β-amyloid-containing plaques, and a progressive association of both microglia overexpressing interleukin-1α and astrocytes overexpressing S100B with neurofibrillary tangle stages has been documented.[49] In C6 glioma cells and primary astrocyte cultures, β-amyloid has been shown to stimulate the synthesis of both S100B mRNA and S100B protein.[50] Although chronic elevation of S100B does not appear to induce formation of β-amyloid plaques in brains of aging transgenic mice,[51] β-amyloid precursor protein (β-APP) and its mRNA have been shown to increase in neuronal cultures exposed to S100B.[52]

In studies using cultured cells, S100B at a high concentration has been shown to induce death of cocultured neurons through nitric oxide release from astrocytes.[53] Lam et al. (2001) reported that S100B stimulates inducible nitric oxide synthase (iNOS) in rat primary cortical astrocytes through a signal transduction pathway that involves activation of the transcription factor, nuclear factor κB (NFκB).[54] In this regard, the receptor for advanced glycation end-products (RAGE) has recently been shown to be a signal transduction receptor for S100/calgranulin (S100A8, -A9, or -A12)–like.[55,56] The fact that the RAGE promoter has functional NFκB sites reinforces the likelihood that RAGE may contribute to the pathogenesis of inflammation.[57] Furthermore, it has been shown that RAGE engagement of S100/calgranulins at inflammatory loci propagates the host response by perpetuating recruitment and activation of cellular effectors.[55,57] Taken together, the above findings suggest a novel paradigm in which RAGE plays a central role in a wide spectrum of chronic inflammatory disorders such as inflammatory bowel disease, Alzheimer's disease, atherosclerosis, rheumatoid arthritis, and the complications of diabetes.[55] In particular, Sasaki et al.[58] have provided immunohistochemical evidence that AGE- and RAGE-positive granules are present in most astrocytes in the brains obtained from patients with Alzheimer's disease, whereas they are rare in control brains. Also, Ma et al.[59] have recently reported that RAGE is expressed in pyramidal cells of the hippocampus after moderate hypoxic-ischemic brain injury in rats.

In the clinical setting, elevations of the cerebral spinal fluid (CSF) content of S100 in various neurological diseases, including stroke, have been reported.[60] The concentrations of S100 protein in the serum[61] or in the CSF[62,63] have been shown to be indicators of the infarct volume and prognosis in acute ischemic brain damage. Since the presumption of the foregoing clinical studies was that the rise in the CSF or the serum concentration of S100B reflects the release of the intracellular S100 proteins upon cell damage, the possibility that S100B may be newly synthesized by astrocytic activation, exerting a detrimental influence on the process of brain damage, was not pursued in those studies.

With the above knowledge as background, we undertook a series of experiments to investigate the possible relationship between the augmented production of S100B by astrocytic activation and the occurrence of delayed infarct expansion after pMCAO in the rat. These studies can be summarized as follows.

17.4 RESULTS OF EXPERIMENTS USING THE RAT pMCAO MODEL

A brief account of our preceding papers[19,20,64,65] is given below with schematic representation of the results obtained.

17.4.1 The Number of TUNEL-Positive Cells in the Peri-Infarct Area

At 24, 48, 72, 120, and 168 h after pMCAO, the number of terminal deoxynucleotidyl transferase-mediated 2′-deoxyuridine 5′-triphosphate-biotin nick end labeling (TUNEL)–positive cells (TPCs) present in the ischemic hemisphere excluding the infarct area was counted using the coronal section encompassing the anterior commissure. TPCs were most numerous in the vicinity of the infarct border. As shown in Figure 17.1, TPCs (designated by bars) were undetectable before 24 h, increased in number from 24 h to 120 h (with a significant difference), and then decreased at 168 h. Since positive staining by TUNEL is considered to be a nonspecific indicator of cell death,[66] the explicit correlation between the increase in the number of TPCs in the peri-infarct area and the occurrence of delayed expansion during the time interval between 24 h and 120 h is indicative of the existence of a causal relationship between the two events. Subsequently, we proceeded to examine the temporal and topographical relationship between the astrocytic production of S100B and the increase of TPCs in the peri-infarct area.

17.4.2 Determination of the S100B Levels in the CSF and the Brain Tissue in the Peri-Infarct Area

The S100B concentration in the CSF obtained by cisternal puncture as well as that in the brain tissue excised from a predetermined position in the peri-infarct area of the rats subjected to pMCAO were determined by enzyme-linked immunosorbent assay (ELISA). As shown in Figure 17.2, the tissue S100B levels in the peri-infarct area transiently decreased immediately after pMCAO. It then precipitously increased around 12 h, peaked at 24 h, and then gradually decreased. Since it was excised from the peri-infarct area, the specimen included variable portions of the infarct. Therefore, the measurement tends to underestimate the actual rise in the S100B content. Notwithstanding, the peak S100B concentrations in the peri-infarct area were severalfold higher than those in the contralateral hemisphere, which were comparable to or even more pronounced than those reported for the brain tissue of patients with advanced Alzheimer's disease.[67] In the infarct area, the S100B levels decreased immediately after pMCAO. Thereafter, it remained at near-zero levels throughout the experiment. The S100B levels in the CSF showed a biphasic change, as shown in Figure 17.3. Immediately after pMCAO, it showed a transient elevation, which represented the release of S100B stored in the ischemic core as well as the peri-infarct area. After a small decline, it rapidly increased

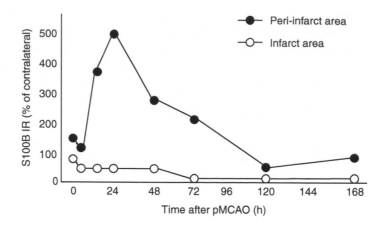

Figure 17.2 The brain tissue levels of S100B after permanent middle cerebral artery occlusion (pMCAO).

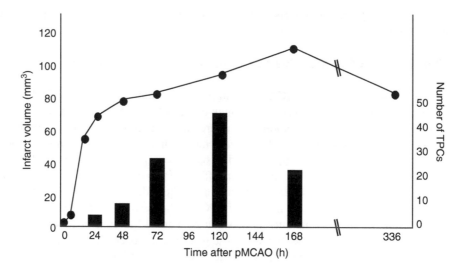

Figure 17.3 The cerebral spinal fluid (CSF) levels of S100B after permanent middle cerebral artery occlusion (pMCAO).

again to reach a much larger peak at 14 h. Relatively high values were maintained for the succeeding 3 days. Importantly, the S100B concentrations in the CSF from the onset of pMCAO until 5 days thereafter far exceeded 10 ng/ml. It seems important to mention here that the protective effect of S100B on endangered neurons was observed in the concentration range below 10 ng/ml.[46] Therefore, the secondary increase in the S100B concentration in the CSF may be harmful rather than protective to the brain. In conjunction with the results of our parallel immunohistochemical studies, which are presented next, the secondary increase in the S100B concentration in the CSF is attributable to the enhanced S100B synthesis by astrocytic activation around the peri-infarct area.

17.4.3 Immunohistochemical Study

As shown in Figure 17.4, an immunohistochemical study using a polyclonal antibody against S100 revealed that there was a paucity of S100-positive astrocytes in the ishchemic hemisphere at 6 h, particularly in the ischemic core, most likely due to depletion of the intracellular store triggered by ischemia. At 24 h, the nuclei of some astrocytes in the vicinity of the peri-infarct border were strongly positive for S100. From 48 h until 96 h, hypertrophy and hyperplasia of astrocytes

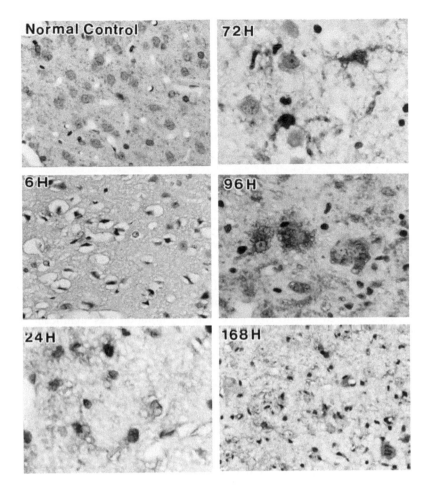

Figure 17.4 *(A color version of this figure follows page 236.)* Immunohistochemical detection of S100 in reactive astrocytes around the infarct border in the rat brain subjected to permanent middle cerebral artery occlusion (pMCAO). (Courtesy of Dr. Fusahiro Ikuta, Niigata Brain Research Institute, Niigata, Japan.)

progressed in the infarct border, accompanied by marked stainability for S100. At 168 h, the stainability for S100 was considerably diminished. S100 was virtually absent in the infarct area. Taken together, the above results provide a clear explanation for the cause of the biphasic elevation of the CSF concentration of S100B. The first transient elevation of S100B represents the release from swollen astrocytes as well as threatened neurons in both the ischemic core and the peri-infarct area, while the second much larger peak is due to the augmented synthesis of the protein by activated astrocytes in the peri-infarct area.

17.4.4 Expression of iNOS in Reactive Astrocytes

Both the augmented astrocytic synthesis of S100B and the appearance of TPCs were shown to occur in the peri-infarct area. Whereas such a topographical concurrence is indicative of a causal relationship, the relatively long peak-to-peak time interval indicates that an additional pathogenic event might intervene between the two events. In this regard, the relevance of iNOS expression is obvious, since S100B has been shown to activate NFκB, leading to the expression of iNOS, as previously described. Iadecola et al.[68] reported that in mice, iNOS mRNA expression in the postischemic brain began between 24 h and 48 h, peaked at 96 h, and subsided 7 days after pMCAO.

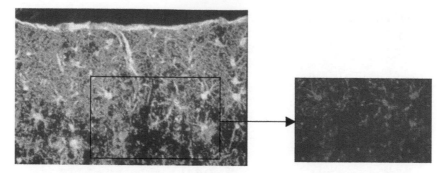

Figure 17.5 *(A color version of this figure follows page 236.)* The left photo shows double immunohistochemical staining of reactive astrocytes in the rat cortex region for S100B (green) and inducible nitric oxide synthesis (iNOS) (red) as examined by fluorescence microscopy. The right photo shows the immunohistochemical staining for iNOS alone. (Courtesy of Dr. Setsuya Fujita, Louis Pasteur Center for Medical Research, Kyoto, Japan.)

Mice lacking the iNOS gene did not express iNOS mRNA or protein after MCAO. Whereas amelioration of ischemic damage as well as the neurological deficits was observed in mice lacking the iNOS gene as compared to wild-type mice, this amelioration was significant at 96 h, not at 24 h, when iNOS is not yet expressed. Delayed treatment with aminoguanidine, a relatively selective inhibitor of iNOS administered at 12 h or 24 h after MCAO in the rat, induced a significant decrease in the neocortical infarct volume.[69] Iadecola et al.[70] detected iNOS immunoreactivity only in polymorphonuclear leukocytes invading the infarct border, not in astrocytes. Nonetheless, Iadecola stated in a paper published later[71] that the cellular localization of iNOS upregulation requires further investigation. Whereas all three NOS isoforms can be expressed by astrocytes, there is no firm evidence for the constitutive expression of NOS in oligodendrocytes or microglia.[72]

Subsequent reports showed that iNOS is primarily induced in reactive astrocytes after transient global ischemia,[73] or pMCAO.[74] Transcriptional activation of iNOS in astrocytes did not occur until 1 day after pMCAO.[75] Using male and female mice in which the iNOS gene was disrupted and their normal littermates as controls, Loihl et al.[75] showed that there was no effect of gender or genotype on the infarct size at 24 h after pMCAO. At 72 h, however, the infarct size was increased in male mice, but not in female mice or in mice of either gender with the gene disruption. The above results were interpreted as indicating that iNOS plays a role in the later development of the infarct in male mice. Also, it was surmised that female mice are either protected against the damaging effects of NO, or protected because iNOS expression/activity is modulated by steroids.

Using the transient MCAO (tMCAO) model in the rat, Tateishi et al.[76] showed that cerebral infarction was minimal on 1 day but progressively increased until 7 days. This delayed infarct expansion was preceded by an enhanced immunoreactivity to S100 and GFAP as well as increases of cyclo-oxygenase-2 and iNOS mRNA expression in the peri-infarct area.[77] The results of our immunohistochemical study with the above model are shown in Figure 17.5 (unpublished data). Three days after tMCAO, the gemistocytic, reactive astrocytes showed a striking increase in the expression of both S100B and iNOS, whose colocalization in astrocytes is apparent in the presented photos.

From the above pieces of evidence, it may be surmised that pMCAO leads to delayed infarct expansion through the following sequence of events:

1. Cell damage in the ischemic core sends signals such as decreased pH, increased extracellular concentration of K^+, Ca^{2+}, or glutamate, and spreading depression to ambient microglia and astrocytes.[78]
2. Microglia and astrocytes in the vicinity of the infarct border that have escaped death are activated and develop the "cytokine cycle,"[3] leading to a precipitous increase in the extracellular concentration of S100B.

3. The high concentration of S100B induces NFκB activation, presumably through binding with RAGE,[55] inducing the expression of iNOS in reactive astrocytes.[54]

4. The resultant production of a large amount of NO induces apoptotic death of astrocytes, neurons, and other cells,[53] and hence delayed infarct expansion ensues.

Although the occurrence of the individual event constituting each event in the above chain of events is supported by ample evidence, direct evidence supporting their linkage is still lacking. In particular, the issue of the central role of S100B in connecting the "cytokine cycle" to the occurrence of cell death requires verification. If at least a significant suppression of the astrocytic S100B production achieved by a suitable method should result in the inhibition of succeeding events, the linkage among the components of the above chain of events would be bolstered. For this purpose, we used a novel pharmacological agent, ONO-2506 ([R]-[–]-2-propyloctanoic acid, ONO Pharmaceutical Co. Ltd.), which has been shown to suppress astrocytic S100B synthesis as follows.

17.5 PROPERTIES AND *IN VITRO* AND *IN VIVO* EFFECTS OF ONO-2506

ONO-2506 was discovered through vigorous screening carried out in search of an agent that possesses an inhibitory action on astrocytic synthesis of S100B. In preliminary experiments, astrocytes obtained from the cerebral cortex of neonatal Wistar rat were cultured for 14 days in the presence or absence of ONO-2506. While the agent did not affect the viability or growth of cultured astrocytes or neurons, it dose-dependently decreased the contents of S100B in cultured astrocytes and nerve growth factor β (NGFβ) in the culture media with an ED_{50} value of around 60 μM for each effect.[79] Using a similar method, the effect of this agent on the mRNA expression of various proteins in cultured astrocytes was examined. In the concentration range of 30–300 μM, ONO-2506 significantly increased the mRNA expression of glutamate transporters (GLT-1 and GLAST) and GABA receptors (GABA$_A$-R β1, GABA$_A$-R β2, and GABA$_A$-R β3). At a similar concentration range, the agent significantly decreased the mRNA expression of S100B and NGFβ and the lipopolysaccharide-stimulated mRNA expression of iNOS and cyclooxygenase-2.[77,79,80] The mRNA expression of glyceraldehyde-3-phosphate dehydrogenase or GFAP was not affected. The molecular mechanism underlying the above wide spectrum of actions of ONO-2506 is unclear at present. Since a variety of proteins, including those mentioned above, are expressed downstream of the S100B binding, the most plausible hypothesis would be that ONO-2506 interferes with the pathway of S100B synthesis in activated astrocytes. It seems plausible that the observed effects on other proteins may simply be owing to the lowering of the intra- and extracellular levels of S100B.

Regarding the cell viability, ONO-2506 exerted no significant effects on the unstimulated neuronal cells cocultured with astrocytes. Glutamate-induced neuronal death was not inhibited by ONO-2506 at doses of 1–300 μM when neurons were cultured alone. When neurons were cocultured with astrocytes, a remarkable protective effect of ONO-2506 became manifest against glutamate- or lipopolysaccharide-induced neuronal death in the concentration range of 30–300 μM.[77,79,80] In addition, ONO-2506 inhibited the augmentation of S100B mRNA expression in cultured astrocytes but did not affect that of GFAP mRNA. In the rat tMCAO model, ONO-2506 markedly inhibited the rise of S100B concentration in the CSF and the cerebral cortex. Administration of ONO-2506 at 6 h after tMCAO significantly inhibited the increase in iNOS and cyclooxygenase-2 mRNA compared with the control levels at 24 h after tMCAO, significantly ameliorating the neurological deficits and reducing the infarct volume at 72 h.[77] The results of the above studies led us to examine the effect of ONO-2506 on the infarct volume and the neurological deficits after pMCAO in the rat. The major findings of this study are reported elsewhere[64] and are briefly described in the following section.

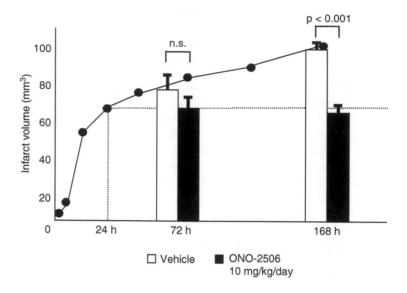

Figure 17.6 Effect of ONO-2506 on the infarct volume at 72 and 168 h after permanent middle cerebral artery occlusion (pMCAO) (bars). The line denotes alternations in the infarct volume after pMCAO as shown in Figure 17.1.

17.5.1 Effects of ONO-2506 on the Infarct Volume after pMCAO

Administration of ONO-2506 started immediately after pMCAO significantly reduced the infarct volume at 168 h but not at 72 h (Figure 17.6). Clearly, the agent did not inhibit the acute infarct expansion during the initial 24 h but almost completely inhibited delayed infarct expansion. In the succeeding experiment, rats were allocated to groups receiving the first drug administration at 24, 48, or 72 h after pMCAO, and the infarct volumes at 168 h were compared. ONO-2506 administration started at 24 h and 48 h significantly decreased the infarct volumes at 168 h by approximately 43% and 35%, respectively. That started at 72 h was without effect. This result clearly showed that the agent has a wide therapeutic time window, which coincides with the peak of augmented astrocytic S100B synthesis after pMCAO. The number of TPCs in the ischemic hemisphere at 72 h after pMCAO was markedly reduced by administration of ONO-2506. Also, the tissue levels of S100B as well as GFAP in the peri-infarct area were significantly decreased by the agent. Since the agent does not affect the GFAP expression in cultured astrocytes, the above result may be interpreted as indicating that the reduction in the tissue levels of S100B led to the inhibition of astrocytic activation in the peri-infarct area, and hence the decrease in the GFAP levels ensued. Thus, administration of ONO-2506 led to a significant inhibition of the astrocytic synthesis of S100B, a generalized inhibition of astrocytic activation, and a reduction in the number of TPCs in the peri-infarct area. Delayed infarct expansion was almost completely inhibited. Importantly, the above temporal profile of infarct inhibition is congruous with those observed in mice lacking iNOS[68] and aminoguanidine-treated mice.[69] Thus, our results bolster the concept that the augmented production of S100B by astrocytic activation plays a pivotal role in the occurrence of delayed infarct expansion after pMCAO, whereby the enhancement of astrocytic expression of iNOS plays a cardinal role in the occurrence of cell death. Conversely, it may be conjectured that ONO-2506 inhibits delayed infarct expansion primarily through its inhibitory action on the astrocytic S100B synthesis.

17.5.2 Effects of ONO-2506 on the Neurological Deficits

Representative data showing the effect of the administration of ONO-2506 started at 24 h after pMCAO (10 mg/kg, i.v., once a day, for 6 days) on the neurological deficits determined by the

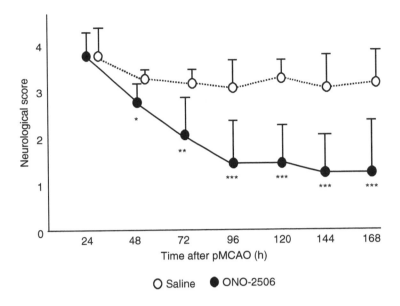

Figure 17.7   Effects of ONO-2506 on the neurological score after permanent middle cerebral artery occlusion (pMCAO).

modified neurological deficits score[81] are shown in Figure 17.7. Surprisingly, a significant improvement in the neurological score as well as the spontaneous activities was manifest as early as the day after the first administration and continued for the succeeding several days. Since the inhibition of infarct expansion does not reach a significant level at 72 h in the present model, the above result clearly indicates that ONO-2506 acts to improve the brain dysfunction through a mechanism that is unrelated or parallel to the inhibition of infarct expansion. In this regard, Dirnagl et al.[82] pointed out that symptoms can regress while the lesion actually expands. This statement is based on the assumption that the neurological deficits reflect injury to the core as well as the penumbra, and as collateral perfusion develops, brain function can be restored within the penumbra. However, since ONO-2506 does not induce any acute alteration in the regional cerebral blood flow within the peri-infarct area (unpublished data), it seems unlikely that the observed improvement of the neurological deficits is primarily induced by the recovery of regional cerebral blood flow. Based on our preceding result showing that ONO-2506 enhances the expression as well as the activities of astrocytic glutamate transporters, we proceeded as described in the following section to examine if the agent might influence the extracellular levels of glutamate in the peri-infarct area.

17.5.3 Effects of ONO-2506 on the Extracellular Levels of Glutamate after tMCAO in the Rat

In the rats subjected to tMCAO, the levels of extracellular glutamate ([Glu]e) in the ischemic cortices in a predetermined location within the peri-infarct area were continuously measured using intracerebral microdialysis. The alterations in the [Glu]e levels in the sham-operated and tMCAO-operated groups with or without drug administration were compared.[65] After tMCAO, the [Glu]e levels transiently increased, returned to normal on reperfusion, and increased again at around 5 h. In the saline-treated group, the [Glu]e levels further increased starting at 15 h, reaching about 280% of the normal level at 24 h. This secondary increase in the [Glu]e levels did not occur in the ONO-2506–treated group. Around 24 h after tMCAO, the basal level of glutamate was markedly increased in the saline-treated group, whereas that in the ONO-2506–treated group was close to normal (Figure 17.8). Then, the glutamate transporter inhibitor, L-*trans*-pyrrolidine-2,4-dicarboxylic acid was infused intracerebrally for 60 min. The agent induced immediate and prominent increases in the [Glu]e levels in the sham-operated

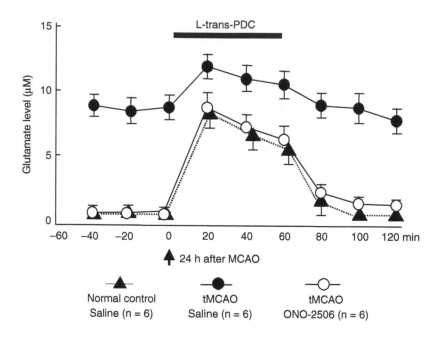

Figure 17.8 Effects of glutamate transporter inhibitor, L-*trans*-pyrrolidine-2,4-dicarboxylic acid (L-*trans*-PDC) on tissue glutamate levels in the normal control rats and the rats subjected to transient middle cerebral artery occlusion (tMCAO) with and without ONO-2506 administration.

and ONO-2506–treated group, but the increase in the saline-treated group was far less marked. This result indicates that the activities of astrocytic glutamate transporters were maintained at near normal levels in the ONO-2506–treated group, whereas they were markedly attenuated in the saline-treated group (Figure 17.8). It may be surmised, therefore, that ONO-2506 prevented the progressive dysfunction of astroglical glutamate transporters in the peri-infarct astrocytes, leading to the maintenance of the near-normal levels of [Glu]e. Corollaries of the above finding are discussed below.

First, it has been shown that cerebral ischemia induces a prompt and delayed rise in the [Glu]e levels in the ischemic core as well as the penumbra,[83–87] which exerts a toxic influence on neurons.[88,89] In the tMCAO model used in this study, the peri-infarct area where the [Glu]e levels were measured was histologically intact at 24 h but was incorporated in the expanding infarct area a few days thereafter. In addition to the putative toxic action of S100B, therefore, the elevated [Glu]e levels may also be involved in the causative mechanisms underlying delayed infarct expansion. It is worth mentioning that N-methyl-D-aspartate receptor activation produces concurrent generation of nitric oxide and reactive oxygen species by NOS,[90] particularly in the state of L-arginine depletion.[91] Therefore, the elevated [Glu]e levels in the peri-infarct area is likely to induce neuronal death by the above mechanism. Second, astrocytes have the ability to modulate neuronal currents and synaptic transmission.[92] The elevation of the [Glu]e levels is considered to participate in the functional suppression in the wide peri-infarct area[93] because glutamate is one of the factors controlling gap-junction–mediated communication between astrocytes.[94] Specifically, GLT-1 plays critical roles in long-term potentiation induction, as well as in short-term potentiation, through regulation of the [Glu]e levels.[95]

17.5.4 Effects of Astrocytic Proteins on Cognitive Functions and Neuronal Survival

In addition to the possible detrimental role of the elevated [Glu]e levels on brain functions described in the previous section, several lines of evidence indicate that S100B and even GFAP may be involved in the modulation of cognitive functions such as learning and memory. While pMCAO in the rat has been shown to cause long-term spatial cognitive impairment,[96] recent evidence suggests that S100B itself may affect spatial learning and memory. Overexpression of S100B has been shown to alter synaptic plasticity and impair spatial learning in transgenic mice, effects attributable to decreased long-term potentiation.[97,98] Conversely, mutant mice devoid of S100B have strengthened synaptic plasticity as identified by enhanced long-term potentiation in the hippocampal CA1 region.[99] Regarding GFAP, GFAP-null astrocytes have been shown to be a better substrate for neuronal survival and neurite outgrowth than wild-type astrocytes.[100] GFAP-null mice displayed enhanced long-term potentiation of both population spike amplitude and excitatory postsynaptic potential slope compared with control mice, suggesting that GFAP is important for astrocyte–neuronal interactions.[101] Conversely, overexpression of human GFAP in astrocytes of transgenic mice provokes fatal encephalopathy.[102]

Furthermore, aging seems to be associated with an inflammatory response and oxidative stress in both the neocortex and the cerebellum, where GFAP is one of the genes that undergoes a twofold increase in expression,[103] which parallels changes seen in human neurodegenerative disorders. Although the increases in GFAP seen in aged astrocytes may be the result of a response to the inflammatory and oxidative state of the aging brain, it is important to note that inflammatory and oxidative responses promote alterations in calcium signaling, which is the primary signaling mechanism by which astrocytes modulate neuronal function and could thus be critical for the progression from the "aged" to the "diseased" brain.[104] Indeed, Badan et al.[105] recently reported that compared with young rats, aged rats showed accelerated astrocytic and microglial reactions that peaked during the first week after tMCAO, which coincided with the stagnation of functional recovery. In a recent review, Cotrina and Nedergaard[104] advocated that the changes exhibited by activated astrocytes, together with the associated modifications in cell surface molecules and extracellular matrix, may be associated with alterations in astrocytic–neuronal interactions that affect synaptic activity and neuronal survival.

Our results with ONO-2506 are in harmony with the above line of thinking. This agent induced significant decreases in astrocytic synthesis of S100B and GFAP as well as the [Glu]e levels in the peri-infarct area. It seems most probable that these effects led in concert to the inhibition of delayed infarct expansion and the prompt improvement of the neurological deficits. Decreased GFAP expression appears to indicate a generalized suppression of astrocytic activation, which in turn indicates the suppression of calcium signaling that triggers astrocytic activation. The key event causing the suppression of calcium signaling would be the inhibition of astrocytic synthesis of S100B, because S100B is known to raise the intracellular levels of calcium and to propagate the host response by perpetuating recruitment and activation of cellular effectors.[47,55,57] Thus, S100B is thought to play a central role in the astrocytic responses in the peri-infarct area, particularly in the subacute phase of permanent focal ischemia, thereby inducing widespread inhibition of neuronal functions. Furthermore, astrocytic activation has been shown to exert harmful effects in phases and loci different from those in the rat pMCAO model, as is described in the following sections.

17.6 CHRONIC NEURODEGENERATION IN REMOTE AREAS AFTER tMCAO OR pMCAO IN THE RAT

Focal cerebral ischemia results in neuronal changes in remote areas that have fiber connections with the ischemic area, such as the substantia nigra[106] and the striatum.[107] Four days after pMCAO in the rat, cell swelling in the pars reticulata took place not in the neurons but in the astrocytes, which is consistent with the high signal intensity seen on T2- and diffusion-weighted images, as well as with the apparent diffusion coefficient reduction.[107] In human stroke patients, secondary degeneration in the ventral nuclei of the thalamus was seen as regions of slightly low signal on proton-density or T2-weighted images, mostly obtained a few weeks after the occurrence of the stroke.[107] The changes in the dorsolateral striatum of the rat from 4 h through 16 weeks after 15-min tMCAO were examined in detail by Fujioka et al.[108] The T1-weighted magnetic resonance imaging signal intensity of the dorsolateral striatum was found to increase from 5 days to 4 weeks, and then subsequently to decrease until 16 weeks. The manganese concentration of the dorsolateral striatum was significantly increased (2.5-fold) from 1 to 4 weeks after ischemia and was associated with induction of manganese-superoxide dismutase and glutamine synthase by astrocytic activation. The neuronal survival ratio in the dorsolateral striatum decreased significantly from 4 h through 16 weeks, accompanied by extracellular β-APP accumulation and chronic glial/inflammatory responses. The author concluded that the observed long-lasting glial and inflammatory responses may play an important role in the slowly maturing striatal degeneration and that the delayed magnetic resonance imaging change of T1 hyperintensity results at least partly from manganese accumulation associated with manganese-superoxide dismutase and glutamine synthase induction by astrocytic activation. The rats with striatal T1 hyperintensity after tMCAO exhibited late behavioral deterioration that was thought to be due to striatal neurodegeneration. The above findings are of utmost importance on account of their apparent relation to the late-onset cognitive decline in stroke patients[109,110] as well as the progression of Alzheimer's disease after stroke.[111] In this regard, it is of interest that the well-documented detrimental effect of the *APOE* ε4 allele on a variety of brain lesions including stroke and Alzheimer's disease appears to be at least partly related to enhancement of astrocytic activation, as is discussed in the following section.

17.7 INCREASED VULNERABILITY TO FOCAL ISCHEMIA IN HUMAN APOLIPOPROTEIN E4 KNOCK-IN MICE

Human apolipoprotein E (apoE) plays a crucial role in lipid and cholesterol transport among various cells, playing a key role in the mobilization and repartitioning of cholesterol and phospholipids during membrane remodeling, repair, and regeneration.[112] Among the three human apoE isoforms encoded by the human *APOE* gene, the apoE4 isoform has been shown to exacerbate brain damage in humans and animals.[113–115] Mori et al.[116] compared the isoform-specific vulnerability conferred by human apoE to ischemic brain damage, using homozygous human apoE2 (2/2)-, apoE3 (3/3)-, or apoE4 (4/4)-knock-in (KI) mice. Twenty-four hours after pMCAO, the infarct volume as well as the neurological scores were significantly more exacerbated in 4/4-KI mice versus 2/2- and 3/3-KI mice, showing that the apoE4 isoform exacerbates acute infarct expansion. In a subsequent study using similar genetically engineered KI mice, whether an apoE isoform-specific exacerbation of delayed infarct expansion occurs after pMCAO was examined.[117] At 1, 3, 5, and 7 days after pMCAO, 4/4-KI mice exhibited significantly larger infarct volumes and worse neurological deficits than 2/2- or 3/3-KI mice, with no significant differences between the latter two groups. Infarct volumes were significantly increased from 1 to 5 days after pMCAO only in 4/4-KI mice. While reactive astrocytosis in the peri-infarct area as evidenced by the GFAP immunoreactive burden was notable in all KI mice, it was significantly augmented in 4/4-KI mice versus 2/2- or 3/3-KI mice (Figure 17.9).

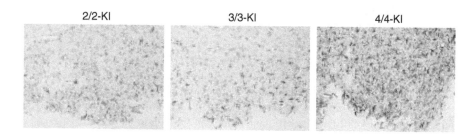

| 2/2-KI | 3/3-KI | 4/4-KI |

Figure 17.9 *(A color version of this figure follows page 236.)* Expression of glial fibrillary acidic protein in the peri-infarct area at 5 days after permanent middle cerebral artery occlusion (pMCAO) in homozygous human apolipoprotein E (apoE)2 (2/2)-, apoE3 (3/3)-, or apoE4 (4/4)-knock-in mice. Note the pronounced astrocytic activation in 4/4-KI mice compared to 2/2- or 3/3-KI mice at 5 days after pMCAO. The infarct border lies in the lowermost part of each figure.

A significant correlation was demonstrated between the GFAP immunoreactive burden and delayed infarct expansion in 4/4-KI mice, but not in 2/2- or 3/3-KI mice. From the above results, it may be surmised that the apoE4 isoform acts to aggravate delayed infarct expansion at least partly through augmentation of reactive astrocytosis and the associated inflammatory responses.

17.8 CONCLUDING REMARKS

After cerebral ischemia, astrocytes change with regard to their shapes and functions, playing different roles in the different phases of infarct evolution. Particularly during the subacute phase of permanent focal cerebral ischemia, activated astrocytes are thought to promote delayed infarct expansion by creating the "cytokine cycle" with activated microglia, wherein the astrocytic synthesis of S100B plays a central role. A similar pathogenic mechanism seems to be in play in the chronic neurodegeneration after focal cerebral ischemia in remote areas such as the thalamus, the striatum, and the substantia nigra.

Insomuch as the augmented astrocytic S100B synthesis is a widely prevalent feature accompanying diverse brain pathology its biological significance with regard to cerebral ischemia has remained elusive. As an adjunct of the "clean-up mechanism" for the infarcted brain tissue,[118] it may serve to kill the still-viable but inept neurons and other damaged cells. If so, the biological significance of S100B expression would be to promote the repair of damaged tissue and prepare for the forthcoming angiogenesis and possible neurogenesis. Alternatively, it may serve to perpetuate inflammatory and immunological reactions as suggested by Hofmann et al.[55] If so, in the absence of pathogens, the S100B-induced inflammatory reaction in cerebral ischemia might be regarded as a cryptic activity of the redundant biological defense mechanisms, hence a phenomenon amenable to therapy. In any event, it seems premature to conclude that the suppression of astrocytic S100B synthesis, even if it could be accomplished by some means, would invariably induce favorable results.

The series of our studies have shown that ONO-2506 inhibits delayed infarct expansion and promptly improves the neurological deficits after focal cerebral ischemia, which is consistent with the current view that activated astrocytes affect synaptic activity and neuronal survival. However, since a number of agents that exhibited beneficial effects in the rodent models of cerebral ischemia have failed to live up to their promise clinically,[119] there is no room for excessive optimism in the future development of the agent. Notwithstanding, we can think of some advantages as well as disadvantages of ONO-2506 in comparison with previous putative brain protective agents such as glutamate inhibitors. First, it is of importance that the agent targets activated astrocytes. Astrocytes outnumber neurons by five- to tenfold in the adult brain,[120] and the ratio of glia to neurons increases through phylogeny.[94] Whether this feature would render the beneficial effect of the agent stronger or weaker in the human brain than in the rodent brain is unpredictable. Second, S100B released from reactive astrocytes has a Janus face, killing the already moribund cells in the vicinity of the infact border in one hand, and promoting the reparative processes such as synaptogenesis on the other hand. Therefore, suppression of S100B synthesis by ONO-2506 would also have its bright and dark sides. Third, the fact that the agent solely inhibits delayed infarct expansion, not the acute, might obviate its clinical efficacy for the following reasons. Being situated in the periphery of the MCA territory, the sensorimotor cortices are only partially damaged after ischemia in the rodent models. By contrast, the sensorimotor cortices as well as the internal capsule are situated in the center of the MCA territory in the human, and hence they tend to be irreversibly damaged immediately after ischemia. Therefore, we could hardly expect that ONO-2506 would exhibit such a marked amelioration of the neurological deficits in man as was observed in the rat. As far as stroke is concerned, therefore, the marked beneficial effect of the agent on sensorimotor functions that was observed in the rodent model would be expected to occur only in those patients who escaped irreversible damage to the sensorimotor cortex immediately after ischemia. By contrast, poststroke dementia and depression, which have recently been attracting attention more than ever, may be regarded an appropriate therapeutic target of the agent. Furthermore, it is tempting to speculate that chronic neurodegerative diseases rather than stroke might also be amenable to treatment by ONO-2506. This line of investigation is underway in our laboratory.

In conclusion, regardless of whether the future development of the agent will end up as a success or failure, it may be stated that ONO-2506 has contributed to partly unraveling the role of glial–neuronal interaction in infarct development and that the role of astrocytic activation in various CNS disorders certainly deserves further exploration.

REFERENCES

1. Plum, F. et al., Glial cells: The silent partners of the working brain, *J. Cereb. Blood Flow Metab.*, 5 (Suppl. 1), 1985.
2. Sheng, J.G. et al., *In vivo* and *in vitro* evidence supporting a role for the inflammatory cytokine interleukin-1 as a driving force in Alzheimer pathogenesis, *Neurobiol. Aging*, 17, 761–766, 1996.
3. Griffin, W.S. et al., Glial-neuronal interactions in Alzheimer's disease: The potential role of a 'cytokine cycle' in disease progression, *Brain Pathol.*, 8, 65–72, 1998.
4. Giulian, D., Microglia and neuronal dysfunction, in *Neuroglia*, Kettenmann, H. and Ransom, B.R., Eds., Oxford University Press, New York, 1995, pp. 671–684.
5. Anderson, M.F. et al., Astrocytes and stroke: Networking for survival?, *Neurochem. Res.*, 28, 293–305, 2003.
6. Eddleston, M. and Mucke, L., Molecular profile of reactive astrocytes—implications for their role in neurologic disease, *Neuroscience*, 54, 15–36, 1993.
7. Ridet, J.L. et al., Reactive astrocytes: Cellular and molecular cues to biological function, *Trends Neurosci.*, 20, 570–577, 1997.
8. Petito, C.K. et al., The two patterns of reactive astrocytosis in postischemic rat brain, *J. Cereb. Blood Flow Metab.*, 10, 850–859, 1990.
9. Schmidt-Kastner, R., Szymas, J., and Hossmann, K.A., Immunohistochemical study of glial reaction and serum-protein extravasation in relation to neuronal damage in rat hippocampus after ischemia, *Neuroscience*, 38, 527–540, 1990.
10. Chen, H. et al., Sequential neuronal and astrocytic changes after transient middle cerebral artery occlusion in the rat, *J. Neurol. Sci.*, 118, 109–116, 1993.
11. Clark, R.K. et al., Development of tissue damage, inflammation and resolution following stroke: An immunohistochemical and quantitative planimetric study, *Brain Res. Bull.*, 31, 565–572, 1993.
12. Garcia, J.H. et al., Progression from ischemic injury to infarct following middle cerebral artery occlusion in the rat, *Am. J. Pathol.*, 142, 623–635, 1993.
13. Yamashita, K. et al., Monitoring the temporal and spatial activation pattern of astrocytes in regional cerebral ischemia using *in situ* hybridization to GFAP mRNA: Comparison with sgp-2 and hsp70 mRNA and the effect of glutamate receptor antagonists, *Brain Res.*, 735, 285–297, 1996.
14. Jabs, R., Bekar, L.K., and Walz, W., Reactive astrogliosis in the injured and postischemic brain, in *Cerebral Ischemia*, Walz, W., Ed., Humana Press, Totowa, 1999, pp. 233–249.
15. Wei, L. et al., Collateral growth and angiogenesis around cortical stroke, *Stroke*, 32, 2179–2184, 2001.
16. Song, H., Stevens, C.F., and Gage, F.H., Astroglia induce neurogenesis from adult neural stem cells, *Nature*, 417, 39–44, 2002.
17. Garcia, J.H. and Kamijyo, Y., Cerebral infarction. Evolution of histopathological changes after occlusion of a middle cerebral artery in primates, *J. Neuropathol. Exp. Neurol.*, 33, 408–421, 1974.
18. Kirino, T., Tamura, A., and Sano, K., Early and late neuronal damage following cerebral ischemia, in *Advance in Behavioral Biology, Mechanism of Cerebral Hypoxia and Stroke*, Somjen, G.J., Ed., Plenum Press, New York, 1988, pp. 23–34.
19. Asano, T. et al., The ameliorative effect of ONO-2506 on the delayed and prolonged expansion of the infarct volume following permanent middle cerebral artery occlusion in rats, *J. Cereb. Blood Flow Metab.*, 119 (Suppl. 1), S64, 1999.

20. Matsui, T. et al., Astrocytic activation and delayed infarct expansion after permanent focal ischemia in rats. Part I: Enhanced astrocytic synthesis of S-100β in the periinfarct area precedes delayed infarct expansion, *J. Cereb. Blood Flow Metab.*, 22, 711–722, 2002.

21. Du, C. et al., Very delayed infarction after mild regional cerebral ischemia: A role for apoptosis?, *J. Cereb. Blood Flow Metab.*, 16, 195–201, 1996.

22. Warach, S. et al., Acute human stroke studied by whole brain echo planar diffusion-weighted magnetic resonance imaging, *Ann. Neurol.*, 37, 231–241, 1995.

23. Baird, A.E. et al., Enlargement of human cerebral ischemic lesion volumes measured by diffusion-weighted magnetic resonance imaging, *Ann. Neurol.*, 41, 581–589, 1997.

24. Beaulieu, C. et al., Longitudinal magnetic resonance imaging study of perfusion and diffusion in stroke: Evolution of lesion volume and correlation with clinical outcome, *Ann. Neurol.*, 46, 568–578, 1999.

25. Swanson, R.A., Farrell, K., and Stein, B.A., Astrocyte energetics, function, and death under conditions of incomplete ischemia: A mechanism of glial death in the penumbra, *Glia*, 21, 142–153, 1997.

26. Kimelberg, H.K., Brain edema, in *Neuroglia*, Kettenmann, H. and Ransom, B.R., Eds., Oxford University Press, New York, 1995, pp. 919–935.

27. Nedergaard, M., Cooper, A.J., and Goldman, S.A., Gap junctions are required for the propagation of spreading depression, *J. Neurobiol.*, 28, 433–444, 1995.

28. Martins-Ferreira, H., Nedergaard, M., and Nicholson, C., Perspectives on spreading depression, *Brain Res. Brain Res. Rev.*, 32, 215–234, 2000.

29. Hossmann, K.A., Viability thresholds and the penumbra of focal ischemia, *Ann. Neurol.*, 36, 557–565, 1994.

30. Liu, D. et al., Astrocytic demise precedes delayed neuronal death in focal ischemic rat brain, *Brain Res. Mol. Brain Res.*, 68, 29–41, 1999.

31. Graham, D.I. and Lantos, P.L., *Greenfield's Neuropathology*, 6th ed., Arnold, London, 1997.

32. Barone, F.C. and Feuerstein, G.Z., Inflammatory mediators and stroke: New opportunities for novel therapeutics, *J. Cereb. Blood Flow Metab.*, 19, 819–834, 1999.

33. Benveniste, E.N., Cytokine production, in *Neuroglia*, Kettenmann, H. and Ransom, B.R., Eds., Oxford University Press, New York, 1995, pp. 700–713.

34. Davies, C.A. et al., An integrated analysis of the progression of cell responses induced by permanent focal middle cerebral artery occlusion in the rat, *Exp. Neurol.*, 154, 199–212, 1998.

35. Rothwell, N.J., Annual review prize lecture cytokines-killers in the brain?, *J. Physiol.*, 514, 3–17, 1999.

36. Donato, R., Functional roles of S100 proteins, calcium-binding proteins of the EF-hand type, *Biochim. Biophys. Acta*, 1450, 191–231, 1999.

37. Moore, B.W., A soluble protein characteristic of the nervous system, *Biochem. Biophys. Res. Commun.*, 19, 739–744, 1965.

38. Zimmer, D.B. et al., The S100 protein family: History, function, and expression, *Brain Res. Bull.*, 37, 417–429, 1995.

39. Heizmann, C.W. and Cox, J.A., New perspectives on S100 proteins: A multi-functional Ca^{2+}-, Zn^{2+}-, and Cu^{2+}-binding protein family, *Biometals*, 11, 383–397, 1998.

40. Barger, S.W., Wolchok, S.R., and Van Eldik, L.J., Disulfide-linked S100β dimers and signal transduction, *Biochim. Biophys. Acta*, 1160, 105–112, 1992.

41. Schäfer, B.W. and Heizmann, C.W., The S100 family of EF-hand calcium-binding proteins: Functions and pathology, *Trends Biochem. Sci.*, 21, 134–140, 1996.

42. Isobe, T., Ishioka, N., and Okuyama, T., Structural relation of two S-100 proteins in bovine brain: Subunit composition of S-100a protein, *Eur. J. Biochem.*, 115, 469–474, 1981.

43. Kligman, D. and Marshak, D.R., Purification and characterization of a neurite extension factor from bovine brain, *Proc. Natl. Acad. Sci. U.S.A.*, 82, 7136–7139, 1985.

44. Van Eldik, L.J. et al., Neurotrophic activity of S-100β in cultures of dorsal root ganglia from embryonic chick and fetal rat, *Brain Res.*, 542, 280–285, 1991.

45. Barger, S.W., Van Eldik, L.J., and Mattson, M.P., S100β protects hippocampal neurons from damage induced by glucose deprivation, *Brain Res.*, 677, 167–170, 1995.

46. Ahlemeyer, B. et al., S-100β protects cultured neurons against glutamate- and staurosporine-induced damage and is involved in the antiapoptotic action of the $5\text{-}HT_{1A}$-receptor agonist, Bay x 3702, *Brain Res.*, 858, 121–128, 1999.

47. Barger, S.W. and Van Eldik, L.J., S100β stimulates calcium fluxes in glial and neuronal cells, *J. Biol. Chem.*, 267, 9689–9694, 1992.

48. Griffin, W.S., Brain interleukin 1 and S-100 immunoreactivity are elevated in Down syndrome and Alzheimer disease, *Proc. Natl. Acad. Sci. U.S.A.*, 86, 7611–7615, 1989.

49. Sheng, J.G., Mrak, R.E., and Griffin, W.S., Glial-neuronal interactions in Alzheimer disease: Progressive association of IL-1α+ microglia and S100β+ astrocytes with neurofibrillary tangle stages, *J. Neuropathol. Exp. Neurol.*, 56, 285–290, 1997.

50. Penã, L.A., Brecher, C.W., and Marshak, D.R., β-Amyloid regulates gene expression of glial trophic substance S100β in C6 glioma and primary astrocyte cultures, *Brain Res. Mol. Brain Res.*, 34, 118–126, 1995.

51. Yao, J., Kitt, C., and Reeves, R.H., Chronic elevation of S100β protein does not alter APP mRNA expression or promote β-amyloid deposition in the brains of aging transgenic mice, *Brain Res.*, 702, 32–36, 1995.

52. Li, Y. et al., S100β increases levels of β-amyloid precursor protein and its encoding mRNA in rat neuronal cultures, *J. Neurochem.*, 71, 1421–1428, 1998.

53. Hu, J., Ferreira, A., and Van Eldik, L.J., S100β induces neuronal cell death through nitric oxide release from astrocytes, *J. Neurochem.*, 69, 2294–2301, 1997.

54. Lam, A.G. et al., Mechanism of glial activation by S100B: Involvement of the transcription factor NFκB, *Neurobiol. Aging*, 22, 765–772, 2001.

55. Hofmann, M.A. et al., RAGE mediates a novel proinflammatory axis: A central cell surface receptor for S100/calgranulin polypeptides, *Cell*, 97, 889–901, 1999.

56. Bucciarelli, L.G. et al., RAGE is a multiligand receptor of the immunoglobulin superfamily: Implications for homeostasis and chronic disease, *Cell Mol. Life Sci.*, 59, 1117–1128, 2002.

57. Schmidt, A.M. et al., The biology of the receptor for advanced glycation end products and its ligands, *Biochim. Biophys. Acta*, 1498, 99–111, 2000.

58. Sasaki, N. et al., Immunohistochemical distribution of the receptor for advanced glycation end products in neurons and astrocytes in Alzheimer's disease, *Brain Res.*, 888, 256–262, 2001.

59. Ma, L. et al., RAGE is expressed in pyramidal cells of the hippocampus following moderate hypoxic-ischemic brain injury in rats, *Brain Res.*, 966, 167–174, 2003.

60. Mokuno, K. et al., Neuron-specific enolase and S-100 protein levels in cerebrospinal fluid of patients with various neurological diseases, *J. Neurol. Sci.*, 60, 443–451, 1983.

61. Missler, U. et al., S-100 protein and neuron-specific enolase concentrations in blood as indicators of infarction volume and prognosis in acute ischemic stroke, *Stroke*, 28, 1956–1960, 1997.

62. Hårdemark, H.G. et al., S-100 protein and neuron-specific enolase in CSF after experimental traumatic or focal ischemic brain damage, *J. Neurosurg.*, 71, 727–731, 1989.

63. Büttner, T. et al., S-100 protein: Serum marker of focal brain damage after ischemic territorial MCA infarction, *Stroke*, 28, 1961–1965, 1997.

64. Tateishi, N. et al., Astrocytic activation and delayed infarct expansion after permanent focal ischemia in rats. Part II: Suppression of astrocytic activation by a novel agent (*R*)-(-)-2-propyloctanoic acid (ONO-2506) leads to mitigation of delayed infarct expansion and early improvement of neurologic deficits, *J. Cereb. Blood Flow Metab.*, 22, 723–734, 2002.

65. Mori, T. et al., Attenuation of a delayed increase in the extracellular glutamate level in the peri-infarct area following focal cerebral ischemia by a novel agent ONO-2506, *Neurochem. Int.*, 45, 381–387, 2004.

66. Ben-Sasson, S.A., Sherman, Y., and Gavrieli, Y., Identification of dying cells — *in situ* staining, *Methods Cell Biol.*, 46, 29–39, 1995.

67. Van Eldik, L.J. and Griffin, W.S., S100β expression in Alzheimer's disease: Relation to neuropathology in brain regions, *Biochim. Biophys. Acta*, 1223, 398–403, 1994.

68. Iadecola, C. et al., Delayed reduction of ischemic brain injury and neurological deficits in mice lacking the inducible nitric oxide synthase gene, *J. Neurosci.*, 17, 9157–9164, 1997.

69. Nagayama, M., Zhang, F., and Iadecola, C., Delayed treatment with aminoguanidine decreases focal cerebral ischemic damage and enhances neurologic recovery in rats, *J. Cereb. Blood Flow Metab.*, 18, 1107–1113, 1998.

70. Iadecola, C. et al., Inducible nitric oxide synthase gene expression in brain following cerebral ischemia, *J. Cereb. Blood Flow Metab.*, 15, 378–384, 1995.

71. Iadecola, C., Bright and dark sides of nitric oxide in ischemic brain injury, *Trends Neurosci.*, 20, 132–139, 1997.

72. Murphy, S., Production of nitric oxide by glial cells: Regulation and potential roles in the CNS, *Glia*, 29, 1–13, 2000.

73. Endoh, M., Maiese, K., and Wagner, J., Expression of the inducible form of nitric oxide synthase by reactive astrocytes after transient global ischemia, *Brain Res.*, 651, 92–100, 1994.

74. Loihl, A.K. and Murphy, S., Expression of nitric oxide synthase-2 in glia associated with CNS pathology, *Prog. Brain Res.*, 118, 253–267, 1998.

75. Loihl, A.K. et al., Expression of nitric oxide synthase, (NOS)-2 following permanent focal ischemia and the role of nitric oxide in infarct generation in male, female, and NOS-2 gene-deficient mice, *Brain Res.*, 830, 155–164, 1999.

76. Tateishi, N. et al., Activation of astrocytes and ischemic damage following the transient focal ischemia, *Nippon Yakurigaku Zasshi*, 112, 103P–107P, 1998.

77. Shimoda, T. et al., ONO-2506, a novel astrocyte modulating agent, suppresses the increase of COX-2 and iNOS mRNA expression in cultured astrocytes and ischemic brain, *Soc. Neurosci. Abstr.*, 24, 384.13, 1998.

78. Kraig, R.P., Lascola, C.D., and Caggiano, A., Glial response to brain ischemia, in *Neuroglia*, Kettermann, H. and Ransom, B.R., Eds., Oxford Press, New York, 1995, pp. 964–976.

79. Shinagawa, R. et al., Modulating effects of ONO-2506 on astrocytic activation in cultured astrocytes from rat cerebrum, *Soc. Neurosci. Abstr.*, 25, 843.10, 1999.

80. Shinagawa, R. et al., ONO-2506 ameliorates neurodegeneration through inhibition of reduction of GLT-1 expression, *Soc. Neurosci. Abstr.*, 24, 384.14, 1998.

81. Bederson, J.B. et al., Rat middle cerebral artery occlusion: Evaluation of the model and development of a neurologic examination, *Stroke*, 17, 472–476, 1986.

82. Dirnagl, U., Iadecola, C., and Moskowitz, M.A., Pathobiology of ischaemic stroke: An integrated view, *Trends Neurosci.*, 22, 391–397, 1999.

83. Benveniste, H. et al., Elevation of the extracellular concentrations of glutamate and aspartate in rat hippocampus during transient cerebral ischemia monitored by intracerebral microdialysis, *J. Neurochem.*, 43, 1369–1374, 1984.

84. Takagi, K. et al., Changes in amino acid neurotransmitters and cerebral blood flow in the ischemic penumbral region following middle cerebral artery occlusion in the rat: Correlation with histopathology, *J. Cereb. Blood Flow Metab.*, 13, 575–585, 1993.

85. Matsumoto, K. et al., Secondary elevation of extracellular neurotransmitter amino acids in the reperfusion phase following focal cerebral ischemia, *J. Cereb. Blood Flow Metab.*, 16, 114–124, 1996.

86. Taguchi, J. et al., Prolonged transient ischemia results in impaired CBF recovery and secondary glutamate accumulation in cats, *J. Cereb. Blood Flow Metab.*, 16, 271–279, 1996.

87. Enblad, P. et al., Middle cerebral artery occlusion and reperfusion in primates monitored by microdialysis and sequential positron emission tomography, *Stroke*, 32, 1574–1580, 2001.

88. Choi, D.W., Glutamate neurotoxicity and diseases of the nervous system, *Neuron*, 1, 623–634, 1988.

89. Olney, J.W., Excitatory transmitter neurotoxicity, *Neurobiol. Aging*, 15, 259–260, 1994.

90. Gunasekar, P.G. et al., NMDA receptor activation produces concurrent generation of nitric oxide and reactive oxygen species: Implication for cell death, *J. Neurochem.*, 65, 2016–2021, 1995.

91. Culcasi, M. et al., Glutamate receptors induce a burst of superoxide via activation of nitric oxide synthase in arginine-depleted neurons, *J. Biol. Chem.*, 269, 12589–12593, 1994.

92. Aschner, M., Neuron-astrocyte interactions: Implications for cellular energetics and antioxidant levels, *Neurotoxicology*, 21, 1101–1107, 2000.

93. Witte, O.W. et al., Functional differentiation of multiple perilesional zones after focal cerebral ischemia, *J. Cereb. Blood Flow Metab.*, 20, 1149–1165, 2000.

94. Haydon, P.G., GLIA: Listening and talking to the synapse, *Nat. Rev. Neurosci.*, 2, 185–193, 2001.

95. Katagiri, H., Tanaka, K., and Manabe, T., Requirement of appropriate glutamate concentrations in the synaptic cleft for hippocampal LTP induction, *Eur. J. Neurosci.*, 14, 547–553, 2001.

96. Okada, M. et al., Long-term spatial cognitive impairment following middle cerebral artery occlusion in rats. A behavioral study, *J. Cereb. Blood Flow Metab.*, 15, 505–512, 1995.

97. Gerlai, R. et al., Overexpression of a calcium-binding protein, S100β, in astrocytes alters synaptic plasticity and impairs spatial learning in transgenic mice, *Learn Mem.*, 2, 26–39, 1995.

98. Roder, J.K., Roder, J.C., and Gerlai, R., Memory and the effect of cold shock in the water maze in S100β transgenic mice, *Physiol. Behav.*, 60, 611–615, 1996.

99. Nishiyama, H. et al., Glial protein S100B modulates long-term neuronal synaptic plasticity, *Proc. Natl. Acad. Sci. U.S.A.*, 99, 4037–4042, 2002.

100. Menet, V. et al., GFAP null astrocytes are a favorable substrate for neuronal survival and neurite growth, *Glia*, 31, 267–272, 2000.

101. McCall, M.A. et al., Targeted deletion in astrocyte intermediate filament (GFAP) alters neuronal physiology, *Proc. Natl. Acad. Sci., U.S.A.*, 93, 6361–6366, 1996.

102. Messing, A. and Brenner, M., GFAP: Functional implications gleaned from studies of genetically engineered mice, *Glia*, 43, 87–90, 2003.

103. Lee, C.K., Weindruch, R., and Prolla, T.A., Gene-expression profile of the ageing brain in mice, *Nat. Genet.*, 25, 294–297, 2000.

104. Cotrina, M.L. and Nedergaard, M., Astrocytes in the aging brain, *J. Neurosci. Res.*, 67, 1–10, 2002.

105. Badan, I. et al., Accelerated glial reactivity to stroke in aged rats correlates with reduced functional recovery, *J. Cereb. Blood Flow Metab.*, 23, 845–854, 2003.

106. Zhao, F. et al., Ultrastructural and MRI study of the substantia nigra evolving exofocal post-ischemic neuronal death in the rat, *Neuropathology*, 22, 91–105, 2002.

107. Nakane, M. et al., Astrocytic swelling in the ipsilateral substantia nigra after occlusion of the middle cerebral artery in rats, *Am. J. Neuroradiol.*, 22, 660–663, 2001.

108. Fujioka, M. et al., Magnetic resonance imaging shows delayed ischemic striatal neurodegeneration, *Ann. Neurol.*, 54, 732–747, 2003.

109. Tatemichi, T.K. et al., Risk of dementia after stroke in a hospitalized cohort: Results of a longitudinal study, *Neurology*, 44, 1885–1891, 1994.

110. Kokmen, E. et al., Dementia after ischemic stroke: A population-based study in Rochester, Minnesota (1960–1984), *Neurology*, 46, 154–159, 1996.

111. Snowdon, D.A. et al., Brain infarction and the clinical expression of Alzheimer disease. The Nun Study, *J.A.M.A.*, 277, 813–817, 1997.

112. Poirier, J., Apolipoprotein E in animal models of CNS injury and in Alzheimer's disease, *Trends Neurosci.*, 17, 525–530, 1994.

113. Nicoll, J.A., Roberts, G.W., and Graham, D.I., Apolipoprotein E ε4 allele is associated with deposition of amyloid β-protein following head injury, *Nat. Med.*, 1, 135–137, 1995.

114. Strittmatter, W.J. and Roses, A.D., Apolipoprotein E and Alzheimer's disease, *Annu. Rev. Neurosci.*, 19, 53–77, 1996.

115. Laskowitz, D.T., Horsburgh, K., and Roses, A.D., Apolipoprotein E and the CNS response to injury, *J. Cereb. Blood Flow Metab.*, 18, 465–471, 1998.

116. Mori, T. et al., Increased vulnerability to focal ischemic brain injury in human apolipoprotein E4 knock-in mice, *J. Neuropathol. Exp. Neurol.*, 62, 280–291, 2003.

117. Mori, T. et al., Aug mented delayed infarct expansion and reactive astrocytosis after permanent focal ischemia in apolipoprotein E4 knock-in mice, *J. Cereb. Blood Flow Metab.*, 24, 646–656, 2004.

118. Manoonkitiwongsa, P.S. et al., Angiogenesis after stroke is correlated with increased numbers of macrophages: The clean-up hypothesis, *J. Cereb. Blood Flow Metab.*, 21, 1223–1231, 2001.

119. De Keyser, J., Sulter, G., and Luiten, P.G., Clinical trials with neuroprotective drugs in acute ischaemic stroke: Are we doing the right thing?, *Trends Neurosci.*, 22, 535–540, 1999.

120. Bignami, A., Glial cells in the central nervous system, in *Discussion in Neuroscience*, Vol. 8, Elsevier, Amsterdam, 1991, pp. 1–45.

CHAPTER 18

Role of Microglia and Astrocytes in Alzheimer's Disease

Peter Schubert and Stefano Ferroni

CONTENTS

18.1 INTRODUCTION

The question of to what extent reactive glial cells contribute to the pathogenesis of neuron damaging brain diseases is a matter of ongoing debate. Microglia as well as astrocytes are the brain endogenous representatives of the general immune system and can be recruited under pathological conditions as immune competent cells. The microglia are highly sensitive sensors for brain insults and, in standby position, are prepared to bring into play the immunological defense mechanisms. The astrocytes stay in the second line for a possible recruitment, if required. Upon activation, the reactive glial cells gain a number of potentially neurotoxic potencies, such as the release of inflammation promoting cytokines and aggressive oxidative radicals. As long as the gradual use of these weapons remains under the strict yin-yang control of a sophisticated intercellular communication network, there is no doubt about the primarily beneficial function of reactive glial cells in defense and repair. However, evidence has accumulated that an escalating pathological glial activation, which involves both microglia and astrocytes, may contribute to secondary nerve cell damage. The latter occurs, for example, after transient brain ischemia, head trauma, and toxic insults and during the course of neurodegenerative diseases.

When trying to delineate basic pathomechanisms, it turns out that oxidative stress, overstimulation of glutamate receptors, and the consecutive excessive loading of neurons with calcium are instrumental in maintaining progressive nerve cell death.[1] They all may be reinforced by potentially harmful functions of reactive microglia and astrocytes. Since the currently available treatments for neurodegenerative diseases are only symptomatic, an elucidation of the glia-linked pathomechanisms could provide a base for the development of a target-specific causal therapy. This would be urgently required for the treatment of the progressively dementing Alzheimer's disease (AD), which hits 10% of people over 65 years and more than 50% of those over 85 years.[2]

18.2 REACTIVE MICROGLIA: NEUROTOXIC VERSUS BENEFICIAL EFFECTS

When considering the possible contribution of glial cell reactions to the pathogenesis of AD, a central role has been attributed to the microglial cells. They are found in the vicinity of aggregated fibrillar β-amyloid (Aβ) plaques that contain damaged neurites, the so-called active neuritic plaques.[3] This histopathological picture has already been described by Alois Alzheimer in his report of the first AD case.[4] A concomitant microglial reaction was not seen in the diffuse amyloid plaques of nondemented elderly people,[5] supporting a relation of fibrillar Aβ neurotoxicity and microglial activation. These neuritic Aβ plaques, the hallmark of AD, are apparently sites of intensive chronic inflammation.[6] This is concluded from their high content of multiple inflammation-related factors, such as proinflammatory cytokines, acute phase proteins, and complement factors (for a review, see Reference 7). Their expression in reactive microglial cells together with an upregulation of respective surface receptors, including the major histocompatibility complex class II (MHC class II) and complement receptors, indicate an advanced state of activation.[8,9] The finding that aggregated Aβ itself is able to activate the classical complement pathway[10] led to the concept that a great part of neuronal damage in AD results from a self-attack that is based on a microglia-linked aggravation of inflammatory responses.[8] According to this concept, initial neuronal death and fibrillar Aβ reinforces complement activation; this enhances the local recruitment of activated microglia inducing a consecutive glial production of inflammatory cytokines that, in turn, exaggerates neuronal death—several vicious circles that cause the so-called bystander damage via an Aβ- and microglia-maintained autotoxic loop.[7,9]

On the other hand, reactive microglia exert beneficial functions in response to more acute brain insults, e.g., providing trophic support and assistance in regeneration and repair (for a review, see Reference 11). The ability of microglia to transform into effective macrophages, which could diminish the burden of the inflammatory irritant Aβ, would be required for interfering with the neuron-damaging autotoxic loop. But the clearance of the Aβ protein is apparently reduced and not sufficient in AD. It may be that the chronic course of this disease modifies the chemical structure of the β-amyloid plaques,[12] which makes them indigestible for microglia and leads to "frustrated phagocytosis." Alternatively, the microglia could be overstimulated by the piling-up Aβ load and brought to an extreme activation state, which favors their autotoxic properties but hinders their capability for phagocytosis. Another explanation has been offered by Streit,[11] i.e., that the microglia has become "diseased" because of aging or genetic and epigenetic risk factors that are associated with chronic neurodegenerative diseases.

There is, however, some recently evolving possibility that such dysfunctional microglial cells, whether diseased or pathologically overdemanded, may regain their capability of Aβ phagocytosis upon treatment. Thus, vaccination of human Aβ–expressing transgenic mice with Aβ was found to reduce the amount of Aβ deposits.[13] Related *in vitro* studies indicated that the anti-Aβ antibodies, which are generated in response to vaccination, promote Aβ phagocytosis by stimulating fc and complement receptors in microglial cells.[14] It should be considered, however, that a further microglial activation upon vaccination might turn out to be counterproductive in human AD, where the chronic presence of aggregated Aβ already maintains increased complement activation and an overstimulation of microglial cells.[15]

Jantzen and coworkers have reported another experimental approach, which could be relevant for the future therapy of AD.[16] They achieved a pronounced reduction of the toxic fibrillar Aβ load in double transgenic (APP/PS-1) mice by treating the animals with a nitric oxide (NO)–releasing derivative of the nonsteroidal anti-inflammatory drug (NSAID) fluobiprofen. The concomitantly observed dramatic increase of microglia, which expressed MHC class II and complement receptors as marker for an advanced state of activation, suggests that the clearance of the Aβ deposits can be attributed to the pharmacological reinforcement of microglial phagocytosis. A similar reduction of Aβ deposits has been obtained in the same strain of mice upon administration of lipopolysaccharide (LPS), an experimentally used potent trigger of microglial activation.[17]

Taken together, the studies on the transgenic mice models indicate that the microglial capability for phagocytosis is related to an elevated activation state, whereas an extreme state of microglial activation, as apparently reached in human AD, no longer allows Aβ phagocytosis but favors autotoxicity. Since the pathological activation of microglia is a graded response,[18] one can expect that its intensity and quality is determined by the complex interplay of harmful and beneficial regulatory pathways. If so, a major therapeutic question would be how to titrate the microglial activation in order to keep it in a range that allows Aβ phagocytosis but does not reinforce neurotoxic pathomechanisms. The above-described promising findings of the Jantzen group may provide a perspective regarding how this question could be answered. Although the mechanism, which is responsible for the observed beneficial stimulation of microglia, is not clear, it seems likely that it is related to the action of nitric oxide. The reason is that fluobiprofen, the NSAID mother compound lacking the NO-releasing group, turned out to be ineffective in stimulating microglial Aβ phago-cytosis.[16] Interestingly, such NO-releasing drugs have also been observed to block the cellular induction of interleukin-1β (IL-1β) by inhibiting the IL-1 converting enzyme.[19] This could mean that the observed beneficial Aβ clearance upon treatment with the NO-releasing fluobiprofen derivative results from a microglial stimulation that leads to a qualitatively quite different activation state than conventionally seen, i.e., lacking the production of IL-1β. This cytokine not only plays a key role in mediating proinflammatory microglial actions that promote the neuron-damaging autotoxic loop but is also the major trigger for the pathological activation of astrocytes. In our opinion, a most critical point is reached for the development of glia-related progressive nerve cell damage when the secondarily recruited astrocytes are forced to give up physiological control functions and form a neurotoxic alliance with the reactive microglia.[20,21]

18.3 SIGNALING UNDERLYING THE ESCALATING GLIA REACTIONS

Neuritic β-amyloid plaques, the sites of progressive nerve cell damage, are not only associated with reactive microglia.[3] They are also surrounded by reactive astrocytes forming an outer shell from where they send glial fibrillar acidic protein (GFAP)–positive processes into the plaque.[22] This inclusion of astrocytes reflects a highly advanced state of pathological glia activation, which occurs in a gradual manner. The upgrading of the glial response is promoted and controlled by a complex intercellular signaling via cytokine-mediated positive and negative feedback loops.

At low activation states, the immunological recruitment remains restricted to the microglia. While the astrocytes are allowed to maintain their important physiological functions, they also participate in the feedback control of reactive microglial functions. The massive production of reactive oxidative intermediates (ROI) is one of the powerful weapons of activated microglia used for defense.[23,24] However, the highly aggressive peroxynitrates, which may add to autotoxic damage, are only formed in the presence of nitric oxide radicals (NO). Although their release could experimentally be achieved in murine microglia, it seems that human microglia is not able to produce significant amounts of NO.[24] On the other hand, some production of reactive nitrogen intermediates upon Aβ stimulation has been observed in microglia obtained from autopsies of AD patients.[25] As long as the glia activation remains confined to the microglia, astrocytes are able to inhibit a potential microglial NO production via the release of the tumor growth factor-β (TGF-β). This is indicated by *in vitro* experiments on rodent microglia.[26]

TGF-β functions as an immune-suppressive agent[27] and inhibited also the production of the tumor necrosis factor-α (TNF-α) in human microglial cells.[28] Although TNF-α may exert beneficial functions, its capacity to exaggerate brain inflammation seems to be an important pathogenic factor in AD. Accordingly, elevated concentrations were found in the cerebrospinal fluid and in the cortex of Alzheimer patients.[29] This cytokine belongs to the group of positive feedback signals, which may promote the pathological glia activation. But the observation that TNF-α itself can induce the expression of the inhibitory TGF-β in glial cells[30] indicates the existence of primarily installed control mechanisms, which limit the risk of an inadequate glial response. Another control is provided by the ambiguous cytokine interleukin-6 (IL-6), which cooperates with TNF-α in mediating brain inflammation but, conversely, may also inhibit the glial production of TNF-α.[31] Again, this negative feedback control is induced by TNF-α itself, which is able stimulate the expression of IL-6 in astrocytes.[32]

On the other hand, TNF-α has the potential to reinforce the free radical as well as the cytokine production via autocrine and paracrine circuits (for a review, see Reference 33). Thus, microglia possess TNF-α receptors (type IP55), which mediate via the activation of NFκB a further increased TNF-α expression in microglia.[34] This autocrine loop is amplified by a paracrine loop that also forces the astrocytes to add to an increased TNF-α production. As a consequence, the threshold may be reached to upgrade the microglial activation state, as indicated by the expression of MHC-class II known to be inducible upon stimulation with TNF-α.[32] Upgraded microglia will release the whole battery of glia-produced inflammatory cytokines at increased amounts. Among those, IL-1β may reach a sufficient level to trigger the secondary recruitment of astrocytes into the phalanx of immune-effective cells. Such a highly upregulated IL-1β production occurs in β-amyloid plaques, where it seems to be involved in a number of functions:

1. It could further increase Aβ production by stimulating the synthesis of APP and the production of chaperones in astrocytes that promote the formation of toxic fibrillar β-amyloid.
2. It may induce the potentiated production of other cytokines in an autocrine/paracrine fashion.
3. It may induce the NO production in astrocytes.
4. It may stimulate via cyclooxygenase activation the production of prostaglandin that is involved in the regulation of glutamate release from astrocytes[33] (see below).

It is still an open question how the glia-mediated autotoxic loop is started in the amyloid plaques. The dangerous cytokine IL-1β has already been detected in diffuse amyloid plaques before the onset of neuronal damage.[35] The major stimulus that turns the plaques into sites of progressive inflammation is thought to be a pathological APP processing, which favors an excessive formation of aggregating β-amyloid.[8] Additional factors, such as brain trauma, intoxication, or ischemia, may promote the glia activation and together with genetic risk factors for AD contribute to the generation of initial lesions. As soon as neurons are damaged, they will release ATP. Extracellular ATP causes a pathological activation of purinergic P2X7 receptors, which is thought to play an important role in the development of AD.[36] There is evidence that this ATP signaling participates in the cytokine regulation and that it heavily stimulates the glial release of TNF-α and IL-1β.[37] This can escalate the glia-mediated inflammation and cause additional nerve cell damage. The latter seems to be further increased by a neurotoxic factor (Ntox), which was released from microglial cells upon stimulation with β-amyloid plaques *in vitro* and was also found in the brains of Alzheimer patients.[38]

Taken together, the glial activation state is largely determined by the regulatory cross talk between reactive microglia and astrocytes. The escalation of the pathological glia activation seems to be driven by a reinforcement of the positive feedback loops with the result that also the astrocytes are recruited as immune-effective cells gaining potentially neurotoxic properties. Glial neurotoxicity is promoted by fibrillar Aβ, which acts as a chronic stimulus.

18.4 ROLE OF ASTROCYTES

The astrocyte reaction seems to be associated with a return to fetal states, regaining functions that are required during brain development for neuronal growth and trophic support.[39] Under pathological conditions, they may assist in repair and regeneration. In addition to such beneficial growth factors,[40] reactive astrocytes express a repertoire of potentially neurotoxic agents, particularly in conjunction with Aβ and associated plaque components (for a detailed review, see Reference 33). They cause a dramatic upregulation of genes encoding for proinflammatory cytokines, chemokines, and inducible enzymes, such as cyclooxygenase-2 (COX2) and iNOS. An additional overexpression of S100β could be responsible for increased neuronal apoptosis, which has been shown to be mediated in rat neuron/astrocyte cocultures by Aβ-conditioned astrocytes.[41] The signal transduction pathways involve the activation of mitogen-associated-protein-kinases (MAPKs) and of the nuclear-factor-κB (NFκB),[42] but the specific mechanisms underlying the Aβ-linked potentiated glial activation are not clear. For astrocytes, the presence of GFAP seems to be necessary.[43]

There is evidence that the astrocytic induction of iNOS by Aβ is partly mediated by IL-1β and TNF-α.[44] In human glial cells, such an upregulated iNOS expression has been shown to occur exclusively in astrocytes[45,46] but not in microglia,[24] indicating that reactive astrocytes are the major source for pathological NO generation. The latter is required to form the aggressive peroxynitrates in conjunction with the massive oxygen radical release from microglia. This underlines that the peroxynitrate-mediated oxidative neuronal damage apparently relies on the deleterious cooperation between reactive microglia and astrocytes.

The dangerous astrocytic iNOS induction could be eventually controlled by TGF-β, which has been shown *in vitro* to suppress NOS-2 gene expression and NO release in astrocytes.[47] TGF-β is present in AD plaques,[48] and its production upon stimulation with Aβ could be evoked in astrocytes of transgenic mice.[49] However, the expression of TGF-β in AD plaques may not be strong enough to counteract the astrocytic iNOS induction. This is suggested by the finding that Aβ-stimulated phagocytes do not produce inhibitory cytokines, such as TGF, which may cause an imbalance between positive and negative feedback signaling.[33] Furthermore, the β-amyloid plaque-associated protein, antichymotrypsin (ACT), is a powerful stimulus for the interaction of astrocytes and Aβ, further strengthening IL-1 production as well as the expression of iNOS.[50]

A highly interesting topic is the evolving essential role of astrocytes in the control of nerve cell activity and synaptic transmission. A pathological alteration of this physiological astrocyte function may turn out to be a major factor for the development of AD. Thus, it has been argued that the β-amyloid–induced impairment of the synaptic function is probably more important for the pathogenesis of AD than nerve cell death.[51] According to the concept of "tripartite synapses,"[52] astrocytes participate actively in the regulation of excitatory synapses by releasing glutamate. This release occurs through exocytosis that is triggered by calcium mobilization from intracellular stores in response to the activation of metabotropic glutamate receptors.[53] Since the astrocytes with their extended processes contact a large number of neurons and synapses, it can be assumed that their calcium-controlled glutamate release could influence the synaptic transmission in extended brain areas. It follows that a pathological alteration of the astrocytic calcium mobilization may aggravate and spread synaptic dysfunction.

Indeed, the signaling paths that regulate the calcium-dependent glutamate release from astrocytes have an Achilles' heel. Thus, the astrocytic glutamate release, elicited upon stimulation with glutamate receptor agonists, could be mimicked by prostaglandin (PGE2), known to act as a potential mediator of inflammatory responses.[54] In detail, costimulation of AMPA and metabotropic glutamate receptors in astrocytes generates an intracellular calcium signal, which activates the phospholipase-A2 causing the release of arachidonic acid. The latter stimulates via COX2 the production of PGE2, which in turn triggers the calcium-dependent glutamate release. However, COX2 is an inducible enzyme, and its mRNA expression was found to be upregulated in astrocytes upon stimulation with the proinflammatory cytokine IL-1β.[55] In view of these findings, an excessive production of IL-1β in AD plaques that overactivates the COX2-dependent astrocytic glutamate release to excitatory synapses can be expected to cause a widespread disturbance of synaptic functions in the AD brain, increasing dementia.

The Achilles' heel, provided by the involvement of PGE2 as a possible trigger for the glutamate release from astrocytes, is also target for another signaling path that acts upstream to PGE2 and enhances its formation. This TNF-α–operated path may serve as a useful instrument for the physiological regulation of synaptic efficacy or turn into a potent pathomechanism.[56] In detail, stimulation of the chemokine receptor CXCR4, which is present in astrocytes, triggers via intracellular calcium mobilization the release of TNF-α. The latter stimulates in an autocrine and paracrine fashion the TNF-α surface receptors in astrocytes, reinforcing their production of PGE2. This causes then an enhanced glutamate exocytosis. The quantity of the evoked astrocytic release of glutamate is determined by the available amount of TNF-α. As long as only the astrocytes contribute to the TNF-α release, the participation of the astrocytes in regulating the glutamate-mediated synaptic transmission will be in the physiological range. However, a simultaneous TNF-α release, elicited also in microglia upon CXCR4 stimulation, would synergistically add to the amount of TNF-α released from astrocytes. As a consequence, the TNF-α–triggered PGE2 formation could be enhanced in astrocytes, forcing a massive glutamate release that exceeds the physiological range. This may add to excitotoxic damage.

18.5 POSSIBLE PATHOMECHANISMS OF NEURONAL DEATH

18.5.1 β-Amyloid–Mediated Neurotoxicity

Toxic fibrillar β-amyloid deposits are the major pathogenetic factor in AD. They act as chronic irritants, which maintain a sustained and escalated reaction of glial cells thought to be involved as amplifier in the autotoxic nerve cell damaging loop. The Aβ-induced glia activation results apparently from a direct interaction with microglia and astrocytes. Their upregulated functions include the production of oxygen radicals by microglia, the induction of iNOS, and NO formation in astrocytes as well as an enhanced production of IL-1β and other interactive proinflammatory cytokines.[33] All these Aβ-potentiated glia functions may contribute to oxidative and excitotoxic nerve cell damage. An interesting recent study shows that Aβ also leads to long-lasting calcium oscillations in astrocytes and to a calcium-dependent impairment of the astrocytic precursor production for glutathione. The observed neurotoxicity is thought to reflect the neuronal dependence on astrocytes for antioxidant support.[57]

The primary trigger for the generation of toxic Aβ is pathological APP generation and processing. Reactive astrocytes may significantly contribute in the following ways:

1. Since they show an upregulated APP expression upon IL-1β stimulation,[58] the big mass of astrocytes may add a large proportion to the potential precursor pool of Aβ.
2. Recent experiments in transgenic mice overexpressing human amyloid precursor protein have shown that reactive astrocytes can selectively express β-secretase, which was also detected in plaque-associated astrocytes of AD patients.[59] Thus, reactive astrocytes seem to possess the instrument for a pathological APP cleaving, which is prerequisite for the formation of aggregated β-amyloid deposits.
3. Reactive astrocytes are the brain-endogenous producers of the chaperones apolipoprotein-E (ApoE) and ACT, which promote the transformation of the amyloid plaques into the neurotoxic fibrillar form.[60,61]

It follows that reactive astrocytes seem to be involved in every step that may turn the physiologically required formation of soluble APP into neurotoxic Aβ.

18.5.2 Oxidative Neuronal Damage

As outlined in the previous section, reactive microglia may produce oxygen radicals and NO. However, the main source for NO seems to be the reactive astrocytes, as concluded from studies on human glial cells. This suggests that a cooperation of both glial cell types is required to produce the two components needed for the formation of peroxynitrates. They are thought to mediate neurotoxicity by β-amyloid (for a recent review, see Reference 62). Such nitro-oxides and their derivatives may cause membrane perturbation and indirect damage by nitrosation reactions known to change the function of membrane transporters and ion channels.[63] There is evidence that oxidative brain damage by NO-related compounds is promoted by the cytokine TNF-α.[64] However, the role of NO and particularly of peroxynitrate in mediating Aβ-linked neuronal oxidative damage still remains circumstantial.

It should be noted that NO is an ambiguous molecular signal that exerts, besides potentially neurotoxic actions, neuroprotection (for a review, see Reference 65). These protective effects of NO are mediated via the activation of the guanylcyclase-generating cGMP. Our own experiments provide an example for this ambiguous NO action. Treatment of neuronal cultures with NO-releasing nitroprusside led to complete neuronal death. This was prevented upon the additional application of a phosphodiesterase inhibitor that increased the cGMP supply provided by the neuroprotective NO path.[20,66]

18.5.3 Excitotoxic Neuronal Damage

Excitotoxic neuronal damage is thought to result from a pathological increase of the extracellular glutamate concentration due to an impaired astrocyte-linked maintenance of the extracellular homeostasis. Excessive glutamate, supported by a membrane-depolarizing increase of extracellular potassium, leads to an overstimulation of the neuronal N-methyl-diaspartate (NMDA) receptors and to an abnormally extended opening of the voltage-controlled NMDA receptor–operated ion channels. This causes in conjunction with an enhanced opening of AMPA receptor–operated ion channels further membrane depolarization and a massive influx of calcium ions into the neurons. The consecutive disturbance of the intracellular calcium homeostasis is the key event causing neuronal death in degenerative diseases.[67]

The main reason for the extracellular glutamate increase could be an impairment or reversal of the cellular reuptake of glutamate, which is thought to be more important than synaptic release.[68] Astrocytes possess the highly effective transporters GLAST and GLT-1, which allows them to take up glutamate against a several thousand–fold concentration gradient (for a review, see Reference 69). Inactivation of the astrocytic transporters has been shown in rat experiments to induce excitotoxic nerve cell death by largely increased extracellular glutamate, whereas a blockade of the neuronal transporters had no effect. This underlines the assumed essential importance of astrocytes for maintaining the extracellular homeostasis.[69] Direct evidence for a pathogenic role of an impaired astrocytic transporter function in the development of AD is missing but suggestive. Thus, a recent study has shown that the glutamate uptake in astrocytes from AD patients was significantly lower than the uptake in astrocytes from nondemented controls.[70] Several pathomechanisms that are operative in AD have been reported to impair the glutamate transporter function, among those the pathological formation of arachidonic acid and oxygen radicals.[71] Overexpression of APP in a transgenic mouse model led to a reduction of the glutamate transporter protein expression and a decreased glutamate uptake,[72] which was also seen in astrocyte cultures upon treatment with Aβ[73] or prostaglandin.[74] Furthermore, specific (inward rectifying) potassium and chloride channels, which are required for extracellular potassium clearance, were absent in cultured astrocytes obtained from newborn rats. But the expression of these channels, which belong to the repertoire of mature astrocytes, could be induced upon long-term treatment with dibutyryl-cAMP, leading to more differentiated phenotypes.[75] The same treatment induced an increased expression of glutamate transporters in astrocytes.[69] These findings suggest that a return of mature astrocytes to a more fetal state, as forced by pathological activation,[39] could be partly responsible for a loss of capabilities that are required to maintain the astrocytic control of the extracellular potassium and glutamate homeostasis.

A pathological alteration of the mechanisms controlling the release of glutamate from astrocytes may significantly contribute to its extracellular rise up to a neurotoxic level. The involvement of the prostaglandin PGE2 in the regulation of the astrocytic glutamate release seems to be a crucial point, where the decision is made whether this primarily physiological mechanism may turn into a pathomechanism. As previously outlined, there are two upstream signaling paths, which regulate the dangerous PGE2 formation and which can be brought in play by cytokines belonging to the upregulated repertoire of the AD pathomechanisms. One is IL-1β, which stimulates the PGE2 production via an increased activation of COX2. The other is TNF-α, which may reach pathologically effective levels when its release from microglia adds to that from astrocytes. Both signaling paths may trigger an excessive PGE2 formation in astrocytes, followed by an excitotoxic release of glutamate.[54,56]

In addition, TNF-α may enhance the efficacy of excitatory synapses by increasing the surface expression of AMPA receptors in neurons.[76] A pathological peroxynitrate formation has been shown to sensitize the neuronal NMDA receptors.[77] Both effects can be expected to increase the vulnerability of neurons to excitotoxic damage.

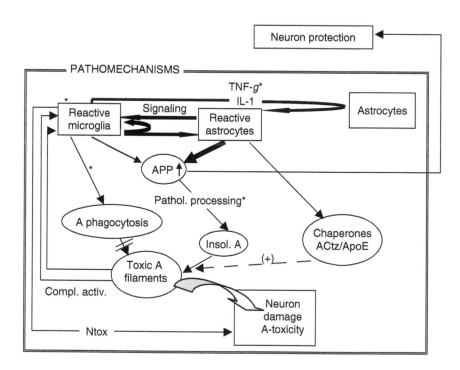

Figure 18.1   Toxic β-amyloid (Aβ) activates microglia and astrocytes, going along with an upregulated signaling via the proinflammatory cytokines IL-1β and TNF-α. The secondarily recruited reactive astrocytes show an increased expression of the amyloid-precursor-protein (APP), which may aggregate to insoluble amyloid plaques (Insol. Aβ) upon pathological APP cleavage. The plaques of Insol. Aβ can be further transformed to toxic Aβ filaments. The transformation is promoted by the chaperones antichymotrypsin (ACT) and apolipoprotein-E (ApoE), produced by reactive astrocytes. This vicious circle, together with the release of a neurotoxic agent (Ntox) from Aβ-stimulated microglia, seems to reinforce an autotoxic loop underlying progressive neuron damage. In principle, reactive microglia could interfere by Aβ phagocytosis.

18.6 SUMMARY AND CONCLUSIONS

The pathological glia activation is a graded response that is submitted to a vigorous control via positive and negative feedback signaling loops. This control provided by microglia- and astrocyte-derived cytokines is overrun in the vicinity of fibrillar β-amyloid plaques and their associated proteins, ApoE and ACT. They act together as chronic irritants, which drive the glia reaction to highly escalated states. Among the upregulated glial repertoire of proinflammatory cytokines is IL-1β, which is involved in a number of vicious circles reinforcing the toxicity of Aβ and of reactive glial cell functions. The secondary recruitment of astrocytes into the phalanx of immune-effective cells seems to be a most powerful pathogenic factor:

1. Reactive astrocytes can be involved in every step that turns the physiologically required formation of soluble APP into toxic fibrillar Aβ.
2. They add in an autocrine/paracrine fashion to an Aβ-triggered excessive production of proinflammatory cytokines.
3. They represent apparently the major source for a pathologically induced NO production that is required to build the aggressive peroxynitrates in conjunction with the massive free oxygen radical release from reactive microglia.
4. Reactive astrocytes may add to a pathological rise of extracellular glutamate by an impairment of their reuptake function or an increased release.

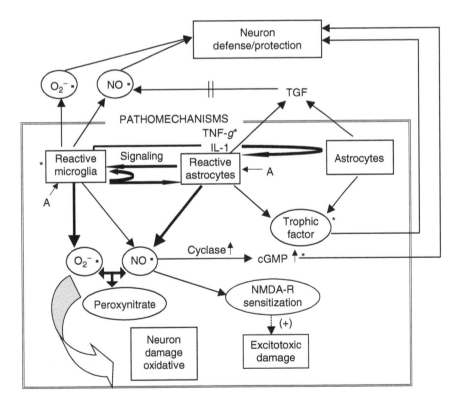

Figure 18.2 In response to escalated glia activation, triggered by toxic β-amyloid (Aβ), reactive microglial cells release massive amounts of oxygen radicals (O2-), whereas reactive astrocytes are the main contributors to nitric oxide (NO). Both radicals together form peroxynitrate, which may promote oxidative neuronal damage. The dangerous NO production is under the negative control of the cytokine tumor growth factor-β (TGF-β), which seems to be overrun under these pathological conditions. NO has an ambiguous function: via activation of the guanyl-cyclase, it increases cyclic GMP (cGMP), which exerts neuroprotection.

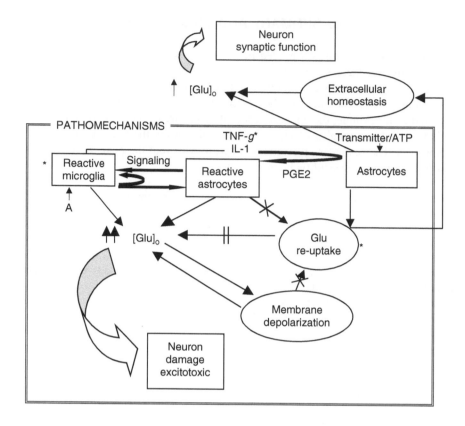

Figure 18.3 An escalated glia activation, triggered by toxic β-amyloid (Aβ), could contribute to a rise of glutamate in the extracellular space ([Glu]o). An excessive Glu release from astrocytes can be triggered by TNF-α from reactive microglia and by upregulated IL-1β. Both cytokines stimulate an increased formation of prostaglandin (PGE2) reinforcing the astrocytic glutamate release. An impairment of the Glu-Re uptake and a pathological membrane depolarization may add to the rise of [Glu]o up to a level that causes excitotoxic damage. Under physiological conditions, astrocytes maintain the extracellular homeostasis and their glutamate release seems to participate in the control of synaptic function.

The emerging active participation of astrocytes in controlling the efficacy of excitatory synapses has an Achilles' heel, since the inflammatory mediator prostaglandin (PGE2) is involved in the regulation of the astrocytic glutamate release. The excessive formation of PGE2 can be stimulated by an Aβ-triggered overproduction of IL-1β as well as by increased TNF-α levels due to an additional TNF-α release from reactive microglia. As a consequence, the physiologically important astrocytic glutamate release may turn into an excitotoxic pathomechanism.

Taken together, as a result of the Aβ-induced escalated pathological glia activation, reactive microglia and astrocytes may form a synergistic alliance that promotes an autotoxic loop and consecutive neuronal damage. The complicated pathomechanisms are illustrated in very simplified diagrams showing the supposed contribution of reactive glial cells to β-amyloid toxicity (Figure 18.1), oxidative neuronal damage (Figure 18.2), and excitotoxic neuronal damage (Figure 18.3). The possible sites of therapeutic interference are indicated. More detailed information is given in a previous review.[21]

REFERENCES

1. Mattson, M.P. et al., Neurodegenerative disorders and ischemic brain disease, *Apoptosis*, 6, 69, 2001.
2. Antuono, P. and Beyer J., The bourdon of dementia. A medical and research perspective. *Theor. Med. Bioeth.*, 20, 3, 1999.

3. Wegiel, J. and Wisniewsky, H.M., The complex of microglial cells and star in three-dimensional reconstruction, *Acta Neuropathol.*, 81, 116, 1990.

4. Alzheimer, A., Über eine eigenartige Erkrankung der Hirnrinde, *Allg. Zeitschr. Psychiatrie Psych.- Gerichtliche Med.*, 64, 146, 1907.

5. Sasaki, A., Yamaguchi, H., Ogawa, A., Sugihara, S., and Nakazato, Y., Microglial activation in early stages of amyloid β protein deposition, *Acta Neuropathol.*, 94, 316, 1997.

6. Eikelenboom, P., Baze, C., van Gool, W.A., Hoozemanns, J.J.M., Rozemuller, J.M., Veerhuis, R., and Williams, A., Neuroinflammation in Alzheimer's disease and prion disease, *Glia*, 40, 232, 2002.

7. Akiyama, H. et al., Inflammation and Alzheimer's disease, *Neurobiol. Aging*, 21, 383, 2000.

8. McGeer, P.L. and McGeer, E.G., The inflammatory response system of brain: implications for therapy of Alzheimer and other neurodegenerative diseases, *Brain Res. Rev.*, 21, 195, 1995.

9. McGeer, P.L. and McGeer, E.G., Inflammation, autotoxicity and Alzheimer disease, *Neurobiol. Aging*, 799, 2001.

10. Rogers, J. et al., Complement activation by β-amyloid in Alzheimer disease, *Proc. Natl. Acad. Sci. U.S.A.*, 89, 1016, 1992.

11. Streit, W.J., Microglia s neuroprotective immunocompetent cells of the CNS, *Glia*, 40, 133, 2002.

12. Kuo, Y.M. et al., Comparative analysis of amyloid-β chemical structure and amyloid plaque morphology of transgenic mouse and Alzheimer disease brains, *J. Biol. Chem.*, 276, 1291, 2001.

13. Schenk, D. et al., Immunization with amyloid-β attenuates Alzheimer-disease-like pathology in the PDAPP mouse, *Nature*, 400, 173, 1999.

14. Webster, S.D. et al., Antibody-mediated phagocytosis of the amyloid-β peptide in microglia is differentially modulated by C1q, *J. Immunol.*, 166, 7496, 2001.

15. McGeer, P.L. and McGeer, P.G., Is there a future for vaccination as a treatment for Alzheimer disease?, *Neurobiol. Aging*, 24, 391, 2003.

16. Jantzen, P.T. et al., Microglial activation and β-amyloid deposit reduction caused by a nitric oxide-releasing non-steroidal anti-inflammatory drug in amyloid precursor protein plus preseniline-1 transgenic mice, *J. Neurosci.*, 22, 2246, 2002.

17. DiCarlo, G. et al., Intrahippocampal LPS injections reduce Aβ load in APP+PS1 transgenic mice, *Neurobiol. Aging*, 22, 1007, 2001.

18. Raivich, G. et al., Neuroglial activation repertoire in the injured brain: graded response, molecular mechanisms, and cues to physiological function, *Brain Res. Rev.*, 30, 77, 1999.

19. Fiorucci, S. et al., IL-1β converting enzyme is a target for nitric oxide–releasing aspirin: new insights in the anti-inflammatory mechanism of nitric oxide–releasing nonsteroidal anti-inflammatory drugs, *J. Immunol.*, 165, 5245, 2000.

20. Schubert, P. et al., Glia-related pathomechanisms in Alzheimer's disease: a therapeutic target?, *Mechanisms Aging Dev.*, 123, 47, 2001.

21. Schubert, P. and Ferroni, S., Pathological glial reactions in neurodegenerative disorders: prospects for future therapeutics, *Expert Rev. Neurotherapeutics*, 13, 279, 2003.

22. Wisniewski, H.H. and Wegiel, J., Spatial relationships between astrocytes and classical plaque components, *Neurobiol. Aging,* 12, 593, 1991.

23. Banati, R.B., Schubert, P., and Kreutzberg, G., Cytotoxicity of microglia, *Glia*, 7, 111, 1993.

24. Colton, C.A. et al., Microglia contribution to oxidative stress in Alzheimer's disease, *Ann. N.Y. Acad. Sci.*, 899, 292, 2000.

25. Lue, L.F. et al., Inflammatory repertoire of Alzheimer's disease and nondemented elderly microglia *in vitro*, *Glia*, 35, 72, 2001.

26. Vincent, V.A.M., Tilders, F.J.H., and van Dam, A.M., Inhibition of endotoxin-induced oxide synthase production in microglial cells by the presence of astroglial cells: a role for transforming growth factor β, *Glia*, 19, 190, 1997.

27. Letterio, J.J. and Roberts, A.B., Regulation of immune response by TGFβ, *Annu. Rev. Immunol.*, 16, 137, 1998.

28. Chao, C.C. et al., Tumor necrosis factor-α production by human fetal microglial cells: regulation by other cytokines, *Dev. Neurosci.*, 17, 97, 1995.

29. Tarkowski, E. et al., Intracerebral production of TNF-α, a local neuroprotective agent in Alzheimer's disease and vascular dementia, *J. Clin. Immunol.*, 19, 223, 1999.

30. Chao, C.C. et al., Tumor necrosis factor-α mediates the release of bioactive transforming growth factor-β in murine microglial cell cultures, *Immunol. Immunopathol.*, 77, 358, 1995.

31. Beneviste, E.N., Inflammatory cytokines within the central nervous system: sources, function and mechanism of action, *Am. J. Physiol.*, 263, C1, 1992.

32. Aschner, M., Astrocytes as mediators of immune and inflammatory responses in the CNS, *Neurotoxicology*, 19, 269, 1998

33. Meda, L., Baron, P., and Scarlato, G., Glial activation in Alzheimer's disease: the role of Aβ and its associated proteins, *Neurobiol. Aging*, 22, 885, 2001.

34. Pocock, J.F. and Liddle, A.C., Microglial signalling cascades in neurodegenerative disease, *Prog. Brain Res.*, 132, 555, 2001.

35. Griffin, W.S. et al., Glial-neuronal interactions in Alzheimer's disease: the potential role of a cytokine cycle in disease progression, *Brain Pathol.*, 8, 65, 1998.

36. Le Feuvre, R., Brough, D., and Rothwell, N., Extracellular ATP and P2X7 receptors in neurodegeneration, *Eur. J. Pharmacol.*, 447, 261, 2002.

37. Liu, J.S.H. et al., Modulation of interleukin-1β and tumor necrosis factor α signaling by P2 purinergic receptors in human fetal astrocytes, *J. Neurosci.*, 20, 5292, 2000.

38. Giullian, D. et al., Senile plaques stimulate microglia to release a neurotoxin found in Alzheimer brain, *Neurochem. Int.*, 27, 119, 1995.

39. Norenberg, M.D., Astrocyte responses to CNS injury, *J. Neuropathol. Exp. Neurol.*, 53, 231, 1994.

40. Rudge, J.S., Astrocyte derived neurotrophic factors, in *Astrocytes: Pharmacology and Function*, Murphy, S., Ed., Academic Press, San Diego, 1993, p. 267.

41. Malchiodi-Albedi, F. et al., Astrocytes contribute to neuronal impairment in βA toxicity increasing apoptosis in rat hippocampal neurons, *Glia*, 34, 68, 2001.

42. Hanada, T. and Yoshimura, A., Regulation of cytokine signaling and inflammation, *Cytokine Growth Factor Rev.*, 13, 413, 2002.

43. Xu, K. et al., Glial fibrillary acidic protein is necessary for mature astrocytes to react to β-amyloid, *Glia*, 25, 390, 1999.

44. Akama, K.T. and van Eldik, L.J., β-Amyloid stimulation of inducible nitric oxide synthase in astrocytes is IL-1β and TNF-dependent, and involves TNα receptor-associated factor, and NFKb-inducing kinase-dependent signaling mechanisms, *J. Biol. Chem.*, 275, 7918, 2000.

45. Lee, S.C. et al., Induction of nitric oxide synthase activity in human astrocytes by IL-1 beta and interferon-gamma, *J. Neuroimmunol.*, 46, 19, 1993.

46. Zhao, M.L. et al., Inducible nitric oxide synthase expression is selectively induced in astrocytes isolated from adult human brain, *Brain Res.*, 813, 402, 1998.

47. Simmons, M.L. and Murphy, S., Roles for protein kinases in the induction of nitric oxide synthase in astrocytes, *Glia*, 11, 227, 1994.

48. Van der Wal, E.A. et al., Transforming growth factor-beta 1 is in plaques in Alzheimer and Down pathologies, *Neuroreport*, 4, 69, 1993.

49. Wyss-Coray, T. et al., Chronic overproduction of transforming growth factor-beta by astrocytes promotes Alzheimer's disease-like microvascular degeneration in transgenic mice, *Am. J. Pathol.*, 156, 139, 2000.

50. Hu, J. and Van Eldik, L.J., Glial-derived proteins activate cultured astrocytes and enhance β amyloid-induced glial activation, *Brain Res.*, 842, 46, 1999.

51. Small, D.H., Mok, S.S., and Bornstein, J.C., Alzheimer's disease and Aβ toxicity: from top to bottom, *Nat. Rev. Neurosci.*, 2, 595 2001.

52. Volterra, A., Magistretti, P.J., and Haydon, P.G., Eds., *The Tripartite Synapse: Glia in Synaptic Transmission*, Oxford University Press, Oxford, 2002.

53. Araque, A. et al., Dynamic signaling between astrocytes and neurons, *Annu. Rev. Physiol.*, 63, 795, 2001.

54. Bezzi, P. et al., Neuron-astrocyte cross-talk during synaptic transmission: physiological and neuro-pathological implications, *Prog. Brain Res.*, 132, 255, 2001.

55. O'Banion, M.K. et al., Interleukin-1β induces prostaglandin G/H synthase2 (cyclooxygenase-2) in primary murine astrocyte cultures, *J. Neurochem.*, 66, 2532, 1996.

56. Bezzi, P. et al., CXCR4-activated astrocyte glutamate release via TNFα: amplification by microglia triggers neurotoxicity, *Nat. Neurosci.*, 4, 702, 2001.

57. Abramov, A.Y. et al., Changes in intracellular calcium and glutathione in astrocytes as the primary mechanism of amyloid neurotoxicity, *J. Neurosci.*, 23, 5088, 2003.

58. Forloni, G. et al., Expression of amyloid precursor protein mRNAs in endothelial, neuronal and glial cells: modulation by interleukin-1, *Mol. Brain Res.*, 16, 128, 1992.

59. Hartlage-Rübsamen, M. et al., Astrocytic expression of the Alzheimer disease β-secretase (BACE1) is stimulus-dependent, *Glia*, 41, 169, 2003.

60. Bowman, B.H. et al., Human APOE protein localized in brains of transgenic mice, *Neurosci. Lett.*, 219, 57, 1996.

61. Das, S. and Potter, H., Expression of the Alzheimer amyloid-promoting factor antichymotrypsin is induced in human astrocytes by IL-1, *Neuron*, 14, 447, 1995.

62. Xie, Z. et al., Peroxynitrate mediates neurotoxicity of amyloid β-peptide 1-42- and lipopolysaccharide-activated microglia, *J. Neurosci.*, 22, 3484, 2002.

63. Trotti, D. et al., Peroxynitrite inhibits glutamate transporter subtypes, *J. Biol. Chem.*, 271, 5976, 1996.

64. Meda, L., Cassatella, M.A., and Szendrei, I., Activation of microglial cells by β-amyloid protein and interferon-γ, *Nature*, 374, 647, 1995.

65. Murphy, S., Production of nitric oxide by glial cells: regulation and potential roles in the CNS, *Glia*, 29, 1, 2000.

66. Ogata, T. et al., Nitric oxide-induced neurotoxicity is inhibited by propentofylline via cyclic GMP elevation, *Neurobiol. Aging*, 19, 255, 1998.

67. LaFerla, F.M., Calcium dyshomeostasis and intracellular signalin in Alzheimer's disease, *Nat. Rev. Neurosci.*, 3, 862, 2002.

68. Attwell, D., Barbour, B., and Szatkowski, M., Nonvesicular release of neurotransmitter, *Neuron*, 11, 401, 1993.

69. Sonnewald, U. and Aschner, M., Pharmacology and toxicology of astrocyte-neuron glutamate transport cycling, *JPET*, 301, 1, 2002.

70. Liang, Z. et al., Effects of estrogen treatment on glutamate uptake in cultured human astrocytes derived from cortex of Alzheimer's disease patients, *J. Neurochem.*, 80, 807, 2002.

71. Volterra, A., Trotti, D., and Racagni, G., Glutamate uptake is inhibited by arachidonic acid and oxygen radicals via two distinct and additive mechanisms, *Mol. Pharmacol.*, 46, 986, 1994.

72. Masliah, E. et al., Abnormal glutamate transporter function in mutant amyloid precursor protein transgenic mice, *Exp. Neurol.*, 163, 381, 2000.

73. Harris, M.E. et al., Amyloid beta peptide inhibits Na-dependent glutamate uptake in rat hippocampal astrocyte cultures, *J. Neurochem.*, 67, 277, 1996.

74. Breitner, J.C.S., Inflammatory processes and anti-inflammatory drugs in Alzheimer's disease: a current appraisal, *Neurobiol. Aging*, 17, 789, 1996.

75. Ferroni, S. et al., Two distinct inwardly rectifying conductances are expressed in long-term dibutyryl-cyclic-AMP treated rat cultured cortical astrocytes, *FEBS Lett.*, 367, 319, 1995.

76. Beatti, E.C. et al., Control of synaptic strength by glial TNF-α, *Science*, 295, 2282, 2002.

77. Hewett, S.J., Csernansky, C.A., and Choi, D.W., Selective potentiation of NMDA induced neuronal injury following induction of astrocytic iNOS, *Neuron*, 134, 487, 1994.

CHAPTER 19

Developmental Effects of Xenobiotics on Thyroid Hormone Regulation in the CNS with Emphasis on Glial Function

Offie P. Soldin

CONTENTS

19.1 INTRODUCTION

The thyroid hormones thyroxine (T_4) and triiodothyronine (T_3) play an essential role in mammalian brain development, maturation, and function. Thyroid hormone effects are regional, temporal, and of a defined duration. The cellular targets, specific receptors, timing, and duration of thyroid hormone effects are still not fully understood. Whereas the thyroid synthesizes both hormones, the main hormone, T_4, acts as a prohormone to the active hormone, T_3. Therefore, collectively these hormones are known as "thyroid hormone." Thyroid gland activity is predominantly regulated by the pituitary glycoprotein hormone, thyroid-stimulating hormone (TSH). Thyroid hormone exerts a negative feedback effect on pituitary secretion of TSH, such that when T_4 concentrations drop, TSH concentrations increase, thus maintaining stable T_3 levels. In the brain, T_3 is derived in large part from 5′ deiodination of T_4 in a pathway closely regulated by developmental and physiologic factors.[1] This chapter deals with the relationship between thyroid hormone regulation and glial function. Before addressing the details of this relationship, it is important to note some general points about thyroid hormone transport, the effect of thyroid hormones on brain development, and the role of glia in neurodevelopment.

T_3 action is mediated through nuclear hormone receptors. Lack of adequate thyroid hormone supply to the developing brain, as in iodine deficiency, congenital hypothyroidism, maternal hypothyroidism, and hypothyroxinemia (low circulating T_4) during pregnancy, especially during the first half of gestation, can result in long-lasting psychoneurological consequences for the fetus.[2] The maternal thyroid is the only source of thyroid hormone for the fetal brain before the onset of fetal thyroid function (beginning in gestation week 12). Transplacental transfer of maternal thyroid hormone to the fetal circulation continues after the onset of fetal thyroid function.[3,4] Thyroid hormone supply to the growing fetus is of both maternal and fetal origin during the second trimester (at the time of neuronal division, migration, and architectural organization) and at later phases of fetal brain development, when there is glial cell multiplication, migration, and myelination.

Iodine is an essential element for T_4 and T_3 synthesis. Epidemiological studies in iodine-sufficient and iodine-deficient areas strongly suggest that early maternal hypothyroxinemia increases the risk of neurodevelopmental deficits of the fetus, even in the absence of clinical hypothyroidism.[5,6] In addition to iodine deficiency, circulating antithyroid antibodies may block fetal thyroid activity, resulting in symptoms similar to severe iodine deficiency. Generally, infants born to hypothyroid mothers appear healthy and without evidence of thyroid dysfunction, provided that there was no severe iodine deficiency *in utero*.[7,8] The controversy is whether these infants suffer from minor deficits.

Thyroid hormone and glia work in concert to maintain normal brain development. Glia appears to be important in T_3 generation from T_4, and thyroid hormone action in brain is influenced by glia–neuron cooperation. This requires that thyroid hormone reach glial cells, implying T_4 transport across the blood–brain barrier (BBB) or across the blood–cerebrospinal fluid (CSF) barrier. Most circulating thyroid hormones are bound to binding proteins—thyroid binding globulin (TBG), transthyretin (TTR), and albumin—and only the free hormones cross the barrier endothelium. The choroid plexus has been shown to be capable of synthesizing TTR.[9,10] Although TTR was proposed as a carrier of T4 in the choroid plexus, TTR-mutant mice have shown no defect in transport.[11] Other transporters such as monocarboxylate transporter 8 (MCT8) may play an important role in thyroid hormone transport.[12,13]

In summary, appropriate brain development requires not only adequate supplies of thyroid hormone, but also their delivery to glia. This chapter focuses on the key effects of thyroid hormones on central nervous system (CNS) development, and specifically on the impact of thyroid hormone and its deficits on glial cell function during development.

19.2 THE EFFECTS OF THYROID HORMONE AND ITS DEFICITS ON NEURODEVELOPMENT

Brain development proceeds through a series of intricately associated events, determined by genetic, as well as epigenetic factors. Thyroid hormone is involved in regulation of brain cell migration, a process important in the formation of cortical layers, as well as differentiation of neurons and oligodendrocytes,[14] astrocytes, and microglia.[15,16] T_3 is an instructive factor in the early steps of oligodendrocyte generation from stem cells and controls the timing of oligodendrocyte precursor cell differentiation.[17,18] In the hypothyroid rat model, hypothyroidism during the postnatal period of myelination onset leads to severe delays in the expression of oligodendrocyte genes, resulting in a reduced number of myelinated axons.[19] Defects in myelination, alterations of cell migration in the cerebral cortex and the cerebellum, and abnormal differentiation of neurons have been noted in hypothyroid rats. Absence of thyroid hormone causes diminished axonal growth and dendritic arborization in the cerebral, visual, and auditory cortex, hippocampus, and cerebellum. In the cerebellum, absence of thyroid hormone also delays proliferation and migration of granule cells from the external to the internal granular layer.

In humans, iodine deficiency resulting in a deficiency in thyroid hormone during gestation leads to neurological deficits, irreversible damage, and severe mental retardation.[2] It is thought that maternal–fetal transfer of thyroid hormone is crucial during the first half of pregnancy before the fetal thyroid is capable of synthesizing its own thyroid hormone. Severe maternal thyroid hormone deficiency is associated with neurological cretins born in areas of severe iodine deficiency. In these cases, CNS damage is irreversible by birth and could have only been prevented if maternal iodine deficiency (and therefore thyroid hormone deficiency) was prevented during the first half of pregnancy.[20,21]

19.2.1 Thyroid Hormone Regulation at the Nuclear Level

T_4 has been implicated in neural cell migration via integrin–laminin interactions and regulation of deiodinase activity in astrocytes.[22] It may also have direct action on factin polymerization.[23,24] The role of T_4 is nongenomic; T_4 delivers T_3 intracellularly through 5′ deiodination. T_3 binds with a greater affinity to specific thyroid hormone receptor (TR) within the nucleus than does T_4. TR is a ligand-modulated transcription factor belonging to the nuclear hormone receptor superfamily. T_3 acts by regulating gene expression after binding to nuclear receptors. T_3 receptor mRNAs are located on neurons as well as on oligodendrocytes. However, although T_3 affects astrocytes, it is not clear yet if the effect is mediated through nuclear TR or some other mechanism. T_3 is also known to have extra nuclear and extra genomic actions (described in 19.3.3).

TRs activate or suppress target gene expression in a hormone-dependent or hormone-independent fashion. The primary action of T_3 on gene expression is mediated through interaction of the TRs with responsive elements located in gene regulatory regions.[25] Triiodothyronine response elements (TREs) are specific sequences of DNA in genes responsive to thyroid hormone. For a review of brain gene expression regulation by thyroid hormone, the correlation between gene expression and physiologic effects, and the likely mechanisms of action of thyroid hormone on brain gene expression, see Bernal et al.[26]

In the absence of T_3, the receptor interacts with corepressors (also a group of nuclear proteins), whereas in the presence of T_3, TR interacts with coactivators.[26] Thus, T_3 determines whether coactivators or corepressors are localized near the initiation site of gene transcription. Conformational changes of chromatin are then directed by acetylating histones (promoted by coactivators), thereby loosening chromatin structures, or by the deacetylation of histones (promoted by corepressors), resulting in chromatin compaction, thus affecting transcription. A negative feedback mechanism between TRs and its own corepressor has been suggested, where the reduction of corepressor levels may represent a control mechanism of TR-mediated gene silencing.[27] Bernal et al. propose a role for unliganded TRs in the pathogenesis of hypothyroidism.[26]

19.2.2 Thyroid Hormone Coordinates Maturational Processes in a Regional and Temporal Pattern

The mammalian brain is a direct target organ of thyroid hormone both during neurodevelopment and in the adult. During development, the role of thyroid hormone is the coordination of ostensibly unrelated maturational processes in a regional and temporal pattern. In addition, thyroid hormone may have extranuclear and extragenomic actions.[28] The physiological and biochemical processes underlying thyroid hormone action in the brain have been reviewed elsewhere.[26,29]

During early embryonic brain development, thyroid hormone has no effects on neural induction, neurulation, and the establishment of polarity or segmentation. At the time of cell migration and the formation of layers, thyroid hormones contribute to cerebral cortex layering and to the establishment of callosal connections. In the cerebellum, thyroid hormones control the rate of migration of granular cells from the external germinal layer to the internal granular layer. It also induces oligodendrocyte differentiation and expression of myelin genes. Brain damage associated with maternal hypothyrodism due to lack of iodine may be prevented, if corrected early enough in pregnancy.

19.2.3 Thyroid Hormone Deficiency Affects Glial Function—The Rat Model

The hypothyroid rat is useful for investigating the effects of thyroid hormone deficiency. Fetuses from pregnant hypothyroid rats display impaired maturation of radial glia cells.[30] The proportion of mature glial cell fibers and glial fibrillary acidic protein (GFAP) in the CA1 region of the hippocampus was significantly decreased in these pups. Similar results were obtained when the rats were treated with a goitrogen.[30]

Thyroid hormone deficiency during late development in the rat resulted in permanent deficits in brain function.[31-33] Delayed maturation of hippocampal and neocortical radial glia has been found in fetuses on gestational day 21 when dams were maintained on a low-iodine diet.[31] In a different study, dams were fed an iodine-deficient diet resulting in maternal hypothyroxinemia (TSH was not measured). Their pups showed an alteration of neuronal migration during neocorticogenesis, possibly as a consequence of glial cell dysfunction.[34] Studies of women who had high TSH and normal T_4 concentrations (mild or "subclinical" hypothyroidism) during gestation week 17[35] or low T_4 alone[36] suggest that these changes are associated with neurodevelopmental deficits, corroborating the effects noted in thyroid deficiency in the rodent model.

19.3 GLIA MEDIATE THYROID HORMONE ACTION

Glia serve as a significant link between the endocrine and nervous systems.[37] Glial nuclear receptors for thyroid and steroid hormones, as well as the metabolism of these hormones and the generation of neuroactive metabolites, make this possible. Furthermore, glial cells synthesize endogenous neuroactive steroids, including pregnenolone and progesterone, from cholesterol.

This section will focus on the mechanisms by which glial cells mediate the action of thyroid hormone on the nervous system. A discussion of thyroid hormone receptors found on glia cells will be followed by considerations of specific effects of thyroid hormone on glial cell proliferation and maturation and effects that are not mediated by thyroid hormone receptors.

19.3.1   The Role of Glia Thyroid Hormone Receptors

Thyroid hormone action is initiated by the binding of T_3 to TRs, which are members of a large superfamily of zinc finger transcription factors.[38] Two genes (*TRα* and *TRβ*) encode four receptor and four nonreceptor proteins. The functional thyroid hormone receptors (isoforms) known to date are $TRα_1$, $TRβ_1$, $TRβ_2$, and $TRβ_3$.

Thyroid hormone receptors are expressed in neurons, oligodendrocytes, and astrocytes.[39–41] The receptor isoforms have functional specificity that depends on the timing and place of their expression. They are ligand-activated transcription factors that modulate gene expression. Activation of both subtypes in glial cells elevates the level of intracellular cyclic AMP (cAMP), leading to alterations in morphology and activity.[42,43] $TRα_1$ accounts for 70–80% of total receptor protein present in the brain. Specificity of actions among the different receptor isoforms is determined primarily by the regional distribution. Other determinants may include the structure of the thyroid hormone responsive elements (TRE), and perhaps heterodimerization partners,[45] but there is less experimental evidence that this may play a role in receptor specificity *in vivo*.

TRα is highly expressed in oligodendrocytes during early development, whereas TRβ expression in these cells is upregulated during late stages of development. Oligodendrocyte progenitor cells express $TRα_1$ and possibly $TRβ_2$, whereas mature oligodendrocytes express $TRα_1$ and $TRβ_1$.[46] Myelin gene expression correlates with a striking increase in $TRβ_1$ expression in the developing brain.[47]

Different receptors are expressed in relation to the maturation state of oligodendrocytes. These receptors may be regulated by thyroid hormone or may change in an hormone-independent way. Transcriptional regulation of myelination[48] and a nuclear protein that binds to $TRβ_1$ forming myelin basic protein-TRE complex have been identified.[49] The expression of this brain nuclear factor is restricted to the perinatal period when myelination is sensitive to T_3. T_3 can function independently of other hormones and serum factors in exerting relatively specific effects on the regulation of myelination.[50,51]

Consistent with T_3 regulation of development, the brain contains abundant T_3 receptors.[52–54] In humans, there is a 40-fold increase in $TRβ_1$ mRNA expression throughout the brain shortly after birth that reaches maximal levels at 10 days postpartum and persists until adulthood. In contrast, $TRα_1$ and *c-erbAα2* mRNA undergo a transient twofold increase and a subsequent decrease to adult levels 2 weeks after birth. Although the effects of thyroid hormones on the development of neurons and oligodendrocytes are well documented, less is known about their effects on astrocytes. Normal astrocytes contain both TRβ subtypes, with greater abundance of $TRβ_2$ receptors.[55] During conditions of reactive gliosis, the $TRβ_1$ subtype increases to a greater proportion,[56] and TR isoforms change with astroglial cell maturation.[57]

19.3.2   Thyroid Hormone Increases Glial Cell Proliferation and Maturation

The differentiation of oligodendrocytes, as well as the degree of myelin synthesis, is increased by thyroid hormone.[57,58] Hyperthyroidism accelerates the myelination process, while a deficiency in thyroid hormone leads to profound and irreversible defects of brain development and maturation.[5,60–63]

T_3 promotes morphological and functional maturation of postmitotic oligodendrocytes.[47,64] Hypothyroidism coordinately and transiently also affects myelin protein gene expression in most brain regions during postnatal development[64]; with advancement in age, the hypothyroid animals show increasing myelin deficits.

An increase in T_3 controls astrocyte number and the maturation of Golgi epithelial cells (Bergmann glia) in the cerebellum. An acceleration in the morphological development of Bergmann glia cells was observed in the hyperthyroid rats, whereas in hypothyroid rats the formation and morphological maturation of these cells were retarded and their final number was increased.[65]

Thyroid hormone also plays a major role in microglial ontogenesis.[16] The density of microglial cells was reduced, as was microglial process formation in the forebrain of developing rats deprived of thyroid hormone from a late fetal stage of life. Conversely, neonatal rat hyperthyroidism increased the density of microglial cells and the growth of microglial processes during the first postnatal week. $TR\alpha_1$ and $TR\beta_1$ were detected in the nuclei of cultured microglial cells, and T_3 was found to promote growth of their processes.

The first postnatal week of life is a critical period for thyroid hormone action on developing cortical microglia. Early transplacental transfer of maternal thyroid hormone[66] influenced the early fetal development of the microglial cells. Thyroid hormone modulated proliferation of microglia, promoting morphological differentiation, growth, and maturation of these cells.[16,67,68] Thyroid hormone control of the number of microglial cells may regulate the availability of microglial-derived growth factors,[69-71] perhaps affecting axonal growth and regeneration.

19.3.3 Nongenomic Transcriptional Regulation of Cell Structure Proteins by Thyroid Hormones

Although thyroid hormones regulate cells at the nuclear level, nongenomic effects on cell structure proteins have also been described.[72] For example, in addition to regulation of actin gene expression at the level of transcription,[73] thyroid hormone modulates actin polymerization through an extra-nuclear action.[74] This regulation could contribute to the effects of thyroid hormone on arborization, axonal transport, and cell–cell contacts during brain development.

Astrocytes are heterogeneous in their ability to respond to T_3, including their capacity to secrete different growth factors or related molecules. In contrast to morphological differentiation, T_3 induces secretion of growth factors responsible for cerebellar astrocyte autocrine proliferation. This effect is accompanied by the reorganization of GFAP, fibronectin, and laminin filaments, suggesting a possible role for T_3 in cerebellar astrocyte adhesion.

In the rat cerebellum, migration of neurons from the external granular layer to the internal granular layer occurs postnatally and is dependent upon the presence of thyroid hormone. Key guidance signals to these migrating neurons are provided by laminin, an extracellular matrix protein that is fixed to the surface of astrocytes.[75] Expression of laminin in the brain is developmentally timed to coincide with neuronal growth spurts.[76] Farwell et al. have shown that T_4 was required for integrin clustering and for the attachment of integrin to laminin in astrocytes.[22]

Leonard et al.[77] have shown that astrocytes lack significant numbers of functional nuclear TRs. The predominant (> 95%) TR in astrocytes is the non-T_3-binding isoform c-$erbA\alpha2$. Small quantities of $TR\alpha_1$[77] and $TR\beta_2$[78] have been identified in cultured astrocytes, although the dominant negative activity of c-$erbA\alpha2$[79] is likely to render these TR isoforms transcriptionally inert.

Additional extranuclear regulatory effects of T_4 in cultured astrocytes include microfilament organization[23,80] and vesicle recycling.[81] In hypothyroids, the appearance of laminin in the molecular layer of the cerebellum is markedly delayed and less abundant in relation to its appearance in the euthyroid cerebellum.[76]

19.4 DEIODINASES IN GLIA AND CONTROL OF BRAIN T_3 CONCENTRATIONS

19.4.1 Deiodinases in the Brain

Brain T_3 concentrations are determined by circulating T_4 and T_3 that cross the BBB and by tissue deiodinase activity. The expression of type II deiodinase (DII) and DIII in the brain is developmentally regulated. DII generates T_3 from T_4 by phenolic ring deiodination. In hypothyroidism, increased DII activity tends to normalize T_3 concentrations even with greatly reduced T_4 concentrations. DIII is necessary for T_3 homeostasis in the CNS. While deiodinases DI and DII generate the active hormone T_3 from T_4, DIII degrades both hormones. DIII inactivates T_4 by tyrosil ring deiodination to generate the inactive reverse triiodothyronine (rT_3). In addition, DIII inactivates T_3 by tyrosil ring deiodination to generate the inactive 3,3' diiodothyronine (T_2). DIII is highly responsive to T_3; its activity has been identified in many regions of the brain paralleling the thyroid status. DIII is specifically expressed in selected nuclei around birth, its focal localization within the hippocampal pyramidal neurons and granule cells of the dentate nucleus.[82]

DII mRNA can be found in different regions of the brain including the hippocampus, hypothalamus, olfactory bulb and anterior olfactory nucleus, caudate-putamen nucleus, the dorsal area of lateral septal nucleus, and the thalamus. Glia, and especially astrocytes and tanycytes, are the primary cells to express DII activity. The first four layers of the cerebral cortex possess the strongest DII expression. Very high DII signal is found in the lateral aspects of the median eminence and the lining of the lower portion of the third ventricle, in an area known to have the specialized glial cells known as ependymal tanycytes.[83,84]

19.4.2 DII in Tanycytes

The highest expression of DII is found in tanycytes. Tanycytes send processes to the adjacent hypothalamus, where they frequently end in blood vessels, and to the median eminence, ending in portal vessels. These cells are thought to be involved in T_4 uptake and its conversion to T_3. Tanycytes transport hormones and other molecules from and to the CSF and the hypothalamus/median eminence and show a much stronger expression of DII compared with the outer cortex layers. Thus, DII in tanycytes could be involved in providing T_3 to the CSF, from which it would reach nearby structures by diffusion, and the portal blood, thus influencing pituitary function.[85]

19.4.3 DII in Astrocytes

Astrocytes express DII[86] and may have an active role in the uptake of T_4 from the blood and in generating T_3 and delivering it to nearby neurons.[26] T_4 regulates DII activity by inactivating actin depolymerization in cAMP-stimulated glial cells, suggesting that an intact actin cytoskeleton is important for downregulation of deiodinase activity.[81,87] T_4 may also influence the downregulation of DII activity in astrocytes by a secondary mechanism, perhaps by targeting lysosomes.[81,87]

19.4.4 The Timing of Oligodendrocyte Development Is Regulated by Thyroid Hormones

Oligodendrocyte precursor cells divide a limited number of times before terminal differentiation. The timing of oligodendrocyte differentiation depends on both intracellular mechanisms and extracellular signals, including thyroid hormone. Ahlgren et al.[88] showed both *in vivo* and *in vitro* that thyroid hormone is required for the normal development of rodent optic nerve oligodendrocytes. Both maternal and fetal thyroid hormone played a part in oligodendrocyte development.

19.4.5 Thyroid Hormone Gene Regulation

Studies of brain-derived neutropic factor have demonstrated that thyroid hormone can regulate MBP expression in a promoter-, developmental-, and region-specific manner.[89] Thyroid hormone also regulates several genes that are involved in a wide range of cellular functions: glutamine synthase, protein kinase C, substrate RC3/neurogranin, prostaglandin D_2 synthase, hairless (a potential transcription factor), and adhesion molecules, as well as matrix proteins such as tenascin and proteins important for neuronal migration. Three phases in the regulation of these particular genes have been identified: a refractory prenatal period, a T_3-responsive period (typically from postnatal days 3–20), and a T_3-independent period.

19.5 ALTERATIONS IN THYROID HORMONE REGULATION

Given the critical role of thyroid hormone in neurodevelopment, it is important to consider various ways in which thyroid hormone regulation can be affected. There are two main ways for such alteration: the first is related to thyroid hormones synthesis and metabolism, the other to direct or indirect interference with thyroid hormone action. Thyroid hormone synthesis is dependent on the nutritional intake of iodine, as well as the effects of goitrogens, substances that either inhibit uptake of iodine by the thyroid or block the synthesis of T_3 and T_4 within the thyroid or the peripheral tissues. Thyroid hormones exert their action at a nuclear level in the brain by regulating the transcription of thyroid hormone–responsive genes. This process is initiated when T_3 binds to thyroid hormone receptors. The consequences of perturbations in this system depend on the nature, concentration, timing, and duration of exposure and the levels of thyroid hormones available.

Analysis of the effects of xenobiotics on thyroid hormone economy or activity is complicated by our limitation in estimating these factors. On the one hand, we do not have knowledge of all thyroid hormone toxicants and their sites and mechanisms of action. On the other, we may be limited by our ability to estimate the effect since our methods of detection may not be specific or sensitive enough, or we may not be measuring at the right time or location.

Throughout pregnancy, maternal–fetal transfer of iodide is the sole source of iodide for fetal thyroid hormone synthesis. Placental transfer of T_4 helps to maintain T_3 concentrations even when the fetus suffers from congenital hypothyroidism and is unable to produce adequate amounts of thyroid hormone. Most often this is a hereditary condition, originating from athyreosis, ectopic, or dysgenetic thyroid, which may result in a permanent defect of thyroid hormone synthesis, or metabolism. It can also occur as a result of prenatal exposure to radioiodine or to other antithyroid drugs. Newborns with congenital hypothyroidism that are treated promptly and adequately with L-T_4 soon after birth can compensate for this deficiency, avoiding mental retardation.[90] In humans, early treatment of congenital hypothyroidism with T_4 reduces intellectual impairment and is the major impetus for neonatal screening for congenital hypothyroidism.

19.5.1 The Role of Iodine

Iodine is an essential component of thyroid hormone. The iodothyronine molecules contain three or four iodine atoms that are covalently bound during iodide organification. The iodination process of thyroglobulin is catalyzed by thyroid peroxidase (TPO) and requires that the thyrocyte concentrate iodide ions from plasma. Iodine availability can be rate limiting, although a variety of physiological mechanisms have evolved to mitigate the consequences of iodine deficiency on thyroid hormone synthesis. In areas of severe iodine deficiency, however, these compensatory mechanisms are often inadequate and lead to iodine deficiency disorders that include hypothyroidism, irreversible mental retardation, goiter, reproductive failure, and increased infant mortality.

There is considerable posttranslational control over the synthesis of thyroid hormone, as a result of the balance between the regulation by TSH and iodide. Since iodide may enter additional metabolic pathways outside the thyroid, over a very wide range of iodide intake the mean hormone levels seem to be unaffected. Under normal conditions, the lack of serum T_4 (due to lack of iodine in the thyroid) is the trigger at the pituitary to increase TSH production.

The sodium iodide symporter (NIS) is a cotransporter responsible for the transport and concentration of iodine. NIS is expressed in the thyroid, the lactating mammary gland, salivary glands, gastric mucosa, and the choroid plexus.[91] NIS has been detected in the human choroid plexus[92,93] and is likely located in the basolateral membrane. Perchlorate, a known inhibitor of iodide uptake by the thyroid NIS, is known to prevent radiolabeled iodide uptake into the brain. Since the NIS is a known site for goitrogen inhibition, the choroid plexus is a possible site for goitrogenic activity.

19.5.2 The Effects of Iodine Deficiency

Iodine deficiency affects millions of people worldwide and is recognized as the most common preventable cause of mental retardation.[94] WHO estimates that some 2.2 billion people are at risk from iodine deficiency in 130 countries.[95] The most severe condition associated with iodine deficiency is neurological cretinism, while milder forms of iodine deficiency may lead to mental retardation.[96] Iodine-deficient mothers are not able to synthesize adequate amounts of thyroid hormone, especially when they are both iodine deficient and selenium deficient; selenium is necessary for deiodinases synthesis.[97] Brain damage occurs during the first trimester, before the onset of thyroid hormone synthesis by the fetus, at a time when the fetus is completely dependent on thyroid hormone supplied by the mother. The resulting neurological cretins are not necessarily hypothyroid, have normal growth, and are not affected by thyroid hormone administration early after birth.[31]

19.5.3 Goitrogens

Goitrogens are compounds that lead to lower thyroid hormone synthesis or activity due to numerous mechanisms of action. They may act by suppressing T_3 and T_4 secretion, resulting in increased TSH production and thyroid gland enlargement (goiter). By blocking hormone synthesis, goitrogens increase the sensitivity of the gland to TSH, which further promotes goitrogenicity. Iodide deficiency is the major cause of endemic goiter and cretinism throughout the world.[98] Several areas of endemic goiter have been attributed to dietary goitrogens, usually acting together with iodine deficiency.

The mechanisms of action of goitrogens and of xenobiotics that affect thyroid hormone action may vary. Some goitrogens act by inhibiting iodide uptake or trapping; others act by interfering with thyroid peroxidase activity, preventing binding to carrier proteins or augmenting urinary iodide loss, which can increase thyroid hormone breakdown or interfere with its metabolism. While some xenobiotics mimic thyroid hormone's effects on expression of genes that are expressed before the onset of fetal thyroid function, there may also be a nongenomic molecular mechanism for perturbations of the thyroid hormone system.[99]

Goitrogens include goitrogenic drugs such as salicylates, lithium, iodides, phenylbutazone, and iodoantipyrine; antithyroidal drugs such as methymazole, propylthiouracil, and perchlorate[100]; food containing goitrogens; and xenobiotics such as cigarette smoke, environmental pollutants, and pesticides. For reviews on goitrogens, please see References 101–103.

19.5.3.1 Dietary Goitrogens

The dietary goitrogens fall into several categories and may contain cyanogenic glucosides such as cassava, sorghum, maize, and millet.[104] Diets high in soybean components or other components that increase fecal bulk may cause excess fecal loss of T_4 and increase the need for the hormone.[105,106] Thioglucosides metabolize an active goitrogenic thioglycoside, L-5-vinyl-2-thio-oxazolidone, to goitrin.[107] Goitrin inhibits oxidation of iodine and blocks its binding to thyroid globulins in the same way as do the thiocarbamides. Thioglucosides also metabolize to thiocyanates and isothiocyanates and are found in the seeds of plants of the genus brassica and the cruciferae, compositae, and unbelliferae.[108] Among the plants containing these compounds are cabbage, kale, brussels sprout, cauliflower, kohlrabi, turnip, rutabaga, mustard, and horseradish. Cattle may ingest these goitrogens and pass them to humans through milk.

19.5.3.2 Environmental and Other Goitrogens

In general, the basic qualitative changes that occur in hypothalamic-pituitary-thyroid adaptation to iodine deficiency or to drugs that interfere with thyroid hormone synthesis are similar in rats and humans. However, there are major species differences in the dose-response relationships of a number of drugs that act on the thyroid. Propylthiouracil (PTU) is a much more potent analog of thiouracil in the rat, but in humans it is less potent. Methimazole (MMI), the most universally employed clinical antithyroid agent, is 10 to 100 times as active as PTU in man but is only equally active in the rat.

19.5.3.3 Polychlorinated Biphenyls

Polychlorinated biphenyls (PCBs) are a class of industrial compounds that consist of paired phenyl rings with various degrees of chlorination. Several human studies claim to illustrate a link between the exposure to PCBs with neurological effects similar to those associated with thyroid hormone deficiencies. A dose-dependent association between PCB exposure and thyroid hormone levels has been described.[109–111] PCBs can enhance thyroid hormone metabolism by the liver, thus increasing their biliary excretion, or interfere with serum proteins that bind and transport thyroid hormone, effects that reduce circulating levels of thyroid hormone.

Exposure to high concentrations of PCBs can reduce circulating levels of thyroid hormone in animals.[112] In a congenitally hypothyroid mouse model (hyt/hyt), abnormalities in both the cognitive and motor systems are exhibited.[113] In this mouse and other animal models of thyroid hormone disorders, delayed somatic and reflexive development are noted, as are permanent deficits in hearing and locomotor and adaptive motor behavior. The hyt/hyt mouse has a mutation in the TSH receptor gene that renders it incapable of transducing the TSH signal in the thyrocyte to produce thyroid hormone. Some behavioral and possibly some biochemical abnormalities in mice exposed to PCBs are similar to those seen in the hyt/hyt mouse.

PCBs can mimic thyroid hormone effects on gene expression before the onset of fetal thyroid function interfering with thyroid hormone signaling without necessarily inhibiting the function of the thyroid gland.[114,115] Epidemiological studies that have attempted to relate PCB body burden to thyroid hormone levels have produced inconsistent results, as have studies aimed at linking PCB exposure to neurodevelopmental effects. This may be due to the difficulties in finding a uniform way to compare exposure levels so that data from all studies can be compared.

Polychlorinated dibenzofurans (PCDFs), produced when PCBs are burned, are equally persistent and toxic. Dioxins may also interfere with thyroid hormone action and thus adversely affect the developing and mature brain. Contaminants that can competitively bind to thyroid hormone receptors include bisphenol A[116] and halogenated bisphenol A derivatives.[117] Bisphenol A can inhibit thyroid hormone ability to regulate gene expression, while bisphenol A derivatives serve as TR antagonists, thus interfering with thyroid hormone signaling. The extent and nature of xenobiotic effects on developmental disruptions in glia, thyroid hormone, and brain require further investigation.

19.6 SUMMARY

Glial cells mediate the action of thyroid hormone on the nervous system. Glial cells express nuclear receptors for both thyroid and steroid hormones and participate in the metabolism of thyroid hormones, resulting in neuroactive metabolites. Brain T_3 concentrations are determined by circulating T_4 and T_3 and deiodinases DI and DII that generate the active hormone T_3 from T_4. Astrocytes and tanycytes are primarily responsible for DII expression. Adverse alterations in thyroid hormone concentrations or responsiveness to thyroid hormone may have significant neurologic sequelae throughout the life cycle. During fetal and early neonatal periods, disorders of thyroid hormone may lead to the development of motor and cognitive disorders. During childhood and adult life, thyroid hormone is required for neuronal maintenance as well as for normal metabolic function. Thyroid hormone effects on astrocytes and oligodendrocytes span from nuclear genomic effects to nongenomic effects on cell structure proteins. Thyroid hormone regulation may be adversely affected if the regulating mechanisms are disturbed by the presence of xenobiotics. Those with an underlying disorder of thyroid hormone homeostasis may be at greater risk for developing cognitive, motor, or metabolic dysfunction upon exposure to xenobiotics that alter thyroid hormone economy. Disruption of thyroid hormone regulation may, however, occur at the level of the receptor, deiodinases, and any points of interaction between the hormones and the glial cells.

ACKNOWLEDGMENTS

I am very grateful to Dr. Juan Bernal for reading the manuscript in depth and making constructive and thoughtful comments and to Dr. Michael Aschner for his support and helpful comments.

REFERENCES

1. Bianco, A.C. et al., Biochemistry, cellular and molecular biology, and physiological roles of the iodothyronine selenodeiodinases, *Endocr. Rev.*, 23, 38, 2002.
2. Porterfield, S.P. and Hendrich, C.E., The role of thyroid hormones in prenatal and neonatal neurological development—current perspectives, *Endocr. Rev.*, 14, 94, 1993.
3. Contempre, B. et al., Detection of thyroid hormones in human embryonic cavities during the first trimester of pregnancy, *J. Clin. Endocrinol. Metab.*, 77, 1719, 1993.
4. Vulsma, T., Gons, M.H., and de Vijlder, J.J., Maternal-fetal transfer of thyroxine in congenital hypothyroidism due to a total organification defect or thyroid agenesis, *N. Engl. J. Med.*, 321, 13, 1989.
5. Morreale de Escobar, G., Obregon, M.J., and Escobar del Rey, F., Is neuropsychological development related to maternal hypothyroidism or to maternal hypothyroxinemia?, *J. Clin. Endocrinol. Metab.*, 85, 3975, 2000.
6. Calvo, R.M. et al., Fetal tissues are exposed to biologically relevant free thyroxine concentrations during early phases of development, *J. Clin. Endocrinol. Metab.*, 87, 1768, 2002.
7. Glinoer, D. and Delange F., The potential repercussions of maternal, fetal, and neonatal hypothyroxinemia on the progeny, *Thyroid*, 10, 871, 2000.
8. Smallridge, R.C. and Ladenson P.W., Hypothyroidism in pregnancy: consequences to neonatal health, *J. Clin. Endocrinol. Metab.*, 86, 2349, 2001.
9. Schreiber, G., The evolution of transthyretin synthesis in the choroid plexus, *Clin. Chem. Lab. Med.*, 40, 1200, 2002.
10. Schreiber, G., Richardson, S.J., and Prapunpoj, P., Structure and expression of the transthyretin gene in the choroid plexus: a model for the study of the mechanism of evolution, *Microsc. Res. Tech.*, 52, 21, 2001.
11. Palha, J.A. et al., Transthyretin regulates thyroid hormone levels in the choroid plexus, but not in the brain parenchyma: study in a transthyretin-null mouse model, *Endocrinology*, 141, 3267, 2000.
12. Dumitrescu, A.M. et al., A novel syndrome combining thyroid and neurological abnormalities is associated with mutations in a monocarboxylate transporter gene, *Am. J. Hum. Genet.*, 74, 168, 2004.
13. Friesema, E.C. et al., Identification of monocarboxylate transporter 8 as a specific thyroid hormone transporter, *J. Biol. Chem.*, 278, 40128, 2003.
14. Munoz, A. and Bernal, J., Biological activities of thyroid hormone receptors, *Eur. J. Endocrinol.*, 137, 433, 1997.
15. Gharami, K. and Das, S., Thyroid hormone-induced morphological differentiation and maturation of astrocytes are mediated through the beta-adrenergic receptor, *J. Neurochem.*, 75, 1962, 2000.
16. Lima, F.R. et al., Regulation of microglial development: a novel role for thyroid hormone, *J. Neurosci.*, 21, 2028, 2001.
17. Ben-Hur, T. et al., Growth and fate of PSA-NCAM+ precursors of the postnatal brain, *J. Neurosci.*, 18, 5777, 1998.
18. Billon, N. et al., Normal timing of oligodendrocyte development depends on thyroid hormone receptor alpha 1 (TRalpha1), *EMBO J.*, 21, 6452, 2002.

19. Guadano Ferraz, A. et al., The development of the anterior commissure in normal and hypothyroid rats, *Brain Res. Dev. Brain Res.*, 81, 293, 1994.

20. Pharoah, P.O., Buttfield, I.H., and Hetzel, B.S., Neurological damage to the fetus resulting from severe iodine deficiency during pregnancy, *Lancet*, 1, 308, 1971.

21. Xue-Yi, C. et al., Timing of vulnerability of the brain to iodine deficiency in endemic cretinism, *N. Engl. J. Med.*, 331, 1739, 1994.

22. Farwell, A.P., Tranter, M.P., and Leonard, J.L., Thyroxine-dependent regulation of integrin-laminin interactions in astrocytes, *Endocrinology*, 136, 3909, 1995.

23. Farwell, A.P. et al., The actin cytoskeleton mediates the hormonally regulated translocation of type II iodothyronine 5'-deiodinase in astrocytes, *J. Biol. Chem.*, 265, 18546, 1990.

24. Siegrist-Kaiser, C.A. et al., Thyroxine-dependent modulation of actin polymerization in cultured astrocytes. A novel, extranuclear action of thyroid hormone, *J. Biol. Chem.*, 265, 5296, 1990.

25. Harvey, C.B. and Williams, G.R., Mechanism of thyroid hormone action, *Thyroid*, 12, 441, 2002.

26. Bernal, J., Guadano-Ferraz, A., and Morte, B., Perspectives in the study of thyroid hormone action on brain development and function, *Thyroid*, 13, 1005, 2003.

27. Tenbaum, S.P. et al., Alien/CSN2 gene expression is regulated by thyroid hormone in rat brain, *Dev. Biol.*, 254, 149, 2003.

28. Davis, P.J. and Davis, F.B., Nongenomic actions of thyroid hormone on the heart, *Thyroid*, 12, 459, 2002.

29. Bernal, J., Action of thyroid hormone in brain, *J. Endocrinol. Invest.*, 25, 268, 2002.

30. Martinez-Galan, J.R. et al., Myelin basic protein immunoreactivity in the internal capsule of neonates from rats on a low iodine intake or on methylmercaptoimidazole (MMI), *Brain Res. Dev. Brain Res.*, 101, 249, 1997.

31. Martinez-Galan, J.R. et al., Early effects of iodine deficiency on radial glial cells of the hippocampus of the rat fetus. A model of neurological cretinism, *J. Clin. Invest.*, 99, 2701, 1997.

32. Morreale de Escobar, G., Obregon, M.J., and Escobar del Rey, F., Is neuropsychological development related to maternal hypothyroidism or to maternal hypothyroxinemia? *J. Clin. Endocrinol. Metab.*, 85, 3975, 2000.

33. Morreale de Escobar, G., The role of thyroid hormone in fetal neurodevelopment, *J. Pediatr. Endocrinol. Metab.*, 14, 1453, 2001.

34. Lavado-Autric, R. et al., Early maternal hypothyroxinemia alters histogenesis and cerebral cortex cytoarchitecture of the progeny, *J. Clin. Invest.*, 111, 1073, 2003.

35. Haddow, J.E. et al., Maternal thyroid deficiency during pregnancy and subsequent neuropsychological development of the child, *N. Engl. J. Med.*, 341, 549, 1999.

36. Pop, V.J. et al., Low maternal free thyroxine concentrations during early pregnancy are associated with impaired psychomotor development in infancy, *Clin. Endocrinol. (Oxf.)*, 50, 149, 1999.

37. Garcia-Segura, L.M., Chowen, J.A., and Naftolin, F., Endocrine glia: roles of glial cells in the brain actions of steroid and thyroid hormones and in the regulation of hormone secretion, *Front. Neuroendocrinol.*, 17, 180, 1996.

38. Brent, G.A., The molecular basis of thyroid hormone action, *N. Engl. J. Med.*, 331, 847, 1994.

39. Carlson, D.J. et al., Immunofluorescent localization of thyroid hormone receptor isoforms in glial cells of rat brain, *Endocrinology*, 135, 1831, 1994.

40. Strait, K.A. et al., Transient stimulation of myelin basic protein gene expression in differentiating cultured oligodendrocytes: a model for 3,5,3'-triiodothyronine-induced brain development, *Endocrinology*, 138, 635, 1997.

41. Carre, J.L. et al., Thyroid hormone receptor isoforms are sequentially expressed in oligodendrocyte lineage cells during rat cerebral development, *J. Neurosci. Res.*, 54, 584, 1998.

42. Kimelberg, H.K., Narumi, S., and Bourke, R.S., Enzymatic and morphological properties of primary rat brain astrocyte cultures, and enzyme development *in vivo*, *Brain Res.*, 153, 55, 1978.

43. Magistretti, P.J. et al., Neurotransmitters regulate energy metabolism in astrocytes: implications for the metabolic trafficking between neural cells, *Dev. Neurosci.*, 15, 306, 1993.

44. Hemmings, S.J. and Shuaib, A., Hypothyroidism-evoked shifts in hippocampal adrenergic receptors: implications to ischemia-induced hippocampal damage, *Mol. Cell. Biochem.*, 185, 161, 1998.

45. Bogazzi, F., Hudson, L.D., and Nikodem, V.M., A novel heterodimerization partner for thyroid hormone receptor. Peroxisome proliferator-activated receptor, *J. Biol. Chem.*, 269, 11683, 1994.

46. Barres, B.A., Lazar, M.A., and Raff, M.C., A novel role for thyroid hormone, glucocorticoids and retinoic acid in timing oligodendrocyte development, *Development*, 120, 1097, 1994.

47. Baas, D. et al., Oligodendrocyte maturation and progenitor cell proliferation are independently regulated by thyroid hormone, *Glia*, 19, 324, 1997.

48. Hudson, L.D., Ko, N., and Kim, J.G., Control of myelin gene expression, in *Glial Cell Development*, Jessen K. and Richardson, W.D., Eds., Bios Scientific, Oxford, U.K., 1994, p. 101.

49. Huo, B., Dozin, B., and Nikodem, V.M., Identification of a nuclear protein from rat developing brain as heterodimerization partner with thyroid hormone receptor-beta, *Endocrinology*, 138, 3283, 1997.

50. Poduslo, S.E., Miller, K., and Pak, C.H., Induction of cerebroside synthesis in oligodendroglia, *Neurochem. Res.*, 15, 739, 1990.

51. Bhat, N.R., Rao, G.S., and Pieringer, R.A., Investigations on myelination *in vitro*. Regulation of sulfolipid synthesis by thyroid hormone in cultures of dissociated brain cells from embryonic mice, *J. Biol. Chem.*, 256, 1167, 1981.

52. Legrand, J., Variations, as a function of age, of the response of the cerebellum to the morphogenetic action of the thyroid in rats, *Arch. Anat. Microsc. Morphol. Exp.*, 56, 291, 1967.

53. Schwartz, H.L. and Oppenheimer, J.H., Ontogenesis of 3,5,3'-triiodothyronine receptors in neonatal rat brain: dissociation between receptor concentration and stimulation of oxygen consumption by 3,5,3'-triiodothyronine, *Endocrinology*, 103, 943, 1978.

54. Perez-Castillo, A. et al., The early ontogenesis of thyroid hormone receptor in the rat fetus, *Endocrinology*, 117, 2457, 1985.

55. Shao, Y. and Sutin J., Expression of adrenergic receptors in individual astrocytes and motor neurons isolated from the adult rat brain, *Glia*, 6, 108, 1992.

56. Shao, Y. and McCarthy, K.D., Plasticity of astrocytes, *Glia*, 11, 147, 1994.

57. Morte, B. et al., Aberrant maturation of astrocytes in thyroid hormone receptor alpha 1 knock out mice reveals an interplay between thyroid hormone receptor isoforms, *Endocrinology*, 145, 1386, 2004.

58. Amat, J.A. et al., Cells of the oligodendrocyte lineage proliferate following cortical stab wounds: an *in vitro* analysis, *Glia*, 22, 64, 1998.

59. Noguchi, T., Effects of growth hormone on cerebral development: morphological studies, *Horm. Res.*, 45, 5, 1996.

60. Rosman, N.P. et al., The effect of thyroid deficiency on myelination of brain. A morphological and biochemical study, *Neurology*, 22, 99, 1972.

61. Balazs, R. et al., The effect of neonatal thyroidectomy on myelination in the rat brain, *Brain Res.*, 15, 219, 1969.

62. Walters, S.N. and Morell, P., Effects of altered thyroid states on myelinogenesis, *J. Neurochem.*, 36, 1792, 1981.

63. Shanker, G., Amur, S.G., and Pieringer, R.A., Investigations on myelinogenesis *in vitro*: a study of the critical period at which thyroid hormone exerts its maximum regulatory effect on the developmental expression of two myelin associated markers in cultured brain cells from embryonic mice, *Neurochem. Res.*, 10, 617, 1985.

64. Ibarrola, N. and Rodriguez-Pena, A., Hypothyroidism coordinately and transiently affects myelin protein gene expression in most rat brain regions during postnatal development, *Brain Res.*, 752, 285, 1997.

65. Clos, J., Legrand, C., and Legrand, J., Effects of thyroid state on the formation and early morphological development of Bergmann glia in the developing rat cerebellum, *Dev. Neurosci.*, 3, 199, 1980.

66. Porterfield, S.P. and Hendrich, C.E., Tissue iodothyronine levels in fetuses of control and hypothyroid rats at 13 and 16 days gestation, *Endocrinology*, 131, 195, 1992.

67. Perry, V.H., Modulation of microglia phenotype, *Neuropathol. Appl. Neurobiol.*, 20, 177, 1994.

68. Mallat, M. et al., New insights into the role of thyroid hormone in the CNS: the microglial track, *Mol. Psychiatry*, 7, 7, 2002.

69. Giulian, D. et al., Interleukin-1 is an astroglial growth factor in the developing brain, *J. Neurosci.*, 8, 709, 1988.

70. Chamak, B., Morandi, V., and Mallat, M., Brain macrophages stimulate neurite growth and regeneration by secreting thrombospondin, *J. Neurosci. Res.*, 38, 221, 1994.

71. Chamak, B., Dobbertin, A., and Mallat, M., Immunohistochemical detection of thrombospondin in microglia in the developing rat brain, *Neuroscience*, 69, 177, 1995.

72. Yen, P.M., Physiological and molecular basis of thyroid hormone action, *Physiol. Rev.*, 81, 1097, 2001.
73. Poddar, R. et al., Regulation of actin and tubulin gene expression by thyroid hormone during rat brain development, *Brain Res. Mol. Brain Res.*, 35, 111, 1996.
74. Siegrist-Kaiser, C.A. et al., Thyroxine-dependent modulation of actin polymerization in cultured astrocytes. A novel, extranuclear action of thyroid hormone, *J. Biol. Chem.*, 265, 5296, 1990.
75. Chiu, A.Y. et al., Laminin and s-laminin are produced and released by astrocytes, Schwann cells, and schwannomas in culture, *Glia*, 4, 11, 1991.
76. Farwell, A.P. and Dubord-Tomasetti, S.A., Thyroid hormone regulates the extracellular organization of laminin on astrocytes, *Endocrinology*, 140, 5014, 1999.
77. Leonard, J.L. et al., Differential expression of thyroid hormone receptor isoforms in neurons and astroglial cells, *Endocrinology*, 135, 548, 1994.
78. Carlson, D.J. et al., Thyroid hormone receptor isoform content in cultured type 1 and type 2 astrocytes, *Endocrinology*, 137, 911, 1996.
79. Farsetti, A. et al., Active repression by thyroid hormone receptor splicing variant alpha 2 requires specific regulatory elements in the context of native triiodothyronine-regulated gene promoters, *Endocrinology*, 138, 4705, 1997.
80. Leonard, J.L. and Farwell, A.P., Thyroid hormone-regulated actin polymerization in brain, *Thyroid*, 7, 147, 1997.
81. Farwell, A.P., DiBenedetto, D.J., and Leonard, J.L., Thyroxine targets different pathways of internalization of type II iodothyronine 5'-deiodinase in astrocytes, *J. Biol. Chem.*, 268, 5055, 1993.
82. Tu, H.M. et al., Regional expression of the type 3 iodothyronine deiodinase messenger ribonucleic acid in the rat central nervous system and its regulation by thyroid hormone, *Endocrinology*, 140, 784, 1999.
83. Guadano-Ferraz, A. et al., The type 2 iodothyronine deiodinase is expressed primarily in glial cells in the neonatal rat brain, *Proc. Natl. Acad. Sci. U.S.A.*, 94, 10391, 1997.
84. Tu, H.M. et al., Regional distribution of type 2 thyroxine deiodinase messenger ribonucleic acid in rat hypothalamus and pituitary and its regulation by thyroid hormone, *Endocrinology*, 138, 3359, 1997.
85. Zhang, L.C. et al., The distributions and signaling directions of the cerebrospinal fluid contacting neurons in the parenchyma of a rat brain, *Brain Res.*, 989, 1, 2003.
86. Bernal, J., Iodine and brain development, *Biofactors*, 10, 271, 1999.
87. Farwell, A.P. et al., Degradation and recycling of the substrate-binding subunit of type II iodothyronine 5'-deiodinase in astrocytes, *J. Biol. Chem.*, 271, 16369, 1996.
88. Ahlgren, S.C. et al., Effects of thyroid hormone on embryonic oligodendrocyte precursor cell development *in vivo* and *in vitro*, *Mol. Cell. Neurosci.*, 9, 420, 1997.
89. Farsetti, A. et al., Molecular basis of thyroid hormone regulation of myelin basic protein gene expression in rodent brain, *J. Biol. Chem.*, 266, 23226, 1991.
90. Illig, R. et al., Mental development in congenital hypothyroidism after neonatal screening, *Arch. Dis. Child.*, 62, 1050, 1987.
91. Jhiang, S.M. et al., An immunohistochemical study of Na+/I- symporter in human thyroid tissues and salivary gland tissues, *Endocrinology*, 139, 4416, 1998.
92. Spitzweg, C. et al., Analysis of human sodium iodide symporter immunoreactivity in human exocrine glands, *J. Clin. Endocrinol. Metab.*, 84, 4178, 1999.
93. Ajjan, R.A. et al., Regulation and tissue distribution of the human sodium iodide symporter gene, *Clin. Endocrinol. (Oxf.)*, 49, 517, 1998.
94. Delange, F. et al., Iodine deficiency in the world: where do we stand at the turn of the century?, *Thyroid*, 11, 437, 2001.
95. Hetzel, B.S., Eliminating iodine deficiency disorders — the role of the International Council in the global partnership, *Bull. World Health Organ.*, 80, 410 (discussion 413), 2002.
96. Glinoer, D., Feto-maternal repercussions of iodine deficiency during pregnancy. An update, *Ann. Endocrinol. (Paris)*, 64, 37, 2003.
97. Dumont, J.E., Corvilain, B., and Contempre, B., The biochemistry of endemic cretinism: roles of iodine and selenium deficiency and goitrogens, *Mol. Cell. Endocrinol.*, 100, 163, 1994.
98. Delange, F., Thilly, C., and Ermans, A.M., Iodine deficiency, a permissive condition in the development of endemic goiter, *J. Clin. Endocrinol. Metab.*, 28, 114, 1968.

99. McEwen, B.S., Gonadal and adrenal steroids regulate neurochemical and structural plasticity of the hippocampus via cellular mechanisms involving NMDA receptors, *Cell. Mol. Neurobiol.*, 16, 103, 1996.

100. Wolff, J., Perchlorate and the thyroid gland, *Pharmacol. Rev.*, 50, 89, 1998.

101. Gaitan, E., Goitrogens in food and water, *Annu. Rev. Nutr.*, 10, 21, 1990.

102. Gaitan, E., Goitrogens, *Baillieres Clin. Endocrinol. Metab.*, 2, 683, 1988.

103. Rao, P.S. and Lakshmy, R., Role of goitrogens in iodine deficiency disorders & brain development, *Indian J. Med. Res.*, 102, 223, 1995.

104. Ermans, A.M. et al., Possible role of cyanide and thiocyanate in the etiology of endemic cretinism, in *Human Development and the Thyroid Gland. Relation to Endemic Cretinism*, Kroc, R.L., Ed., Plenum Press, New York, 1972, p. 455.

105. Van Wyk, J.J. et al., The effects of a soybean product on thyroid functions in humans, *Pediatrics*, 24, 752, 1959.

106. Yamada, T., Effect of fecal loss of thyroxine on pituitary-thyroid feedback control in the rat, *Endocrinology*, 82, 327, 1968.

107. Langer, P. and Michajlovskij, N., Studies on the antithyroid activity of naturally occurring L-5-vinyl-2-thiooxazolidone and its urinary metabolite in rats, *Acta Endocrinol. (Copenh.)*, 62, 21, 1969.

108. Langer, P. and Greer, M.A., Antithyroid activity of some naturally occurring isothiocyanates *in vitro*, *Metabolism*, 17, 596, 1968.

109. Hagmar, L. et al., Plasma concentrations of persistent organochlorines in relation to thyrotropin and thyroid hormone levels in women, *Int. Arch. Occup. Environ. Health*, 74, 184, 2001.

110. Sala, M. et al., Association between serum concentrations of hexachlorobenzene and polychlorobiphenyls with thyroid hormone and liver enzymes in a sample of the general population, *Occup. Environ. Med.*, 58, 172, 2001.

111. Persky, V. et al., The effects of PCB exposure and fish consumption on endogenous hormones, *Environ. Health Perspect.*, 109, 1275, 2001.

112. Goldey, E.S. and Crofton, K.M., Thyroxine replacement attenuates hypothyroxinemia, hearing loss, and motor deficits following developmental exposure to Aroclor 1254 in rats, *Toxicol. Sci.*, 45, 94, 1998.

113. Sher, E.S. et al., The effects of thyroid hormone level and action in developing brain: are these targets for the actions of polychlorinated biphenyls and dioxins?, *Toxicol. Ind. Health*, 14, 121, 1998.

114. Zoeller, R.T., Dowling, A.L., and Vas, A.A., Developmental exposure to polychlorinated biphenyls exerts thyroid hormone-like effects on the expression of RC3/neurogranin and myelin basic protein messenger ribonucleic acids in the developing rat brain, *Endocrinology*, 141, 181, 2000.

115. Inglefield, J.R. et al., Identification of calcium-dependent and -independent signaling pathways involved in polychlorinated biphenyl-induced cyclic AMP-responsive element-binding protein phosphorylation in developing cortical neurons, *Neuroscience*, 115, 559, 2002.

116. Moriyama, K. et al., Thyroid hormone action is disrupted by bisphenol A as an antagonist, *J. Clin. Endocrinol. Metab.*, 87, 5185, 2002.

117. Kitamura, S. et al., Thyroid hormonal activity of the flame retardants tetrabromobisphenol A and tetrachlorobisphenol A, *Biochem. Biophys. Res. Commun.*, 293, 554, 2002.

CHAPTER 20

Astrocytes in Ammonia Neurotoxicity: A Target, a Mediator, and a Shield

Jan Albrecht

CONTENTS

20.1 INTRODUCTION

A chapter under a similar title was contributed to the previous edition of *The Role of Glia in Neurotoxicity* in 1996.[1] The present text will primarily focus on the key findings and ideas pertinent to ammonia-induced changes in the metabolism and function of astrocytes that have emerged in the last decade. The leitmotif will be elucidation of the indirect nature of ammonia neurotoxicity, i.e., that the impaired function of neurons (disturbed neural transmission) is associated with, and in most aspects ensues, changes that occur in astrocytes. The stage-setting Section 20.2 will offer brief characteristics of the epidemiology of hyperammonemic conditions, their major neurological symptoms and causes. Section 20.3 will dwell on the various aspects of ammonia-induced alterations in astrocytic metabolism. It will be argued that bioenergetic failure is the essence of these alterations and that many of them are related to the synthesis and accumulation of glutamine, the major ammonia metabolite in the CNS. Section 20.4 will describe the postulated sequence of events that lead to astrocytic swelling, the major contributing factor to hyperammonemic brain edema. Section 20.5 will focus on the proven or easily predictable consequences of astrocytic changes for the functioning of the neurotransmitter systems. Section 20.6 will describe changes in the neuroprotective functions of astrocytes. Perspectives and issues that deserve attention in future studies will be dealt with in Section 20.7.

The chapter does not pretend to cover all the aspects of ammonia-induced changes in astrocytes. It concentrates on those that have a bearing on the astrocytic–neuronal interaction and can be linked to distinct symptoms of CNS dysfunction related to hyperammonemia.

20.2 AMMONIA IN NEUROLOGICAL DISORDERS: EPIDEMIOLOGY AND SYMPTOMS

Ammonia is a well-documented pathogenic factor in a number of neurological disorders. Major disease units in which increased blood or brain ammonia has a discernible impact on the patient's neurological status are listed in Table 20.1. These conditions are collectively coined *hyperammonemic encephalopathies*. Hyperammonemia is most frequently associated with liver damage. Ammonia not metabolized in the liver enters the brain in excess, and the ensuing cerebral symptoms are defined as hepatic encephalopathy (HE).

Available data illustrating the epidemiology of hyperammonemic encephalopathies are incomplete and indirect; exhaustive analyses have dealt with liver cirrhosis. Although the involvement of ammonia-induced brain damage in these cases cannot be predicted with adequate precision, numbers derived from rough estimations are strikingly high. According to a report of the WHO Regional Office issued in June 2000 and accounting for EU countries, liver cirrhosis is diagnosed in 14 cases/100,000, and the death rate of liver cirrhotic patients amounts to ~ 60,000/year. Available data do not directly pertain to ammonia as a cause of death. However, it is estimated that 50–75% of cases of liver cirrhosis are associated with periodical occurrence of hyperammonemia and HE symptoms. By comparison, the number of car accidents in EU does not exceed 40,000/year.

Irrespective of whether acute or chronic, HE and other forms of hyperammonemia are associated with a broad spectrum of neuropsychiatric symptoms, which according to their advancement have been defined as stages I–IV. For detailed description of the mental state and electrophysiologic and biochemical parameters characterizing these different stages, the reader may be referred to recent reviews.[2-4] Briefly, euphoria, irritability, and overexcitation in stage I evolve through gradual decrease of motor activity and lethargy in stages II and III to coma in stage IV. Aggravation of neuropsychiatric symptoms is accompanied by gradual slowing of the EEG pattern and by the development of cerebral edema, which is a major cause of death in acute forms of HE. Collectively,

Table 20.1 Hyperammonemic Encephalopathies

Acute hepatic encephalopathy (viral or toxic liver injury)
Chronic hepatic encephalopathy (cirrhosis, portacaval shunt)
Reye's syndrome
Inborn urea cycle insufficiencies
Uremic encephalopathy
Diabetic encephalopathy
Hypoglycemic coma

the emerging picture is that of a progressing imbalance between inhibitory and excitatory transmission in disfavor of the latter, the degree of the imbalance being fairly well correlated with the increase of blood ammonia.[2] As will be pointed out later, astrocytic swelling is overwhelmingly involved in ammonia-induced cerebral edema, and astrocytic dysfunction largely contributes to the disturbances of neurotransmission.

As described in the previous edition,[1] advanced stages of hyperammonemia are hallmarked by the appearance in brain of transformed astrocytes, the so-called Alzheimer type II cells. However, these cells occur in modest quantities and their metabolic and functional characteristics are largely unknown, making it impossible to implicate them in any defined manifestations of hyperammonemic encephalopathy.

20.3 ASTROCYTE AS A METABOLIC TARGET OF AMMONIA

As mentioned elsewhere in the book, topographic positioning of astrocytes makes them act as a natural barrier for all the toxins that pass from the capillary bed to the neurons and thus a primary target of its neurotoxic action. In this sense, ammonia is no exception. What makes the astrocytes particularly vulnerable to ammonia is that they are the brain compartment responsible for cerebral ammonia metabolism.

20.3.1 Glutamine Synthesis in Astrocytes and the Glutamine–Glutamate Cycle: A Key Route of Ammonia Metabolism in the Brain

The brain has an incomplete urea cycle, and therefore the glutamine synthetase (GS)–mediated amidation of glutamate (Glu) to glutamine (Gln) is the only efficient way in which blood-borne ammonia is trapped and detoxicated.[5,6] Gln synthesis is catalyzed by glutamine synthetase, which is an astroglia-specific enzyme.[7] The thus-formed Gln is transferred to neurons where it becomes substrate in the synthesis of amino acid neurotransmitters glutamate (Glu) and GABA.[8,9] Glu released from neurons reacts with Glu receptors on the postsynaptic membrane. A major proportion of neurotransmitter Glu is transferred back to astrocytes and serves to trap ammonia, giving rise to Gln, thereby closing the Glu-Gln cycle. The major steps of the cycle are illustrated in Figure 20.1. As will be discussed later, overloading astrocytes with ammonia leads to a complex chain of metabolic derangements within and independently of the cycle, many of which involve excessive accumulation of Gln.

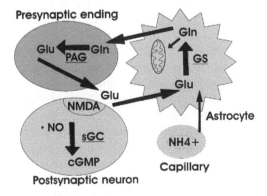

Figure 20.1 Coupling of blood-borne ammonia metabolism to the glutamine–glutamate cycle. For description, see Section 20.3.1. NH_4^+, ammonium cation; Glu, glutamate; Gln, glutamine; GS, glutamine synthetase; PAG, phosphate-activated glutaminase; NMDA, *N*-methyl-*D*-aspartate receptor (only this class of ionotropic Glu receptors is indicated); NO, nitric oxide; sGC, soluble guanyl cyclase; cGMP, cyclic guanosyl-3′-5′-monophosphate.

20.3.2 Bioenergetic Failure

Bioenergetic failure, the primary ammonia-induced derangement of astrocytic metabolism and an underlying cause of astrocytic dysfunction, is to be seen as a cumulative effect of a number of interrelated processes. These include (1) increased ATP consumption, (2) decreased ATP production, and (3) mitochondrial damage, manifested by mitochondrial permeability transition (MPT) and swelling.

Increased ATP consumption in astrocytes exposed to excess ammonia has two major causes: increased Gln synthesis and enhanced activity of the Na/K ATPase. Both events represent the early response to acute ammonia overload *in vivo* and *in vitro*.[10-13] A large body of evidence suggests that ammonia impairs oxidative metabolism in astrocytes. Added to astrocytes in culture, ammonia inhibits oxidation of a plethora of substrates including pyruvate, branched-chain amino acids, and glutamate. These effects, measured as reduction of CO_2 production from,[14-18] or of oxygen consumption supported by, the respective substrates,[19] are consistent with the inhibition by ammonia of the tricarboxylic acid cycle enzymes in various treatment paradigms and CNS preparations. Thus, treatment with ammonia of cerebral nonsynaptic mitochondria, a fraction enriched in astrocytic mitochondria, inhibited the individual components of the α-ketoglutarate dehydrogenase complex[20-22] and the malate–aspartate shuttle enzymes.[23] Increased conversion of Glu to Gln in astrocytes[10-12] depletes the cells of a key energy source. Accordingly, ammonia-treated astrocytes show morphological changes including swelling of matrix.[24] More recent studies have shown that treatment of astrocytes with ammonia leads to a cyclosporin A (CsA)–inhibitable collapse of the inner mitochondrial membrane potential ($_m$) and MPT.[25,26] Of note, bioenergetic failure also results from the direct action of accumulated Gln on mitochondria. This issue will be discussed in more detail in Section 20.4, in the context of the mechanism underlying ammonia-induced astrocytic swelling.

One other aspect of astrocytic energy metabolism pertains to the breakdown of glycogen, a noradrenaline-stimulated, cAMP-dependent process that is thought to fulfill part of the glucose demand of adjacent neurons (reviewed in Reference 27). Treatment of cultured astrocytes with ammonia reduces both noradrenaline-dependent generation of cAMP[28] and cAMP-mediated glycogen breakdown,[29] indicating limitation of the neurosupportive potency of astrocytes.

20.3.3 Oxidative and Nitrosative Stress

A considerable body of evidence suggests that hyperammonemia *in vivo* and treatment of cerebral preparations with ammonia *in vitro* evoke both oxidative and nitrosative stress in CNS cells. Increased production of reactive oxygen species has been recorded in the brains of rats subjected to acute ammonia treatment[30] and in striatal microdialysates of rats in which ammonia was administered via the microdialysis probe.[31] Recently, Murthy et al. demonstrated production of free radicals in ammonia-treated astrocytes in culture.[32] Most interestingly, this effect was suppressed by MSO, supporting the role of Gln synthesis in the ammonia-induced mitochondrial impairment (see also Section 20.4). Ammonia may also promote oxidative stress by upregulating heme oxygenase-1 expression, as demonstrated both in cultured astrocytes and in the brain *in vivo*.[33]

Increased NO production *in vivo* in ammonia-treated mice or rats is monitored as enhanced accumulation in the brain microdialysates of cGMP, which results from soluble guanyl cyclase activation by NO.[31,34] However, the relative contribution of neurons and astrocytes to the ammonia-dependent formation of NO, the degree to which NO synthesis overlaps with the generation of other superoxides, and the underlying mechanisms have not been delineated in detail. For a long time, enhanced generation of NO has been interpreted to reflect overstimulation of NMDA receptors in neurons. Indeed, ammonia toxicity in acutely hyperammonemic mice was attenuated by previous administration of an inhibitor of nitric oxide synthase (NOS), nitroarginine,[35] and by NMDA receptor antagonists including MK-801.[36] The existence of an NMDA receptor–independent component of

NO synthesis has been suspected on the basis of the observation that the effect of nitroarginine was qualitatively different from MK-801.[35] At this stage, however, interaction of ammonia with either the NMDA-dependent or -independent component of NO synthesis has not been considered to take place in astrocytes. While functional NMDA receptors have been documented in astrocytes *in situ*,[37] their presence in cultured astrocytes has long been questioned.[38] In turn, the NMDA receptor–uncoupled NO synthesis has been ascribed to synaptosomes, the fraction where increased uptake of the NO precursor, arginine, was measured in hyperammonemic rats.[39–41] Eventually, refinements of culturing and recording techniques have led to a convincing demonstration of NMDA receptor–mediated electrophysiological and metabolic events in astrocytes in culture. Ammonia was found to induce NO synthase in cultured astrocytes, and the process appeared sensitive to NMDA receptor blockade.[42] Activation of the NMDA/NO pathway in astrocytes promoted protein tyrosine nitration (PTN)[43] and inhibited GS activity,[42,43] supporting the involvement of PTN in tonic regulation of ammonia detoxication in astrocytes. Collectively, the results point to the concerted contribution of astrocytes and neurons to the oxidative and nitrosative stress evoked by ammonia.

20.3.4 Peripheral Benzodiazepine Receptor: Mediator of Ammonia-Induced Oxidative Stress and Mitochondrial Dysfunction in Astrocytes

Peripheral benzodiazepine receptor (PBR) is a hydrophobic transmembrane protein that is located in the outer mitochondrial membrane and is a component of the permeability transition pore (PTP), whose opening leads to MPT (reviewed in Reference 44). In the CNS, PBR are mainly localized in astrocytes.[45] PBR concentrations have been found increased in the autopsy brain samples of HE patients[46] and in rat brain in experimentally induced hyperammonemia.[47] Benzodiazepines acting at PBR in astrocytes induce the formation of free radicals[48] and PTN.[49] Nitration by a model benzodiazepine, diazepam, is mediated by NMDA receptors and involves free radicals (perhaps generated in association with increased heme oxygenase activity) but bypasses NOS activity, indicating a mechanism different from that underlying the effect of ammonia. PTN also occurs in diazepam-treated rats *in vivo*[43] (for recent review, see Reference 49).

20.4 MECHANISMS OF AMMONIA-INDUCED ASTROCYTIC SWELLING

Cerebral edema, the major complication and cause of intracranial hypertension and death in patients with acute hyperammonemia, is primarily due to astrocytic swelling.[50] Cumulative effects of mitochondrial dysfunction and Gln synthesis have been held responsible for the ammonia-induced astrocytic gain of volume or water. A number of studies have pointed to a good correlation between the elevation of brain Gln and brain water content in acutely hyperammonemic animals.[51,52] However, the mechanism by which Gln promotes water accumulation remains a matter of hot debate.

In its original version, the "Gln theory" attributed water accumulation in astrocytes to the osmotic action of glutamine.[53] This interpretation stems from studies of the past decade that measured the pathophysiologic effects of inhibition of Gln synthesis by MSO. In rats made hyperammonemic by ammonium acetate infusion, where an increase of brain water content was associated with more than a tripling of brain Gln, pretreatment with MSO has led to normalization of cerebral Gln and prevented intracerebral accumulation of water.[54] MSO treatment substantially reduced brain swelling and the increase of intracranial pressure in rats in which acute hyperammonemia was produced by continuous infusion of ammonia.[55] Treatment of hyperammonemic rats with MSO attenuated a number of ultrastructural manifestations of astrocytic swelling, including a decrease in cytoplasmic density and an increase of nuclear circumference.[56] Ammonia infusion into rats increased the extracellular potassium activity in the brain, which was ascribed to the shrinkage of the extracellular space in association with astrocytic swelling; this effect was likewise attenuated by MSO treatment.[57] However, as outlined below, recent data question the role of Gln as an edema-producing osmolyte.

Mild hypothermia proved successful as a treatment modality for brain ischemia, countering the consequences of decreased energy supply by reducing energy consumption.[59] In portacaval-shunted rats infused with ammonia, mild hypothermia delayed the development of brain edema without affecting the ammonia-induced rise of the cerebral glutamine.[60] Development of cerebral edema in acute hyperammonemia was associated with increased synthesis of Gln and other glucose-derived metabolites, lactate, and alanine, which in conjunction with a decrease of glutamate and aspartate content indicated cerebral mitochondrial energy failure.[61] Mild hypothermia attenuated brain edema and other pathophysiological HE symptoms in this model and normalized lactate and alanine synthesis and glutamate and aspartate levels but did not reduce glutamine synthesis.[62] Hence, bioenergetic failure plus osmotic effects of lactate and alanine appeared to play a more crucial role in the development of astrocytic swelling and brain edema than the osmotic action of Gln. However, there is evidence to suggest that Gln accumulating in excess contributes to ammonia-induced mitochondrial damage. Glutamine at pathophysiologically relevant (low millimolar) concentrations induced MPT in nonsynaptic mitochondria[63] and cultured astrocytes,[64] and dissipation of $\Delta\psi_m$ in ammonia-treated astrocytes in culture was prevented by coaddition of a GS inhibitor, methionine sulfoximine (MSO).[25] In this context, the fact that excess glutamine did not counter amelioration by hypothermia of ammonia-induced brain edema in the *in vivo* model may be related to a decreased demand for energy production by mitochondria or to decreased entry of Gln through the mitochondrial membrane. Gln transport in mitochondria is an active process that is likely to become stimulated by hyperammonemia.[65]

Our own recent study revealed a Glu-mediated, excitotoxic component of ammonia-induced cell swelling in cerebral cortical slices; the increase of cell volume was reduced by NMDA receptor antagonists, nitric oxide synthase blockers, and free-radical scavengers.[66] Of note, the reduction was not associated with a significant decrease of the cerebral Gln content. Other factors inferred in predisposing to ammonia-induced astrocytic swelling include interaction of endogenous ligands with upregulated PBR receptors[67] and increased aquaporin-4 protein expression,[68] which is likely to promote transmembrane movement of osmotically obligated water.[69] It is thus clear that Gln-dependent and independent mechanisms cooperate in inducing hyperammonemic brain edema.[70]

20.5 CONTRIBUTION OF ASTROCYTES TO THE AMMONIA-INDUCED CHANGES IN NEURAL TRANSMISSION

20.5.1 Glutaminergic Transmission

One of the major functions of astrocytes is synaptic clearance of neuronally released Glu and compensation for its continuous nonvesicular release from intracellular compartments.[71] Acute hyperamonemia is characterized by increased extracellular accumulation of Glu,[72,73] and this phenomenon has been implicated in the earlier discussed excitotoxic aspects of ammonia neurotoxicity. Ammonia has been shown to inhibit Glu uptake in astrocytes in culture[74] and in astroglia-enriched fractions bulk-isolated from adult rat brain.[75] Decreased expression of astrocytic Glu transporters, GLT1 and GLAST, has been demonstrated in *in vivo* hyperammonemic models[76,77] and in cultured astrocytes upon prolonged treatment with ammonium ions.[77–79] By contrast, chronic hyperammonemic conditions are often associated with Glu depletion in the brain.[80] An *in vivo* study using the NMR technique revealed a decreased incorporation into glutamate of a label derived from [2-^{13}C]acetate, a precursor of Gln synthesis in astrocytes, in portacaval-shunted rats.[81] This observation linked Glu depletion to the impaired operation of the Glu-Gln cycle along the astrocyte–neuron axis.

20.5.2 GABAergic Transmission

Hyperammonemia is associated with an increased GABAergic tone, and astrocytes have been implicated in two mechanisms underlying this phenomenon. One is associated with the increased PBR density; PBRs control the synthesis of pregnenolone-derived neurosteroids, some of which are positive modulators of the GABA(A)-benzodiazepine receptor complex.[82,83] In two models of hyperammonemia in mice, the increase of PBR binding was coupled to increased synthesis of pregnenolone and its neuroactive derivatives,[84] and pretreatment of healthy mice with a PBR antagonist increased their resistance to subsequent injection of a toxic dose of ammonium acetate.[85] Ammonia promotes GABAergic transmission directly by increasing the GABA-induced chloride current[86] and enhances the binding of agonist ligands to the GABAA receptor complex.[87] Enhanced release of GABA was measured in astrocytes isolated from animals in which hyperammonemia was associated with toxic liver damage.[88] Increase of astrocyte-derived GABA in the synaptic cleft is expected to further enhance the GABAergic tone already stimulated by ammonia.

20.6 AMMONIA AND ASTROGLIA-DERIVED NEUROMODULATORS: FOCUS ON NEUROPROTECTION

Astrocytes synthesize (or store) and release two molecules that interact with nerve cell receptors and, as such, may be defined as gliotransmitters. The nonproteinaceous sulfur amino acid taurine is a weak GABAA and glycine receptor ligand,[89] and the tryptophan metabolite kynurenic acid is a broad-spectrum antagonist of ionotropic glutamate receptors, showing a strong affinity toward the glycine site of the N-methyl-D-aspartate (NMDA) receptor.[90] Indirect evidence suggests that ammonia alters the synthesis and release of either compound, which is likely to alter their neuroprotective potential.

20.6.1 Taurine

Taurine is a nonproteinaceous sulfur amino acid showing neuroprotective activities associated with its agonistic properties toward the GABAA receptor but also with its ability to act as an antioxidant and cell membrane stabilizer.[91] Our own recent studies have provided strong evidence that taurine protects CNS cells from damage evoked by acute ammonia treatment. Taurine applied intracerebrally limited the ammonia-induced accumulation in the microdialysates of rat striatum of cGMP and free radicals,[31] as well as dopamine and its metabolite, dihydroxyphenylacetic acid.[92] Taurine also counteracted ammonia-induced swelling of cerebral cortical slices, and the effect was antagonized by a GABAA receptor antagonist, bicucculine.[66]

Endogenous, astroglia-derived taurine may likewise offer protection against ammonia neurotoxicity. Acute ammonia treatment of astrocytes in vitro is a potent stimulus of taurine release. Short-term incubation with ammonia concentrations recorded in brain in acute HE evoked a massive release of newly loaded taurine from cultured astrocytes[93] and Müller glial cells[94–96] and of endogenous taurine from cerebrocortical slices.[97,98] The magnitude of stimulation of taurine release by 5 mM ammonia in Müller glia was similar to the release elicited by 65 mM KCl.[94,95] The mechanism of ammonia-induced taurine release differed from that normally evoked by conditions associated with ionic or osmotic stress. It was osmoresistant, not inhibited by the ion cotransport inhibitor furosemide, calcium-independent, and coupled to cAMP synthesis.[95] The release evoked by ammonia in a glia-derived cell line (C6) involved activation of the taurine transporter, TauT.[99] Endogenous taurine release from cerebrocortical slices was only partly inhibited by furosemide, was poorly correlated with ammonia-induced cell volume increase,[97] and was inhibited by a moderate increase of K^+ concentration in the medium.[98] The stimulatory effect of ammonia appears to be amino acid specific: taurine was the only endogenous neuroactive amino acid efficiently released from ammonia-treated cerebrocortical slices,[97] and treatment of cultured astrocytes with ammonia actually decreased the basal efflux of radiolabeled Glu surrogate, D-aspartic acid.[93]

In vivo evidence likewise supports the role of taurine in the neuromodulatory response to ammonia. Microdialysis studies revealed that administration of ammonia to rats either i.p.[100] or intracerebrally by microdialysis[101] evokes taurine accumulation in the extracellular space. In accordance with *in vitro* observations, furosemide did not significantly affect the stimulation by ammonia *in vivo*.[101] Hence, ammonia-induced taurine release is not simply an osmoregulatory or cell volume–regulatory response normally observed following ionic or osmotic stress in a variety of pathophysiological conditions. Increased efflux of Tau and its eventual redistribution is likely to promote neuroprotection or neural inhibition.

Chronic hyperammonemic conditions, including HE, are only rarely associated with brain edema. A view prevails that intracerebral water accumulation related to metabolic changes evoked by ammonia in the acute phase (cf. Section 20.4) is compensated by efflux of metabolically inert osmolytes, including taurine. Indeed, decreased cerebral taurine content was observed in a chronic HE model[102] and in HE patients (reviewed in Reference 103). In accordance with this view, increased spontaneous efflux of newly loaded taurine was noted in striatal or cerebrocortical slices derived from rats with prolonged hyperammonemia or HE induced by thioacetamide, respectively,[104] and the increased efflux correlated with decreased taurine content in the particular structures and models.[53]

In vitro evidence suggests that the loss of taurine from the brain is associated with, and perhaps triggered by, its increased efflux from astrocytes. Exposure of astrocytes[105] or Mller glial cells[96] for 24 h to 1 mM ammonia increased the spontaneous efflux of radiolabeled taurine. Prolonged exposure to ammonia reduced the taurine content of astrocytes[105] and F98 glioma cells.[106] Both effects were absent in cultured neurons, confirming the relatively higher vulnerability of astrocytes to prolonged exposure to ammonia.[105]

On the other hand, the decreased taurine content appears to contribute to a vicious circle of events: it renders the brain more vulnerable to an extra metabolic stress, manifesting a loss of osmoregulatory properties. In cultured astrocytes, increased spontaneous release of taurine was accompanied by a decreased ability to promote extra taurine release in response to high K+ or hypoosmolar media.[96,104,106] Also, a decreased potassium-evoked release of endogenous taurine was noted in cortical slices derived from rats with thioacetamide-induced HE.[98]

20.6.2 Kynurenic Acid

Kynurenic acid (KYNA) is thought to serve a neuroprotective role in neurological disorders associated with excitotoxicity.[107] The KYNA-synthesizing enzyme, kynurenate aminotransferase (KAT), is predominantly located in astrocytes.[108] Therefore, changes of KYNA synthesis may be interpreted to reflect interference with the neuroprotective response of astrocytes. Acute treatment with ammonia *in vivo* or *in vitro* inhibited KYNA synthesis in cerebral cortical slices[109] and in a glial cell line,[110] which is believed to exacerbate the neuroexcitatory effects of ammonia. Hyperammonemia in a toxic liver failure model was accompanied by stage-dependent changes in the cerebral KYNA synthesis that matched the evolution of symptoms from excitation to depression. The synthesis was found decreased at an early stage characterized by metabolic and bioelectric activation of the CNS but was elevated at a period in which the rats showed advanced symptoms of neural inhibition.[111]

20.7 CONCLUSIONS AND FORTHCOMING ISSUES

The last decade of intensive investigations positively verified most of the earlier documented, or purported, ammonia-induced derangements in astrocytic metabolism and function and provided further support for their impact on the astrocytic–neuronal interactions and pathophysiological symptoms of hyperammonemic encephalopathies as a whole. Impaired energy metabolism, excessive glutamine synthesis, upregulation of benzodiazepine receptors, inefficient clearance or excessive release of neuroactive amino acids, and altered intercellular taurine movements have all remained on stage. Their collective or independent contributions to cerebral edema or impaired neurotransmission appear beyond doubt. The key role of Gln, whose excessive synthesis and accumulation is a hallmark of hyperammonemic encephalopathies, has been partly reevaluated in its essence. The process of Gln synthesis itself as a cause of energy depletion and metabolic and amino acid imbalance appears to be more critical than the osmotic effects of the amino acid *per se*. Upregulation of peripheral benzodiazepine receptors (PBR) has been demonstrated to link cerebral edema to the promotion of inhibitory neurotransmission. Kynurenic acid, the natural NMDA receptor antagonist, is a newcomer to the scene whose impact on the ammonia-induced disturbances in neurotransmission deserves further, more direct elucidation.

One aspect of the Gln-Glu cycle that has remained an unexplored "white spot" in the area of ammonia neurotoxicity is the role of Gln transport in determining the availability of Gln synthesized in astrocytes for the metabolic reactions in neurons. Like in other mammalian tissues, Gln transport in the CNS is controlled by transport systems differing in their topographic, cell-type, and substrate specificity.[112] Of these, the system N transporter SNAT3 is predominantly located in astrocytes.[113] A recent study by Rae et al.[114] using 1-^{13}C glucose as a metabolic precursor revealed that a number of steps within the glutamine-glutamate cycle in cerebral cortical slices, glutamate synthesis included, are inhibited by histidine, the competitor of SNAT3. It is thus tempting to speculate that SNAT3-mediated efflux of newly synthesized glutamine may be impaired in astrocytes affected by ammonia. One other forthcoming issue is that of glutathione. Glutathione is a neuroprotectant whose synthesis in the neurons depends on astroglial GSH export.[115] A single report has pointed to the enhanced glutathione synthesis and outtransport in ammonia-treated astrocytes.[116] Further investigations are expected to provide an analysis of the consequences of increased glutathione availability for the integrity and survival of astrocytes and adjacent neurons under hyperammonemic stress.

In conclusion, the studies conducted so far revealed a plethora of metabolic pathways and functions in astrocytes that are vulnerable to ammonia. Still, even a most severe and durable hyperammonemia leaves the structural integrity of neurons virtually untouched and their functional impairment reversible. Hence, although a target of toxic ammonia action, astrocytes remain an effective antiammonia shield for neurons.

REFERENCES

1. Albrecht, J., Astrocytes and ammonia neurotoxicity, in *The Role of Glia in Neurotoxicity*, Aschner, M., and Kimelberg, H.K., Eds., CRC Press, Boca Raton, 1996, p. 137.
2. Conn, H.O. and Bircher, J., *Hepatic Encephalopathy: Syndromes and Therapies*, Medi-Ed Press, Bloomington, 1994.
3. Jones, E.A. and Weissenborn, K., Neurology and the liver, *J. Neurol. Neurosurg. Psychiatry*, 63, 279, 1997.
4. Albrecht, J. and Jones, E.A., Hepatic encephalopathy: Molecular mechanisms underlying the clinical syndrome, *J. Neurol. Sci.*, 170, 138, 1999.
5. Berl, S. et al., Metabolic compartments *in vivo*. Ammonia and glutamic acid metabolism in brain and liver, *J. Biol. Chem.*, 237, 2562, 1962.
6. Farrow, N.A. et al., A ∂^5N-n.m.r. study of cerebral, hepatic and renal nitrogen metabolism in hyperammonemic rats, *Biochem. J.*, 270, 473, 1990.
7. Norenberg, M.D. and Martinez-Hernandez, A., Fine structural localization of glutamine synthetase in astrocytes of rat brain, *Brain Res.*, 61, 303, 1979.
8. Westergaard, N., Sonnewald, U., and Schousboe, A., Metabolic trafficking between neurons and astrocytes: The glutamate/glutamine cycle revisited, *Dev. Neurosci.*, 17, 203, 1995.

9. Sonnewald, U., Westergaard, N., and Schousboe, A., Glutamate transport and metabolism in astrocytes, *Glia*, 21, 56, 1997.

10. Waniewski, R.A., Physiological levels of ammonia regulate glutamine synthesis from extracellular glutamate in astrocyte cultures, *J. Neurochem.*, 58, 167, 1992.

11. Farinelli, S.E., and Nicklas, W.J., Glutamate metabolism in rat cortical astrocyte cultures, *J. Neurochem.*, 58, 1905, 1992.

12. Huang, R., Effects of chronic exposure to ammonia on glutamate and glutamine interconversion and compartmentation in homogeneous primary cultures of mouse astrocytes, *Neurochem. Res.*, 19, 257, 1994.

13. Albrecht, J., Wysmyk-Cybula, U., and Rafaowska, U., Na$^+$/K$^+$-ATPase activity and GABA uptake in astroglial cell-enriched fractions and synaptosomes derived from rats in the early stage of experimental hepatogenic encephalopathy, *Acta Neurol. Scand.*, 72, 317, 1985.

14. Hertz, L. et al., Some metabolic effects of ammonia on astrocytes and neurons in primary cultures, *Neurochem. Pathol.*, 6, 97, 1987.

15. Lai, J.C.K. et al., Differential effects of ammonia and -methylene-DL-aspartate on the metabolism of glutamate and related amino acids by astrocytes and neurones in primary cultures, *Neurochem. Res.*, 14, 377, 1989.

16. Yu, A.C.H., Schousboe, A., and Hertz, L., Influence of pathological concentrations of ammonia on metabolic fate of ∂^4C-labeled glutamate in astrocytes in primary cultures, *J. Neurochem.*, 42, 594, 1984.

17. Rao, V.L.R. and Murthy, Ch.R.K., Hyperammonemic alterations in the metabolism of glutamate and aspartate in rat cerebellar astrocytes, *Neurosci. Lett.*, 138, 107, 1992.

18. Wysmyk-Cybula, U., Faff-Michalak, L., and Albrecht, J., Effects of acute hepatic encephalopathy and *in vitro* treatment with ammonia on glutamate oxidation in bulk-isolated astrocytes and mitochondria of the rat brain, *Acta Neurobiol. Exp.*, 51, 165, 1991.

19. Albrecht, J., Wysmyk-Cybula, U., and Rafaowska, U., Cerebral oxygen consumption in experimental hepatic encephalopathy: Different responses in astrocytes, neurons and synaptosomes, *Exp. Neurol.*, 97, 418, 1987.

20. Lai, J.C.K. and Cooper, A.J.L., Brain a -ketoglutarate dehydrogenase complex: kinetic properties, regional distribution, and effect of inhibitors, *J. Neurochem.*, 47, 1376, 1986.

21. Faff-Michalak, L., Wysmyk-Cybula, U., and Albrecht, J., Different responses of rat cerebral 2-oxoglutarate dehydrogenase activity to ammonia and hepatic encephalopathy in synaptic and nonsynaptic mitochondria, *Neurochem. Int.*, 19, 573, 1991.

22. Faff-Michalak, L. and Albrecht, J., The two catalytic components of the 2-oxoglutarate dehydrogenase complex in rat cerebral synaptic and nonsynaptic mitochondria. Comparison of the response to *in vitro* treatment with ammonia, hyperammonemia and hepatic encephalopathy, *Neurochem. Res.*, 18, 119, 1993.

23. Faff-Michalak, L. and Albrecht, J., Aspartate aminotransferase, malate dehydrogenase, and pyruvate-carboxylase activities in rat cerebral synaptic and nonsynaptic mitochondria: Effects of *in vitro* treatment with ammonia, hyperammonemia and hepatic encephalopathy, *Metabol. Brain Dis.*, 6, 187, 1991.

24. Gregorios, J.B., Mozes, L.W., and Norenberg, M.D., Morphologic effects of ammonia on primary astrocyte cultures. II. Electron microscopic studies, *J. Neuropathol. Exp. Neurol.*, 44, 404, 1985.

25. Bai, G. et al., Ammonia induces the mitochondrial permeability transition in primary cultures of rat astrocytes, *J. Neurosci. Res.*, 66, 981, 2001.

26. Rama Rao, K.V. et al., Suppression of ammonia-induced astrocyte swelling by cyclosporin A, *J. Neurosci. Res.*, 74, 891, 2003.

27. Fillenz, M. et al., The role of astrocytes and noradrenaline in neuronal glucose metabolism, *Acta Physiol. Scand.*, 167, 275, 1999.

28. Liskowsky, D.R., Norenberg, M.D., and Norenberg, L.O.B., Effect of ammonia on cyclic AMP production in primary astrocyte cultures, *Brain Res.*, 386, 386, 1986.

29. Dombro, R.S., Hudson, D.G., and Norenberg, M.D., The action of ammonia on astrocytic glycogen and glycogenolysis, *Mol. Chem. Neuropathol.*, 19, 259, 1993.

30. Kosenko, E. et al., Superoxide production and antioxidant enzymes in ammonia intoxication in rats, *Free Rad. Res.*, 27, 637, 1997.

31. Hilgier, W. et al., Taurine reduces ammonia- and N-methyl-D-aspartate-induced accumulation of cyclic GMP and hydroxyl radicals in microdialysates of the rat striatum, *Eur. J. Pharmacol.*, 468, 21, 2003.

32. Murthy, C.R. et al., Ammonia-induced production of free radicals in primary cultures of rat astrocytes, *J. Neurosci. Res.*, 66, 282, 2001.

33. Warskulat, U. et al., Ammonia-induced heme oxygenase-1 expression in cultured rat astrocytes and rat brain *in vivo*, *Glia*, 40, 324, 2002.

34. Hermenegildo, C., Monfort, P., and Felipo, V., Activation of N-methyl-D-aspartate receptors in rat brain *in vivo* following acute ammonia intoxication: characterization by *in vivo* brain microdialysis, *Hepatology*, 31, 709, 2000.

35. Kosenko, E. et al., Nitroarginine, an inhibitor of nitric oxide synthetase, attenuates ammonia toxicity and ammonia-induced alterations in brain metabolism, *Neurochem. Res.*, 4, 451, 1995.

36. Hermenegildo, C. et al., NMDA receptor antagonists prevent acute ammonia toxicity in mice, *Neurochem. Res.*, 10, 1237, 1996.

37. Kimelberg, H.K. et al., Freshly isolated astrocyte (FIA) preparations: a useful single cell system for studying astrocyte properties, *J. Neurosci. Res.*, 61, 577, 2000.

38. Pearce, B. et al. Astrocyte glutamate receptor activation-promotes inositol phospholipid turnover and calcium flux. *Neurosci. Lett.*

39. Albrecht, J., Hilgier, W., and Rafa•owska, V., Activation of arginine metabolism to glvtamate in rat brain synaptosomes in thioa cetamide-induced hepatic encephalopathy: an adaptive response? *J. Neurosci. Res.*, 25, 125, 1990.

40. Rao, V.L.R., Audet, R.M., and Butterworth, R.F., Increased nitric oxide synthase activities and L-[³H] arginine uptake in brain following portacaval anastomosis, *J. Neurochem.*, 65, 677, 1995.

41. Rao, V.L.R., Audet, R.M., and Butterworth, R.F., Portacaval shunting and hyperammonemia stimulate the uptake of L-[³H] arginine but not of L-[³H] nitroarginine into rat brain synaptosomes, *J. Neurochem.*, 68, 337, 1997.

42. Schliess, F. et al., Ammonia induces MK-801-sensitive nitration and phosphorylation of protein tyrosine residues in rat astrocytes, *FASEB J.*, 16, 739, 2002.

43. Gorg, B. et al., Benzodiazepine-induced protein tyrosine nitration in rat astrocytes, *Hepatology*, 37, 334, 2003.

44. Norenberg, M.D., Itzhak, Y., and Bender, A.S. The peripheral benzodiazepine receptor and neurosteroids in hepatic encephalopathy, in *Cirrhosis, Hyperammonemia and Hepatic Encephalopathy*, Felipo, V., Ed., Plenum Press, New York, 1997, p. 95.

45. Itzhak, Y., Baker, L., and Norenberg, M.D., Characterization of the peripheral-type benzodiazepine receptors in cultured astrocytes: evidence for multiplicity, *Glia*, 9, 211, 1993.

46. Lavoie, J., Pomier-Layrargues, G., and Butterworth, R.F., Increased densities of "peripheral-type" benzodiazepine receptors in autopsied brain tissue from cirrhotic patients with hepatic encephalopathy, *Hepatology*, 11, 874, 1990.

47. Itzhak, Y. et al., Acute liver failure and hyperammonemia increase peripheral-type benzodiazepine receptor binding and pregnenolone synthesis in mouse brain, *Brain Res.*, 705, 345, 1995.

48. Jayakumar, A.R., Panickar, K.S., and Norenberg, M.D., Effects on free radical generation by ligands of the peripheral benzodiazepine receptor in cultured neural cells, *J. Neurochem.*, 83, 1226, 2002.

49. Norenberg, M.D., Oxidative and nitrosative stress in ammonia neurotoxicity, *Hepatology*, 37, 245, 2003.

50. Traber, P. et al., Electron microscopic evaluation of brain edema in rabbits with galactosamine-induced fulminant hepatic failure, *Hepatology*, 7, 1257, 1987.

51. Swain, M., Butterworth, R.F., and Blei, A., Ammonia and related amino acids in the pathogenesis of brain edema in acute ischemic liver failure, *Hepatology*, 15, 449, 1992.

52. Hilgier, W. and Olson, J.E., Brain ion and amino acid contents during edema development in hepatic encephalopathy. *J. Neurochem.*, 62, 197, 1994.

53. Brusilow, S.W., Hepatic encephalopathy (letter), *N. Engl. J. Med.*, 314, 786, 1986.

54. Takahashi, H. et al., Inhibition of brain glutamine accumulation prevents cerebral edema in hyperammonemic rats, *Am. J. Physiol.*, 261, H825, 1991.

55. Master, S., Gottstein, J., and Blei, A.T., Cerebral blood flow and the development of ammonia-induced brain edema in rats after portacaval anastomosis, *Hepatology*, 30, 876, 1999.

56. Willard-Mack, C.L. et al., Inhibition of glutamine synthetase reduces ammonia-induced astrocyte swelling in the rat, *Neuroscience*, 71, 589, 1996.

57. Sugimoto, H. et al., Methionine sulfoximine, a glutamine synthetase inhibitor, attenuates increased extracellular potassium activity during acute hyperammonemia, *J. Cereb. Blood Flow Metab.*, 17, 44, 1997.

58. Norenberg, M.D. and Bender, A.S., Astrocyte swelling in liver failure: role of glutamine and benzo-diazepines, *Acta Neurochir.*, 60, 24, 1994.

59. Dietrich, W.D. and Kuluz, J.W., New research in the field of stroke: therapeutic hypothermia after cardiac arrest, *Stroke*, 34, 1051, 2003.

60. Cordoba, J. et al., Mild hypothermia modifies ammonia-induced brain edema in rats after portacaval anastomosis, *Gastroenterology*, 116, 686, 1999.

61. Zwingmann, C. et al., Selective increase of brain lactate synthesis in experimental acute liver failure: results of a [^{1}H-^{13}C] nuclear magnetic resonance study, *Hepatology*, 37, 420, 2003.

62. Chatauret, N. et al., Protective effects of mild hypothermia on brain glucose metabolism in rats with acute liver failure: a ^{1}H/^{13}C-NMR study, *Gastroenterology*, 125, 815, 2003.

63. Ziemi•ska, E. et al., Induction of permeability transition and swelling of rat brain mitochondria by glutamine, *Neurotoxicology*, 21, 295, 2000.

64. Rama Rao, K.V., Jayakumar, A.R., and Norenberg, M.D., Induction of the mitochondrial permeability transition in cultured astrocytes by glutamine, *Neurochem. Int.*, 43, 517, 2003.

65. Doli•ska, M., Hilgier, W., and Albrecht, J., Ammonia stimulates glutamine uptake to nonsynaptic cerebral mitochondria of the rat, *Neurosci. Lett.*, 11, 175, 1996.

66. Zieli•ska, M., Law, R.O., and Albrecht, J., Excitotoxic mechanism of cell swelling in rat cerebral cortical slices treated acutely with ammonia, *Neurochem. Int.*, 43, 299, 2003.

67. Bender, A.S. and Norenberg, M.D., Effect of benzodiazepines and neurosteroids on ammonia-induced swelling in cultured astrocytes, *J. Neurosci. Res.*, 54, 673, 1998.

68. Rama Rao, K.V. et al., Increased aquaporin-4 expression in ammonia-treated cultured astrocytes, *Neuroreport*, 14, 2379, 2003.

69. Badaut, J., Lasbennes, F., Magistretti, P.J., and Regli, L., Aquaporins in brain: distribution, physiology, and pathophysiology, *J. Cereb. Blood Flow Metab.*, 22, 367, 2002.

70. Albrecht, J., Glucose-derived osmolytes and energy impairment in brain edema accompanying liver failure: the role of glutamine reevaluated, *Gastroenterology*, 125, 976, 2003.

71. Jabaudon, D. et al., Inhibition of uptake unmasks rapid extracellular turnover of glutamate of nonvesicular origin, *Proc. Natl. Acad. Sci. U.S.A.*, 96, 8733, 1999.

72. Michalak, A. et al., Neuroactive amino acids and glutamate (NMDA) receptors in frontal cortex of rats with experimental acute liver failure, *Hepatology*, 24, 908, 1996.

73. Vogels, B.A.P.M. et al., Memantine, a noncompetitive NMDA receptor antagonist improves hyper-ammonemia-induced encephalopathy and acute hepatic encephalopathy in rats, *Hepatology*, 25, 4, 820, 1997.

74. Bender, A.S. and Norenberg, M.D., Effects of ammonia on L-glutamate uptake in cultured astrocytes, *Neurochem. Res.*, 21, 5, 567, 1996.

75. Albrecht, J. et al., Astrocytes in acute hepatic encephalopathy: Metabolic properties and transport function, in *Biochemical Pathology of Astrocytes*, Norenberg, M.D., Hertz, L., and Schousboe, A., Eds., Alan R. Liss, New York, 1988, p. 465.

76. Norenberg, M.D. et al., The glial glutamate transporter in hyperammonemia and hepatic encephalopathy: relation to energy metabolism and glutamatergic neurotransmission, *Glia*, 21, 124, 1997.

77. Knecht, K. et al., Decreased glutamate transporter (GLT-1) expression in frontal cortex of rats with acute liver failure, *Neurosci. Lett.*, 229, 201, 1997.

78. Zhou, B.G. and Norenberg, M.D., Ammonia downregulates GLAST mRNA glutamate transporter in rat astrocyte cultures, *Neurosci. Lett.*, 276, 145, 1999.

79. Chan, H. et al., Effects of ammonia on glutamate transporter (GLAST) protein and mRNA in cultured rat cortical astrocytes, *Neurochem. Int.*, 37, 243, 2000.

80. Lavoie, J. et al., Amino acid changes in autopsied brain tissue from cirrhotic patients with hepatic encephalopathy, *J. Neurochem.*, 49, 692, 1987.

81. Sonnewald, U., Therrien, G., and Butterworth, R.F., Portacaval anastomosis results in altered neuron-astrocytic metabolic trafficking of amino acids: evidence from 13C-NMR studies, *J. Neurochem.*, 67, 1711, 1996.

82. Majewska, M.D., Neurosteroids: endogenous bimodal modulators of the GABA-A receptor: mechanism of action and physiological significance, *Prog. Neurobiol.*, 38, 370, 1992.

83. Mensah-Nyagan, A.G. et al., Neurosteroids: expression of steroidogenic enzymes and regulation of steroid biosynthesis in the central nervous system, *Pharmacol. Rev.*, 51, 63, 1999.

84. Itzhak, Y. et al., Acute liver failure and hyperammonemia increase peripheral-type benzodiazepine receptor binding and pregnenolone synthesis in mouse brain, *Brain Res.*, 705, 345, 1995.

85. Itzhak, Y. and Norenberg, M.D., Attenuation of ammonia toxicity in mice by PK 11195 and pregnenolone sulfate, *Neurosci. Lett.*, 182, 251, 1994.

86. Takahashi, K. et al., Ammonia potentiates GABAA response in dissociated rat cortical neurons, *Neurosci. Lett.*, 151, 51, 1993.

87. Ha, J.H. and Basile, A.S., Modulation of ligand binding to components of the GABAA receptor complex by ammonia: implications for the pathogenesis of hyperammonemic syndromes, *Brain Res.*, 720, 35, 1996.

88. Albrecht, J. and Rafa•owska, U., Enhanced potassium-stimulated γ-aminobutyric acid release by astrocytes derived from rats with early hepatogenic encephalopathy, *J. Neurochem.*, 49, 9, 1987.

89. Chapkova, A.N. et al., Long-lasting enhancement of cortocostriatal neurotransmission by taurine, *Eur. J. Neurosci.*, 16, 1523, 2002.

90. Stone, T.W., Neuropharmacology of quinolinic and kynurenic acids, *Pharmacol. Rev.*, 45, 309, 1993.

91. Saransaari, P. and Oja, S.S., Taurine and neural cell damage, *Amino Acids*, 19, 509, 2000.

92. Anderzhanova, E. et al., Changes in the striatal extracellular levels of dopamine and dihydroxyphenylacetic acid evoked by ammonia and N-methyl-D-aspartate: modulation by taurine, *Brain Res.*, 977, 290, 2003.

93. Albrecht, J., Bender, A.S., and Norenberg, M.D., Ammonia stimulates the release of taurine from cultured astrocytes, *Brain Res.*, 660, 228, 1994.

94. Faff-Michalak, L. et al., K$^+$-, Hypoosmolarity- and NH$_4$$^+$-induced taurine release from cultured rabbit Müller cells: Role of Na$^+$ and Cl$^-$ ions and relation to cell volume changes, *Glia*, 10, 114, 1994.

95. Faff-Michalak, L., Reichenbach, A., and Albrecht, J., Ammonia-induced taurine release from cultured rabbit Muller cells is an osmoresistant process mediated by intracellular accumulation of cyclic AMP, *J. Neurosci. Res.*, 46, 231, 1996.

96. Faff, L., Reichenbach, A., and Albrecht, J., Two modes of stimulation by ammonia of taurine release from cultured rabbit Muller cells, *Neurochem. Int.*, 31, 301, 1997.

97. Zielińska, M. et al., Effects of ammonia *in vitro* on endogenous taurine efflux and cell volume in rat cerebrocortical minislices: influence of inhibitors of volume-sensitive amino acid transport, *Neuroscience*, 91, 631, 1999.

98. Zielińska, M. et al., Effects of ammonia and hepatic failure on the net efflux of endogenous glutamate, aspartate and taurine from rat cerebrocortical slices: modulation by elevated K$^+$ concentrations, *Neurochem. Int.*, 41, 87, 2002.

99. Zielińska, M., Zabocka, B., and Albrecht, J., Effect of ammonia on taurine transport in C6 glioma cells, *Adv. Exp. Med. Biol.*, 526, 463, 2003.

100. Raghavendra Rao, V.L., Audet, R.M., and Butterworth, R.F., Selective alterations of extracellular brain amino acids in relation to function in experimental portal-systemic encephalopathy: results of an *in vivo* microdialysis study, *J. Neurochem.*, 65, 1221, 1995.

101. Zielińska, M. et al., Ammonia-induced extracellular accumulation of taurine in the rat striatum *in vivo*: role of ionotropic glutamate receptors, *Neurochem. Res.*, 27, 37, 2002.

102. Cordoba, J., Gottstein, J., and Blei, A.T., Glutamine, myo-inositol, and organic brain osmolytes after portacaval anastomosis in the rat: implications for ammonia-induced brain edema, *Hepatology*, 24, 919, 1996.

103. Häussinger, D. et al., Hepatic encephalopathy in chronic liver disease: a clinical manifestation of astrocyte swelling and low-grade cerebral edema, *J. Hepatol.*, 32, 1035, 2000.

104. Hilgier, W., Olson, J.E., and Albrecht, J., Relation of taurine transport and brain edema in rats with simple hyperammonemia or liver failure, *J. Neurosci. Res.*, 45, 69, 1996.

105. Wysmyk, U. et al., Long-term treatment with ammonia affects the content and release of taurine in cultured cerebellar astrocytes and granule neurons, *Neurochem. Int.*, 24, 317, 1994.

106. Zwingmann, C. et al., Effects of ammonia exposition on glioma cells: Changes in cell volume and organic osmolytes studied by diffusion-weighted and high-resolution NMR spectroscopy, *Dev. Neurosci.*, 2, 2463, 2000.

107. Schwarcz, R. and Pellicciari, R., Manipulation of brain kynurenines: glial targets, neuronal effects, and clinical opportunities, *J. Pharmacol. Exp. Ther.*, 303, 1, 2002.

108. Du, F. et al., Localization of kynurenine aminotransferase immunoreactivity in the rat hippocampus, *J. Comp. Neurol.*, 321, 477, 1992.

109. Saran, T. et al., Acute ammonia treatment *in vitro* and *in vivo* inhibits the synthesis of a neuroprotectant kynurenic acid in rat cerebral cortical slices, *Brain Res.*, 787, 348, 1998.

110. Kocki, T. et al., Regulation of kynurenic acid synthesis in C6 glioma cells, *J. Neurosci. Res.*, 68, 622, 2002.

111. Saran, T. et al., Kynurenic acid synthesis in cerebral cortical slices of rats with progressing symptoms of thioacetamide-induced hepatic encephalopathy, *J. Neurosci. Res.*, 75, 436, 2004.

112. Chaudhry, F.A., Reimer, R.J., and Edwards, R.H., The glutamine commute: take the N line and transfer to the A, *J. Cell Biol.*, 157, 349, 2002.

113. Boulland, J.L. et al., Cell-specific expression of the glutamine transporter SN1 suggests differences in dependence on the glutamine cycle, *Eur. J. Neurosci.*, 15, 1615, 2002.

114. Rae, C. et al., Inhibition of glutamine transport depletes glutamate and GABA neurotransmitter pools: further evidence for metabolic compartmentation, *J. Neurochem.*, 85, 503, 2003.

115. Dringen, R., Pfeiffer, B., and Hamprecht, B., Synthesis of the antioxidant glutathione in neurons: Supply by astrocytes of CysGly as precursor for neuronal glutathione, *J. Neurosci.*, 19, 562, 1999.

116. Murthy, C.R. et al., Elevation of glutathione levels by ammonium ions in primary cultures of rat astrocytes, *Neurochem. Int.*, 37, 255, 2000.

CHAPTER 21

Alcohol and Glia in the Developing Brain

Lucio G. Costa, Kevin Yagle, Annabella Vitalone, and Marina Guizzetti

CONTENTS

21.1 INTRODUCTION

Research on the deleterious effects of alcohol on the nervous system has focused primarily on neurons. Yet it has been increasingly pointed out that glial cells may play an important role in the effects of ethanol in the developing and adult brain. As comprehensive reviews have summarized earlier research on the effects of alcohol on glial cells, both *in vitro* and *in vivo*,[1–4] this chapter will focus on selected aspects of the interactions of alcohol with astroglia, with an emphasis on astrocytes and on the developmental neurotoxicity of ethanol.

It is well established that exposure of the fetus to alcohol may lead to fetal alcohol syndrome (FAS), whose principal features include central nervous system (CNS) dysfunctions (mental retardation, microencephaly, brain malformations), pre- and postnatal growth deficiency, and a characteristic facial dysmorphology.[5] The CNS deficits of FAS are of most concern, as they are long-lasting and persistent even if other symptoms (growth retardation, facial characteristics) have subsided.[6–8] FAS is now considered a leading cause of mental retardation in the general population.[9] Additionally, FAS is believed to represent only the "tip of the iceberg," as lower doses of alcohol may be associated with neuropsychiatric and intellectual deficits, without the full-blown FAS with facial dysmorphogenesis or other major malformations.[10] This condition is often referred to as alcohol-related neurodevelopmental disorders (ARND).

A large number of studies have been conducted in laboratory animals (primarily rodents but also nonhuman primates) to gain an understanding of the characteristics and mechanisms of alcohol teratogenicity, and these findings have been summarized in several books and reviews.[10–13] In rodents, *in vivo* studies have utilized both prenatal and postnatal exposures to mimic human exposure during the first and early-second trimester of pregnancy and during the late-second and third trimester and early postnatal life, respectively. As the development of the nervous system differs across species,[14] and since within one species, different brain regions and cell types develop at different times, it has become apparent that the timing of exposure to ethanol is extremely important for specific adverse effects to be manifested. Additionally, the blood alcohol concentration at a given time, rather than the total amount of alcohol ingested, has been shown to be a key determinant of ethanol's developmental effects. Thus, even occasional binge-drinking by a pregnant woman is seen as most damaging for the fetus.[15]

When ethanol is administered during the brain growth spurt (the first 2 postnatal weeks in the rat, characterized by proliferation of glial cells and maturation of neurons, which develop dendrites, axons, and synaptic contacts), the most striking effect observed is microencephaly,[16] which is also seen in 80% of FAS children and appears to be irreversible in both animals and humans.[7,8,17,18]

Most studies aimed at exploring the neurologic sequence resulting from alcohol abuse during pregnancy have focused on neuronal susceptibility. Pre- and postnatal exposure to ethanol can cause loss of neurons, particularly in the hippocampus and the cerebellum.[19,20] Recently, it has been shown that a single postnatal administration of ethanol on postnatal day 7 in the rat can cause widespread apoptotic neuronal death, which has been ascribed to its ability to inhibit glutamate NMDA receptors and to activate GABA-A receptors.[21,22]

The extent to which glial cells are vulnerable to ethanol has been less thoroughly investigated, though evidence has accumulated indicating that developmental exposure to ethanol can significantly affect glial cells.[3,20,23–25] Most effects of alcohol on the biochemistry, metabolism, morphology, and differentiation of glial cells have been reviewed in the first edition of this book.[3] Thus, we will focus on the effects of alcohol on glial cell proliferation and survival, as well as some more recent aspects of its effects on glial cell biochemistry.

21.2 ETHANOL AND GLIAL CELL PROLIFERATION

In the past several years, substantial research efforts have been devoted to study the effects of ethanol on proliferation of glial cells, particularly astrocytes, stimulated by a variety of mitogens. The initial rationale behind these studies lies in the fact that there is *in vivo* evidence that developmental exposure to ethanol in experimental animals as well as in humans causes a reduction in glial cell number.[18,26,27] As mentioned earlier, one of the main findings in individuals diagnosed with FAS is the presence of microcephaly and microencephaly. Animal models of FAS have shown that microencephaly is observed when ethanol is given during the first 2 postnatal weeks in the rat, a period corresponding to the third trimester of pregnancy in humans.[16,28] This period of brain development corresponds to the so-called brain growth spurt, which is characterized by proliferation of glial cells and synaptogenesis.[14] As glial cells represent some 80–90% of total brain cells, it is plausible to hypothesize that microencephaly may involve, in addition to loss of neurons, loss of glial cells.[29]

Most research on ethanol and glial cell proliferation has been carried out *in vitro*, utilizing primary cell cultures or established glial cell lines. Different studies have examined the effects of ethanol on proliferating glial cells kept in serum or in synchronized (serum-deprived) cell upon stimulation by a variety of mitogens.[29] In an early study carried out in proliferating glial cells prepared from 15-day-old chick embryos, high concentrations of ethanol (217 and 434 mM) were found to cause a 20–40% reduction in DNA content.[30] Lower concentrations of ethanol (22 and 109 mM) were devoid of effect, as confirmed by subsequent studies in rat mixed glial primary cultures[31] and in rat fetal astrocytes.[32] On the other hand, another study found a decrease in DNA content in astrocytes from newborn mice cultured in the presence of 45 mM ethanol.[33] Similarly, two additional studies indicated that ethanol (44, 109, and 217 mM) inhibited the growth of rat cerebral glial cultures following exposure for 7 days or longer.[34,35] Inhibition of ^{3}H-thymidine incorporation into DNA by incubations for various times with 100–200 mM ethanol in rat cortical astrocytes was reported in some, but not all, studies.[36–39] The proliferation of astroglia from adult human cerebrum, measured by bromodeoxyuridine incorporation into DNA, was also inhibited by ethanol (22, 45, and 109 mM).[40] In transformed cell lines, ethanol (30–120 mM) was found to inhibit proliferation of C6 rat glioma cells,[41] though this was not confirmed in another study.[42] More recently, however, the same authors reported that ethanol (80 mM) inhibits serum-stimulated proliferation of rat cortical astrocytes and C6 astrocytoma cells.[43] Altogether, these finding indicate that ethanol can inhibit proliferation of astroglial cells *in vitro*, though in several studies only concentrations of ethanol > 100 mM were effective. As stated, in these studies cells were cultured in medium containing serum and concomitantly exposed to ethanol. Under these conditions, cells are expected to be divided among the G_0/G_1, S, and G_2/M phases of the cell cycle.

Since ethanol has been suggested to inhibit cell proliferation by causing a block in the G_0/G_1 phase,[44] a more sensitive way to investigate the effect of ethanol on cell proliferation is to expose cells synchronized in the G_0/G_1 phase to ethanol while they are trying to reenter the cell cycle. This situation mimics that found during the brain growth spurt, when glial cells start actively proliferating after a period of quiescence. Indeed, more recent studies have examined the effect of ethanol on the proliferation of glial cells, serum-deprived for 24–48 h and thus synchronized in the G_0/G_1 phase, followed by the addition of the mitogens together with ethanol. Overall, ethanol appears to be a potent inhibitor of stimulated glial cell proliferation (Table 21.1). The effects of ethanol toward DNA synthesis stimulated by acetylcholine analogs, serum, and insulin-like growth factor I (IGF-I) have been studied mainly from a mechanistic viewpoint and will be discussed in more detail. First, however, its effects toward other mitogens are briefly summarized. In a study by Luo and Miller,[42] proliferation of C6 rat glioma cells induced by basic fibroblast growth factor (bFGF) was found to be almost completely inhibited by 80 mM ethanol, and the IC_{50} was about 20 mM, and this was confirmed by a subsequent study by the same investigators.[43] These results differ from those of de Vito et al.,[38] who reported a lack of effect of ethanol (50 mM) on DNA synthesis in rat cortical astrocytes induced by FGF. This latter study also reported that proliferation of astrocytes induced by prolactin was inhibited by ethanol with an IC_{50} of 25 mM.[38] Ethanol (24-h coincubation) did not affect proliferation induced by insulin or platelet-derived growth factor-BB (PDGF-BB).[38,45,46] On the other hand, with longer incubation times (3 days), ethanol affected PDGF-BB- and in particular PDGF-AA-induced DNA synthesis.[47] Ethanol could also inhibit proliferation of

Table 21.1 Inhibition by Ethanol of Proliferation of Synchronized Astroglial Cell Stimulated by Various Mitogens

Mitogen	Effective Ethanol Concentration (mM)	Ref.
IGF-1	10	46, 49
Thrombin	25–50	Guizzetti and Costa, unpublished
EGF	10–20	46, 49
Prolactin	10–25	38
Muscarinic agonists	10–25	45, 51
FGF	No effect	38
bFGF	20–80	42
Interleukin-1	10–25	45
Insulin	No effect	45
	25	37
PDGF-BB	No effect	38, 45, 46
	40	42
	60	63
PDGF-AA	80	47
Phorbol ester	10–25	45
Serum	20	48
	250	45
	20	63

astrocytes induced by interleukin-1[45] and by phorbol esters[45,48] and of C6 rat glioma cells by epidermal growth factor (EGF).[46,49] This latter effect, however, was shown to be due to induction of IGF-1 receptors by EGF and a subsequent mitogenic action mediated by an IGF-I/IGF-I receptor autocrine pathway.[46]

The inhibitory effect of ethanol on proliferation of astroglial cells induced by acetylcholine analogs, serum, and IGF-I has been studied in more detail, and potential underlying mechanisms have been unraveled. Through activation of the M_3 subtype of cholinergic muscarinic receptors, the cholinergic agonist carbachol causes proliferation of rat cortical astrocytes, human astrocytoma cells, and human fetal astrocytes.[50,51] Ethanol has been shown to inhibit muscarinic receptor-induced DNA synthesis at concentrations of 10–100 mM, with flow cytometry experiments suggesting a block in the G_0/G_1 phase of the cell cycle.[45,51] Studies aimed at elucidating the intracellular signaling pathways involved in the mitogenic action of muscarinic agonists have indicated that two pathways are of relevance (Figure 21.1). One involves activation of phospholipase C, mobilization of intracellular calcium, activation of protein kinase C (PKC) ε, and activation of mitogen-activated protein kinase (MAPK).[52–54] The other pathway involves activation of phospholipase D (PLD), with formation of phosphatidic acid (PA), and activation of phosphatidylinositol-3-kinase (PI-3K), followed by activation of PKC ζ, p70S6 kinase, and nuclear factor–kB.[55–58] Though both pathways can be inhibited by high (> 100 mM) concentrations of ethanol, only the second one is sensitive to the same ethanol concentrations (10–100 mM) that are capable of inhibiting cell proliferation.[51,54,57,59–61] Furthermore, activation of PI-3K is affected only by relatively high (100 mM) ethanol concentrations, suggesting that the primary target for ethanol is PLD.[61] Indeed, through a transphosphatidylation reaction,[62] ethanol, as well as other alcohols such as 1-butanol, causes inhibition of PLD, which leads to reduced formation of PA and reduced sequential activation of PKC ζ and p70S6 kinase (Figure 21.1).[58]

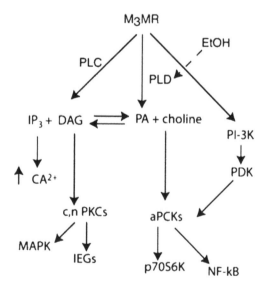

Figure 21.1 Schematic representation of the intracellular signal-transduction pathways activated by muscarinic M_3 receptors in astroglial cells and involved in its mitogenic action. Also indicated is the proposed site of action of ethanol, involved in its inhibition of muscarinic agonist-induced cell proliferation. Abbreviations: M_3MR, muscarinic M_3 receptor; PLC, phospholipase C; PLD, phospholipase D; PI-3 K, phosphatidylinositol-3-kinase; PA, phosphatidic acid; IP3, inositol-1,4,5-triphosphate; DAG, diacylglycerol; a, c, n PKC, atypical, classical, and novel protein kinase C; p70S6K, p70S6 kinase; NF-kB, nuclear factor-kB; MAPK, mitogen-activated protein kinase; IEGs, immediate-early-genes; PDK, phosphoinositide-dependent kinase; EtOH, ethanol.

A similar signaling pathway appears to be involved in ethanol's inhibition of serum-induced proliferation of rat cortical astrocytes following 24-h serum deprivation.[63,64] Indeed ethanol (20–200 mM) was found to cause significant inhibition of serum-induced DNA synthesis and PA formation, and similar effects were seen with 1-butanol, which also undergoes a transphosphatidylation reaction.[63] Additionally, ethanol had only minimal effects on PA-induced glial cell proliferation, suggesting that its primary target is upstream of PA, i.e., at the level of PLD.[58,64]

Ethanol also inhibits proliferation of astroglial cells induced by IGF-I, at concentrations ranging from 10 to 100 mM.[49] Upon binding IGF-I, the receptor undergoes tyrosine autophosphorylation and phosphorylates other cellular protein substrates. Such autophosphorylation at tyrosine residues is required for its mitogenic signal.[65] Ethanol appears to inhibit the enzymatic activity of the tyrosine kinase domain of the IGF-I receptor, thereby blocking upstream the cascade of intracellular events leading to increased DNA synthesis.[66,67] This mechanism, however, may not be operational in all cell types.[68]

Inhibition of astroglial cell growth appears to be due to ethanol itself and not to its conversion to acetaldehyde. Indeed, astrocytes lack alcohol dehydrogenase,[69] and at concentrations devoid of cytotoxicity acetaldehyde did not inhibit DNA synthesis in human astrocytoma cells.[45] Furthermore, 4-methylpyrazole, an inhibitor of alcohol dehydrogenase, failed to affect the inhibiting effect of ethanol on cell proliferation.[41,45] Thus, ethanol appears to be able to inhibit proliferation of astrocytes or glioma cells without the need for metabolic activation.

21.3 ETHANOL AND GLIAL CELL DEATH

As seen in the case of neurons, *in vivo* or *in vitro* exposure of astroglial cells to ethanol can result in cell death. Holownia et al.[70] observed that incubation of rat cortical astroglial-oligodendroglial cells with ethanol (50 or 100 mM for 7 days) caused significant cell death, which they attributed to necrotic rather than apoptotic mechanisms. *In vivo* exposure to ethanol during gestation and lactation in rats was found to trigger necrotic and apoptotic cell death in both neurons and astrocytes in the developing cerebral cortex; of all dead cells, approximately 40% were GFAP positive (i.e., astrocytes), and of these more than 90% were apoptotic.[71] In rat cortical astrocytes *in vitro*, ethanol (50–200 mM) was found to induce apoptotic cell death, which involved stimulation of the sphingomyelinase-ceramide pathway.[72] Similarly, incubation of rat astrocytes with ethanol (50–200 mM) was found to induce apoptosis (Figure 21.2) (Yagle and Costa, unpublished observations).

The exact mechanism by which ethanol causes apoptotic cell death in astrocytes is not clear, but it may be related to its ability to induce oxidative stress in these cells.[73–75] Indeed, there is evidence that free-radical species and oxidative stress can activate the sphingomyelin-ceramide pathway and cause apoptotic cell death.[76] Supporting evidence for this hypothesis is that antioxidants, such as vitamin C or idebenone, antagonize the toxic effects of ethanol in glial cells.[75,77] Activation of the sphingomyelin-ceramide pathway has also been suggested to explain the increase in the susceptibility of astrocytes to TNFα-induced cell death caused by ethanol.[78] An additional effect of ethanol that may be involved in astroglial cell death is represented by its action on the cytoskeleton.[81–83] Ethanol (10–100 mM) has been shown to disrupt the actin cytoskeletal organization in rat cortical astrocytes by causing a Rho-dependent actin reorganization from stress fibers to actin rings,[83] which could represent an initial step in an apoptotic process.[84]

As in the case of ethanol-induced inhibition of glial cell proliferation, induction of apoptotic cell death may be due to ethanol itself rather than to its metabolism to acetaldehyde. Indeed, while acetaldehyde at high micromolar concentrations can cause apoptosis in certain cell types,[79] the amount formed upon incubation of astrocytes with ethanol is quite low (in the nanomolar range), possibly due to P450-mediated metabolism.[80]

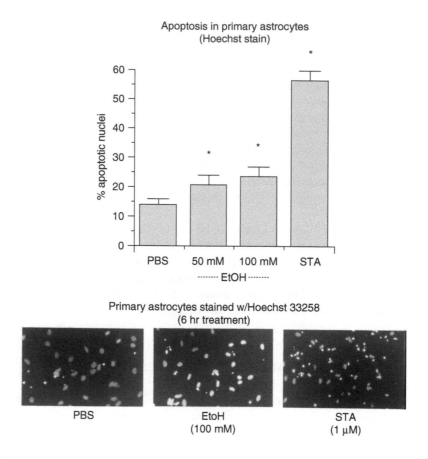

Figure 21.2   Apoptosis induced by ethanol in cultured rat cortical astrocytes. Cells were incubated with ethanol
(50 or 100 m*M*) or with the positive control staurosporine (1 µ*M*) for 6 h. Apoptosis was assessed
by Hoechst 33258 fluorescent staining. (Top) Quantification of apoptotic nuclei in various treatment
groups. (Bottom) Hoechst 33258 fluorescent staining of astrocytes exposed for 6 h to PBS, 100
m*M* ethanol or 1 µ*M* staurosporine. Abbreviations: PBS, phosphate buffer saline; STA, staurosporine; EtOH, ethanol (Yagle and Costa, unpublished observations). *Significantly different from
control, $p < 0.05$.

21.4 RELEVANCE OF ETHANOL–GLIA INTERACTIONS
IN FETAL ALCOHOL SYNDROME

It has been stated that "conceptualizations in neural development have tended to be neuronocen-
tric,"[85] indicating that developmental events have often been viewed from the standpoint of neurons.
Yet, it is now evident that glia plays a most relevant role in neuronal proliferation, migration,
differentiation, and survival.[85] Furthermore, evidence is accumulating indicating that glia exert a
most relevant control on synapse formation.[86,87] Thus, a reduced number of glial cells, caused by
ethanol-induced inhibition of glial cell proliferation or by ethanol-stimulated glial cell apoptotic
death, is expected to have profound consequences for the development of the brain. Abnormalities
in glial development are indeed suspected of contributing to the adverse effects of ethanol in the
developing brain.[23] Evidence exists of abnormal glial migration in humans with FAS, as well as in
primates and rats exposed to ethanol during development.[88,89] A reduction of glial cell number has
also been reported in rat models of FAS.[26,27] A loss of midline S100β-positive glial cells following
prenatal exposure of rats to ethanol has been reported,[90] and a recent study in mice indicates that
such deficits are also caused by moderate alcohol exposures (approximately 10–30 m*M*), suggesting

an exquisite susceptibility of such cells to alcohol's effects and a possible rodent model for ARND.[91] In children affected by FAS, hypoplasia of the corpus callosum and anterior commissure, two areas originally formed by neuroglial cells, has been reported.[92] Additionally, as stated earlier, the findings that microencephaly is strongly associated with ethanol exposure during the brain growth spurt,[16] a period characterized by rapid glial cell proliferation and maturation, also suggests a potential effect of ethanol on proliferation, growth, and maturation of glia.

This chapter has focused on the effects of ethanol on astroglial cell proliferation and survival, as several other reviews have summarized its effects on other metabolic aspects of astrocytes.[3,4] Two aspects, however, deserve, in our opinion, some further considerations: the effects of ethanol on trophic factors and on cholesterol homeostasis. Glial cells synthesize and release multiple neurotrophic factors and express receptors for such factors.[93,94] Such neurotrophic factors—for example, nerve growth factor, brain-derived neurotrophic factor, or fibroblast growth factors—exert most relevant effects on developing neurons as well as on glia itself. Conceivably, an interaction of ethanol with the synthesis or the release of such growth/trophic factors should have a profound effect on brain development. Though evidence for such a possibility is emerging, as indicated by several studies,[95–97] further studies should be directed at investigating the extent of the interaction of ethanol with glial cell–derived neurotrophic factors and the underlying mechanisms.

As mentioned earlier, glial cells control the extent of synapse formation,[86] and such synaptogenic activity is mediated by cholesterol.[98] In the central nervous system, the cholesterogenic activity is localized predominantly within glial cells,[99] and the developing brain appears to rely almost completely on local synthesis of cholesterol.[100] It is known that genetic defects in cholesterol biosynthesis, such as in Smith-Lemli-Opitz syndrome, lead to microcephaly, agenesis of the corpus callosum, mental retardation, behavioral disorders, and facial dismorphology,[101] which are also hallmarks of FAS. Though the effects of ethanol on lipids and lipoproteins have been extensively studied in relation to the cardiovascular system, very limited information exist on its potential effects on cholesterol homeostasis in the developing brain. Given the relevance of glial-derived cholesterol for cell proliferation and synapse development,[98,102] exploring such a possibility seems to be warranted.

In conclusion, ethanol exerts a number of effects on glial cells in the developing brain. Effects leading to a decreased number of astroglia, such as inhibition of proliferation and induction of apoptotic cell death, together with other actions leading to impairment of growth factor synthesis or release and, potentially, of cholesterol homeostasis, would severely disrupt the normal development of the brain, most notably of neurons, which are highly dependent on glia for maturation and survival. It would thus seem that investigations of the interactions of ethanol with developing glia and their role in the CNS effects of FAS still represent a fruitful area for further studies.

ACKNOWLEDGMENTS

Research by the authors was supported by grants from the National Institutes of Health (AA08154 and ES07033) and the Alcohol and Drug Abuse Institute at the University of Washington.

REFERENCES

1. Davies, D.L., The response of astrocytes to ethanol exposure, in *Alcohol Neurobiology: Brain Development and Hormone Regulation*, CRC Press, Boca Raton, 1992, p. 69.
2. Lancaster, F.E., Ed., *Alcohol and Glial Cells*, U.S. Dept. of Health and Human Services, National Institute on Alcohol Abuse and Alcoholism, Research Monograph 27, 1994, p. 250.
3. Snyder, A.K., Responses of glia to alcohol, in *The Role of Glia in Neurotoxicity*, Aschner, M., and Kimelberg, H.K., Eds., CRC Press, Boca Raton, 1996, p. 111.

4. Guerri, C. and Renau-Piqueras, J., Alcohol, astroglia and brain development, *Mol. Neurobiol.*, 15, 65, 1997.

5. Streissguth, A.P., Landesman-Dwyer, S., Martin, J.C., and Smith, D.W., Teratogenic effects of alcohol in humans and laboratory animals, *Science*, 209, 353, 1980.

6. Streissguth, A.P., Aase, J.M., Clarren, S.K., Randels, S.P., LaDue, R.A., and Smith, D.F., Fetal alcohol syndrome in adolescents and adults, *J.A.M.A.*, 265, 1961, 1991.

7. Lemoine, P. and Lemoine, P.H., Outcome of children of alcoholic mothers (study of 105 cases followed to adult age) and various prophylactic findings, *Ann. Pediatr. (Paris)*, 391, 226, 1992.

8. Spohr, H.S., Willms, J., and Steinhausen, H.C., Prenatal alcohol exposure and long-term developmental consequences, *Lancet*, 341, 907, 1993.

9. Abel, E.L. and Sokol, F.J., Fetal alcohol syndrome is now leading cause of mental retardation, *Lancet*, II, 1222, 1986.

10. Stratton, K., Howe, C., and Battaglia, F., Eds., *Fetal Alcohol Syndrome: Diagnosis, Epidemiology, Prevention and Treatment*, National Academy Press, Washington, D.C., 1996, p. 213.

11. West, J.R., Ed., *Alcohol and Brain Development*, Oxford University Press, Oxford, 1986, p. 440.

12. Zagon, I.S. and Slotkin, T.A., Eds., *Maternal Substance Abuse and the Developing Nervous System*, Academic Press, San Diego, 1992, p. 367.

13. Watson, R.R., Ed., *Alcohol and Neurobiology: Brain Development and Hormone Regulation*, CRC Press, Boca Raton, 1992, p. 396.

14. Dobbing, J. and Sands, J., Comparative aspects of the brain growth spurt, *Early Hum. Dev.*, 3, 79, 1979.

15. Maier, S.E. and West, J.R., Drinking patterns and alcohol-related birth defects, *Alcohol Res. Health*, 25, 168, 2001.

16. Samson, H.H., Microencephaly and fetal alcohol syndrome: human and animal studies, in *Alcohol and Brain Development*, West, J.R., Ed., Oxford Press, Oxford, 1986, p. 167.

17. Balduini, W. and Costa, L.G., Effects of ethanol on muscarinic receptor-stimulated phosphoinositide metabolism during brain development, *J. Pharmacol. Exp. Ther.*, 250, 541, 1989.

18. Swayze, V.W., Johnson, V., Hanson, J.W., Piven, J., Sato, Y., Giedd, J.N., Mosnik, D., and Andreasen, N.C., Magnetic resonance imaging of brain anomalies in fetal alcohol syndrome, *Pediatrics*, 99, 232, 1997.

19. Miller, M.W., Generation of neurons in the rat dentate gyrus and hippocampus: effect of prenatal and postnatal treatment with ethanol, *Alcohol. Clin. Exp. Res.*, 19, 1500, 1995.

20. Phillips, S.C., Alcohol and histology of the developing cerebellum, in *Alcohol and Brain Development*, West, J.R., Ed., Oxford Press, Oxford, 1986, p. 204.

21. Ikonomidou, C., Bittigau, P., Ishimaru, M.J., Wozniak, D.F., Koch, C., Genz, K., Price, M.T., Stefovska, V., Hörster, F., Tenkova, T., Dikranian, K., and Olney, J.W., Ethanol-induced apoptotic neurodegeneration and fetal alcohol syndrome, *Science*, 287, 1056, 2000.

22. Ikonomidou, C., Bittigau, P., Koch, C., Genz, K., Hörster, Fl, Feldenhoff-Mueser, U., Tenkova, T., Dikranian, K., and Olney, J.W., Neurotransmitters and apoptosis in the developing brain, *Biochem. Pharmacol.*, 62, 401, 2001.

23. Phillips, D.E., Effects of alcohol on the development of glial cells and myelin, in *Alcohol and Neurobiology: Brain Development and Hormone Regulation*, Watson, R.R., Ed., CRC Press, Boca Raton, 1992, p. 83.

24. Ledig, M. and Tholey, G., Fetal alcohol exposure and glial cell development, in *Alcohol and Glial Cells*, Lancaster, F.E., Ed., NIH-NIAAA, Bethesda, MD, 1994, p. 117.

25. Guerri, C., Pascual, M., and Renau-Piqueras, J., Glia and fetal alcohol syndrome, *Neurotoxicology*, 22, 593, 2001.

26. Miller, M.W. and Potempa, G., Numbers of neurons and glia in mature rat somatosensory cortex: effects of prenatal exposure to ethanol, *J. Comp. Neurol.*, 293, 92, 1990.

27. Perez-Torrero, E., Duran, P., Granados, L., Gutierrez-Ospina, G., Cintra, L., and Diaz-Cintra, S., Effect of acute prenatal ethanol exposure on Bergman glia cells early postnatal development, *Brain Res.*, 746, 305, 1997.

28. Maier, S.E., Chen, W-J.A., Miller, J.A., and West, J.R., Fetal alcohol exposure and temporal vulnerability: regional differences in alcohol-induced microencephaly as a function of the timing of binge-like alcohol exposure during rat brain development, *Alcohol. Clin. Exp. Res.*, 21, 1418, 1997.

29. Guizzetti, M., Catlin, M.C., and Costa, L.G., Effects of ethanol on glial cell proliferation: relevance to the fetal alcohol syndrome, *Front. Biosci.*, 2, e93, 1997.

30. Davies, D.L. and Vernadakis, A., Effects of ethanol on cultured astroglial cells: proliferation and glutamine synthetase, *Dev. Brain Res.*, 16, 27, 1984.

31. Bass, T. and Volpe, J.J., Ethanol in clinically relevant concentrations enhances expression of oligo-dendroglial differentiation but has no effect on astrocytic differentiation or DNA synthesis in primary cultures, *Dev. Neurosci.*, 11, 52, 1988.

32. Lockhorst, D.K. and Druse, M.J., Effects of ethanol on cultured fetal astroglia, *Alcohol. Clin. Exp. Res.*, 17, 810, 1993.

33. Kennedy, L.A. and Mukerji, S., Ethanol neurotoxicity: 1. Direct effects on replicating astrocytes, *Neurobehav. Toxicol. Teratol.*, 8, 11, 1986.

34. Davies, D.L. and Ross, T.M., Long-term ethanol exposure markedly changes the cellular composition of cerebral glial cultures, *Dev. Brain Res.*, 62, 151, 1991.

35. Davies, D.L. and Cox, W.E., Delayed growth and maturation of astrocytic cultures following exposure to ethanol: electron microscopic observations, *Brain Res.*, 547, 53, 1991.

36. Guerri, C., Saez, R., Sancho-Tello, M., Martin, E., and Renau-Piqueras, J., Ethanol alters astrocyte development: a study of critical periods using primary cultures, *Neurochem. Res.*, 15, 559, 1990.

37. Snyder, A.K., Singh, S., and Ehmann, S., Effects of ethanol on DNA, RNA and protein synthesis in rat astrocyte cultures, *Alcohol. Clin. Exp. Res.*, 16, 295, 1992.

38. de Vito, W.J., Stone, S., and Mori, K., Low concentrations of ethanol inhibit prolactin-induced mitogenesis and cytokine expression in cultured astrocytes, *Endocrinology*, 138, 922, 1997.

39. Barret, L., Soubeyran, A., Usson, Y., Eysseric, H., and Saxod, R., Characterization of the morphological variations of astrocytes in culture following ethanol exposure, *Neurotoxicology*, 17, 487, 1996.

40. Kane, C.J.M., Bessy, A., Boop, F.A., and Davies, D.C., Proliferation of astroglia from the adult human cerebrum is inhibited by ethanol *in vitro*, *Brain Res.*, 731, 39, 1996.

41. Isenberg, K., Zhou, X., and Moore, B.W., Ethanol inhibits C6 cell growth: fetal alcohol syndrome model, *Alcohol. Clin. Exp. Res.*, 16, 695, 1992.

42. Luo, J. and Miller, M.W., Ethanol inhibits basic fibroblast growth factor–mediated proliferation of C6 astrocytoma cells, *J. Neurochem.*, 67, 1448, 1996.

43. Miller, M.W. and Luo, J., Effects of ethanol and basic fibroblast growth factor on the transforming growth factor β1 regulated proliferation of cortical astrocytes and C6 astrocytoma cells, *Alcohol. Clin. Exp. Res.*, 26, 671, 2002.

44. Mikami, K., Haseba, T. and Ohno, Y., Ethanol induces transient arrest of cell division (G2+M block) followed by G0/G1 block: dose effect of short- and longer-term ethanol exposure on cell cycle and functions, *Alcohol Alcohol.*, 32, 145, 1997.

45. Guizzetti, M. and Costa, L.G., Inhibition of muscarinic receptor-stimulated glial cell proliferation by ethanol, *J. Neurochem.*, 67, 2236, 1996.

46. Resnicoff, M., Cui, S., Copula, D., Hoek, J.B., and Rubin, R., Ethanol-induced inhibition of cell proliferation is modulated by insulin-like growth factor-I receptor levels, *Alcohol. Clin. Exp. Res.*, 20, 961, 1996.

47. Luo, J. and Miller, M.W., Platelet-derived growth factor-mediated signal transduction underlying astrocyte proliferation: site of ethanol action, *J. Neurosci.*, 19, 10014, 1999.

48. Kötter, K., Jin, S., and Klein, J., Inhibition of astroglial cell proliferation by alcohols: interference with the protein kinase C—phospholipase D signaling pathways, *Int. J. Dev. Neurosci.*, 18, 825, 2000.

49. Resnicoff, M., Rubini, M., Baserga, R., and Rubin, R., Ethanol inhibits insulin-like growth factor-1-mediated signaling and proliferation of C6 rat glioblastoma cells, *Lab. Invest.*, 71, 657, 1994.

50. Guizzetti, M., Costa, P., Peters, J., and Costa, L.G., Acetylcholine as a mitogen: muscarinic receptor-mediated proliferation of rat astrocytes and human astrocytoma cells, *Eur. J. Pharmacol.*, 297, 265, 1996.

51. Guizzetti, M., Möller, T., and Costa, L.G., Ethanol inhibits muscarinic receptor-mediated DNA synthesis and signal transduction in human fetal astrocytes, *Neurosci. Lett.*, 344, 68, 2003.

52. Guizzetti, M., Wei, M., and Costa, L.G., The role of protein kinase C α and ε isozymes in DNA synthesis induced by muscarinic receptors in a glial cell line, *Eur. J. Pharmacol.*, 359, 223, 1992.

53. Catlin, M.C., Kavanagh, T.J., and Costa, L.G., Muscarinic receptor-induced calcium responses in astroglia, *Cytometry*, 41, 123, 2000.

54. Yagle, K., Lu, H., Guizzetti, M., Möller, T., and Costa, L.G., Activation of mitogen-activated protein kinase by muscarinic receptors in astroglial cells: role in DNA synthesis and effect of ethanol, *Glia*, 35, 111, 2001.

55. Guizzetti, M. and Costa, L.G., Possible role of protein kinase C zeta in muscarinic receptor-induced proliferation of astrocytoma cells, *Biochem. Pharmacol.*, 60, 1457, 2000.

56. Guizzetti, M. and Costa, L.G., Activation of phosphatidylinositol-3-kinase by muscarinic receptors in astrocytoma cells, *Neuroreport*, 12, 1639, 2001.

57. Guizzetti, M., Bordi, F., Dieguez-Acuna, F.J., Vitalone, A., Media, F., Woods, J.S., and Costa, L.G., Nuclear factor kB activation by muscarinic receptors in astroglial cells: effect of ethanol, *Neuroscience*, 120, 941, 2003.

58. Guizzetti, M., Thompson, B.D., Kim, Y., VanDeMark, K., and Costa, L.G., Role of phospholipase D signaling in ethanol-induced inhibition of carbachol-stimulated DNA synthesis of 132 1N1 astrologytome cells. *J. Neurochem.* 90, 646, 2004.

59. Catlin, M.C., Guizzetti, M., and Costa, L.G., Effect of ethanol on muscarinic receptor-induced calcium responses in astroglia, *J. Neurosci. Res.*, 60, 345, 2000.

60. Guizzetti, M. and Costa, L.G., Muscarinic receptors, protein kinase C isozymes and proliferation of astroglial cells: effects of ethanol, *Neurotoxicology*, 21, 1117, 2000.

61. Guizzetti, M. and Costa, L.G., Effect of ethanol on protein kinase C ζ and p70S6 kinase activation by carbachol: a possible mechanism for ethanol-induced inhibition of glial cell proliferation, *J. Neurochem.*, 82, 38, 2002.

62. Gustavsson, L., Phosphatidylethanol formation: specific effects of ethanol mediated by phospholipase D, *Alcohol Alcohol.*, 30, 391, 1995.

63. Kötter, K. and Klein, J., Ethanol inhibits astroglial cell proliferation by disruption of phospholipase D-mediated signaling, *J. Neurochem.*, 73, 2517, 1999.

64. Schatter, B., Walev, I., and Klein, J., Mitogenic effects of phospholipase D and phosphatidic acid in transiently permeabilized astrocytes: effects of ethanol, *J. Neurochem.*, 87, 95, 2003.

65. Kato, H., Faria, T.N., Stannard, B., Roberts, C.T., and LeRoith, D., Role of tyrosine kinase activity in signal transduction by the insulin-like growth factor-I (IGF-I) receptor. Characterization of kinase-deficient IGF-I receptors and the action of an IGF-I mimetic antibody (alpha IR-3), *J. Biol. Chem.*, 268, 2655, 1993.

66. Resnicoff, M., Sell, C., Ambrose, D., Baserga, R., and Rubin, R., Ethanol inhibits the autophosphorylation of the insulin-like growth factor 1 (IGF-1) receptor and the IGF-1-mediated proliferation of 3T3 cells, *J. Biol. Chem.*, 268, 21777, 1993.

67. Seiler, A.E.M., Ross, B.N., Green J.S., and Rubin, R., Differential effects of ethanol on insulin-like growth factor-I receptor signaling, *Alcohol. Clin. Exp. Res.*, 24, 140, 2000.

68. Hallak, H., Seiler, A.E.M., Green, J.S., Henderson, A., Ross, B.N., and Rubin, R., Inhibition of insulin-like growth factor-I signaling by ethanol in neuronal cells, *Alcohol. Clin. Esp. Res.*, 25, 1058, 2001.

69. Iborra, F.J., Renau-Piqueras, J., Boleda, M.D., Guerri, C., and Pares, X., Immunocytochemical and biochemical demonstration of formaldehyde dehydrogenase (Class III alcohol dehydrogenase) in the nucleus, *J. Histochem. Cytochem.*, 40, 1865, 1992.

70. Holownia, A., Ledig, M., and Menez, J.-F., Ethanol-induced cell death in cultured rat astroglia, *Neurotoxicol. Teratol.*, 19, 141, 1997.

71. Climent, E., Pascual, M., Renau-Piqueras, J., and Guerri, C., Ethanol exposure enhances cell death in the developing cerebral cortex: role of brain-derived neurotrophic factor and its signaling pathway, *J. Neurosci. Res.*, 68, 213, 2002.

72. Pascual, M., Valles, S.L., Renau-Piqueras, J., and Guerri, C., Ceramide pathways modulate ethanol-induced cell death in astrocytes, *J. Neurochem.*, 87, 1535, 2003.

73. Montoliu, C., Sancho-Tello, M., Azorin, J., Burgal, M., Valles, S., Renau-Piqueras, J., and Guerri, C., Ethanol increases cytochrome P4502E1 and induces oxidative stress in astrocytes, *J. Neurochem.*, 65, 2561, 1995.

74. Fonseca, L.L., Alves, P.M., Carrondo, M.J.T., and Santos, H., Effect of ethanol on the metabolism of primary astrocytes studied by ^{13}C- and ^{31}P-NMR spectroscopy, *J. Neurosci. Res.*, 66, 803, 2001.

75. Muscoli, C., Fresta, M., Cardile, V., Palumbo, M., Renis, M., Puglisi, G., Paolino, D., Nistico, S., Rotiroti, D., and Mollace, V., Ethanol-induced injury in rat primary cortical astrocytes involves oxidative stress: effect of idebenone, *Neurosci. Lett.*, 329, 21, 2002.

76. Andrieu-Abadie, N., Gouaze, V., Salvayre, R., and Levade, T., Ceramide in apoptosis signaling: relationship with oxidative stress, *Free Rad. Biol. Med.*, 31, 717, 2001.

77. Sanchez-Moreno, C., Paniagua, M., Madrid, A., and Martin, A., Protective effect of vitamin C against the ethanol mediated toxic effects on human brain glial cells, *J. Nutr. Biochem.*, 14, 606, 2003.

78. deVito, W.J., Stone, S., and Shamgochian, M., Ethanol increases the neurotoxic effect of tumor necrosis factor-α in cultured rat astrocytes, *Alcohol. Clin. Esp. Res.*, 24, 82, 2000.

79. Zimmerman, B.T., Crawford, G.D., Dahl, R., Simon, F.R., and Mapoles, J.E., Mechanisms of acetaldehyde-mediated growth inhibition: delayed cell cycle progression and induction of apoptosis, *Alcohol. Clin. Exp. Res.*, 19, 434, 1995.

80. Eysseric, H., Goutheir, B., Soubeyran, A., Bessard, G., Saxod, R., and Barret, L., Characterization of the production of acetaldehyde by astrocytes in culture after ethanol exposure, *Alcohol. Clin. Exp. Res.*, 21, 1018, 1997.

81. Saez, R., Burgal, M., Renau-Piqueras, J., Marques, A., and Guerri, C., Evolution of several cytoskeletal proteins of astrocytes in primary culture: effect of prenatal alcohol exposure, *Neurochem. Res.*, 16, 737, 1991.

82. Tomas, M., Lazaro-Dieguez, F., Duran, J.M., Marin, P., Renau-Piqueras, J., and Egea, G., Protective effects of lysophosphatidic acid (LPA) on chronic ethanol-induced injuries to the cytoskeleton and glucose uptake in rat astrocytes, *J. Neurochem.*, 87, 220, 2003.

83. Guasch, R.M., Tomas, M., Minambres, R., Valles, S., Renau-Piqueras, J., and Guerri, C., RhoA and lysophosphatidic acid are involved in the actin cytoskeleton reorganization of astrocytes exposed to ethanol, *J. Neurosci. Res.*, 72, 487, 2003.

84. Mills, J.C., Stone, N.L., and Pittman, R.N., Extranuclear apoptosis: the role of the cytoplasm in the execution phase, *J. Cell Biol.*, 146, 703, 1999.

85. Lemke, G.K., Glial control of neuronal development, *Annu. Rev. Neurosci.*, 24, 87, 2001.

86. Ullian, E.M., Sapperstein, S.K., Christopherson, S., and Barres, B.A., Control of synapse number by glia, *Science*, 291, 657, 2001.

87. Pfrieger, F.W., Role of glia in synapse development, *Curr. Opin. Neurobiol.*, 12, 486, 2002.

88. Clarren, S.K., Neuropathology in fetal alcohol syndrome, in *Alcohol and Brain Development*, West, J.R., Ed., Oxford University Press, Oxford, 1986, p. 158.

89. Miller, M.W., and Robertson, S., Prenatal exposure to ethanol alters the postnatal development and transformation of radical glia to astrocytes in the cortex, *J. Comp. Neurol.*, 337, 252, 1993.

90. Eriksen, J.L., Gillespie, R.A., and Druse, M.J., Effects of *in utero* ethanol exposure and maternal treatment with a 5-HT(1A) agonist on S100B-containing glial cells, *Brain Res. Dev. Brain Res.*, 121, 133, 2000.

91. Zhou, F.C., Sari, Y., Powrozek, T., Goodlett, C.R., and Li, T.-K., Moderate alcohol exposure compromises natural tube midline development in prenatal brain, *Dev. Brain Res.*, 144, 43, 2003.

92. Riley, E.P., Mattson, S.M., Sowell, E.R., Jernigan, T.L., Sobel, T.F., and Jones, K.L., Abnormalities of the corpus callosum in children prenatally exposed to alcohol, *Alcohol. Clin. Exp. Res.*, 19, 1198, 1995.

93. Müller, H.W., Junghaus, U., and Koppler, J., Astroglial neurotrophic and neurite-promoting factors, *Pharmacol. Ther.*, 65, 1, 1995.

94. Althaus, H.H. and Richter-Landsberg, C., Glial cells as targets and producers of neurotrophins, *Int. Rev. Cytol.*, 197, 203, 2000.

95. Valles, S., Lindo, L., Montoliu, E., Renau-Piqueras, J., and Guerri, C., Prenatal exposure to ethanol induces changes in the nerve growth factor and its receptor in proliferating astrocytes in primary culture, *Brain Res.*, 656, 281, 1994.

96. Kim, J.A. and Druse, M.J., Deficiency of essential neurotrophic factors in conditioned media produced by ethanol-exposed cortical astrocytes, *Brain Res. Dev. Brain Res.*, 96, 1, 1996.

97. Heaton, M.B., Mitchell, J.J., Paiva, M., and Walker, D.W., Ethanol induced alterations in the expression of neurotrophic factors in the developing rat central nervous system, *Dev. Brain Res.*, 121, 97, 2000.

98. Mauch, D.H., Nägler, K., Schumacher, S., Gäritz, C., Müller, E.C., Otto, A., and Pfrieger, F.W., CNS synaptogenesis promoted by glia-derived cholesterol, *Science*, 194, 1354, 2001.

99. Volpe, J.J. and Hennessy, S.W., Cholesterol bioshythesis and 3-hydroxy-3-methyl-glutaryl coenzyme A reductase in cultured glial and neuronal cells. Regulation by lipoprotein and certain free sterols, *Biochem. Biophys. Acta*, 486, 408, 1977.

100. Jurevics, H.A., Kidwai, F.Z., and Morell, P., Sources of cholesterol during development of the rat fetus and fetal organs, *J. Lipid Res.*, 38, 723, 1997.

101. Nowaczyk, M.J., Whelan, D.T., Heschke, T.W., and Hill, R.E., Smith-Lemli-Opitz syndrome, a treatable inherited error of metabolism causing mental retardation, *Can. Med. Assoc. J.*, 161, 165, 1999.

102. Langan, T.J. and Slater, M.C., Quiescent astroglia in long-term primary cultures re-enter the cell cycle and require a non-sterol isoprenoid in late G1, *Brain Res.*, 548, 9, 1991.

CHAPTER 22

Zinc Homeostatic Proteins and Glia

William F. Silverman

CONTENTS

22.1 FREE ZINC IN THE CNS

The distribution and role of zinc ions in the CNS have been studied for many years, particularly with regard to neurons. Zinc is present and essential in the brain, and indeed, throughout the body, in two main forms: tightly bound to metalloproteins, including many enzymes and zinc finger transcription factors, and free or loosely bound. The latter, histochemically reactive or chelatable zinc, represents only a fraction of the total in any given region, and its function and regulation are poorly understood. Typically, this "free" or chelatable Zn^{2+} is present in vesicles within cells, and in the brain, appears largely restricted in a subset of glutamatergic neurons (i.e., "gluzinergic[29]") located principally in the telencephalon. This distribution is the basis of the well-known pattern obtained by the so-called Danscher-Timms or autometallographic methods, which highlight the very dense concentration of Zn^{2+} (and indirectly, glutamate) in the mossy fibers of the dentate gyrus of the hippocampus, amygdala, and neocortex among other areas (Figure 22.1). More precisely, zinc is present in glutamate-containing synaptic vesicles and released during neuronal activity,[28,29] though its physiological function is still debated. What *is* known is that extracellular Zn^{2+} can modulate the activity of several ion channels, most notably those activated by NMDA, AMPA, and GABA receptors.[11,51,91,95] Chelatable, or more specifically, synaptic zinc, has also been shown to participate in the experimental induction of LTP.[49]

In addition to its physiological actions, zinc, under various pathological conditions, e.g., in ischemia or following seizure or closed head trauma, is released at glutamatergic synapses at very high concentrations and permeates cells via a number of routes, including NMDA and voltage-gated Ca^{2+} channels, into neurons.[27,56,94] The subsequent rise in neuronal intracellular zinc (Zn^{2+}_i) is considered a key factor in the resulting neuronal death linked to these syndromes. The specificity

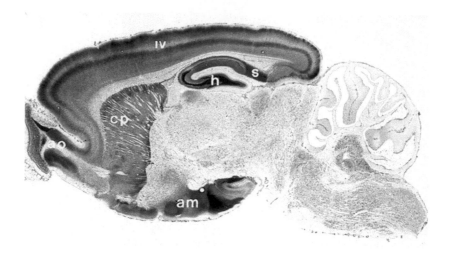

Figure 22.1   Autometallographic (AMG) staining of zinc in a sagital section of mouse brain. Note the dense zinc distribution (gray/black) in cortex, hippocampus, striatum, amygdala, and olfactory bulb. (Modified from Frederickson, C.L. et al., *J. Nutr.*, 130, 1471S, 2000. With permission.)

of this effect has been demonstrated by selective chelation of Zn^{2+} (i.e., as opposed to calcium ions), which reduced the number of neurons that died following an ischemic episode.[44] Enhanced sensitivity of GABA receptors to Zn^{2+} has also been implicated in producing the conditions conducive to the onset of epileptic seizures.[15] In addition, Zn^{2+} inhibits specific elements of the cellular cascade leading to apoptosis.[8,100] Thus, apoptosis occurs in cells cultured in Zn-free medium or following treatment with a membrane-permeable zinc chelator.[10,99] A recent article suggests, however, that zinc can induce oxidative stress, leading to the apoptotic cell death of dopaminergic neurons, and may be involved in the etiology of Parkinson's disease.[50] In addition, zinc, or more correctly, perturbations of Zn-regulation, has recently been linked to Alzheimer's disease, where it is present and involved in the formation of β-amyloid plaques.[41,89] Thus, accumulation of spontaneously produced plaques in a mouse model of Alzheimer's disease was reported in a recent study to be blocked by systemic treatment with clioquinol (CQ), a membrane-permeable chelator.[9]

In peripheral organs, diverse if unproved functions for free or loosely bound zinc have been advanced. Thus, as in the brain, there is growing evidence that chelatable zinc regulates apoptosis, but numerous other activities are also believed to be mediated. In the pancreas, for example, free zinc is found almost exclusively in the beta cells of the Islets of Langerhans,[64] where it is present in secretory granules as crystals containing hexamers of insulin with two or more zinc atoms (see introduction in Zalewski et al.[99]). It is believed that Zn^{2+} ions render insulin insoluble,[22] thereby enabling its storage in the cell. In the testes, zinc is said to be particularly important in the development and function of sperm. Zinc deficiency results in hypogonadism, inhibition of spermatogenesis, and defects in the morphology of spermatozoa, which are themselves rich in chelatable zinc.[88] The effects of excess zinc on cells in various organs of the body, as in the brain, are catastrophic (see Frederickson et al.[26] for review).

From the above description, it is clear that cells face a dilemma. On the one hand, zinc is necessary for normal development and function. On the other hand, it is highly toxic to these cells. Zinc is a small, hydrophilic, highly charged species, which cannot cross biological membranes by passive diffusion. Therefore, specialized mechanisms are required for both its release and its uptake. Indeed, the distribution of free Zn^{2+} across the plasma membrane resembles that of Ca^{2+} insofar as it is distributed as a large gradient, exceeding 3 to 4 orders of magnitude.[4,85] Maintenance of this steep electrochemical gradient requires an active transport system. A number of conserved proteins are known to bind zinc inside cells. The best known of these zinc homeostatic proteins (ZHPs) are metallothioneins (for review, see References 3 and 16), but several other proteins actively buffer zinc, including cellular glutathione and S-100. In addition, a group of transporter proteins in mammals, termed ZnTs, i.e., zinc transporters, have been characterized recently.[65] McMahon and Cousins[60] have extensively reviewed the isolation and characterization of ZnT proteins with regard to neurons. Little information currently exists on the existence within or the effect on this family of proteins in glia. Much more, however, is known about MTs in glia.

22.2 METALLOTHIONEINS

Metallothioneins (MTs) are small cysteine-rich molecules with seven transmembrane domains (see Hidalgo et al.[37] for review) that bind Zn^{2+} with high affinity and large capacity. Several isoforms of MT have been identified, with metallothionein I and II (MT I/II) expressed by neuroglia, particularly astrocytes,[3] and MT III expressed by neurons[23] or glia.[38] Expression of MTs is markedly upregulated following injury or inflammation[32,36,78] or by exposure to zinc.[21]

Metallothioneins are increasingly thought of as a potential intracellular reservoir for zinc. Indeed, transgenic mice missing the Zn^{2+} transporter, ZnT-3, appear to lack vesicular Zn^{2+} yet following seizures, accumulate chelatable Zn^{2+} in hippocampal pyramidal neurons,[14] suggesting that strong synaptic activity can mobilize Zn^{2+} from intracellular sites of sequestration. Similarly, zinc is released from MT following an ischemic episode.[26] Recent studies have shown that Zn^{2+} is released from MTs in response to the intracellular signaling molecule, nitric oxide,[61] and to disulfide-containing oxidants, such as oxidized glutathione (GSSG) or 2,2-dithiodipyridine (DTDP).[57,58] This same mechanism was shown to operate in glia over-expressing MT I/II.[55] From this, it has been suggested that zinc thus sequestered might be mobilized following changes in the redox potential of the cell. Mobilization of intracellular Zn^{2+} by disulfide oxidants has been demonstrated in cultured neurons.[1] It is believed that the redox sensitivity of the cysteine-rich MT to Zn^{2+}, normally a redox-inert metal, is the basis for this phenomenon. Thus, MT is a critical link between cellular redox state and metal ion homeostasis.

Mitochondria constitute another possible site of releasable intracellular Zn^{2+}.[55,83] The existence of a mitochondrial zinc uptake transporter that imports zinc from metalloprotein-bound cytosolic zinc[35] provides a possible mechanism for the accumulation of significant amounts of Zn^{2+} from the sub-nanomolar cytoplasmic concentration of this ion. Yet accumulation of zinc results in compromised mitochondrial function or death due to loss of mitochondrial membrane potential, generation of reactive oxygen species, and release of proapoptotic factors.[81,82] This suggests that, here too, Zn^{2+} is normally buffered to prevent, for example, direct contact between zinc ions and proteins of the electron transport chain. Though the exact mechanism by which this effect is normally prevented is currently unknown, Ye et al.[97] have demonstrated that MT can move into mitochondria and bind zinc ions there. Thus, the means exists for sequestering mitochondrial Zn^{2+} for rapid release to restore homeostasis of intracellular Zn^{2+}.

22.3 ZINC TRANSPORTERS

These proteins are encoded by a conserved family of genes called SLC30. Most (though see Reference 42) are believed to possess six transmembrane domains with both termini on the cytoplasmic side. The first of the 9 ZnT family members so far identified,[65] ZnT-1, is widely distributed throughout the CNS[80] and in all peripheral organs examined to date,[66] in cell plasma membranes as well as cytoplasm. The expression of ZnT-1 has been shown previously to be responsive to zinc.[60,92,62] The mechanism by which ZnT-1 expression is triggered involves the MTF, or metal response element binding transcription factor,[68] and an intracellular receptor for Zn^{2+} (see Reference 2 for review). Thus, Zn^{2+} permeates the cell and binds to and activates MTF, which induces expression of ZnT-1.[2]

A neuroprotective role has been proposed for ZnT-1 on the basis of its ability to vitiate/attenuate the effects of moderate free zinc on baby hamster kidney cells.[67] This effect was observed as well in cultured glial cells (Nolte et al., personal communication) and *in vivo*, with respect to the effect of free Zn^{2+} on neurons in the ischemic gerbil brain.[92] Indeed, protection against toxic accumulation of free zinc is clearly a high priority for neurons as well as for many other cell types. Thus, in the testis, where damage from zinc (and other heavy metals, e.g., cadmium) is known to occur, both ZnTs and MTs are expressed (see below), though not apparently, in the same cells. The work of Palmiter and of Tsuda et al. previously cited, led to the dogma that ZnT-1 is a zinc extruder, though this conclusion has not been proved. What is known is that expression of ZnT-1 mRNA is upregulated by increased zinc load[47,60] and that Zn^{2+} appears to regulate ZnT-1 during development.[62]

22.4 ZINC, ZINC HOMEOSTATIC PROTEINS, AND GLIA

To date, virtually all of the published work carried out on ZnT proteins has focused on neurons. This is likely due to the fact that chelatable Zn^{2+} is restricted to vesicles in glutamatergic synaptic terminals.[27] The first localization of a ZnT, moreover, was for ZnT-3, which was shown to colocalize with zinc in these same synaptic vesicles and is apparently required for the entry of zinc.[66] While the physiological influence and the toxic effects of high concentrations of Zn^{2+} on neurons are well established,[11] much less is known about the effects of zinc on glial cells. Today it is axiomatic that glial cells, which are estimated to outnumber neurons by as much as 9 to 1, interact extensively with neurons and play a key role not only in brain ion homeostasis but also in synapse formation and transmission.[75,86] Although zinc is toxic to cultured neurons at lower concentrations than astrocytes,[19,90] levels of zinc that are toxic to astrocytes are commonly reached in the brain during excitotoxic episodes,[11,95] and CNS ischemia can induce massive glial cell death.[11,39] It appears that at least in some cases, this occurs via the classic apoptotic pathway.[39]

The discovery that astrocytes express ion channels and neurotransmitter receptors, including glutamate, GABA, 5-HT, muscarinic, nicotinic, adrenergic, and others (see Reference 94 for review), raises the possibility, even likelihood, that these cells possess zinc homeostatic proteins in addition to MT. While most interest in ion interactions with glia have centered on Ca^{2+}, a growing literature attests to significant interest in the role of zinc in glial function. Numerous studies have reported glial death subsequent to a focal ischemic episode (see Reference 40 for review). Ischemia causes release of Zn^{2+} at high concentrations, which rapidly enters cells through numerous routes, including voltage-gated Ca^{2+} channels, NMDA, and AMPA kainate channels,[28,29,59,84] where it has been implicated in cell death. The specificity of this effect was demonstrated previously by selective chelation of Zn^{2+} (as opposed to calcium ions), which reduced the number of neurons that died following an ischemic episode.[43]

Generally, it appears that oligodendrocytes are more sensitive to ischemia and to hypoglycemia than astrocytes, and both are less sensitive than neurons,[6,18,52] though there is variation depending on the region examined.[98] The reason that glia in general are less sensitive to ischemia is not known, but their ability to utilize stored glycogen and anaerobic energy production pathways[98] as well as lower metabolic requirements may help them overcome transient reductions in oxygen tension or changes in redox potential. It is also possible that glia possess more effective mechanisms for preventing catastrophic changes in Ca or Zn, which can initiate or enhance cell death cascades or promote production of reactive oxygen species (ROS). The best-known glial protein involved in defense against damage by free radicals is glutathione (GSH), and indeed, glia have been shown to possess a higher GSH content than neurons.[6,54,76] As mentioned previously, MTs, which buffer metal ions and are capable of scavenging oxygen radicals, are therefore also likely to be a factor. A very different mechanism was recently described[20] in which a diffusible peptide, NAP, interacting with tubulin, protects astrocytes against zinc toxicity by inducing microtubule reorganization. Why this would favor glia over neurons, in which microtubules are also the primary component of the neuronal cytoskeleton, is not clear. Nevertheless, the very different mechanism by which cellular protection against the toxic effects of zinc is achieved is instructive.

Our group has observed that ZnT-1, which is expressed by most if not all neurons, is conspicuously absent from astrocytes in the brain, though it is present in oligodendrocytes.[80] In preliminary studies from this lab (Figure 22.2), ZnT-1 is present in cerebral white matter and colocalized with the oligodendrocyte marker, *Rip*.[30] In contrast, ZnT-1 failed to colocalize with either GFAP, vimentin, or MT I/II antisera, which specifically label astrocytes. The situation appears to be different in cell culture, however, where astrocytes, in addition to oligodendrocytes, have been observed to be immunoreactive for ZnT-1 (Figure 22.3) (Nolte et al., personal communication). Prior exposure to nonlethal levels of Zn^{2+}, moreover, boosts ZnT-1 expression in astrocytes. The result is a reduced net zinc influx, mediated by ZnT-1 possibly via its influence on the L-type calcium channel (LTCC), which increases the resistance of these cells to toxic levels of zinc. Indeed, ZnT-1 was shown to significantly reduce influx of both Zn^{2+} and Ca^{2+} in several cell types, including astrocytes, PC-12 cells, and HEK 293 cells heterologously expressing ZnT-1 and the alpha-beta subunits of the L-type Ca^{2+} channel (Segal et al., personal communication). The presence of LTCC in both cultured and endogenous astrocytes has been demonstrated in many studies.[12,53] Against the notion that LTCC closure protects cells from potentially lethal zinc (or calcium) influx, a recent paper has described the protective effects of potassium/depolarization on glia in the presence of high concentrations of extracellular Zn^{2+} [85]. The protective effects of zinc priming or exposure to sublethal concentrations suggests that astrocytes respond to low levels of extracellular zinc by activating a mechanism that protects the cells from the toxic effects of zinc at higher concentrations. This is consistent, moreover, with studies showing that overexpression of ZnT-1 confers resistance to zinc toxicity in a number of cell lines.[43,66,92]

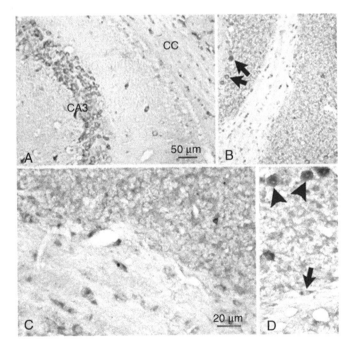

Figure 22.2 ZnT-1–immunoreactive oligodendrocytes *in situ*. (A) ZnT-1–immunoreactive neurons in the CA3 region of the hippocampus and in the corpus callosum (cc). (B) Oligodendrocytes in white matter. Arrows point to Purkinje cell perikarya. (C) High magnification of the white matter in B, showing ZnT-1 labeling of oligodendrocytes soma. (D) Additional detail of area in B demonstrating the Purkinje cell labeling with ZnT-1 (arrowheads) and a sole oligodendrocyte (arrow). (From Sekler and Silverman, unpublished observations.)

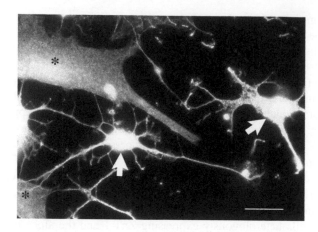

Figure 22.3   Mixed glial cell cultures were prepared from cortex of newborn Wistar rats. Cells were cultured in Dulbecco's modified Eagle's medium supplemented with 10% fetal calf serum. After 10 days in culture, microglial cells were removed by shaking. The remaining monolayer was trypsinized and plated on PLL-coated glass coverslips at low density. Immunocytochemistry was performed 3 days after subcultivation. The ZnT-1 antibody, prepared using a peptide derived from the C-terminal of ZnT-1,[77] was incubated for 3 hours, followed by anti-rabbit IgG coupled to Alexa 594 (Molecular Probes). Both astrocytes (asterisks) and oligodendrocyte precursors (arrows) exhibit a slightly punctate labeling of the cytoplasmic membrane and cellular processes. Note that the oligodendrocytes are more prominently labeled. Magnification bar: 10 μm. (Courtesy of Christiane Nolte.)

But Zn^{2+} will induce the expression of many proteins, not just ZnT-1. One way to circumvent this problem is to knock out the gene—an option precluded by the fact that ZnT-1 KO mice die prior to birth.[47] Another possibility is to employ a more modulated approach, such as antisense gene knockdown or the more advanced siRNA gene silencing method to reduce the expression level of ZnT-1. This latter approach appears to be advancing rapidly toward application, employing a viral vector, and should prove valuable in evaluating the physiological role of this protein both *in vitro* and in the whole animal.

The precise manner in which zinc exerts its negative influence on cells is still under investigation. Zinc has been shown to bind to and inhibit glutathione reductase and peroxidase, the major enzymes responsible for cellular antioxidative defense mechanisms.[87] Zinc has also been reported to reduce cellular glutathione (GSH) in glia. Together with increased zinc-induced ROS production, apoptotic or necrotic cell death is the expected result. Restoration of GSH prevents this result,[77] presumably by blocking the interaction of Zn with mitochondria, which has been shown to disrupt complex II of the electron transport chain and subsequent generation of ROS.[19,83] This pathway appears to be an early event in zinc-induced astrocyte (but not neuronal) death. Others have reported that lethal reduction in GSH also involves impairment in Ca^{2+} homeostasis.[73] In sum, it appears that ZnT-1 is expressed by astrocytes, though under restricted conditions, e.g., following ischemia or in response to oxidative stress. Little evidence exists at present regarding glial expression of other zinc transporters, though transient ZnT-3 immunoreactivity has been reported in a subset of astrocytes in the early postnatal mouse.[93] The possibility that reactive astrocytes could, under the appropriate circumstances, reexpress ZnT-3 should be assessed. This appears to be the case for other proteins, e.g., the intermediate filament nestin, which labels radial glia, an astrocytes precursor, during development and reactive astrocytes in the mature organism.[5,13,33]

In addition to ischemia, zinc may be a factor in a number of other neuropathological conditions involving glia, e.g., multiple sclerosis (MS) and experimental autoimmune encephalomyelitis (EAE). Both MS and EAE cause demyelination and neurodegeneration as well as death of oligodendrocytes.[45] Laboratory studies have demonstrated profound protective effects of MT I and II on the progression and recovery from EAE[71,72] and from other conditions (see Reference 3 for review).

This is likely to be related to the demonstrated efficiency of MT I and II as antioxidant proteins and scavengers of ROS.[16,61] In addition, MT I/II have been shown to reduce proinflammatory cytokines, including TNF alpha and interleukin-6,[70] and stimulate the activation of astrocytes,[7,31,71] all of which could contribute to MT-mediated protection/recovery from EAE. The effects of MT in EAE also highlight its previously described antiapoptotic actions on both neurons and oligodendrocytes. The mechanism by which this aspect of MT activity is manifested is currently not known, though any of the aforementioned actions in addition to its well-known control of intracellular Zn concentrations could be involved. What *is* known is that a Zn-deficient diet causes numerous deleterious effects including degeneration of myelin and proliferation of glial cells.[34] A role for other ZHPs, e.g., ZnT-1, in these cells, therefore, hardly seems fanciful. Indeed, using a zinc chelator, clioquinol, which has a profound effect on tissue zinc concentrations,[63] we recently observed that following several days of treatment with the agent, both ZnT-1 and MT I/II are markedly downregulated in neurons in the mouse cerebellum (Nitzan and Silverman, unpublished observations) but not in hippocampus or olfactory bulb. We theorize that the effect of zinc chelation on ZHPs in the synaptic zinc-poor cerebellum is related to the actions of the chelator on less rapidly replenished internal zinc pools there (see above) in contrast to the plethora of synaptically delivered Zn^{2+} in the hippocampus and olfactory bulb. In the former area, the prolonged drop in cytosolic zinc results in degradation of ZHPs and reduction in new synthesis. In the latter, the rapidly restored supply of Zn^{2+} prevents this cellular shift. That astrocytes were equally affected by CQ suggests that neuronal and glial ZHPs operate by similar mechanisms.

The distribution of zinc-regulating proteins in nonneural tissues can be instructive with regard to the function of these proteins in glia. In the CD-1 mouse testes, we have observed that ZnT-1 is present principally in two populations of cells, the testosterone-producing Leydig cells in the interstitial tissue and the Sertoli cells in the seminiferous tubules (Silverman et al., unpublished observations). Metallothionein I/II in these same mice is notably absent in these two cell types but is expressed by germ cells in the tubule wall. It may be significant that in addition to MT I/II, a number of other glial-specific genes are expressed in the testes, and specifically in Leydig and Sertoli cells.[17] The former cells are widely considered to be embryologically related to neural cells, including glia. The latter are believed to be of mesenchymal origin. Among the proteins produced by both glial and testicular cells are GFAP, vimentin, S-100 protein, galactocerebroside (GalC), cyclic 2′,3′ nucleotide 3′ phosphodiesterase (CNPase), A2B5-antigen (A2B5), and O_4 antigen (O_4).[17] Vimentin, interestingly, is only transiently an astrocyte marker. Similarly, its expression in Leydig cells ceases during prenatal development, to be replaced by GFAP.[25,74] This relationship is reversed in Sertoli cells, which express vimentin strongly in the postnatal mouse, while GFAP is very low.[17]

The nature of the relationship between neuroglia and testicular cells is unclear at present. Nevertheless, there are obvious parallels. Oligodendrocytes, for example, which are ZnT-1 positive and MT negative, surround and nourish neuron processes, while ZnT-1–positive and MT-negative Sertoli (also called sustentacular) cells do much the same for the germ cells of the testes. Astrocytes and Leydig cells each produce and export a wide variety of bioactive substances. Although the astrocytes *in vivo* do not normally express ZnT-1, they are clearly capable of this as shown in glia cultures (see above). It has also been suggested that Leydig cells are, in part, responsible for the blood–testis barrier, much as astrocytes are held to have a role in the blood–brain barrier. Furthermore, both glia and Leydig cells produce and utilize NO,[46,69] as do Sertoli cells, apparently to regulate tight junctions.[48] It is interesting to speculate on the role of ZnT-1 in the Sertoli cells, in light of the well-known destructive effects of cadmium on the integrity of the blood–testis barrier and the ability of zinc to prevent this.[23]

From this brief survey, it is clear that characterizing the various ZHPs that operate in both neurons and neuroglia is an important goal for those studying the neurobiology of zinc. Glial cells face many of the same challenges as neurons, including sensitivity to ischemic attack on energy production by free radicals, changes in redox potential, and dynamic changes in intracellular zinc concentrations in addition to others specific to these cells. Undoubtedly, the rapid pace of developments in our understanding of the biology of transition metals will shed new light on the role played by ZHPs in the once-ignored but now increasingly well-appreciated glial population of the mammalian CNS.

ACKNOWLEDGMENTS

The author would like to thank Dr. Israel Sekler for his critical reading of the manuscript. Thanks also to Dr. Yuval Nitzan-Bebe and Ms. Vered Elgezar who performed the studies from my laboratory described here.

REFERENCES

1. Aizenman, E. et al., Induction of neuronal apoptosis by thiol oxidation: putative role of intracellular zinc release, *J. Neurochem.*, 75, 1878, 2000.
2. Andrews, G.K., Cellular zinc sensors: MTF-1 regulation of gene expression, *Biometals*, 14, 223, 2001.
3. Aschner, M., The functional significance of brain metallothioneins, *FASEB J.*, 10, 1129, 1996.
4. Atar, D. et al., Excitation-transcription coupling mediated by zinc influx through voltage-dependent calcium channels, *J. Biol. Chem.*, 270, 2473, 1985.
5. Baldwin, S.A. and Scheff S.W., Intermediate filament change in astrocytes following mild cortical contusion, *Glia*, 16, 266, 1996.
6. Bolanos, J.P. et al., Effect of peroxynitrite on the mitochondrial respiratory chain: differential susceptibility of neurones and astrocytes in primary culture, *J. Neurochem.*, 64, 1965, 1995.
7. Carrasco, J. et al., Metallothioneins are upregulated in symptomatic mice with astrocyte-targeted expression of tumor necrosis factor-alpha, *Exp. Neurol.*, 163, 46, 2000.
8. Chang, I. et al., Pyruvate inhibits zinc-mediated pancreatic islet cell death and diabetes, *Diabetologia*, 46, 1220, 2003.
9. Cherny, R.A. et al., Treatment with a copper-zinc chelator markedly and rapidly inhibits beta-amyloid accumulation in Alzheimer's disease transgenic mice, *Neuron*, 30, 665, 2001.
10. Chimienti, F. et al., Role of cellular zinc in programmed cell death: temporal relationship between zinc depletion, activation of caspases, and cleavage of Sp family transcription factors, *Biochem. Pharmacol.*, 62, 51, 2001.
11. Choi, D.W. and Koh, J.Y., Zinc and brain injury, *Annu. Rev. Neurosci.*, 21, 347, 1998.
12. Chung, Y.H. et al., Enhanced expression of L-type Ca2+ channels in reactive astrocytes after ischemic injury in rats, *Neurosci. Lett.*, 302, 93, 2001.
13. Clarke, S.R. et al., Reactive astrocytes express the embryonic intermediate neurofilament nestin, *Neuroreport*, 5, 1885, 1994.
14. Cole, T.B. et al., Seizures and neuronal damage in mice lacking vesicular zinc, *Epilepsy Res.*, 39, 153, 2000.
15. Coulter, D.A., Epilepsy-associated plasticity in gamma-aminobutyric acid receptor expression, function, and inhibitory synaptic properties, *Int. Rev. Neurobiol.*, 45, 237, 2001.
16. Coyle, P. et al., Metallothionein: the multipurpose protein, *Cell. Mol. Life Sci.*, 59, 627, 2001.
17. Davidoff, M.S. et al., Leydig cells of the human testis possess astrocyte and oligodendrocyte marker molecules, *Acta Histochem.*, 104, 39, 2002.
18. Dewar, D., Underhill, S.M., and Goldberg M.P., Oligodendrocytes and ischemic brain injury, *J. Cereb. Blood Flow Metab.*, 23, 263, 2003.
19. Dineley, K.E. et al., Astrocytes are more resistant than neurons to the cytotoxic effects of increased [Zn(2+)](i), *Neurobiol Dis.*, 7, 310, 2000.
20. Divinski, I., Mittelman, L., and Gozes, I. A femtomolar acting octapeptide interacts with tubulin and protects astrocytes against zinc intoxication. *J. Biol. Chem.*, 279, 28531, 2004.
21. Ebadi, M. et al., Expression and regulation of brain metallothionein, *Neurochem. Int.*, 27, 1, 1995.
22. Emdin, S.O. et al., Role of zinc in insulin biosynthesis. Some possible zinc-insulin interactions in the pancreatic B-cell, *Diabetologia*, 19, 174, 1980.

23. Erickson, J.C. et al., Disruption of the metallothionein-III gene in mice: analysis of brain zinc, behavior, and neuron vulnerability to metals, aging, and seizures, *J. Neurosci.*, 17, 1271, 1997.

24. Espevik, T. et al., Effects of cadmium on survival and morphology of cultured rat Sertoli cells, *J. Reprod. Fertil.*, 65, 489, 1982.

25. Franke, W.W., Grund, C., and Schmid, E., Intermediate-sized filaments present in Sertoli cells are of the vimentin type, *Eur. J. Cell. Biol.*, 19, 269, 1979.

26. Frederickson, C.J. et al., Nitric oxide causes apparent release of zinc from presynaptic boutons, *Neuroscience*, 115, 471, 2002.

27. Frederickson, C.J. and Bush, A.I., Synaptically released zinc: physiological functions and pathological effects, *Biometals*, 14, 353, 2001.

28. Frederickson, C.J. et al., Importance of zinc in the central nervous system: the zinc-containing neuron, *J. Nutr.*, 130, 1471S, 2000.

29. Frederickson, C.J., Hernandez, M.D., and McGinty, J.F., Translocation of zinc may contribute to seizure-induced death of neurons, *Brain Res.*, 480, 317, 1989.

30. Friedman, B. et al., *In situ* demonstration of mature oligodendrocytes and their processes: an immunocytochemical study with a new monoclonal antibody, rip, *Glia*, 2, 380, 1989.

31. Gaither, L.A., and Eide, D.J., Eukaryotic zinc transporters and their regulation, *Biometals*, 14, 251, 2001.

32. Giralt, M. et al., Metallothionein-1+2 protect the CNS after a focal brain injury, *Exp. Neurol.*, 73, 114, 2002.

33. Goldman, S., Glia as neural progenitor cells, *Trends Neurosci.*, 26, 590, 2003.

34. Gong, H. and Amemiya, T., Optic nerve changes in zinc-deficient rats, *Exp. Eye Res.*, 72, 363, 2001.

35. Guan, Z. et al., Kinetic identification of a mitochondrial zinc uptake transport process in prostate cells, *J. Inorg. Biochem.*, 97, 199, 2003.

36. Hamer, D.H., Metallothionein, *Annu. Rev. Biochem.*, 55, 913, 1986.

37. Hidalgo, J. et al., Metallothionein expression and oxidative stress in the brain, *Methods Enzymol.*, 348, 238, 2002.

38. Hozumi, I., Inuzuka, T., and Tsuji, S., Brain injury and growth inhibitory factor (GIF) — a minireview, *Neurochem. Res.*, 23, 319, 1998.

39. Hyun, H.J. et al., Depletion of intracellular zinc induces macromolecule synthesis- and caspase-dependent apoptosis of cultured retinal cells, *Brain Res.*, 869, 39, 2000.

40. Juurlink, B.H., Response of glial cells to ischemia: roles of reactive oxygen species and glutathione, *Neurosci. Biobehav. Rev.*, 21, 151, 1997.

41. Kaiser, J., Alzheimer's: could there be a zinc link?, *Science*, 265, 1365, 1994.

42. Kambe, T. et al., Cloning and characterization of a novel mammalian zinc transporter, zinc transporter 5, abundantly expressed in pancreatic beta cells, *J. Biol. Chem.*, 277, 19049, 2002.

43. Kim, Y.H. et al., Zn2+ entry produces oxidative neuronal necrosis in cortical cell cultures, *Neuroscience*, 89, 175, 1999.

44. Koh, J.Y. et al., The role of zinc in selective neuronal death after transient global cerebral ischemia, *Science*, 272, 1013, 1996.

45. Kornek, B. et al., Multiple sclerosis and chronic autoimmune encephalomyelitis: a comparative quantitative study of axonal injury in active, inactive, and remyelinated lesions, *Am. J. Pathol.*, 157, 267, 2000.

46. Kugler, P. and Drenckhahn, D., Astrocytes and Bergmann glia as an important site of nitric oxide synthase I, *Glia*, 16, 165, 1996.

47. Langmade, S.J. et al., The transcription factor MTF-1 mediates metal regulation of the mouse ZnT1 gene, *J. Biol. Chem.*, 275, 3480, 2000.

48. Lee, N.P. and Yan Cheng, C. Regulation of Sertoli cell tight junction dynamics in the rat testis via the nitric oxide synthase/soluble guanylate cyclase/3',5'-cyclic guanosine monophosphate/protein kinase G signaling pathway: an *in vitro* study, *Endocrinology*, 144, 3114, 2003.

49. Li, Y., Hough, C.J., Frederickson, C.J., and Sarvey, J.M., Induction of mossy fiber —> Ca3 long-term potentiation requires translocation of synaptically released Zn2+, *J. Neurosci.*, 21, 8015, 2001.

50. Lin, A.M. et al., Zinc-induced apoptosis in substantia nigra of rat brain: neuroprotection by vitamin D3, *Free Radic. Biol. Med.*, 34, 1416, 2003.

51. Lin, D.D., Cohen, A.S., and Coulter, D.A., Zinc-induced augmentation of excitatory synaptic currents and glutamate receptor responses in hippocampal CA3 neurons, *J. Neurophysiol.*, 85, 1185, 2001.

52. Liu, D. et al., Astrocytic demise precedes delayed neuronal death in focal ischemic rat brain, *Mol. Brain Res.*, 68, 29, 1999.

53. MacVicar, B.A. et al., Modulation of intracellular Ca++ in cultured astrocytes by influx through voltage-activated Ca++ channels, *Glia*, 4, 448, 1991.

54. Makar, T.K. et al., Vitamin E, ascorbate, glutathione, glutathione disulfide, and enzymes of glutathione metabolism in cultures of chick astrocytes and neurons: evidence that astrocytes play an important role in antioxidative processes in the brain, *J. Neurochem.*, 62, 45, 1994.

55. Malaiyandi, L.M., Dineley, K.E., and Reynolds, I.J. Divergent concequences arise from metallothionein overexpression in astrocytes: zinc buffering and oxidant-induced zinc release. *Glia*, 45, 346, 2004.

56. Manev, H. et al., Characterization of zinc-induced neuronal death in primary cultures of rat cerebellar granule cells, *Exp. Neurol.*, 146, 171, 1997.

57. Maret, W., Oxidative metal release from metallothionein via zinc-thiol/disulfide interchange, *Proc. Natl. Acad. Sci. U.S.A.*, 91, 237, 1994.

58. Maret, W., The function of zinc metallothionein: a link between cellular zinc and redox state, *J. Nutr.*, 30, 1455S, 2000.

59. Marin, P. et al., Routes of zinc entry in mouse cortical neurons: role in zinc-induced neurotoxicity, *Eur. J. Neurosci.*, 12, 8, 2000.

60. McMahon, R.J. and Cousins, R.J., Regulation of the zinc transporter ZnT-1 by dietary zinc, *Proc. Natl. Acad. Sci. U.S.A.*, 95, 4841, 1998.

61. Molinero, A. et al., Effect of nitric oxide synthesis inhibition on mouse liver and brain metallothionein expression, *Neurochem. Int.*, 33, 559, 1998.

62. Nitzan, Y.B. et al., Postnatal regulation of ZnT-1 expression in the mouse brain, *Dev. Brain Res.*, 137, 149, 2002.

63. Nitzan, Y.B. et al., Clioquinol effects on tissue chelatable zinc in mice, *J. Mol. Med.*, 81, 637, 2003.

64. Okamoto, K. and Kawanishi, H., Submicroscopic histochemical demonstration of intracellular reactive zinc in beta cells of pancreatic islets, *Endocrinol. Jpn.*, 13, 305, 1966.

65. Palmiter, R.D. and Huang, L., Efflux and compartmentalization of zinc by members of the SLC30 family of solute carriers, *Pflugers Arch.*, May14 [Epub ahead of print], 2003.

66. Palmiter, R.D. et al., ZnT-3, a putative transporter of zinc into synaptic vesicles, *Proc. Natl. Acad. Sci. U.S.A.*, 93, 14934, 1996.

67. Palmiter, R.D. and Findley, S.D., Cloning and functional characterization of a mammalian zinc transporter that confers resistance to zinc, *EMBO J.*, 14, 639, 1995.

68. Palmiter, R.D., Regulation of metallothionein genes by heavy metals appears to be mediated by a zinc-sensitive inhibitor that interacts with a constitutively active transcription factor, MTF-1, *Proc. Natl. Acad. Sci. U.S.A.*, 91, 1219, 1994.

69. Park, S.K. et al., Modulation of inducible nitric oxide synthase expression in astroglial cells, *Neuropharmacology*, 33, 1419, 1994.

70. Penkowa, M. and Hidalgo, J., Treatment with metallothionein prevents demyelination and axonal damage and increases oligodendrocyte precursors and tissue repair during experimental autoimmune encephalomyelitis, *J. Neurosci. Res.*, 72, 574, 2003.

71. Penkowa, M. and Hidalgo, J., Metallothionein treatment reduces proinflammatory cytokines IL-6 and TNF-alpha and apoptotic cell death during experimental autoimmune encephalomyelitis (EAE), *Exp. Neurol.*, 170, 1, 2001.

72. Penkowa, M. and Hidalgo, J., Metallothionein I+II expression and their role in experimental autoimmune encephalomyelitis, *Glia*, 32, 247, 2000.

73. Pereira, C.F. and Oliveira, C.R., Oxidative glutamate toxicity involves mitochondrial dysfunction and perturbation of intracellular Ca2+ homeostasis, *Neurosci. Res.*, 37, 227, 2000.

74. Pixley, S.K. and de Vellis, J., Transition between immature radial glia and mature astrocytes studied with a monoclonal antibody to vimentin, *Brain Res.*, 317, 201, 1984.

75. Ransom, B., Behar, T., and Nedergaard, M., New roles for astrocytes (stars at last), *Trends Neurosci.*, 26, 520, 2003.

76. Raps, S.P. et al., Glutathione is present in high concentrations in cultured astrocytes but not in cultured neurons, *Brain Res.*, 493, 398, 1989.

77. Ryu, R. et al., Depletion of intracellular glutathione mediates zinc-induced cell death in rat primary astrocytes, *Exp. Brain Res.*, 143, 257, 2002.

78. Sato, M. and Bremner, I., Oxygen free radicals and metallothionein, *Free Rad. Biol. Med.*, 14, 325, 1993.

79. Segel, D., Ohana, T., Besser, L., Hershfinkel, M., Moran, A., and Sekler, I. A role for ZnT-1 in regulating cellular cation influx. *Biochem. and Biophys. Res. Comm.* 323, 1144, 2004.

80. Sekler, I. et al., Distribution of the zinc transporter ZnT-1 in comparison with chelatable zinc in the mouse brain, *J. Comp. Neurol.*, 447, 201, 2002.

81. Sensi, S.L. et al., Modulation of mitochondrial function by endogenous Zn2+ pools, *Proc. Natl. Acad. Sci. U.S.A.*, 100, 6157, 2003.

82. Sensi, S.L., Yin, H.Z., and Weiss, J.H., AMPA/kainate receptor-triggered Zn2+ entry into cortical neurons induces mitochondrial Zn2+ uptake and persistent mitochondrial dysfunction, *Eur. J. Neurosci.*, 12, 3813, 2000.

83. Sensi, S.L. et al., Preferential Zn2+ influx through Ca2+-permeable AMPA/kainate channels triggers prolonged mitochondrial superoxide production, *Proc. Natl. Acad. Sci. U.S.A.*, 96, 2414, 1999.

84. Sensi, S.L. et al., Measurement of intracellular free zinc in living cortical neurons: routes of entry, *J. Neurosci.*, 17, 9554, 1997.

85. Scheline, C.T., Takata, T., Ying, H., Canzoniero, L.M., Yang, A., Yu, S.P., and Choi, D.W. Potassium atttenuates zinc-induced death of cultured cortical astrocytes. *Glia*, 46, 18, 2004.

86. Slezak, M. and Pfrieger, F.W., New roles for astrocytes: regulation of CNS synaptogenesis, *Trends Neurosci.*, 26, 531, 2003.

87. Splittgerber, A.G. and Tappel, A.L., Inhibition of glutathione peroxidase by cadmium and other metal ions, *Arch. Biochem. Biophys.*, 197, 534, 1979.

88. Stoltenberg, M. et al., Autometallographic demonstration of zinc ions in rat sperm cells, *Mol. Hum. Reprod.*, 3, 763, 2001.

89. Suh, S.W. et al., Histochemically-reactive zinc in amyloid plaques, angiopathy, and degenerating neurons of Alzheimer's diseased brains, *Brain Res.*, 852, 274, 2000.

90. Swanson, R.A. and Sharp, F.R., Zinc toxicity and induction of the 72 kD heat shock protein in primary astrocyte culture, *Glia*, 6, 198, 1992.

91. Takeda, A., Zinc homeostasis and functions of zinc in the brain, *Biometals*, 14, 343, 2001.

92. Tsuda, M. et al., Expression of zinc transporter gene, ZnT-1, is induced after transient forebrain ischemia in the gerbil, *J. Neurosci.*, 17, 6678, 1997.

93. Valente, T. and Auladell, C., Developmental expression of ZnT3 in mouse brain: correlation between the vesicular zinc transporter protein and chelatable vesicular zinc (CVZ) cells. Glial and neuronal CVZ cells interact, *Mol. Cell. Neurosci.*, 21, 189, 2002.

94. Verkhratsky, A. and Steinhauser, C., Ion channels in glial cells, *Brain Res. Rev.*, 32, 380, 2000.

95. Vogt, K. et al., The actions of synaptically released zinc at hippocampal mossy fiber synapses, *Neuron*, 26, 187, 2000.

96. Wang, Z. et al., Inhibitory zinc-enriched terminals in the mouse cerebellum: double-immunohistochemistry for zinc transporter 3 and glutamate decarboxylase, *Neurosci. Lett.*, 321, 37, 2002.

97. Ye, B., Maret, W., and Vallee, B.L., Zinc metallothionein imported into liver mitochondria modulates respiration, *Proc. Natl. Acad. Sci. U.S.A.*, 98, 2317, 2001.

98. Xu, L., Sapolsky, R.M., and Giffard, R.G., Differential sensitivity of murine astrocytes and neurons from different brain regions to injury, *Exp. Neurol.*, 169, 416, 2001.

99. Zalewski, P.D. et al., Video image analysis of labile zinc in viable pancreatic islet cells using a specific fluorescent probe for zinc, *J. Histochem. Cytochem.*, 42, 877, 1994.

100. Zalewski, P.D., Forbes, I.J., and Betts, W.H., Correlation of apoptosis with change in intracellular labile Zn(II) using zinquin [(2-methyl-8-p-toluenesulphonamido-6-quinolyloxy)acetic acid], a new specific fluorescent probe for Zn(II), *Biochem. J.*, 296, 403, 1993.

CHAPTER 23

Iron and Glial Toxicity

James R. Connor, Xuesheng Zhang, and Poonlarp Cheepsunthorn

CONTENTS

23.1 INTRODUCTION

Oxidative stress associated with cell death in progressive neurological disease often involves the loss of metal homeostasis, resulting in increased generation of free radicals and loss of mitochondrial function. Radical formation can lead to lipid peroxidation and ultimately loss of plasma membrane integrity. Iron is a transition metal that plays an important functional role in many biological processes. However, even in a bound form, iron is dangerous owing to its ability to increase oxidative stress through conversion of H_2O_2 to more reactive ROS such as the hydroxyl radical (Fenton reaction). There is evidence that iron can be mobilized and redistributed from cell storage pools after brain insult. Some of this redistribution is not only intracellular but also intercellular, involving glial cells. The redistribution of iron may result in increased cellular vulnerability of some glial cells and dysfunction of others that ultimately promotes neurodegeneration and therefore may exacerbate the disease process. There are also important cellular functions and interactions that are iron dependent. Therefore, iron homeostasis is key to normal neurological function. The role of glial cells in maintaining brain iron homeostasis and the consequence of loss of homeostasis is discussed in this chapter.

23.2 BRAIN INJURY AND IRON ACCUMULATION IN GLIA

A common consequence of almost all brain injury is a reactive gliosis. This reactive gliosis is consistently associated with iron accumulation by astrocytes and microglia and a loss of iron staining in oligodendrocytes. This latter statement appears true regardless of whether the damage is directly associated with white matter as in periventricular leukomalacia[1] or in a predominantly gray matter region as in Parkinson's disease.[2] Evidence shows that even in gray matter regions the predominant iron positive cells are oligodendrocytes.[3-6] In the premature fetus or newborn human infant, acute lesions of white matter range from an early reactive gliosis with relative preservation of axon cylinders to frank infarctions. The lesions are typically seen in subcortical and periventricular white matter of the cerebral hemispheres as well as in the centrum ovale and corpus callosum.[7] Reactive white matter gliosis is a frequent postmortem finding in both premature and full-term infants, suggesting that multiple causes (inflammation, hypoxia, ischemia) underlie the process. However, why this region is relatively more vulnerable is not known, and we propose that the relatively high iron content of the cells in this region may underlie the vulnerability. This proposition is supported by the observations that, unlike gray matter infarction, the lesions in white matter do not follow a vascular distribution suggesting an intrinsic predisposition of cells to damage. Further support for this proposition is found in the literature that oligodendrocytes are the primary targets in periventricular leukomalacia in the premature infant.[8]

23.3 OLIGODENDROCYTES AND IRON

The expression of ferritin and transferrin mRNAs is an early event in differentiating oligodendrocytes.[3,9] The expression of transferrin mRNA and transferrin protein in the brain is almost entirely due to oligodendrocytes.[3] Thus, there is a unique relationship between oligodendrocytes and expression of iron-related proteins in the brain. In the brain, iron concentrations in the white matter consistently exceed those of corresponding gray matter,[10] and microscopic analyses by our laboratory and others have established that the principal cells in the brain that stain following iron histochemistry are white matter oligodendrocytes.[6,11,12] Convincing evidence has been provided that iron acquisition by oligodendrocytes is essential for myelination.[13] Iron is required for the biosynthesis of cholesterol and lipids that are abundant and key components of myelin. HMG-CoA reductase, which catalyzes the NADPH-dependent reduction of HMG-CoA to mevalonate, is also an iron-requiring enzyme. This is the first committed reaction in cholesterol synthesis and the major step that regulates the overall pathway. HMG-CoA reductase is enriched in oligodendrocytes relative to other cells.[13] There are additional iron-requiring steps in the synthesis of cholesterol. For example, the demyelination associated with tellurium toxicity is thought to result from blockage of an iron-requiring step in cholesterol biosynthesis.[14] A relatively high iron requirement by oligodendrocytes is also consistent with the reported relatively high rate of oxidative metabolism for these cells.[15] Specific iron-requiring enzymes involved in maintaining a high rate of metabolic and biosynthetic activity, such as glucose-6-phosphate dehydrogenase, dioxygenase, succinic dehydrogenase, and NADH dehydrogenase, as well as the cytochrome oxidase system, are all elevated in oligodendrocytes relative to other cells in brain.[16] Both lipid saturase and desaturase enzymes involved in myelin synthesis and turnover are iron requiring.[13] Thus, iron is involved in myelination, both directly through synthetic pathways and indirectly as an essential cofactor for the activity of enzymes in oligodendrocytes, most of which are elevated compared with other cells in the brain. It is not surprising that peak iron uptake into the brain coincides with the onset of myelination and that we have found that the developmental pattern of ferritin binding in the brain follows the pattern for myelinogenesis both spatially and temporally.[17]

23.3.1 What Role Does Iron Play in the Vulnerability of Oligodendrocytes to Physiological Stresses?

The vulnerability of oligodendrocytes to hypoxic/ischemic insult has been demonstrated, but the cause of the vulnerability has not been elucidated. Oligodendrocytes may have a poorer ability to scavenge free radicals than astrocytes.[18] Oligodendrocytes are susceptible to oxidative stress but cannot be protected by standard antioxidants such as vitamin E, superoxide dismutase, or glutathione.[19] Recently, vitamin K has been reported as providing protection to oligodendrocytes against free-radical accumulation and cell death.[20]

There is compelling evidence that iron mismanagement by oligodendrocytes contributes to the pathogenesis of multiple sclerosis[21–23] and other demyelinating disorders such as Pelizaeous-Merzbacher,[24,25] central pontine myelinolysis,[26] and progressive rubella panencephalitis.[27] In animal models of dysmyelinating disorders (EAE), iron chelation therapy decreases both the severity of the histopathology and clinical symptoms[28] and an iron-deficient diet protects against EAE.[29] A recent study involving targeted disruption of the gene encoding a cytoplasmic mRNA binding protein known as iron regulatory protein 2 (IRP2) reported that disruption of this gene was associated with significant accumulation of iron and an increase in ubiquitin staining in white matter oligodendrocytes. The mice developed a movement disorder characterized by ataxia, tremor, and bradykinesia.[30] IRP2 is responsible for the coordinate posttranscriptional regulation of transferrin receptor and ferritin mRNAs, thus regulating the amount of intracellular iron.[31]

23.3.2 Evidence that Cytokine-Mediated Toxicity in Oligodendrocytes Is Iron Mediated

Inflammatory cytokines such as TNF-α, IFN-γ, and IL-1β are under intense investigation as causative or major contributors to oligodendrocytic cell death in multiple sclerosis.[32,33] Not coincidentally, these cytokines also impact iron metabolism.[34,35] We have shown that IL-1β increases the labile iron pool (LIP) of astrocytoma cells followed by an induction of ferritin synthesis.[36] The timing of the increase in the LIP and ferritin synthesis may be critical in determining whether a cell will undergo oxidative stress, because increases in the LIP can predispose cells to oxidative stress[37] but an increase in ferritin could be protective. Indeed, expression of H-ferritin is an early event in oligodendrocyte maturation,[38] and limiting expression of H-ferritin by genetic manipulation decreases myelin production (unpublished observation). There is compelling evidence from *in vivo* and *in vitro* studies that strongly supports the involvement of TNF-α and IFN-γ in inflammatory demyelinating conditions. TNF-α and IFN-γ have both been shown to selectively damage oligodendrocytes and myelin *in vitro*.[39,40] TNF-α is toxic to CG-4 and O2A oligodendrocyte cultures, and this toxic effect can be blocked by ciliary neurotrophic factor (CNTF) or insulin-like growth factor I (IGF-I), cytokines associated with oligodendroglial development and repair.[41,42] TNF-α and IFN-γ are reportedly elevated in lesion sites and in the CSF of MS patients,[43–45] and administration of IFN-γ to MS patients worsens clinical severity of the disease.[46] Overexpression of TNF-α or IFN-γ in the brains of transgenic mice results in severe CNS hypomyelination during development and early death of the affected animals.[47,48] TNF-α transgenic mice can be rescued by a monoclonal antibody against TNF-α. Injection of TNF-α into the vitreous humor induces demyelination of the optic nerve in mice.[49] Similarly, injection of TNF-α or IFN-γ into rat lumbosacral spinal cord produces an inflammatory response similar to EAE.[50]

Evidence for a relationship between iron metabolism and cytokines is equally compelling. IL-1β and TNF-α both increase serum ferritin and increase synthesis of H-ferritin.[51] These cytokines and γ interferon (IFN) decreased the amount of iron uptake in a human monocytic line[52] and decreased circulating Tf receptor expression.[53] Iron administration can reverse the cytokine effect on ferritin synthesis.[54] Thus, the evidence strongly supports our hypothesis that iron status will influence the response of oligodendrocytes to cytokines. Oligodendrocytes express H-ferritin, the iron-sequestering protein, and hypoxia induces H-ferritin expression.[55] This is strong evidence that iron is involved in the oxidative stress process in oligodendrocytes.[56] Ferritin expression parallels iron accumulation in oligodendrocytes,[57] but the timing of this relationship may be critical from a cell protection viewpoint. The intracellular iron required for synthesis of myelin and metabolic activities of the cell will be found in the labile iron pool, not sequestered in ferritin. It is the iron in the labile iron pool that will be involved primarily in redox reactivations that will generate reactive oxygen species.[58] Ferritin synthesis follows changes in the labile iron pool as we have shown[36] using cytokine exposure as a model. Furthermore, hypoxia may impact the mRNA binding proteins responsible for promoting ferritin synthesis, but this area has not been actively explored in oligodendrocytes.

23.4 A RELATIONSHIP BETWEEN IRON, MICROGLIA, AND OLIGODENDROCYTES

During development, iron and ferritin are initially found in microglia, and then as myelination is initiated, iron, ferritin, and transferrin mRNA are found within oligodendrocytes.[3,9,57,59] The relationship between oligodendrocytic maturation and iron accumulation has also been examined. If oligodendrocytes are present but not functional (myelin-deficient rats) brain iron uptake continues[60] but accumulates in astrocytes and microglia.[11] If oligodendrocytes reach maturity and produce myelin, even if the myelin is altered (shiverer mouse model), iron accumulates in oligodendrocytes.[12,61] Following neonatal brain injury, iron-laden microglia are observed in white matter tracts in which there are no iron-positive oligodendrocytes. As myelination proceeds, iron- and ferritin-positive oligodendrocytes replace the iron-positive microglia. We have proposed that under normal conditions, iron-loaded microglia release trophic factors that provide a supportive environment for oligodendrocytes. However, there is evidence that activated microglia release soluble mediators responsible for killing oligodendrocytes.[62,63] Therefore, it appears that when microglia are stimulated, their trophic tendencies can be reversed. We have proposed a model system in which to study this concept and further propose that the iron status of microglia impacts on the expression of the soluble factors. This dual effect is consistent with the nature of iron because of its ability to not only interact with oxygen to promote metabolic activities but also promote free-radical species.

It is not inconsistent to propose that the toxicity of iron can be cell type dependent. Iron acquisition is used by macrophages to promote their bactericidal and tumoricidal activities.[64,65] Activation of microglia can cause release of iron from ferritin that was mediated by superoxide production.[66] Also, brain injury can induce heme oxygenase 1 (HO-1) in microglia,[67] which can release iron.[68] If ferritin is present in sufficient quantities, the iron may be detoxified or ferritin may have to first be synthesized depending on the amount of iron released. These latter studies provide compelling evidence of how activation of microglia could change the dynamics of intercellular interaction, when iron, rather than being sequestered by microglia, suddenly is released in a toxic form. The cumulative evidence also suggests that microglia are more resistant to oxidative stress, whereas other cells are very vulnerable to oxidative stress.

23.4.1 Iron and Microglia

Activation of microglia has been described within the central nervous system (CNS) following a wide range of pathological stimuli. Microglial activation is considered a key in the defense of the CNS against infection and injury. Activated microglia are involved in inflammation in neural tissue by secreting inflammatory cytokines: tumor necrosis factor (TNF), interleukin (IL)-1, and IL-6.[69–71] Furthermore, they can amplify the inflammation by recruiting peripheral immune cells to the sites of injury. Activated microglia release nitric oxide, reactive oxygen species, and excitatory amino acids[72–74] that may contribute to progressive damage of oligodendrocytes. In addition, recent studies have demonstrated that microglia exposed to Aβ peptides or the envelope glycoprotein of HIV-1 release neurotoxins.[75–77] Thus, the evidence for a significant role for microglia in brain injury is compelling. However, a major unresolved issue is the identification of factors involved in activating and maintaining the activation of microglia.

One common observation in progressive activation of microglia is that the process is generally accompanied by intracellular accumulation of iron. There are a number of examples of this in both human and animal studies. In Alzheimer's disease, iron-loaded microglia are associated with compacted amyloid deposition.[78–80] Iron-enriched microglia are commonly observed in Parkinson's disease.[81–83] In an animal model of Parkinson's disease (6-OHDA), the temporal relationship between iron accumulation, ferritin expression in microglia, and dopaminergic cell death led the authors to conclude that the primary cause of cell death in this model was most likely the result of accumulation of iron-laden microglia.[84] In multiple sclerosis, iron is found in activated microglia associated with the MS plaques.[82,83] Iron-loaded microglia are also seen in demyelinated regions associated with HIV infection. At least one-third of the patients with HIV infection have excessive accumulation of perivascular iron-loaded microglia. Immunocytochemistry of the iron storage protein ferritin demonstrated that there is an increase in ferritin-positive, activated microglia in the cerebral cortex in HIV dementia.[85] Multinucleated giant cells formed by the fusion of infected microglia, which are characteristics of this disease, also accumulate iron. High expression level of proinflammatory cytokines such as TNF-α and IL-1β is observed in the infected brain. In animal models of hypoxia/ischemia, iron-enriched microglia are present in the damaged areas.[86,87] In the kainite model of epilepsy, rats fed an iron-deficient diet suffered less damage throughout the brain and had less microgliosis compared with those on a control diet, whereas rats fed an iron-supplemented diet had increased brain damage and microgliosis.[88]

Thus, activation of microglia may be an important factor that links abnormal brain iron metabolism to the pathogenesis of neurodegeneration. Very few studies have addressed the issue that iron accumulation in microglia influences their function. Iron increases the transcript expression of inducible nitric oxide synthase in macrophages[89] and nitric oxide (NO) production in mouse BV2 cells.[90] Iron status also affected the production of NO and TNF-α in a rat microglial cell line.[91] Iron can exert its biological effect by functioning as a cofactor for the activation of nuclear transcription factor κB, because the chelation of iron is shown to inactivate nuclear transcription κB, resulting in a decreased cytokine gene expression in macrophages.[92] The expression of ferritin in activated microglia due to an increase in cellular iron could also inhibit NFkB-mediated TNF-α expression in a manner analogous to that of iron chelator.

Understanding the role of iron in activated microglia under pathological conditions should provide insight into the mechanisms of microglial activation and the potential toxicity of microglia to oligodendrocytes. Furthermore, demonstrating that iron loading contributes to microglial activity could provide different perspectives for those interested in targeting activated microglia or their by-products for therapeutic intervention in neuroinflammation and other diseases.

23.5 IRON AND ASTROCYTES

Astrocytes produce trophic factors, regulate neurotransmitter and ion concentrations, and remove toxins and debris from the extracellular space of the CNS, thereby maintaining an extracellular milieu that is optimally suited for neuronal function. Astrocytes protect neurons from various external insults, including reactive oxygen species (ROS).[93,94] Consequently, astrocytic functional impairments, as well as physiological reactions of astrocytes to injury, have the potential to induce or exacerbate neuronal dysfunction.

A deeper understanding of the multiple aspects of astrocyte functioning *in vivo* and the modulation by inflammatory molecules is needed to successfully control homeostasis and to reduce neurotoxicity in CNS disorders. A proposed model has been published,[95] and the data are reviewed in this section.

23.5.1 Iron, Oxidative Stress, and Excitotoxicity in Astrocytes

Alterations in astroglial cell function during oxidative stress may play a key role in the pathophysiology of brain injury. Glutamate, the main excitatory neurotransmitter from mammalian brain, is also a potent neurotoxin, as high extracellular levels of glutamate can induce neuronal death through overstimulation of glutamate receptors, a mechanism referred to as excitotoxicity. There is increasing evidence that excitotoxicity and oxidative stress are interdependent phenomena.[96] Oxidative stress is implied in the progression of neuronal death,[97] and there have been several reports showing degeneration of astrocytes, apparently preceding neuronal loss and reactive gliosis, in experimental animal models of these pathologies.[98–100] Iron accumulation occurs in astrocytes under a variety of experimental injury models, and proteins involved in cellular iron acquisition are increased in astrocytes in a kainate injury model.[101] Astrocytes are uniquely positioned to interact with iron as it passes from the blood into the brain.[102] In a disorder known as aceruloplasminemia, a disorder of iron metabolism caused by the lack of ferroxidase activity of ceruloplasmin, iron accumulates in astrocytes.[103] The mechanism of this iron accumulation likely relates to a dysfunction in a glycosylphosphatidylinositol-anchored form of ceruloplasmin in astrocytes, which appears to be required for iron efflux from astrocytes.[104]

A number of studies have demonstrated that astrocytes are vulnerable to oxidative stress and that accumulation of iron promotes oxidative stress in these cells.[105] MPTP induces HO-1 expression in striatal astrocytes,[106] and HO-1 is also induced in astrocytes by hypoxia-ischemia.[107] As mentioned, HO-1 could result in release of free iron, which would induce generation of reactive oxygen species via the Fenton reaction. Therefore, the production of ferritin in astrocytes would be required to protect these cells from iron-induced oxidative stress.[108] HO-1 induction may result in iron deposition in astrocytic mitochondria,[109] leading to mitochondrial insufficiency. This mitochondrial iron deposition in astrocytes may be a normal consequence of aging that could be related to age-related neurodegenerative disorders.[110]

Studies by our laboratory support the idea that the mechanism by which oxidative stress increases vulnerability of astrocytes appears to involve mitochondrial dysfunction.[111,112] Mitochondria are highly sensitive to oxidative stress, and ROS production by mitochondria can in turn disturb mitochondrial function, causing a vicious cycle. Loss of mitochondrial function contributes directly to diseases such as Alzheimer's disease (AD), Parkinson's disease (PD), and Huntington's disease (HD).[113] Selective vulnerability of astrocyte mitochondria to mitochondrial inhibitors has been described in the 3-nitropropionic acid (3-NPA) model of HD[114] with astrocytic mitochondrial swelling and disruption of cristea occurring prior to neuronal damage.[114] The observation that damage to astrocyte mitochondria precedes damage to neurons is consistent with the idea that loss of astrocyte protective functions could directly mediate death of neurons in neurodegenerative disease. The idea that oxidative stress to astrocytes could precede or promote neuronal cell death is also supported by evidence in Parkinson's disease. In this disease, therapeutic approaches include inhibition of monoamine oxidase B (MAO-B), an astrocytic enzyme, and administration of glial-derived neural trophic factor (GDNF), which is normally provided by astrocytes.

Robb and colleagues[111,112] also have provided evidence that the events associated with peroxide-induced death of astrocytes involves generation of superoxide at the site of mitochondria, loss of mitochondrial membrane potential, and depletion of ATP. These events are iron mediated, with iron loading exacerbating and iron chelation reducing oxidative stress. Iron chelation maintained the mitochondrial membrane potential, prevented peroxide-induced elevations in superoxide levels, and preserved ATP levels. Increases in intracellular calcium are thought to mediate mitochondrial dysfunction and cell death in neurons.[111,112] Although increased intracellular calcium occurred after oxidative stress to astrocytes, the calcium increase was not necessary for collapse of mitochondrial membrane potential. Indeed, when astrocytes were oxidatively stressed in the absence of extracellular calcium, cell death was enhanced, mitochondrial membrane potential collapsed at an earlier time point, and superoxide levels increased. Additionally, our data do not support opening of the mitochondrial permeability transition pore as part of the mechanism of peroxide-induced oxidative stress of astrocytes. These studies led to the conclusion that the increase in intracellular calcium following peroxide exposure does not mediate astrocytic death and may even provide a protective function. The vulnerability of astrocytes and their mitochondria to oxidative stress correlates more closely with iron availability than with increased intracellular calcium.[95,111]

23.6 CONCLUSION

The evidence is overwhelming that iron accumulation in glial cells is a process that promotes their development (oligodendrocytes) yet increases their vulnerability to oxidative stress (all types of glia). There is growing evidence that the iron accumulation in astrocytes will result in dysfunction of this cell type that may result in secondary neuronal degeneration due to loss of astrocytic protective activities. Microglia appear particularly intriguing because they may acquire iron to promote growth and development of oligodendrocytes, yet when they are exposed to inflammatory agents, they may use the iron to kill surrounding cells. Clearly, management of iron homeostasis is an important function of glial cells, and understanding the dynamics of this homeostatic role may be crucial to promoting healthy brain development and minimizing the effects of brain injury.

REFERENCES

1. Inder, T.E. et al., Periventricular white matter injury in the premature infant is followed by reduced cerebral cortical gray matter volume at term, *Ann. Neurol.*, 46, 755–760, 1999.
2. Berg, D. et al., Brain iron pathways and their relevance to Parkinson's disease, *J. Neurochem.* 79, 225–236, 2001.
3. Connor, J.R. and Menzies, S.L., Relationship of iron to oligodendrocytes and myelination, *Glia*, 17, 83–89, 1996.
4. Dwork, A.J., Schon, E.A., and Herbert, J., Nonidentical distribution of transferrin and iron in rat brain, *Neuroscience*, 27, 333–335, 1988.
5. Hill, J.M., The distribution of iron in the brain, in *Topics in Neurochemistry and Neuropharmacology*, Youdim, M.B.H., Lowenberg, W., and Tipton, K.R., Eds., 1989. Taylor & Francis, New York, New York.
6. Benkovic, S. and Connor, J.R., Ferritin, transferrin and iron in normal and aged rat brains, *J. Comp. Neurol.*, 337, 1, 1993.
7. Tourbah, A. et al., Magnetic resonance imaging using FLAIR pulse sequence in white matter diseases, *J. Neuroradiol.*, 23, 217–222, 1996.
8. Volpe, J.J., Brain injury in the premature infant — from pathogenesis to prevention, *Brain Dev.*, 19, 519–534, 1997.
9. Connor, J.R., Iron acquisition and expression of iron regulatory proteins in the developing brain: manipulation by ethanol exposure, iron deprivation and cellular dysfunction, *Dev. Neurosci.*, 16, 233–47, 1994.
10. Rajan, K.S., Colburn, R.W., and Davis, J.M., Distribution of metal ions in the subcellular fractions of several rat brain areas, *Life Sci.*, 18, 423, 1976.
11. Connor, J.R. and Menzies, S.L., Altered cellular distribution of iron in the central nervous system of myelin deficient rats, *Neuroscience*, 34, 265–271, 1990.
12. Levine, S.M., Oligodendrocytes and myelin sheaths in normal, quaking and shiverer brains are enriched in iron, *J. Neurosci. Res.*, 29, 413–419, 1991.
13. Beard, J.L., Wiesinger, J.A., and Connor, J.R., Pre- and postweaning iron deficiency alters myelination in Sprague-Dawley rats, *Dev. Neurosci.*, 25, 308–315, 2003.
14. Wagner-Recio, M., Toews, A.D., and Morell, P., Tellurium blocks cholesterol synthesis by inhibiting squalene metabolism: preferential vulnerability to this metabolic block leads to peripheral nervous system demyelination, *J. Neurochem.*, 57, 1891, 1991.
15. Hamberger, A., Oxidation of tricarboxylic acid cycle intermediates by nerve cell bodies and glial cells, *J. Neurochem.*, 8, 31–35, 1961.

16. Cammer, W., Oligodendrocyte associated enzymes, in *Oligodendroglia*, Norton, W.T., Ed., Plenum Press, New York, 1984, pp. 199–232.

17. Hulet, S.W., Menzies, S., and Connor, J.R., Ferritin binding in the developing mouse brain follows a pattern similar to myelination and is unaffected by the jimpy mutation, *Dev. Neurosci.*, 24, 208–213, 2002.

18. Husain, J. and Juurlink, B.H., Oligodendroglial precursor cell susceptibility to hypoxia is related to poor ability to cope with reactive oxygen species, *Brain Res.*, 698, 86–94, 1995.

19. Kim, Y.S and Kim, S.U., Oligodendroglial cell death induced by oxygen radicals and its protection by catalase, *J. Neurosci. Res.*, 29, 100–106, 1991.

20. Li, J. et al., Novel role of vitamin k in preventing oxidative injury to developing oligodendrocytes and neurons, *J. Neurosci.*, 23, 5816–5826, 2003.

21. Craelius, W. et al., Iron deposits surrounding multiple sclerosis plaques, *Arch. Pathol. Lab. Med.*, 106, 397, 1982.

22. Esiri, M.M., Taylor, C.R., and Mason, D.Y., Applications of an immunoperoxidase method to a study of the central nervous system: Preliminary findings in a study of human formalin-fixed material, *Neuropathol. Appl. Neurobiol.*, 2, 233–246, 1976.

23. Drayer, B. et al., Reduced signal intensity on MR images of thalamus and putamen in multiple sclerosis: increased iron content?, *AJNR*, 8, 413–419, 1987.

24. Koeppen, A.H., Barron, K.D., Csiza, C.K., and Greenfield, E.A., Comparative immunocytochemistry of Pelizaeus-Merzbacher disease, the jimpy mouse, and the myelin-deficient rat, *J. Neurol. Sci.*, 84, 315–327, 1988.

25. Jaeken, J. et al., Sialic acid-deficient serum and cerebrospinal fluid transferrin in a newly recognized genetic syndrome, *Clin. Chem. Acta*, 144, 245–247, 1984.

26. Gocht, A. and Lohler, J., Changes in glial cell markers in recent and old demyelinated lesions in central pontine myelinolysis, *Acta Neuropathol.*, 80, 46–58, 1990.

27. Valk, J., *Magnetic Resonance of Myelin, Myelination and Myelin Disorders*, Springer-Verlag, Vienna, Austria, 1989, pp. 225–268.

28. Bowern, N. et al., Inhibition of autoimmune neuropathological process by treatment with an iron-chelating agent, *J. Exp. Med.*, 160, 1532–1543, 1984.

29. Grant, S.M. et al., Iron-deficient mice fail to develop autoimmune encephalomyelitis, *J. Nutr.*, 133, 2635–2638, 2003.

30. LaVaute, T. et al., Target deletion of the gene encoding iron regulatory protein-2 causes misregulation of iron metabolism and neurodegenerative disease in mice, *Nat. Genet.*, 27, 209–214, 2001.

31. Klausner, R.D., Rouault, T.A., and Harford, J.B., Regulating the fate of mRNA: the control of cellular iron metabolism, *Cell*, 72, 19–28, 1993.

32. Merrill, J.E. and Benveniste, E.N., Cytokines in inflammatory brain lesions: helpful and harmful, *Trends Neurosci.*, 19, 331–338, 1996.

33. Leeden, R.W. and Chakraborty, G., Cytokines, signal transduction, and inflammatory demyelination: review and hypothesis, *Neurochem. Res.*, 23, 277–289, 1998.

34. Chenais, B., Morjani, H., and Drapier, J.C., Impact of endogenous nitric oxide on microglial cell energy metabolism and labile iron pool, *J. Neurochem.*, 81, 615–623, 2002.

35. Miller, L.L. et al., Iron-independent induction of ferritin H chain by tumor necrosis factor, *Proc. Natl. Acad. Sci. U.S.A.*, 88, 4946–4950, 1991.

36. Pinero, D.J., Hu, J., Cook, B., and Connor, J.R., Interleukin-1beta increases binding of the iron regulatory protein and the synthesis of ferritin by increasing the labile iron pool, *Biochim. Biophys. Acta*, 1497, 279–288, 2000.

37. Halliwell, B., Gutteridge, J.M.C., and Cross, C.E., Free radicals, antioxidants and human disease: Where are we now?, *J. Lab. Clin. Med.*, 119, 598–620, 1992.

38. Miller, L.L. et al., Iron-independent induction of ferritin H chain by tumor necrosis factor, *Proc. Natl. Acad. Sci. U.S.A.*, 88, 4946–4950, 1991.

39. Selmaj, K.W. and Raine, C.S., Tumor necrosis factor mediates myelin and oligodendrocyte damage *in vitro*, *Ann. Neurol.*, 23, 339–346, 1988.

40. Baerwald, K.D. and Popko, B., Developing and mature oligodendrocytes respond differently to the immune cytokine interferon-gamma, *J. Neurosci. Res.*, 52, 230–239, 1998.

41. Ye, P. and D'Ercole, A.J., Insulin-like growth factor I protects oligodendrocytes from tumor necrosis factor-α-induced injury, *Endocrinology*, 140, 3063–3072, 1999.

42. Louis, J.C. et al., CG-4 a new bipotential glial cell line from rat brain, is capable of differentiating *in vitro* into either mature oligodendrocytes or type-2 astrocytes, *J. Neurosci. Res.*, 31, 193–204, 1992.

43. Levine, S.M. et al., Ferritin, transferrin and iron concentrations in the cerebrospinal fluid of multiple sclerosis patients, *Brain Res.*, 821, 511–515, 1999.

44. Merrill, J.E. and Benveniste, E.N., Cytokines in inflammatory brain lesions: helpful and harmful, *Trends Neurosci.*, 19, 331–338, 1996.

45. Traugott, U. and Lebon, P., Demonstration of alpha, beta, and gamma interferon in active chronic multiple sclerosis lesions, *Ann. N.Y. Acad. Sci.*, 540, 309–311, 1988.

46. Panitch, H.S. et al., Treatment of multiple sclerosis with gamma interferon: exacerbations associated with activation of the immune system, *Neurology*, 37, 1097–1102, 1987.

47. Probert, L. et al., Spontaneous inflammatory demyelinating disease in transgenic mice showing central nervous system-specific expression of tumor necrosis factor α, *Proc. Natl. Acad. Sci. U.S.A.*, 92, 11294–11298, 1995.

48. Corbin, J.G. et al., Targeted CNS expression of interferon-gamma in transgenic mice leads to hypo-myelination, reactive gliosis, and abnormal cerebellar development, *Mol. Cell. Neurosci.*, 7, 354–370, 1996.

49. Butt, A.M. and Jenkins, H.G., Morphological changes in oligodendrocytes in the intact mouse optic nerve following intravitreal injection of tumor necrosis factor, *J. Immunol.*, 51, 27–33, 1994.

50. Simmons, R.D. and Willenborg, D.O., Direct injection of cytokines into the spinal cord causes autoimmune encephalomyelitis-like inflammation, *J. Neurol. Sci.*, 100, 37–42, 1990.

51. Smirnov, I.M. et al., Effects of TNF-α and IL-1β on iron metabolism by A459 cells and influence on cytotoxicity, *Am. J. Physiol.*, 277, L257–L263, 1999.

52. Fahmy, M. and Young, S.P., Modulation of iron metabolism in monocyte cell line U937 by inflammatory cytokines: changes in transferrin uptake, iron handling and ferritin mRNA, *Biochem. J.*, 296, 175–181, 1993.

53. Feelders, R.A. et al., Regulation of iron metabolism in the acute-phase response: interferon γ and tumour necrosis factor α induce hypoferraemia, ferritin production and a decrease in circulating transferrin receptors in cancer patients, *Eur. J. Clin. Invest.*, 28, 520–527, 1998.

54. Miller, L.L. et al., Iron-independent induction of ferritin H chain by tumor necrosis factor, *Proc. Natl. Acad. Sci. U.S.A.*, 88, 4946–4950, 1991.

55. Qi, Y. and Dawson, G., Hypoxia specifically and reversibly induces the synthesis of ferritin in oligodendrocytes and human oligodendrogliomas, *J. Neurochem.*, 63, 1485–1490, 1994.

56. Goldbaum, O. and Richter-Landsberg, C., Stress proteins in oligodendrocytes: differential effects of heat shock and oxidative stress, *J. Neurochem.*, 78, 1233–1242, 2001.

57. Cheepsunthorn, P., Palmer, C., and Connor, J.R., Cellular distribution of ferritin subunits in postnatal rat brain, *J. Comp. Neurol.*, 400, 73–86, 1998.

58. Kruszewski, M., Labile iron pool: the main determinant of cellular response to oxidative stress, *Mutat. Res.*, 531, 81–92, 2003.

59. Connor, J.R. et al., A histochemical study of iron-positive cells in the developing brain, *J. Comp. Neurol.*, 355, 111–123, 1995.

60. Gocht, A. et al., Iron uptake in the brain of the myelin deficient rat, *Neurosci. Lett.*, 154, 187–190, 1993.

61. Connor, J.R. et al., Transferrin and iron in the shiverer mouse brain, *J. Neurosci. Res.*, 36, 501–507, 1993.

62. Lehnardt, S. et al., The toll-like receptor TLR4 is necessary for lipopolysaccharide-induced oligoden-drocyte injury in the CNS, *J. Neurosci.*, 22, 2478–2486, 2002.

63. Zhang, S.C., Goetz, B.D., and Duncan, I.D., Suppression of activated microglia promotes survival and function of transplanted oligodendroglial progenitors, *Glia*, 41, 191–198, 2003.

64. Caccavo, D. et al., Antimicrobial and immunoregulatory functions of lactoferrin and its potential therapeutic application, *J. Endotoxin Res.*, 8, 403–17, 2002.

65. Green, R., Esparza, I., and Schreiber, R., Iron inhibits the nonspecific tumoricidal activity of mac-rophages. A possible contributory mechanism for neoplasia in hemochromatosis, *Ann. N.Y. Acad. Sci.*, 526, 301–309, 1988.

66. Yoshida, T. et al., Activated microglia cause iron-dependent lipid peroxidation in the presence of ferritin, *Neuroreport*, 9, 1929–1933, 1998.

67. Turner, C.P. et al., Heme oxygenase-1 is induced in glia throughout brain by subarachnoid hemoglobin, *J. Cereb. Blood Flow Metab.*, 18, 257–273, 1998.

68. Schipper, H.M. et al., Mitochondrial iron sequestration in dopamine-challenged astroglia: role of heme oxygenase-1 and the permeability transition pore, *J. Neurochem.*, 72, 1802–1811, 1999.

69. Tham, C.S. et al., Microglial activation state and lysophospholipid acid receptor expression, *Int. J. Dev. Neurosci.*, 21, 431–443, 2003.

70. Dickson, D.W. et al., Microglia in human disease, with an emphasis on acquired immune deficiency syndrome, *Lab. Invest.*, 64, 135–156, 1991.

71. Kreutzberg, G.W., Microglia: a sensor for pathological events in the CNS, *Trends Neurosci.*, 19, 312–318, 1996.

72. Giulian, D. et al., Interleukin 1 of the central nervous system is produced by ameboid microglia, *J. Exp. Med.*, 164, 594–604, 1986.

73. Piani, D. et al., Murine brain macrophages induced NMDA receptor mediated neurotoxicity *in vitro* by secreting glutamate, *Neurosci. Lett.*, 133, 159–162, 1991.

74. Boje, K.M. and Arora, P.K., Microglial-produced nitric oxide and reactive nitrogen oxides mediate neuronal cell death, *Brain Res.*, 587, 250–256, 1992.

75. Giulian, D. et al., The envelope glycoprotein of human immunodeficiency virus type 1 stimulates release of neurotoxins from monocytes, *Proc. Natl. Acad. Sci. U.S.A.*, 90, 2769–2773, 1993.

76. Giulian, D. et al., Study of receptor-mediated neurotoxins released by HIV-1-infected mononuclear phagocytes found in human brain, *J. Neurosci.*, 16, 3139–3153, 1996.

77. Perini, G. et al., Role of p75 neurotrophin receptor in the neurotoxicity by beta-amyloid peptides and synergistic effect of inflammatory cytokines, *J. Exp. Med.*, 195, 907–918, 2002.

78. Kaneko, Y. et al., Ferritin immunohistochemistry as a marker for microglia, *Acta Neuropathol. (Berl.)*, 79, 129–136, 1989.

79. Grundke-Iqbal, I. et al., Ferritin is a component of the neuritic (senile) plaque in Alzheimer dementia, *Acta Neuropathol. (Berl.)*, 81, 105–110, 1990.

80. Connor, J.R. et al., A histochemical study of iron, transferrin, and ferritin in Alzheimer's diseased brains, *J. Neurosci. Res.*, 31, 75–83, 1992.

81. Kaneko, Y. et al., Ferritin immunohistochemistry as a marker for microglia, *Acta Neuropathol. (Berl.)*, 79, 129–136, 1989.

82. Craelius, W. et al., Iron deposits surrounding multiple sclerosis plaques, *Arch. Pathol. Lab. Med.*, 106, 397–399, 1982.

83. LeVine, S.M., Iron deposits in multiple sclerosis and Alzheimer's disease brains, *Brain Res.*, 760, 298–303, 1997.

84. He, Y., Lee, T., and Leong, S.K., Time course of dopaminergic cell death and changes in iron, ferritin and transferrin levels in the rat substantia nigra after 6-hydroxydopamine (6-OHDA) lesioning, *Free Rad. Res.*, 31, 103–112, 1999.

85. Gelman, B.B. et al., Siderotic cerebral macrophages in the acquired immunodeficiency syndrome, *Arch. Pathol. Lab. Med.*, 116, 509–516, 1992.

86. Bidmon, H.J. et al., Heme oxygenase-1 (HSP-32) and heme oxygenase-2 induction in neurons and glial cells of cerebral regions and its relation to iron accumulation after focal cortical photothrombosis, *Exp. Neurol.*, 168, 1–22, 2001.

87. Cheepsunthorn, P. et al., Hypoxic/ischemic insult alters ferritin expression and myelination in neonatal rat brains, *J. Comp. Neurol.*, 431, 382–396, 2001.

88. Shoham, S. and Youdim, M.B., Iron involvement in neural damage and microgliosis in models of neurodegenerative diseases, *Cell. Mol. Biol. (Noisy-le-grand)*, 46, 743–760, 2000.

89. Weiss, G. et al., Iron regulates nitric oxide synthase activity by controlling nuclear transcription, *J. Exp. Med.*, 180, 969–976, 1994.

90. Saleppico, S. et al., Iron regulates microglial cell-mediated secretory and effector functions, *Cell Immunol.*, 170, 251–259, 1996.

91. Cheepsunthorn, P. et al., Characterization of a novel brain-derived microglial cell line isolated from neonatal rat brain, *Glia*, 35, 53–62, 2001.

92. Xiong, S. et al., Signaling role of intracellular iron in NF-kappaB activation, *J. Biol. Chem.*, 278, 17646–17654, 2003.

93. Desagher, S., Glowinski, J., and Premont, J., Astrocytes protect neurons from hydrogen peroxide toxicity, *J. Neurosci.*, 16, 2553–2562, 1996.

94. Tanaka, J. et al., Astrocytes prevent neuronal death induced by reactive oxygen and nitrogen species, *Glia*, 28, 85–96, 1999.

95. Robb, S.J. and Connor, J.R., Oxidative stress-induced cell damage in the CNS: A proposal for a final common pathway, in *Metals and Oxidative Damage in Neurological Disorders*, Connor, J.R., Ed., Plenum Press, New York, 1997, pp. 341–351.

96. Trotti, D., Danbolt, N.C., and Volterra, A., Glutamate transporters are oxidant-vulnerable: a molecular link between oxidative and excitotoxic neurodegeneration?, *Trends Pharmacol. Sci.*, 19, 328–334, 1998.

97. Coyle, J.T. and Puttfarcken, P., Oxidative stress, glutamate, and neurodegenerative disorders, *Science*, 262, 689–695, 1993.

98. Bruijn, L.I. et al., ALS-linked SOD1 mutant G85R mediates damage to astrocytes and promotes rapidly progressive disease with SOD1-containing inclusions, *Neuron*, 18, 327–338, 1997.

99. Martin, L.J. et al., Hypoxia-ischemia causes abnormalities in glutamate transporters and death of astroglia and neurons in newborn striatum, *Ann. Neurol.*, 42, 335–348, 1997.

100. Liu, D. et al., Astrocytic demise precedes delayed neuronal death in focal ischemic rat brain, *Brain Res. Mol. Brain Res.*, 68, 29–41, 1999.

101. Wang, X.S., Ong, W.Y., and Connor, J.R., A light and electron microscopic study of divalent metal transporter-1 distribution in the rat hippocampus, after kainate-induced neuronal injury, *Exp. Neurol.*, 177, 193–201, 2002.

102. Malecki, E.A. et al., Existing and emerging mechanisms for transport of iron and manganese to the brain, *J. Neurosci. Res.*, 56, 113–122, 1999.

103. Miyajima, H., Aceruloplasminemia, an iron metabolic disorder, *Neuropathology*, 23, 345–350, 2003.

104. Jeong, S.Y. and David, S., Glycosylphosphatidylinositol-anchored ceruloplasmin is required for iron efflux from cells in the central nervous system, *J. Biol. Chem.*, 278, 27144–27148, 2003.

105. Robb, S.J. and Connor, J.R., An *in vitro* model for analysis of oxidative death in primary mouse astrocytes, *Brain Res.*, 788, 125–132, 1998.

106. Fernandez-Gonzalez, A., Perez-Otano, I., and Morgan, J.I., MPTP selectively induces haem oxygenase-1 expression in striatal astrocytes, *Eur. J. Neurosci.*, 12, 1573–1583, 2000.

107. Bidmon, H.J. et al., Heme oxygenase-1 (HSP-32) and heme oxygenase-2 induction in neurons and glial cells of cerebral regions and its relation to iron accumulation after focal cortical photothrombosis, *Exp. Neurol.*, 168, 1–22, 2001.

108. Regan, R.F. et al., Ferritin induction protects cortical astrocytes from heme-mediated oxidative injury, *Neuroscience*, 113, 985–994, 2002.

109. Ham, D. and Schipper, H.M., Heme oxygenase-1 induction and mitochondrial iron sequestration in astroglia exposed to amyloid peptides, *Cell. Mol. Biol. (Noisy-le-grand)*, 46, 587–596, 2000.

110. Schipper, H.M. et al., Astrocyte mitochondria: a substrate for iron deposition in the aging rat substantia nigra, *Exp. Neurol.*, 152, 188–196, 1998.

111. Robb, S.J. et al., Influence of calcium and iron on cell death and mitochondrial function in oxidatively stressed astrocytes, *J. Neurosci. Res.*, 55, 674–686, 1999.

112. Robb, S.J. et al., Influence of nitric oxide on cellular and mitochondrial integrity in oxidatively stressed astrocytes, *J. Neurosci. Res.*, 56, 166–176, 1999.

113. Beal, M.F., Mitochondria, free radicals, and neurodegeneration, *Curr. Opin. Neurobiol.*, 6, 661–666, 1996.

114. Nishino, H. et al., Acute 3-nitropropionic acid intoxication induces striatal astrocytic cell death and dysfunction of the blood-brain barrier: involvement of dopamine toxicity, *Neurosci. Res.*, 27, 343–355, 1997.

CHAPTER 24

Astroglia and Methylmercury Neurotoxicity

Gouri Shanker, Tore Syversen, Judy L. Aschner, and Michael Aschner

CONTENTS

24.1 INTRODUCTION

All sources of environmental mercury represent a risk to human health as its conversion by microorganisms in waterways results in MeHg accumulation in the food chain. Excessive MeHg ingestion from a diet high in fish is associated with neurological dysfunction.[1-3] More recent studies in human populations are in agreement with the earlier findings that maternal mercury exposure during pregnancy is related to neurological as well as neuropsychological deficits in the offspring noticeable at 6 to 7 years of age.[4-6] Another recent report points to the detrimental effects of MeHg on neurogenesis.[7] However, in spite of all these findings, the issue remains far from settled as indicated by other studies[8-10] where no definitive linkage was found between MeHg and neurodevelopmental deficits in children at 66 months of age. Thus, there remains a need to understand the mechanisms and consequences of MeHg exposure on CNS function.

Among the various cell types in the brain, astrocytes occupy approximately 50% of the CNS volume.[11] The "foot" processes of these cells are in close proximity to synapses, axonal tracts, nodes of Ranvier, and capillaries. Critically important roles of astrocytes during early brain development include the synthesis and elaboration of cues for neuronal migration and supply of neurotrophic factors necessary for neuronal division and differentiation. Astrocytes express a wide variety of receptors and uptake systems for neurotransmitters, functions which were formerly thought to be exclusively neuronal, thus preventing the buildup of highly damaging levels of neurotransmitters (e.g., glutamate) in the extracellular fluid.[12] In addition, astrocytes maintain optimal extracellular ion concentrations, control extracellular pH, and maintain the high electrical resistance (tightness) of the blood–brain barrier (BBB).[13]

24.2 ASTROCYTES, MeHg, AND GSH

24.2.1 MeHg and Astrocytes

Although the detrimental effects of MeHg in the CNS are not confined to only astrocytes, a large volume of studies points to their important role in mediating MeHg neurotoxicity.

1. Astrocytes represent a preferential cellular site for MeHg accumulation.[14–17]
2. MeHg is known to induce swelling of astrocytes.[14]
3. MeHg selectively inhibits astrocytic uptake of cystine and cysteine, thereby detrimentally affecting cellular redox status and reducing GSH content.[18–21]
4. Neurons become more vulnerable to excitotoxic injury due to increased extracellular glutamate concentrations as a result of MeHg-dependent inhibition of glutamate (and aspartate) uptake and stimulation of its efflux from astrocytes.[22–26]
5. MeHg-triggered neuronal dysfunction is the result of disturbances in astrocytes,[27] and furthermore, the *in vitro* combined application of nontoxic concentrations of mercury and glutamate leads to the appearance of the typical neuronal lesions found with excitotoxic stimulation.[28]
6. MeHg exposure results in the activation of cytosolic phospholipase A_2 ($cPLA_2$) in astrocytes leading to arachidonic acid (AA) release and further inhibition of glutamate transporters feeding forward on an unimpeded cytotoxic cycle.[21,29]
7. MeHg causes an increase in free-radical species contributing to astrocytic damage.[30,31]

Taken together, the results point to the MeHg-induced disruption in astrocytes of EAA homeostasis and the involvement of glutamate-mediated excitotoxic mechanism in MeHg neurotoxicity (see Figure 24.1 for details).

24.2.2 GSH Biosynthesis in Brain and Its Role in the Maintenance of Redox Homeostasis

GSH (a tripeptide comprising L-glutamate, L-cysteine, and glycine) functions as a major antioxidant in mammalian cell systems and composes almost 90% of intracellular nonprotein thiols.[32] Some of its important functions include the maintenance of optimal intracellular redox homeostasis, thus protecting cells against oxidative stress and damage, and conjugation and elimination of a variety of toxic molecules, including MeHg.[33,34] The levels of GSH are known to be lower in neurons than in astrocytes,[33,35] thereby increasing the vulnerability of neurons to increased intracellular reactive oxygen species (ROS) formed as a result of MeHg exposure.[36,37]

The cysteine supplied to the CNS through the vascular system is quickly oxidized to cystine in the extracellular fluid (ECF) by oxygen.[38] The intracellular concentrations of GSH are to a large extent dependent on the availability of cysteine, the rate-limiting precursor for its synthesis.[39] *In vitro* studies have shown that it is difficult to maintain adequate intracellular GSH levels in neurons

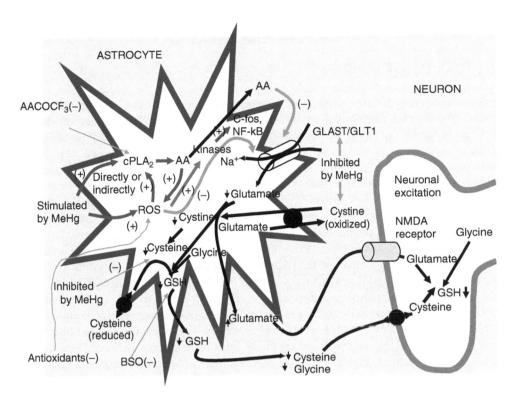

Figure 24.1 *(A color version of this figure follows page 236.)* A schematic astrocyte/neuronal model depicting various MeHg-initiated processes resulting in neurotoxicity. Please refer to text for details. Abbreviations: AA, arachidonic acid; GSH, glutathione; BSO, buthionine sulfoximine; ROS, reactive oxygen species; cPLA$_2$, cytosolic phospholipase A$_2$; AACOCF3, arachidonyl trifluoromethyl ketone; GLAST/GLT-1, astrocytic glutamate transporters in rodents; NF-kB, nuclear factor kappa B.

unless they are cocultured with astrocytes.[40] Astrocytes release cysteine for neuronal GSH synthesis[41] and also release the dipeptide cysteine-glycine (CysGly).[42] The GSH of astrocytic origin serves as a substrate for the astrocytic ectoenzyme γ-glutamyltranspeptidase (γ-GT).[42] The product CysGly thus generated serves as a precursor for neuronal GSH synthesis either via direct uptake of this dipeptide or via cysteine and glycine in the ECF (Figure 24.1). This metabolic interaction between neurons and astrocytes provides multiple targets for interfering with GSH homeostasis. Toxic substances such as MeHg, which disturb cystine and cysteine uptake, GSH release, γ-GT activity, or utilization of CysGly or cysteine by neurons, would thus likely lead to lowered neuronal GSH levels, increasing their susceptibility to oxidative stress.

24.2.3 Association between MeHg Neurotoxicity and GSH Levels

MeHg is a strong neurotoxic molecule whose mechanisms of action are only partially understood. A number of studies have indicated an important role for GSH in the modulation of MeHg neurotoxicity.[33,43] A significant association between GSH concentrations and MeHg toxicity was observed in a MeHg-resistant cell line.[44] The levels of GSH in these cells are fourfold higher compared with nonresistant cells, which resulted in higher MeHg efflux and lower intracellular retention of MeHg. The toxicity of MeHg can be increased by lowering intracellular GSH levels,[33] or decreased by increasing intracellular GSH levels.[25,45] Possible explanations for this protective effect of GSH may include direct conjugation of GSH with MeHg followed by efflux of this conjugate, GSH acting as an intracellular buffer for MeHg, thus reducing the net MeHg concentration available for interaction with other sensitive macromolecules, or GSH acting as a ROS scavenger.

24.3 INHIBITORY EFFECTS OF MeHg ON EAA, CYSTINE, AND CYSTEINE TRANSPORT

24.3.1 EAA Uptake Inhibition

The EAA glutamate is one of the most abundant neurotransmitters found in the brain. While necessary for normal functioning of the CNS, increased extracellular levels of glutamate are neurotoxic. A number of studies have shown that MeHg potently and specifically inhibits the uptake of EAA in cultured astrocytes[16,17,22–24,26] (Figure 24.1). Evidence that mercurial neurotoxicity is at least in part mediated by glutamate includes the observations that (1) the toxic actions of MeHg are prevented[46] by N-methyl-D-aspartate (NMDA) receptor antagonists, (2) neurons are unaffected by MeHg in the absence of glutamate,[27] and (3) nontoxic levels of glutamate and mercury produce neurotoxicity when applied together.[28] MeHg-induced inhibition of ^{3}H-aspartate/glutamate was found to be significantly attenuated by the antioxidant catalase. This suggests that MeHg-induced inhibition of EAAT1 (astrocytic excitatory amino acid transporter) function is due to, among other things, the excess H_2O_2 formation in cultured astrocytes.[26] Prolonged exposure to inhibitors of glutamate (i.e., MeHg and threo-hydroxyaspartate [THA]) significantly decreased EAAT1 mRNA levels, suggesting that transporter expression is related to function. The study points to MeHg-induced overproduction of ROS, notably H_2O_2, as one of the mechanisms for inhibition of glutamate transport and transporter expression in cultured astrocytes.[26] In a recently reported study,[47] the effect of oral ingestion of MeHg on glutamate uptake by mice brain cortical slices and also the protective effect of ebselen (a lipid-soluble seleno-organic compound able to detoxify peroxides and peroxynitrite) in this process were examined. The results demonstrated a MeHg-induced and dose-dependent decrease in glutamate uptake and a reversal of this effect with ebselen. This suggests that the *in vivo* inhibitory effect of MeHg on glutamate uptake could most probably be related to overproduction of H_2O_2, consistent with the observations of Allen et al.[26]; the protective effect of ebselen could be related to its ability to detoxify this peroxide.

24.3.2 Cystine Uptake Inhibition

Adequate cystine transport is essential for the maintenance of optimal intracellular cystine concentrations.[48,49] A number of reports have suggested that neurons rely almost exclusively on GSH of astrocytic origin[35,50] to supply the precursor molecules for neuronal GSH synthesis, either as cysteine[41] or CysGly.[42] These conclusions are based on observations showing sparse cystine transport by neurons, coupled with their inability to maintain optimal intracellular GSH levels in medium containing only cystine.[35,50] More recent studies, however, demonstrate that cultured hippocampal neurons are capable of readily transporting cystine,[18] consistent with earlier findings.[51,52] Thus, it appears that at least some neuronal cell types (hippocampal) in culture can transport cystine and maintain optimal GSH levels even in the absence of astrocytic support. Recent structural analysis of mammalian and human sodium-independent cystine transporter Xc (cystine-glutamate heteroexchange transporter) has shown that both contain heavy- and light-chain subunits.[53,54] The light-chain xCT (system Xc transporter-related protein) confers structural specificity, whereas the cell surface antigen 4F2 (4F2hc, also known as CD98) composes the heavy chain. Human glioma cells expressing two variants of light chain have been described in the spinal cord and brain.[55] Exchange of cystine and glutamate occurs with this transporter, with cystine moving inward (where it gets reduced to cysteine) and glutamate moving out down its concentration gradient (see Figure 24.1). The Xc system represents a major cystine transport route in fetal brain cells[35] but plays a minor role in synaptosomes.[56] In contrast, the sodium-dependent X_{AG} transporter (a family comprising five subtypes of proteins) represents a major pathway for cystine accumulation in synaptosomes[56] and also a significant mode (30–50%) for its uptake in fetal brain cells.[35] Recent work shows that both cultured cerebral astrocytes as well as hippocampal neurons readily take up cystine by multiple mechanisms. The transport of cystine in neurons was found to occur through the sodium-independent system Xc and the multifunctional ectoenzyme/amino acid transporter γ-glutamyl transpeptidase (GGT), while in astrocytes, along with these uptake systems, it also occurred through the sodium-dependent glutamate/aspartate transporter system (X_{AG}),[18] in agreement with previously reported observations.[57] It was found that although MeHg failed to affect cystine uptake in neurons, it inhibited its transport in astrocytes,[18] and this differential sensitivity could be fully accounted for by the MeHg-induced inhibition of astrocytic cystine accumulation via the X_{AG} transporter.

24.3.3 Cysteine Uptake Inhibition

The amino acid cysteine represents the rate-limiting substrate for the synthesis of GSH. Both cerebral astrocytes as well as cerebral and hippocampal neurons can readily accumulate cysteine as shown in recent studies.[19,20,58] Approximately 80–90% of cysteine uptake is sodium dependent (with X_{AG} system as the major and ASC system, the neutral amino acids transporter system, as the minor contributor) in all these cell types, similar to the observations in rat alveolar type II cells as well as in bovine pulmonary artery endothelial cells (BPAEC).[59] MeHg caused a significant concentration-dependent inhibition of cysteine transport in astrocytes (Figure 24.1) but not in neurons, by inhibition of both X_{AG} and the ASC transport systems.[60] A significant concentration-dependent inhibition of cysteine (and also of cystine) uptake in astrocytes was also observed with glutamate, suggesting that the latter may, at least in part, be mediating this MeHg response. The lack of inhibitory effect on neurons cannot be fully explained, but some of the possible mechanisms include (1) a comparatively more efficient transport system for MeHg in astrocytes compared with neurons, consistent with preferential astrocytic accumulation of MeHg,[14,15] and (2) indirect inhibition of uptake in astrocytes secondary to reduced uptake and increased efflux of glutamate,[25,26] with glutamate acting as a surrogate inhibitor of cysteine/cystine transport.

24.4 ROLE OF cPLA$_2$ IN MeHg NEUROTOXICITY

Activation of cPLA$_2$ stimulates the breakdown of membrane phospholipids, releasing AA with subsequent formation of arachidonate metabolites (e.g., prostaglandins, leukotrienes, platelet-activating factor).[61] AA and its metabolites are known to influence activation of voltage-dependent and ligand-gated channels[62,63] and potentiate ischemic damage.[64] Enhanced cPLA$_2$ activity is commonly noted in various neurodegenerative processes.[65–67] MeHg has also been shown to stimulate cPLA$_2$ in neurons.[68,69] The ability of MeHg to enhance AA release from astrocytes and increase cPLA$_2$ mRNA and protein expression (Figure 24.1) was also recently demonstrated.[29] In this study, MeHg was shown to cause a dose-dependent increase in ^{3}H-AA release from neonatal rat primary cerebral astrocytes. This effect was reversed upon treatment with the specific cPLA$_2$ inhibitor, arachidonyl trifluoromethyl ketone (AACOCF$_3$), suggesting direct stimulation of cPLA$_2$ by MeHg (see Figure 24.1). This enhanced astrocytic ^{3}H-AA release following MeHg treatment was not associated with increased incorporation of ^{3}H-AA into putative substrates of cPLA$_2$ given that MeHg did not alter ^{3}H-AA incorporation into total or individual thin-layer chromatographic (TLC)–separated phospholipids.[29] Besides MeHg, glutamate has also been shown to enhance cPLA$_2$ activity.[70] In summary, these studies suggests that cPLA$_2$ activation and AA release play an important role in MeHg-induced neurotoxicity, and the latter is mediated, at least in part, by an excitotoxic mechanism. MeHg-dependent AA release will trigger a cycle of events starting with interference and inhibition of glutamate transporter function (Figure 24.1) and increased extracellular glutamate accumulation and its effects on cPLA$_2$ activation. The net result of this relentless excitotoxic cycle is damage to juxtaposed neurons.

24.5 ROLE OF ROS IN MeHg NEUROTOXICITY

Oxidative stress has been implicated in a variety of neurodegenerative diseases (e.g., Alzheimer's and Parkinson's disease)[65,71] and non-neurodegenerative conditions (i.e., rheumatoid arthritis, cancer, diabetes mellitus, and atherosclerosis)[72–74] and in metal-induced neurotoxicity.[31,75–78] The balance between oxidative and reductive cellular processes is known to be severely affected in these disorders.[74,79] There is a strong association between oxidative stress and the accumulation of high levels of toxic reactive species, such as reactive oxygen species (ROS), reactive nitrogen species (RNS), and reactive nitrogen oxygen species (RNOS), as well as unbound metal ions.[74,78,80] Typical ROS include oxygen radicals such as superoxide radical (O_2^{*-}) and hydroxyl radical ($*OH$), as well as nonradical derivatives of oxygen including hydrogen peroxide (H_2O_2). Examples of RNS include nitric oxide (NO) radical, and that of RNOS include the highly reactive oxidant species peroxynitrite (ONOO-), a product of the reaction between NO and O_2^{*-}. These reactive species are highly oxidizing and damaging to cellular redox-sensitive proteins, enzymes, and DNA, and they also cause membrane peroxidation. ROS are reported to mediate MeHg-dependent neurotoxicity in multiple experimental models. For example, MeHg stimulates ROS formation *in vivo* (rodent cerebellum) and *in vitro* (isolated rat brain synaptosomes),[36] as well as in cerebellar neuronal cells, a hypothalamic neuronal cell line, and mixed reaggregating cell cultures.[37,45,81–83] Additionally, increased ROS formation (contributing to mitochondrial dysfunction) has been observed in (1) mitochondria isolated from MeHg-injected rat brain,[76] (2) mitochondria isolated *in vitro* from rat brain and then exposed to MeHg,[84] and (3) mitochondria from Hg- and glutamate-exposed astrocytes and neurons.[85,86] Involvement of immediate early genes (e.g., c-fos) and the closely related transcription factor NF-kB (nuclear factor kappa B) in the modulation of oxidative stress has also been recently suggested[87,88] (Figure 24.1). Oxidative stress also results in increased lipid peroxidation leading to the formation of a toxic product 4-hydroxy-2-nonenal (HNE) associated with the inhibition of mitochondrial respiration.[89] The mechanism involves inhibition of α-ketoglutarate dehydrogenase (KGDH) and pyruvate dehydrogenase (PDH). MeHg-dependent overproduction of ROS is mediated, at least in part, by glutamate, since it can be attenuated by NMDA receptor antagonists.[45] The source of glutamate is likely to be astrocytic given the stimulatory effect of MeHg on astrocytic glutamate release and the inhibition of astrocytic glutamate uptake by this metal.[26,33]

24.5.1 Biochemical Studies

In a recent report, Sanfeliu et al.[30] examined and observed the protective effects of antioxidants against MeHg neurotoxicity using cell viability assays in cultured astrocytes, neurons, and neuroblastoma cells. Recent studies[31] with cultured neonatal rat cerebral astrocytes have examined in detail the MeHg-induced ROS formation and the effect of various antioxidants and other agents in this process. Some of these results are presented in Table 24.1. As can be seen from the table, there was a MeHg-concentration-dependent increase in ROS formation, which was attenuated in the presence of trolox, a non-thiol-containing antioxidant (Figure 24.1). In addition, AACOCF$_3$ (a cPLA$_2$ specific inhibitor) significantly decreased MeHg-induced astrocytic ROS formation while BSO (buthionine-L-sulfoxamine, a GSH synthesis inhibitor) exacerbated MeHg-induced ROS generation, thus indicating a role for both cPLA$_2$ and GSH in mediating and attenuating MeHg's effects, respectively.

24.5.2 Confocal Microscopic Studies

Detection and assessment of intracellular ROS changes using redox-sensitive fluorescent probes and laser-scanning confocal microscopy represent a powerful and sensitive technique to investigate the role of oxidative stress in cellular dyshomeostasis. Recently, using neurons and C6 glioma cells, the effect of toxic metals, such as zinc and MeHg, on intracellular oxidized states, including

Table 24.1 ROS Formation in Variously Treated Astrocytes

Additions	DCF Fluorescence (% Control) (Mean ± SEM)
Control	100 ± 3
5 μM MeHg	158 ± 24
10 μM MeHg	171 ± 14*
20 μM MeHg	278 ± 40***
10 μM MeHg + BSO (100 μM)	231 ± 23** ****
10 μM MeHg + AACOCF$_3$ (20 μM)	112 ± 17**
10 μM MeHg + 0.5 mM Trolox	110 ± 14**

Note: Cultured neonatal rat cerebral astrocytes (4–6 weeks) were loaded for 30 min with 50 μM 2′,7′-dichlorodihydrofluorescein diacetate (H$_2$DCFDA) dye at 37°C. Next, the cells were treated with MeHg (0, 5, 10, 20 μM), and the fluorescence intensity was measured after 60 min at 37°C with a multiwell fluorescence plate reader set at 485-nm excitation and 530-nm emission wavelength. In experiments with antioxidant trolox (6-hydroxy-2,5,7,8-tetramethylchroman-2-carboxylic acid), after cell loading with dye and a subsequent 10-min preincubation with 0.5 mM trolox, MeHg (10 μM) was added and fluorescence measured after 60 min. In additional experiments, after preincubation of cells at 37°C either with AACOCF$_3$ (arachidonyl trifluoromethyl ketone) for 3 h or with BSO (buthionine-L-sulfoxamine) for 24 h, cells were loaded with dye, which was followed by treatment with MeHg (10 μM) for 60 min. Then fluorescence was measured as described before.

*$p < 0.05$ versus control.
**$p < 0.05$ versus 10 μM MeHg.
***$p < 0.01$ versus control.
****$p < 0.001$ versus control; $n = 5 - 10$.

oxidative DNA damage, was assessed by monitoring fluorescent changes by confocal microscopy in the presence of redox-sensitive probes.[90–92] In C6 glioma cells, MeHg exposure induced both ROS formation as well as DNA oxidative damage.[92] There are hardly any studies with primary astrocytes in which the effect of MeHg on ROS formation has been examined with laser-scanning confocal microscopy. Recently, we have carried out these investigations with rat neonatal cerebral astrocytes employing two redox-sensitive fluorescent probes. One of them is 5,6-chloromethyl-2′,7′-dichlorohydrofluorescein diacetate (CM-H$_2$DCFDA). This is a general redox probe that reacts with intracellular ROS in various subcellular locations. A second probe to measure ROS is reduced chloromethyl-X-rosamine (CM-H$_2$XRos alias reduced mitotracker red), a mitochondria-specific fluorescent dye. The nonfluorescent CM-H$_2$DCFDA passively diffuses into cells, where its acetate groups are cleaved by intracellular esterases and its thiol-reactive chromethyl groups react with intracellular GSH and other thiols. Subsequent oxidation yields a fluorescent adduct.[93] The intensity of fluorescence reflects the intracellular oxidized states.[94,95] Reduced mitotracker red CM-H$_2$XRos, a mitochondria-specific probe, does not fluoresce until it enters an actively respiring cell, where it is oxidized to the fluorescent mitochondrion-selective probe and then is sequestered in the mitochondria.[96] Using the CM-H$_2$DCFDA redox probe, the intracellular ROS production in MeHg-exposed (30 min) cerebral astrocytes was assessed by confocal laser scanning microscopy (Figure 24.2). The images (A = control and B = MeHg treatment), taken 25 minutes after dye exposure, show a clear increase in fluorescent intensity in MeHg-exposed cells, which was found to be significant upon quantitative analysis (Figure 24.3). Additionally, production of mitochondrial ROS was examined by confocal laser scanning microscopy in the presence of mitotracker red dye CM-H$_2$XRos. Astrocytes were pretreated for 30 minutes at 37°C with 10 μM MeHg. After the addition of 250 nM CM-H$_2$XRos, the fluorescence images were taken at 1-minute intervals using a Zeiss inverted microscope equipped with a HE/NE-1 laser, at an excitation wavelength of 543 nm and emission wavelength of 570–600 nm. Figure 24.4 shows (A = control and B = MeHg treatment) images taken after 25 minutes of dye addition and demonstrates an increase in fluorescence intensity with MeHg, which upon quantitative analysis was found to be significant (Figure 24.5).

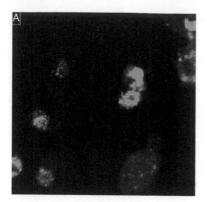
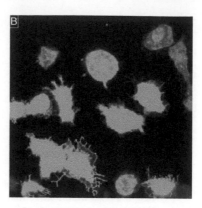

Figure 24.2   *(A color version of this figure follows page 236.)* ROS formation in rat primary cerebral astrocytes exposed to 10 μ*M* MeHg as assessed by changes in DCF fluorescence. After growing the neonatal rat cerebral astrocytes (4–6 weeks) in poly-*d*-lysine–coated Corning 35-mm dishes (MatTek Co., Ashland, MA), they were treated with 10 μ*M* MeHg for 30 min at 37˚C. Subsequent to the addition of 10 μ*M* CM-H$_2$DCFDA dye, fluorescence images were recorded at 1-min intervals using a laser-scanning (Zeiss LSM 510, inverted Axiovert 100M) microscope, at excitation wavelength of 488 nm (argon laser) and emission wavelength of 515 nm. The above pictures (A = control and B = MeHg treated) were taken at 25 min after dye addition.

In summary, the results obtained with both CM-H$_2$DCFDA and the mitochondria-specific mitotracker red dye, CM-H$_2$XRos, indicate enhanced formation of ROS in MeHg-exposed astrocytes. Additionally, the time-series experiments carried out in MeHg-exposed astrocytes to monitor intracellular ROS formation in the presence of both of the above-mentioned redox fluorescent probes showed that the initial increase in fluorescence first appeared with the mitotracker red dye. This suggests that mitochondria most likely represent the initial primary site of ROS formation.

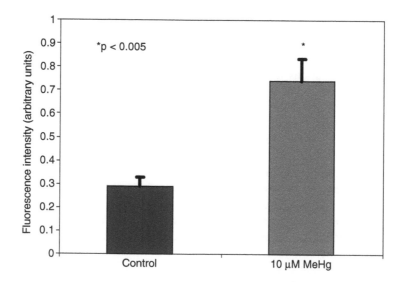

Figure 24.3   The MeHg-induced ROS formation assessed by changes in DCF fluorescence. The images obtained in Figure 24.1 were quantitatively analyzed for changes in fluorescence intensities within cells using the Zeiss LSM software. For each treatment, cellular fluorescence was analyzed from two to three culture dishes of different culture batches (typical total number of cells per dish used for fluorescence analysis was 5–15). The statistical analysis was done with the student's *t*-test comparing control and MeHg-treated samples. The results are presented as mean ± SEM and indicate MeHg-induced increase in fluorescent intensity normalized to the corresponding cross-sectional area for each cell/region of interest.

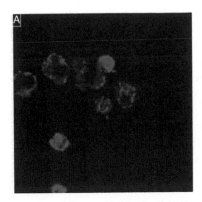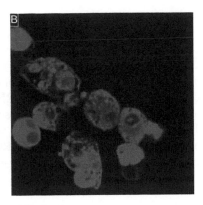

Figure 24.4 *(A color version of this figure follows page 236.)* Mitochondrial ROS formation in cerebral astrocytes exposed to 10 μ*M* MeHg as assessed by changes in fluorescence of the oxidized form of X-rosamine mitotracker dye. Neonatal rat cerebral astrocytes, grown in poly-*d*-lysine–coated Corning 35-mm dishes, were treated for 30 min at 37°C with 10 μ*M* MeHg. After the addition of 250 n*M* CM-H$_2$XRos, the fluorescence images were taken at 1-min intervals using a laser-scanning confocal microscope equipped with a rhodamine laser (543 / nm excitation and 570–600-nm emission). The indicated pictures (A = control and B = MeHg treated) were taken 25 min after dye addition.

This is consistent with observations in MeHg-treated C6 glioma cells and in manganese-treated pheochromocytoma (PC12) cells.[92,97] Thus, these studies point to the important etiologic role played by ROS in MeHg-induced neurotoxic damage. Additionally, within the mitochondrial organelle, various studies have pointed to the mitochondrial electron transport chain complexes as the most vulnerable sites for free-radical-induced damage. Both NO, O_2^{*-}, and $ONOO^-$ species are damaging to the activities of various mitochondrial complexes in GSH-depleted astrocytes or in neurons with relative paucity of GSH.[98–102] Recent elegant coculturing studies have shown that a sustained exposure of neurons to NO- and ONOO-generating astrocytes (following treatment with LPS and cytokines) was found to cause considerable damage to the neuronal mitochondrial respiratory chain complexes.[99,101] In addition to this, another mechanism by which $NO/ONOO^-$ radicals have been reported to damage neuronal mitochondria is through the involvement of NMDA receptors.[103,104] In these

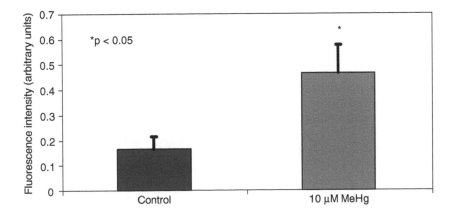

Figure 24.5 The MeHg-induced ROS production as assessed by changes in fluorescence of the oxidized form of the mitotracker dye. The images obtained in Figure 24.4 were quantitatively analyzed for changes in fluorescence intensities as described in detail in the legend for Figure 24.3. The statistical analysis was carried out with the student's *t*-test comparing control and MeHg-treated cells analyzed at the same time point. The results are presented as mean ± SEM and show an increase in fluorescent intensity with MeHg treatment.

coculture studies (with astrocytes and neurons), NO generated as a result of prolonged exposure of astrocytes to cytokines was found to damage complexes II, III, and IV of neighboring neurons in coculture, resulting in neuronal death. However, the NMDA receptor antagonist, MK-801, prevented the damage, suggesting involvement of glutamate. The authors postulate that astrocyte-derived NO stimulated release of neuronal glutamate, which, in turn, activated neuronal NMDA receptors and stimulated further formation of RNS via neuronal nitric oxide synthases, leading to mitochondrial damage and neuronal death. These results, coupled with the reported observation of neuronal GSH depletion (20%) by glutamate,[103] suggest mitochondrial dysfunction and impairment of antioxidant status as contributing factors to ROS/RNS-mediated neurotoxicity. It is quite likely that similar mechanisms may operate in MeHg-induced neurotoxicity involving ROS/RNS.

24.6 CONCLUSIONS

From the studies presented in this chapter, it becomes abundantly clear that there is no single uniform mechanism that can explain the plethora of effects observed in MeHg neurotoxicity. As discussed below, expression and function of a wide variety of molecules (including some described in this chapter) are altered in response to this heavy metal, contributing to altered cell homeostasis. Our studies on astrocytes showed the following:

1. MeHg inhibits cystine and cysteine uptake, thus contributing to decreased GSH levels and altered redox status.
2. Astrocytic $cPLA_2$ has also been suggested as a key molecular switch, whose stimulation directly or indirectly by MeHg will lead to release of AA, a modulator of EAA release.
3. Interference with glutamate transporter function (by both AA as well as MeHg) will result in increased extracellular glutamate levels.
4. Increased extracellular glutamate also results from swelling[105] as well as direct interaction with the cystine transporter and will result ultimately in decreased cysteine supply for neuronal GSH synthesis.
5. Superimposed on these functional disturbances are a number of other cellular events that will accentuate the effects of MeHg, such as induction of ROS formation in astrocytes. The latter will prevent the removal of extracellular glutamate by inhibiting the high-affinity glutamate transporters. Decreased intracellular GSH levels (as a result of MeHg-dependent cystine/cysteine uptake inhibition or following BSO treatment) will further increase the toxic buildup of ROS.
6. In addition, the involvement of glutamate in MeHg-dependent ROS formation has been demonstrated by the inhibition of this effect with NMDA receptor antagonists.[45]
7. Interestingly, our recent observations of the inhibition of MeHg-stimulated ROS formation by $AACOCF_3$ (an inhibitor of $cPLA_2$) suggests a link between $cPLA_2$ inhibition and ROS formation.

In summary, these above-described key events can be synthesized into the following tentative model. A number of interconnected molecular events triggered by MeHg-dependent $cPLA_2$ activation, AA and ROS formation, glutamate transporter inhibition, and excessive extracellular glutamate accumulation and its effect on $cPLA_2$ activation will ultimately result in higher toxic levels of ROS and glutamate.[26,70,106–107] Maintenance of adequate intracellular GSH levels in astrocytes is essential for their protection against oxidative stress. A decrease in intracellular GSH (resulting from cystine/cysteine uptake inhibition) will exacerbate the MeHg-induced ROS formation. The overall outcome of excessive ROS and glutamate accumulation leads to an amplifying destructive cycle of both oxidative and excitotoxic injury, eventually ending in astrocytic dysfunction and compromised ability to maintain optimal control over the extracellular environment, thus indirectly leading to neuronal demise.

It is clear that additional studies are needed to clearly understand the modes of action of this potent toxicant. Future therapeutic counterstrategies aimed at reducing or eliminating the toxic and damaging effects of this heavy metal should take into consideration inhibitors of both oxidative and excitotoxic stress, obviously keeping in mind their ability to cross the blood–brain barrier and to cause minimal side effects.

ACKNOWLEDGMENTS

Preparation of this review was supported by NIEHS Grant #07331 awarded to MA.

REFERENCES

1. Kjellstrom, T., Kennedy, P., Wallis, S., Stewart, A., Friberg, L., Lind, B., Witherspoon, T., and Mantell, C., Physical and mental development of children with prenatal exposure to mercury from fish. Stage II: interviews and psychological tests at age 6, *National Swedish Environ. Protect. Board Report*, Solna, Sweden, 1989, p. 3642.
2. Grandjean, P., Weihe, P., White, R.F., Debes, F., Araki, S., Yokoyama, Murata, K., Sorenson, N., Dahl, R., and Jorgensen, P.J., Cognitive deficit in 7-year old children with prenatal exposure to methylmercury, *Neurotoxicol. Teratol.*, 19, 417–428, 1997.
3. Grandjean, P., Budtz-jorgensen, E., White, R.F., Jorgensen, P.J., Weihe, P., Debes, F., and Keiding, N., Methylmercury exposure biomarkers as indicators of neurotoxicity in children aged 7 years, *Am. J. Epidemiol.*, 150, 301–305, 1999.
4. Steuerwald, U., Weihe, P., Jorgenson, P.J., Bjerve, K., Brock, J., Heinzow, B., Budtz-jorgensen, E., and Grandjean, P., Maternal seafood diet, methylmercury exposure, and neonatal neurologic function, *J. Pediatr.*, 136, 599–605, 2000.
5. Cordier, S., Garel, M., Mandereau, L., Morcel, H., Doineau, P., Gosme-Segure, S., Josse, D., White, R., and Amiel-Tison, C., Neurodevelopmental investigations among methylmercury-exposed children in French Guyana, *Environ. Res.*, 89, 1–11, 2002.
6. Grandjean, P., White, R.F., Weine, P., and Jorgensen, P.J., Neurotoxic risk caused by stable and variable exposure to methylmercury from seafood, *Ambul. Pediatr.*, 3, 18–23, 2003.
7. Faustman, E.M., Ponce, R.A., Ou, Y.C., Mendoza, M.A.C., Lewandowski, T., and Kavanagh, T., Investigations of methylmercury-induced alterations in neurogenesis, *Environ. Health Perspect.*, 110 (Suppl. 5), 859–864, 2002.
8. Palumbo, D.R., Cox, C., Davidson, P.W., Myers, G.J., Choi, A., Shamlaye, C., Sloane-Reeves, J., Cermichiari, E., and Clarkson, T.W., Association between prenatal exposure to methylmercury and cognitive functioning in Seychellois children: a reanalysis of the McCarthy scales of children's ability from the main cohort study, *Environ. Res.*, 84, 81–88, 2000.
9. Davidson, P.W., Kost, J., Myers, G.J., Cox, C., Clarkson, T.W., and Shamlaye, C.F., Methylmercury and neurodevelopment: reanalysis of the Seychelles child development study outcomes at 66 months of age, *JAMA*, 285, 1291–1293, 2001.
10. Myers, G.J., Davidson, P.W., Cox, C., Shamlaye, C.F., Palumbo, D., Cernichiari, E., Sloane-Reeves, J., Wilding, G.E., Kosti, J., Huang, L.S., and Clarkson, T.W., Prenatal methylmercury exposure from ocean fish consumption in the Seychelles child development study, *Lancet*, 361, 1686–1692, 2003.
11. Chen, Y. and Swanson, R.A., Astrocytes and brain injury, *J. Blood Flow Metab.*, 23, 137–149, 2003.
12. Aschner, M. and Kimelberg, H.K., Eds., in *The Role of Glia in Neurotoxicity*, CRC Press, Boca Raton, FL, 1996.
13. Aschner, M., Allen, J.W., Kimelberg, H.K., LoPachin, R.M., and Streit, W.J., Glial cells in neurotoxicity development, *Annu. Rev. Pharmacol. Toxicol.*, 39, 151–173, 1999.
14. Garman, R.H., Weiss, B., and Evans, H.L., Alkylmercurial encephalopathy in the monkey: a histopathologic and autoradiographic study, *Acta Neuropathol. (Berl.)*, 32, 61–74, 1975.
15. Charleston, J.S., Body, R.C., Bolander, R.P., Mottet, N.K., Vahter, M.E., and Burbacher, T.M., Changes in the number of astrocytes and microglia in the thalamus of the monkey Macaca fascicularis following long-term subclinical methylmercury exposure, *Neurotoxicology*, 17, 27–38, 1996.
16. Aschner, M., Methylmercury in astrocytes—what possible significance?, *Neurotoxicology*, 17, 93–106, 1996.

17. Aschner, M., Yao, C.P., Allen, J.W., and Tan, K.H., Methylmercury alters glutamate transport in astrocytes, *Neurochem. Int.*, 37, 199–206, 2000.

18. Allen, J.W., Shanker, G., and Aschner, M., Methylmercury inhibits the *in vitro* uptake of the glutathione precursor, cystine, in astrocytes, but not in neurons, *Brain Res.*, 694, 131–140, 2001.

19. Shanker, G., Allen, J.W., Mutkus, L.A., and Aschner, M., The uptake of cysteine in cultured primary astrocytes and neurons, *Brain Res.*, 902, 156–163, 2001.

20. Shanker, G. and Aschner, M., Identification and characterization of uptake system for cystine and cysteine in cultured astrocytes and neurons: evidence for methylmercury-targeted disruption of astrocytic transport, *J. Neurosci. Res.*, 66, 998–1002, 2001.

21. Shanker, G., Syversen, T., and Aschner, M., Astrocyte-mediated methylmercury neurotoxicity, *Biol. Trace Elem. Res.*, 95, 1–10, 2003.

22. Brookes, N. and Kristt, D.A., Inhibition of amino acid transport and protein synthesis by $HgCl_2$ and methylmercury in astrocytes, selectivity and reversibility, *J. Neurochem.*, 53, 1228–1237, 1989.

23. Aschner, M., Du, Y.L., Gannon, M., and Kimelberg, H.K., Methylmercury-induced alterations in excitatory amino acid transporter in rat primary astrocytic cultures, *Brain Res.*, 602, 181–186, 1993.

24. Dave, V., Mullaney, K.J., Godorie, S., Kimelberg, H.K., and Aschner, M., Astrocytes as mediators of methylmercury neurotoxicity: effects on D-aspartate and serotonin uptake, *Dev. Neurosci.*, 16, 222–231, 1994.

25. Mullaney, K.J., Fehm, M.N., Vitarella, D., Wagoner, D.E., Jr., and Aschner, M., The role of –SH groups in methylmercuric chloride induced D-aspartate and rubidium release from rat primary astrocytic cultures, *Brain Res.*, 641, 1–9, 1994.

26. Allen, J.W., Mutkus, L.M., and Aschner, M., Methylmercury-mediated inhibition of ^{3}H-D-aspartate transport in cultured astrocytes is reversed by the antioxidant catalase, *Brain Res.*, 902, 92–100, 2001.

27. Brookes, N., *In vitro* evidence for the role of glutamate in CNS toxicity of mercury, *Toxicology*, 76, 245–256, 1992.

28. Matyja, E. and Albrecht, J., Ultrastructural evidence that mercuric chloride lowers the threshold for glutamate neurotoxicity in an organotypic culture of rat cerebellum, *Neurosci. Lett.*, 158, 155–158, 1993.

29. Shanker, G., Mutkus, L.A., Walker, S.J., and Aschner, M., Methylmercury enhances arachidonic acid release and cytosolic phospholipase A_2 expression in primary cultures of neonatal astrocytes, *Mol. Brain Res.*, 106, 1–11, 2002.

30. Sanfeliu, C., Sebastia, J., and Kim, S.U., Methylmercury neurotoxicity in cultures of human neurons, astrocytes, neuroblastoma cells, *Neurotoxicology*, 22, 317–327, 2001.

31. Shanker, G. and Aschner, M., Methylmercury-induced reactive oxygen species formation in neonatal cerebral astrocytic cultures is attenuated by antioxidants, *Mol. Brain Res.*, 110, 85–91, 2003.

32. Anderson, M.E. and Meister, A., Transport and direct utilization of gamma-glutamylcyst(e)ine for glutathione synthesis, *Proc. Natl. Acad. Sci. U.S.A.*, 80, 707–711, 1983.

33. Aschner, M., Mullaney, K.J., Wagoner, D., Lash, L.H., and Kimelberg, H.K., Intracellular glutathione (GSH) levels modulate mercuric chloride (MC)- and methylmercuric chloride (MeHgCl)-induced amino acid release from neonatal rat primary astrocytes cultures, *Brain Res.*, 664, 133–140, 1994.

34. Stohs, S.J. and Bagchi, D., Oxidative mechanisms in the toxicity of metal ions, *Free Rad. Biol. Med.*, 18, 321–336, 1995.

35. Sagara, J., Miura, K., and Bannai, S., Cystine uptake and glutathione level in fetal brain cells in primary culture and suspension, *J. Neurochem.*, 61, 1667–1671, 1993.

36. Ali, S.F., LeBel, C.P., and Bondy, S.C., Reactive oxygen species formation as a biomarker of methylmercury and trimethyltin neurotoxicity, *Neurotoxicology*, 13, 637–648, 1992.

37. Sorg, O., Schilter, B., Honnegger, P., and Monnet-Tschudi, F., Increased vulnerability of neurons and glial cells to low concentrations of methylmercury in a prooxidant situation, *Acta Neuropathol. (Berl.)*, 96, 621–627, 1998.

38. Wade, L.A. and Brady, H.M., Cysteine and cystine transport at the blood-brain barrier, *J. Neurochem.*, 37, 730–734, 1981.

39. Bannai, S. and Tateishi, N., Role of membrane transport in metabolism and function of glutathione in mammals, *J. Membr. Biol.*, 89, 1–8, 1986.

40. Dichter, M.A., Rat cortical neurons in cell culture: culture methods, cell morphology, electrophysiology and synapse formation, *Brain Res.*, 149, 279–293, 1978.

41. Wang, X.F. and Cyander, M.S., Astrocytes provide cysteine in neurons by releasing glutathione, *J. Neurochem.*, 74, 1434–1442, 2000.

42. Dringen, R., Pfeiffer, B., and Hamprecht, B., Synthesis of the antioxidant glutathione in neurons: supply by astrocytes of CysGly as precursor for neuronal glutathione, *J. Neurosci.*, 19, 562–569, 1999.

43. Choi, B.H., Yee, S., and Robles, M., The effects of glutathione glycoside in methylmercury poisoning, *Toxicol. Appl. Pharmacol.*, 141, 357–364, 1996.

44. Miura, K. and Clarkson, T.W., Reduced methylmercury accumulation in a methylmercury-resistant rat pheochromocytoma PC12 cell line, *Toxicol Appl. Pharmacol.*, 118, 39–45, 1993.

45. Park, S.T., Lim, K.T., Chung, Y.T., and Kim, S.U., Methylmercury-induced neurotoxicity in cerebral neuron culture is blocked by antioxidants and NMDA receptor antagonists, *Neurotoxicology*, 17, 37–46, 1996.

46. Kim, S.U., Park, S.T., Lim, K.T., and Chung, Y.T., Methylmercury-induced neurotoxicity in cerebral neuron culture is blocked by antioxidants and NMDA receptor antagonists, *Neurotoxicology*, 17, 37–45, 1996.

47. Farina, M., Frizzo, M.E., Soares, F.A., Schwalm, F.D., Dietrich, M.O., Zeni, G., Rocha, J.B., and Souza, D.B., Ebselen protects against methylmercury-induced inhibition of glutamate uptake by cortical slices from adult mice, *Toxicol. Lett.*, 144, 351–353, 2003.

48. Bannai, S., Transport of cystine and cysteine in mammalian cells, *Biochim. Biophys. Acta*, 779, 289–306, 1984.

49. O'Connor, E., Devesa, A., Garcia, C., Puertes, I.R., Pellin, A., and Vina, J.R., Biosynthesis and maintenance of GSH in primary astrocyte cultures: role of L-cystine and ascorbate, *Brain Res.*, 680, 157–163, 1995.

50. Sagara, J., Miura, K., and Bannai, S., Maintenance of neuronal glutathione by glial cells, *J. Neurochem.*, 61, 1672–1676, 1993.

51. Murphy, T.H., Schnaar, R.L., and Coyle, J.T., Immature cortical neurons are uniquely sensitive to glutamate toxicity by inhibition of cystine uptake, *FASEB J.*, 4, 1624–1633, 1990.

52. Sagara, J. and Schubert D., The activation of metabotropic receptor protects nerve cells from oxidative stress, *J. Neurosci.*, 18, 6662–6671, 1998.

53. Sato, H., Tamba, M., Ishii, T., and Bannai, S., Cloning and expression of a plasma membrane cystine/glutamate exchange transporter composed of two distinct proteins, *J. Biol. Chem.*, 274, 11455–11458, 1999.

54. Sato, H., Tamba, M., Kuriyama-Matsumura, K., Okuno, S., and Bannai, S., Molecular cloning and expression of human xCT, the light chain of amino acid transport system Xc-, *Antioxid. Redox Signal*, 2, 665–671, 2000.

55. Kim, J.Y., Kanai, Y., Chairoungdua, A., Cha, S.H., Matsuo, H., Kim, D.K., Inatomi, J., Sawa, H., Ida, Y., and Endou, H., Human cystine/glutamate transporter: cDNA cloning and upregulation by oxidative stress in glioma cells, *Biochim. Biophys. Acta*, 1512, 335–344, 2001.

56. Flynn J. and McBean, G.J., Kinetic and pharmacological analysis of L-[^{35}S]-cystine transport into rat brain synaptosomes, *Neurochem. Int.*, 36, 513–521, 2000.

57. Bender, A.S., Reichelt, W., and Norenberg, M.D., Characterization of cystine uptake in cultured astrocytes, *Neurochem. Int.*, 37, 269–276, 2000.

58. Chen, Y. and Swanson, R.A., The glutamate transporters EAAT$_2$ and EAAT$_3$ mediate cysteine uptake in cortical neuron cultures, *J. Neurochem.*, 84, 1332–1339, 2003.

59. Bukowski, D.M., Deneke, S.M., Lawrence, R.A., and Jenkinson, S.G., A non-inducible cystine transport system in rat alveolar type II cells, *Am. J. Physiol.*, 268, 121–126, 1995.

60. Shanker, G., Allen, J.W., Mutkus, L.A., and Aschner, M., Methyl-mercury inhibits cysteine uptake in cultured primary astrocytes, but not in neurons, *Brain Res.*, 914, 159–165, 2001.

61. Kishimoto, K., Matsumura, K., Kataoka, Y., Morii, H., and Watanabe, Y., Localization of cytosolic phospholipase A$_2$ messenger RNA mainly in neurons in the rat brain, *Neuroscience*, 92, 1061–1077, 1999.

62. Ordway, R.W., Singer, J.J., and Walsh, J.V., Jr., Direct regulation of ion-channels by fatty acids, *Trends Neurosci.*, 14, 96–100, 1991.

63. Miller, B., Sarantis, M., Traynelis, S.F., and Attwell, D., Potentiation of NMDA receptor currents by arachidonic acid, *Nature*, 355, 722–725, 1992.

64. Matsumura, K., Cao, C., Watanabe, Y., and Watanabe, Y., Prostaglandin system in the brain: sites of biosynthesis and sites of action under normal and hyperthermic states, *Prog. Brain Res.*, 115, 275–295, 1998.

65. Lukiw, W.J. and Bazan, N.G., Neuro-inflammatory signaling upregulation in Alzheimer's disease, *Neurochem. Res.*, 25, 1173–1184, 2000.

66. Sapirstein, A. and Bonventre, J.V., Phospholipase A_2 in ischemic and toxic brain injury, *Neurochem. Res.*, 25, 745–753, 2000.

67. Arai, K., Ikegaya, Y., Nakatani, Y., Kudo, I., Nishiyama, N., and Matsuki, N., Phospholipase A_2 mediates ischemic injury in the hippocampus: a regional difference of neuronal vulnerability, *Eur. J. Neurosci.*, 13, 2319–2323, 2001.

68. Verity, M.A., Sarafian, T., Pacifici, E.H.K., and Sevanian, A., Phospholipase A_2 stimulation by methylmercury in neuron culture, *J. Neurochem.*, 62, 705–714, 1994.

69. Shanker, G., Hampson, R.E., and Aschner, M., Methylmercury stimulates arachidonic acid release and cytosolic phospholipase A_2 expression in primary neuronal cultures, *Neurotoxicology*, in press, 2003.

70. Kim, D.K., Rordorf, G., Nemenoff, R.A., Koroshetz, W.J., and Bonventre, J.V., Glutamate stably enhances the utility of the two cytosolic forms of phospholipase A_2 in brain cortical cultures, *Biochem. J.*, 310, 83–90, 1995.

71. Kitamura, Y., Shimohama, S., Ota, T., Matsuoka, Y., Nobura, Y., and Taniguchi, T., Alterations of transcription factor NF-kappa B and STAT 1 in Alzheimer's disease brains, *Neurosci. Lett.*, 237, 17–20, 1997.

72. Kehrer J.P. and Smith, C.V., Free radicals in biology: sources, reactivities, and roles in the etiology of human diseases, in *Natural Antioxidants in Human Health and Disease*, Frei, B., Ed., Academic Press, San Diego, 1994, pp. 25–62.

73. Halliwell, B., Free radicals and antioxidants: a personal view, *Nutr. Rev.*, 52, 253–265, 1994.

74. Betteridge, D.J., What is oxidative stress?, *Metabolism*, 49, 3–8, 2000.

75. LeBel, C.P., Ishiropoulos, H., and Bondy, S.C., Evaluation of the probe 2′, 7′-dichlorofluorescin as an indicator of reactive oxygen species formation and oxidative stress, *Chem. Res. Toxicol.*, 5, 227–231, 1992.

76. Yee, S. and Choi, B.H., Oxidative stress in neurotoxic effects of methylmercury poisoning, *Neurotoxicology*, 17, 17–26, 1996.

77. Almazan, G., Liu, H.-N., Khorchild, A., Sunderrajan, S., Martinez-Bermudez, A.K., and Chemtoz, S., Exposure of developing oligodendrocytes to cadmium causes HSP 72 induction, free radical generation, reduction in glutathione levels, and cell death, *Free Rad. Biol. Med.*, 29, 858–869, 2000.

78. Bush, A.I., Metals and neuroscience, *Curr. Opin. Chem. Biol.*, 4, 184–191, 2000.

79. Sayre, L.M., Perry, G., and Smith, M.A., Redox metals and neurodegenerative disease, *Curr. Opin. Chem. Biol.*, 3, 220–225, 1999.

80. Davis, K.L., Martin, E., Turko, I.V., and Murad, F., Novel effects of nitric oxide, *Annu. Rev. Pharmacol. Toxicol.*, 41, 203–236, 2001.

81. Sarafian, T.A., Vartavarian, L., Kane, D.J., Bredeson, D.E., and Verity, M.A., Bcl-2 expression decreases methylmercury-induced free radical generation and cell killing in a neural cell line, *Toxicol. Lett.*, 74, 149–155, 1994.

82. Sarafian, T.A., Methylmercury-induced generation of free radicals: Biological implications, *Metal Ions Biol. Syst.*, 36, 415–444, 1999.

83. Gasso, S., Cristofol, R.M., Selema, G., Rosa, R., Rodriguez-Farre, E., and Sanfeliu, C., Antioxidant compounds and Ca^{2+} pathway blockers differentially protect against methylmercury and mercuric chloride neurotoxicity, *J. Neurosci. Res.*, 66, 133–145, 2001.

84. Myhre, O. and Fonnum, F., The effect of aliphatic, naphthenic, and aromatic hydrocarbons on production of reactive oxygen species and reactive nitrogen species in rat brain synaptosome fraction: the involvement of calcium, nitric oxide synthase, mitochondria, and phospholipase A_2, *Biochem. Pharmacol.*, 62, 119–128, 2001.

85. Dugan, L.L., Sensi, S.L., Canzoniero, L.M., Haudran, S.D., Rothman, S.M., Lin, T.S., Goldberg, M.P., and Choi, D.W., Mitochondrial production of reactive oxygen species in cortical neurons following exposure to N-methyl-D-aspartate, *J. Neurosci.*, 151, 6377–6388, 1995.

86. Brawer, J.R., McCarthy, G.F., Gornitsky, M., Frankel, D., Mehindale, K., and Schipper, H.M., Mercuric chloride induces a stress response in cultured astrocytes characterized by mitochondrial uptake of iron, *Neurotoxicology*, 19, 767–776, 1998.

87. Crawford, D.R., Suzuki, T., and Davies, K.J.A., Redox regulation of gene expression, in *Antioxidants and Redox Regulation of Genes*, Sen, C.K., Sies, H., and Baeuerle, P.A., Eds., Academic Press, San Diego, CA, 2000, pp. 21–45.

88. Christman, J.W., Blackwell, T.S., and Jurlink, B.H., Redox regulation of nuclear factor kappa B: therapeutic potential for attenuating inflammatory response, *Brain Pathol.*, 10, 153–162, 2000.

89. Humphries, K.M. and Szweda, L.I., Selective inactivation of a-ketoglutarate dehydrogenase and pyruvate dehydrogenase: reaction of lipoic acid with 4-hydroxy–nonenal, *Biochemistry*, 37, 15835–15841, 1998.

90. Sensi, S.L., Yin, H.Z., and Weiss, J.H., Glutamate triggers preferential Zn^{2+} flux through Ca^{2+} permeable AMPA channels and consequent ROS production, *NeuroReport*, 10, 1723–1727, 1999.

91. Gotz, M.E., Koutsilieri, E., Riederer, P., Ceccatelli, S., and Dare, E., Methylmercury induces neurite degeneration in primary culture of dopaminergic mesencephalic cells, *J. Neural Transm.*, 109, 597–605, 2002.

92. Belletti, S., Orlandini, G., Vettori, M.V., Mutti, A., Uggeri, J., Scandroglio, R., Alinovi, R., and Gatti, R., Time course assessment of methylmercury effects on C6 glioma cells: submicromolar concentrations induce oxidative DNA damage and apoptosis, *J. Neurosci. Res.*, 70, 703–711, 2002.

93. Liu, S.X., Athar, M., Lippai, I., Waldren, C., and Hei, T.K., Induction of oxyradicals by arsenic: implication for mechanism of genotoxicity, *Proc. Natl. Acad. Sci. U.S.A.*, 98, 1643–1648, 2001.

94. Takeuchi, T., Nakajima, M., and Morimoto, K., Relationship between the intracellular reactive oxygen species and the induction of oxidative DNA damage in human neutrophil-like cells, *Carcinogenesis*, 17, 1543–1548, 1996.

95. Bagchi, M., Kuszynski, C., Patterson, E.B., Tang, L., Bagchi, D., Saani, T., and Stohs, S.J., *In vitro* free radical production in human oral keratinocytes induced by smokeless tobacco extract, *in vitro Toxicology*, 10, 263–274, 1997.

96. Krieg, T., Landsberger, M., Alexeyev, M.F., Felix, S.B., Cohen, M.V., and Downey, J., Activation of Akt is essential for acetylcholine to trigger generation of oxygen free radicals, *Cardiovasc. Res.*, 58, 196–202, 2003.

97. Anantharam, V., Kitazawa, M., Wagner, J., Kaul, S., and Kanthasamy, A.G., Caspase-3 dependent proteolytic cleavage of protein kinase Cδ is essential for oxidative stress-mediated dopaminergic cell death after exposure to methylcyclopentadienyl manganese tricarbonyl, *J. Neurosci.*, 22, 1738–1751, 2002.

98. Bolanos, J.P., Heales, S.J., Land, J.M., and Clark, J.B., Effect of peroxynitrite on the mitochondrial respiratory chain: differential susceptibility of neurons and astrocytes in primary culture, *J. Neurochem.*, 64, 1965–1972, 1995.

99. Bolanos, J.P., Heales, S.J., Peuchen, S., Barker, J.E., Land, J.M., and Clark, J.B., Nitric-oxide mediated mitochondrial damage: a potential neuroprotective role for glutathione, *Free Rad. Biol. Med.*, 21, 995–1001, 1996.

100. Barker, J.E., Heales, S.J., Cassidy, A., Bolanos, J.P., Land, J.M., and Clark, J.B., Depletion of brain glutathione results in a decrease of glutathione reductase activity; an enzyme susceptible to oxidative damage, *Brain Res.*, 716, 118–122, 1996.

101. Stewart, V.C., Sharpe, M.A., Clark, J.B., and Heales, S.J., Astrocyte-derived nitric oxide causes both reversible and irreversible damage to the neuronal mitochondrial respiratory chain, *J. Neurochem.*, 75, 694–700, 2000.

102. Heales, S.J. and Bolanos, J.P., Impairment of brain mitochondrial function by reactive nitrogen species: the role of glutathione in dictating susceptibility, *Neurochem. Int.*, 40, 469–474, 2002.

103. Almeida, A., Heales, S.J., Bolanos, J.P., and Medina, J.M., Glutamate neurotoxicity is associated with nitric oxide-mediated mitochondrial dysfunction and glutathione depletion, *Brain Res.*, 790, 209–216, 1998.

104. Stewart, V.C., Heslegrave, A.J., Brown, G.C., Clark, J.B., and Heales, S.J., Nitric oxide-dependent damage to neuronal mitochondria involves the NMDA receptor, *Eur. J. Neurosci.*, 15, 458–464, 2002.

105. Aschner, M., Eberle, N., Miller, K., and Kimelberg, H.K., Interaction of methylmercury with rat primary astrocyte cultures: effects on rubidium uptake and efflux and induction of swelling, *Brain Res.*, 530, 245–250, 1990.

106. Yamashita, T., Ando, Y., Sakashita, N., Hirayama, K., Tanaka, Y., Tashima, K., Uchino, M., and Ando, M., Role of nitric oxide in the cerebellar degeneration during methylmercury intoxication, *Biochim. Biophys. Acta*, 1334, 303–311, 1997.

107. Aschner, M., Astrocytic swelling, phospholipase A_2, glutathione and glutamate: interactions in methylmercury induced neurotoxicity, *Cell. Mol. Biol.*, 46, 843–854, 2000.

CHAPTER 25

Manganese Dynamics, Distribution, and Neurotoxicity

Vanessa A. Fitsanakis, Stephanie J. Garcia, and Michael Aschner

CONTENTS

25.1 INTRODUCTION

Manganese (Mn) is the tenth most abundant element on Earth and comprises 0.1% of the Earth's crust, making it ubiquitous in soil, air, water, and food. Mn exists physiologically as Mn(II), Mn(III), and Mn(IV) in both animals and humans (Takeda, 2003). Cotzias summarized early work recognizing the nutritional importance of Mn for central nervous system (CNS) function (Cotzias, 1958). As an essential trace element in mammals, it is necessary for normal development and life functions, and both Mn deficiency and excess can lead to health problems. Mn is a component or cofactor for several enzymes involved in the metabolism of fats and proteins, as well as antioxidants such as superoxide dismutase (SOD) and glutamine synthetase (GS). It is involved in immune function, regulation of blood sugars, production of cellular energy, reproduction, digestion, bone growth, carbohydrate metabolism, and blood clotting (Aschner, 2000). Recent evidence implicates Mn in the stellate process formation of astrocytes (Liao and Chen, 2001), although the relevance to normal *in vivo* development of astrocytes has yet to be determined. Although Mn deficiency is uncommon, it may result in birth defects, poor bone formation, impaired fertility, and an increased susceptibility to seizures (Aschner, 2000; Aschner et al., 2002). On the other hand, exposure to high levels of respirable Mn (< 5 $\mu g/m^3$) in adults results in symptoms similar to Parkinson's disease and is characterized initially by psychological disturbances such as manic or compulsive behaviors, weakness, or apathy. Latter symptoms include extrapyramidal motor system symptoms such as tremors, ataxia, bradykinesia, and dystonia (Gerber et al., 2002).

Mn overexposure in children, and consequent neurobehavioral indices, are not well established (Aschner, 2000); however, one study shows possible associations between high Mn in drinking water and cognitive deficits in children of an aboriginal population of Australia (Cawte, 1985). Other studies have evaluated neurological risks in children associated with increased Mn concentrations in both water and food. Children (11–13 years old) who drank water for 3 years or more at average Mn concentration of 0.241 ± 0.051 mg/l performed more poorly on neurobehavioral tests compared with children who drank water with an average Mn concentration that did not exceed 0.040 ± 0.012 mg/l (He et al., 1994). The children with increased Mn exposure also performed more poorly in school, and their blood chemistry (neurotransmitter levels) was altered (Zhang et al., 1995). The He and Zhang studies are poorly controlled for confounding variables, such as potential exposure to other metals, duration and amount of Mn uptake from flour fertilized with sewage, and health and nutritional status. Thus, they fail to establish a heightened neurological risk for chronic consumption of Mn (at 0.241 ± 0.051 mg/l) in drinking water.

Mn is consumed in the diet, especially from fruits, vegetables, grains, nuts, tea, and several spices (Roth and Garrick, 2003). Mn-containing compounds include inorganic salts, such as $MnCl_2$, $MnSO_4$, MnO_2, and Mn_3O_4, and organic compounds, such as the fungicide Maneb and methylcyclopentadienyl Mn tricarbonyl (MMT), which has replaced lead as an antiknock agent in gasoline in Canada and parts of the United States (Aschner, 2000). The best-characterized examples of Mn overexposure are from occupational settings involving mining, welding, and steel factory work. Moreover, there is an increasing awareness of potential Mn toxicity in patients receiving total parenteral nutrition (TPN) (Spencer, 1999). Whereas only 1–5% of ingested Mn is absorbed through the gastrointestinal tract (Davis et al., 1993), intravenous TPN bypasses intestinal absorption and Mn retention is nearly 100%; moreover, excretion is minimal. It is noteworthy to point out that patients with liver damage and gastrointestinal disorders and preterm infants most commonly receive TPN, patients already at risk for Mn accumulation due to inefficient elimination.

The focus of this chapter is to review the role of Mn in the CNS, particularly with emphasis on glia, in relation to brain physiology and function. Specifically, this chapter will discuss the following: the transport of Mn into and out of the CNS; transport dynamics and distribution in neurons and glia; Mn interactions with specific enzymes; and possible mechanisms for Mn toxicity, including Mn involvement in neurochemical components and oxidative stress. Finally, this chapter will shed some light on a burgeoning area of Mn research involving magnetic resonance (MR) technology.

25.2 CURRENT QUESTIONS

25.2.1 At and Across the Blood–Brain Barrier

The blood–brain barrier (BBB) consists of tightly joined capillary endothelial cells surrounded by astrocytic foot processes. Astrocytes enhance the activity and protein content of endothelial cell MnSOD, a prerequisite for normal BBB function (Schroeter et al., 1999). In this way, astrocytes are important in the integrity and function of the BBB, which protects the brain from foreign or toxic substances. Because the rate and extent of Mn transport across the BBB modulates its neurotoxicity (Aschner, 2000), to understand this phenomenon, it is important to discuss how Mn gets into and out of the brain.

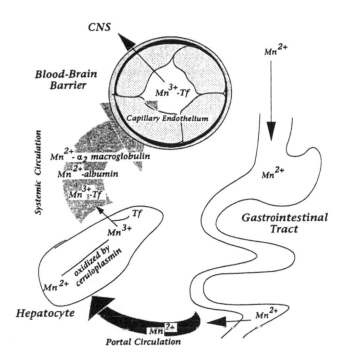

Figure 25.1 Proposed metabolism and transport of Mn in mammals; Tf = transferrin. (From Aschner, M. and Aschner, J.L., *Neurosci. Biobehav. Rev.*, 15, 333, 1991. With permission.)

The overall process from ingestion to intrabrain availability is summarized in Figure 25.1 (Aschner and Aschner, 1991). Dietary Mn(II) is absorbed from the gut, possibly by divalent metal transporter1 (DMT1), and transported to the liver (Malecki et al., 1999). Hepatocytes convert ingested Mn(II) to Mn(III), via ceruloplasmin, which then binds to transferrin. While most Mn is excreted into the bile from hepatocytes, some moves through the systemic circulation as Mn(III)-transferrin, Mn(II)-albumin, and Mn(II)-α_2-macroglobulin. Mn(III)-transferrin is involved in Mn transport across the BBB via receptor-mediated endocytosis. Once in the brain, astrocytes also convert Mn(II) to Mn(III) via ceruloplasmin (Malecki et al., 1999). Transferrin receptor is located on various cell types including luminal and abluminal membranes of CNS capillary endothelial cells (Aschner and Aschner, 1991), although transferrin itself is mainly secreted only from oligodendrocytes (Connor et al., 1990; Connor, 1994). Under normal plasma levels, Mn enters the CNS across the capillary endothelium, whereas during elevated plasma Mn conditions, Mn enters across the choroid plexus. Using high resolution radiography, Takeda found that ^{54}Mn was initially concentrated in the choroid plexus and then gradually taken up into the brain from the CSF; levels were especially high in the dentate gyrus and hippocampus and lower in the cerebral cortex (Takeda et al., 1994).

Evidence for transferrin-mediated Mn transport across the BBB comes from studies using fast protein liquid chromatography (Davidsson et al., 1989) and gel exclusion chromatography (Aschner and Aschner, 1990) following exposure of Mn to rats. Mn(III)-transferrin uptake by receptor-mediated endocytosis also occurs in cultured neuroblastoma cells (Suarez and Eriksson, 1993). Further, in cultured brain endothelial cells, competition studies with ^{59}Fe-transferrin showed that excess Mn attenuated iron (Fe) uptake and vice versa (Aschner et al., 2002). These results, coupled with the finding that Mn uptake across the BBB is inhibited by Fe(III)-dextran, indicate that Fe homeostasis plays a role in regulating Mn transport across the BBB (Aschner and Aschner, 1990).

Mn crosses the BBB via active transferrin and non-transferrin-mediated mechanisms (Aschner and Aschner, 1990; Aschner and Gannon, 1994). This is supported by studies demonstrating Mn accumulation in the CNS of mutant mice with only 1% of the normal transferrin levels. Cultured glial cells from hypotransferrinemic mice take up Mn via non-transferrin-mediated mechanisms (Takeda et al., 1998). Additionally, Mn(II) accumulation in rat brains exhibits saturation kinetics by a separate mechanism from Mn(III)-transferrin uptake (Aschner and Gannon, 1994). Further evidence for both active and passive transport mechanisms comes from studies measuring ^{54}Mn(II) uptake into brain and choroid plexus in which free Mn(II) ions are rapidly taken up and binding to blood proteins restricts Mn(II) influx (Rabin et al., 1993). The non-transferrin-mediated transport of Mn(III) across the BBB is unknown (Takeda, 2003); however, several transport systems have been reported for Mn(II) uptake, including the following: calcium (Ca) channels, Na/Ca exchanger, active Ca uniporter, and Na/Mg antiporter (Takeda, 2003). As previously mentioned, DMT1, the gene for a nonspecific transmembrane protein capable of transporting divalent Fe, Mn, Cd, Co, Ni, Cu, and Pb, has been identified in the brain (Gunshin et al., 1997).

DMT1 is expressed in several tissues throughout the body, including the brain, where it is present in both brain capillary endothelial cells and choroidal epithelial cells (Takeda, 2003). Evidence also suggests DMT1 is expressed *in vivo* in astrocytes and in primary astrocyte cultures (Burdo et al., 1999; Williams et al., 2000; Lis et al., 2004), though this is contradicted by Gunshin's reported lack of DMT1 in adult rat astrocytes (Gunshin et al., 1997). Mn(II) has a relatively high affinity for DMT1 and competes with Fe(II) for both binding and subsequent transport (Roth and Garrick, 2003). Mechanisms that regulate Fe status may influence uptake and toxicity of Mn (Roth and Garrick, 2003). During Fe deficiency, duodenal DMT1 expression is increased, enhancing intestinal Mn absorption (Mena et al., 1969; Oates et al., 2000; Moos et al., 2002). By comparison, brain Mn levels are elevated in Fe-deficient rats (Erikson et al., 2002). The function and properties of DMT1 have been characterized in the Belgrade rat and the microcytic mouse, both of which have mutations in DMT1 that render it inactive (Roth and Garrick, 2003). Both Mn and Fe uptake are decreased in the brains of the Belgrade rats, suggesting that DMT1 is critical for the transport of both metals (Roth and Garrick, 2003).

Recent work has begun to explore efflux kinetics of Mn. Using the *in situ* brain perfusion technique, brain influx and efflux of Mn(II) ion and Mn citrate were determined. Brain influx of both Mn species was greater than that attributable to diffusion, based on calculated estimates, suggesting carrier-mediated influx. On the other hand, brain Mn efflux was not more rapid than that predicted from diffusion, suggesting non-carrier-mediated efflux (Crossgrove et al., 2003; Yokel et al., 2003).

25.2.2 Mn Levels in the Brain

The concentration of Mn in brain tissue averages ~ 1–2 µg/g dry weight (Prohaska, 1987), but can range from 1 to 10 µg/g wet weight (Chan et al., 1992). Mn is distributed heterogeneously in both the human and rat brain, with the highest values reported for palladium and putamen in humans and hypothalamus in rats (Prohaska, 1987). Regional concentrations of brain Mn, therefore, vary across species. Using neutron activation analysis, brain regional Mn concentration averaged 0.4 µg/g wet weight (Erikson et al., 2002). Total Mn(II) in the cerebral cortex has been estimated at 2 µ*M* using electron spin resonance spectroscopy (Wedler et al., 1982), although in cases of Mn toxicity, the metal concentrates in the basal ganglia (Malecki et al., 1999). Intracellular concentrations of Mn and various other metals have been determined in cultured chick neurons, glia, and cell media via dispersive x-ray fluorescence and atomic absorption spectroscopy. Both neurons and glia accumulate metals; Mn accumulates up to 800 times more in neurons (380 µ*M*) and 150 times in glia (75 µ*M*) compared with the cell media (0.5 µ*M*). This may indicate that the intracellular levels of Mn are higher than assumed from measurements of Mn in whole tissue samples (Tholey et al., 1988; Tholey et al., 1988). Mn concentrations in the brain seem to plateau at two- to threefold of "normal" levels even in severe or prolonged low-level Mn exposure (Subhash and Padmashree, 1991; Dorman et al., 2001) and may accumulate up to fivefold following exposure throughout development and into adulthood (Lai et al., 1992). The average Mn concentration in normal whole blood ranges from 7 to 12 µg/l (Barceloux, 1999; Aschner, 2000).

The U.S. Environmental Protection Agency (EPA) has set the Mn inhalation reference concentration (RfC) at 0.05 μm Mn/m³ for both environmental and occupational exposures. The reference dose (RfD) for food intake is 10 mg/d (Barceloux, 1999; Gerber et al., 2002), while the average daily food intake of Mn for adult humans is between 2 and 10 mg (Greger, 1998; Aschner, 2000). In cases of occupational Mn exposure known to induce manganism, workers were exposed to 20–30 mg Mn/m³, but chronic exposure to air levels from 1 to 5 mg/m³ can lead to subclinical alterations (Barceloux, 1999).

It is noteworthy that the Food and Nutrition Board of the National Research Council (NRC) has not established a Recommended Daily Allowance for Mn because too little information is available regarding the dietary requirements for Mn. In a recent publication (2002), the NRC states the following: "There were insufficient data to set an Estimated Average Requirement (EAR) for Mn. An Adequate Intake (AI) was set based on median intakes reported from the Food and Drug Administration Total Diet Study. The AI for adult men and women is 2.3 and 1.8 mg/day, respectively. A Tolerable Upper Limit Level (UL) of 11 mg/day was set for adult based on a no-observed-adverse-effect-level of Western diets." The adequate intake for infants 0–6 and 7–12 months is set at 0.003 and 0.6 mg/day, respectively. The UL (defined as the maximum level of daily nutrient intake that is likely to pose no risk of adverse effects) for the same age groups was not determined (Council, 2002).

25.2.3 Dynamics and Distribution in Glia

In order to understand the effects of Mn on various cell populations, it is necessary to discuss both cellular influx and efflux mechanisms, as well as intracellular amounts of the bound and free metal. This will lay the groundwork concerning the interaction of Mn with various proteins.

Although several laboratories have used both primary and immortalized cell cultures to determine the kinetics of Mn influx, these experiments have been conducted exclusively in cultured astrocytes. As it is generally accepted that astrocytes do not to have transferrin receptors, there is some controversy as to how the Mn is taken up by these cells. That astrocytes can transport Mn intracellularly, however, has been demonstrated by several labs. Studies with astrocytes isolated from rat (Aschner et al., 1992) and chick (Wedler et al., 1989) determined that Mn uptake was not only saturable, but competitively and noncompetitively inhibited by Ca(II). Studies by Wedler et al. demonstrate that Mn uptake follows first-order kinetics, with saturation at $K_m = 18$ μM and $V_{max} = 100$ nmol Mn/min/mg protein (Wedler et al., 1989). Although the kinetic parameters for rat astrocytes are considerably different, with a $K_m = 0.3$ μM and $V_{max} = 0.30$ nmol Mn/min/mg protein (Aschner et al., 1992), the reasons for this are not entirely clear. Perhaps differences in transport kinetics exist across species, although this has not been pursued further in the literature.

Little is known about the uptake kinetics in microglia, and even less is known concerning oligodendrocytes. Takeda et al. compared Mn uptake among different glial populations (Takeda et al., 1998), although no kinetics were reported. It is interesting to note that this group observed a greater Mn uptake in microglia compared with astrocytes, the traditional Mn "sinks," or oligodendrocytes, cells normally associated with Fe uptake. Some labs have examined the ability of oligodendrocytes to both take up and efflux Mn (Golub et al., 1996; Takeda et al., 1998), but again no kinetic parameters were reported. Thus, the focus of glial studies remains on the role of astrocytes in Mn neurotoxicity.

The uptake of Mn by astrocytes appears to be rather stable over time; however, Mn efflux seems to be a biphasic process (Figure 25.2). Efflux experiments involving chick astrocytes indicated that about 30–50% of intracellular ⁵⁴Mn is released within 20–30 minutes, while the remaining metal is released more slowly, over a period of hours (Wedler et al., 1989). The efflux is not a static process, though. Experiments demonstrate that the amount of ⁵⁴Mn released in the initial rapid phase can be increased to almost 100% (from the 30–50% previously mentioned) if extracellular Mn is increased to concentrations approaching 12.8 μM. A similar efflux phenomenon is observed in rat astrocytes as well (Aschner et al., 1992), where the rate, but not the extent of Mn release, which appears to plateau at 70%, increases. The reason for the difference between chick and rat

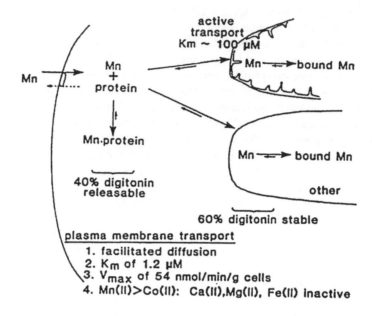

Figure 25.2 Mn(II) economy of brain glial cells, by analogy to rat hepatocytes. (From Wedler, F.C., Ley, B.W., and Grippo, A.A., *Neurochem. Res.*, 14, 1129, 1989. With permission.)

astrocytes is again unknown. It is hypothesized that a pool of tightly bound Mn must exist in rat, but not chick, glia. More work is needed to determine whether differences in Mn kinetics in astrocytes derived from other species are observed, as this could provide more consistency among labs for Mn toxicity studies in astrocytes.

In order to determine whether Mn efflux is an energy-dependent process, others have addressed this question in the presence or absence of ATP synthesis inhibitors. Results suggested that Mn release was not due to facilitated diffusion (relying on ATP) but likely involved a carrier/transporter mechanism across the astrocytic membrane that was ATP independent (Wedler et al., 1989). This conclusion is supported by data from rat hepatocytes (Schramm and Brandt, 1986) and isolated brain mitochondria (Konji et al., 1985), suggesting that Mn transport actually involves the Ca(II) uniporter system (Gavin et al., 1990). Co-use of this uniporter by Mn fits with observations that a significant amount of Mn entering the cell is eventually stored in the mitochondria.

Indeed, the propensity of Mn to concentrate in the mitochondria was reported as early as 1955 (Maynard and Cotzias, 1955) and has been confirmed since (Schramm and Brandt, 1986; Liccione and Maines, 1988; Brouillet et al., 1993). The data indicate that approximately 50–70% of cellular Mn is indeed associated in a bound state with the mitochondria. The remaining 30–50% of Mn(II) is found in the cytosol in a "free" form, with concentrations approaching 1 μM. This amount accounts for about 5% of the total 20 μM Mn in the cytosol, the majority of which presumably binds to specific cytosolic proteins. Perhaps it is this cytosolic fraction of Mn that is responsible for the rapidly effluxing Mn observed in kinetic experiments, while the mitochondrial fraction represents the slowly effluxing, or tightly bound, Mn.

Other physiologically relevant metal ions, such as Fe, Cu, Zn, and Al, have been studied in the context of their effect on Mn uptake and efflux (Malhotra et al., 1984; Takeda et al., 1994; Takeda et al., 1998; Roth and Garrick, 2003; Zatta et al., 2003). In general, these metals appear to alter uptake but not efflux dynamics of Mn. Furthermore, these metals influence the transport of Mn without direct competition for Mn(II) sites. This is in contrast to interactions observed between Ca and Mn.

Initial studies examining the dynamics of Mn(II) transport indicated that Ca(II) ions were an integral part of uptake and efflux in both rat and chick astrocytes (Wedler et al., 1989; Aschner et al., 1992). Indeed, Ca(II) appeared to be a competitive inhibitor of Mn(II) uptake, affecting initial velocity and attenuating net Mn(II) uptake. It is thought that the effect on velocity is due to direct competition for transport binding sites at the level of both facilitated diffusion (at the plasma membrane), and active transport (Ca uniporter) of the mitochondrial membrane. The effect of Ca(II) on the extent of Mn(II) uptake appears to involve competition for binding sites within the mitochondria, since Mn(II) ions alter Ca(II) efflux (Gavin et al., 1990). A more detailed review of the interactions of Ca(II) and Mn(II) was written by Gavin et al. (1999). In general, it appears that the presence of both internal and external Ca(II) inhibits the initial velocity and extent of ^{54}Mn(II) uptake. The presence of external Ca(II) increases the rate and extent of ^{54}Mn(II) efflux, while internal or equilibrated Ca(II) inhibits Mn(II) efflux.

25.2.4 Manganese Interactions with Glutamine Synthetase

Schramm discussed several useful criteria for determining whether an enzyme is actually regulated by Mn (Schramm, 1982). These include the following:

1. The total amount of Mn(II) present in the cell or tissue is equal to or greater than the net amount of target enzyme.
2. The concentration of free Mn(II) is sufficient to bind a significant fraction of the target enzyme.
3. The affinity of the target enzyme for Mn(II) is high compared to both the amount of free Mn(II) present and the enzyme's affinity for Mg(II) ions, which are known to compete with Mn(II) for binding sites.
4. The magnitude of changes in concentrations of Mn(II) and enzyme-Mn(II) complexes, due to intracellular signals, changes in enzyme affinity, or concentration of binding sites, are sufficient to account for changes in enzyme activity needed for regulation.

Many enzymes are known to depend on Mn for function and regulation. This includes, but is not limited to, the following: glutamine synthetase, hexokinase, mitochondrial superoxide dismutase, pyruvate carboxylase, phosphoenolpyruvate (PEP) carboxykinase, ATPase, protein (Tyr, Ser) kinases, DNA and RNA polymerases, protein phosphatases, endonuclease, nucleotide (ATP, GTP) cyclase, and ligases. Although many of these enzymes are ubiquitous, glutamine synthetase in CNS is specific to astrocytes and will be discussed further.

Glutamine synthetase (GS) is a key enzyme in the glutamate–glutamine–γ-aminobutyric acid cycle, a fundamental pathway involving both neurons and astrocytes. This cycle has been extensively reviewed by others (Wedler and Denman, 1984; Stadtman, 1990; Weber et al., 2002) and is shown in schematic in Figure 25.3. It is estimated that between 60 and 70% of the receptors in the CNS detect either glutamate (Glu) or γ-aminobutyric acid (GABA), indicating the importance of GS not only in recycling Glu, but also in providing a necessary pool of glutamine (Gln) for subsequent production of Glu and GABA. As Glu is the major excitatory neurotransmitter, and GABA is the major inhibitory neurotransmitter in the brain, it is of critical importance that these two neurochemicals be tightly regulated.

The Glu-Gln-GABA cycle is initiated when glutamatergic neurons release L-Glu into the synaptic cleft, resulting in excitation of nearby neurons and a lowering membrane potential. If Glu is not quickly removed from the extracellular space, it results in continued stimulation of local neurons (termed *excitotoxicity*), rapidly depleting the cell's storehouse of ATP. As astrocytes have high-affinity transporters for L-Glu, they quickly remove it from the synaptic cleft, concentrating it in their cytoplasm where GS is located. Astrocytes also have a high affinity for ammonia, another potentially neurotoxic substance, and can concentrate it in their cytoplasm to metabolize Gln from Glu. Gln, which does not stimulate neurons, then diffuses across the astrocytic cellular membrane where it is taken up by both glutamatergic and GABAergic neurons to synthesize their respective neurotransmitters.

Early experiments involving Mn(II) and Mg(II) indicated that micromolar quantities of Mn(II) could upregulate GS activity even in the presence of millimolar concentrations of Mg(II) (Wedler

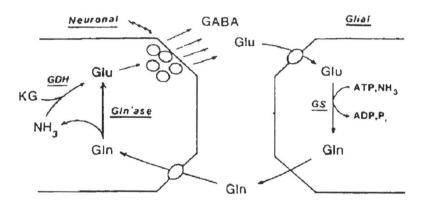

Figure 25.3 Glutamate-glutamine cycle for neurons and glia. (From Wedler, F.C. and Toms, R., *Manganese in Metabolism and Enzyme Function*, Schramm, V.L. and Wedler, F.C., Eds., Academic Press, Orlando, FL, 1986, chap. 13. With permission.)

et al., 1982). This suggested that Mn(II) was involved in the regulation of GS. Although the initial data was later determined to be an artifact of the isolation procedure (Wedler and Tom, 1986), subsequent experiments indicated that GS had both a dissociation constant for Mn in the micromolar range (Schramm, 1982; Wedler and Tom, 1986) and multiple metal binding sites (Maurizi et al., 1987). This latter data was confirmed in *Escherichia coli*, where the high-affinity metal site partly stabilizes interactions of the GS subunits, allowing for L-Glu binding, while lower-affinity binding sites appears to be involved in binding ATP (Ginsburg, 1972; Villafrance and Nowak, 1992). Although the above suggest that Mn is involved at the active site, it is possible that Mn acts at a remote site in a strictly regulatory capacity.

Based on reports that Mn is bound to GS, it is estimated that the total concentration of Mn in the cytoplasm of glia is about 20–30 μM. ESR measurements suggest that 5% of this Mn exists in a "free form" (Wedler and Ley, 1994), while the total concentration of Mn-containing sites on GS (calculated at eight subunits per enzyme) is close to 0.52 μM (Wedler et al., 1982). This latter concentration is, in reality, probably significantly higher if the cytoplasmic volume of astrocytes is taken into account. This results in GS being a substantial source of bound, cytoplasmic Mn. Other experiments have determined that approximately 30% of GS subunits contained bound Mn(II) (Wedler and Ley, 1994), leading to the conclusion that not only is the amount of Mn bound to GS within reasonable limits, but a significant fraction of GS is indeed associated with Mn *in vivo*.

As GS is the principal glial cytosolic enzyme activated by Mn, further research is needed to characterize the effects of neurotoxic levels of Mn on the function of GS, since conditions inhibiting the ability of GS to detoxify ammonia or remove L-Glu from the synaptic cleft could leave vulnerable or dead neuronal populations.

25.2.5 Neurochemical Components

Although Mn deficiency or excess can disrupt a slew of physiological parameters, neurochemical consequences of Mn imbalance are the best characterized. Several reviews have discussed the involvement of Mn in biochemical processes in mammalian cells (Schramm and Wedler, 1986; Prohaska, 1987; Aschner and Aschner, 1991; Wedler, 1993; Wedler, 1994; Wedler, 1994; Aschner et al., 1999; Aschner, 2000; Hazell, 2002; Erikson and Aschner, 2003; Roth and Garrick, 2003; Takeda, 2003).

Mn alters neurotransmitters, most notably, dopamine (Tran et al., 2002a, 2002b), Glu (Takeda et al., 2002), and GABA (Lai et al., 1984; Lipe et al., 1999; Erikson et al., 2002). Although Mn-related neurotransmitter effects are primarily associated with neurons, some may relate to glial function. These include neurotransmitter release (Drapeau and Nachshen, 1984), monoamine metabolism

(Kontur and Fechter, 1985; Subhash and Padmashree, 1991), and uptake of the catecholamines dopamine and norepinephrine (Lai et al., 1982; Chandra et al., 1984; Lai et al., 1984; Hussain et al., 1987).

The response of brain receptors to elevated Mn are age-, dose-, and time-dependent. Normal, physiological levels of Mn modulate receptor function, generally by altering the specificity of ligand binding at the receptor binding site. Additionally, Mn is involved in catalysis of adenylate cyclase and guanylate cyclase, as well as specific protein kinases and phosphatases that regulate activity of the receptor or intracellular enzymes. Mn also stabilizes agonist- and antagonist-specific receptor forms and subunit interactions. Mn interacts with several receptor systems, both ion-gated channels and G-protein coupled receptors, including the following: dopaminergic (Seth and Chandra, 1984; Butterworth et al., 1995; Husain et al., 2001; Kessler et al., 2003), glutaminergic (Cano et al., 1997; Hazell and Norenberg, 1997; Erikson and Aschner, 2002), GABAergic (Hazell et al., 1999; Hazell et al., 2003), and to a lesser extent, serotonergic (Battaglia et al., 1984; Seth and Chandra, 1984), cholinergic (Donaldson and LaBella, 1984; Seth and Chandra, 1984), and opioid (Jauzac et al., 1984; Zajac and Roques, 1985; Szucs et al., 1987) receptors.

25.2.5.1 Dopamine

Because Mn neurotoxicity produces extrapyramidal motor symptoms similar to Parkinson's disease, research efforts have focused on the involvement of dopamine. Most studies in animals and humans concur that Mn exposure leads to decreased dopamine levels, as well as decreases in receptor or transporter (DAT) expression concomitant with elevated Mn levels in the brain. For example, rats treated with $MnCl_2$ had increased L-tyrosine hydroxylase activity that persisted for up to 6 months in the striatum, declining by 8 months. This lends credence to an association between the initial psychiatric phase of manganism, speculated to coincide with overactivity of dopamine neurons, and the later neurological phase, associated with reduced striatal dopamine (Bonilla, 1980). Correspondingly, decreased striatal dopamine concentration was found in neonatal rats given supplemental Mn during the lactation period. This decrease was linked to neurocognitive deficits (Tran et al., 2002a, 2002b).

Although there is general agreement that Mn leads to decreased dopamine receptor binding, this has not been proven; the response seems to vary with age, exposure level, and duration, and it is possibly species-dependent. The autopsy of patients on long-term total parenteral nutrition (TPN) who also exhibited extrapyramidal motor deficits during life revealed two- to sevenfold increases in Mn in the globus pallidus and a concomitant decrease in dopamine D_2 binding sites (Butterworth et al., 1995). Likewise, a case study of a patient in the latter stages of chronic manganism found decreased D_2 receptor density using ^{18}F-methylspiperone positron emission tomography (PET) (Kessler et al., 2003). On the other hand, rats exposed to Mn exhibited an increase in striatal dopamine receptors (Seth and Chandra, 1984; Husain et al., 2001). Receptor response is age-dependent, as the same Mn exposure level in neonatal rats resulted in decreased dopamine receptor binding sites (Seth and Chandra, 1984).

A recent study investigated the function of presynaptic dopaminergic terminals by using a cocaine analog that binds to the DAT. Although there was a slight decrease in DAT in the putamen of patients with a history of Mn exposure and concurrent bradykinetic-rigid syndrome compared with normal controls, data from Huang et al. indicate that the presynaptic dopaminergic terminals are not the main target of chronic Mn intoxication (Huang et al., 2003).

25.2.5.2 Glutamate

A proximate mechanism for Mn neurotoxicity involves altered Glu metabolism, which can lead to excitotoxicity. $MnCl_2$-exposed primary cultured astrocytes take up less Glu, which may be due to attenuated Glu:aspartate transporter (GLAST) function (Hazell and Norenberg, 1997; Erikson and Aschner, 2002). If astrocytes can no longer efficiently soak up Glu, extracellular levels may increase, leading to excitotoxicity in neighboring neurons (Erikson and Aschner, 2003). As a potential protective mechanism, Glu receptor binding sites may be downregulated during $MnCl_2$ exposure in mice; alternatively, the decrease in Glu receptor binding sites may be indicative of neuronal loss associated with excitotoxicity (Cano et al., 1997).

Takada provides an elegant review regarding Mn action in synaptic neurotransmission (Takeda, 2003). Using *in vivo* microdialysis, the level of [54]Mn in the extracellular space of the hippocampus was increased alongside Glu, suggesting that Mn may be released from glutamatergic neuron terminals (Takeda et al., 2002). Mn also can act on voltage-dependent Ca channels and NMDA receptors, although the functional significance remains to be elucidated (Takeda, 2003). It is possible that Mn released from glutamatergic neurons suppresses neighboring neuronal activity via the inhibition of voltage-dependent Ca channels (Takeda et al., 2002). More about the relationship between Mn and excitotoxicity, as it relates to energy depletion and mitochondrial inhibition, is discussed in Section 25.3.

On the other hand, nutritional Mn deficiency, though not common, can contribute to epilepsy or susceptibility to seizures due to altered metabolism of L-Glu and the role of GS activity, potentially dependent on Mn (Dupont and Tanaka, 1985; Carl et al., 1993). The exact mechanism for this phenomenon is unknown, however.

25.2.5.3 GABA

Although work traditionally has focused on the involvement of dopamine, Mn-induced neurode-generation primarily involves GABAergic neurons within the globus pallidus, rather than dopam-inergic neurons within the nigra-striatal pathway as in Parkinson's disease (Olanow et al., 1996; Pal, 1999; Verity, 1999; Roth and Garrick, 2003). Mn exposures in rodent models produce incon-sistent alterations in GABA; however, a general pattern may be that lower (6 mg/kg/d) and short-term exposures result in decreased levels of GABA (Lai et al., 1984), whereas higher (20 mg/kg/d) and long-term exposures result in elevated GABA levels (Shukla and Singhal, 1984; Lipe et al., 1999; Erikson and Aschner, 2003). Accordingly, a relationship may exist between the severity of Mn exposure and GABA levels. In a pre-Parkinsonism rat model generated by 6-hydroxydopamine (6-OHDA)–induced dopamine depletion, Mn treatment produced deficits in motor function con-current with increases in striatal GABA (~ 20%) but did not affect dopamine. This suggests that changes in GABAergic and glutamatergic neurotransmission in the globus pallidus likely precede those in striatal dopamine.

Though not formally a GABA receptor, peripheral-type benzodiazepine receptor (PTBR) binding sites were upregulated following Mn treatment of cultured astrocytes (Hazell et al., 1999; Hazell et al., 2003). PTBRs are localized primarily to the mitochondria of astrocytes and are involved in oxidative metabolism, mitochondrial proliferation, and neurosteroid synthesis (Hazell, 2002). Because certain neurosteroids modulate $GABA_A$, NMDA/Glu, glycine, and AMPA/kainate receptors, astrocytic PTBRs may alter neuronal excitability indirectly via neurosteroid synthesis (Hazell, 2002).

In all, Mn affects motor function, and it appears that the GABAergic system may be more sensitive than the dopaminergic system to increased Mn levels, especially in the striatum. Addi-tionally, aberrant Glu metabolism may be a mechanism for enhanced excitotoxicity. Further studies are necessary to clarify the mechanisms involved.

25.3 MANGANESE, MITOCHONDRIAL INHIBITION, AND REACTIVE OXYGEN SPECIES PRODUCTION

Mitochondrial function depends on the reduction of various substrates (Glu and malate for complex I, succinate for complex II, ascorbate for complex IV, for example) in the presence of specific inner mitochondrial membrane enzymes. The translocation of protons from the matrix to the intermem-branous space sets up a gradient that facilitates ATP synthesis from ADP and P_i.

As mentioned previously, intracellular Mn accumulates to a large degree in the mitochondrial matrix (Maynard and Cotzias, 1955; Liccione and Maines, 1988), probably via Ca transporters (Chance, 1965; Gavin et al., 1990; Gavin et al., 1999). Several labs have estimated the concentration of Mn in the mitochondrial matrix (Puskin and Gunter, 1972; Puskin and Gunter, 1973), but these range from 10 nM to 10 mM, depending on how much Mn was in the original system (Brown and Taylor, 1999). Under normal conditions, it is thought that the Mn concentration in the intramitochondrial space is around 50 nM, sufficiently small to prevent interference with normal mitochondrial functioning (Brown and Taylor, 1999).

Since Mn accumulates in the mitochondria, it is important to examine whether Mn can inhibit mitochondrial respiration. Mitochondria consume about 90% of the cell's oxygen and, even under nonpathological conditions, represent a significant source of reactive oxygen species (ROS) production (Turrens and Boveris, 1980; Liu et al., 2002). Should the electron transport chain be inhibited, the production of ROS increases significantly. Many labs have examined whether Mn can inhibit mitochondrial respiration, providing a potential mechanism for Mn neurotoxicity. It is known that Mn can inhibit respiration at complex I (Gavin et al., 1992; Galvani et al., 1995), complex II (Husain et al., 1976; Gavin et al., 1992; Galvani et al., 1995; Malecki, 2001; Tomas-Camardiel et al., 2002), and perhaps even complex III and IV (Husain et al., 1976). Others have demonstrated that Mn-containing compounds, such as the gasoline additive MMT or the fungicide Maneb, also inhibit mitochondrial respiration (Autissier et al., 1977; Zhang et al., 2003). Since mitochondria are ubiquitous, not only are neurons susceptible to increases in oxidative stress, but glia also could experience increased ROS due to mitochondrial inhibition. Although astrocytes are a source of antioxidants, both neurons and glia alike may be compromised if ATP levels are decreased or if oxidative stress increases.

Mitochondrial inhibition generally is considered an indirect mechanism by which compounds induce oxidative stress. Data exist suggesting that Mn, or Mn-containing compounds, also can directly produce ROS. For example, both types of compounds can interact directly with catechols to produce both semiquinones, through one-electron oxidation (Fitsanakis et al., 2002), and quinones, through two-electron oxidation (Kalyanaraman et al., 1985; Archibald and Tyree, 1987; Shen and Dryhurst, 1998). Although many of these experiments were done with neuronal cultures, it is likely that Mn could catalyze catechol oxidation extracellularly in the synaptic cleft, leading to the exposure of both neurons and glia to increased concentrations of ROS. As oxygen is usually the recipient of electrons in the oxidation of catechols, this results in production not only of ROS but also of highly redox active quinone species (Graham, 1978; Schweigert et al., 2001).

Mn also can potentiate the formation of nitric oxide (NO), an ROS capable of reacting with other oxygen species to form peroxynitrite. Chang and Liu report that Mn potentiates formation of nitric oxide in activated, but not naïve, microglia without cytotoxic changes (Chang and Liu, 1999). Others have confirmed an increase in the production of ROS, or a modulation of antioxidant status, in astrocytes exposed to Mn (Spranger et al., 1998; Chen and Liao, 2002). Interestingly, the work by Chang and Liu demonstrate that only Mn, and not other metals such as Zn, Cu, Co, Fe, or Ni, induced large amounts of nitrite production in microglia cultures (Chang and Liu, 1999), suggesting that Mn may be novel in its ability to produce nitric oxide in microglia. Although studies have examined the propensity of Mn to increase ROS formation in neuronal cultures, more work needs to be done with various glial populations to verify that similar observations hold true in these cell types as well.

Inhibition of mitochondrial respiration has a dual consequence for the affected cells. On the one hand, there is an increase in oxidative stress; on the other, reduced ATP levels leave the cell vulnerable to excitotoxicity due to its inability to maintain a normal polarized state. In the latter scenario, slightly depolarized neurons become more susceptible to excitation from basal levels of Glu. This excitotoxicity has been documented in studies involving injection of Mn into rat striatum (Brouillet et al., 1993), although this finding has not been repeated. As discussed earlier, astrocytes are the major cell type responsible for the reuptake of Glu. If astrocytes were unable to clear this excitatory neurotransmitter from the synaptic cleft, due to reduction in their ATP levels, this could have the net result of inducing excitotoxicity in the surrounding neuronal populations. Indeed, reduced Glu clearance following Mn exposure has been documented (Hazell and Norenberg, 1997; Erikson and Aschner, 2002; Erikson et al., 2002), although the reasons for this are not entirely clear.

Although production of ROS has mainly been studied in the context of neurons, the data is highly suggestive that glial cells, especially microglia and astrocytes, contribute to the production of reactive nitrogen and oxygen species; however, the general effect of ROS on glial populations needs to be further examined. Additionally, more work needs to focus on what effects increased levels of excitatory amino acids, namely glutamate, may have on glial cell populations.

Oxidation of important cell components by Mn(III) has been suggested as a cause of the toxic effects of Mn, given its strong pro-oxidant activity. Thus, determination of the oxidation states of Mn could help to identify the dominant mechanism of Mn-induced CNS damage. Gunter et al. (2004) have recently utilized X-ray Absorbance Near Edge Structure (XANES) spectroscopy, characterizing the oxidation state of Mn in mitochondria isolated from brain (Gunter et al., 2004). Results from these studies showed that Mn existed primarily as a combination of Mn(II) complexes, and there was minimal evidence for Mn(III). This suggests that Mn complexes are most likely to form Mn(II)-ATP, and that if oxidative stress is indeed a dominant mechanism in mediating Mn neurotoxicity, most likely it is not related to the existence of Mn(III) in the cells. Similar studies are currently being carried out in primary astrocyte cultures.

25.4 MANGANESE AND MAGNETIC RESONANCE STUDIES

Although noninvasive procedures, such as x-ray, computed tomography (CT), magnetic resonance (MR), and positron emission technology (PET), have been available in clinical settings, only recently have these latter technologies become more widely utilized in the research community. Of these techniques, MR is uniquely capable of imaging Mn due to the metal's paramagnetic properties. Theoretically, then, increases in Mn deposition could be visualized *in vivo* as regions of hyperintensity on T_1-weighted images or as regions of hypointensity on T_2-weighted images. Recent MR work has also focused on establishing correlations of T_1-relaxation times, rather than image intensity, with blood-Mn concentrations (Kim et al., 1999; Gallez et al., 2001).

Although early studies with Mn and MR focused mainly on the metal's potential as a tissue-specific contrast agent (Kreft et al., 1993; Baba et al., 1994; Ni et al., 1997; Brurok et al., 1999; Weinmann et al., 2003), more recent studies have focused on the propensity of Mn to be released during neuronal activation (Lin and Koretsky, 1997; Morita et al., 2002; Natt et al., 2002; Pautler and Koretsky, 2002; Ryu et al., 2002; Saleem et al., 2002; Watanabe et al., 2002; Leergaard et al., 2003). Indeed, these studies have led to intricate detailing of neuronal pathways in various animal species. Clinically, MR has been used to document brain-Mn deposition in patients with chronic liver failure (Barron et al., 1994; Hernandez et al., 2002; Park et al., 2003), in those receiving total parenteral nutrition (Fell et al., 1996; Chaki et al., 1998; Nagatomo et al., 1999; Reimund et al., 2000), and in mining populations (Dietz et al., 2001; Hernandez et al., 2002). For now, the use of MR imaging to determine Mn deposition in toxicological models, especially trying to differentiate neuronal versus glial targets, is still underdeveloped. This is due to several factors: the inability of imaging to distinguish among various cell populations, the difficulty of correlating changes in image intensity with blood or brain Mn concentration, and the high concentration of Mn that must be present in order to visualize areas of hyperintensity. The development of better MR markers for various cell types, new methods of quantification, and more refined scanning protocols will make imaging more suited to answering questions associated with Mn neurotoxicity.

A relatively new development in the area of MR technology is MR spectroscopy (MRS). This technique allows for the measurement of both concentrations and synthetic rates of individual metabolites or neurochemicals within a narrowly defined region. In an excellent review concerning the potential of this technology, Novotny et al. (2003) describe how MRS can detect such relevant neurochemicals as GABA, Glu, Gln, glycine, histidine, phenylalanine, and taurine (Novotny et al., 2003). MRS appears to be uniquely poised to address the questions of whether GABA or Glu changes precede dopamine changes and to what extent Glu uptake is inhibited *in vivo*. For example, since MRS can determine metabolic rates related to Glu metabolism (Gruetter et al., 1994; Bluml et al., 2001; Novotny et al., 2003), which intimately involves astrocytes, it is possible that MRS could be used to determine the effects of Mn on astrocyte populations. This is an area of research ripe for further exploitation in the area of Mn neurotoxicity.

25.5 CONCLUSIONS AND FUTURE DIRECTIONS

While the evidence is clear that Mn neurotoxicity selectively targets the basal ganglia and induces neuronal cell loss, the progression of degeneration has yet to be clarified. Neurons may be affected first and compromised further when mitochondria in astrocytes are inhibited. Although neuronal degeneration clearly results from excess Mn in the brain, ignoring the impact of glial cells, particularly astrocytes, in the equation fails to consider the scope of Mn neurotoxicity. Indeed, this review begins to delve into the dynamic neuron–astrocyte interactions that can go awry following excess Mn exposure. Alterations in astrocyte gene expression, failure of energy metabolism, increased production of reactive oxygen and nitrogen species, and increased extracellular Glu leading to excitotoxicity could all contribute to cell death.

Mn plays a crucial role during all life stages in animals and humans. To gain a clearer picture of the mechanisms of neurotoxicity, it is important to understand Mn transport and regulation in the brain. Although our knowledge of Mn transport and distribution has increased considerably in recent years, a host of questions have yet to be addressed adequately. Questions to ponder include the following: To what extent do novel transporters facilitate Mn entry into and efflux from the brain? Once in the brain, how is Mn regulated at a cellular level? What roles do microglia and oligodendrocytes play in storing and utilizing Mn? Are microglia important in the etiology and development of Mn-induced neurotoxicity? Which neurotransmitters play a defining role in the neurotoxic consequences of excess Mn levels? To what degree does an increase in oxidative stress, or conversely, a decrease in antioxidant defenses, affect the observed cell death in manganism? The current work in the field is diligently pursuing these and other questions to further our understanding of how Mn interacts with various cell populations and proteins in the brain.

ACKNOWLEDGMENTS

Preparation of this review was supported in part by grant numbers ES10563 *(to MA)* and 1-F32-ES012768-01 from the National Institute of Environmental Health Sciences (NIEHS) (SJG). Its contents are the sole responsibility of the authors and do not necessarily represent the official views of the NIEHS or the NIH. The authors wish to acknowledge Dr. Frederick C. Wedler; his chapter in the first edition of this book (1996) was integral in the preparation of this review.

ABBREVIATIONS

ADP — Adenosine diphosphate
ATP — Adenosine triphosphate
AI — Adequate intake
AMPA — α-Amino-3-hydroxy-5-methyl-ioxyzole-propionic acid
BBB — Blood–brain barrier

Ca — Calcium
CNS — Central nervous system
CT — Computed tomography
DNA — Deoxyribonucleic acid
DMT1 — Divalent metal transporter1
DAT — Dopamine transporter
EPA — Environmental Protection Agency
EAR — Estimated average requirement
ESR — Electron spin resonance
Fe — Iron
GABA — γ-Aminobutyric acid
Glu — Glutamate
GLAST — Glutamate aspartate transporter
Gln — Glutamine
GS — Glutamine synthetase
GTP — Guanosine triphosphate
6-OHDA — 6-hydroxydopamine
MR — Magnetic resonance
MRS — Magnetic resonance spectroscopy
Mn — Manganese
MnSOD — Manganese superoxide dismutase
MMT — Methylcyclopentadienyl Mn tricarbonyl
NMDA — *N*-methyl-*D*-aspartate
NRC — National Research Council
PEP — Phosphoenolpyruvate
PTBR — Peripheral-type benzodiazepine receptors
PET — Positron emission tomography
ROS — Reactive oxygen species
RfC — Reference concentration
RfD — Reference dose
RNA — Ribonucleic acid
SOD — Superoxide dismutase
TPN — Total parenteral nutrition
UL — Upper limit level

REFERENCES

Archibald, F. and Tyree, C. (1987). Manganese poisoning and the attack of trivalent manganese upon catecholamines, *Arch. Biochem. Biophys.*, 256, 638–650.

Aschner, M. (2000). Manganese in health and disease: From transport to neurotoxicity, in *Handbook of Neurotoxicology*, Massaro, E., Ed., Humana Press, Totowa, NJ, pp. 195–209.

Aschner, M. and Aschner, J. (1990). Manganese transport across the blood-brain barrier: Relationship to iron homeostasis, *Brain Res. Bull.*, 24, 857–860.

Aschner, M. and Aschner, J. (1991). Manganese neurotoxicity: Cellular effects and blood-brain barrier transport, *Neurosci. Biobehav. Rev.*, 15, 333–340.

Aschner, M. and Gannon, M. (1994). Manganese (Mn) transport across the rat blood-brain barrier: Saturable and transferrin-dependent transport mechanisms, *Brain Res. Bull.*, 33, 345–349.

Aschner, M., Gannon, M., and Kimelberg, H. (1992). Manganese uptake and efflux in cultured rat astrocytes, *J. Neurochem.*, 58, 730–735.

Aschner, M., Shanker, G., Erikson, K., Yang, J., and Mutkus, L.A. (2002). The uptake of manganese in brain endothelial cultures, *Neurotoxicology*, 23, 165–168.

Aschner, M., Vrana, K.E., and Zheng, W. (1999). Manganese uptake and distribution in the central nervous system (CNS), *Neurotoxicology*, 20, 173–180, 1999.

Autissier, N., Dumas, P., Brosseau, J., and Loireau, A. (1977). Action du manganese methylcyclopentadienyle tricarbonyle (MMT) sur les mitochondries I. Effects du MMT, *in vitro*, sur la phosphorylation oxydative des mitochondries hepatiques de rats, *Toxicology*, 7, 115–122.

Baba, Y., Lerch, M.M., Tanimoto, A., Kreft, B.P., Saluja, A.K., Zhao, L., Chen, J., Steer, M.L., and Stark, D.D. (1994). Enhancement of pancreas and diagnosis of pancreatitis using manganese dipyridoxyl diphosphate, *Invest. Rad.*, 29 (Suppl. 2), S300–S301.

Barceloux, D. (1999). Manganese, *J. Toxicol. Clin. Toxicol.*, 37, 293–307.

Barron, T.F., Devenyi, A.G., and Mamourian, A.C. (1994). Symptomatic manganese neurotoxicity in a patient with chronic liver disease: Correlation of clinical symptoms with MRI findings, *Ped. Neurol.*, 10, 145–148.

Battaglia, G., Shannon, M., and Titeler, M. (1984). Guanyl nucleotide and divalent cation regulation of cortical S_2 serotonin receptors, *J. Neurochem.*, 43, 1213–1219.

Bluml, S., Moreno, A., Hwang, J., and Ross, B. (2001). 1-(^{13}C) glucose magnetic resonance spectroscopy of pediatric and adult brain disorders, *NMR Biomed.*, 14, 19–32.

Bonilla, E. (1980). L-tyrosine hydroxylase activity in the rat brain after chronic oral administration of manganese chloride, *Neurobehav. Toxicol.*, 2, 37–41.

Brouillet, E.P., Shinobu, L., McGarvey, U., Hochberg, F., and Beal, M.F. (1993). Manganese injection into the rat striatum produces excitotoxic lesions by impairing energy metabolism, *Exper. Neurol.*, 120, 89–94.

Brown, S. and Taylor, N. (1999). Could mitochondrial dysfunction play a role in manganese toxicity?, *Environ. Tox. Pharm.*, 7, 49–57.

Brurok, H., Ardenkjaer-Larsen, J.H., Hansson, G., Skarra, S., Berg, K., Karlsson, J.O.G., Laursen, I., and Jynge, P. (1999). Manganese dipyridoxyl diphosphate: MRI contrast agent with antioxidative and cardioprotective properties?: *In vitro* and *ex vivo* assessments, *Biochem. Biophys. Res. Commun.*, 254, 768–772.

Burdo, J., Martin, J., Menzies, S., Dolan, K., Romano, M., Fletcher, R., Garrick, M., Garrick, L., and JR C. (1999). Cellular distribution of iron in the brain of the Belgrade rat, *Neuroscience*, 93, 1189–1196.

Butterworth, R., Spahr, L., Fontaine, S., and Layrargues, G. (1995). Manganese toxicity, dopaminergic dysfunction and hepatic encephalopathy, *Metab. Brain Dis.*, 10, 259–267.

Cano, G., Suarez-Roca, H., and Bonilla, E. (1997). Alterations of excitatory amino acid receptors in the brain of manganese-treated mice, *Mol. Chem. Neuropathol.*, 30, 41–52.

Carl, G., Blackwell, L., Barnett, F., Thompson, L., Rissinger, C., Olin, K., Critchfield, J., Keen, C., and Gallagher, B. (1993). Manganese and epilepsy: Brain glutamine synthetase and liver arginase activities in genetically epilepsy prone and chronically seizured rats, *Epilepsia*, 34, 441–446.

Cawte, J. (1985). Psychiatric sequelae of manganese exposure in the adult, foetal and neonatal nervous systems, *Aust. N.Z. J. Psychiatry*, 19, 211–217.

Chaki, H., Matsuda, A., Yamamoto, K., Kokuba, Y., Kataoka, M., Fujibayashi, Y., Matsuda, T., and Yamamoto, K. (1998). Significance of magnetic resonance image and blood manganese measurement for the assessment of brain manganese during total parenteral nutrition in rats, *Biol. Trace Elem. Res.*, 63, 37–50.

Chan, A., Minski, M., Lim, L., and Lai, J. (1992). Changes in brain regional manganese and magnesium levels during postnatal development: modulations by chronic manganese administration, *Metab. Brain Dis.*, 7, 21–33.

Chance, B. (1965). The energy-linked reaction of calcium with mitochondria, *J. Biol. Chem.*, 240, 2729–2748.

Chandra, S., Murthy, R., Husain, T., and SK, B. (1984). Effect of interaction of heavy metals on (Na⁺ -K⁺) ATPase and the uptake of ^{3}H-DA and ^{3}H-NA in rat brain synaptosomes, *Acta Pharmacol. Toxicol. (Copenh.)*, 54, 210–213.

Chang, J.Y. and Liu, L.Z. (1999). Manganese potentiates nitric oxide production by microglia, *Molec. Brain Res.*, 68, 22–28.

Chen, C.J. and Liao, S.L. (2002). Oxidative stress involves in astrocytic alterations induced by manganese, *Exp. Neurol.*, 175, 216–225.

Connor, J. (1994). Iron acquisition and expression of iron regulatory proteins in the developing brain: manipulation by ethanol exposure, iron deprivation and cellular dysfunction, *Dev. Neurosci.*, 16, 233–247.

Connor, J., Menzies, S., St. Martin, S., and Mufson, E. (1990). Cellular distribution of transferrin, ferritin, and iron in normal and aged human brains, *J. Neurosci. Res.*, 27, 595–611.

Cotzias, G. (1958). Manganese in health and disease, *Physiol. Rev.*, 38, 503–532.

Crossgrove, J.S., Allen, D.D., Bukaveckas, B.L., Rhineheimer, S.S., and Yokel, R.A. (2003). Manganese distribution across the blood-brain barrier: I. Evidence for carrier-mediated influx of manganese citrate as well as manganese and manganese transferrin, *Neurotoxicology*, 24, 3–13.

Davidsson, L., Lonnerdal, B., Sandstrom, B., Kunz, C., and Keen, C. (1989). Identification of transferrin as the major plasma carrier protein for manganese introduced orally or intravenously or after *in vitro* addition in the rat, *J. Nutr.*, 119, 1461–1464.

Davis, C., Zech, L., and Greger, J. (1993). Manganese metabolism in rats: An improved methodology for assessing gut endogenous losses, *Proc. Soc. Exp. Biol. Med.*, 202, 103–108.

Dietz, M.C., Ihrig, A., Wrazidlo, W., Bader, M., Jansen, O., and Triebig, G. (2001). Results of magnetic resonance imaging in long-term manganese dioxide-exposed workers, *Environ. Res.*, 85, 37–40.

Donaldson, J. and LaBella, F. (1984). The effects of manganese on the cholinergic receptor *in vivo* and *in vitro* may be mediated through modulation of free radicals, *Neurotoxicology*, 5, 105–112.

Dorman, D., Struve, M., James, R., Marshall, M., Parkinson, C., and Wong, B. (2001). Influence of particle solubility on the delivery of inhaled manganese to the rat brain: manganese sulfate and manganese tetroxide pharmacokinetics following repeated (14-day) exposure, *Toxicol. Appl. Pharmacol.*, 170, 79–87.

Drapeau, P. and Nachshen, D. (1984). Manganese fluxes and manganese-dependent neurotransmitter release in presynaptic nerve endings isolated from rat brain, *J. Physiol.*, 348, 493–510.

Dupont, C. and Tanaka, Y. (1985). Blood manganese levels in children with convulsive disorder, *Biochem. Med.*, 33, 246–255.

Erikson, K. and Aschner, M. (2002). Manganese causes differential regulation of glutamate transporter (GLAST) taurine transporter and metallothionein in cultured rat astrocytes, *Neurotoxicology*, 23, 595–602.

Erikson, K., Shihabi, Z.K., Aschner, J., and Aschner, M. (2002). Manganese accumulates in iron-deficient rat brain regions in a heterogeneous fashion and is associated with neurochemical alterations, *Biol. Trace Elem. Res.*, 87, 143–156.

Erikson, K.M. and Aschner, M. (2003). Manganese neurotoxicity and glutamate-GABA interaction, *Neurochem. Int.*, 43, 475–480.

Fell, J.M.E., Meadows, N., Khan, K., Long, S.G., Milla, P.J., Reynolds, A.P., Quaghebeur, G., and Taylor, W.J. (1996). Manganese toxicity in children receiving long-term parenteral nutrition, *Lancet*, 347, 1218–1221.

Fitsanakis, V.A., Amarnath, V., Moore, J.T., Montine, K.S., Zhang, J., and Montine, T.J. (2002). Catalysis of catechol oxidation by metal-dithiocarbamate complexes in pesticides, *Free Rad. Bio. Med.*, 33, 1714–1723.

Gallez, B., Demeure, R., Baudelet, C., Abdelouahab, N., Beghein, N., Jordan, B., Geurts, M., and Roels, H.A. (2001). Noninvasive quantification of manganese deposits in the rat brain by local measurement of NMR proton T_1 relaxation times, *Neurotoxicology*, 22, 387–392.

Galvani, P., Fumagalli, P., and Santagostino, A. (1995). Vulnerability of mitochondrial complex I in PC12 cells exposed to manganese, *Eur. J. Pharm.*, 293, 377–383.

Gavin, C.E., Gunter, K.K., and Gunter, T.E. (1990). Manganese and calcium efflux kinetics in brain mitochondria, *Biochem. J.*, 266, 329–334.

Gavin, C.E., Gunter, K.K., and Gunter, T.E. (1992). Mn^{+2} sequestration by mitochondria and inhibition of oxidative phosphorylation., *Toxicol. Appl. Pharm.*, 115, 1–5.

Gavin, C.E., Gunter, K.K., and Gunter, T.E. (1999). Manganese and calcium transport in mitochondria: implications for manganese toxicity, *Neurotoxicology*, 20, 445–454.

Gerber, G.B., Leonard, A., and Hantson, P. (2002). Carcinogenicity, mutagenicity and teratogenicity of manganese compounds, *Crit. Rev. Oncol. Hemat.*, 42, 25–34.

Ginsburg, A. (1972). Glutamine synthetase of *Escherichia coli:* Some physical and chemical properties, in *Advances in Enzymology*, Anfinson, C., Edsall, J., and Richards, F., Eds., Academic Press, New York, chap. 24.

Golub, M.S., Han, B., and Keen, C.L. (1996). Aluminum alters iron and manganese uptake and regulation of surface transferrin receptors in primary rat oligodendrocyte cultures, *Brain Res.*, 719, 72–77.

Graham, D.G. (1978). Oxidative pathways for catecholamines in the genesis of neuromelanin and cytotoxic quinones, *Mol. Pharmacol.*, 14, 633–643.

Greger, J. (1998). Dietary standards for manganese: overlap between nutritional and toxicological studies, *J. Nutr.*, 128 (Suppl. 2), 368S–371S.

Gruetter, R., Novotny, E., Boulware, S., Mason, G., Rothman, D., Shulman, G., Prichard, J., and Shulman, R. (1994). Localized ^{13}C NMR spectroscopy in the human brain of amino acid labeling from D-[1-^{13}C]glucose, *J. Neurochem.*, 63, 1377–1385.

Gunshin, H., Mackenzie, B., Berger, U., Gunshin, Y., Romero, M., Boron, W., Nussberger, S., Gollan, J., and Hediger, M. (1997). Cloning and characterization of a mammalian proton-coupled metal-ion transporter, *Nature*, 388, 482–488.

Gunter, T., Miller, L., Gavin, C., Eliseev, R., Salter, J., Buntinas, L., Alexandrov, A., Hammond, S., and Gunter, K. (2004). Determination of the oxidation states of manganese in brain, liver, and heart mitochondria, *J. Neurochem.*, 88, 266–280.

Hazell, A. and Norenberg, M. (1997). Manganese decreases glutamate uptake in cultured astrocytes, *Neurochem. Res.*, 22, 1443–1447.

Hazell, A.S. (2002). Astrocytes and manganese neurotoxicity, *Neurochem. Int.*, 41, 271–277.

Hazell, A.S., Desjardins, P., and Butterworth, R.F. (1999). Chronic exposure of rat primary astrocyte cultures to manganese results in increased binding sites for the 'peripheral-type' benzodiazepine receptor ligand 3H-PK 11195, *Neurosci. Lett.*, 271, 5–8.

Hazell, A.S. and Norenberg, M. (1997). Manganese decreases glutamate uptake in cultured astrocytes, *Neurochem. Res.*, 22, 1443–1447.

Hazell, A.S., Normandin, L., Nguyen, B., and Kennedy, G. (2003). Up-regulation of 'peripheral-type' benzodiazepine receptors in the globus pallidus in a sub-acute rat model of manganese neurotoxicity, *Neurosci. Lett.*, 349, 13–16.

He, P., Liu, D., and Zhang, G. (1994). Effects of high-level-manganese sewage irrigation on children's neurobehavior, *Zhonghua Yu Fang Yi Xue Za Zhi*, 28, 216–218.

Hernandez, E.H., Valentini, M.C., and Discalzi, G. (2002). T_1-weighted hyperintensity in basal ganglia at brain magnetic resonance imaging: Are different pathologies sharing a common mechanism?, *Neurotoxicology*, 23, 669–674.

Huang, C., Weng, Y., Lu, C., Chu, N., and Yen, T. (2003). Dopamine transporter binding in chronic manganese intoxication, *J. Neurol.*, 250, 1335–1339.

Husain, M., Khanna, V., Roy, A., Tandon, R., Pradeep, S., and Set, P.K. (2001). Platelet dopamine receptors and oxidative stress parameters as markers of manganese toxicity, *Hum. Exp. Toxicol.*, 20, 631–636.

Husain, R., Seth, P.K., and Chandra, S.V. (1976). Early inhibition of succinic dehydrogenase by manganese in rat gonads, *Bull. Environ. Contam. Toxicol.*, 16, 118–121.

Husain, T., Ali, M., and Chandra, S. (1987). The combined effect of Pb^{2+} and Mn^{2+} on monoamine uptake and Na^+, K^+-ATPase in striatal synaptosomes, *J. Appl. Toxicol.*, 7, 277–280.

Jauzac, P., Frances, B., Puget, A., and Meunier, J. (1984). Divalent cations: Do they stabilize the agonist (12S) form of the mu opioid receptor?, *Neuropeptides*, 5, 125–128.

Kalyanaraman, B., Felix, C., and Sealy, R. (1985). Semiquinone anion radicals of catechol(amine)s, catechol estrogens, and their metal ion complexes, *Environ. Health Perspect.*, 64, 185–198.

Kessler, K., Wunderlich, G., Hefter, H., and Seitz, R. (2003). Secondary progressive chronic manganism associated with markedly decreased striatal D_2 receptor density, *Mov. Disord.*, 18, 217–218.

Kim, S.H., Chang, K.H., Chi, J.G., Cheong, H.K., Kim, J.Y., Kim, Y.M., and Han, M.H. (1999). Sequential change of MR signal intensity of the brain after manganese administration in rabbits: Correlation with manganese concentration and histopathologic findings, *Invest. Rad.*, 34, 383–393.

Konji, V., Montag, A., Sandri, G., Nordenbrand, K., and Ernster, L. (1985). Transport of Ca^{2+} and Mn^{2+} by mitochondria from rat liver, heart and brain, *Biochimie*, 67, 1241–1250.

Kontur, P. and Fechter, L. (1985). Brain manganese, catecholamine turnover, and the development of startle in rats prenatally exposed to manganese, *Teratology*, 32, 1–11.

Kreft, B.P., Baba, Y., Tanimoto, A., Finn, J.P., and Stark, D.D. (1993). Orally administered manganese chloride: Enhanced detection of hepatic tumors in rats, *Radiology*, 186, 543–548.

Lai, J., Chan, A., Leung, T., Minski, M., and Lim, L. (1992). Neurochemical changes in rats chronically treated with a high concentration of manganese chloride, *Neurochem. Res.*, 17, 841–847.

Lai, J., Leung, T., Guest, J., Davison, A., and Lim, L. (1982). The effects of chronic manganese chloride treatment expressed as age-dependent, transient changes in rat brain synaptosomal uptake of amines, *J. Neurochem.*, 38, 844–847.

Lai, J., Leung, T., and Lim, L. (1984). Differences in the neurotoxic effects of manganese during development and aging: some observations on brain regional neurotransmitter and non-neurotransmitter metabolism in a developmental rat model of chronic manganese encephalopathy, *Neurotoxicology*, 5, 37–47.

Leergaard, T.B., Bjaalie, J.G., Devor, A., Wald, L.L., and Dale, A.M. (2003). *In vivo* tracing of major rat brain pathways using manganese-enhanced magnetic resonance imaging and three-dimensional digital atlasing, *Neuroimage*, 20, 1591–1600.

Liao, S. and Chen, C. (2001). Manganese stimulates stellation of cultured rat cortical astrocytes, *Neuroreport*, 12, 3877–3881.

Liccione, J. and Maines, M. (1988). Selective vulnerability of glutathione metabolism and cellular defense mechanisms in rat striatum to manganese, *J. Pharmacol. Exp. Ther.*, 247, 156–161.

Lin, Y.J. and Koretsky, A.P. (1997). Manganese ion enhanced T_1-weighted MRI during brain activation: An approach to direct imaging of brain function, *Magn. Reson. Med.*, 38, 378–388.

Lipe, G., Duhart, H., Newport, G., Slikker, W.J., and Ali, S. (1999). Effect of manganese on the concentration of amino acids in different regions of the rat brain, *J. Environ. Sci. Health B*, 34, 119–132.

Lis, A., Barone, T., Paradkar, P., Plunkett, R., and Roth, J. (2004). Expression and localization of different forms of DMT1 in normal and tumor astroglial cells, *Brain Res. Mol. Brain Res.*, 122, 62–70.

Liu, Y., Fiskum, G., and Schubert, D. (2002). Generation of reactive oxygen species by the mitochondrial electron transport chain, *J. Neurochem.*, 80, 780–787.

Malecki, E.A. (2001). Manganese toxicity is associated with mitochondrial dysfunction and DNA fragmentation in rat primary striatal neurons, *Brain Res. Bull.*, 55, 225–228.

Malecki, E.A., Devenyi, A.G., Beard, J.L., and Connor, J.R. (1999). Existing and emerging mechanisms for transport of iron and manganese to the brain, *J. Neurosci. Res.*, 56, 113–122.

Malhotra, K.M., Murthy, R.C., Srivastava, R.S., and Chandra, S.V. (1984). Concurrent exposure of lead and manganese to iron-deficient rats: Effect on lipid peroxidation and contents of some metals in the brain, *J. Appl. Toxicol.*, 4, 22–25.

Maurizi, M., Pinkofsky, H., and Ginsburg, A. (1987). ADP, chloride ion, and metal ion binding to bovine brain glutamine synthetase, *Biochemistry*, 26, 5023–5031.

Maynard, L. and Cotzias, G. (1955). Partition of Mn among organs and organelles of the rat, *J. Biol. Chem.*, 214, 789–795.

Mena, I., Horiuchi, K., Burke, K., and Cotzias, G. (1969). Chronic manganese poisoning: Individual susceptibility and absorption of iron, *Neurology*, 19, 1000–1006.

Moos, T., Trinder, D., and Morgan, E. (2002). Effect of iron status on DMT1 expression in duodenal enterocytes from beta$_2$-microglobulin knockout mice, *Am. J. Physiol. Gastrointest. Liver Physiol.*, 283, G687–G694.

Morita, H., Ogino, T., Seo, Y., Fujiki, N., Tanaka, K., Takamata, A., Nakamura, S., and Murakami, M. (2002). Detection of hypothalamic activation by manganese ion contrasted T_1-weighted magnetic resonance imaging in rats, *Neurosci. Lett.*, 326, 101–104.

Nagatomo, S., Umehara, F., Hanada, K., Nobuhara, Y., Takenaga, S., Arimura, K., and Osame, M. (1999). Manganese intoxication during total parenteral nutrition: Report of two cases and review of the literature, *J. Neurol. Sci.*, 162, 102–105.

National Research Council (2002). Manganese, in *Dietary Reference Intakes for Vitamin A, Vitamin K, Arsenic, Boron, Chromium, Copper, Iodine, Iron, Manganese, Molybdenum, Nickel, Silicon, Vanadium and Zinc*, Panel on Micronutrients, and the Standing Committee on the Scientific Evaluation of Dietary Reference Intakes, pp. 394–419.

Natt, O., Watanabe, T., Boretius, S., Radulovic, J., Frahm, J., and Michaelis, T. (2002). High-resolution 3D MRI of mouse brain reveals small cerebral structures *in vivo*, *J. Neurosci. Meth.*, 120, 203–209.

Ni, Y., Petre, C., Bosmans, H., Miao, Y., Grant, D., Baert, A.L., and Marchal, G. (1997). Comparison of manganese biodistribution and MR contrast enhancement in rats after intravenous injection of MnDPDP and MnCl$_2$, *Acta Radiol.*, 38, 700–707.

Novotny, E., Fulbright, R., Pearl, P., Gibson, K., and Rothman, D. (2003). Magnetic resonance spectroscopy of neurotransmitters in human brain, *Ann. Neruol.*, 54 (Suppl. 6), S25–S31.

Oates, P., Thomas, C., Freitas, E., Callow, M., and Morgan, E. (2000). Gene expression of divalent metal transporter 1 and transferrin receptor in duodenum of Belgrade rats, *Am. J. Physiol. Gastrointest. Liver Physiol.*, 278, G930–G936.

Olanow, C., Good, P., Shinotoh, H., Hewitt, K., Vingerhoets, F., Snow, B., Beal, M., Calne, D., and Perl, D. (1996). Manganese intoxication in the rhesus monkey: a clinical, imaging, pathologic, and biochemical study, *Neurology*, 46, 492–498.

Pal, P.K., Samii, A., and Calne, D.B. (1999). Manganese neurotoxicity: A review of clinical features, imaging and pathology, *Neurotoxicology*, 20, 227–238.

Park, N.H., Park, J.K., Choi, Y., Yoo, C., Lee, C.-R., Lee, H., Kim, H.K., Kim, S.-R., Jeong, T.-H., Park, J., Yoon, C.S., and Kim, Y. (2003). Whole blood manganese correlates with high signal intensities on T_1-weighted MRI in patients with liver cirrhosis, *Neurotoxicology*.

Pautler, R.G. and Koretsky, A.P. (2002). Tracing odor-induced activation in the olfactory bulbs of mice using manganese-enhanced magnetic resonance imaging, *Neuroimage*, 16, 441–448.

Prohaska, J. (1987). Functions of trace elements in brain metabolism, *Physiol. Rev.*, 67, 858–901.

Puskin, J. and Gunter, T. (1973). Ion and pH gradients across the transport membrane of mitochondria following Mn^{++} uptake in the presence of acetate, *Biochem. Biophys. Res. Commun.*, 51, 797–803.

Puskin, J.S. and Gunter, T.E. (1972). Evidence for the transport of manganous ion against an activity gradient by mitochondria, *Biochimica et Biophysica Acta (BBA) — Bioenergetics*, 275, 302–307.

Rabin, O., Hegedus, L., Bourre, J., and Smith, Q. (1993). Rapid brain uptake of manganese(II) across the blood-brain barrier, *J. Neurochem.*, 61, 509–517.

Reimund, J., Dietemann, J., Warter, J., Baumann, R., and Duclos, B. (2000). Factors associated with hyper manganesemia in patients receiving home parenteral nutrition, *Clin. Nutr.*, 19, 343–348.

Roth, J.A. and Garrick, M.D. (2003). Iron interactions and other biological reactions mediating the physiological and toxic actions of manganese, *Biochem. Pharm.*, 66, 1–13.

Ryu, S., Brown, S.L., Kolozsvary, A., Ewing, J.R., and Kim, J.H. (2002). Noninvasive detection of radiation-induced optic neuropathy by manganese-enhanced MRI, *Rad. Res.*, 157, 500–505, 2002.

Saleem, K.S., Pauls, J.M., Augath, M., Trinath, T., Prause, B.A., Hashikawa, T., and Logothetis, N.K. (2002). Magnetic resonance imaging of neuronal connections in the Macaque monkey, *Neuron*, 34, 685–700.

Schramm, V. (1982). Metabolic regulation: Could Mn(II) be involved?, *Trends Biochem. Sci.*, 7, 369–371.

Schramm, V. and Brandt, M. (1986). The manganese (II) economy of rat hepatocytes, *Fed. Proc.*, 45, 2817–2820.

Schramm, V. and Wedler, F. (1986). *Manganese in Metabolism and Enzyme Function*, Academic Press, Orlando, FL.

Schroeter, M.L., Mertsch, K., Giese, H., Muller, S., Sporbert, A., Hickel, B., and Blasig, I.E. (1999). Astrocytes enhance radical defense in capillary endothelial cells constituting the blood-brain barrier, *FEBS Lett.*, 449, 241–244.

Schweigert, N., Zehnder, A.J.B., and Eggen, R.I.L. (2001). Chemical properties of catechols and their molecular modes of toxic action in cells, from microorganisms to mammals, *Environ. Microbiol.*, 3, 81–91.

Seth, P. and Chandra, S. (1984). Neurotransmitters and neurotransmitter receptors in developing and adult rats during manganese poisoning, *Neurotoxicology*, 5, 67–76.

Shen, X.-M. and Dryhurst, G. (1998). Iron- and manganese-catalyzed autoxidation of dopamine in the presence of L-cysteine: Possible insights into iron- and manganese-mediated dopaminergic neurotoxicity, *Chem. Res. Toxicol.*, 11, 824–837.

Shukla, G. and Singhal, R. (1984). The present status of biological effects of toxic metals in the environment: lead, cadmium, and manganese, *Can. J. Physiol. Pharmacol.*, 62, 1015–1031.

Spencer, A. (1999). Whole blood manganese levels in pregnancy and the neonate, *Nutrition*, 15, 731–734.

Spranger, M., Schwab, S., Desiderate, S., Bonmann, E., Krieger, D., and Fandrey, J. (1998). Manganese augments nitric oxide synthesis in murine astrocytes: A new pathogenetic mechanism in manganism?, *Exp. Neurol.*, 149, 277–283.

Stadtman, E. (1990). Discovery of glutamine synthetase cascade, *Methods Enzymol.*, 182, 793–809.

Suarez, N. and Eriksson, H. (1993). Receptor-mediated endocytosis of a manganese complex of transferrin into neuroblastoma (SHSY5Y) cells in culture, *J. Neurochem.*, 61, 127–131.

Subhash, M. and Padmashree, T. (1991). Effect of manganese on biogenic amine metabolism in regions of the rat brain, *Food Chem. Toxicol.*, 29, 579–582.

Szucs, M., Spain, J., Oetting, G., Moudy, A., and Coscia, C. (1987). Guanine nucleotide and cation regulation of mu, delta, and kappa opioid receptor binding: Evidence for differential postnatal development in rat brain, *J. Neurochem.*, 48, 1165–1170.

Takeda, A. (2003). Manganese action in brain function, *Brain Res. Brain Res. Rev*, 41, 79–87.

Takeda, A., Akiyama, T., Sawashita, J., and Okada, S. (1994). Brain uptake of trace metals, zinc and manganese, in rats, *Brain Res.*, 640, 341–344.

Takeda, A., Devenyi, A.G., and Connor, J.R. (1998). Evidence for non-transferrin-mediated uptake and release of iron and manganese in glial cell cultures from hypotransferrinemic mice, *J. Neurosci. Res.*, 51, 454–462.

Takeda, A., Sawashita, J., and Okada, S. (1994). Localization in rat brain of the trace metals, zinc and manganese, after intracerebroventricular injection, *Brain Res.*, 658, 252–254.

Takeda, A., Sotogaku, N., and Oku, N. (2002). Manganese influences the levels of neurotransmitters in synapses in rat brain, *Neuroscience*, 114, 669–674.

Tholey, G., Ledig, M., Kopp, P., Sargentini-Maier, L., Leroy, M., Grippo, A., and Wedler, F. (1988). Levels and sub-cellular distribution of physiologically important metal ions in neuronal cells cultured from chick embryo cerebral cortex, *Neurochem. Res.*, 13, 1163–1167.

Tholey, G., Ledig, M., Mandel, P., Sargentini, L., Frivold, A., Leroy, M., Grippo, A., and Wedler, F. (1988). Concentrations of physiologically important metal ions in glial cells cultured from chick cerebral cortex, *Neurochem. Res.*, 13, 45–50.

Tomas-Camardiel, M., Herrera, A.J., Venero, J.L., Cruz Sanchez-Hidalgo, M., Cano, J., and Machado, A. (2002). Differential regulation of glutamic acid decarboxylase mRNA and tyrosine hydroxylase mRNA expression in the aged manganese-treated rats, *Molec. Brain Res.*, 103, 116–129.

Tran, T.T., Chowanadisai, W., Crinella, F.M., Chicz-DeMet, A., and Lonnerdal, B. (2002a). Effect of high dietary manganese intake of neonatal rats on tissue mineral accumulation, striatal dopamine levels, and neurodevelopmental status, *Neurotoxicology*, 23, 635–643.

Tran, T.T., Chowanadisai, W., Lonnerdal, B., Le, L., Parker, M., Chicz-Demet, A., and Crinella, F.M. (2002b). Effects of neonatal dietary manganese exposure on brain dopamine levels and neurocognitive functions, *Neurotoxicology*, 23, 645–651.

Turrens, J. and Boveris, A. (1980). Generation of superoxide anion by the NADH dehydrogenase of bovine heart mitochondria, *Biochem. J.*, 191, 421–427.

Verity, M.A. (1999). Manganese neurotoxicity: A mechanistic hypothesis, *Neurotoxicology*, 20, 489–498.

Villafrance, J. and Nowak, T. (1992). Metal ions at enzyme active sites, in *The Enzymes*, Vol. XX, Sigman, D., Ed., Academic Press, New York, chap. 2.

Watanabe, T., Natt, O., Boretius, S., Frahm, J., and Michaelis, T. (2002). *In vivo* MRI staining of mouse brain after subcutaneous application of $MnCl_2$, *Magn. Reson. Med.*, 48, 852–859.

Weber, S., Dorman, D.C., Lash, L.H., Erikson, K., Vrana, K.E., and Aschner, M. (2002). Effects of manganese (Mn) on the developing rat brain: oxidative-stress related endpoints, *Neurotoxicology*, 23, 169–175.

Wedler, F. (1993). Magnesium and manganese enzymes, in *Inorganic Reactions and Methods*, Vol. 16, Zuckerman, J. and Norman, A., Eds., VCH Publishers, Deerfield Beach, FL, chap. 1.

Wedler, F. (1994a). Biochemical and nutritional role of manganese: An overview, in *Manganese in Health and Disease*, Klimis-Tavantzis, D., Ed., CRC Press, Boca Raton, FL, chap. 1.

Wedler, F. (1994b). Biological significance of manganese in mammalian systems, in *Progress in Medicinal Chemistry*, Vol. 30, Ellis, G. and Luscombe, D., Eds., Elsevier, Amsterdam, chap. 3.

Wedler, F. and Denman, R. (1984). Glutamine synthetase: the major Mn(II) enzyme in mammalian brain, *Cur. Top Cell Regul.*, 24, 153–169.

Wedler, F., Denman, R., and Roby, W. (1982). Glutamine synthetase from ovine brain is a manganese(II) enzyme, *Biochemistry*, 21, 6389–6396.

Wedler, F. and Ley, B. (1994). Kinetic, ESR, and trapping evidence for *in vivo* binding of Mn(II) to glutamine synthetase in brain cells, *Neurochem. Res.*, 19, 139–144.

Wedler, F. and Tom, R. (1986). Interaction of Mn(II) with mammalian glutamine synthesis, in *Manganese in Metabolism and Enzyme Function*, Schramm, V. and Wedler, F., Eds., Academic Press, New York, pp. 221–228.

Wedler, F.C., Ley, B.W., and Grippo, A.A. (1989). Manganese (II) dynamics and distribution in glial cells cultured from chick cerebral cortex, *Neurochem. Res.*, 14, 1129–1135.

Weinmann, H.-J., Ebert, W., Misselwitz, B., and Schmitt-Willich, H. (2003). Tissue-specific MR contrast agents, *Eur. J. Radiol.*, 46, 33–44.

Williams, K., Wilson, M., and Bressler, J. (2000). Regulation and developmental expression of the divalent metal-ion transporter in the rat brain, *Cell Mol. Biol. (Noisy-le-grand)*, 46, 563–571.

Yokel, R.A., Crossgrove, J.S., and Bukaveckas, B.L. (2003). Manganese distribution across the blood-brain barrier: II. Manganese efflux from the brain does not appear to be carrier mediated, *Neurotoxicology*, 24, 15–22.

Zajac, J. and Roques, B. (1985). Differences in binding properties of mu and delta opioid receptor subtypes from rat brain: kinetic analysis and effects of ions and nucleotides, *J. Neurochem.*, 44, 1605–1614.

Zatta, P., Lucchini, R., van Rensburg, S.J., and Taylor, A. (2003). The role of metals in neurodegenerative processes: aluminum, manganese, and zinc, *Brain Res. Bull.*, 62, 15–28.

Zhang, G., Liu, D., and He, P. (1995). Effects of manganese on learning abilities in school children, *Zhonghua Yu Fang Yi Xue Za Zhi*, 29, 156–158.

Zhang, J., Fitsanakis, V., Gu, G., Jing, D., Ao, M., Amarnath, V., and Montine, T. (2003). Manganese ethylene-*bis*-dithiocarbamate and selective dopaminergic neurodegeneration in rat: A link through mitochondrial dysfunction, *J. Neurochem.*, 84, 336–346.

CHAPTER 26

Astroglia and Lead Neurotoxicity

Evelyn Tiffany-Castiglioni and Yongchang Qian

CONTENTS

26.1 INTRODUCTION

Lead (Pb) is an archetypal neurotoxic environmental contaminant because it is pervasive, metabolically nonessential, and toxic through multiple mechanisms. Furthermore, Pb produces sublethal functional impairment at low exposures and is more toxic to the immature than mature brain. In these respects it is similar to other common neurotoxicants, including mercurials, polyaromatic hydrocarbons, and ethanol. Lead's pervasiveness in the environment is characterized by its presence in all ecological and biological niches of modern Earth (Patterson and Settle, 1993). Though Pb is not a prominent contaminant of the food chain, it is a contaminant of water (from lead pipes and solder), soil (from Pb-containing paint dust and fuel emissions), and dwellings (from Pb-containing paint). Like the other neurotoxicants mentioned, Pb has no known essential role in metabolism. Instead, Pb is thought to exploit metabolic pathways used by other metals and thereby alter normal cell function. That Pb acts by multiple mechanisms at the cellular and molecular levels will be a major theme of this chapter, with emphasis on its affects on astrocytes, which accumulate Pb in the central nervous system (CNS). Other environmental neurotoxicants also act upon numerous targets, and indeed only a few, such as tetrodotoxin, are known to have a single mechanism of action (Lai et al., 2004).

0-8493-1794-0/05/$0.00+$1.50
© 2005 by CRC Press LLC

The neurotoxicity of low-level Pb exposure can be appreciated in several key observations. The brain accumulates and retains Pb (Bradbury and Deane, 1993). Thus, brain tissues take up and cannot readily clear Pb in the Pb-exposed vertebrate. Human exposure occurs throughout life, prenatally, postnatally, and in adulthood. The most sensitive developmental period for Pb exposure of the human brain appears to be the first 2 years of life. This age corresponds to physiologic immaturity of the brain and gut that increase Pb accumulation and to behaviors, such as mouthing and crawling, that increase exposures (Charney et al., 1980; Bellinger et al., 1986). Furthermore, neuronal circuitry develops after birth and Pb is toxic to synapses, the function of which is critical in the process of circuit development (Nihei and Guilarte, 2001). Most Pb neurotoxicity studies in humans have focused on deleterious effects in children, in whom the blood lead level of health concern is 10 µg/dl or higher (Centers for Disease Control, 1991). In children, reduced IQ scores (0.25–0.5 units per 1 µg/dl above 10 µg/dl), reduced attention span, and increased aggression are associated with blood lead levels above 10 µg/dl (Bellinger et al., 1992; Stiles and Bellinger, 1993; Schwartz, 1994; Needleman et al., 1996; Banks et al., 1997; Tong et al., 2000). New epidemiologic studies show that Pb levels below 10 µg/dl are also associated with an IQ decline estimated at 0.74 units per µg/dl in the range of 1 to 10 µg/dl (Bellinger and Needleman, 2003; Canfield et al., 2003). Furthermore, delayed pubertal development in girls has been reported at 3 µg/dl and appears to result from disruptions of the neuroendocrine system (Selevan et al., 2003). Although Pb is well known as a developmental neurotoxicant, the idea is gaining support that Pb exposure is also associated with adult neurologic diseases and disorders. Epidemiological evidence indicates an increased risk for gliomas in adult workers exposed to Pb (Anttila et al., 1996; Vainio, 1997). Long-term occupational exposure (> 20 years) to Pb in combination with Cu or Mn is associated with Parkinson's disease (Gorell et al., 1997, 1999). Furthermore, developmental exposure to neurotoxicants may plausibly form a fetal basis for adult neurodegenerative diseases, though direct links between such diseases and specific toxic agents have not yet been established. Thus, despite decades of attention to the problem of Pb neurotoxicity, Pb continues to be a threat to human health.

Much attention has been paid to astrocytes in Pb neurotoxicity because they resist gross pathologic effects, except at very high exposures, and because they accumulate and store Pb (Holtzman et al., 1997; Tiffany-Castiglioni et al., 1989). However, all cell types in the CNS show alterations in response to Pb exposure, some of which appear to be direct responses and others secondary reactions to the direct effects of Pb upon other cells. This review will consider three emerging topics on astroglia exposed to low and moderate levels of Pb: gene responses, stress responses, and the changing responses of astroglia to Pb across the spectrum of development and aging. The term *astrocytes* will be used to refer to fully mature cells *in vivo*, whereas *astroglia* will refer to immature cells or cells of unknown stages of differentiation, including astroblasts and astrocytes in culture. The known interactions of Pb with astrocytes have been reviewed earlier from several perspectives. The reader is referred to Holtzman et al. (1984, 1987) and Tiffany-Castiglioni and colleagues (Tiffany-Castiglioni et al., 1989, 1996a; Tiffany-Castiglioni, 1993) for sequential historical perspectives, including the "lead sink hypothesis." Additional reviews are available on astrocytes as metal-storing cells (Tiffany-Castiglioni and Qian, 2001), cell-type specific responses to Pb *in vitro* (Tiffany-Castiglioni and Qian, 2004), and the endoplasmic reticulum as a site of toxic action of Pb in astroglia (Qian and Tiffany-Castiglioni, 2003).

A comparative summary of studies on Pb and glial cell lines, cultured astroglia, and astrocytes *in vivo* is provided in Table 26.1. The table includes most of the studies of Pb effects on astroglia and astroglia-like cells published since 1990, an arbitrarily selected date that corresponds with what we view as the "intensification stage" of Pb neurotoxicity research *in vitro* (Tiffany-Castiglioni, 1993). In the intensification stage, investigators began to identify subtle effects on cells at submicromolar concentrations of Pb and draw parallels to the low doses that are deleterious to the nervous system *in vivo*. Though some of the results reported in Table 26.1 are conflicting, a consensus view is that Pb treatment decreases the activity of the astroglial enzyme glutamine synthetase (GS) (Sierra et al., 1989; Engle and Volpe, 1990; Sierra and Tiffany-Castiglioni, 1991; Zurich et al., 2002), alters

Table 26.1 Effects of Pb on Astroglia-like Cell Lines, Astroglia in Culture, and Astrocytes *In Vivo* Reported 1990–2004

Model Type	Approach	Results	Ref.
Cell Lines			
1321 N1 human astrocytoma	5–100 µM Pb acetate 24 or 48 h in serum-free medium; tritiated thymidine incorporation; flow cytometry; immunoblots for PKC isozymes in cell fractions	Stimulation of DNA synthesis by 5 µM and higher; mechanism involved protein kinase C alpha (PKCα) at 50 and 100 µM Pb; potentiation of carbachol-induced DNA synthesis by 5 µM Pb; stimulation of cell cycle progression at 20 µM	Lu et al., 2001
A172 human glioblastoma	10 µM Pb for 3–4 d in medium with serum; expression of glial markers identified by immunocytochemistry	Modulation of glial cell differentiation by exposure to Pb and Cd; decreased expression of glial fibrillary acidic protein (GFAP) and S-100; increased expression of transferrin receptor	Stark et al., 1992
C6 rat glioma	0.03–10 µM Pb acetate for 3–16 h; cell fractions separated by centrifugation and PKC in cell fractions detected by Western blot analysis	Increased membrane-bound and decreased cytosolic pools of PKC	Laterra et al., 1992
C6 rat glioma	1 or 10 µM Pb nitrate for several minutes in medium with serum; Ca^{2+} imaging in live cells with indo-1 by interactive laser cytometry	Pb enters cells via a cation channel activated by depletion of intracellular Ca^{2+}	Kerper and Hinkle, 1997
C6 rat glioma	10 µM Pb acetate daily up to 50 min in serum-free medium; Cu measurements by ^{67}Cu	Blockage of Cu efflux apparently by inhibition of ATP7a (Menkes protein)	Qian et al., 1995, 1999
C6 rat glioma	1 µM Pb acetate for 1 week in medium with serum; total proteins fractioned by gel chromatography; Pb in fractions measurement by atomic absorption spectroscopy	Over 90% of Pb is bound to high molecular weight fraction (~ 66 kD)	Qian et al., 1999
C6 rat glioma	Total proteins extracted from untreated C6 cells and poured over a Pb affinity column; proteins identified by immunoblot and amino acid sequencing	Pb is bound to 78 kD glucose-regulated protein (GRP78)	Qian et al., 2000
C6 rat glioma	1 µM Pb acetate daily for 1 week in medium with serum; mRNA detected by Northern blot; protein identified by 2-D gel, immunoblot, and amino acid sequencing	Induction of glucose regulated protein 78 kD (GRP78) at both mRNA and protein levels	Qian et al., 2000, 2001
C6 rat glioma	1 µM Pb acetate daily for 1 week in medium with serum; transcription factor detected by gel shift	Activation of redox-regulated transcription factor	Qian et al., 2001
C6 rat glioma	100–600 µM Pb nitrate 24 h in medium with serum; Pb-resistant and wild-type cells; suppression subtractive hybridization (SSH) between mRNAs	Upregulation of gene for heat-shock protein HSP90	Li and Rossman, 2001

(Continued)

Table 26.1 Effects of Pb on Astroglia-like Cell Lines, Astroglia in Culture, and Astrocytes *In Vivo* Reported 1990–2004 (*Continued*)

Model Type	Approach	Results	Ref.
Cell Lines			
C6 rat glioma	1–16 μM Pb acetate in serum-free medium with 0.1% BSA for 48 h; tumor necrosis factor alpha (TNF-α) or interleukin (IL)-1β for 24 h; cytokine plus Pb; metalloproteinase MMP-9 activity in culture medium measured by gelatin zymography	Enhancement of MMP-9 secretion by Pb-exposed cells cotreated with TNF-α or IL-1β	Lahat et al., 2002
C6 rat glioma	0.5–10 μM Pb acetate in medium with low serum for 3 h followed by 24 h treatment with combinations of TNF-α, interferon (IFN)-γ, and IL-1β; nitric oxide (NO) measured by Griess reaction; inducible NO synthase (iNOS) levels measured by immunoblot	Inhibition of cytokine-induced NO production; suppression of iNOS protein levels	Garber and Heiman, 2002
Immortalized rat astroglia	10 μM Pb acetate for 24 h; cDNA microarray analysis of Pb-induced gene expression; confirmation by RT-PCR and immunoblot	Differential regulation of 44 genes of 418 tested in Pb-exposed cells; upregulation of annexin A5 gene product expression by Pb was confirmed by RT-PCR and immunoblot	Bouton et al., 2001
U-373MG human astroglioma	0.1, 1.0, or 10 μM Pb acetate 24 or 48 h in medium with serum; gene expression detected by RT-PCR	Increased expression of tumor necrosis factor alpha (TNF-α); decreased expression of IL-1β, IL-6, and GABA transaminase	Liu et al., 2000
U-373MG human astrocytoma	0.1, 1.0, or 10 μM Pb acetate 24 or 48 h in medium with serum; RT-PCR and bioassay for tumor necrosis factor alpha (TNF-α); merocyanine and propidium iodide staining for apoptosis	Increased expression of TNF-α; no additive or synergistic effect with lipopolysaccharide; no effect on apoptosis	Cheng et al., 2002
Primary Cell Cultures			
Primary astroglial cultures from neopallium of neonatal cf.-1 mouse	1–300 μM Pb chloride or other metals for 48 h; metallothionein (MT) protein measured by Cd-hemoglobin assay	No effect of Pb on MT protein levels	Kramer et al., 1996
Primary astroglial cultures from cortex of 16-d C57BL mouse embryos	10 μM Pb acetate up to 2 h; AP-1 DNA binding activity detected by gel shift with quantitative analysis of phosphorimages; Ref-1 determined by nuclei isolation and immunoblot	Increased AP-1 DNA binding activity after 1 h and return to control level after 2 h; increased AP-1 site binding protein c-jun mRNA after 15 min, peaking at 30 min and returning to control level after 1 h; Ref-1 accumulation in nucleus after 30 min, peaking at 60 min and returning to control level after 2 h	Scortegagna and Hanbauer, 2000
Astroglial-enriched mixed primary cultures from neonatal rat	2.5 μM Pb acetate in homogenates of untreated cultures; ion exchange chromatography of ^{14}C-glutamate	70% decrease of glutamine synthetase activity	Engle and Volpe, 1990

Table 26.1 Effects of Pb on Astroglia-like Cell Lines, Astroglia in Culture, and Astrocytes *In Vivo* Reported 1990–2004 (*Continued*)

Model Type	Approach	Results	Ref.
Primary Cell Cultures			
13-d first passage cultures from neonatal rat cortical primary astroglial cultures	Cells treated three times per week for 1–3 weeks with 0.25–1 μM Pb acetate in medium with serum; 0.1nM-10μM Pb acetate added to untreated homogenates; ion exchange chromatography of ^{14}C-glutamate	Concentration-dependent reduction of glutamine synthetase activity with 0.25–1 μM long-term Pb exposure; decreased activity in acutely treated cell homogenates at Pb levels as low as 0.1 nM	Sierra and Tiffany-Castiglioni, 1991
Primary astroglial cultures from cortex of 1–2-day-old rat pups	10 μM Pb acetate for 15 min in buffer; HPLC; Ca^{2+} imaging in live cells with fura-2 by interactive laser cytometry	Increased IP$_3$ generation; no effect on Ca^{2+} transients either through influx through plasma membrane or release from intracellular Ca^{2+} stores	Dave et al., 1993
Neonatal rat cortical primary astroglial cultures	5 and 50 μM Pb acetate for 24 h in medium; SDS-PAGE and 2-D gel of total ^{3}H-labeled proteins	*De novo* synthesis of stress proteins, including HSP70	Opanashuk and Finkelstein, 1995b
Primary astroglial cultures from cortex of 1–2-day-old rat pups	1 μM Pb acetate daily up to 14 d in medium with serum; atomic absorption spectroscopy	Transient fourfold increase in total Ca levels in 7-d cultures and twofold increase in 21-d cultures, with normal levels reestablished by day 14; greater accumulation of Pb by 21-d than 7-d cells	Tiffany-Castiglioni et al., 1996a
Primary or first passage astroglial cultures from cortex of 1–2-day-old rat pups	0.1 and 1 μM Pb acetate daily up to 9 d in medium with serum; quantitative imaging with fluorescent probes for glutathione and mitochondrial membrane potential	Transient depletion and subsequent elevation of intracellular glutathione levels; delayed loss of mitochondrial membrane potential	Legare et al., 1993; Tiffany-Castiglioni et al., 1996b
Primary or first passage astroglial cultures from cortex of 1–2-day-old rat pups	1 μM Pb acetate daily in medium with 10% fetal bovine serum; Pb measured by atomic absorption spectroscopy on days 1, 4, and 7 of treatment	Loss of Pb over time from astroglial cultures in the presence of continued exposure	Tiffany-Castiglioni et al., 1996a; Lindahl et al., 1999
Astroglial-enriched rat cerebellar cultures from first- and second-generation rats exposed to Pb	Gestational exposure to Pb via 0.25% Pb acetate in dams' drinking water; continued exposure throughout postnatal development and breeding; continued exposure of F2 generation; cells cultured from F1 and F2 generation pups; cells cultured from untreated rats were exposed to 1 or 5 μM Pb acetate for 8 d; HPLC measurements of glutamate uptake	No effect of Pb on glutamate uptake; prevention by Pb of the blockage of glutamate uptake by *L-trans*-pyrollodine-2,4-dicarboxilic acid (PDC), a competitive inhibitor of glutamate uptake, in cultures from unexposed animals treated *in vitro*; enhanced prevention by Pb of PDC blockage in cultures from Pb-treated F1 and F2 rats	Yi and Lim, 1998
Primary astroglial cultures from cortex of 1–2-day-old rat pups	1 μM Pb acetate daily for 1 week in medium with serum; Ca^{2+} imaging in live cells with indo-1 by interactive laser cytometry	Pb enters cells via an L-type Ca^{2+} channel	Legare et al., 1998
14-d astroglial-enriched first passage cultures from neonatal rat cerebellum	10, 100, or 1000 μM Pb acetate in buffer for 10 min; ^{3}H-glutamate or ^{45}Ca uptake measured after 10 min incubation; cellular Pb content measured by atomic absorption spectroscopy	No effect of acute Pb treatment at levels tested on glutamate uptake; stimulation of Ca^{2+} uptake with acute exposure to 10 μM Ca plus 1000 μM Pb or 100 μM Ca plus 100 μM Pb; no effect of Ca on Pb uptake	Raunio and Tähti, 2001

Table 26.1 Effects of Pb on Astroglia-like Cell Lines, Astroglia in Culture, and Astrocytes *In Vivo* Reported 1990–2004 (*Continued*)

Model Type	Approach	Results	Ref.
Primary Cell Cultures			
Primary astroglial cultures from cortex of 1–2 day-old rat pups	1 µM Pb acetate daily up to 7 d in medium with serum; quantitative RT-PCR and Northern blot analysis	Induction of glucose regulated protein 78 kD (GRP78) at mRNA levels	Qian and Tiffany-Castiglioni, 2003
Systems with Heterogeneous Cell Interactions			
Reaggregate cultures of dispersed cells from fetal rat telencephalon	1 µM Pb acetate daily in serum-free medium on day 25–35 of culture; Pb measured by atomic absorption spectroscopy on day 35	Neuron-enriched cultures took up more Pb than glial-enriched cultures; high K+ in neuronal cultures was a confounding factor	Zurich et al., 1998
14-d primary neonatal rat astroglia fed conditioned medium from other cells	1 µM Pb acetate daily in medium with 10% fetal bovine serum; Pb measured by atomic absorption spectroscopy on days 1, 4, and 7 of treatment	Stimulation of intracellular Pb accumulation in astroglia by conditioned medium from SY5Y cells but not from primary rat astroglia and mouse cerebrovascular endothelial cells	Lindahl et al., 1999
Cocultures of rat astroglial primaries and SH-SY5Y human neuroblastoma cells in a Millipore® semipermeable membrane system	1 µM Pb acetate daily in medium with 10% fetal bovine serum; Pb measured by atomic absorption spectroscopy on days 1, 4, and 7 of treatment	Selective accumulation of Pb by astroglia compared with SY5Y cells from shared medium containing Pb; stimulation of intracellular Pb accumulation in astroglia by coculture with SY5Y cells but not astroglia	Lindahl et al., 1999
Reaggregate cultures of dispersed cells from fetal rat telencephalon	1, 10, or 100 µM Pb acetate daily for 10 d in serum-free medium on days 5–15 or 25–35 of culture; cells harvested immediately or after 11-d recovery; Pb measured by atomic absorption spectroscopy; biochemical assays for enzymes; immunocytochemistry for GFAP	Immature cultures accumulate and retain more Pb than differentiated cultures; glutamine synthetase activity is reduced in immature cultures, and in a more delayed fashion in differentiated cultures; GFAP amounts were reduced in immature cultures; microglia strongly activated	Zurich et al., 2002
Pb-Treated Animals			
Rhesus monkeys exposed to chronic low and moderate Pb levels (ok)	Prenatal and postnatal exposure to Pb acetate in diet, 350 (1.85 µM) or 600 ppm (2.88 µM) 9 y; protein detected by immunohistochemistry after Pb-free diet 32 months	Increased GFAP-positive radial glia and star-shaped vimentin-positive astroglia in hippocampus in a dose-related manner	Buchheim et al., 1994
Adult male rats exposed to low and moderate Pb levels	4, 8, or 12.5 mg/kg 5 d per week for 4 weeks; 4-week recovery; GFAP measured	Region-specific alteration in GFAP: elevated in frontal cortex, occipital cortex, striatum, and hippocampus, but not thalamus, cerebellum, or brain stem	VandenBerg et al., 1996
Young rats exposed to moderate Pb levels	Exposure from birth to postnatal day 20 via 0.02% Pb acetate in dams' drinking water	Decrease in gene expression of GFAP day 30 to day 50 in cerebellum	Zawia and Harry, 1996

Table 26.1 Effects of Pb on Astroglia-like Cell Lines, Astroglia in Culture, and Astrocytes *In Vivo* Reported 1990–2004 (*Continued*)

Model Type	Approach	Results	Ref.
Pb-Treated Animals			
Rats exposed to low Pb levels during early development, early development and adult hood, or postweaning	Exposure from conception to postnatal day 16 or 100 or from postnatal day 16 to 100 to 750 ppm Pb acetate in diet; qualitative histologic examination of brain tissues; immunohistochemistry and *in situ* hybridization for GFAP expression; *in situ* hybridization for vimentin	Increase in gliosis and mRNA levels of GFAP on day 100 in hippocampus of rats exposed postweaning, but not rats exposed earlier in development; no effect on vimentin	Peters at al., 1994; Stoltenburg-Didinger et al., 1996
Young rats exposed to moderately high Pb levels	15-month exposure from postnatal day 1 or 7 via lactation with 1% Pb acetate in dams' drinking water, followed by direct exposure to 1% Pb acetate in drinking water after weaning at postnatal day 25; GFAP and vimentin detected by immunohistochemistry, light microscopy, and computer-assisted image analysis	Transient increase of immunoreactivity for GFAP in hippocampus and cerebellum after 2–3 months; change in GFAP expression to vimentin in hypertrophied astrocytes of hippocampus after 12 months	Selvín-Testa et al., 1995
Young Long-Evans rats exposed to moderate Pb levels	Exposure from birth to postnatal day 15 via 0.02% Pb acetate in dams' drinking water; AP-1 DNA binding activity measured by gel shift; Jun protein detected by immunoblot	Significant increase in AP-1 DNA binding activity on postnatal day 3, not on day 9 and 15, in hippocampus and cortex; Jun protein level unchanged	Pennypacker et al., 1997
Young rats exposed to moderately high Pb levels	Exposure from prenatal to postnatal day 160 via gestation and lactation with 1% Pb acetate in dams' drinking water, followed by direct exposure to 1% Pb acetate in drinking water after weaning at postnatal day 25, or exposure starting at postnatal day 1; proteins detected by immunohistochemistry, transmission electron microscopy, and computer-assisted image analysis	Increase in HSP70 from postnatal day 21 to 45 in hippocampus of prenatally exposed group, decreasing to control level after postnatal day 60; elevation of GFAP and vimentin levels from postnatal day 60 to 90 in cerebral cortex and hippocampus in both prenatal and postnatal exposure groups; recovery of vimentin level after PND120 in hippocampus and cortex only in prenatally exposed group	Selvín-Testa et al., 1997
Adult rats acutely exposed to Pb	25 mg Pb acetate per kg body weight i.p. daily for 3 d; immunoblot of brain regions for GFAP	Increased GFAP expression in hippocampus and cerebral cortex, but not cerebellum	Struyñska et al., 2001
Adult male B6 mice acutely exposed to Pb	12.5 mg Pb acetate per kg body weight i.p.; killed 24 h later; immunohistochemistry for TNF-α; TUNEL assay for apoptosis	Increased expression of TNF-α protein; no effect on apoptosis	Cheng et al., 2002
Middle-aged (15-month-old) male rats	Exposure to water containing 200 mg Pb acetate per l for 2 months followed by intracerebroventricular injection of streptozotocin (STZ); killed after 1 month; additional Pb exposure for analysis of brain by immunocytochemistry and immunoblot	Blood Pb levels 10.8 µg/dl in both Pb-exposed groups; brain Pb doubles in Pb+STZ group compared with Pb alone; astroglial hypertrophy and sporadically increases staining for GRP94 protein in animals given Pb and STZ, an energy-depleting diabetogenic agent, but not STZ alone	Yun et al., 2000

the developmental pattern of astrocyte cytoskeletal protein glial fibrillary acidic protein (GFAP) (Buchheim et al., 1994; Peters et al., 1994; Stoltenburg-Didinger et al., 1996; VandenBerg et al., 1996; Zawia and Harry, 1996; Selvín-Testa et al., 1997; Stružyňska et al., 2001), and stimulates stress responses (Legare et al., 1993; Opanashuk et al., 1995a, 1995b; Selvín-Testa et al., 1997; Yun et al., 2000; Cheng et al., 2002; Qian and Tiffany-Castiglioni, 2003).

In Table 26.1, an attempt is made to categorize the level of Pb exposure used, as well as its duration. Pb concentrations used to treat cultures are listed and the presence or absence of serum in the medium is noted, if known. Serum has nonlinear effects on the apparent bioavailability of Pb to cultured cells, a topic that has received no systematic attention experimentally (Tiffany-Castiglioni, 1993). Glia in culture accumulate more Pb in serum-free than serum-supplemented medium (Qian et al., 1995), hypothetically because serum decreases either its bioavailability or suppresses cellular uptake or retention mechanisms. In addition, other components of medium precipitate Pb (Rowles et al., 1989). Thus it is difficult to arrive at a rule of thumb regarding whether a Pb concentration tested in culture is low, medium, or high. In general, however, throughout this chapter, concentrations ≤ 1.0 μM (about 20 µg/dl) will be considered low, 10 μM moderate, and ≥ 100 μM high. Likewise, doses given to animals are noted in Table 26.1. Qualifying remarks, such as low, moderate, or high exposure, are given for *in vivo* studies, based on blood Pb values reported in the study or typically associated with that dose if not reported, or on other effects reported in the study, such as weight loss, which characterizes high exposure.

26.2 EFFECTS OF LEAD ON GENE AND PROTEIN EXPRESSION

26.2.1 Regulation of Gene Expression

Theoretically, Pb may affect gene expression by at least two indirect mechanisms, and in either case the initial effect may be amplified through cell-signaling cascades. One mechanism is the postulated ability of Pb to substitute for physiologic metals such as Zn^{2+} and Ca^{2+} at their binding sites on peptides. In addition, Pb may disrupt cell homeostatic mechanisms that buffer against oxidative stress. Bouton and Pesvner (2000) provide a very useful review of Pb effects on gene expression that includes a "network diagram" of potential Pb targets (see Figure 1 of their article). Targets may be transcriptional regulators of gene expression, such as calmodulin, protein kinase C (PKC), and stress proteins, or they may be transcription factors, such as activator protein-1 (AP-1), zinc-finger protein Sp1, and nuclear factor kappa B (NF-kB). PKC has been shown to be activated by Pb in a number of studies (e.g., Laterra et al., 1992; Long et al., 1994; Tomsig and Suszkiw, 1995; Lu et al., 2001; Deng and Poretz, 2002). The transcriptional regulators act upon the transcription factors, which in turn activate or inhibit the expression of target genes. Thus, if Pb alters regulatory activity at either level, amplification of the response may occur. Two examples are given in Figure 26.1, AP-1 in astroglia and Sp1 in oligodendroglia. In the model diagrammed, pathways are summarized by which Pb may alter glial differentiation via actions upon PKC that are amplified through AP-1 or Sp1, or by a direct effect on Sp1. Dashed arrows indicate postulated pathways that have not been tested directly, and solid arrows indicate pathways that are supported by experimental evidence in cell culture models.

The promoter region of the *GFAP* gene contains a consensus AP-1 sequence, which is the binding site for dimers of fos, jun, and fos-like families of transcription factors (Masood et al., 1993; Pennypacker et al., 1994). In Pb-treated neonatal rats, AP-1 binding to DNA is significantly increased on postnatal day 3 in hippocampus and cortex, but not in the frontal lobe and brainstem (Pennypacker et al., 1997). However, the level of the jun protein, which appears to modulate expression of developmentally important genes, is unchanged. Within 60 minutes of exposure, Pb (10 μM) also elicits an increase in AP-1 DNA-binding activity in primary cultures of rat astroglia and the nuclear accumulation of redox factor-1 (Ref-1) (Scortegagna and Hanbauer, 2000).

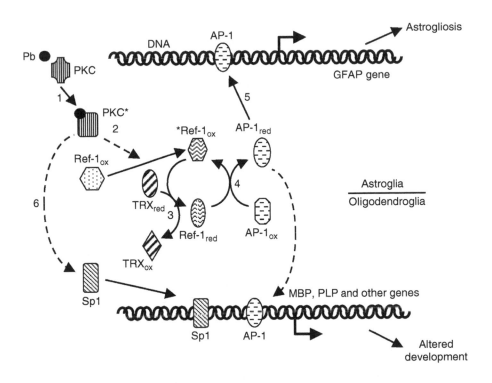

Figure 26.1 Modulation of gene expression by lead (Pb) in astroglia and oligodendroglia. This diagram presents a model whereby Pb may hypothetically alter the differentiation of astroglia or oligodendroglia or their responsiveness to stress by actions upon transcriptional regulators, such as protein kinase C (PKC), and transcription factors, such as activator protein-1 (AP-1) and the zinc-finger protein Sp1. AP-1 is a transcription factor that binds to the glial fibrillary acid protein (GFAP) gene promoter region in astroglia. Sp1 is a component of a transcriptional complex for gene regulation and targets genes for myelin basic protein (MBP) and proteolipid protein (PLP) in oligodendroglia, as well as other genes in other cell types. In the model diagrammed, the following possible effects of Pb are shown: (1) Pb activates PKC to PKC*, (2) PKC* phosphorylates oxidized redox factor-1 (Ref-1$_{ox}$ to *Ref-1$_{ox}$), (3) *Ref-1$_{ox}$ is reduced by reduced thioredoxin (TRX$_{red}$), (4) Ref-1$_{red}$ reduces oxidized AP-1 (AP-1$_{ox}$ to AP-1$_{red}$), (5) AP-1$_{red}$ binds in concert with other transcriptional factors to a targeted gene promoter and induces targeted gene expression, and (6) PKC* stimulates Sp1 binding to DNA by a separate mechanism. Pathways for which experimental data exist are depicted by solid arrows, and pathways for which little or no data yet exist are indicted by dashed arrows.

In Figure 26.1, Pb-activated PKC is shown to phosphorylate oxidized redox factor-1 (Ref-1$_{ox}$). Reduced thioredoxin (TRX$_{red}$) converts oxidized redox factor-1 (Ref-1$_{ox}$) into reduced Ref-1$_{red}$, which in turn reduces oxidized activator protein-1 (AP-1) to reduced activator protein-1 (AP-1$_{red}$) (Hirota et al., 1997; Wei et al., 2000). This form of AP-1 can bind directly to the promoter region of *GFAP* in astroglia (Sarid, 1991; Masood et al., 1993; Pennypacker et al., 1994) and probably also binds to the promoter region of *MBP* that contains AP-1 binding elements in oligodendroglia (Miskimins and Miskimins, 2001). These findings suggest that Pb may directly regulate *GFAP* gene expression by modulating AP-1 DNA binding activity. GFAP is an astroglial-specific intermediate filament protein that is expressed in mature astrocytes and that forms prominent bundles in astrogliosis (Eng et al., 2000). The linkage between GFAP regulation by Pb and Pb-induced gliosis has not been established, but GFAP induction hypothetically could be a mechanism for gliosis observed in some studies of Pb-exposed rats, as will be discussed later in this chapter (Peters et al., 1994; Stoltenburg-Didinger et al., 1996).

Though existing work on Sp1 pertains to oligodendroglia rather than astroglia, it is of interest to consider because it highlights the difficulties of predicting the effects of altered gene expression on subsequent development. A diagram of a possible interaction of Pb with Sp1 leading to upregulation of the genes for myelin basic protein (*MPB*) and proteolipid protein (*PLP*) expression is therefore included in Figure 26.1. Sp1 is a component of a transcriptional complex that participates in the regulation of genes rich in GC elements. *MBP* and *PLP* genes contain multiple Sp1 promoter regions, and Sp1 plays a critical role during oligodendrocyte development in the human brain (Henson et al., 1992). In the cerebellum of rats exposed to moderate Pb levels in the dams' water from birth to weaning, the peak developmental period of increased binding of the Sp1 consensus sequence to nuclear extracts is shifted from postnatal days 20–30 to days 5–10. This shift corresponds temporally with the premature expression of mRNAs for MBP and PLP (Zawia and Harry, 1996; Zawia et al., 1998). These findings suggest transiently accelerated oligodendroglial development. However, the investigators also found that cerebellar mRNA levels for MBP are depressed after weaning and into adulthood by Pb treatment, which suggests delayed development. Delayed development is consistent with a loss of myelin markers observed by Deng and Poretz (2001b) in rats exposed to moderately high Pb levels during gestation and with *in vitro* evidence that Pb blocks the developmental transition of oligodendroglial progenitor cells from antigenically distinct early to late stages (Deng and Poretz, 2001a, 2002). The developmental period of Pb exposure probably is a key factor in responses of oligodendroglial lineage cells to Pb exposure. Further investigation is needed to clearly identify target genes and gene products that are affected by Pb during each developmental event, as well as their window of vulnerability.

PKC activation is required for the delay of differentiation by oligodendroglial progenitor cells in culture (Deng and Poretz, 2002), which strengthens the circumstantial case that PKC is involved in the process of gene regulation depicted in Figure 26.1. PKC may thus activate Sp1 binding to MBP and PLP promoter regions in oligodendroglia. Evidence that this process occurs in neural cells has been reported for the rat PC12 pheochromocytoma cell line, in which PKC and mitogen-activated protein kinase (MAPK) participate in Pb-induced Sp1 activity. In PC12 cells exposed to Pb (0.1 μM) and nerve growth factor (NGF), administration of staurosporine, a PKC inhibitor, attenuates NGF and Pb-induced Sp1 DNA-binding, while PD 98059, a MAPK inhibitor, depletes both basal and induced Sp1 DNA-binding (Atkins et al., 2003). However, direct evidence is lacking in support of this mechanism in Pb-treated oligodendroglia.

26.2.2 Genes in Development

The effects of Pb on gene expression are important in two respects. First, brain development is controlled by genes, and the expression of these genes is programmed and coordinated in time and space. Either precocious or delayed gene expression induced by environmental chemicals may establish a physiologic substrate for developmental and degenerative disorders by subtle alterations in CNS structure and function. With the mapping of the human genome and clues provided by genetic polymorphisms, the genetic details of nervous system development and neurodegenerative disorders are beginning to come into focus. Astroglia outnumber neurons in the mammalian brain by about 5–10 to 1. Their functions are to support neurons, provide them with nutrients, and protect them from metabolic waste products and harmful chemicals. Recent evidence clearly demonstrates that astroglia play active roles in CNS development and function, including enhancement of synaptogenesis, regulation of synaptic plasticity, and induction of neurogenesis and neuronal differentiation (Carmignoto, 2000; Araque et al., 2001; Mauch et al., 2001; Ullian et al., 2001; Song et al., 2002). One can speculate that the alteration of gene expression in astroglia would modulate their functions not only as a "security screen" for neurons but also as "technical support" for neuronal functions. Pb, which is preferentially accumulated in astroglia, alters gene expression and therefore astroglial biology. Two gene products that have been studied by several groups are GS and GFAP, which serve as informative indicators of astroglial function and structure, respectively.

26.2.2.1 Glutamine Synthetase (GS)

Glutamine synthetase (GS, glutamate ammonia ligase, E.C. 6.3.1.2) is a critical enzyme of the glutamine-glutamate cycle in brain that amidates the excitatory neurotransmitter glutamate to glutamine in the presence of ATP (Benjamin and Quastel, 1975). Its almost exclusive localization to astroglia (Norenberg and Martinez-Hernandez, 1979) makes GS a subject of considerable interest because it could be a functional biomarker for astroglial toxicity. Furthermore, GS activity (Caldani et al., 1982) and gene expression (Mearow et al., 1989) serve as developmental markers for astroglia. Astroglia possess two excitatory amino acid transporters, EAAT1 and EAAT2 (GLAST and GLT-1), that remove most of the glutamate released by neurons (Rothstein et al., 1996). GS detoxifies ammonia and produces glutamine that is taken up by neighboring neurons for synthesis into glutamate or γ-amino-butyric acid (GABA) (Schousboe et al., 1992, 1993; Suarez et al., 2002). GS also provides glutamate-derived metabolites, such as citrate, that are used by neurons for energy. Inhibition of GS may therefore lead to ammonia toxicity (Benjamin and Quastel, 1975), depletion of glutamate stores in neurons (Rothstein and Tabakoff, 1984), and depletion of neuronal energy (Cohen, 1997).

Several studies listed in Table 26.1 report the alteration of GS activity by Pb in cell culture systems and *in vivo*. The reduction of activity has been shown to occur in cytosolic fractions of untreated cultured astroglia at micromolar and nanomolar Pb levels (Engle and Volpe, 1990; Sierra and Tiffany-Castiglioni, 1991), in several cell culture models at micromolar levels or lower (Sierra and Tiffany-Castiglioni, 1991; Liu et al., 2001; Zurich et al., 2001), and in the spinal cord of fetal guinea pigs and their dams (Sierra et al., 1989), though not in postnatal rat cerebellum, in which GS activity is increased (Cookman et al., 1988). GS is the only toxicologic endpoint in astroglia that has been shown to respond similarly to Pb exposure in cell homogenates, cultured cells, and the intact animal, and therefore offers a robust biomarker for Pb neurotoxicity studies. The lack of complete consensus among studies will be addressed later in this section. The study by Zurich et al. (2002) is of particular interest because it reports developmental differences between cultures of two ages.

Zurich et al. (2002) carried out a study with aggregating brain cell cultures consisting of tissue-like spheres of mixed cell types suspended in culture by constant rotation. In this tissue culture system, aggregates form spontaneously among dissociated cells from the fetal rat telencephalon, and investigators can assess characteristics of interacting astroglia, microglia, and neurons (Zurich et al., 2004). The investigators compared the responses of proliferating immature cultures treated from days 5–15 in culture with Pb acetate to differentiated cultures treated from days 25–35. In each case, some cultures were assessed at the end of the 10-d Pb exposure period and others after an additional 11-d posttreatment recovery period. This discussion will focus on results obtained from treatment with 1 μM Pb, though higher concentrations were also tested that were moderately or highly cytotoxic. The study showed that with 1 μM treatment, Pb content (nmol/mg protein) of the cultures, as measured by atomic absorption spectroscopy, is similar in immature and differentiated cultures and shows a reduction of at least 75% after recovery. The basis for the observed decrease in tissue Pb concentration, such as cell proliferation, loss of Pb-containing cells, or extrusion of Pb from the cells, was not addressed. However, Lindahl et al. (1999) have likewise reported loss of Pb from astroglial cultures not owing to loss of cells from the culture or dilution of total Pb in the culture from cell proliferation. Zurich et al. (2002) measured both the specific activity of GS and GFAP immunoreactivity. In agreement with other cell culture studies, they found that GS activity decreases by treatment with 1 μM Pb and extended this finding to both immature and differentiated cultures, as well as postrecovery cultures. Of further interest is that GFAP protein levels, as measured by immunoblots, are not decreased at this Pb level, except in the immature cultures following recovery. This finding suggests indirectly that astroglia are not selectively lost from the cultures. The staining intensity of GFAP was qualitatively examined in cryostat-cut sections of the cultures. Results suggest a mild gliosis at the end of the 10-d Pb treatment in the immature cultures after the 11-d recovery period. Whereas GFAP protein levels had not increased, as shown by immunoblot, this finding could reflect bundling or aggregation of GFAP. Increased numbers and clustering of microglia could be observed by immunohistochemical staining in areas of astrogliosis. The authors additionally measured specific activities of two neuronal enzymes in their cultures: choline acetyltransferase (ChAT) and glutamic acid decarboxylase (GAD). These enzymes showed no change with 1 μM Pb treatment except for a slight decrease in ChAT activity in immature cultures after the 11-d recovery period, the same treatment group in which GFAP levels were slightly decreased. These findings together suggest that GS is a very sensitive marker of astroglial responsiveness to Pb exposure.

The finding by Zurich et al. (2002) that an astroglial biomarker is more sensitive than neuronal biomarkers to Pb exposure is consistent with a study of rhesus monkeys postnatally exposed to Pb for 9 years in a diet containing 600 mg/kg, in which a remarkable decrease in the astroglial protein S-100 is observed in the hippocampus. In this study, levels of neuronal Ca-binding proteins, parvalbumin, calbindin D28K, and calretinin, remain unchanged (Noack et al., 1996), again indicating a greater sensitivity of astroglial than neuronal biomarkers to Pb exposure, a conclusion that is consistent with the notion that astroglia are a "lead sink" (Tiffany-Castiglioni et al., 1989).

In contrast to the above studies, Cookman et al. (1988) reported that rats exposed postnatally to moderately low Pb levels (400 mg Pb chloride/l to produce blood Pb levels of 17–24 µg/dl) show increased cerebellar GS activity, which suggests stimulation of precocious differentiation by Pb. This conclusion is supported by the investigators' *in vitro* study, in which rat C6 glioma and cerebellar astrocytes exposed to 1 to 100 μM Pb chloride for 3 days extend cytoplasmic processes, some of which resemble glial end-feet (Cookman et al., 1988). This phenotypic induction is cell-type specific, as it is not observed in neuro-2a neuroblastoma and chick spinal cord explants. In view of our earlier discussion of possible dual stimulatory and inhibitory effects of Pb on the same glial genes though transcription factors such as AP-1 and Sp1, future experiments should systematically address the stage of development during which Pb exposure is tested in order to more fully understand its long-term effects.

26.2.2.2 *Glial Fibrillary Acidic Protein (GFAP)*

The consensus view from *in vivo* studies is that GFAP expression is altered in animals exposed to moderate and high Pb levels and astrogliosis occurs under some conditions, although the types of responses observed are not consistent across studies. Astrogliosis is a unique morphological response in the CNS akin to scar formation in connective tissue. A feature of astrogliosis is an increase in GFAP level or its bundling into fibrous aggregates in the cell. Gliosis does not occur in purified astroglial cultures treated with Pb (Holtzman et al., 1987), though it does occur in mixed cultures in which microglia are present, as discussed earlier (Zurich et al., 2002). *In vivo*, variability in Pb responses appears to arise from differences in experimental designs used, including brain region studied, age during Pb exposure, dose and duration of exposure, and posttreatment recovery period. The findings of various studies are listed in Table 26.1. A general conclusion from these studies is that the susceptibility of the brain to astrogliosis after moderately high Pb exposure changes depending on the period of development during which the brain was first challenged (Peters et al., 1994; Buchheim et al., 1994; Selvín-Testa et al., 1995, 1997; Strużyñska et al., 2001).

Two findings of particular interest with regard to Pb effects on GFAP will be highlighted. The first is that GFAP gene expression in brain regions differs in response to moderate Pb exposure. In rats exposed from birth to postnatal day 20 via 0.02% Pb acetate in dams' drinking water, the peak expression of GFAP mRNA is delayed from postnatal day 15 to 20 in the cortex and accelerated from postnatal day 25 to 20 in the hippocampus (Harry et al., 1996). Furthermore, GFAP gene expression is decreased in the cerebellum of rat pups after postnatal day 20 (Zawia and Harry, 1996). Precocious GFAP mRNA expression in the hippocampus is consistent with the observation that AP-1 binding to DNA is increased in Pb-exposed neonatal rats on postnatal day 3, not on days 9 and 15, in hippocampus and cortex (Pennypacker et al., 1997). The existence of an AP-1 site in the *GFAP* promoter region, as discussed earlier, implicates Pb in regulating *GFAP* gene expression by modulating AP-1 DNA binding activity (Sarid, 1991; Pennypacker et al., 1994). In another study of prolonged Pb exposure in rats to moderately high Pb levels, three phases of responsiveness with regard to levels of GFAP and vimentin were found. In this model, consisting of a 15-month exposure from postnatal day 1 or 7 via lactation with 1% Pb acetate in the dams' drinking water, followed by direct exposure to 1% Pb acetate in drinking water after weaning at postnatal day 25, immunohistochemical staining for GFAP increases in hippocampal and cerebellar astrocytes during the first 2–3 months of Pb exposure and then declines to control levels during the next 4–12 months. However, a recurrence of increased GFAP staining and an inappropriate appearance of the immature glial protein vimentin are observed in cerebral cortex after 14 months of Pb exposure (Selvín-Testa et al., 1995).

In order to understand the involvement of astroglial development in Pb neurotoxicity through the measurement of the expression of astroglial marker genes, some fundamental issues must be taken into consideration. Astroglia from different brain regions of the brain have different time courses of development (Harry et al., 1996). Thus, their vulnerability to Pb dose and exposure duration will be different. Another consideration is that astroglial responses to Pb exposure change depending on the developmental stage in which they occur. This subject has already been explored with regard to GFAP. An additional aspect of the same issue is that astroglia of different developmental stages may carry different burdens of Pb. The ability of astrocytes to sequester Pb from neurons develops at about postnatal day 20–22 in the cerebral cortex of rats exposed acutely to high Pb levels (Holtzman et al., 1984). This phenomenon has been reproduced in cell culture at low Pb levels, where it seems to be dependent on neuronal induction of the Pb-sequestering property in astroglia (Lindahl et al., 1999). The subject of tolerance to intracellular accumulated Pb will be further discussed in the next section on stress proteins.

Responses to Pb exposure at the transcriptional level provide some additional appreciation of underlying developmental processes of Pb neurotoxicity. AP-1 has already been mentioned. Another candidate transcription factor by which Pb might modulate *GFAP* gene expression is NF-*k*B, a critical mediator in signal pathway of stress response. NF-*k*B is activated in C6 rat glioma cells exposed to 1 μM Pb acetate daily for 1 week (Qian et al., 2001), and the *GFAP* promoter region contains an NF-*k*B binding site (Krohn et al., 1999). However, mutations in the NF-*k*B binding site of the *GFAP* promoter region do not block induction by transforming growth factor-1β (TGF-1β) or inhibition by interleukin-1β (IL-1β) (Krohn et al., 1999). Whereas TGF-1β and IL-1β are involved in the activation of astrogliosis (Duguid et al., 1989; Steward et al., 1990; Logan et al., 1992), it is unlikely that NF-*k*B modulates *GFAP* gene expression in astrogliosis. However, the activation of NF-*k*B by Pb exposure occurs in neurons (Ramesh et al., 1999; Ramesh and Jadhav, 2001) and is therefore still of interest in mechanisms of Pb neurotoxicity.

26.2.3 Stress-Responsive Genes

Several studies listed in Table 26.1 provide evidence that Pb induces stress responses in astroglia. For example, *de novo* biosynthesis of stress proteins, including heat-shock protein 70 (HSP70), occurs in rat astroglial cultures after short-term Pb exposure (Opanashuk and Finkelstein, 1995a, 1995b). The transient induction of HSP70 protein also occurs in brain tissue of post-weanling rats exposed to a moderately high Pb dose during gestation and postnatally (Selvín-Testa et al., 1997). Increased mRNA and protein levels of the endoplasmic reticulum stress protein glucose-regulated protein 78 (GRP78) occur in C6 cells (Qian et al., 2001; Qian and Tiffany-Castiglioni, 2003). GRP94, another glucose-regulated protein, is upregulated in the astrocytes of chronically Pb-exposed middle-aged rats when they receive an intracerebroventricular (i.c.v.) injection of strepto-zotocin (STZ) to inhibit glucose metabolism and energy formation (Yun et al., 2000). This finding further implicates the GRP family of proteins in Pb neurotoxicity *in vivo*, because GRP78 and

GRP94 are regulated by glucose (Shiu et al., 1977). Some of the stress responses induced by Pb may be related to oxidative stress, as cultured astroglia exposed to low Pb levels show transient depletion of intracellular GSH levels (Legare et al., 1993), loss of mitochondrial membrane potential (Legare et al., 1993), and increased binding of the reactive oxygen species (ROS) activated transcription factors AP-1 (Scortegagna and Hanbauer, 2000) and NF-kB to DNA (Qian et al., 2001). The source of oxidative stress may be transiently increased copper ion levels through the blockage by Pb of the Cu transporter ATP7a (Rowles et al., 1989; Qian et al., 1995; Qian et al., 1999).

Although the studies cited above report the induction of stress proteins in Pb-treated astroglia, no attempt has yet been made to link tolerance of intracellular accumulated Pb in astroglia to stress protein induction. However, an *in vivo* study by Selvín-Testa and colleagues (1997) provides a temporal context in which to consider both stress responses and gliotic responses of astroglia to moderately high Pb doses. In rats exposed to Pb during gestation, lactation, and postweaning postnatal, HSP70 immunoreactivity increases from postnatal day 21 to 45 in astroglia and nonpyramidal neurons. Astroglial hypertrophy is observed at the same time. GFAP and vimentin immunoreactivity increase after postnatal day 45, when HSP70 induction subsides. This time course indicates that Pb exposure takes several weeks to induce HSP70 expression; induction of HSP70 occurs before the accumulation or aggregation of intermediate filament proteins (Selvín-Testa et al., 1997). New questions arising from this study are whether astroglia between postnatal days 21 and 45 are more sensitive to Pb than other developmental stages, whether threshold level of intracellular Pb must be reached to trigger HSP70 induction, and whether astroglia have other mechanisms to adapt to and tolerate intracellular Pb accumulation, such as Pb-binding proteins. Pb-binding proteins are hypothesized to provide a Pb-buffering capability in the cytoplasm. However, no data are available as yet that demonstrate this capacity or link their induction with Pb exposure (Fowler, 1998). In this section, one stress protein induced by Pb, GRP78, which coincidentally binds strongly to Pb, will be discussed, as well as the cytokine TNF-α, which is implicated in stress responses in the brain.

26.2.3.1 *Glucose-Regulated Protein 78 (GRP78)*

GRP78 is a molecular chaperone localized to the endoplasmic reticulum (ER) that can be upregulated by chemicals that induce oxidative stress or cause an accumulation of unfolded or misfolded proteins to induce the "unfolded protein response" (Little et al., 1994). GRP78 is involved in protein folding and trafficking and in calcium homeostasis (Little et al., 1994; McCormick et al., 1997; Liu et al., 1998). In eukaryotic cells, GRP78 binds transiently to proteins traversing through the ER and facilitates their folding, assembly, and transport (Simons et al., 1995; Hendershot et al., 1996; Hamman et al., 1998). We have reported interactions between Pb and GRP78 in C6 rat glioma cells and cultured astroglia, including upregulation of GRP78 by Pb and binding of Pb to GRP78. GRP78 gene and protein expression is elevated in rat C6 glioma cells exposed to 1 μM Pb acetate for 1 week, as observed by Northern blot analysis and immunodetection and confirmed by amino acid sequencing of the protein (Qian et al., 2000; Qian et al., 2001). Induction of GRP78 is also observed in rat primary astroglial cultures exposed to 1 μM Pb for 1 week (Qian and Tiffany-Castiglioni, 2003). In addition, Pb specifically binds to GRP78 *in vitro*, which was demonstrated by use of a Pb-affinity column (Qian et al., 2000). An illustration of the Pb-affinity column and immunoblot methods used to detect binding of Pb to GRP78 is given in Figure 26.2A and B. Pb binding to GRP78 is also supported by the observation that more than 90% of glia-associated Pb is localized to a protein fraction of C6 cells with molecular weights larger than 66 kD (Qian et al., 1999). In addition, we have found that Pb induces GRP78 aggregation intracellularly. A diagram of the experimental method used to detect the GRP78 aggregation and fluorescence micrographs of control and Pb-treated cells expressing a GRP78/green fluorescence protein chimera are presented in Figure 26.3A and B (Qian et al., unpublished data).

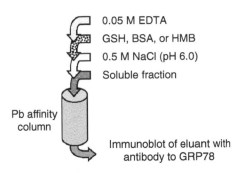

0.05 M EDTA
GSH, BSA, or HMB
0.5 M NaCl (pH 6.0)
Soluble fraction

Pb affinity
column

Immunoblot of eluant with
antibody to GRP78

Figure 26.2 Specificity of binding by lead to GRP78. (A) Diagram of the experimental method used to detect binding of cytosolic proteins to a Pb affinity column. The Pb affinity column was prepared from a proprietary Ni column from which the Ni was removed and replaced with Pb. A high-molecular-weight cytosolic fraction of C6 cells in which moieties that bind Pb were thought to reside was poured through the column, followed by successive washes of buffered saline, buffer containing a potentially competing Pb-binding moiety, and EDTA. The eluant was run on an immunoblot gel and probed with an antibody to GRP78. Three moieties were tested for their abilities to dislodge any bound protein in the cytosol from the Pb column: glutathione (GSH), bovine serum albumin (BSA), and the heavy-metal-binding (HMB) region of the Menkes protein, ATP7a. (B) Immunoblot of eluants from the Pb-affinity column. GRP78 protein was detected in eluants from all EDTA washes, but not from the buffered saline or GSH washes, indicating that GRP78 was strongly bound by the Pb-affinity column. A band corresponding to BSA was found in the BSA eluant, but no band corresponding to GRP78, indicating cross-reactivity of the antibody with BSA. A band corresponding to GRP78 was detected in the HMB eluant, indicating that HMB partially dislodged GRP78 from the Pb-affinity column and therefore also binds Pb. (Figure 26.2B from Qian, Y. et al., *Toxicol. Appl. Pharmacol.*, 163, 260, 2000. With permission.)

In Pb-exposed astroglia, Pb binding to GRP78 may provide a transient defense mechanism to prevent cells from damage induced by the free Pb ion until it is deposited in other unknown storage sites in the cell. The idea that GRP78 itself is a Pb-storage protein is biologically implausible because of its critical chaperone functions. However, we propose that GRP78 is involved in the transport of Pb to other storage sites (Qian and Tiffany-Castiglioni, 2003).

Thus GRP78 may contribute to astroglial tolerance of intracellular Pb. On the other hand, whereas Pb induces GRP78 expression, this event could also be pathologic, as GRP78 is overexpressed in malignant human breast lesions (Fernandez et al., 2000).

The mechanism of regulation of *GRP78* gene expression by Pb is unknown. However, several putative xenobiotic-regulated sequences, including the NF-kB consensus sequence, have been identified in the *GRP78* gene promoter region (Resendez et al., 1988; Roy et al., 1996). We have observed that activation of NF-kB DNA binding activity is accompanied by the induction of GRP78 by Pb treatment in rat C6 glioma cells (1–10 μM for 3 h) (Qian et al., 2001). In addition, the parallel increase of NF-kB mRNA levels and GRP78 gene expression occurs in cultured rat astroglia treated with 1 μM Pb (Qian and Tiffany-Castiglioni, 2003). Our current view is that Pb binding to GRP78, mimicking the unfolded protein response (UPR), is probably an early step in regulating GRP78 expression (Forman et al., 2003; Qian and Tiffany-Castiglioni, 2003).

26.2.3.2 *Tumor Necrosis Factor-α (TNF-α)*

Cytokines are a group of small polypeptides that play important roles in immune responses in the CNS. An example is tumor necrosis factor-α (TNF-α), a proinflammatory cytokine that not only acts on astroglia but is also produced by them in response to specific signals (John et al., 2003). Available data suggest that TNF-α plays both promoting and inhibiting roles in the nervous system because of the existence of two distinguishable signaling pathways mediated through TNF receptors 1 and 2 (TNFR1 and TNFR2) (John et al., 2003). TNF-α is involved in the induction of astrogliosis (Ait-Ikhlef et al., 1999; Harry et al., 2002). Two findings support the idea that TNF-α can mediate

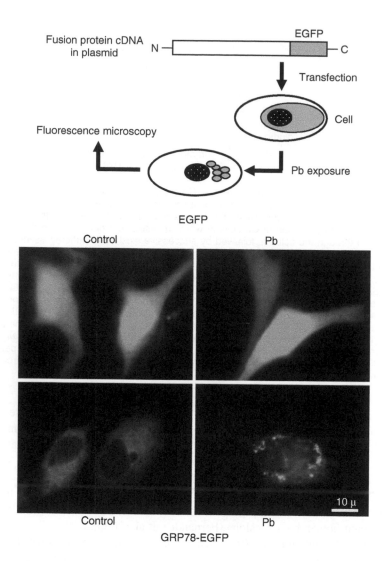

Figure 26.3   *(A color version of part b of this figure follows page 236.)* Lead-induced GRP78 aggregation in human CCF-STTG1 astrocytoma cells. (A) Diagram of an immunofluorescence approach used to examine the cellular distribution of GRP78 in Pb-treated cells. Cells were transiently transfected with a cDNA construct of GRP78 and enhanced green fluorescence protein (EGFP) in order to induce cells to express the chimeric fluorescent GRP78 protein (GRP78-EGFP). Fluorescence was detected by digital microscopic imaging. (B) Representative micrographs of transfected control cells and cells that had been exposed to 5 μM Pb in medium for 10 h. The clumping of the fluorescent chimeric protein is observed only in the Pb-treated cells transfected with GRP78-EGFP, suggesting that Pb causes GRP78 aggregation. The clumping does not appear in Pb-treated cells transfected with EGFP only.

Pb-induced astrogliosis. First, Pb upregulates TNF-α gene expression in human U-373MG glioma cells (Liu et al., 2000). Second, in rat C6 glioma cells, Pb enhances TNF-α-induced gene expression of the matrix metalloproteinases, an important family of proteases that are involved in both physiological and pathological processes of the CNS (Lahat et al., 2002). Furthermore, the elevation of TNF-α induced by Pb does not trigger cell apoptosis in human U-373MG glioma cells (Cheng et al., 2002), implying that Pb-induced TNF-α plays a promoting role in the activation of astrogliosis.

26.3 Conclusions and Future Directions

This review considers gene responses and stress responses of astroglia exposed to low and moderate levels of Pb, particularly as they change over the time course of astroglial development and aging. One of the most striking developmental differences between immature and mature astroglia is their capacity for the accumulation and retention of Pb. The main ideas presented in this chapter and possible future directions for research are as follows:

1. Pb, which is preferentially accumulated in astroglia, alters gene expression and therefore astroglial biology. The alteration of gene expression in astroglia would modulate their functions not only as a "security screen" for neurons but also as "technical support" for neuronal functions.
2. Targets for Pb-induced astroglial responses may be transcriptional regulators of gene expression, such as PKC and stress proteins, such as GRP78, or they may be transcription factors, such as AP-1.
3. Two gene products that have been studied most extensively are glutamine synthetase (GS) and glial fibrillary acidic protein (GFAP), which serve as informative indicators of astroglial function and structure, respectively. The localization of GS to astroglia and its essential metabolic roles in the brain make it a subject of considerable interest because it could be a functional biomarker for astroglial toxicity. GFAP induction by AP-1 may be a mechanism for gliosis observed in some studies of Pb-exposed rats.
4. Because dual stimulatory and inhibitory effects of Pb on the same glial genes are possible though transcription factors such as AP-1 and Sp1, future experiments should systematically address the stage of development during which Pb exposure is tested in order to more fully understand its long-term effects.
5. Stress proteins are induced in Pb-treated astroglia and may be linked to tolerance of astroglia to intracellular accumulated Pb.

ACKNOWLEDGMENTS

The authors' work is supported by National Institutes of Health Grants P42 ES04917, P30 ES09106, and T32 ES07273.

REFERENCES

Ait-Ikhlef, A., Hantaz-Ambroise, D., Jacque, C., Belkadi, L., and Rieger, F. (1999). Astrocyte proliferation induced by wobbler astrocyte conditioned medium is blocked by tumor necrosis factor-alpha (TNF-alpha) and interleukin-1beta (IL-1beta) neutralizing antibodies *in vitro*, *Cell Mol. Biol.*, *(Noisy-le-grand)*, 45, 393.

Anttila, A., Heikkila, P., Nykyri, E., Kauppinen, T., Pukkala, E., Hernberg, S., and Hemminki, K. (1996). Risk of nervous system cancer among workers exposed to lead, *J. Occup. Environ. Med.*, 38, 131.

Araque, A., Carmignoto, G., and Haydon, P.G. (2001). Dynamic signaling between astrocytes and neurons, *Annu. Rev. Physiol.*, 63, 795.

Atkins, D.S., Basha, M.R., and Zawia, N.H. (2003). Intracellular signaling pathways involved in mediating the effects of lead on the transcription factor Sp1, *Int. J. Dev. Neurosci.*, 21, 235.

Banks, E.C., Ferretti, L.E., and Shucard, D.W. (1997). Effects of low level lead exposure on cognitive function in children: a review of behavioral, neuropsychological and biological evidence, *Neurotoxicology*, 18, 237.

Bellinger, D., Leviton, A., Rabinowitz, M., Needleman, H., and Waternaux, C. (1986). Correlates of low-level lead exposure in urban children at 2 years of age, *Pediatrics*, 77, 826.

Bellinger, D.C. and Needleman, H.L. (2003). Intellectual impairment and blood lead levels, *N. Engl. J. Med.*, 349, 500; author reply 500.

Bellinger, D.C., Stiles, K.M., and Needleman, H.L. (1992). Low-level lead exposure, intelligence and academic achievement: a long-term follow-up study, *Pediatrics*, 90, 855.

Benjamin, A.M. and Quastel, J.H. (1975). Metabolism of amino acids and ammonia in rat brain cortex slices *in vitro*: a possible role of ammonia in brain function, *J. Neurochem.*, 25, 197.

Bouton, C.M., Hossain, M.A., Frelin, L.P., Laterra, J., and Pevsner, J. (2001). Microarray analysis of differential gene expression in lead-exposed astrocytes, *Toxicol. Appl. Pharmacol.*, 176, 34.

Bouton, C.M. and Pevsner, J. (2000). Effects of lead on gene expression, *Neurotoxicology*, 21, 1045.

Bradbury, M.W.B. and Deane, R. (1993). Permeability of the blood-brain barrier to lead, *Neurotoxicology*, 14, 131.

Buchheim, K., Noack, S., Stoltenburg, G., Lilienthal, H., and Winneke, G. (1994). Developmental delay of astrocytes in hippocampus of rhesus-monkeys reflects the effect of prenatal and postnatal chronic low-level lead-exposure, *Neurotoxicology*, 15, 665.

Caldani, M., Rolland, B., Fages, C., and Tardy, M. (1982). Glutamine synthetase activity during mouse brain development, *Experientia*, 38, 1199.

Canfield, R.L., Henderson, C.R., Jr., Cory-Slechta, D.A., Cox, C., Jusko, T.A., and Lanphear, B.P. (2003). Intellectual impairment in children with blood lead concentrations below 10 microg per deciliter, *N. Engl. J. Med.*, 348, 1517.

Carmignoto, G. (2000). Reciprocal communication systems between astrocytes and neurones, *Prog. Neurobiol.*, 62, 561.

Centers for Disease Control (1991). Preventing lead poisoning in young children: a statement by the Centers for Disease Control, Atlanta.

Charney, E., Sayre, J., and Coulter, M. (1980). Increased lead absorption in inner city children: where does the lead come from?, *Pediatrics*, 65, 226.

Cheng, Y.-J., Yang, B.-C., Hsieh, W.-C., Huang, B.-M., and Liu, M.-Y. (2002). Enhancement of TNF-α expression does not trigger apoptosis upon exposure of glial cells to lead and lipopolysaccharide, *Toxicology*, 178, 183.

Cohen, D.M. (1997). Inhibition of glutamine synthetase induces critical energy threshold for neuronal survival, *Ann. N.Y. Acad. Sci.*, 826, 456.

Cookman, G.R., Hemmens, S.E., Keane, G.J., King, W.B., and Regan, C.M. (1988). Chronic low level lead exposure precociously induces rat glial development *in vitro* and *in vivo*, *Neurosci. Lett.*, 86, 33.

Dave, V., Vitarella, D., Aschner, J.L., Fletcher, P., Kimelberg, H.K., and Aschner, M. (1993). Lead increases inositol 1,4,5-trisphosphate levels but does not interfere with calcium transients in primary rat astrocytes, *Brain Res.*, 618, 9.

Deng, W. and Poretz, R.D. (2001a). Lead exposure affects levels of galactolipid metabolic enzymes in the developing rat brain, *Toxicol. Appl. Pharmacol.*, 172, 98.

Deng, W. and Poretz, R.D. (2001b). Lead alters the developmental profile of the galactolipid metabolic enzymes in cultured oligodendrocyte lineage cells, *Neurotoxicology*, 22, 429.

Deng, W. and Poretz, R.D. (2002). Protein kinase C activation is required for the lead-induced inhibition of proliferation and differentiation of cultured oligodendroglial progenitor cells, *Brain Res.*, 929, 87.

Duguid, J.R., Bohmont, C.W., Liu, N.G., and Tourtellotte, W.W. (1989). Changes in brain gene expression shared by scrapie and Alzheimer disease, *Proc. Natl. Acad. Sci. U.S.A.*, 86, 7260.

Eng, L.F., Ghirnikar, R.S., and Lee, Y.L. (2000). Glial fibrillary acidic protein: GFAP-thirty-one years (1969–2000), *Neurochem. Res.*, 25, 1439.

Engle, M.J. and Volpe, J.J. (1990). Glutamine synthetase activity of developing astrocytes is inhibited *in vitro* by very low concentrations of lead, *Dev. Brain Res.*, 55, 283.

Fernandez, P.M., Tabbara, S.O., Jacobs, L.K., Manning, F.C., Tsangaris, T.N., Schwartz, A.M., Kennedy, K.A., and Patierno, S.R. (2000). Overexpression of the glucose-regulated stress gene GRP78 in malignant but not benign human breast lesions, *Breast Cancer Res. Treat.*, 59, 15.

Forman, M.S., Lee, V.M., and Trojanowski, J.Q. (2003). 'Unfolding' pathways in neurodegenerative disease, *Trends Neurosci.*, 26, 407.

Fowler, B.A. (1998). Roles of lead-binding proteins in mediating lead bioavailability, *Environ. Health Perspect.*, 106 (Suppl. 6), 1585.

Garber, M.M. and Heiman, A.S. (2002). The *in vitro* effects of Pb acetate on NO production by C6 glial cells, *Toxicol. In Vitro*, 16, 499.

Gorell, J.M., Johnson, C.C., Rybicki, B.A., Peterson, E.L., Kortsha, G.X., Brown, G.G., and Richardson, R.J. (1999). Occupational exposure to manganese, copper, lead, iron, mercury and zinc and the risk of Parkinson's disease, *Neurotoxicology*, 20, 239.

Gorell, J.M., Rybicki, B.A., Johnson, C.C., and Peterson, E.L. (1997). Occupational metal exposures and the risk of Parkinson's disease, *Neuroepidemiology*, 18, 303.

Hamman, B.D., Hendershot, L.M., and Johnson, A.E. (1998). BiP maintains the permeability barrier of the ER membrane by sealing the lumenal end of the translocon pore before and early in translocation, *Cell*, 92, 747.

Harry, G.J., Schmitt, T.J., Gong, Z., Brown, H., Zawia, N., and Evans, H.L. (1996). Lead-induced alterations of glial fibrillary acidic protein (GFAP) in the developing rat brain, *Toxicol. Appl. Pharmacol.*, 139, 84.

Harry, G.J., Tyler, K., d'Hellencourt, C.L., Tilson, H.A., and Maier, W.E. (2002). Morphological alterations and elevations in tumor necrosis factor-α, interleukin (IL)-1α, and IL-6 in mixed glia cultures following exposure to trimethyltin: Modulation by proinflammatory cytokine recombinant proteins and neutralizing antibodies, *Toxicol. Appl. Pharm.*, 180, 205.

Hendershot, L., Wei, J., Gaut, J., Melnick, J., Aviel, S., and Argon, Y. (1996). Inhibition of immunoglobulin folding and secretion by dominant negative BiP ATPase mutants, *Proc. Natl. Acad. Sci. U.S.A.*, 93, 5269.

Henson, J., Saffer, J., and Furneaux, H. (1992). The transcription factor Sp1 binds to the JC virus promoter and is selectively expressed in glial cells in human brain, *Ann. Neurol.*, 32, 72.

Hirota, K., Matsui, M., Iwata, S., Nishiyama, A., Mori, K., and Yodoi, J. (1997). AP-1 transcriptional activity is regulated by a direct association between thioredoxin and Ref-1, *Proc. Natl. Acad. Sci. U.S.A.*, 94, 3633.

Holtzman, D., DeVries, C., Nguyen, H., Olson, J., and Bensch, K. (1984). Maturation of resistance to lead encephalopathy: cellular and subcellular mechanisms, *Neurotoxicology*, 5, 97.

Holtzman, D., Olson, J.E., DeVries, C., and Bensch, K. (1987). Lead toxicity in primary cultured cerebral astrocytes and cerebellar granular neurons, *Toxicol. Appl. Pharmacol.*, 89, 211.

John, G.R., Lee, S.C., and Brosnan, C.F. (2003). Cytokines: powerful regulators of glial cell activation, *Neuroscientist*, 9, 10.

Kerper, L.E. and Hinkle, P.M. (1997). Cellular uptake of lead is activated by depletion of intracellular calcium stores, *J. Biol. Chem.*, 272, 8346.

Kramer, K.K., Liu, J., Choudhuri, S., and Klaassen, C.D. (1996). Induction of metallothionein mRNA and protein in murine astrocyte cultures, *Toxicol. Appl. Pharmacol.*, 136, 94.

Krohn, K., Rozovsky, I., Wals, P., Teter, B., Anderson, C.P., and Finch, C.E. (1999). Glial fibrillary acidic protein transcription responses to transforming growth factor-beta1 and interleukin-1beta are mediated by a nuclear factor-1-like site in the near-upstream promoter, *J. Neurochem.*, 72, 1353.

Lahat, N., Shapiro, S., Froom, P., Kristal-Boneh, E., Inspector, M., and Miller, A. (2002). Inorganic lead enhances cytokine-induced elevation of matrix metalloproteinase MMP-9 expression in glial cells, *J. Neuroimmunol.*, 132, 123.

Lai, J., Porreca, F., Hunter, J.C., and Gold, M.S. (2004). Voltage-gated sodium channels and hyperalgesia, *Annu. Rev. Pharmacol. Toxicol.*, 44, 371.

Laterra, J., Bressler, J.P., Indurti, R.R., Belloni-Olivi, L., and Goldstein, G.W. (1992). Inhibition of astroglia-induced endothelial differentiation by inorganic lead: a role for protein kinase C, *Proc. Natl. Acad. Sci. U.S.A.*, 89, 10748.

Legare, M.E., Barhoumi, R., Burghardt, R.C., and Tiffany-Castiglioni, E. (1993). Low-level lead exposure in cultured astroglia: identification of cellular targets with vital fluorescent probes, *Neurotoxicology*, 14, 267.

Legare, M.E., Barhoumi, R., Hebert, E., Bratton, G.R., Burghardt, R.C., and Tiffany-Castiglioni, E. (1998). Analysis of Pb^{2+} entry into cultured astroglia, *Toxicol. Sci.*, 46, 90.

Li, P. and Rossman, T.G. (2001). Genes upregulated in lead-resistant glioma cells reveal possible targets for lead-induced developmental neurotoxicity, *Toxicol. Sci.*, 64, 90, 2001.

Lindahl, L., Bird, L., Legare, M.E., Mikeska, G., Bratton, G.R., and Tiffany-Castiglioni, E. (1999). Differential ability of astroglia and neuronal cells to accumulate lead: Dependence on cell type and on degree of differentiation, *Toxicol. Sci.*, 50, 236.

Little, E., Ramakrishnan, M., Roy, B., Gazit, G., and Lee, A.S. (1994). The glucose-regulated proteins (GRP78 and GRP94): functions, gene regulation, and applications, *Crit. Rev. Eukaryot. Gene Expr.*, 4, 1.

Liu, H., Miller, E., van de Water, B, and Stevens, J.L. (1998). Endoplasmic reticulum stress proteins block oxidant-induced Ca^{2+} increases and cell death, *J. Biol. Chem.*, 273, 12858.

Liu, M.Y., Hsieh, W.C., and Yang, B.C. (2000). *In vitro* aberrant gene expression as the indicator of lead-induced neurotoxicity in U-373MG cells, *Toxicology*, 147, 59.

Logan, A., Frautschy, S.A., Gonzalez, A.M., Sporn, M.B., and Baird, A. (1992). Enhanced expression of transforming growth factor beta 1 in the rat brain after a localized cerebral injury, *Brain Res.*, 587, 216.

Long, G.J., Rosen, J.F., and Schanne, F.A. (1994). Lead activation of protein kinase C from rat brain. Determination of free calcium, lead, and zinc by 19F NMR, *J. Biol. Chem.*, 269, 834.

Lu, H., Guizzetti, M., and Costa, L.G. (2001). Inorganic lead stimulates DNA synthesis in human astrocytoma cells: role of protein kinase Cα, *J. Neurochem.*, 78, 590.

Masood, K., Besnard, F., Su, Y., and Brenner, M. (1993). Analysis of a segment of the human glial fibrillary acidic protein gene that directs astrocyte-specific transcription, *J. Neurochem.*, 61, 160.

Mauch, D.H., Nagler, K., Schumacher, S., Goritz, C., Muller, E.C., Otto, A., and Pfrieger, F.W. (2001). CNS synaptogenesis promoted by glia-derived cholesterol, *Science*, 294, 1354.

McCormick, T.S., McColl, K.S., and Distelhorst, C.W. (1997). Mouse lymphoma cells destined to undergo apoptosis in response to thapsigargin treatment fail to generate a calcium-mediated grp78/grp94 stress response, *J. Biol. Chem.*, 272, 6087.

Mearow, K.M., Mill, J.F., and Vitkovic, L. (1989). The ontogeny and localization of glutamine synthetase gene expression in rat brain, *Brain Res. Mol. Brain Res.*, 6, 223.

Miskimins, R. and Miskimins, W.K. (2001). A role for an AP-1-like site in the expression of the myelin basic protein gene during differentiation, *Int. J. Dev. Neurosci.*, 19, 85.

Needleman, H.L., Riess, J.A., Tobin, M.J., Biesecker, G.E., and Greenhouse, J.B. (1996). Bone lead levels and delinquent behavior, *JAMA*, 275, 363.

Nihei, M.K. and Guilarte, T.R. (2001). Molecular changes in glutamatergic synapses induced by Pb2+: association with deficits of LTP and spatial learning, *Neurotoxicology*, 22, 635.

Noack, S., Lilienthal, H., Winneke, G., and Stoltenburg-Didinger, G. (1996). Immunohistochemical localization of neuronal and glial calcium-binding proteins in hippocampus of chronically low level lead exposed rhesus monkeys, *Neurotoxicology*, 17, 679.

Norenberg, M.D. and Martinez-Hernandez, A. (1979). Fine structural localization of glutamine synthetase in astrocytes of rat brain, *Brain Res.*, 161, 303.

Opanashuk, L.A. and Finkelstein, J.N. (1995a). Relationship of lead-induced proteins to stress response proteins in astroglial cells, *J. Neurosci. Res.*, 42, 623.

Opanashuk, L.A. and Finkelstein, J.N. (1995b). Induction of newly synthesized proteins in astroglial cells exposed to lead, *Toxicol. Appl. Pharmacol.*, 131, 21.

Patterson, C. and Settle, D. (1993). New mechanisms in lead biodynamics at ultra-low levels, *Neurotoxicology*, 14, 291.

Pennypacker, K.R., Thai, L., Hong, J.S., and McMillian, M.K. (1994). Prolonged expression of AP-1 transcription factors in the rat hippocampus after systemic kainate treatment, *J. Neurosci.*, 14, 3998.

Pennypacker, K.R., Xiao, Y., Xu, R.H., and Harry, G.J. (1997). Lead-induced developmental changes in AP-1 DNA binding in rat brain, *Int. J. Dev. Neurosci.*, 15, 321.

Peters, B., Stoltenburg, G., Hummel, M., Herbst, H., Altmann, L., and Wiegand, H. (1994). Effects of chronic low level lead exposure on the expression of GFAP and vimentin mRNA in the rat brain hippocampus analysed by *in situ* hybridization, *Neurotoxicology*, 15, 685.

Qian, Y. and Tiffany-Castiglioni, E. (2003). Lead-induced endoplasmic reticulum (ER) stress responses in the nervous system, *Neurochem. Res.*, 28,153.

Qian, Y., Falahatpsheh, M.H., Zheng, Y., Ramos, K.S., and Tiffany-Castiglioni, E. (2001). Induction of 78 kD glucose-regulated protein (GRP 78) expression and redox-regulated transcription factor activity by lead and mercury in C6 rat glioma cells, *Neurotox. Res.*, 3, 581.

Qian, Y., Harris, E.D., Zheng, Y., and Tiffany-Castiglioni, E. (2000). Lead targets GRP78, a molecular chaperone, in C6 rat glioma cells, *Toxicol. Appl. Pharmacol.*, 163, 260.

Qian, Y., Mikeska, G., Harris, E.D., Bratton, G.R., and Tiffany-Castiglioni, E. (1999). Effect of lead exposure and accumulation on copper homeostasis in cultured C6 rat glioma cells, *Toxicol. Appl. Pharmacol.*, 158, 41.

Qian, Y., Tiffany-Castiglioni, E., and Harris, E.D. (1995). Copper transport and kinetics in cultured C6 rat glioma cells, *Am. J. Physiol.*, 269, C892.

Ramesh, G.T. and Jadhav, A.L. (2001). Levels of protein kinase C and nitric oxide synthase activity in rats exposed to sub chronic low level lead, *Mol. Cell. Biochem.*, 223, 27.

Ramesh, G.T., Manna, S.K., Aggarwal, B.B., and Jadhav, A.L. (1999). Lead activates nuclear transcription factor-kappaB, activator protein-1, and amino-terminal c-Jun kinase in pheochromocytoma cells, *Toxicol. Appl. Pharmacol.*, 155, 280.

Raunio, S. and Tähti, H. (2001). Glutamate and calcium uptake in astrocytes after acute lead exposure, *Chemosphere*, 44, 355.

Resendez, E., Jr., Wooden, S.K., and Lee, A.S. (1988). Identification of highly conserved regulatory domains and protein-binding sites in the promoters of the rat and human genes encoding the stress-inducible 78-kilodalton glucose-regulated protein, *Mol. Cell. Biol.*, 8, 4579.

Rothstein, J.D., Dykes-Hoberg, M., Pardo, C.A., Bristol, L.A., Jin, L., Kuncl, R.W., Kanai, Y., Hediger, M.A., Wang, Y., Schielke, J.P., and Welty, D.F. (1996). Knockout of glutamate transporters reveals a major role for astroglial transport in excitotoxicity and clearance of glutamate, *Neuron*, 16, 675.

Rothstein, J.D. and Tabakoff, B. (1984). Alteration of striatal glutamate release after glutamine synthetase inhibition, *J. Neurochem.*, 43, 1438.

Rowles, T.K., Womac, C., Bratton, G.R., and Tiffany-Castiglioni, E. (1989). Interaction of lead and zinc in cultured astroglia, *Metab. Brain Dis.*, 4, 187.

Roy, B., Li, W.W., and Lee, A.S. (1996). Calcium-sensitive transcriptional activation of the proximal CCAAT regulatory element of the grp78/BiP promoter by the human nuclear factor CBF/NF-Y, *J. Biol. Chem.*, 271, 28995.

Sarid, J. (1991). Identification of a cis-acting positive regulatory element of the glial fibrillary acidic protein gene, *J. Neurosci. Res.*, 28, 217.

Schousboe, A., Westergaard, N., Sonnewald, U., Petersen, S.B., Huang, R., Peng, L., and Hertz, L. (1993). Glutamate and glutamine metabolism and compartmentation in astrocytes, *Dev. Neurosci.*, 15, 359.

Schousboe, A., Westergaard, N., Sonnewald, U., Petersen, S.B., Yu, A.C., and Hertz, L. (1992). Regulatory role of astrocytes for neuronal biosynthesis and homeostasis of glutamate and GABA, *Prog. Brain Res.*, 94, 199.

Schwartz, J. (1994). Low-level lead exposure and children's IQ: a meta-analysis and search for threshold, *Environ. Res.*, 65, 42.

Scortegagna, M. and Hanbauer, I. (2000). Increase AP-1 binding activity and nuclear REF-1 accumulation in lead-exposed primary cultures of astroglia, *Neurochem. Res.*, 25, 861.

Selevan, S.G., Rice, D.C., Hogan, K.A., Euling, S.Y., Pfahles-Hutchens, A., and Bethel, J. (2003). Blood lead concentration and delayed puberty in girls, *N. Engl. J. Med.*, 348, 1527.

Selvín-Testa, A., Capani, F., Loidl, C.F., Lopez, E.M., and Pecci-Saavedra, J. (1997). Prenatal and postnatal lead exposure induces 70 kDa heat shock protein in young rat brain prior to changes in astrocyte cytoskeleton, *Neurotoxicology*, 18, 805.

Selvín-Testa, A., Loidl, C.F., Lopez, E.M., Capani, F., Lopez-Costa, J.J., and Pecci-Saavedra, J. (1995). Prolonged lead exposure modifies astrocyte cytoskeletal proteins in the rat brain, *Neurotoxicology*, 16, 389.

Shiu, R.P., Pouyssegur, J., and Pastan, I. (1977). Glucose depletion accounts for the induction of two trans-formation-sensitive membrane proteinsin Rous sarcoma virus-transformed chick embryo fibroblasts, *Proc. Natl. Acad. Sci. U.S.A.*, 74, 3840.

Sierra, E.M. and Tiffany-Castiglioni, E. (1991). Reduction of glutamine synthetase activity in astroglia exposed in culture to low levels of inorganic lead, *Toxicology*, 65, 295.

Sierra, E.M., Rowles, T.K., Martin, J., Bratton, G.R., Womac, C., and Tiffany-Castiglioni, E. (1989). Low level lead neurotoxicity in a pregnant guinea pig model: Neuroglial enzyme activities and brain trace metal concentrations, *Toxicology*, 59, 81.

Simons, J.F., Ferro-Novick, S., Rose, M.D., and Helenius, A. (1995). BiP/Kar2p serves as a molecular chaperone during carboxypeptidase Y folding in yeast, *J. Cell Biol.*, 130, 41.

Song, H., Stevens, C.F., and Gage, F.H. (2002). Astroglia induce neurogenesis from adult neural stem cells, *Nature*, 417, 39.

Stark, M.,. Wolff, J.E.A., and Korbmacher, A. (1992). Modulation of glial cell differentiation by exposure to lead and cadmium, *Neurotoxicol. Teratol.*, 14, 247.

Steward, O., Torre, E.R., Phillips, L.L., and Trimmer, P.A. (1990). The process of reinnervation in the dentate gyrus of adult rats: time course of increases in mRNA for glial fibrillary acidic protein, *J. Neurosci.*, 10, 2373.

Stiles, K.M. and Bellinger, D.C. (1993). Neuropsychological correlates of low-level lead exposure in school-age children: a prospective study, *Neurotoxicol. Teratol.*, 15, 27.

Stoltenburg-Didinger, G., Pünder, I., Peters, B., Marcinkowski, M., Herbst, H., Winneke, G., and Wiegand H. (1996). Glial fibrillary acidic protein and RNA expression in adult rat hippocampus following low-level lead exposure during development, *Histochem. Cell Biol.*, 105, 431.

Stružyńska, L., Bubko, I., Walski, M., and Rafałowska, U. (2001). Astroglial reaction during the early phase of acute lead toxicity in the adult rat brain, *Toxicology*, 165, 121.

Suarez, I., Bodega, G., and Fernandez, B. (2002). Glutamine synthetase in brain: effect of ammonia, *Neurochem. Int.*, 41, 123.

Tiffany-Castiglioni, E. (1993). Cell culture models for lead toxicity in neuronal and glial cells, *Neurotoxicology*, 14, 513.

Tiffany-Castiglioni, E. and Qian, Y. (2001). Astroglia as metal depots: molecular mechanisms for metal accumulation, storage and release, *Neurotoxicology*, 22, 577.

Tiffany-Castiglioni, E. and Qian, Y. (2004). Cell-type specific responses of the nervous system to lead, in *In Vitro Neurotoxicology: Principles and Challenges*, Hollinger, M., Ed., Humana Press, Inc., Totowa, NJ, p. 151.

Tiffany-Castiglioni, E., Legare, M.E., Schneider, L.A., Hanneman, W.H., Zenger, E., and Hong, S.J. (1996a). Astroglia and neurotoxicity, in *The Role of Glia in Neurotoxicity*, Aschner, M. and Kimelberg, H.K., Eds., CRC Press, Boca Raton, p. 175.

Tiffany-Castiglioni, E., Legare, M.E., Schneider, L.A., Harris, E.D., Barhoumi, R., Zmudzki, J., Qian, Y., and Burghardt, R.C. (1996b). Heavy metal effects on glia, in *Methods in Neurosciences*, Vol. 30, Regino Perez-Polo, J., Ed., Academic Press, New York, p. 135.

Tiffany-Castiglioni, E., Sierra, E.M., Wu, J.-N., and Rowles, T.K. (1989). Lead toxicity in neuroglia, *Neurotoxicology*, 10, 383.

Tomsig, J.L. and Suszkiw, J.B. (1995). Multisite interactions between Pb2+ and protein kinase C and its role in norepinephrine release from bovine adrenal chromaffin cells, *J. Neurochem.*, 64, 2667.

Tong, S., von Schirnding, Y.E., and Prapamontol, T. (2000). Environmental lead exposure: a public health problem of global dimensions, *Bull. World Health Org.*, 78, 1068.

Ullian, E.M., Sapperstein, S.K., Christopherson, K.S., and Barres, B.A. (2001). Control of synapse number by glia, *Science*, 291, 657.

Vainio, H. (1997). Lead and cancer—association or causation?, *Scand. J. Work Environ. Health*, 23, 1, 1997.

VandenBerg, K.J., Lammers, J.H.C.M., Hoogendijk, E.M.G., and Kulig, B.M. (1996). Changes in regional brain GFAP levels and behavioral functioning following subchronic lead acetate exposure in adult rats, *Neurotoxicology*, 17, 725.

Wei, S.J., Botero, A., Hirota, K., Bradbury, C.M., Markovina, S., Laszlo, A., Spitz, D.R., Goswami, P.C., Yodoi, J., and Gius, D. (2000). Thioredoxin nuclear translocation and interaction with redox factor-1 activates the activator protein-1 transcription factor in response to ionizing radiation, *Cancer Res.*, 60, 6688.

Yi, E.Y. and Lim, D.K. (1998). Effects of chronic lead exposure on glutamate release and uptake in cerebellar cells of rat pups, *Arch. Pharm. Res.*, 21, 113.

Yun, S.W., Gartner, U., Arendt, T., and Hoyer, S. (2000). Increase in vulnerability of middle-aged rat brain to lead by cerebral energy depletion, *Brain Res. Bull.*, 52, 371.

Zawia, N.H. and Harry, G.J. (1996). Developmental exposure to lead interferes with glial and neuronal differential gene expression in the rat cerebellum, *Toxicol. Appl. Pharmacol.*, 138, 43.

Zawia, N.H., Sharan, R., Brydie, M., Oyama, T., and Crumpton, T. (1998). Sp1 as a target site for metal-induced perturbations of transcriptional regulation of developmental brain gene expression, *Dev. Brain Res.*, 107, 291.

Zurich, M.G., Eskes, C., Honegger, P., Berode, M., and Monnet-Tschudi, F. (2002). Maturation-dependent neurotoxicity of lead acetate *in vitro*: implication of glial reactions, *J. Neurosci. Res.*, 70, 108.

Zurich, M.G., Monnet-Tschudi, F., Bérode, M., and Honegger, P. (1998). Lead acetate toxicity *in vitro*: Dependence on the cell composition of the cultures, *Toxicol. In Vitro*, 12, 191.

Zurich, M.G., Monnet-Tschudi, F., Costa, L.G., Schilter, B., and Honegger, P. (2004). Aggregation brain cell cultures for neurotoxicological studies, in *In Vitro Neurotoxicology: Principles and Challenges*, Hollinger, M., Ed., Humana Press, Inc., Totowa, NJ, p. 243.

CHAPTER 27

Role of Glia in MPTP Toxicity and Parkinson's Disease

Lucio G. Costa and Michael Aschner

CONTENTS

27.1 INTRODUCTION

1-methyl-4-phenyl-1,2,3,6-tetrahydropyridine (MPTP) is an analog of the opiate analgesic meperidine (Demerol®). MPTP surfaced on the illegal markets of San Francisco in the early 1980s. It was produced in a "clandestine" fashion and was referred to as a "synthetic heroin." Heroin addicts consuming MPTP rapidly developed a permanent neurological disorder that was clinically indistinguishable from nonidiopathic Parkinson's disease (Langston et al., 1982). Subsequent experiments have confirmed the propensity of MPTP to preferentially damage nigrostriatal dopamine-containing neurons. While tragically debilitating to its abusers, the availability of MPTP offered an exciting model for the disease, triggering resurgence in research on the etiology of Parkinson's disease and potential modalities for its treatment. The results from a number of these studies suggest that astrocytes play a pivotal role in the neurodegenerative changes associated with exposure to MPTP. MPTP has been shown to damage dopaminergic neurons in the nigrostriatal pathway in humans (Speciale, 2002) as well as experimental animals (Heikkila et al., 1984). Given these properties, it has proven to be an important tool for modeling various aspects of dopamine (DA) neuron degeneration occurring in idiopathic Parkinson's disease (Speciale, 2002).

27.2 MPTP AND ASTROCYTES

MPTP represents a classical compound that is metabolized within astrocytes to a reactive intermediate, 1-methyl-4-phenylpyridinium ion (MPP$^+$) with subsequent propensity to selectively destroy nigrostriatal dopaminergic neurons (Burns et al., 1983; Heikkila et al., 1984; Langston et al., 1984).

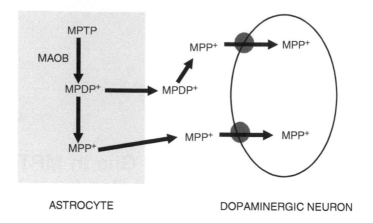

Figure 27.1 Bioactivation of MPTP to MPP+ in estrocytes and uptake of MPP+ in dopaminergic nevrons.

The fully oxidized pyridinium metabolite (MPP+) is likely to be the mediator of MPTP neurotoxicity and is apparently able to damage neuronal cells after being formed within and released from astrocytes (Di Monte et al., 1996). Of note is that MPP+, despite its changed chemical structure, can cross the plasma membrane into the extracellular space after being formed within astrocytes (Di Monte et al., 1992a). Furthermore, there is some evidence that MPP+ can also be formed extracellularly, presumably via autoxidation of the 1-methyl-4-phenyl-2,3-dihydropyridinium intermediate (MPDP+), after this latter compound has been generated within astrocytes and has crossed astrocyte membranes (Di Monte et al., 1992a) (Figure 27.1).

Astrocytes *in situ* are MAO-B immunocytochemically positive (Levitt et al., 1982). This raises the possibility that these glial cells might participate in conversion of MPTP to MPP+. Indeed, evidence from primary cell cultures has demonstrated the time-dependent production of MPP+ by astrocytes exposed to MPTP (Ransom et al., 1987). That neuroglia mediate an essential step in MPTP toxicity is supported by *in vivo* studies with the astroglial-selective toxicant, α-aminoadipic acid (α-AA), a six-carbon chemical analog of glutamate (Takada et al., 1990). When MPTP is microinjected into the substantia nigra pars compacta (SNc), tremendous loss of nigral neurons is observed as indicated by fluorescent retrograde axonal tracing. In contrast, coinjection of MPTP plus α-AA into the SNc is associated with reduced neuronopathy. Upon reactive gliosis, the protective effect of α-AA is curtailed, which suggests that once repopulated with astrocytes, MPTP is again oxidized to MPP+.

Based on these studies, a heuristic model of MPTP toxicity has been proposed, which involves diffusion of pro-neurotoxicant into the brain where it is taken up by MAO-B–positive astrocytes and neurons and subsequently metabolized to the reactive neurotoxic metabolite, MPP+. MPP+ is eventually released from glial and neuronal sites and is subsequently concentrated in nigrostriatal neurons via dopamine uptake. Accumulation of MPP+ produces nerve cell death presumably via inhibition of mitochondrial NADH oxidation and, consequently, of oxidative phosphorylation.

Because of their role in the bioactivation of MPTP, astrocytes would be expected to be targets of the cytotoxic properties of MPP+. However, degeneration of glial cells does not appear to be an obvious neuropathological feature of MPTP exposure; rather, astrogliosis has been found following administration of MPTP (see below). As *in vitro* studies have shown that astrocytes in culture are sensitive to the toxic effects of MPTP, due primarily to inhibition of mitochondrial complex I and ATP depletion (Di Monte et al., 1992b), the lack of glial toxicity observed *in vivo* may be ascribed to the following factors: (1) MPP+ may not reach intracellular concentrations great enough to cause astrocytic death, or (2) toxic intracellular concentrations of MPP+ may not persist long enough to induce irreversible cell damage (Di Monte et al., 1992b).

The situation is clearly different in dopaminergic neurons. The MAO-derived metabolite of MPTP, MPP$^+$, requires uptake by the dopamine transporter (VMAT2) to gain access to dopaminergic neurons (Figure 27.1). Evidence for this comes from studies using selective dopamine uptake inhibitors (Javitch et al., 1985; Melamed et al., 1985), as well as data showing that MPTP toxicity is prevented in mice genetically engineered to lack the dopamine transporter (Gainetdinov et al., 1997). Dopaminergic neurons are particularly vulnerable to MPTP toxicity because of their ability to accumulate MPP$^+$ (via dopamine uptake) and to retain it for a prolonged period of time.

Thus far, two pathways of MPP$^+$ formation have been identified in astrocytes: one that is dependent upon the activity of monoamine oxidase (MAO) and one that is related to the presence of transition metals (Di Monte et al., 1996). Increased glutamatergic drive to basal ganglia output nuclei has also been considered a likely contributor to the pathogenesis of MPTP-induced symptoms. Since astrocytes efficiently transport MPTP and glutamate intracellularly (see above), increased excitatory "tone" and impairment in glutamate uptake in astrocytes have also been implicated in the etiology of MPTP-induced neurotoxicity. The ability of MPTP to affect D-aspartate (a nonmetabolizable analog of L-glutamate) uptake in astrocytes was recently demonstrated by Hazell et al. (1997). The effect was shown to be dependent upon the conversion of MPTP to MPP$^+$. Another study has investigated the cellular and molecular mechanisms underlying the restorative actions of basic fibroblast growth factor (FGF-2) and the changes within astroglial cells in the MPTP-lesioned striatum. Specifically, striatal expression and regulation of connexin-43 (cx43), the principal gap junction protein of astroglial cells, along with the expression of GFAP, FGF-2, and functional coupling were studied (Rufer et al., 1996). These authors report that MPTP alters the expression and protein levels of cx43, providing for another possible mechanism for MPTP-induced neurotoxicity and the direct involvement of astrocytes in this process.

In addition to the long-lasting effects of MPTP on DA neurotransmitter levels, administration of the compound also results in neurodegeneration as evidenced by astrogliosis. Astrocyte hypertrophy is a universal response to a variety of nervous system insults, and changes in GFAP, the major astrocyte intermediate filament protein, can be used to assess reactive gliosis. MPTP exposure is associated with robust time-dependent increases in striatal GFAP that peak at 48 h posttreatment (Reinhard et al., 1988; O'Callaghan et al., 1990). More recent studies have suggested that both neuronal and astroglial responses upon MPTP exposure might be mediated via serotonergic effects (Lullen et al., 2003). Astrocytic activation has also been suggested in studies where S-100 protein was shown selectively to be expressed by astrocytes, but not by microglia, after MPTP treatment (Muramatsu et al., 2003).

Fernandez-Gonzalez et al. (2000) have investigated the expression of the heme oxygenase isoform 1 (HO-1) in a mouse model for Parkinson's disease (PD). The authors report that MPTP triggered a rapid and persistent increase in HO-1 mRNA exclusively in the mouse striatum. *In situ* hybridization and immunohistochemistry showed HO-1 to be localized to striatal astrocytes. The induction of HO-1 by MPTP was blocked by selegiline and GBR-12909, indicating that the protoxin had to be metabolized by monoamine oxidase B and taken up by dopaminergic neurons to exert its action in astrocytes. MPTP did not alter the expression of other enzymes of heme synthesis or degradation, nor were the levels of mRNA for heme or iron-binding proteins changed. Thus, expression of HO-1 was not part of a cellular program involving heme biosynthesis or homeostasis. In addition, heat-shock proteins (Hsp) were not induced by MPTP. Fernandez-Gonzalez et al. (2000) concluded that MPTP elicited a selective transcriptional response in striatal astrocytes and that this response is likely mediated by molecules released from affected dopaminergic nerve terminals in the striatum acting upon neighboring astrocytes.

On the basis of the observation that different strains of mice are differentially sensitive to MPTP neurotoxicity (Hamre et al., 1999), Smeyne et al. (2001) have proposed that sensitivity to MPTP is conferred by glia. Such modulation does not appear to involve the conversion of MPTP to MPP$^+$, but rather different, still undefined, facilitative or protective factors, which may include inducible nitric oxide synthase and reduced glutathione (GSH) levels, respectively (Smeyne et al., 2001).

27.3 GLIA AND PARKINSON'S DISEASE

As in the case of MPTP administration (see previous section), loss of dopaminergic neurons in postmortem Parkinsonian brain is associated with a significant glial reaction (Forno et al., 1992). Specifically, a mild degree of astrogliosis is observed, along with a more robust microglial response (McGeer et al., 1988; Mirza et al., 2000). Activation of microglial cells is also observed in the MPTP mouse model (Czlonkowska et al., 1986); in MPTP-treated animals, however, the astrocytic reaction is more pronounced than that seen in Parkinsonian brain (Langston et al., 1999).

Glial response to injury may have beneficial effects, which in the case of Parkinson's disease could attenuate degeneration (Teisman et al., 2003). For example, glial cells can produce and release trophic factors, such as GDNF (glial-derived neurotrophic factor) or BDNF (brain-derived neurotrophic factor), that may attenuate dopaminergic neuron degeneration (Kordower et al., 2000; Batchelor et al., 1999). Alternatively, glial cells may protect dopaminergic neurons against degeneration by scavenging toxic compounds released by the dying neurons (Teisman et al., 2003). Both actions of glial cells may thus have neuroprotective roles in Parkinson's disease.

On the other hand, other lines of evidence support the contention that glial cells, and particularly activated microglial cells, could have a deleterious role in Parkinson's disease. Activated microglia can produce a variety of noxious compounds including reactive oxygen species (ROS), reactive nitrogen species (RNS), proinflammatory prostaglandins, and cytokines (Teiman et al., 2003). While the role of ROS has been debated for some time as a cardinal feature of the oxidative stress hypothesis for Parkinson's disease (Jenner, 2003), recent attention has been devoted to RNS, given the idea that nitric oxide–mediated nitrating stress could be pivotal in the pathogenesis of Parkinson's disease (Giasson et al., 2000). Prostaglandins and their synthesizing enzymes such as cyclooxygenase-2 (COX-2) are also receiving increasing attention, as they are emerging as important determinants of cytotoxicity associated with inflammation (O'Banion, 1999). It is of interest, in this regard, that pharmacological inhibition of both COX-2 and COX-1 attenuates MPTP neurotoxicity in mice (Teisman and Ferger, 2001). Another group of glial-derived compounds that may inflict damage in Parkinson's disease is that of cytokines, such as tumor necrosis factor-α (TNF-α) and interleukin-1β (IL-1β). Such cytokines may act on glial cells themselves (providing, for example, further microglial activation), as well as on dopaminergic neurons, where they may trigger intracellular death-related signaling pathways (Teisman et al., 2003). Altogether, these detrimental actions of glial cells may play a role in propagating the neurodegenerative process; if so, they may be targeted by therapeutic interventions to slow or halt the progression of neurodegeneration in Parkinson's disease.

REFERENCES

Batchelor, P.E., Liberatore, G.T., Wong, J.Y., Porriti, M.J., Frerichs, F., Donnan, G.A., and Howell, D.W. (1999). Activated macrophages and microglia induce dopaminergic sprouting in the injured striatum and express brain-derived neurotrophic factor and glial cell line-derived neurotrophic factor, *J. Neurosci.*, 19, 1708–1716.

Burns, R.S., Chiueh, C.C., Markey, S.P., Ebert, M.H., Jacobowitz, D.M. et al. (1983). A primate model of parkinsonism: Selective destruction of dopaminergic neurons in the pars compacta of the substantia nigra by N-methyl-4-phenyl-1, 2, 3, 6-tetrahydropyridine, *Proc. Nat. Acad. Sci. U.S.A.*, 80, 4546–4550.

Czlonkowska, A., Kohutnicka, M., Kurkowska-Jastrzebska, I., and Czlonkowski, A. (1996). Microglial reaction in MPTP (1-methyl-4-phenyl- 1,2,3,6-tetrahydopyridine) induced Parkinson's disease mice model, *Neurodegeneration*, 5, 137–143.

Di Monte, D.A., Royland, J.E., Irwin, I., Langston, J.W. (1996). Astrocytes as the site for bioactivation of neurotoxins, *Neurotoxicology*, 17, 697–703.

Di Monte, D.A., Wu, E.Y., Irwin, I., Delanney, L.E., and Langston, J.W. (1992a). Production and disposition of 1-methyl-4-phenylpyridinium in primary cultures of mouse astrocytes, *Glia*, 5, 48–55.

Di Monte, D.A., Wu, E.Y., Irwin, I., Delanney, L.E., and Langston, J.W. (1992b). Toxicity of 1-methyl-4-phenyl-1,2,3,6-tetrahydopyridine in primary cultures of mouse astrocytes, *J. Pharmacol. Exp. Ther.*, 261, 44–49.

Fernandez-Gonzalez, A., Perez-Otano, I., and Morgan, J.I. (2000). MPTP selectively induces haem oxygenase-1 expression in striatal astrocytes, *Eur. J. Neurosci.*, 12, 1573–1583.

Forno, L.S., Delanney, L.E., Irwin, I., Di Monte, D.A., and Langston, J.W. (1992). Astrocytes and Parkinson's disease, *Prog. Brain Res.*, 94, 429–436.

Gainetdinov, R.R., Fumagalli, F., Jones, S.R., and Caron, M.G. (1997). Dopamine transporter is required for *in vivo* MPTP neurotoxicity: evidence from mice lacking the transporter, *J. Neurochem.*, 69, 1322–1325.

Giasson, B.I., Duda, J.E., Murray, I.V., Chen, Q., Souza, J.M. et al. (2000). Oxidative damage linked to neurodegeration by selective alpha-synuclein nitration in synucleinopathy lesions, *Science*, 290, 985–989.

Hamre, K., Tharp, R., Pon, K., Xiong, X., and Smeyne, R.J. (1999). Differential strain susceptibility following 1-methyl-4-phenyl-1,2,3,6-tetrahydopyridine (MPTP) administration acts in an autosomal dominant fashion: quantitative analysis in seven strain of *Mus musculus*, *Brain Res.*, 828, 91–103.

Hazell, A.S., Itzhak, Y., Liu, H., and Norenberg, M.D. (1997). 1-Methyl-4-phenyl-1,2,3,6-tetrahydopyridine (MPTP) decreases glutamate uptake in cultured astrocytes, *J. Neurochem.*, 68, 2216–2219.

Heikkila, R.E., Hess, A., and Duvoisin, R.C. (1984). Dopaminergic neurotoxicity of 1-methyl-4-phenyl-1, 2, 5, 6-tetrahydropyridine in mice, *Science*, 224, 1451–1453.

Javitch, J.A., D'Amato, R.J., Strittmatter, S.M., and Snyder, S.H. (1985). Parkinsonism-inducing neurotoxin, N-methyl-4-phenyl-1,2,3,6-tetrahydopyridine: uptake of the metabolite N-methyl-4-phenylpyridine by dopamine neurons explains selective toxicity, *Proc. Natl. Acad. Sci. U.S.A.*, 82, 2173–2177.

Jenner, P. (2003). Oxidative stress in Parkinson's disease, *Ann. Neurol.*, 53 (Suppl. 3), 528–538.

Kordower, J.H., Palfi, S., Chen, E.Y., Ma, S.Y., Sendera, T., Cochran, E.J. et al. (1999). Clinicopathological findings following intraventricular glial-derived neurotrophic factor treatment in a patient with Parkinson's disease, *Ann. Neurol.*, 46, 419–424.

Langston, J.W., Ballard, P., Tetrud, J.W., and Irwin, I. (1982). Chronic parkinsonism in humans due to a product of meperidine-analog synthesis, *Science*, 219, 979–980.

Langston, J.W., Forno, L.S., Rebert, C.S., and Irwin, I. (1984). Selective nigral toxicity after systemic administration of 1-methyl-4-phenyl-1,2,3,5,6-tetrahydopyridine (MPTP) in the squirrel monkey, *Brain Res.*, 292, 390–394.

Langston, J.W., Forno, L.S., Tetoud, J., Reeves, A.G., Kaplan, J.A., and Karluk, D. (1999). Evidence of active nerve cell degeneration in the substantia nigra of humans years after 1-methyl-4-phenyl-1,2,3,6-tetrahydopyridine exposure, *Ann. Neurol.*, 46, 598–605.

Luellen, B.A., Miller, D.B., Chisnell, A.C., Murphy, D.L., O'Callaghan, J.P., and Andrews, A.M. (2003). Neuronal and astroglial responses to the serotonin and norepinephrine neurotoxin: 1-methyl-4-(2'-aminophenyl)-1,2,3,6-tetrahydopyridine, *J. Pharmacol. Exp. Ther.*, 307, 923–931.

McGeer, P.L., Itagai, S., Boyes, B.E., and McGeer, E.G. (1988). Reactive microglia are positive for HLA-DR in the substantia nigra of Parkinson's and Alzheimer's disease brains, *Neurology*, 38, 1285–1291.

Melamed, E., Rosenthal, J., Cohen, O., Globus, M., and Uzzan, A. (1985). Dopamine but not norepinephrine or serotonin uptake inhibitors protect mice against neurotoxicity of MPTP, *Eur. J. Pharmacol.*, 116, 179–181.

Mirza, B., Hadberg, H., Thomsen, P., and Moos, T. (2000). The absence of reactive astrocytosis is indicative of a unique inflammatory process in Parkinson's disease, *Neuroscience*, 95, 425–432.

Muramatsu, Y., Kurosaki, R., Watanabe, H., Michimata, M., Matsubara, M., Imai, Y., and Araki, T. (2003). Expression of S-100 protein is related to neuronal damage in MPTP-treated mice, *Glia*, 42, 307–313.

O'Banion, M.K. (1999). Cyclooxygenase –2: molecular biology, pharmacology and neurobiology, *Crit. Rev. Neurobiol.*, 13, 45–82.

O'Callaghan, J.P., Miller, D.B., and Reinhard, J.J. (1990). Characterization of the origins of astrocyte response to injury using the dopaminergic neurotoxicant, 1-methyl-4-phenyl-1,2,3,6-tetrahydopyridine, *Brain Res.*, 521, 73–80.

Ransom, B.R., Kunis, D.M., Irwin, I., and Langston, J.W. (1987). Astrocytes convert the parkinsonism inducing neurotoxin, MPTP, to its active metabolite, MPP$^+$, *Neurosci. Lett.*, 75, 323–328.

Reinhard, J.J., Miller, D.B., and O'Callaghan, J.P. (1988). The neurotoxicant MPTP (1-methyl-4-phenyl-1,2,3,6-tetrahydopyridine) increases glial fibrillary acidic protein and decreases dopamine levels of the mouse striatum: evidence for glial response to injury, *Neurosci. Lett.*, 95, 246–251.

Rufer, M., Wirth, S.B., Hofer, A., Dermietzel, R., Pastor, A. et al. (1996). Regulation of connexin-43, GFAP, and FGF-2 is not accompanied by changes in astroglial coupling in MPTP-lesioned, FGF-2-treated parkinsonian mice, *J. Neurosci. Res.*, 46, 606–617.

Smeyne, M., Goloubeva, O., and Smeyne, R.J. (2001). Strain-dependent susceptibility to MPTP and MPP+-induced Parkinsonism is determined by glia, *Glia*, 34, 73–80.

Speciale, S.G. (2002). MPTP: insights into parkinsonian neurodegeneration, *Neurotoxicol. Teratol.*, 24, 607–620.

Takada, M., Li, Z.K., and Hattori, T. (1990). Astroglial ablation prevents MPTP-induced nigrostriatal neuronal death, *Brain Res.*, 509, 55–61.

Teisman, P. and Ferger, B. (2001). Inhibition of the cyclooxygenase isoenzymes Cox-1 and Cox-2 provide neuroprotection in the MPTP-mouse model of Parkinson's disease, *Synapse*, 39, 167–174.

Teisman, P., Tieu, K., Cohen, O., Choi, D.K., Wu, D.C., Marks, D., Vila, M., Jackson-Lewis, V., Przedborski, S. (2003). Pathogenic role of glial cells in Parkinson's disease, *Mov. Disord.*, 18, 121–129.

Index

A

P